TRANSITIONING TO A PROSPEROUS, RESILIENT AND CARBON-FREE ECONOMY

This book is a comprehensive manual for policy-makers addressing the issues around human-caused climate change, which threatens communities with increasing extreme weather events, sea-level rise and declining habitability of some regions due to desertification or inundation. The book looks at both mitigation of greenhouse gas emissions and global warming and adaption to changing conditions as the climate changes. It encourages the early adoption of climate change measures that this can be achieved while maintaining prosperity. The book takes a sector-by-sector approach, starting with energy and includes cities, industry, natural resources and agriculture, enabling practitioners to focus on actions relevant to their field. It uses case studies across a range of countries, and various industries, to illustrate the opportunities available. Blending technological insights with economics and energy policy, the book presents the tools decision makers need to achieve rapid decarbonisation, while unlocking and maintaining productivity, profit and growth.

KENNETH G. H. BALDWIN is Director of the Australian National University (ANU) Energy Change Institute (ECI) and an ANU Public Policy Fellow. He is also Director of the ANU ECI Grand Challenge Zero-Carbon Energy for the Asia-Pacific. Professor Baldwin won the 2004 Australian Government Eureka Prize for Promoting Understanding of Science for initiating and championing 'Science meets Parliament'. He received the Australian Optical Society medal and the Barry Inglis Medal of the National Measurement Institute.

MARK HOWDEN is Director of the Australian National University Institute for Climate Energy and Disaster Solutions. He has researched climate impacts, adaptation, and emission reduction for food security, energy, water resources and urban systems. He has partnered with many industry, community, and policy groups. A major contributor to the Intergovernmental Panel on Climate Change (IPCC) since 1991, he is now a Vice-Chair of the IPCC. He shared the 2007 Nobel Prize with Al Gore and IPCC colleagues.

MICHAEL H. SMITH has been a Research Fellow or Adjunct Senior Lecturer at the Australian National University since 2006. He has co-authored or contributed to over 200 publications on climate change, including *Climate, Energy and Water:*

Managing Trade-offs, Seizing Opportunities (Cambridge University Press, 2015), *Factor Five* (Routledge, 2010), *Cents and Sustainability* (Routledge, 2010), *Whole System Design* (Routledge, 2009) and *Natural Advantage of Nations* (Routledge, 2005), and has served as a policy adviser to national and subnational governments.

KAREN HUSSEY is an Honorary Professor with the Centre for Policy Futures at the University of Queensland. As an academic, Karen's research and teaching focused on the policy, institutional and governance arrangements needed to address issues including climate change, water resource management, waste management, biodiversity conservation and the 'trade–environment' nexus. Karen currently works in the Queensland Government.

PETER J. DAWSON is a consultant and writer and former senior officer for the Australian Government. He has worked on international trade, industry development, technology transfer and government procurement policy. He has consulted internationally on Small and Medium Enterprise development for the World Bank and other agencies, mainly in Indonesia.

TRANSITIONING TO A PROSPEROUS, RESILIENT AND CARBON-FREE ECONOMY

A Guide for Decision Makers

Edited by

KENNETH G. H. BALDWIN

Australian National University, Canberra

MARK HOWDEN

Australian National University Institute for Climate Energy and Disaster Solutions, Canberra

MICHAEL H. SMITH

Australian National University, Canberra

KAREN HUSSEY

University of Queensland

PETER J. DAWSON

P J Dawson & Associates

CAMBRIDGE
UNIVERSITY PRESS

CAMBRIDGE
UNIVERSITY PRESS

University Printing House, Cambridge CB2 8BS, United Kingdom

One Liberty Plaza, 20th Floor, New York, NY 10006, USA

477 Williamstown Road, Port Melbourne, VIC 3207, Australia

314–321, 3rd Floor, Plot 3, Splendor Forum, Jasola District Centre, New Delhi – 110025, India

103 Penang Road, #05–06/07, Visioncrest Commercial, Singapore 238467

Cambridge University Press is part of the University of Cambridge.

It furthers the University's mission by disseminating knowledge in the pursuit of education, learning, and research at the highest international levels of excellence.

www.cambridge.org
Information on this title: www.cambridge.org/9781107118348
DOI: 10.1017/9781316389553

© Cambridge University Press 2021

First published 2021

Printed in the United Kingdom by TJ Books Limited, Padstow Cornwall

A catalogue record for this publication is available from the British Library.

ISBN 978-1-107-11834-8 Hardback

In memory of founding editor Michael Raupach

Contents

The colour plates can be found between pages 388 and 389

Figures

Tables

Contributors

Philip Adams
Monash University

Eshan Ahuja
Australian National University

Xuemei Bai
Australian National University

Kenneth G. H. Baldwin
Australian National University

Timothy M. Baynes
CSIRO

Fiona J. Beck
Australian National University

Pablo Berrutti
Altiorem

Lachlan Blackhall
Australian National University

Andrew Blakers
Australian National University

Nicolette Boele
Responsible Investment Association of Australasia

Tim Capon
CSIRO

Ken Coghill
Monash University

Peter J. Dawson
P J Dawson & Associates

Amandine Denis-Ryan
Climate Works Australia

Retno Gumilang Dewi
Bandung Institute of Technology

Nathan Fabian
Responsible Investment Association of Australasia

Thomas Faunce
Australian National University

Evan Franklin
Australian National University

Scott Ferraro
Climate Works Australia

Paul Gauché
Heliogen Inc

Alexey M. Glushenkov
Australian National University

David Gourlay
Australian National University

Paul Graham
CSIRO

Ali Hasanbeigi
Global Efficiency Intelligence

Steve Hatfield-Dodds
Australian National University

Jane Hodgkinson
CSIRO

Mark Howden
Australian National University

Karen Hussey
University of Queensland

Tony Irwin
Australian National University

Andy Jones
Climate Works Australia

Frank Jotzo
Australian National University

Niina Kauto
Climate Works Australia

Heather Keith
Griffith University

Rob Kelly
Climate Works Australia

Abel Kinyondo
University of Dar es Salaam

Sarah Levy
Climate Works Australia

Keith Lovegrove
ITP Thermal Pty Ltd

Michelle Lyons
Australian National University

Andrew Macintosh
Australian National University

Brendan Mackey
Griffith University

Mark Mehos
National Renewable Energy Laboratory

Lynette Molyneaux
University of Queensland

Cristina Neesham
Newcastle University (UK)

Peter Newton
Swinburne University of Technology

Barbara Norman
University of Canberra

David Osmond
Windlab

Alan Pears
RMIT University

Jamie Pittock
Australian National University

John Pye
Australian National University

Hedda Ransan-Cooper
Australian National University

Chris Ryan
Australian National University

Ucok W. R. Siagian
Bandung Institute of Technology

Anna Skarbek
Climate Works Australia

Michael H. Smith
Australian National University

Thomas Smith
Monash University

Mark Stafford Smith
CSIRO

Peter Stasinopoulos
RMIT University

Nathan Steggel
Windlab

Andrew Stuchbery
Australian National University

Bjorn Sturmberg
Australian National University

Keith Sue
James Cook University

John Thwaites
Climate Works Australia

Andrea Turner
University of Technology Sydney

Mahesh B. Venkataraman
Australian National University

Robert Webb
Australian National University

Stuart White
University of Technology Sydney

Russell Wise
CSIRO

Foreword

MALCOLM TURNBULL

Former Prime Minister of Australia

Another title for this book could be: *No More Excuses: How We Can Slash Our Emissions, Save Our Planet and Pay Less for Electricity.*

The brutal physical consequences of global warming are confronting us every day – from an eerily temperate arctic to the worst bushfires in our history, only the wickedly and wilfully blind can ignore the urgency for decisive action to reduce our emissions and slow the relentless heating of our planet.

Global warming has always been a wicked problem because it calls on present generations to take action to prevent adverse consequences for generations yet unborn. Or that's how it was seen 20 years ago. Now the adverse consequences are upon us, and our children and grandchildren's generations are marching in the street to demand immediate action.

And yet, because of a toxic alliance between right-wing populist politics, their amplifiers in the media and the fossil fuel lobby, we have seen climate action delayed and frustrated, and the science that demands it denied.

Right at the heart of the political, or policy, problem has been the need to persuade today's voters to pay more for energy in order to protect the planet for future generations. Well, that was the economic argument we used to face. But today, thanks to extraordinary improvements in the technology of renewable generation and storage, we can now say with confidence that, with the right planning, we can rapidly transition to a world where energy is both much greener and cheaper.

It doesn't often happen, but right now, thanks to science, we can have our cake and eat it too. But we cannot delay. All of the political manoeuvres and debate can delay action to address global warming, but they cannot delay its consequences.

The more than 60 authors of this book demonstrate with comprehensive research, backed by practical examples, that we can transition to a low-carbon economy in order to avoid catastrophic damage to our planet, and at the same time support prosperous and adaptive societies. This book has the facts and figures that will refute the arguments of those who say 'we can't afford it'. It is, in short, a pragmatic handbook on how to achieve the transition to a decarbonised world that is both prosperous and resilient.

Although it covers many technical areas, the book is designed to be accessible to a wide audience including non-technical readers. The book examines the policy environments and the challenges of bringing about the necessary transition. It describes the main technologies available to deliver clean energy, including solar photovoltaics and wind which are now the cheapest sources of energy in most parts of the world. The chapters on wind energy, solar photovoltaics and storage in particular demonstrate how and why the costs have dramatically declined, in large measure thanks to the contribution of Australian scientists and researchers.

I was pleased to see the detailed discussion of pumped storage hydro, including the two large-scale Australian projects: Snowy Hydro 2.0 and Battery of the Nation. As Prime Minister of Australia, I had the opportunity to be very effectively educated in pumped storage hydro by one of the authors of this book, Andrew Blakers. This enabled me to conceive the national pumped hydro agenda in general and these two major projects in particular.

While the message of the authors is overwhelmingly positive and encouraging for those who support stronger immediate action to cut emissions, it is not unrealistic. The discussion of the opportunities for green hydrogen is very objective and canvases the particular challenges of storing and transporting hydrogen. On the very encouraging side, the authors make the good point that desalinating seawater for the purpose of electrolysis is a negligible additional cost to the process.

The question of how countries can transition to low-carbon economies and the likely outcomes for societies and economies were explored in the UN Deep Decarbonisation Pathways Projects. The book reviews these studies and their outcomes in a chapter covering 16 countries. It applies these studies to Australia, to examine how economies can navigate the transition with continuous growth, and emerge with greater and more sustainable prosperity than would be possible under 'business-as-usual'. This is a strong message based on deep research and is pitched in a way that can be understood by key decision makers in government, industry and the wider community.

The book also examines key sectors of the economy and how they are both contributing to and being affected by climate change. These sectors include: cities and their components; land use, forestry and agriculture; transport; mining, oil and gas; and industry and manufacturing. In many cases, measures to reduce emissions also have co-benefits. Replacing gas heating with electric heat pumps not only reduces emissions but also delivers cost savings. Electrification of mines using electrically powered conveyors to replace diesel trucks not only reduces greenhouse emissions but also avoids air pollution. The chapters on land use, forests and agriculture explore in depth the complex problems of achieving sustainability in these ecosystems, which are highly involved in carbon cycles affected by deforestation and changes in land use, but also critical to the health of people and the planet.

The book adopts an optimistic tone but also acknowledges the challenges and barriers to transitioning to low-carbon economies. It discusses the ways in which resources can be misallocated as a consequence of poor governance in various ways, including in embedded subsidies for fossil fuels which can distort markets at the expense of renewables.

This is not a book which addresses the poisonous politics of climate change – there wouldn't be enough room for the science and engineering if it was – but there is a brief discussion of the ways in which vested interests can use existing provisions in free trade agreements, such as investor–state dispute settlement clauses to block renewables sup-ported by state environmental policies.

The Intergovernmental Panel on Climate Change (IPCC) considers that to have a reasonable chance of avoiding catastrophic effects from climate change it is necessary to limit global warming to 2 °C, requiring substantial and rapid reductions in greenhouse gas emissions and adaptation to the climate change impacts we cannot avoid. This is the imperative behind this handbook for policy-makers and practitioners that will enable a prosperous and resilient transition to future decarbonised economies. It also emphasises that now is the time for new and effective solutions.

Every politician and policy adviser who cares about climate action should read this book. It provides so many of the practical answers we need if we are to effect this transition, as we must. Just as the existential challenge we face is grounded in the laws of physics, so must our response be equally scientific. We need evidence-based policy founded on engineering and economics – not ideology and idiocy.

Introduction

This book comes at a critical point in the transition to a low-carbon future. The global scientific consensus evidenced in the most recent Intergovernmental Panel on Climate Change (IPCC) report warns of the need to make purposeful reductions in global greenhouse gas (GHG) emissions urgently over the next decade, trending to net zero emissions globally by 2050, to avoid the worst forms of dangerous climate change. Yet, global GHG emissions have risen again in 2019, at the end of the hottest decade on record, and temperature rises continue largely unabated. The good actions of some countries, regions and cities are to be commended but do not yet add up to a solution for our planet. How do we turn this around?

Profound changes are called for to reduce GHG emissions and, as this book points out, these changes are achievable. Not only can these measures slow and eventually stop dangerous climate change and its impacts, but they can also unlock new sources of economic growth, enabling this transition to be achieved in ways that build a solid foundation for economic prosperity in the twenty-first century, while delivering huge benefits for economies, communities and the natural environment. Carbon-free energy generation can entirely replace coal, oil and (fossil) natural gas over a decade or two, reducing not only GHG emissions, but also the dangerous air pollution that affects millions of people in many countries, including some of the most populous and the most impoverished, thus improving public health.

Huge opportunities for profitable businesses and well-paid jobs beckon in this new environment. Economists show how the transition to carbon-free economies will drive increases in GDP, if properly handled. Reputable economic modelling shows that transition to low-carbon economies will bring more rapid growth than 'business as usual'. Two highly regarded studies have been the OECD's *Investing in Climate, Investing in Growth* (2017), and the Global Commission on Climate and the Economy's *Unlocking the Inclusive Growth Story of the 21st Century: Accelerating Climate Action in Urgent Times* (2018). These have followed the landmark Stern Review, *The Economics of Climate Change* (2006). But 'proper handling' means that the transition must be managed in a way that ensures those employed in carbon-intensive industries are able to secure new and valued jobs and lead satisfactory lifestyles in a decarbonised economy.

The outlook for the alternative future – that is, taking no or insufficient action to drive the transition to a carbon-free economy – is grim. Higher temperatures that can cause heatwaves, droughts and wildfires, as well as other extreme weather events, will disrupt economies, reduce productivity and displace populations. Sea-level rise will threaten coastal cities, industries, infrastructure and communities. The costs of climate impacts will rise, as will the cost of adapting to those impacts. In this uncertain future, humanity will have passed over the great benefits of a carbon-free economy and will face new adversities that may include lower real incomes, fewer options for development and less comfortable lifestyles.

More seriously, the displacement of populations may precipitate conflict as people in the most affected areas struggle to survive amid rising temperatures and sea levels, more severe droughts and water shortages, deepening poverty and food insecurity. Such displacement may reach tipping points beyond which established laws and social arrangements catastrophically fail, precipitating millions of desperate refugees to look for new homes.

The challenges are daunting but are surmountable with existing technologies and systems. The priority must be a fair transformation that simultaneously responds to the urgency of the challenge, and the need to be sensitive to populations' aspirations for the future. The fundamental change that is required is not without precedent, but the urgency with which it must be achieved is, and this will make the transformation significantly more difficult.

The Aim of This Book

This book originates in a global environment with a cacophony of voices offering differing solutions and conflicting priorities for addressing climate change. It is intended to clarify the policy reform and complementary actions needed for the transformation, and be a practical 'how-to' guide for decision makers to achieve the commitments made in the Paris Climate Change Agreement:[1] to keep global warming as close as possible to 1.5 °C and well under 2 °C to avoid the worst of dangerous climate change.

Throughout the book, the authors have highlighted climate strategies and responses that can achieve both mitigation and adaptation outcomes, an approach that provides many potential co-benefits, including enhanced prosperity and well-being. The book is written to be accessible to non-experts, to provide a source of wide-ranging information for the community at large about how the world can manage the transformation to a prosperous, resilient, carbon-free economy.

How the Book Is Organised

The book incorporates studies of each economic sector, indicating sectoral challenges and the available strategies for both climate mitigation and adaptation. The political and

---

[1] *Paris Agreement Under the United Nations Framework Convention on Climate Change*, opened for signature 16 February 2016. Available at: https://unfccc.int/process-and-meetings/the-paris-agreement/the-paris-agreement.

institutional frameworks within which action on climate change must proceed are described, and the barriers to change are discussed.

Decarbonising the energy and electricity sectors is essential for the rapid mitigation of climate change, and enables emissions reductions in the buildings, transport, industry and agricultural sectors through electrification. Chapter 1 discusses the complexity and challenges inherent in developing climate policy through the lens of the energy sector. It provides a comprehensive analysis of the technological and institutional complexities of decarbonising the energy sector, and explores strategic options based on different political and administrative models. More public-sector-driven models such as those in France and Germany are contrasted with the market-based models of the USA, the UK and Australia. It also discusses the decarbonisation of the electricity sector as a 'litmus test' for decarbonisation of economies. Barriers to changing to new dominant technologies are also explored.

Chapters 2–8 then lay out the main technologies available for the transition to low-carbon energy: wind, solar photovoltaics (PV), solar thermal, nuclear and hydropower, as well as energy storage and the role of hydrogen.

The next section reviews two decarbonisation studies, projecting the effects of deployment of the new technologies on climate change and economies. Chapter 9 compares 16 countries but focuses particularly on Australia, noting that Australia can maintain 2.4% annual GDP growth while transitioning to a low-carbon economy without problematic structural change. In Chapter 10, a country study of an emerging giant economy – Indonesia – shows how developing countries are addressing climate change, including a cameo example of how renewable energy is being encouraged in India.

The next group of chapters (Chapters 11–16) take the reader from the global and national levels to the sectoral and city scales. They explore how cities, and precincts within them, can mitigate GHG emissions while adapting to changes in the climate that cannot be avoided. Chapter 13 considers issues around water: its changing patterns of availability, its relationship to food production and the need for more efficient use and conservation, especially in regions experiencing reduced rainfall as a consequence of climate change. This is followed by a discussion of transport and industry – key sectors with very significant GHG emissions. Opportunities for electrification based on renewable energy are presented: for example, the replacement of gas heating with heat pumps that are more efficient, cost-effective and emissions-free. In transport, electrification can extend to sea and air transport as electric and hybrid models are developed.

The next theme is land use, forests and agriculture, covered in Chapters 17–19. The natural carbon cycle is explained and pressure on land resources for human exploitation, deforestation, the need to conserve carbon-rich native forests, and mitigation strategies are explored. Agriculture, with its related value chains, is a major source of emissions, but emissions reduction is challenging, and population growth with associated increased demand for food implies even greater challenges in the future. Nevertheless, there are possible win–wins that can be achieved.

Transitioning the mining, metals, oil, gas and petrochemicals industries is then reviewed in Chapter 20, noting that mining has a major role in producing the raw materials necessary for the transformation, while reducing its own emissions. This can be achieved by electrification of mines, for example by replacing diesel trucks with electrically powered conveyor systems. Oil and gas will have diminishing roles in economies but can reduce emissions from their operations, including through controlling fugitive emissions and flaring.

The last section of the book revisits themes from the first chapter and looks at the issues posed by current institutional arrangements and trading systems. Chapter 21 discusses the role that international trade will play in the global transition to low-carbon economies. In particular, this chapter examines the impact of investor–state dispute settlement clauses in international agreements, which enable corporations to sue states over perceived discriminatory action, which can include environmental regulation. Conflicts arising from global trade arrangements that have focused on open markets and 'competitive economies' – sometimes to the detriment of national action on important environmental matters including climate change – are also analysed in the trade chapter, which then discusses the opportunities presented by the Paris Climate Change Agreement, and offers some recommendations to successfully support the low-carbon agenda.

The need for good governance and the elimination of corruption is discussed in Chapter 22, while Chapter 23 looks at how the transformation can be financed, noting that financial institutions are not only increasingly taking into account climate risk in their investment decisions and that climate-friendly investments generally show superior returns, but also that financing the operation and expansion of fossil fuel industries continues to take a large share of available funds. New and encouraging developments in the finance sector that support environmental, social and governance-focused enterprises are highlighted.

The final chapter, Chapter 24, discusses bottom-up social movements that can help drive the transformation to carbon-free and resilient economies.

Reasons for Optimism

According to many reports, global negotiations at the 25th Conference of the Parties (COP 25) in Madrid ended with disappointment on many fronts, highlighting the growing gap between the stronger action needed to avert climate disaster and the sluggish responses of most major economies. In addition, the International Energy Agency (IEA) *World Energy Outlook 2019* reports (IEA 2019):

Current country commitments, the Nationally Determined Contributions (NDCs) made under the Paris Agreement and domestic energy policy plans fail to bring about the rapid, far-reaching changes required to avert dangerous and irreversible changes in the global climate system. These are assessed in our Stated Policies Scenario and lead to total global energy-related CO_2 emissions growing steadily from today's levels before plateauing around 36 Gt after the mid-2040s. This trajectory is consistent with limiting the temperature increase to below 2.7 °C above pre-industrial averages with a 50% probability(or below 3.2 °C with a 66% probability).

So, are there reasons for optimism? The answer is yes. The widespread adoption of renewable energy has far outstripped earlier predictions, and the steep fall in costs of solar PV technology, wind-powered energy and batteries, as well as energy- and fuel-efficient technologies, such as LED (light-emitting diode) lighting, heat pumps, electric vehicles, e-buses and e-bikes, have greatly surprised and are continuing. There have also been significant advances in the development of new business and financing models, as outlined in Chapter 23, which can be complemented by proven and effective climate change policies at the national, subnational and city levels (as covered in Chapters 1 and 11 to 23). And, as discussed above, implementing effective climate change policy reform can enable significant economic and employment growth – a fact that has become more evident in recent years and which lies at the heart of many of the economic stimulus measures being developed to support post-COVID-19 economic recovery.

Notwithstanding some failings of COP 25, there is a growing consensus among communities that climate change is real and action is needed urgently. This is reflected in growing concern about policy stasis at the global level and in some countries. The community consensus needs to translate into political will and effective government policies that can overcome barriers to both climate mitigation and adaptation.

Communities also have to be convinced that life in a transformed economy will be good; even better than the present. Studies indicate that deep decarbonisation processes are consistent with continuing economic growth. Change is often hard and frequently resisted until the new paradigm becomes 'normal'. For example, the adoption of electric vehicles has been particularly slow, but many predict that there will soon be a 'tipping point' and rapid take-up will follow as the cost savings in operation are realised.

Ensuring that the transition is just and fair is critical, and structural adjustment packages will be needed to retrain and support affected workers. Some governments, notably Germany, are already developing transition plans for vulnerable communities where former industries such as coal mining and coal-fired generation are shutting down. The move away from coal by major miners, BHP and Rio Tinto, and direction of investment away from fossil fuels by some financial institutions, show growing momentum behind decarbonisation, but on close inspection those shifts are, in the main, modest, conditional and not yet mainstream.

The broad consensus among populations and businesses about the need to act urgently on climate change is growing, as the benefits of carbon-free energy generation and decarbonising the wider economy become ever more obvious, along with the need for immediate and strategic climate adaptation. Yet as consensus emerges around the 'why' of decarbonisation and climate adaptation, decision makers and political leaders are still grappling with the 'how' – particularly, how best to achieve climate change mitigation and adaptation goals while simultaneously growing jobs and economies.

This book responds to that challenge.

References

Global Commission on the Economy and Climate (2018). *Unlocking the Inclusive Growth Story of the 21st Century: Accelerating Climate Action in Urgent Times*. Washington, DC: The Global Commission on the Economy and Climate. Available at: http:// newclimateeconomy.report/2018/.

IEA (International Energy Agency) (2019). *World Energy Outlook 2019*. Paris: International Energy Agency.

OECD (Organisation for Economic Co-operation and Development) (2017). *Investing in Climate, Investing in Growth*. Paris: OECD. Available at: www.oecd.org/economy/taking-action-on-climate-change-will-boost-economic-growth.htm.

Stern, N. (2006). *The Stern Review: The Economics of Climate Change*. Cambridge: Cambridge University Press.

1

Policy Frameworks and Institutions for Decarbonisation: The Energy Sector as 'Litmus Test'

LYNETTE MOLYNEAUX AND KEITH SUE

Executive Summary

As a global community, we have known about the threats associated with the burning of fossil fuels for several decades, and yet greenhouse gas emissions have continued to increase. This is as a result of the perceived economic imperative of cheap energy from fossil fuels, complex technological and institutional structures that govern investment and supply of energy, the power of fossil fuel vested interests to direct the political discourse towards inaction, and a global decarbonisation framework which is effectively uncoupled from global trade rules. Building a successful policy framework requires a nod to each of these complexities. In this chapter we look to the energy sector to explain how these complexities have affected the success of decarbonisation plans and what can be done to mitigate against their hindrance of success.

We discuss how context matters and that there is no single framework that works. Each country or region needs to account for the specifics of their energy system, institutional structures, public expectations, level of development and political context in policy formulation. The chapter therefore gives consideration to the tools that can be included within the policy framework that will suit the expectations of a diversity of public views on the need to decarbonise. We conclude with a discussion on the challenges that need to be confronted to achieve intended outcomes.

1.1 Why Climate and Energy Policy Is on the Critical Path to Decarbonisation

The combustion of fossil fuels produces 73% of annual global greenhouse gas (GHG) emissions. The largest sectoral consumer of fossil fuel is electricity and heat generation, which contributes 30% to GHG emissions, followed by industrial, construction and mining processes, which contribute 32% to annual totals, and transport, which contributes 16% (CAIT 2019). Agriculture contributes 12%, land use, land-use change and forestry contribute 7% and waste, 3%. Climate policy therefore has to accommodate varying scope and scale to reach all sectoral processes which result in GHG emissions. It is an economy-wide challenge, with almost no sector or industry immune from the need to reduce emissions.

Inevitably, then, designing and implementing comprehensive, effective climate policy involves hard political choices: prioritising between options which could result in

7

suboptimal outcomes, and attempting to resolve inequitable distributional consequences. The stakes are high and mobilisation of power by fossil fuel regimes is large, such that the temptation for politicians to defer the big decisions is strong (Meadowcroft 2011). Politicians may reference uneven global decarbonisation action, which results in decreased international competitiveness and general concerns about increasing costs for voters as justification for limiting climate and energy policy. The inertia from national governments despite commitments made in the Paris Agreement[1] is testament to the strength of these arguments.

This chapter discusses the complexity and challenges inherent in developing climate policy through the lens of the energy sector. Emissions from the generation of electricity form the major contribution to most countries' carbon budgets because coal, a carbon-intensive fuel, geographically dispersed around the world, has been the preferred, and, until recently, most affordable, fuel for electricity generation. Electricity generation is the largest contributor to GHG emissions, and the Climate Change Authority in the UK concluded that 'any path to an 80% reduction by 2050 requires that electricity generation is almost entirely decarbonised by 2030' (UK CCC 2008: 173). Similarly, the European Commission's *Roadmap for Moving to a Competitive Low Carbon Economy in 2050* finds that '[e]lectricity will play a central role in the low carbon economy' (European Commission 2011: 6). Decarbonisation plans therefore commonly focus on electricity, with additional policy measures to address emissions in the other sectors, including regulation to address energy efficiency, taxes and land clearing. As is clear from our analysis, many of the bottlenecks experienced in the energy sector are not unique to that sector, and indeed there are many similarities with decarbonisation efforts in the transport, agriculture and built environment sectors – such is the case with profound structural change.

The chapter begins by focusing on the complexities associated with electricity supply and other industry and transport sectors, then moves on to the tools required for a coherent policy framework and finishes with a detailed discussion on the challenges to confront in decarbonising the energy sector.

1.1.1 Electricity Generation

The sector has complex technical and institutional characteristics, which require powerful policy frameworks to regulate and facilitate change. A discussion of these complexities follows.

1.1.1.1 Technological Complexity

Large power plants are constructed to gain economies of scale and reduce the unit cost of electricity. Capital cost for a 1000-MW power plant ranges from USD 0.7–1.3 billion for a combined-cycle gas turbine, USD 3.0–6.2 billion for a coal-fired power station, to USD 6.9–12.2 billion for a nuclear power plant (Lazard 2019). Greater utilisation of power

---

[1] *Paris Agreement Under the United Nations Framework Convention on Climate Change*, opened for signature 16 February 2016. Available at: https://unfccc.int/process-and-meetings/the-paris-agreement/the-paris-agreement.

plants can be gained by transferring electricity between consumption regions as demand varies using electricity networks. The scale of investment requires detailed planning and stakeholder buy-in, with large risks for investors and financiers who in turn seek secure returns from long periods of operation. The power grid connects every generator to every consumer, supplying power to meet demand, all and every day, making operation and management extremely complex.

Technological innovation in the electricity sector is increasing this complexity. Energy dispatched from variable sources or discharged from stored energy needs to be integrated with existing, less flexible technologies to ensure grid stability. Demand-side innovation has been slow to emerge because of the challenges associated with motivating individual actors to respond dynamically.

1.1.1.2 Institutional Complexity

Governance and Regulation

Where electricity systems have not been liberalised, they are generally operated as vertically integrated utilities, with governance structures dictated by state actors. Utilities provide advice to politicians on operational and strategic requirements, and politicians direct utilities according to political commitments.

From the mid-1990s, liberalisation of the electricity sector has led to the dismantling of vertically integrated utilities, which involved: introducing competition among generators to supply electricity to a centralised electricity market, and among retailers to supply electricity to consumers; and the regulation of network organisations with respect to returns and investment. While many countries and states have pursued liberalisation, implementations stalled after flaws were exposed in California's electricity market in 2000. Changes to market designs resolved the flaws, but the consequence was that national/regional electricity sectors are often characterised by a mix of public and private, vertically integrated utilities and independent power producers and retailers, functioning alongside each other. Governance arrangements have sought to encompass a broad range of responsibilities, including the establishment of legislative frameworks, the definition and implementation of operational standards and rules, and the regulation of industry participants. Electricity sectors are often complicated by a 'mixed-model energy governance structure' where the regions/states are responsible for the provision of electricity but the sector is subject to national/federal electricity governance, oversight and planning (Osofsky and Wiseman 2014; Warren et al. 2016).

Institutions and Path Dependence

In large technical systems such as electricity networks, technological infrastructures and the institutional frameworks which support them are tightly bound and co-evolutionary (Künneke 2008). Dominant design theorists argue that such systems develop along a predictable pathway (Nelson 1995) and culminate in the emergence of a dominant technology class (Anderson and Tushman 1990). Upon establishment as the dominant design, the technology class is reinforced by institutional biases to gain precedence over other, potentially superior, designs (Arthur 1989). An extension of this theory of path dependency by Gregory Unruh (2000) moves beyond the genesis of an industry and the competition to establish any given technology as the dominant design; instead addressing the techno-institutional relationship in a mature system. Unruh specifically focuses on the electricity industry in the context of emission-intensive generation and argues that the development of technological trajectories in conjunction with the institutional environment establishes a positive feedback between institutions and a particular technology class, leading to a 'lock out' of emerging technologies – which he describes as a techno-institutional complex. It has been argued that the propensity for new technologies to be 'locked out' of a system due to ideological inertia is further reinforced by the disparity between the rapid development of technological pathways and the comparatively slow pace of regulatory and institutional change (Williamson 1998; Künneke 2008). It follows that these two aspects form an internal ideological force acting to maintain the status quo in technical systems such as electricity networks.

Due to the inherent bias towards this dominant technology class, it could be argued that technologies external to the current framework, such as renewable forms of electricity generation, energy storage, energy efficiency technologies and demand management, may struggle to integrate. Industries operating in flexible and adaptable environments may embrace such technologies – termed disruptive technologies – and radically change the industrial paradigm through a discontinuous transition to a new dominant design (Christensen 1997). Rapidly integrating large-scale technological disruption into the supply of electricity, in a potentially rigid and unresponsive institutional environment, may lead to detrimental operation at both a technical and market level.

This techno-institutional complexity means that the challenges confronting policy-makers in their endeavour to develop more sustainable electricity systems are manifold. These challenges transcend simple policy formulation, and require a detailed appreciation for competing sociopolitical objectives and institutional inertia (Figure 1.1). Those embarking on electricity market reform for decarbonisation must recognise that institutional innovation is required to support the entry of new technologies and business models, but the risks they may pose to the existing system must be managed. Ideally, the design of the policy framework should maximise technological flexibility, while managing policy objectives.

1.1.1.3 Competing Objectives of Electricity Policy

In considering specific policy frameworks to drive the decarbonisation of the electricity sector, we must first recognise that there are a range of other policy objectives which must be managed and prioritised in a coherent manner. These objectives underpin existing sectoral institutional structures, and the broader energy policy frameworks. While this may appear to be quite obvious, the tensions between competing objectives can be easily overlooked in the heat of policy formulation; that is, until one is confronted with a suboptimal mix of unanticipated outcomes. Any discussion surrounding effective sectoral decarbonisation policies must, then, acknowledge the breadth of policy objectives relevant

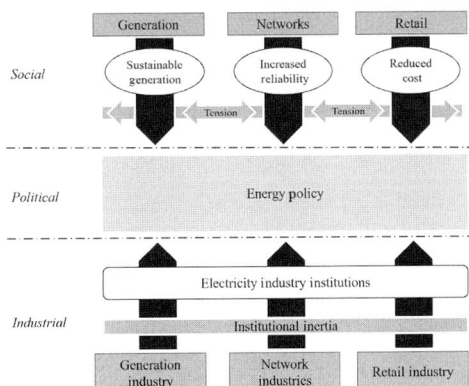

Figure 1.1 Social interactions in electricity markets.
Source: Courtesy of the authors. Note, the sociopolitical objectives (white circles) influence policy at both the federal and jurisdictional level. These objectives can be conflicting, as shown here. An additional tension exists between the inherent institutional inertia and desired social change (Sue et al. 2014).

to electricity provision and market design, and highlight the limitations that conflicting objects can impose.

Context Matters

The stark contrast between electricity provision in the developed and developing world highlights the contextual nature of objectives. Developed nations exhibiting well-developed electricity industries have traditionally been focused on reliability of service and energy security. In contrast, developing nations are clearly more directed towards availability and affordability of electricity supply. The push towards more sustainable electricity provision in the previous two decades interacts with these two dynamics in contrasting ways. For developed countries, the focus has been largely placed on the integration of renewable energy (RE) systems into existing systems. In contrast, for developing countries, availability of electricity supply is affected by the remoteness of communities and, because household demand exists for only a few hours a day, and efficient coal, nuclear and some gas power stations have minimum stable operating levels for all 24 hours per day, industry demand is required in conjunction with household demand, to make electricity technically and economically feasible for households and small commercial consumers.

In many cases, decentralised RE systems can be the primary vehicle for access to electricity in remote locations. A recent International Energy Agency (IEA) report finds that decentralised, renewable energy is the least-cost way to provide power to more than half of the population gaining access by 2030 (IEA 2019a).

Availability and Access

Although considerable policy focus to date has been placed on lifting regions out of energy poverty, access to electricity in many developing regions remains poor (Table 1.1). This is

Table 1.1. *Access to electricity in developing regions*

Region	Population without electricity (millions)	Electrification rate (%)	Urban electrification rate (%)	Rural electrification rate (%)
Developing countries	860	86	95	77
Africa	595	54	79	35
North Africa	*1*	*>99*	*>99*	*>99*
Sub-Saharan Africa	*594*	*45*	*74*	*26*
Developing Asia	230	94	98	91
China	*3*	*>99*	*>99*	*>99*
India	*74*	*95*	*>99*	*92*
Indonesia	*5*	*98*	*>99*	*96*
Other South East Asia	*46*	*90*	*97*	*83*
Other developing Asia	*103*	*79*	*89*	*73*
Central/South America	16	97	99	88
Middle East	19	93	98	78
World	**862**	**89**	**96**	**79**

Note: Access shown as absolute figures and percentage levels, as assessed by the IEA in 2018.
Source: Adapted from IEA (2019a).

most pronounced in sub-Saharan Africa, where 74% of the rural population has no access to electricity. Achieving 100% electrification requires investment of USD 40 billion a year (IEA 2019a).

The deployment of low-carbon technologies has been seen as a key pillar to achieve these goals. In attempting to balance supply availability and environmental impact through RE systems, we are confronted by issues of affordability and access. Thus, requirements must be tempered by the financial resources of consumers to pay for availability and service levels.

Security of Supply

Security objectives focus on constraints to meet the required demand. Typically, these are framed across two dimensions: resource availability and the capacity of the electricity system to couple supply and demand. Resource availability has traditionally been the cornerstone issue for national supply security concerns. For example, until 2010, the current mix of fuel sources for the electricity industry in the USA grew largely out of the oil crises throughout the 1970s, where investment in coal power generation was required in preference to other fuel sources (US DoE 1978). A more contemporary example is that of

Germany which has, in recent years, attempted to become more energy self-sufficient to decrease its dependence on imported gas (Joas et al. 2016). Internal to the electricity network security of supply is the effective coupling of supply and demand. Under the current paradigm of liberalised electricity markets, supply is largely driven by forecast demand over the short and long term, and by long-term predictions for infrastructure investment. Forecasting errors can result in affordability concerns. This can occur even in well-developed countries where planning processes are normally robust, as exemplified by the long-term forecasting difficulties in the Australian National Electricity Market, where the Australian Energy Market Operator has consistently overestimated demand since 2008 (Simshauser 2019).

Resilience

Definitions of resilience vary but, in the main, an electricity system is resilient when it is able to adjust to significant change and still maintain function. Engineers seek to design systems that are fail-safe, but the cost associated does not meet affordability requirements. Electricity systems are therefore designed to tolerate predictable faults, outages and weather events. This form of resilience can be thought of as *engineering resilience*, as it is the design of the system to adapt to predictable changes (Molyneaux et al. 2016). Energy *system resilience* requires the ability to adapt to more than technical faults and constraints. Unpredictable change may come in the form of higher global fuel prices, environmental policies or technological innovation. Examples of change pertinent to decarbonisation include carbon constraints or prices, a large take-up of air conditioning or electric vehicles and consumer investment in solar photovoltaic (PV) panels. Factors that are effective in improving energy system resilience include spare capacity in fuel sources, spare capacity in generation capacity and diversity in fuel sources for generation. Diversity in fuel sources is not simply a matter of pursuing strategy to encourage fuel switching, because fossil fuel prices are often correlated. For instance, in the USA in the 1970s, oil price escalation led to natural gas and coal price escalation, all of which led to electricity price escalation, which had severe consequences for manufacturing. Technologies that use solar, wind, geothermal and hydro resources are all independent of each other and fossil fuel prices, making a combination of these fuel sources highly desirable for energy system resilience. Systems that have more diverse fuel sources, reliant on unpredictable weather conditions, require interconnection and energy storage to secure supply from areas with different weather and demand patterns. Strategies of diversity and spare capacity require rules to direct planning, operation and actions by electricity sector participants. A lack of guidance by rule-makers will result in electricity systems vulnerable to a high dependence on any one fuel source because investors will seek the current cheapest technology. Fuel source diversity acts as insurance against future fuel shortages or escalating fuel prices to enhance system resilience and tariff certainty.

Balancing Objectives

Priorities shift as countries develop. In developing countries, access and affordability are prioritised above all objectives. As economies mature and industrial and commercial demand grows to underwrite the cost of electricity generation, priority shifts to security of supply. Industrialised countries' prioritisation of security of supply is related to the business imperative for secure, affordable supply to facilitate competitive advantage. In the development of policy frameworks for energy and climate, industry and business demands for affordable, secure supply have taken precedence over requirements for emissions reduction (Meadowcroft 2011). The discourse has not included the potential for future threats from carbon constraints or climate change. The challenge for policy frameworks is to prioritise the objectives of emissions reductions and resilience in the short term to gain energy security and affordability in the long term.

1.1.2 Manufacturing, Construction and Mining Processes

The manufacturing, construction and mining sectors release emissions through burning fuel for heating and cooling, through chemical reactions and from leaks from industrial processes or equipment. These sectors contribute 32% of GHGs and are discussed in more detail in Chapters 16, 12 and 20, respectively. Decarbonisation of electricity supply, either by on-site generation or access to grid power from renewable resources, can reduce manufacturing emissions. In the longer term, the replacement of gas for industrial processes may be possible, using concentrated solar thermal in some locations or hydrogen and its derivatives. The sector chapters also emphasise the potential for improved energy efficiency that can reduce energy consumption and costs. The possible emergence of direct reduction of metal ores with renewable energy utilising 'green' hydrogen (produced by water splitting) could have highly favourable prospects for some economies while reducing emissions. A resurgence in mining based on the raw materials needed for modern technologies such as lithium is predicted. Such mining would desirably be as carbon-free as possible, notably through renewals-based electrification. Material switching, such as to more wood in construction and to new technologies to replace Portland cement, also hold promise for emissions reduction in construction.

For detail on the technology options for these sectors, readers should turn to the appropriate chapters. Climate policy for these sectors is subject to a different set of complexities to those encountered in the electricity supply industry.

1.1.2.1 Complexity of Global Competitiveness

The post-war rise of Japan as a global manufacturer initiated intense international competition to supply product to the world's consumers. The growth of new manufacturing sectors required low production costs. In this tightly contested space, national industries that are large employers seek preferential treatment to facilitate global competitiveness through reduced taxation, removal of international tariff barriers and forgiveness from climate policy costs. An unintended consequence of the Kyoto Protocol's focus on equity for developing nations is carbon subsidies for internationally traded goods. For a detailed discussion on the challenges of integrating climate change policy with international trade, refer to Chapter 21.

1.1.2.2 Complexity in Assigning Responsibility for International Emissions

As China has developed to be the world's pre-eminent manufacturer, it has become the world's largest GHG emitter. Much of China's production (along with other developing nations in Asia) is exported without transfer of responsibility to the consuming nation for the GHG emissions embodied in the products. Marine and air transports consume fossil fuels that emit GHGs in international waters and space, but are not accounted for in national inventories or targets. These result in inaccurate accounting for GHG emissions and increase the complexity of policy to reduce emissions at a national level.

The original framework envisaged in the Kyoto Protocol included the concept of internationally linked emissions trading schemes incorporating flexible mechanisms like the Clean Development Mechanism (CDM). These schemes were aimed at encouraging cross-border emissions trading and developed countries' investment in developing countries' emissions reduction projects in order to meet national emissions reduction targets. The complexity of establishing global mechanisms of this nature has inhibited their potential.

1.1.2.3 Mining and the Resource Curse

The resource curse is frequently associated with the unexpected economic disadvantage that results from the development of lucrative extractive industries. It is primarily expected in developing countries, which lack the institutional, legal and political infrastructure to protect against political capture by extractive industries to the detriment of other sectors of the economy (Sachs and Warner 1997; Barbier 2003). Developed countries may, however, also fall prey to another aspect of the resource curse, known as 'Dutch disease'. In this form of the malaise, the extractive industries drive rising exchange rates that render manufacturing or agricultural products less competitive internationally. Examples of Dutch disease include the impact of natural gas exports on the Netherlands in the 1960s, North Sea oil on the United Kingdom in the 1980s and coal and natural gas exports on Australia from the mid-2000s (Goodman 2008).

The prospect of reduced coal, natural gas or oil extraction to mitigate against climate change is particularly challenging for national and subnational states already experiencing the consequences of the resource curse or Dutch disease. These states may be particularly constrained from restructuring due to heavy reliance on the proceeds and employment from fossil fuel mining (Goodman 2008; Jakob and Steckel 2014; Baer 2016).

1.1.3 Transport

Motor vehicles, aircraft and marine vessels release emissions through burning fuel. In transport as in manufacturing and processing, electrification and other fuel switching – to hydrogen fuel cell or synthetic fuels – are key strategies for emissions reduction for passenger cars and road freight. Challenges are posed in the aviation and marine transport sectors but electric and hybrid aeroplanes are already in development. Chapter 15 emphasises the desirability of lifestyle adjustments with the related changes to cityscapes to encourage bicycling and walking. The advantages of compact cities are emphasised in Chapter 11.

Climate policy to address emissions in transport tends to rely heavily on innovations; for example, electric vehicles or hydrogen fuel cell vehicles. There are however multiple opportunities for emissions reduction from investing in public transport and encouraging greater use of public transport. Also, shared vehicle ownership presents opportunities for significant emissions reductions. Further discussion on this can be found in the chapters on transport and cities.

1.2 Developing Coherent Policy Frameworks for Decarbonisation

At the highest level, a robust array of non-political meta-governance and governance structures is required to facilitate the transformation of the energy sector to new low-carbon technologies. This integration of new and existing institutional structures to support decarbonisation is critical to ensure effective implementation and acceptance of policy pathways. To facilitate the ability to adapt, rules should avoid being prescriptive and planning needs to place the highest priority on decarbonisation options.

1.2.1 Creating Incentives for Investment

1.2.1.1 Pricing Carbon

Economists have long argued that market-related polices will be the most efficient mechanism for influencing investment decisions. Currently, low-carbon plant and industrial processes are more expensive than existing plant and processes reliant on fossil fuels because there are no costs associated with carbon emissions. Pricing of carbon emissions creates behavioural change by consumers, which leads to reduced consumption of plant and processes reliant on fossil fuels.

Pricing can take the form of a carbon tax (CT) or the trading of permits to emit carbon through an Emissions Trading System (ETS). With a CT, the state or regulator decides on the price of emissions and emitters pay the state for emissions generated. With an ETS, the state or regulator decides on the reductions in emissions over a specified time period, and then auctions the available stock of emissions for emitters to purchase to meet their liability. Penalties result if emitters emit more than the quantity of permits purchased.

There are multiple implementations of both ETSs and CTs around the world, but in the main ETSs have struggled to achieve significant emissions reductions because of difficulties in predicting appropriate emissions targets without compromising national economies. The surplus supply of permits leads to permit prices too low to induce investment. Carbon taxes have also not been very successful, due to political concerns and public antipathy to taxation and increasing tariffs.

Australia's climate policy provides an illustration of the difficulties of implementing carbon prices. There have been two attempts at pricing carbon. In 2009, the Carbon Pollution Reduction Scheme (CPRS), effectively an ETS, was introduced and passed in the House of Representatives but failed multiple times in the Senate. As a result of the divisive debate around climate policy, Malcolm Turnbull, the leader of the Opposition Liberal–National Coalition, lost the leadership to Tony Abbott. Because of falling popularity as a result of multiple issues, but including climate policy, Kevin Rudd, the Prime Minister, lost the leadership of the Australian Labor Party to Julia Gillard. In 2011, Julia Gillard's government passed the *Securing a clean energy future: Implementing the Australian Government's climate change plan*, which included a CT for top polluters from 1 July 2012 until 30 June 2015, transforming to an ETS, with 50% of permits purchased from domestic sources, and a price ceiling and floor for the first 3 years. Tony Abbott, reflecting the view of some sectors, called carbon pricing 'a so-called market in the non-delivery of an invisible substance to no-one' (Metherell and Hawley 2013) and claimed: 'The thing about the carbon tax is that it will clean out people's wallets and it will wipe out jobs big time.' (Parliament of Australia 2011: 10990). At the next election, the Liberal–National Coalition won government. Tony Abbott had campaigned using the slogan 'Axe the Tax', and repealed the Clean Energy Legislation in 2014.

> **Key areas to consider when implementing pricing mechanisms** are: who sets long- and short-term target reduction or carbon price; what sectors to include; how to educate the public and how to reduce political interference with respect to setting the cap or the tax.

1.2.1.2 Feed-In Tariffs

Many countries have pursued feed-in tariffs (FITs) as a means to incentivise investment in RE. The most successful country in this regard has been Germany. Certainty for investors was gained through prioritised access to the electricity grid for RE generators and fixed tariffs for RE owners. Investment in RE from private citizens, farmers and cooperative programmes has contributed to broad public support for RE projects. After investment in 38-GW (gigawatt) wind and 40.5-GW PV panels between 2002 and 2016, there is debate about the cost of that incentive scheme to German consumers. For now, it appears that the majority of consumers are not unhappy with significant investment in energy from wind and sunshine (German AEE 2019), and the manufacturing capacity to sustain that investment. Germany continues to encourage investment in RE, although FITs are now only available for small installations and are adjusted in line with achievement of targets to avoid large overshoots in investment like those experienced in 2010–11. For larger installations, an auction mechanism for tariffs dictates investment decisions. Feed-in tariffs have been successful at incentivising investment in RE although there are questions around the cost associated with this policy mechanism.

> **Key areas to consider when implementing FITs** are: who sets long- and short-term FITs; how to ensure flexibility in FITs to reflect achievement of targets; how to ensure FITs don't produce regressive outcomes that drive inequity among households.

1.2.1.3 Regulating Renewable Energy Targets

Some countries have avoided the political challenges of pricing carbon, and the costs associated with FITs, and have regulated to ensure that desired renewable energy targets (RETs) are met. These policies have been popular in the USA and Australia because although they are not perceived to be economically efficient, they don't require significant government intervention, and are in line the public's preferences for RE.

Responsibilities for the institutional framework include: the establishment of rules associated with eligibility of sources to qualify as generators of electricity from RE; the establishment of rules associated with the surrender of certificates by liable entities to meet the RE target; the issue of tradable certificates to RE generators for the electricity produced; and the management of the surrendered tradable certificates against the liability of whole-sale purchasers of electricity.

While Australia does have a federal RET scheme, it has been subjected to a fractious political debate and a review that stifled investment for more than a year, and culminated in a reduced target in 2015. Responding to public preferences for RE, many state governments have announced RETs independent of the federal scheme.

Key areas to consider when implementing RETs include: who sets long- and short-term targets; how to reduce political interference in the setting of targets and how to encourage regionalisation of targets for geographical spread and security of supply.

1.2.1.4 Support for Financing

An overarching challenge in the deployment of low-carbon plants and processes is in accessing appropriate finance. Low-carbon plants are typically characterised by large, upfront capital costs and relatively low operational expenses when compared with more traditional supply technologies. Due to a lack of familiarity with the characteristics of low-carbon investment, perceived and actual finance risks are high. In recognition of this information gap, development banks are established to invest in low-carbon projects deemed too risky for traditional financing institutions. Examples include the Green Investment Group (GIG) in the UK and the Clean Energy Finance Corporation (CEFC) in Australia. Funds sourced from these institutions help to lower the cost of capital for new energy-related technology and improve incentives for investment in decarbonisation.

Originating as state-based institutions subject to revision in ownership, scope and focus, these low-carbon development banks can be hampered by the ideological and technological preferences of the incumbent political parties. Australia's CEFC has periodically faced pressure by politicians to reassign its mandate and enable investment in coal plants (Chan 2017; Coorey 2019). The UK's GIG originated as the Green Investment Bank (GIB) in 2012 but was sold to the Australia-based Macquarie Group in 2017, which may have altered its appetite for risk and investment in decarbonising technologies (UK GIG 2019: 3).

> **Key areas to consider when establishing financing support** include: coordination with institutions tasked with decarbonisation responsibilities to develop products suitable to meet targets; and the implementation of strategies to reduce political interference in development bank mandates.

1.2.1.5 Facilitating Financing from Novel Sources

Significant support for investment in low-carbon technologies in the community and business sectors in the USA, Europe and Australia has harnessed funds from a broad spectrum of small to large investors. Indeed, community financing in Germany and Denmark has done much of the heavy lifting of investment in RE in those countries. Where individuals within a community decide to invest in RE as a cooperative venture, investment may be made by local private investors on a for- or not-for-profit basis. Returns for investment can be delivered to investors through: credits to investors' electricity bills; proportionate income from sales of electricity; proportionate income from sales of RE certificates if the country where the investment is made has some form of RE target; or a combination of the above. In the USA, where there is the potential for the application of investment tax credits (ITCs) to earnings, the business structure of the community entity can improve returns by reducing income tax liability. Consequently, there are numerous combinations of investor structures and business models in the USA to maximise returns. Institutional frameworks for community RE investment need to ensure that electricity rules allow: access to the grid and provide rules for the flow of returns to the investors from retailers or generators; that investment rules accommodate groups of small investors who seek to fund RE; and that, where applicable, tax incentives are transparent.

Crowdfunding, defined as 'the collective effort by people who network and pool their money together, usually via the internet, in order to invest in and support efforts initiated by other people or organizations' (Ordanini et al. 2011), is a novel business model which has the potential to tap into small investors with interests in RE and could result in disruption in both the energy and financial sectors. There has been some formalisation of the institutional structure, along the lines of the Jumpstart Our Business Start-ups (JOBS) Act in the USA[2] to legalise certain forms of equity crowdfunding. Australia has also legislated to facilitate crowdfunding (Parliament of Australia 2016) while seeking to protect small, immature investors.

Corporate power purchase agreements (PPAs) are a further novel approach to underwriting investment in RE. Where large commercial energy users seek to hedge against increasing power prices, long-term (anywhere from 10 to 25 years) PPAs are signed. A recent focus has been for large corporations to sign PPAs for RE to underwrite project investment.

---

[2] *Jumpstart Our Business Startups Act*, Pub L No 112–106, 126 Stat 306 (2012). Available at: www.govtrack.us/congress/bills/112/hr3606/text.

1.2.1.6 Regulating Options for Renewal of Ageing Electricity Generation Fleets

The investment in coal-fired power from the mid-1970s to 1980s has resulted in coal generator fleets now ready for replacement in the USA, Europe and Australia. The US Clean Power Plan (CPP), announced in 2015, set emissions benchmarks for power stations too low for the operation of coal-fired power stations (US EPA 2015). The CPP never took effect (Scobie 2016) but the low price of natural gas and RE, and the intention of the CPP, resulted in reduced coal plant capacity as retiring plants were replaced by lower emitting plants as shown in Figure 1.2.

A 15% reduction in carbon dioxide (CO_2) emissions from electricity generation was achieved from 2014 to 2017 through this fleet renewal process, as seen in Figure 1.3. The legal challenge to and repeal of the CPP, and instatement of the Affordable Clean Energy (ACE) rule on 19 June 2019 (US EPA 2019), which relaxed limits on GHGs from power plants, halted emissions reductions.

China too has pursued efficiency gains through fleet renewal. In the 'large substitute for small' programme, 76.8 GW of small inefficient coal plants were decommissioned and replaced by large efficient plants (IEA 2014). Nevertheless, while efficiency programmes have slowed China's growth in carbon emissions, it has not facilitated any absolute decline in them. With a relatively young coal power fleet (the average age of the coal fleet is 11.1

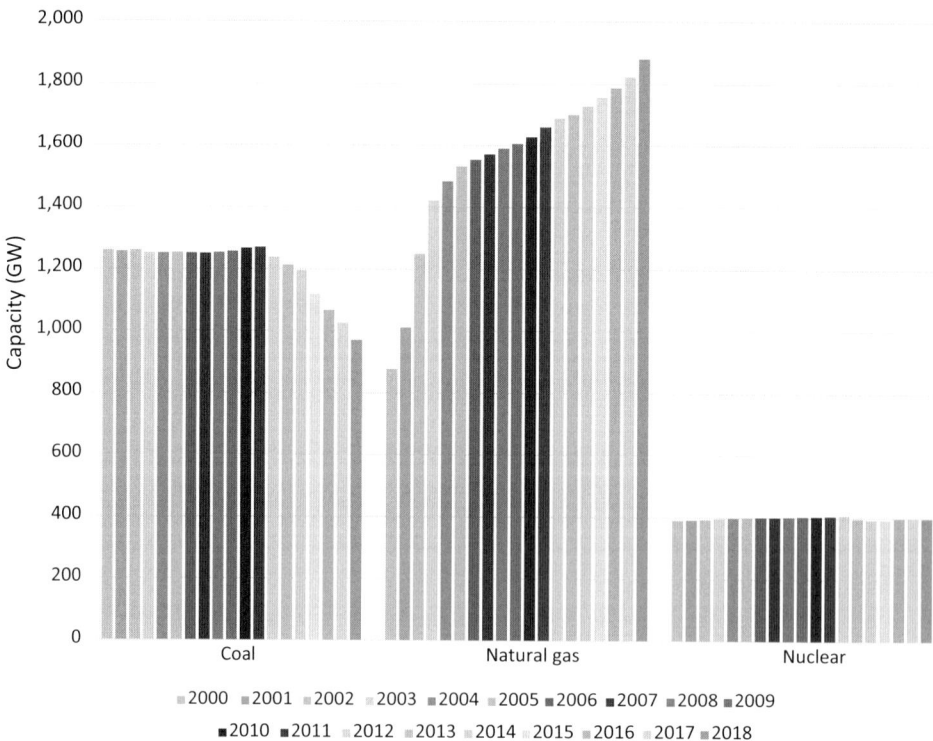

Figure 1.2 US generator capacity by fossil fuel source.

Source: US EIA (2019a). For a colour version of this figure, please see the colour plate section.

Carbon dioxide (CO_2) from heat and power

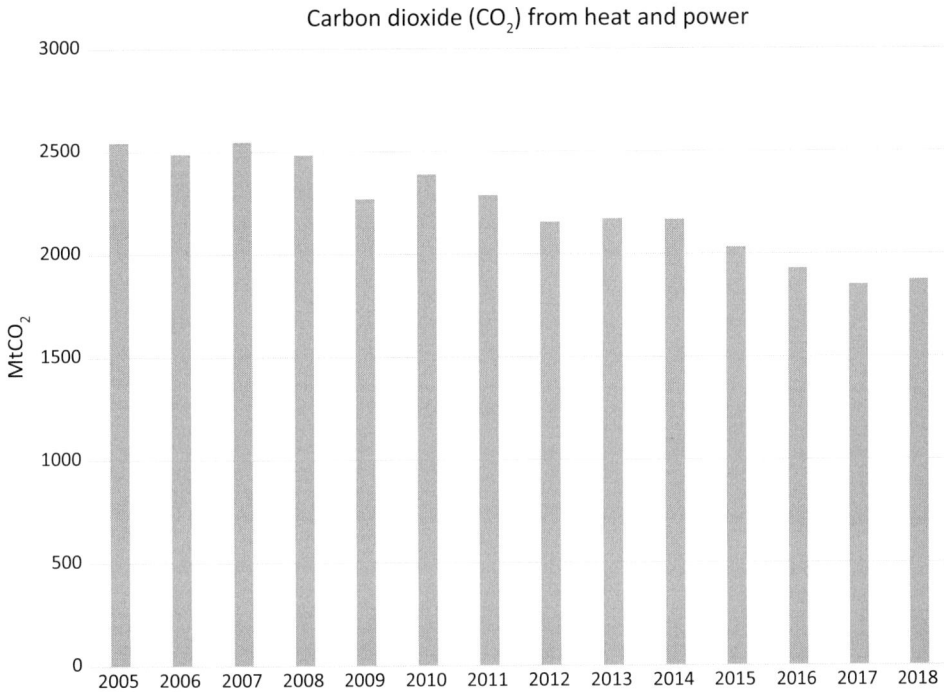

Figure 1.3 US CO_2 emissions from power generation.
Source: US EIA (2019b).

years (Tong et al. 2019)), there is now less opportunity to reap efficiency gains and emissions reductions from plant renewals.

Efficiency gains and fuel switching are positive outcomes but decarbonisation requires a faster transition than a simple switch between fossil fuels. Figure 1.4 shows the extraordinary challenge required of China if the world is to limit CO_2 emissions and contain the consequences of climate change as modelled by the IEA.

Key areas to consider when regulating fleet renewal include: coordination with institutions tasked with decarbonisation responsibilities to develop suitable targets associated with fleet renewal.

1.2.1.7 Developing New Industries That Advance Decarbonisation

While the price of low-carbon technologies may have reduced dramatically over the last decade, there is still a view that market forces alone will not be able to drive the implementation of these technologies and industries at the rate required for decarbonisation. For this reason there is still a significant role for the state in supporting low-carbon industries to facilitate the growth of low-carbon economies (Moe 2010).

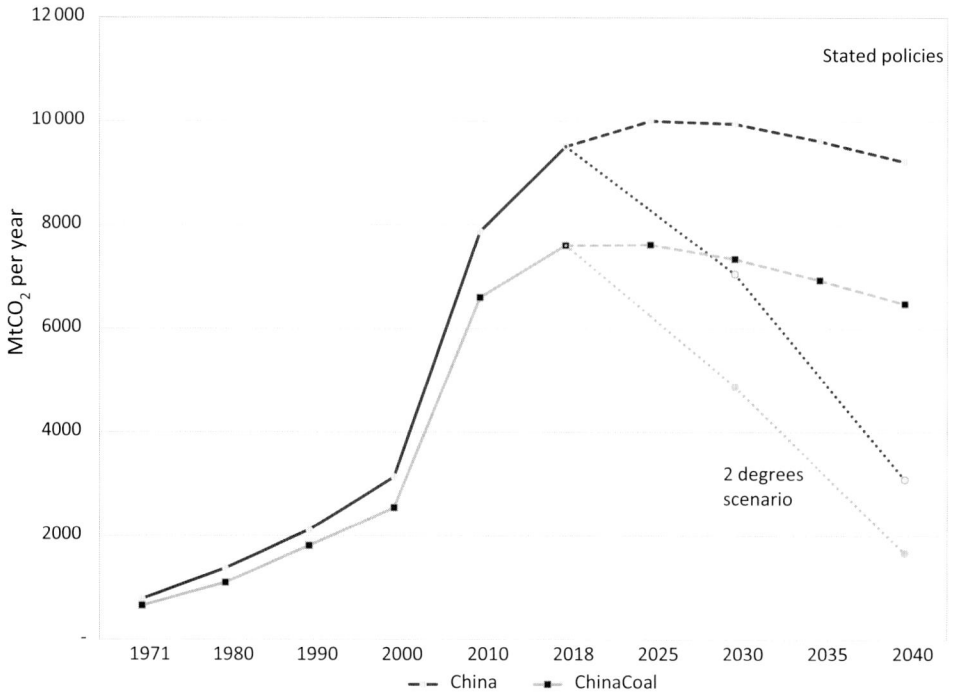

Figure 1.4 China CO_2 emissions from energy combustion: actuals 1970–2018 and projections 2025–2040. Note, the dashed lines show projections of emissions based on current stated policies. Dotted lines show projections of emissions based on policies required to limit global warming to below 2 °C.
Sources: IEA (2019b, 2019c, 2019d).

France, Germany and Denmark provide excellent examples of the development benefits of advancing investment in decarbonisation technology. In France's case, a nuclear power industry grew out of a regulated response to energy security in the 1970s, which created employment and in turn facilitated exports through the application of knowledge gained in the process of nuclear deployment to international projects. In Denmark's case, a wind power industry grew out of carbon price and FITs to foster energy self-sufficiency, created employment and in turn facilitated exports through the manufacture and deployment of new technology. In Germany's case, a PV industry grew out of FITs in response to a requirement for decarbonisation and energy security, which created employment and in turn facilitated exports through the manufacture and deployment of new technology in Europe. The European Commission also took the opportunity to boost economic recovery and GHG reductions by funding the European Energy Programme for Recovery in 2009, in the wake of the financing constraints that followed the Lehman Brothers collapse (European Commission 2009).

1.2.1.8 Research into New Technology

Institutions that facilitate research remain the best mechanism for incentivising greater levels of research into new electricity system-related technology. After the global financial crisis in 2008, the US government made available USD 1.5 billion in research funding for energy storage to benefit both the US motor vehicle sector and the electricity sector. Germany, in its 6th Energy Research Programme, allocated approximately EUR 3.5 billion for funding the research and development of energy technologies between 2011 and 2014. Other countries make available public money for research in universities and technical institutions. Australia established the Australian Renewable Energy Agency in 2011 for seed funding for new and promising technologies, although from 2014 its funding has been progressively eroded to alleviate government budget constraints (St John 2016). Government commitment to funding of appropriate research into energy technologies remains an important pathway to rapid decarbonisation but there has to be the political will to adequately fund research, and public support for the funding to be made available.

Key areas to consider when allocating funds for research include: funding allocation mechanisms and how to raise adequate funding for research.

1.2.2 Creating Incentives to Change Consumer Behaviour

1.2.2.1 Efficiency

Market, information, organisational and institutional failures have led to firms operating at suboptimal energy efficiency levels. Unexploited potential energy efficiencies are often referred to as the 'energy efficiency gap' (Hirst and Brown 1990). There are several reasons for this seemingly irrational behaviour, but the most common include market failures and barriers like: shareholder expectations with respect to returns, which constrain investment in energy efficiency; low priority of energy issues, which leads to inadequate skills within firms to identify opportunities or keep up with innovations; and subsidies and unpriced costs such as air pollution (Brown and Sovacool 2011). Increasing efficiency in electricity consumption reduces demand for electricity and thus in the investment required for decarbonisation, suggesting an important role for government intervention.

Studies point to significant energy consumption reduction by relatively small changes to behaviour and investment. This is discussed in detail in the industry chapter (Chapter 16) but, in summary: new buildings can be designed and existing buildings retrofitted for greater energy efficiency; energy efficiency can be engineered into industrial processes and investment in LED (light-emitting diode) lighting and combined heat and power can reduce electricity demand.

Efficiency programmes assist companies to identify and implement energy efficiency opportunities. Of these programmes the following are amongst the most effective:

Energy Price Increases

In line with economic theory, an increase in price decreases consumption, theoretically through energy efficiency measures taken to counteract increases in energy costs. It has been estimated that, in China, electricity tariff subsidies have led to heightened electricity consumption (IEA 2014). If subsidies were removed, efficiency measures would follow (Wang and Lin 2017). A study of industrial electricity consumption in South Korea found that electricity reductions of between 1.3% and 10.5% (depending on the price elasticity of demand) could be achieved by increasing the electricity price to average Organisation for Economic Co-operation and Development (OECD) levels (Han and Yun 2015).

While price may be the most effective tool for improving efficiency, it is not popular with consumers and causes problems for politicians. Since 2007, electricity prices in Australia have increased significantly, driven initially and primarily by heavy investment in networks, but also by FITs, the CT and larger profits for retailers. These increases have helped reduce consumption but have been extremely unpopular. Price increases in Britain since 2003 have also led to reductions in electricity consumption but the price rises have been accompanied by public distrust of energy companies.

Energy Efficiency Standards

Energy efficiency standards for buildings, transport vehicles, and energy-using industrial equipment and appliances play a significant role in improving energy efficiency, providing a clear role for governments to intervene. There are, broadly, two important approaches: minimum energy performance standards and top-runner or best available technology standards. Both provide incentives for the ratcheting up of energy efficiency performance standards. These are further discussed in the industry chapter (Chapter 16).

White Certificate Schemes

White certificates are used to certify energy reduction attained. Under these schemes, firms producing, supplying or distributing oil, gas or electricity are required to use measures to achieve predefined energy efficiency levels. If defined targets are not met, penalties are incurred. Certificates are issued to the entity on receipt of evidence of energy savings. In most of these schemes, certificates can be traded or used to meet energy-saving obligations.

1.2.2.2 Demand Management

It is estimated that demand-side initiatives could save AUD 4–12 billion over 10 years in Australia (AEMC 2012), with associated emissions abatement. The most powerful of demand response tools is time-of-use (TOU) tariffs for electricity. With the increasing concentration of electric goods in homes, residential demand has increased relative to commercial and industrial demand such that peaks in electricity demand occur during the late afternoon and early evening. Time-of-use tariffs reflect the cost of delivering electricity when demand is increased but require advanced metering infrastructure to record electricity consumption so that tariffs can be applied appropriately. Where new meters and tariffs are implemented and consumers are not advised about the costs, benefits and required behaviour change, consumer anger can result, as was the Victorian experience in Australia (King 2015).

1.2.2.3 Communicating Benefits of Low-Emission Options to the Public

With large centralised generation and networks for distribution dominating existing electricity sector technology, it is imperative for the public to be informed about the potential for new technologies and the benefits experienced from policy action. Good policy frameworks should include funding for government agencies that are able to communicate with the public. In Australia, the Australian Renewable Energy Agency (ARENA) is tasked with that responsibility, while globally the International Renewable Energy Agency (IRENA), the International Atomic Energy Agency (IAEA) and the Global Carbon Capture and Storage Institute (GCCSI) all participate with the International Energy Agency (IEA) as intergovernmental agencies tasked with communicating the benefits of low-emission options to the public.

In support of its energy transition policy ('Energiewende'), Germany has several additional research and policy institutes that conduct research into, and communicate, the potential and outcomes of decarbonisation policies. Examples include: the Fraunhofer Institute, which has published research findings on the cost benefits of RE policies in Germany (German BMU 2009) and facts about photovoltaics in Germany (Wirth 2017); the Renewable Energies Agency, which has published findings on public acceptance of RE in Germany (German AEE 2016, 2019); and the German Aerospace Centre, which has published research on possible energy system scenarios for Germany (Pregger et al. 2013) and, in collaboration with Australian researchers, has published on how climate targets can be achieved (Teske 2019).

1.2.3 Adapting Institutional Frameworks

1.2.3.1 Requirements for a Decarbonisation Framework for Liberalised Electricity Systems

Merit order dispatch and wholesale electricity markets have evolved to discover lowest-cost generation of energy from fossil fuels. The arrival of new technologies has highlighted the inadequacies of markets established purely for the dispatch of fossil fuels. Renewable energy has no or very low marginal costs and thus is dispatched in preference to energy generated from fossil fuels, which creates technical problems for generators that have must-run properties. Distributed generation in the form of rooftop solar panels needs to be integrated into the electricity market and yet each household/consumer is too small to participate in an electricity market. With this level of change facing the electricity sector, there have been questions raised as to the efficacy of electricity markets to deliver investment and integration of new technologies with different characteristics into a decarbonised electricity sector (UK CCC 2009; Rhys 2010; Thomas 2016). Some jurisdictions have already initiated changes. Germany's FIT legislation provided preferential dispatch and fixed-term contracts for RE generators in 1991. Britain's Energy Act 2013 tasked OFGEM with establishing long-term contracts for difference (CFDs) to provide stable returns for investment in low-carbon generation, to provide a capacity market for supply security and to ensure the availability of long-term contracts for independent RE generators (Thomas

2016). The state of South Australia has declared an intention to commission investment to counteract the shortcomings of the electricity market and federal energy policy which have resulted in reduced grid stability from high levels of generation from wind and solar power in that state (AAP 2017).

As an alternative to increasingly complex market manipulations, one proposal suggests a central agency to take responsibility for delivering both the low-carbon emissions target and adequate investment in infrastructure (Rhys 2010, 2014). As there is no or low marginal cost associated with low-carbon electricity, the concept of a central buyer tasked with awarding preferential supply arrangements at project sign-off through an auction mechanism to guarantee competition among technologies, abatement and investors, has its merits. German investment in large-scale RE is currently awarded using this framework. As variable and intermittent sources of energy supply higher proportions of electricity, backup reserve, supply balancing and demand response assume higher importance in securing supply. Firm capacity markets (as being discussed in Europe) (Eurelectric 2015), balancing markets (as being pursued by the western USA) (Lenhart et al. 2016) and demand response markets (Wang and Brown, 2014; Mahmoudi et al. 2017) therefore should be developed to replace or evolve from more traditional wholesale markets to become the primary short- and medium-term market mechanisms to respond to supply gaps, voltage instability or demand fluctuations.

There have been calls for more inclusive network planning frameworks to replace traditional processes. One of the key areas that has gained traction is *integrated resources planning* (Tellus 2000), which takes a holistic approach incorporating both supply- and demand-side options. By internalising the consideration of demand-side options, this process allows for the incorporation of environmental and social objectives, while potentially promoting more cost-effective network solutions. These gains are achievable due to greater contestability of service provision. There has been considerable work done to date on the development of these processes in markets such as California (CPUC 2019), Australia (AEMO 2019) and the UK (UK OFGEM 2016).

1.2.3.2 Requirements for Regulated Vertically Integrated Electricity Sectors

Without the complexity of governance and operation of liberalised electricity systems, decarbonisation requirements can be regulated and the vertically integrated electricity sector can be required to comply with decarbonisation goals. A good example of decarbonisation achieved through regulated vertically integrated electricity systems is the decision by France taken in the 1970s to deploy nuclear power. While many other nations were considering the benefits of nuclear power in the late 1960s and 1970s, none other than France was able to implement a wholesale transformation of the electricity sector. What facilitated France's transition was a combination of political will, informed by a highly technocratic and powerful administrative system, effectively executed through the Ministry of Industry, which was responsible for both the atomic regulatory body and the state-owned power utility. What was particularly significant was that despite a lack of bipartisanship on the benefits of nuclear power during elections, French presidents from both sides of the political spectrum continued deployment after being elected. The French approach to the politics of nuclear power deployment differed remarkably from the USA, where a non-interventionist, free market approach ensured that policies were more concerned with formalistic regulatory management of utilities 'from the outside' with little ability to 'impose its will on a myriad of electric utilities' (Jasper 2014). The French example indicates that there are benefits to vertically integrated, publicly managed institutional frameworks when considering decarbonisation of the electricity sector.

1.2.3.3 New Models for Remote Regions

Access to electricity remains a challenge for remote regions for a variety of reasons, including the cost of infrastructure required to supply to remote locations, the ability of consumers to pay for electricity and the industrial demand to underwrite the cost for residential and small business consumers. Models for electrification in remote regions are emerging and reflect the context of electricity provision for agrarian villagers with little ability to pay for electricity.

Addressing energy poverty in rural areas therefore requires subsidisation, new micro-payment models, and local RE generated for village microgrids, without a requirement for a developed country-style liberalised electricity market. Funding for the capital investment needs to come from investors in microfinance, development, aid and industries aligned with remote locations like mobile telephony and information technology, because income in remote rural areas is insufficient to pay for large infrastructure. In pursuit of the roadmap to achieve decarbonisation while enhancing productivity and jobs growth, the deployment of village-based electricity infrastructure using the investment models mentioned above can encourage the creation of jobs to support and administer the systems within the remote areas themselves, and in so doing creates an industry sector alongside the electrification programme. This removes the binary nature of options for rural electrification – of either emissions from electricity generation or no electricity access – and thus reduces the tensions between sociopolitical objectives.

The institutional framework for rural electrification therefore requires structures or agencies to identify: demand for electrification; supply of different classes of investors and funds; appropriate technologies and sources of RE; business models for implementation; operational support and administration requirements; and training opportunities for local individuals to be able to operate and support business models.

1.2.3.4 Addressing the Inconsistencies in International Accounting and Responsibility for Carbon Emissions

The Kyoto Protocol was created as a framework of consensus on the individual ability of nations to achieve emissions reductions. It has not been successful. This does not mean that individual countries seeking to act decisively should not continue to implement climate policy that is cognisant of the risks of climate change. Indeed, the EU has taken such a stance, seeking much greater carbon emissions reductions than other nation states or economic blocs, in line with their concerns about the consequences of climate change. This is discussed in detail in Chapter 21 but, in summary, the inconsistencies that arise from uneven action to address carbon emissions can be addressed by nation states or economic blocs through border adjustments and trade agreements that account for the costs of embedded carbon content. This is not a perfect outcome, but it does help to advance emissions reductions as nation states are subjected to the political exigencies associated with their responsibilities to reduce emissions.

1.3 The Challenges to Confront

We have discussed the linkages between broad policy objectives and the instruments used to implement those objectives. We have explicitly recognised the importance of the contextual environment for the formation of new industrial configurations which are enabled through the emergence of new technologies. We will now explore the current challenges being presented to institutional environments as a result of new technologies aimed at decarbonisation.

1.3.1 Integrating the Old and the New

The management of the transition from centralised to decentralised systems is challenging. It requires careful and pragmatic planning and management of the grid. The German grid operators have found that improved weather forecasting has significantly improved the management of fast shifts in supply. The Australian Energy Market Operator has found that the ability to direct centralised generators to supply in times of supply constraint is more effective than awaiting market offers for supply. The overarching requirement though is for grid and market operators to proactively forecast, plan and manage according to the situation at hand. A prescriptive rule-based electricity sector will not be adaptive nor will it deliver a secure supply of electricity.

Decarbonisation and supply security are public concerns and if tariffs are to be kept affordable then incumbents will need to be supported through the transition. Managing incumbents will be as much about end-of-life asset management and integration with new technologies as it will be about designing markets to ensure that there is sufficient competition amongst incumbents to reduce rent-seeking behaviour.

The consensus on the need for action that emerged from the 2015 United Nations Conference of the Parties (COP 21) has set a level of expectation that action at all levels of society is necessary. It is noteworthy that some business and private actors are now acting independently of regulatory frameworks on mechanisms like corporate PPAs, direct investment in RE and new funding mechanisms. There is a role for investment to proceed outside of current mechanisms and regulation to pursue decarbonisation. Facilitating the ability of investors and system managers to operate independently of regulatory frameworks will significantly enhance the adaptation of the system.

> **Key areas to consider when integrating the old and the new** are: management of variable supply and demand; and coordinating market structures.

1.3.2 Managing the Politics

There are 'boundaries to economic rationality and calculation, so that the latter are allowed to operate only in certain circumstances. The most precise methods for making policy recommendations eventually run into the realities of political resistance and conflict' (Jasper 2014: 3).

And yet it is important to remember that history shows 'only states that managed to prevent vested interests from gaining control over economic policy-making stood much chance of pursuing policies of structural change' (Moe 2010).

1.3.2.1 Political Involvement

Despite the complexity of the decisions facing politicians, there is evidence of political roadblocks being applied to climate policies (Bayulgen and Ladewig 2017). These political roadblocks in the USA reflect historic tepid public mood on the need for climate policy action (at about 50% support) (Pew Research Center 2016). In Australia, only 55% of Australians think that governments are not doing enough on climate change, although there are significant differences across the political spectrum (Merzian et al. 2019). Overcoming political obstacles to decarbonisation needs a combination of determination, persuasion and horse-trading between political actors (Compston and Bailey 2012) to convince a sceptical public of the long-term benefits.

1.3.2.2 Lessons from Historical Roll-Outs of New Technology

There are a number of lessons to be learned from the institutional frameworks that facilitated the roll-out of nuclear power in France and curtailed it in the USA in the 1970s. Of importance here is that US politicians used their ideological frameworks to twist the energy policy debate and frame it as a free market versus government intervention decision (Jasper 2014). The US predisposition towards free market ideology burdened potential investors and financiers with significant risks associated with the choice of technology, the costs of operation and waste disposal, public resistance and public safety in the event of accidents. These were too large to attract significant levels of investment in the US nuclear power industry. By comparison, the French state assumed responsibility for these risks and thus formulated energy policy which drove strategic state objectives of energy security despite public resistance.

By comparison, Germany has a history of strong and effective policy frameworks to secure environmental benefits by pursuing technologies rather than ideologies. In 1983, regulation to limit emissions of sulphur dioxide (SO_2) from all power stations through the implementation of flue gas desulphurisation technology, despite concerns raised by industry, electricity and coal-mining sectors, achieved 93% reduction in emissions within 12 years, faster and more effectively than the cap-and-trade policy mechanism employed in the USA, which achieved a 51% reduction in emissions from 1995 to 2010 (Harrington et al. 2004). Legislation to encourage investment in new technology since 1991 has had consistent bipartisan support resulting in significant investment. Initial responsibility for RE policy was assigned to the ministry of the environment, which avoided confrontation with industry actors with political power and contrasting priorities. Jobs created in the RE sector tended to be in small- and medium-sized enterprises and in regions that were rural or economically less developed. Coupled with RE investors, which included private citizens, farmers and cooperative programmes, this technology-led approach helped to create a positive view of the benefits of RE and public support for the decarbonisation framework.

The lesson from the politics of nuclear power in France and the USA, and the technocratic approaches preferred in France and Germany, is that political ideological world views should be removed from the formulation of policy frameworks.

1.3.2.3 Lessons on Gaining Independence from Political Interference

To illustrate the advantages of institutional frameworks that seek independence from politics, it is pertinent to look to the frameworks that emerged after the damaging levels of inflation created during the 1970s and 1980s. In many jurisdictions, national reserve banks have been given the independence and responsibility to direct monetary policy to seek financial system stability, currency stability, full employment and economic prosperity. These objectives allowed technocrats in national reserve banks, removed from politics, to focus on long-term economic growth and politicians in national governments to accept responsibility for disciplined fiscal policy to achieve medium-term stability (Reserve Bank of Australia 2013).

A similar structure may be beneficial for decarbonisation. A Climate Change Authority (CCA) was established in Australia in 2011 but not endowed with the responsibility to manage the achievement of targets. Consequently, it was perceived more as an advocate of climate policy than as the body responsible for achieving targets. For effective pathways and roadmaps to decarbonisation, an independent authority, populated with technocrats, could be given the responsibility to meet a long-term decarbonisation goal. The most effective tools would be applied to meet that goal, with the provisos of minimising impact on employment and economic stability. This framework could facilitate stable policy and investment through greater levels of certainty and predictability and provide a credible pathway to decarbonisation.

1.3.2.4 The Influence of Powerful Vested Interests

Climate policy requires investment in new technologies and a fundamental change in the harvesting of energy sources. There are thus enormous benefits to be gained by investors in new technologies, but some of the most powerful industries dependent on mining, transporting and marketing fossil fuels for their ongoing hegemony face decline and irrelevance in the process of decarbonisation. Fossil fuel companies and their associated lobbyists have significant access to bureaucrats, politicians and regulators, which provide opportunities for a revolving door between industry, the public service and politicians. A good example of this emerged after investigations into the Deepwater Horizon oil spill highlighted how the cosy relationships between industry and the regulator led to 'principal capture' (Neill and Morris 2012).

Research into Appalachian coal regions shows that they are dependent on the coal industry for economic opportunity, their cultural identity as rural Americans and their sense of worth. As demand for coal in the USA has declined since 2009, the coal industry has framed environmentalism as the cause of economic decline in coal regions. Fear of job losses as a result of mine closures has resulted in significant electoral anger against climate policy and the election of politicians opposed to decarbonisation (Lewin 2017). Other research finds that communities that host coal mines tend to have high poverty, poor health outcomes (Hendryx 2011) and reduced economic development compared to non-coal-mining communities. As coal miners are paid relatively high wages for low-skilled work, there is a disincentive to improve education, which increases risk of unemployment as the sector mechanises or contracts (Douglas and Walker 2017). Policies intended to promote employment from coal extraction will come at the price of lower long-term income growth. Policy-makers are advised to use the current income from fossil fuel extraction to promote economic diversification to prepare for adaptation (Douglas and Walker 2017).

Thus, decarbonising is not the techno-economic win–win solution as it is commonly framed by engineers, economists and policy-makers. In reality, decarbonisation is characterised by power structures, control of resources and political economy, all of which will result in a deeply political struggle (Healy and Barry 2017). There is increasing reference to a 'just transition' for workers who bear the burden of decarbonisation policies (Mayer 2018), but it is important to recognise that there needs to be a transition 'to' something tangible, not a theoretical shift to some unknown new opportunity. For this reason, it is essential that climate/energy policy interacts closely with economic development to channel employment away from fossil fuel extraction to specific new opportunities also located in rural settings. Proposals like Ross Garnaut's development of Australian renewable energy resources to underpin a low-cost manufacturing hub (Garnaut 2019) and the Asian Renewable Energy Hub (Asian Renewable Energy Hub 2018) proposed to be developed by a consortium including InterContinental Energy, Macquarie Capital and Vestas (cwpRenewables 2018), are evidence of the potential of this interaction of climate/energy policy and economic development.

1.3.3 Pathways and Roadmaps to Decarbonisation Resilience

> There are many pathways to deep decarbonization, and they do not require major technological breakthroughs.
>
> *(The White House 2016: 30)*

Context is crucial. A country that is reliant on coal for electricity generation and export income will pursue a different pathway to one that has few energy resources and historically has been reliant on imports of energy for economic development. Equally, a country that has little electricity infrastructure will consider a different roadmap to one that has an ageing, inefficient fleet of high-emission generators. Countries where politics is driven by a conceptual framework of free markets will choose different policy mixes to those more comfortable with the welfare state. Pathways and roadmaps will be determined by context and the elements of climate policy, identified in Sections 1.2.1 and 1.2.2 above, applied as appropriate.

In particular, every nation, city, region and industry needs to address public commitment to elements of climate policy that are acceptable. Public antipathy to carbon pricing led to its demise in Australia. Irrespective of its economic credentials, forcing the public to accept a carbon price when there is antipathy to such a device will lead to a roadblock to decarbonisation. Instead, if the pursuit of higher levels of RE has broad public support, then it is more effective to pursue an RET. Decarbonisation pathways require a framework of policy measures that are supported by and acceptable to the public. In the absence of a more technocratic approach to decarbonisation, a framework based on popular policy measures will: garner political support and facilitate the prompt roll-out of mechanisms to guide the process of both mitigation and adaptation; ensure that decarbonisation follows a consistent pathway, avoiding political interference which leads to competing and contradictory policy agendas and a stop–go approach to policies; develop a pathway based on a diverse combination of policy measures that together will accelerate investment and behaviour modification to mitigate against and adapt to climate change; and facilitate adaptation to changing circumstances using a multiplicity of policy tools, thereby avoiding reliance on any one single tool and allowing the introduction of new tools into the framework to address gaps or changes in requirement. This democratisation of decarbonisation policy could provide a more powerful and durable framework to achieve targets in the requisite time frames. So too, strong climate policies need to recognise and tackle the inevitable impact on some communities from decarbonisation. The disenfranchisement felt in America's industrial heartland saw the Trump Administration retreat from the global climate leadership and policies of the Obama Administration – arguably because not enough was done to provide the necessary social support through what is in essence a form of profound structural adjustment.

References

AAP (Australian Associated Press) (2017). SA to go it alone on power quest. *SBS Korean*. 10 February. Available at: www.sbs.com.au/language/english/sa-to-go-it-alone-on-power-quest.

AEMC (Australian Energy Market Commission) (2012). *Power of Choice Review: Giving Consumers Options in the Way They Use Electricity*. Sydney: Australian Energy Market Commission.

AEMO (Australian Energy Market Operator) (2019). *Integrated System Plan (ISP)*. Melbourne: Australian Energy Market Operator. Available at: www.aemo.com.au/Electricity/National-Electricity-Market-NEM/Planning-and-forecasting/Integrated-System-Plan.

Anderson, P. and Tushman, M. L. (1990). Technological discontinuities and dominant designs: A cyclical model of technological change. *Administrative Science Quarterly*, 35, 604–633.

Arthur, W. B. (1989). Competing technologies, increasing returns, and lock-in by historical events. *The Economic Journal*, 99, 116–131.

Asian Renewable Energy Hub (2018). Renewable Energy at Oil and Gas Scale. Available at: https://asianrehub.com/.

Baer, H. A. (2016). The nexus of the coal industry and the state in Australia: Historical dimensions and contemporary challenges. *Energy Policy*, 99, 194–202.

Barbier, E. B. (2003). The role of natural resources in economic development. *Australian Economic Papers*, 42, 253–272.

Bayulgen, O. and Ladewig, J. W. (2017). Vetoing the future: Political constraints and renewable energy. *Environmental Politics*, 26, 49–70.

Brown, M. A. and Sovacool, B. (2011). Barriers to the diffusion of climate-friendly technologies. *International Journal of Technology Transfer and Commercialization*, 10, 43–62.

CAIT Climate Data Explorer (2019). *Country Greenhouse Gas Emissions*. Washington, DC: World Resources Institute. Available at: http://cait.wri.org.

Chan, G. (2017). Coalition says it may change Clean Energy Finance Corporation rules to fund coal plants. *The Guardian*. 19 February.

Christensen, C. M. (1997). *The Innovator's Dilemma: How New Technologies Cause Great Firms to Fail*. Boston, MA: Harvard Business School Press.

Compston, H. and Bailey, I. (2012). *Climate Clever: How Governments Can Tackle Climate Change (and Still Win Elections)*. Florence, KY: Taylor and Francis.

Coorey, P. (2019). Coalition eyes funding for carbon capture. *Australian Financial Review*. 18 November.

CPUC (California Public Utilities Commission) (2019). *Integrated Resource Plan and Long Term Procurement Plan (IRP-LTPP)*. San Francisco, CA: California Public Utilities Commission. Available at: www.cpuc.ca.gov/irp/.

cwpRenewables (2018). Asian Renewable Energy Hub. Available at: https://cwprenewables.com/projects/asian-renewable-energy-hub/.

Douglas, S. and Walker, A. (2017). Coal mining and the resource curse in the Eastern United States. *Journal of Regional Science*, 57, 568–590.

Eurelectric (2015). A Reference Model for European Capacity Markets. Eurelectric position paper. Available at: www.eurelectric.org/media/1918/a_reference_model_for_european_capacity_markets-2015-030-0145-01-e.pdf.

European Commission (2009). *European Energy Programme for Recovery*. Brussels: European Commission. Available at: http://ec.europa.eu/energy/eepr/projects/.

European Commission (2011). *A Roadmap for Moving to a Competitive Low Carbon Economy in 2050*. Brussels: European Commission. Available at: www.europarl.europa.eu/meetdocs/2009_2014/documents/com/com_com(2011)0112_/com_com(2011)0112_en.pdf [English translation].

Garnaut, R. (2019). *Superpower: Australia's Low-Carbon Opportunity*. Melbourne: La Trobe University Press.

German AEE (Renewable Energies Agency) (2016). Acceptance of renewable energy in Germany. Available at: www.unendlich-viel-energie.de/english/acceptance-of-renewable-energy-in-germany.

German AEE (2019). Important for the fight against climate change: Citizens want more renewable energies. Available (in German) at: www.unendlich-viel-energie.de/akzeptanzumfrage-2019.

German BMU (Federal Ministry of the Environment, Nature Conservation and Nuclear Safety) (2009). *Cost and Benefit Effects of Renewable Energy Expansion in the Renewable and Heat Sectors*. Berlin: German Federal Ministry of the Environment, Nature Conservation and Nuclear Safety. Available at: www.ctc-n.org/resources/cost-and-benefit-effects-renewable-energy-expansion-german-power-and-heat-market.

Goodman, J. (2008). The minerals boom and Australia's resource curse. *Journal of Australian Political Economy*, 61, 201–219.

Han, J. and Yun, S.-J. (2015). An analysis of the electricity consumption reduction potential of electric motors in the South Korean manufacturing sector. *Energy Efficiency*, 8, 1035–1047.

Harrington, W., Morgenstern, R. D. and Sterner, T., eds. (2004). *Choosing Environmental Policy: Comparing Instruments and Outcomes in the United States and Europe*. Washington, DC: Resources for the Future.

Healy, N. and Barry, J. (2017). Politicizing energy justice and energy system transitions: Fossil fuel divestment and a 'just transition'. *Energy Policy*, 108, 451–459.

Hendryx, M. (2011). Poverty and mortality disparities in Central Appalachia: Mountaintop mining and environmental justice. *Journal of Health Disparities Research and Practice*, 4, 44–53.

Hirst, E. and Brown, M. (1990). Closing the efficiency gap: Barriers to the efficient use of energy. *Resources, Conservation and Recycling*, 3, 267–281.

IEA (International Energy Agency) (2014). *Emissions Reduction through Upgrade of Coal-Fired Power Plants: Learning from Chinese Experience*. Paris: OECD.

IEA (2019a). SDG7*: Data and Projections: Access to Affordable, Reliable, Sustainable and Modern Energy for All*. Paris: International Energy Agency. Available at: www.iea.org/reports/sdg7-data-and-projections/access-to-electricity#abstract.

IEA (2019b). World energy balances. *IEA World Energy Statistics and Balances* [database]. Available at: https://doi.org/10.1787/data-00512-en.

IEA (2019c). CO_2 emissions by product and flow. *IEA CO_2 Emissions from Fuel Combustion Statistics* [database]. Available at: https://doi.org/10.1787/data-00430-en.

IEA (2019d). *World Energy Outlook 2019*. Paris: International Energy Agency.

Jakob, M. and Steckel, J. C. (2014). How climate change mitigation could harm development in poor countries. *WIREs Climate Change*, 5, 161–168.

Jasper, J. M. (2014). *Nuclear Politics: Energy and the State in the United States, Sweden, and France*. Princeton: Princeton University Press.

Joas, F., Pahle, M., Flachsland, C. and Joas, A. (2016). Which goals are driving the Energiewende? Making sense of the German Energy Transformation. *Energy Policy*, 95, 42–51.

King, S. (2015). Smart meters, dumb policy: The Victorian experience. *The Conversation*. 17 September. Available at: https://theconversation.com/smart-meters-dumb-policy-the-victorian-experience-47685.

Künneke, R. W. (2008). Institutional reform and technological practice: The case of electricity. *Industrial and Corporate Change*, 17, 233–265.

Lazard (2019). Levelized cost of energy analysis: Version 13. *Lazard*. 7 November. Available at: www.lazard.com/perspective/lcoe2019.

Lenhart, S., Nelson-Marsh, N., Wilson, E. J. and Solan, D. (2016). Electricity governance and the Western energy imbalance market in the United States: The necessity of interorganizational collaboration. *Energy Research & Social Science*, 19, 94–107.

Lewin, P. G. (2017). 'Coal is not just a job, it's a way of life': The cultural politics of coal production in Central Appalachia. *Social Problems*, 66, 51–68.

Mahmoudi, N., Heydarian-Forushani, E., Shafie-Khah, M., Saha, T. K., Golshan, M. E. H. and Siano, P. (2017). A bottom-up approach for demand response aggregators' participation in electricity markets. *Electric Power Systems Research*, 143, 121–129.

Mayer, A. (2018). A just transition for coal miners? Community identity and support from local policy actors. *Environmental Innovation and Societal Transitions*, 28, 1–13.

Meadowcroft, J. (2011). Engaging with the politics of sustainability transitions. *Environmental Innovation and Societal Transitions*, 1, 70–75.

Merzian, R., Quicke, A., Bennett, E., Campbell, R. and Swann, T. (2019). *Climate of the Nation*. Research report. Canberra: The Australia Institute. Available at: https://australiainstitute.org.au/report/climate-of-the-nation-2019/.

Metherell, L. and Hawley, S. (2013). Tony Abbott says ETS a 'market in an invisible substance'; Labor denies scrapping carbon price will leave $6bn budget hole. *ABC News*. 15 July. Available at: www.abc.net.au/news/2013-07-15/abbott-dismisses-ets-as-market-in-an-invisible-substance/4820564.

Moe, E. (2010). Energy, industry and politics: Energy, vested interests, and long-term economic growth and development. *Energy*, 35, 1730–1740.

Molyneaux, L., Brown, C., Wagner, L. and Foster, J. (2016). Measuring resilience in energy systems: Insights from a range of disciplines. *Renewable and Sustainable Energy Reviews*, 59, 1068–1079.

Neill, K. A. and Morris, J. C. (2012). A tangled web of principals and agents: Examining the Deepwater Horizon oil spill through a principal-agent lens. *Politics and Policy*, 37, 1047–1072.

Nelson, R. R. (1995). Recent evolutionary theorizing about economic change. *Journal of Economic Literature*, 33, 48–90.

Ordanini, A., Miceli, L., Pizzetti, M. and Parasuraman, A. (2022). Crowd-funding: Transforming customers into investors through innovative service platforms. *Journal of Service Management*, 22, 443–470.

Osofsky, H. and Wiseman, H. (2014). Hybrid energy governance. *University of Illinois Law Review*, 2014, 1–66.

Parliament of Australia (2011). Commonwealth Parliamentary Debates, House of Representatives. No. 14, 2011. Wednesday, 21 September. Canberra: Hansard.

Parliament of Australia (2016). Corporations Amendment (Crowd-Sourced Funding) Bill: Explanatory Memorandum. Canberra: Parliament of Australia. Available at: www.aph.gov.au/Parliamentary_Business/Bills_Legislation/Bills_Search_Results/Result?bId=r5766.

Pew Research Center (2016). The Politics of Climate [survey]. Available at: www.pewresearch.org/science/2016/10/04/the-politics-of-climate/.

Pregger, T., Nitsch, J. and Naegler, T. (2013). Long-term scenarios and strategies for the deployment of renewable energies in Germany. *Energy Policy*, 59, 350–360.

Reserve Bank of Australia (2013). Statement on the conduct of monetary policy. *Reserve Bank of Australia*. 24 October. Available at: www.rba.gov.au/monetary-policy/framework/stmt-conduct-mp-6-24102013.html.

Rhys, J. (2010). Reforming UK electricity markets. *Oxford Energy Forum*, 81, 20–23. Available at: www.oxfordenergy.org/wpcms/wp-content/uploads/2011/02/OEF-81.pdf.

Rhys, J. (2014). Back to the CEGB? Greater central control of UK energy may be inevitable. *The Conversation*. 9 May. Available at: https://theconversation.com/back-to-the-cegb-greater-central-control-of-uk-energy-may-be-inevitable-26474.

Sachs, J. D. and Warner, A. M. (1997). Fundamental sources of long-run growth. *The American Economic Review*, 87, 184–188.

Scobie, C. (2016). Supreme Court stays EPA's clean power plan. *American Bar Association*. 17 February. Available at: www.americanbar.org/groups/litigation/com mittees/environmental-energy/practice/2016/021716-energy-supreme-court-stays-epas-clean-power-plan/.

Simshauser, P. (2019). Missing money, missing policy and resource adequacy in Australia's National Electricity Market. *Utilities Policy*, 60.

St John, A. (2016). What's happening with ARENA? *Flagpost*. 19 September. Available at: www.aph.gov.au/About_Parliament/Parliamentary_Departments/Parliamentary_ Library/FlagPost/2016/September/ARENA-changes.

Sue, K., Macgill, I. and Hussey, K. (2014). Distributed energy storage in Australia: Quantifying potential benefits, exposing institutional challenges. *Energy Research & Social Science*, 3, 16–29.

Tellus (2000). *Best Practices Guide: Integrated Resource Planning for Electricity.* Washington, DC: United States Agency for International Development.

Teske, S. (2019). *Achieving the Paris Climate Agreement Goals: Global and Regional 100% Renewable Energy Scenarios with Non-Energy GHG Pathways for +1.5 °C and +2 °C.* Switzerland: Springer Nature.

The White House (2016). *United States Mid-Century Strategy for Deep Decarbonization.* Washington, DC: The White House. Available at: https://unfccc.int/files/focus/long-term_strategies/application/pdf/mid_century_strategy_report-final_red.pdf.

Thomas, S. (2016). A perspective on the rise and fall of the energy regulator in Britain. *Utilities Policy*, 39, 41–49.

Tong, D., Zhang, Q., Zheng, Y., Caldeira, K., Shearer, C., Hong, C., Qin, Y. and Davis, S. J. (2019). Committed emissions from existing energy infrastructure jeopardize 1.5°C climate target. *Nature*, 572, 373–377.

UK CCC (Committee on Climate Change) (2008). *Building a Low-Carbon Economy: The UK's Contribution to Tackling Climate Change.* London: UK Committee on Climate Change.

UK CCC (2009). *Meeting Carbon Budgets: The Need for a Step Change.* London: UK Committee on Climate Change.

UK GIG (Green Investment Group) (2019). *Progress Report: Accelerating the Transition to a Greener Global Economy.* Sydney: UK Green Investment Group. Available at: https://greeninvestmentgroup.com/media/230066/gig-progress-report-2019_final-a4 .pdf.

UK OFGEM (Office of Gas and Electricity Markets) (2016). Integrated Transmission Planning and Regulation. *UK Office of Gas and Electricity Markets*. Available at: www.ofgem.gov.uk/electricity/transmission-networks/integrated-transmission-plan ning-and-regulation.

Unruh, G. C. (2000). Understanding carbon lock-in. *Energy Policy*, 28, 817–830.

US DOE (Department of Energy) (1978). Power Plant and Industrial Fuel Use Act. *energy. gov*. Available at: https://energy.gov/oe/services/electricity-policy-coordination-and-implementation/other-regulatory-efforts/power-plant.

US EIA (Energy Information Administration) (2019a). Existing nameplate and net summer capacity by energy source, producer type and state (EIA-860) [data resource]. *US Energy Information Administration*. Available at: www.eia.gov/electricity/data/state/.

US EIA (2019b). Table 9.1 emissions from energy consumption at power plants [data resource]. *US Energy Information Administration*. Available at: www.eia.gov/electri city/data.php#elecenv.

US EPA (Environmental Protection Agency) (2015). Overview of the Clean Power Plan: Cutting carbon pollution from power plants. *United States Environmental Protection*

Agency. Available at: https://archive.epa.gov/epa/cleanpowerplan/fact-sheet-overview-clean-power-plan.html.

US EPA (2019). Affordable Clean Energy Rule. *United States Environmental Protection Agency*. Available at: www.epa.gov/stationary-sources-air-pollution/affordable-clean-energy-rule.

Wang, X. and Lin, B. (2017). Impacts of residential electricity subsidy reform in China. *Energy Efficiency*, 10, 499–511.

Wang, Y. and Brown, M. A. (2014). Policy drivers for improving electricity end-use efficiency in the USA: An economic–engineering analysis. *Energy Efficiency*, 7, 517–546.

Warren, B., Christoff, P. and Green, D. (2016). Australia's sustainable energy transition: The disjointed politics of decarbonisation. *Environmental Innovation and Societal Transitions*, 21, 1–12.

Williamson, O. E. (1998). Transaction cost economics: How it works; where it is headed. *De Economist*, 146, 23–58.

Wirth, H. (2017). *Recent Facts about Photovoltaics in Germany*. Freiburg: Fraunhofer Institute.

Technologies for Decarbonising the Electricity Sector

2

Wind Energy

NATHAN STEGGEL AND DAVID OSMOND

Executive Summary

Wind energy technologies are a proven, established and important source of electricity generation. Engineers and financiers view wind energy as a mature technology and a low-risk investment. Wind energy provides substantial portions of the electricity supply across a diverse number of locations due to its flexibility on land and sea.

Through the use of strong renewable energy targets and effective policy reform, many jurisdictions are rapidly increasing wind power to decarbonise their electricity grids. States, provinces and countries where wind generation exceeds 15% of total electrical requirements include: parts of the USA (Iowa, Kansas, Oklahoma, North and South Dakota, Maine, Minnesota, Idaho, Colorado, Texas), Denmark, Portugal, Ireland, Scotland, Spain, Germany, Tamil Nadu (India) and Inner Mongolia (China). The state of South Australia generated approximately 40% of its electricity from wind in 2018.

To assist decision makers to better understand the potential of wind power to help achieve a low-carbon, resilient future, the chapter provides an introduction to key developments in the technology and economics of wind power. Extensive research and development has led to the building of increasingly tall turbines with large swept areas and refined blade design that can harvest greatly increased amounts of energy, thus reducing unit costs of generation.

Historically, offshore wind has been considerably more expensive than onshore. However, the cost for offshore wind has fallen rapidly in recent years. Significant research and development advances are expected to push pricing down further and to open up vast additional resources through floating offshore technologies.

Where good wind resources are available, new wind energy developments can provide electricity more cost-effectively than new fossil-fuel-based power stations and in many cases compete with power stations that have fully amortised their set-up costs.

2.1 Wind Technology: History and Current Status

The extraction of useful energy from the wind has a long and important history in human development. Sailing boats are known to have been used by Ancient Egyptians around 3000 BCE and windmills for grinding corn and pumping water were first used in the Middle East around the eighth century CE before spreading to Europe, India and China.

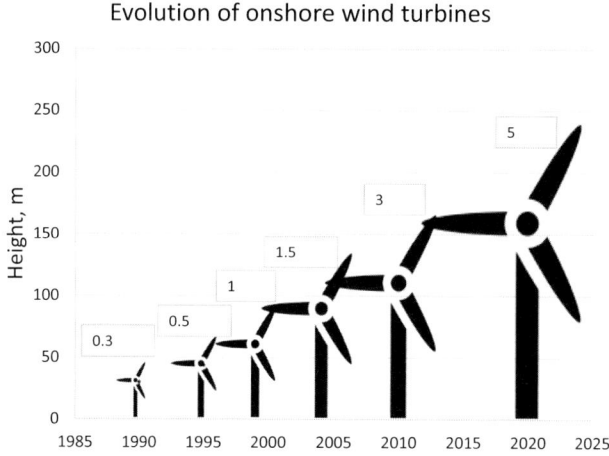

Figure 2.1 Evolution of size and capacity of onshore wind turbines.
Source: Courtesy of the authors.

Wind power, along with hydropower, played a critical role in providing energy to drive many early industrial processes developed in the Middle Ages that helped underpin the first industrial revolution (Ahuja 2015). At their peak during the early industrial revolution, the total amount of wind-powered mills in Europe is estimated to have been around 100 000, compared to some 200 000 waterwheels (Hills 1996).

Converting wind into electrical energy was first undertaken in 1887 in Scotland, but it is Danish scientist and inventor Poul la Cour who is perhaps best recognised as the early pioneer of wind turbines used for electrical generation. La Cour was also concerned with the storage of energy, and from 1895 the electricity from la Cour's wind turbine at Askov Folk High School was used to generate hydrogen and oxygen through the process of electrolysis. The hydrogen was used by the gas light in the school for lighting until 1902.

Small windmills were used extensively across countries such as the USA and Australia to power irrigation pumps. In the USA, by 1900 more than six million windmills were erected throughout the countryside, providing energy for farmers to pump water. Significant advances in wind power technology from the 1890s to the 1940s paved the way for the modern large-scale wind power technologies generating significant amounts of electricity. For instance, in Yalta, by 1931, a precursor to the modern horizontal wind generator was used, generating 100 kW (kilowatts). Wind turbines of the megawatt (MW) scale were developed in the 1980s and 1990s. In the last 30 years, there has been significant progress in increasing the capacity of wind turbines to provide electricity. This factor, in addition to reduction in costs from economies of scale, has seen wind power costs reduce by 40-fold during this period (Figure 2.1).

2.2 Current Technology

The vast majority of modern large-scale wind turbines have three bladed rotors that rotate upstream of the tower and generator. A basic schematic of a modern turbine is provided in Figure 2.2:

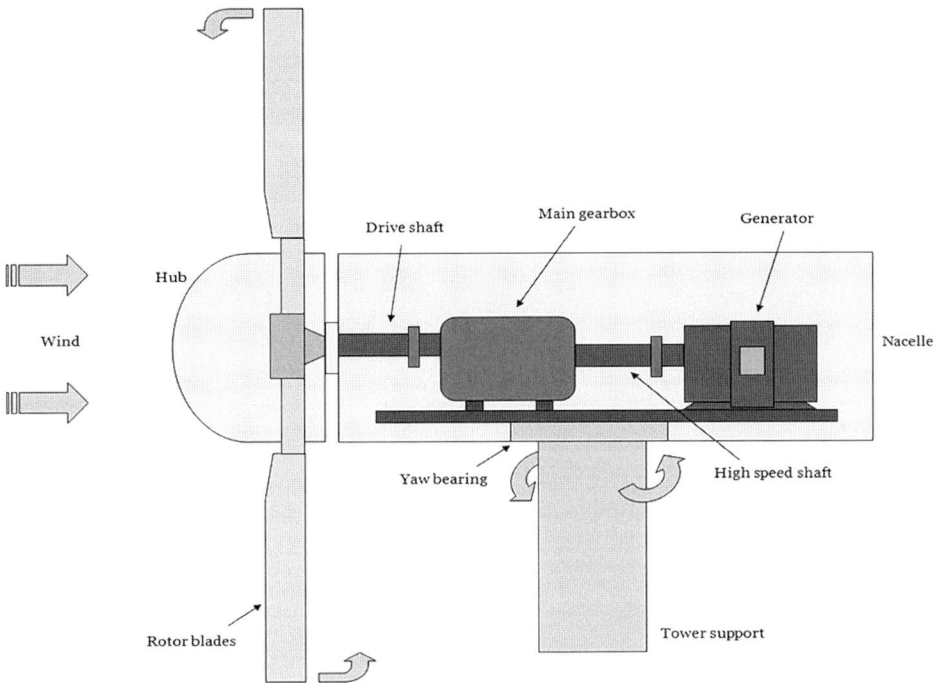

Figure 2.2 Basic schematic of a modern wind turbine.
Source: Courtesy of the authors.

- the wind exerts a turning torque on the rotor, which in turn spins the drive shaft;
- a gearbox increases the rotation rate of a second drive shaft, which then generates electricity in the generator; and
- yaw motors act to ensure the turbine is always facing into the wind, while the turbine blades have adjustable pitch angles to ensure optimal aerodynamic efficiency.

A small but growing proportion of modern turbines now use direct drive technology, which means they do not have a gearbox, and typically use permanent magnets instead of electromagnets in the generator.

Modern turbines generally fit into one of three turbine classes. Class I turbines are designed for very windy sites, Class II turbines for medium-speed sites, and Class III turbines for low wind speed sites. As fewer very windy locations for wind farm sites remain undeveloped, developers are looking increasingly at less windy sites, and turbine manufacturers have responded by rapidly improving the technology and resulting economics for the Class II and III turbines.

Figure 2.3 indicates a typical power curve for a 135-m-diameter turbine in each turbine class. It can be seen that the power curves are very similar up to a wind speed of about 8 m/s, but at higher speeds the power from the Class III turbine levels off as it reaches rated power. The same figure also shows the turbine efficiency, defined as the energy produced by the turbine relative to the kinetic energy of the incoming air. All three turbine class curves reach a peak efficiency of about 45%; however, the Class III turbine has been

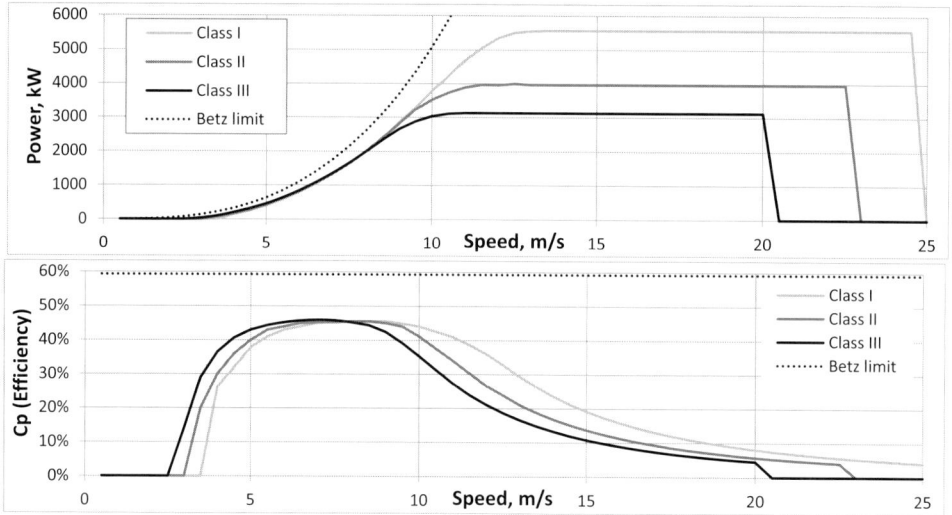

Figure 2.3 Generic turbine power curve (top) and efficiency (bottom) for a 135-m-diameter turbine in each of the turbine classes.
Source: Courtesy of the authors.

optimised to reach peak efficiency at lower wind speeds in comparison to the Class I or II turbines. Also shown is the Betz limit, which is approximately 59% and is the theoretical limit of turbine efficiency in a free wind stream.

At low wind speed sites, stresses from wind loading are reduced, which has helped turbine manufacturers design Class III turbines with very long blades. These long blades allow the turbine to capture more wind, and hence to generate more power. However, this comes at the expense of the turbines generally having to be spaced further apart in order to reduce wake-induced loadings.

The specific power of a wind turbine is the ratio of the generator size to the swept area of the rotor. Class III turbines have reduced generator capacities relative to the swept area of the rotor, and therefore have a smaller specific power ratio than a Class II or I turbine. A smaller specific power ratio reduces loading on the turbine at high wind speeds.

The capacity factor of a turbine is defined as the average turbine output normalised by the maximum capacity of the turbine. The smaller specific power ratio of the Class III turbine means that it reaches a given capacity factor at a lower wind speed than a Class I turbine. This means that even at relatively low wind speed locations, the capacity factor of a modern Class III turbine is often high compared to Class I turbines from an earlier era. This is also helped by modern turbines having much taller hub heights than their earlier counterparts. The evidence for this can be seen in Figure 2.4, which indicates the capacity factor of US wind farms in 2017 by commercial operation date. It illustrates that newer wind farm sites using more modern turbine technology tend to have higher capacity factors than older sites, despite many of the more recent sites having lower wind speeds than earlier sites.

Many European and some Asian countries are finding it increasingly difficult to find sufficient wind farm sites on land and are therefore exploring offshore locations. Offshore

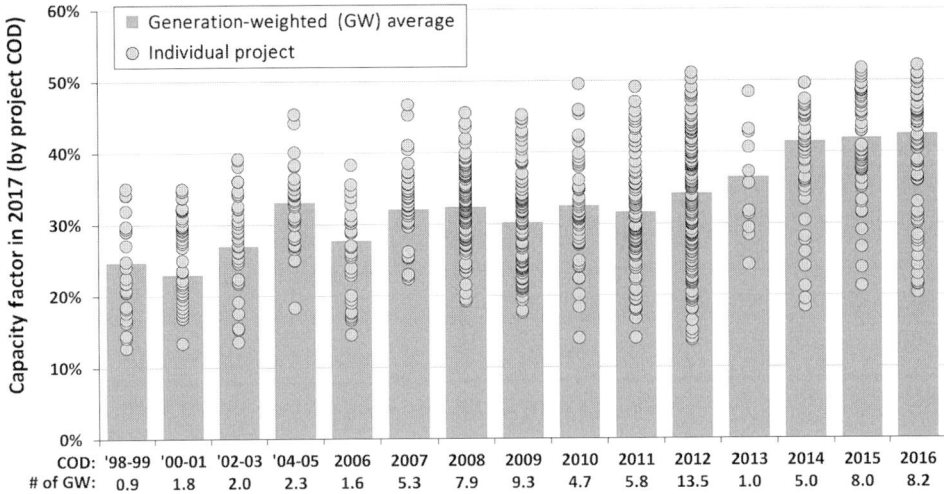

Figure 2.4 Calendar year 2017 capacity factors by commercial operation date (COD) for US wind farms. Also shown along the x axis is the installed capacity of wind that reached COD in each year.

Source: 2017 Wind Technologies Market Report, US Department of Energy (2017). For a colour version of this figure, please see the colour plate section.

sites typically have higher wind speeds, lower turbulence levels and can be optimally arranged in arrays. The capital costs of building in seawater mean that the economics of offshore wind is less competitive than onshore, however this is partly countered by the higher capacity factors achieved by offshore projects. In recent years the cost of offshore projects has fallen dramatically, and as a result, the International Renewable Energy Agency (IRENA) expects offshore wind to increase from 23 GW (gigawatts) in 2018 to almost 1000 GW by 2050 (IRENA 2019). This now means that many nations need to review and potentially include offshore wind farm resources in their planning for future renewable energy investment and reverse auctions.

Currently, offshore wind is most popular in northern Europe where space limitations have reduced the number of viable onshore locations. It is however possible that future improvements in offshore construction techniques, together with technology advances such as floating platforms, may further improve offshore economics and see offshore wind become competitive in more countries.

2.3 Wind Assessment and Development Process

A wind farm developer must consider a large variety of factors when prospecting for a new site. These include:

- the quality of the wind resource;
- the distance to nearby transmission, its level of spare capacity and the cost of connection;
- the terrain complexity and transport considerations for construction;

- land permitting requirements;
- environmental restrictions; and
- community attitudes and considerations, such as the number and distance of surrounding residences, and whether the development would negatively impact current land use in the region such as tourism, agriculture and lifestyle.

If consideration of the above 'development risks' does not present any 'fatal flaws', then the developer may choose to continue progressing the development. This typically involves agreements with the landowners and community, direct measurement of the wind resource and many environmental and planning studies.

2.3.1 Wind Measurement

The economics of each project depends critically on the wind resource, and for this reason it is important to obtain accurate predictions of the wind across the site. As the wind can vary substantially from year to year, it is important to commence measurements of the wind as early as possible so as to maximise the length of available data. It is common for a developer to initially obtain 12 months of data, which will enable a preliminary though moderately accurate estimate of the wind resource, before deciding whether to spend more money on additional monitoring at other potential turbine locations.

Historically, it has been common practice to use masts with cup anemometers to measure the wind speed, in conjunction with wind vanes to measure the wind direction. More recently, remote sensing using LIDAR (light detection and ranging) or SODAR (sonic detection and ranging) have also been used. These instruments use laser or sound pulses to measure the wind speed and direction. They have improved portability and they can also measure at higher elevations than conventional masts. However, acceptance of the validity or accuracy of the data varies within the industry, and often these instruments are used to supplement mast-based data rather than used as a complete replacement. Regardless of the type of instrument used, it is important that recommended industry standard procedures for wind monitoring are followed, so as to minimise uncertainty in the modelled wind resource.

In addition to on-site monitoring, developers usually also seek external longer-term data sets in order to (long-term) adjust the on-site data. Weather station data from a nearby location are often used, though the use of reanalysis[1] data is now a common alternative or addition. Whatever the source of the data, the idea is to obtain a good correlation between the on-site measurement data and the longer-term external data, and then to apply those correlations to the long-term data. The end goal of the long-term adjustment process is to reduce uncertainty in the wind resource.

---

[1] Reanalysis data sets are global data sets that cover the Earth's surface with a grid, typically with a temporal resolution of 6 hours. The data sets are compiled by atmospheric models that embed data from global weather stations, weather balloons, ocean buoys and shipping data.

2.3.2 Energy Assessment

Software using computational fluid dynamic models is used to predict the wind resource away from the monitoring locations, to all the potential turbine locations. Following this, optimisation software is used along with potential turbine power curves to create the best possible turbine layout. The optimisation software incorporates turbine wake models, and aims to maximise the energy prediction of the resultant layout, keeping in mind various constraints such as land inclusion or exclusion zones and minimum turbine separation requirements, as well as potential construction costs.

As more information on the wind resource becomes available, the developer will be able to make a more accurate optimisation of the expected turbine layout. As this progresses, they will continue discussions with any additional landowners, as required, and also start working through the various environmental, land, grid and community approval processes.

2.3.3 Uncertainty and Financial Due Diligence

It is common practice for the wind farm to be funded through a combination of equity and debt. Those providing the debt typically want to be very confident that the revenue from the operating wind farm will be sufficient to cover the interest and capital repayment expenses each year over the life of the loan. This means that in addition to assessing the most likely energy production of the wind farm, assessments must also be made of the uncertainty in those predictions. Uncertainty can be influenced by many factors, including:

- the accuracy of instruments used to measure the wind speed;
- the vertical distance between the height of measurements and the turbine hub height;
- the length of the on-site data monitoring period;
- the degree of correlation with any external wind data sources and their data length;
- the complexity of the site; and
- the average distance between monitoring locations and turbine locations.

The task of the developer is to weigh the benefits that come with reduced uncertainty against the increased expenses that are associated with minimising this uncertainty.

2.4 Financing Wind Farm Projects: Factors That Influence the Price of Wind Energy

The key drivers that impact the price of delivered energy from a wind farm (or any other form of electricity generation) are:

- cost of capital (debt and equity or tax equity in the USA);
- operational period or expected lifetime of asset (typically 20–30 years);
- strength of resource (often expressed in terms of net capacity factor);
- capex (capital expenditure) – cost of development and construction (includes financing and transmission connection);

- opex (operational expenditure) – operations costs (includes all warranty, maintenance and operations together with asset management costs); and
- guaranteed revenue and the period for which revenue is guaranteed (or expected if it is a merchant plant).

Each of these drivers is explored in more detail below before a summary of the costs of wind in various global markets is provided.

2.4.1 Cost of Capital

In a project finance arrangement, the cost of both debt and equity will depend on a number of regional and project-specific factors. These include: local debt rates; local risk premiums or margins (typically higher in emerging economies); strength and creditworthiness of counterparties (experienced equity investors and turbine manufacturers with strong balance sheets provide greater comfort to lenders); and, perhaps most importantly, the amount and creditworthiness of revenue that the project has contracted.

2.4.2 Asset Lifetime

There are many wind farms around the world that have been operational for over 20 years and modern wind farms are expected to last at least 20 years. Turbine manufacturers will undertake loading assessments prior to construction and often provide each wind turbine with a design life of 25–30 years.

2.4.3 Net Capacity Factor (NCF)

The net capacity factor (NCF) is dependent on the strength and quality of the wind resource, the turbine power curve and the energy losses. Larger blade diameters and hub heights mean that most sites would expect to have an NCF in the range of 30–50%. This in itself represents a substantial reduction in costs, as in the late 1990s typical NCFs were around 15–25%.

2.4.4 Capex and Opex

The total construction cost (capex) of each wind farm depends primarily on the turbine costs, the balance of plant costs including roads, collector network and control building and the costs of connecting the wind farm onto the local transmission system.

The operating costs (opex) for a wind farm are largely driven by the costs of scheduled and unscheduled maintenance of the wind turbines and associated infrastructure. As turbine sizes have increased, the costs of scheduled maintenance have not increased substantively on a per turbine basis. This has led to a reduction in opex on a dollar per megawatt-hour

(MWh) basis. The risk of component failures that result in unscheduled maintenance is often offset through the purchase of an extended warranty from the manufacturer.

2.4.5 Revenue

Fixing or guaranteeing the revenue for a period that exceeds the term of the debt contract for the project is the preferred mechanism for ensuring that projects are 'bankable'. In merchant markets, where large volumes of electricity are bought and sold and the price fluctuates with supply and demand, it is possible to 'hedge' a significant portion of the electricity expected to be generated by the wind farm and finance the project against the hedged portion.

2.4.6 Comparing the Cost of Wind in Global Markets

The price of delivered energy varies significantly with the above variables and from market to market. The (US) National Renewable Energy Laboratory (NREL) has estimated the range of LCOE (levelised cost of energy) for onshore wind energy in 2017 in the USA (NREL 2017). They calculated an average number of USD 47/MWh and a range of USD 42–65/MWh with the key sensitivities presented in Figure 2.5.

It is useful to compare[2] this number with recent published power purchase agreement rates in various global markets.

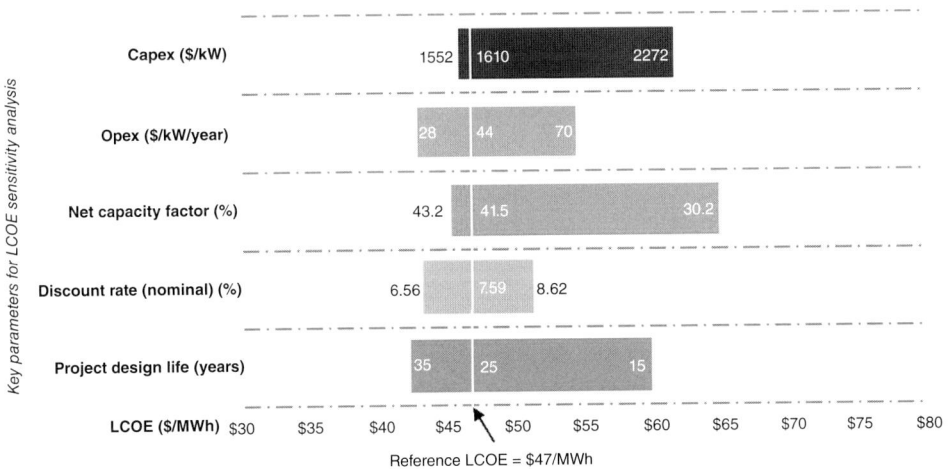

Figure 2.5 Levelised cost of energy of wind in USA.
Source: Reproduced from NREL (2017). For a colour version of this figure, please see the colour plate section.

---

[2] Exchange rate assumptions for the comparison were 1 USD = 0.8 AUD = 0.084 ZAR = 0.336 BRL (2015) = 0.499 BRL (2013). BRL is Brazilian real, AUD is Australian dollar and ZAR is South African rand.

- Australia: Publicly released numbers from the 2015 and 2016 Australian Capital Territory Wind Auctions (ACT Government 2016) are in the range of AUD 73–92/MWh with no indexation for the 20-year life of the offtake agreement; once indexation and conversion to US dollars is undertaken these prices are equivalent to a range of around USD 47–58/MWh.
- South Africa: Four bidding rounds have taken place for renewables in South Africa's Renewable Energy Independent Power Procurement Programme (IPP Renewables n.d.). Round-four announcements were made in April 2015 with wind prices ranging from around USD 47/MWh to USD 56/MWh.
- Mexico: Mexico has held three rounds of power auctions. In round two (2016), prices averaged USD 36/MWh, while in round three (2017) prices had dropped to USD 21/MWh (Mora 2017).
- Peru: In February 2016, the Peruvian energy watchdog secured wind energy at prices of around USD 37/MWh (Hristova 2016).
- Morocco: A wind tender for 850 MW of wind generation attracted bids with an average price of USD 30 (Yaneva 2016).

Although these markets are quite distinct, the pricing is relatively consistent in a US dollar context and at or below the lower end of the NREL LCOE estimates.

Offshore wind has historically been considerably more expensive than onshore, due to the complexities of building large structures at sea. However, costs for offshore wind farms have fallen significantly between 2015 and 2018, achieving a LCOE in 2018 that was previously predicted for 2030. These have fallen to USD 0.13/kWh in 2018 (IRENA 2019), and are predicted to fall further, to USD 0.05–0.09/kWh by 2030, based on current trends and prices awarded in auctions in 2016–2018 (IRENA 2019). According to IRENA (2019), offshore wind is now an attractive proposition to provide clean, low-cost electricity that can compete head-to-head with fossil fuels without financial support in certain European markets. By 2030 it is expected to be competitive with fossil fuels in many countries across the world.

2.4.7 Economic Cost Benefits of Wind

A number of studies have been conducted in recent years on the cost of increasing the amount of renewable energy generation in Australia. A 2018 study conducted by RepuTex compared two emission target scenarios, one of which would result in 42% of generation being derived from renewable energy sources in 2030 associated with a 26% reduction in emissions, and the other would see 50% of generation from renewables associated with a 45% reduction in emissions (RepuTex Energy 2018). The modelling found that wholesale electricity prices would be lower under the more ambitious target with higher level of renewables. Moreover, pricing would be lower in 2030 than in 2017, the start year for the model. In both scenarios, the big drop in wholesale pricing between 2018 and 2020 was due to a large increase in the amount of wind generation displacing higher-cost generation.

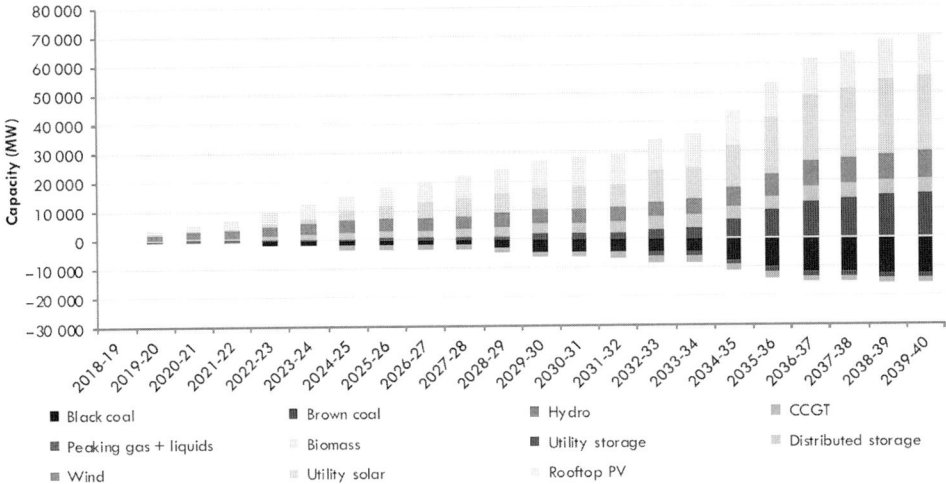

Figure 2.6 Relative change in installed capacity in the neutral scenario.
Source: AEMO (2018). For a colour version of this figure, please see the colour plate section.

The results of modelling conducted by RepuTex are consistent with those found by the Australian Energy Market Operator, AEMO. In its 2018 *Integrated System Plan*, AEMO showed the results of a cost-based engineering optimisation plan to forecast the national electricity market over the next 20 years (AEMO 2018). Under all modelled scenarios, coal was mostly phased out over that 20-year period and was replaced primarily by a mixture of wind and solar, supplemented with storage, as seen in Figure 2.6.

2.5 Transitioning to High Wind and Renewables Penetration

There are today many markets around the world where wind energy makes up a significant portion of local electricity supply. Denmark and Lithuania were the leading countries in 2018, with 46% and 33% of generation coming from wind power, respectively (BP 2019). As a result, there are multiple utilities and network operators with real experience of operating their systems and servicing their customers in situations where wind makes up a significant portion of electricity generation. This experience together with various studies (Cochran et al. 2012) indicate that wind can be integrated without major technical implications up to levels of 20–40%, dependent on the network and resource characteristics. Research is now beginning to focus on the implications of very high renewables penetration levels (up to 100%). In this section we explore a few regions of the world that either already have high levels of wind penetration or where there is clear momentum towards such transition.

2.5.1 Denmark

In 2018, Denmark generated 46% of its electricity from wind energy. This is a remarkable statistic and even more so with reference to Figure 2.7. In this figure the fraction of demand

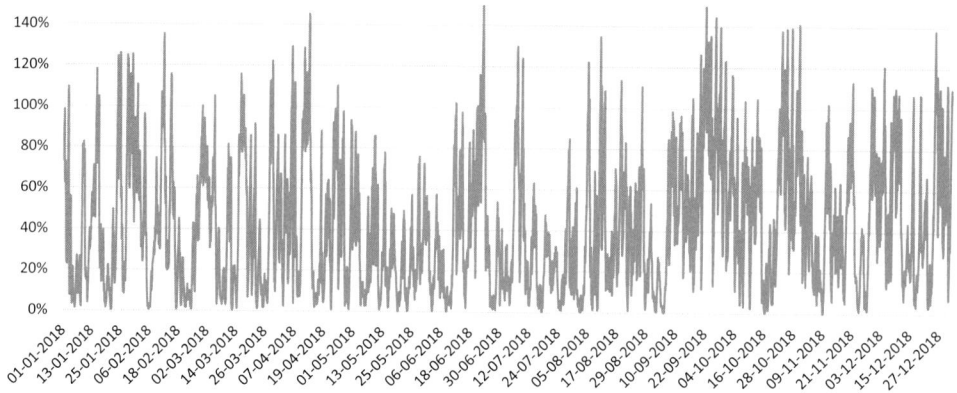

Figure 2.7 Fraction of Danish demand from wind generation, hourly data, December 2014.
Source: Based on Chabot (2016). For a colour version of this figure, please see the colour plate section.

supplied by wind generation is plotted for 2018. The obvious question that this graph poses is how does Denmark balance supply and demand given the fluctuating nature of the wind resource?

There are three key factors that allow Denmark to have such high levels of wind on their electricity system:

(1) integration into the neighbouring electrical networks of Norway, Sweden and Germany – the import and export of electricity is correlated with the level of wind generation;

(2) substantive levels of CHP (combined heat and power) plants that provide district-level heating. These CHP plants can provide some balancing to the network by varying their output when wind output changes; and

(3) day-ahead forecasting and scheduling: forecasts of the level of wind that will be available for electricity generation are now integral to the balancing process. Utilities and networks will therefore plan how much thermal generation will be required, and the level of imports and exports, hours and days in advance.

Denmark has an interim target to achieve 50% wind energy on its network by 2020 and appears likely to reach this target. In addition, they have an overall vision to become completely fossil-fuel-free by 2050.

2.5.2 USA

The USA has seen very large quantities of wind energy introduced to its varied electricity networks since the early 2000s. Even with the recent shale gas revolution and resulting historically low natural gas pricing, wind continues to compete economically as the lowest cost of new energy generation available.

Table 2.1. *Top 10 US states by wind energy as percentage of electricity supply (AWEA n.d.)*

Rank/State	Percentage	Rank/State	Percentage
1. Iowa	37%	6. Maine	20%
2. Kansas	36%	7. Minnesota	18%
3. Oklahoma	32%	8. Colorado	18%
4. South Dakota	30%	9. Idaho	15%
5. North Dakota	27%	10. Texas	15%

It is not surprising to observe, therefore, multiple states where wind energy contributes a significant portion of electricity supply. The top 10 as of 2017 are tabulated in Table 2.1.

California, the second largest energy user in the USA after Texas, has an ambitious goal of achieving 50% renewables by 2026, 60% by 2030 and 100% carbon-free energy by 2045. Around 7% of California's electricity production came from 5.6 GW of installed wind capacity in 2017 (AWEA n.d.). Reaching the 50% target is expected to see at least a doubling of wind production in California, with the balance coming from solar, existing hydro and biomass. There is some potential for large imports of wind power from wind-rich states such as Wyoming and New Mexico, which may end up being more cost-effective than building wind energy within California.

The National Renewable Energy Laboratory (NREL) has undertaken several recent studies (NREL 2012) to explore technical issues of integrating very high proportions of renewables into the US electricity system and has published the *Renewable Electricity Futures Study*. The central conclusion of this study is that:

renewable electricity generation from technologies that are commercially available today, in combination with a more flexible electric system, is more than adequate to supply 80% of total U.S electricity generation in 2050 while meeting electricity demand on an hourly basis in every region of the United States. *(NREL 2012: iii)*

2.5.3 Australia

Australia's wind resource is well distributed and has a technical potential that significantly exceeds total generation requirements. This is reflected in Figure 2.8 and by the fact that wind has been installed in all corners of Australia, from Mount Emerald in Far North Queensland to Woolnorth in north-western Tasmania and Collgar in the Western Australian wheat belt.

By the start of 2019 there was 5679 MW of wind generating capacity in operation across the country (Clean Energy Council 2019), but an investment boom resulted in another 5.69 GW of capacity under construction or due to start soon. These projects will likely generate around 36 TWh of electricity, and on their own will easily exceed the 2020 large-scale renewable energy target of 33 TWh. Wind generation of 36 TWh would represent

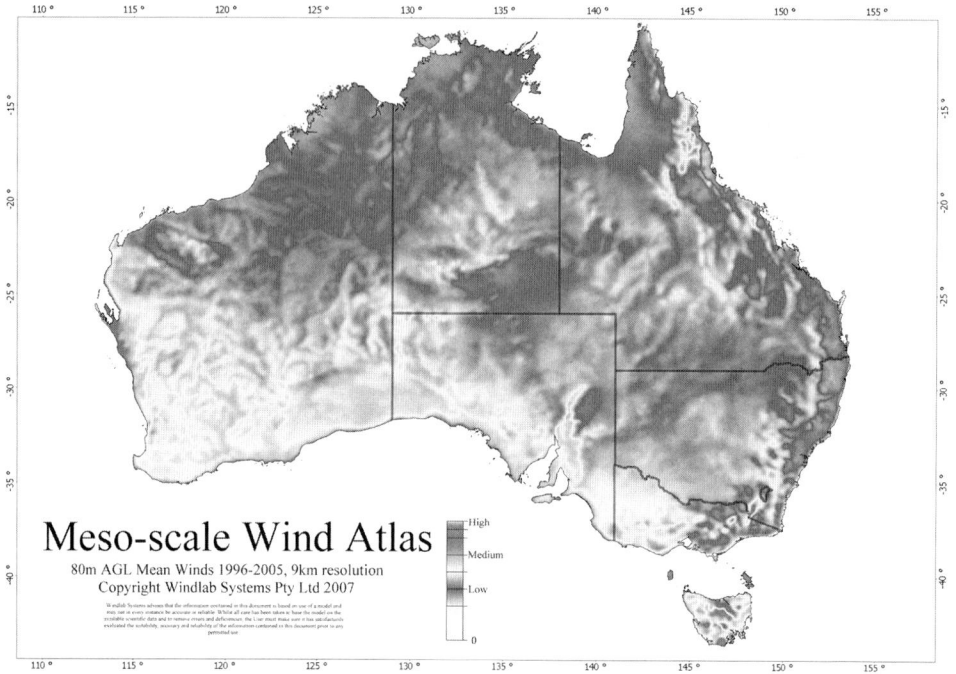

Figure 2.8 Australian Wind Atlas.
Source: Courtesy of the authors. For a colour version of this figure, please see the colour plate section.

approximately 14% of Australia's electricity demand. This is not unexpected, as Australia is a large, windy country. China produced around ~240 TWh of wind electricity in 2016, approximately the equivalent of Australia's entire electricity generation output.

The state of South Australia is relatively weakly interconnected with the rest of the National Electricity Market, but has already achieved a 40% level of wind integration (OpenNEM n.d.). Most of that generation was built over the course of a decade, a period which saw the closure of all its coal generators, which had previously provided about a third of all generation. The transition also saw a reduction of imports from the neighbouring state of Victoria, with no trend in gas generation.

2.5.4 United Kingdom

By the end of 2018 there was around 22 GW of wind energy installed in the UK (RenewableUK n.d.), generating approximately 17% of the country's electrical needs (Figure 2.9). The UK has agreed to a 2020 renewable electricity target of 30%. Regionally, more ambitious targets are in place. The Scottish government has a target of 100% renewables equivalent generation by 2020, with wind expected to be dominant in this mix.

The UK has been an early adopter in the installation of offshore wind energy, around 8.2 GW was installed by the end of 2018. Much of the UK's anticipated renewables growth

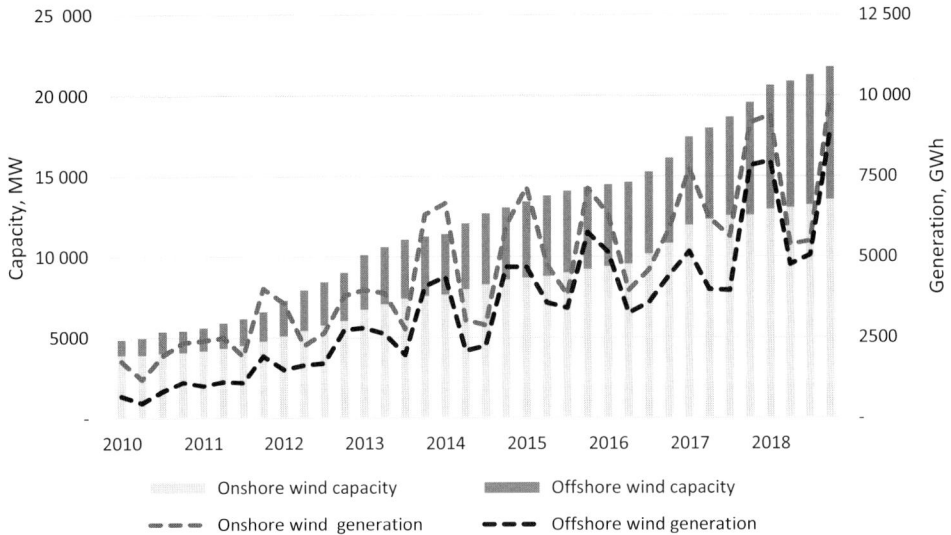

Figure 2.9 UK wind energy capacity and generation by quarter, 2000–2018.
Source: UK DBEIS (2013).

will be in offshore wind, and learnings from the UK are expected to allow new offshore markets, predominantly in Europe and Asia where land constraints reduce the opportunities for onshore wind, to ramp up by the turn of the decade.

Offshore wind auctions in 2017 saw prices fall over 30% from 2012 levels. Developers believe that a new generation of even bigger turbines mean they can achieve further cost reductions. The sector is perceived by investors to be much less risky than it was 5 years ago, bringing down the cost of capital.

2.5.5 South Africa

In South Africa, the Renewable Energy Independent Power Producer Procurement Programme (REIPPPP) commenced in 2010 and the first four rounds have seen 34 wind projects awarded Preferred Bidder status (RSA Department of Energy 2018). A total of 6328 MW of renewable energy has been procured from independent power producers (of which 53% is wind and 36% is solar photovoltaics), committing USD 20.5 billion in total investment since the programme's inception (Eberhard and Naude 2017). The country enjoys some of the best wind and solar resources in the world, with significant land area in the interior of the country suitable for the development of wind farms (Wright, Ireland et al. 2017).

As of late 2018, 3357 MW of wind energy was operational and connected to the national grid, supplying around 3% of total installed capacity (Eberhard and Naude 2017). Note that coal, as an energy generation source, contributed 36.8 GW (74%) to the total generation mix (Wright, Bischof-Niemz et al. 2017). As environmental impera-tives become more important, it is envisioned that decommissioned coal will struggle to be replaced with new coal generation. Significant reductions in tariffs for renewable energy have been observed as ensuing rounds of the REIPPPP have become increasingly competitive; average wind costs in the latest round four were USD 0.05/kWh (Eberhard and Naude 2017).

The rate of investment into the REIPPPP had slowed since 2015, due to delays in the procurement program associated with a much debated and politically complex update to the Integrated Resource Plan (IRP). However, a new IRP was finally gazetted by the Minister of Mineral Resources and Energy in October 2019, which sets an exciting pathway for the procurement of new renewables (predominantly wind and solar photovoltaics) over the period to 2030. It requires that 1600 MW of wind is procured per annum from 2022 to 2030, and estimates that wind energy will provide around 18% of South African electricity by 2030. The policy framework was confirmed when REIPPPP Bid Window 5 was launched in March 2021, and if the IRP is maintained in its current form around 15 GW of new wind energy will be procured and constructed over the next 10 years.

2.6 Future Wind Scenario

In previous sections we have noted that there is no technical barrier to wind penetration levels exceeding 30% in certain markets. Given the favourable economics of wind and the growing requirements to mitigate against carbon, it is useful to explore whether a 20% global wind scenario by 2040 is feasible and what that would mean in terms of annual global investments. The assumptions and projections associated with our 20% scenario are summarised in Table 2.2.

Around 56 GW of wind power was installed globally each year from 2014 to 2017. Therefore, the global install rate needs to increase to realise the 20% scenario. The International Energy Agency's (IEA) 2013 *Technology Roadmap: Wind Energy* (IEA 2013) suggests a lower average install rate of around 65 GW per annum for the period up to 2050, with wind supplying about 18% of global electrical needs in 2050.

Table 2.2. *Assumptions and projections associated with 20% global wind energy generation scenario*

Global installed wind capacity (2017)	~539 GW (REN21 2018)
Projected electricity consumption (2040)	~36 000 TWh/yr[3]
20% by wind generation scenario (2040)	~7200 TWh/yr
Installed wind capacity required (2040)	~2200 GW
Linear annual install rate to 2040	~72 GW/yr
Average capacity factor of new projects	~42%
Annual investment	~USD 100 billion/yr

---

[3] The International Energy Agency (IEA 2018) projects that global electricity consumption will grow from around 25 570 TWh/yr in 2017 to around 36 000 TWh/yr by 2040 under its 'New Policies' scenario.

The IEA *Roadmap* and perhaps even the 20% scenario above can be considered conservative for a number of reasons:

- IEA projections are consistently known to be conservative and there is a long history of upgrading the projections on an annual basis;
- the latest *Technology Roadmap: Wind Energy* was published in 2013, and thus can be considered quite out of date;
- there is known to be a strong pipeline of approved and under-development projects in most mature and early-stage development markets; in many markets these projects can comfortably exceed a 20% penetration level;
- wind is competing on price alone in many markets and with further technology improvements and industry experience the price will become more competitive; and
- an NCF of 42% is assumed for future wind projects in the above scenario: as was noted, higher hub heights, longer blades and more efficient turbines means this is highly likely to be an underestimate, further reducing the required level of installations necessary to achieve 20% wind penetration.

2.7 Conclusion

There no longer appears to be any significant impediment to wind becoming one of the dominant energy sources for the global electricity sector before mid-century. Economics alone make wind a go-to fuel source for new-build in markets with demand growth.

This summary has been largely silent on policy direction as wind continues to grow and mature at a rate that surprises the market analysts. In markets without demand growth the authors consider the most important policy direction to be a strong commitment to decarbonise. Such a commitment could come either through a timetable for decommissioning old thermal power stations or through a strong carbon price. With such policy in place the transition to low-carbon electricity could accelerate further.

References

ACT (Australian Capital Territory) Government (2016). *Canberra 100% Renewable: Leading Innovation with 100% Renewable Energy by 2020*. Canberra: Australian Capital Territory Government. Available at: www.environment.act.gov.au/__data/assets/pdf_file/0007/987991/100-Renewal-Energy-Tri-fold-ACCESS.pdf.

Ahuja, E. (2015). Energy change: Insights from the 1st and 2nd Industrial Revolutions and recent developments to help achieve the next low carbon industrial revolution. Masters Thesis, The Australian National University. Available at: www.academia.edu/27596477/Ahuja_E_2015_Energy_Change_-Insights_from_the_1st_and_2nd_Industrial_Revolutions_and_Recent_Developments_to_help_achieve_the_next_Low_Carbon_Industrial_Revolution._ANU_ECI_Masters_Thesis_advanced_._Supervisor_Dr_Michael_H_Smith.

AEMO (Australian Energy Market Operator) (2018). *Integrated System Plan: For the National Electricity Market*. Australian Energy Market Operator. Available at:

https://aemo.com.au/-/media/files/electricity/nem/planning_and_forecasting/isp/2018/inte grated-system-plan-2018_final.pdf.

AWEA (American Wind Energy Association) (n.d.). State facts sheets. *American Wind Energy Association*. Available at: www.awea.org/resources/fact-sheets/state-facts-sheets.

BP (2019). *BP Statistical Review of World Energy 2019*. London: BP. Available at: www .bp.com/en/global/corporate/energy-economics/statistical-review-of-world-energy .html.

Chabot, B. (2016). Backing up wind and nuclear power. *Erneuerbare Energien*. Available at: www.erneuerbareenergien.de/backing-up-wind-and-nuclear-power/150/437/86412/.

Clean Energy Council (2019). *Clean Energy Australia Report 2019*. Clean Energy Council. Available at: https://assets.cleanenergycouncil.org.au/documents/resources/ reports/clean-energy-australia/clean-energy-australia-report-2019.pdf.

Cochran, J., Bird, L., Heeter, J. and Arent, D. J. (2012). *Integrating Variable Renewable Energy in Electric Power Markets: Best Practices from International Experience*. Springfield, VA: US Department of Energy and US Department of Commerce. Available at: www.nrel.gov/docs/fy12osti/53732.pdf.

Eberhard, A. and Naude, R. (2017). *The South African Renewable Energy IPP Procurement Programme: Review, Lessons Learned & Proposals to Reduce Transaction Costs*. Cape Town: Graduate School of Business, University of Cape Town. Available at: www.gsb.uct.ac.za/files/EberhardNaude_REIPPPPReview_ 2017_1_1.pdf.

Hills, R. (1996). *Power from Wind: A History of Windmill Technology*. Cambridge: Cambridge University Press. Available at: www.cambridge.org/au/academic/sub jects/general-science/history-science/power-wind-history-windmill-technology.

Hristova. D. (2016). Peru shortlists 13 winners in renewables auction. *Renewables Now*. 17 February. Available at: https://renewablesnow.com/news/peru-shortlists-13-winners- in-renewables-auction-513576/.

IEA (International Energy Agency) (2013). *Technology Roadmap: Wind Energy*. Paris: International Energy Agency. Available at: www.energie-nachrichten.info/file/News/ 2013/2013-10/Wind_2013_Roadmap.pdf.

IEA (2018). *World Energy Outlook 2018*. Paris: International Energy Agency. Available at: www.iea.org/reports/world-energy-outlook-2018/electricity.

IPP Renewables (n.d.). BW4 preferred bidder announcement. *IPP Projects*. 16 April. Available at: https://ipp-projects.co.za/PressCentre.

IRENA (International Renewable Energy Agency) (2019). *Future of Wind: Deployment, Investment, Technology, Grid Integration and Socio-economic Aspects (A Global Energy Transformation Paper)*. Abu Dhabi: International Renewable Energy Agency. Available at: www.irena.org/-/media/Files/IRENA/Agency/Publication/ 2019/Oct/IRENA_Future_of_wind_2019.pdf.

Mora, A. (2017). Mexico's third long-term electricity auction: The results and the com-parison. *Mexico Business Publishing*. 29 November. Available at: www .renewableenergymexico.com/mexicos-third-long-term-electricity-auction-the-results- and-the-comparison/ (link discontinued).

NREL (National Renewable Energy Laboratory) (2012). *Renewable Electricity Futures Study*. Edited by M. M. Hand, S. Baldwin, E. DeMeo et al. 4 vols. NREL/TP- 6A20–52409. Golden, CO: National Renewable Energy Laboratory. Available at: www.nrel.gov/analysis/re-futures.html.

NREL (2017). *Cost of Wind Energy Review*. Technical Report NREL/TP-6A20–72167. Golden, CO: National Renewable Energy Laboratory. Available at: www.nrel.gov/ docs/fy18osti/72167.pdf.

OpenNEM (n.d.). *OpenNEM Project*. Created by McConnell, D., Holmes à Court, S. and Tan, S. Available at: https://opennem.org.au/energy/nem.

REN21 (2018). *Renewables 2018: Global Status Report*. Paris: REN21. Available at: www.ren21.net/wp-content/uploads/2019/05/GSR2018_Full-Report_English.pdf.

RenewableUK (n.d.). Wind energy statistics. *renewableUK*. Available at: www.renewableuk.com/page/UKWEDhome/Wind-Energy-Statistics.htm.

RepuTex Energy (2018). *The Impact of the NEG on Emissions and Electricity Prices by 2030: Modelling for Greenpeace Australia Pacific*. Melbourne: RepuTex Australia. Available at: www.reputex.com/wp-content/uploads/2018/07/REPUTEX_Modelling-of-the-National-Energy-Guarantee_0718_26-45.pdf.

RSA (Republic of South Africa) Department of Energy (2018). 2018 Draft Integrated Resource Plan. Available at: www.energy.gov.za/IRP/irp-update-draft-report-2018.html.

UK DBEIS (Department for Business, Energy and Industrial Strategy) (2013). National statistics: Energy trends: UK renewables. *Gov.uk*. 9 January. Available at: www.gov.uk/government/statistics/energy-trends-section-6-renewables.

US Department of Energy (2017). *2017 Wind Technologies Market Report*. US Department of Energy. Available at: www.energy.gov/eere/wind/downloads/2017-wind-technologies-market-report.

Wright, J. G., Bischof-Niemz, S. T., Calitz, J. R., Mushwana, C. and van Heerden, R. (2017). Future wind deployment scenarios for South Africa. Paper presented at WindAc Conference, Cape Town, South Africa, 14–16 November. Available at: http://hdl.handle.net/10204/10070.

Wright, J., Ireland, G., Hartley, F., Merven, B., Burton, J., Ahjum, F. Mccall, B., Caetano, T. and Arndt, C. (2017). *The Developing Energy Landscape in South Africa: Technical Report*. Cape Town: University of Cape Town Energy Research Centre, CSIR (Council for Scientific and Industrial Research) and IFPRI (International Food Policy Research Institute).

Yaneva, M. (2016). Morocco's wind power price goes as low as USD 30/MWh. *Renewables Now*. 19 January. Available at: https://renewablesnow.com/news/moroccos-wind-power-price-goes-as-low-as-usd-30-mwh-509642/.

3

Solar Photovoltaics

ANDREW BLAKERS

Executive Summary

Solar energy is vast, ubiquitous, non-polluting and indefinitely sustainable. It is an ideal energy solution. And it is here and now: solar photovoltaic (PV) technology has recently become the dominant global energy technology in terms of new generation capacity being added each year, with wind energy in second spot.

Worldwide, PV technology accounts for 42% of annual net new generation capacity additions. Wind accounts for another 33%, hydro 7%, fossil and nuclear combined, 17%; and all other renewable technologies combined, 1%.

The costs of PV energy production have dropped rapidly, by over three-quarters since 2010. As the chapter explains, further price cuts are likely in the next decade. PV electricity is now less expensive than domestic and commercial retail electricity from the grid throughout much of the world and has reached cost-competitiveness with wholesale conventional electricity.

The materials required to construct PV systems are available in abundance, principally silicon – the second most abundant element on Earth after oxygen – but also materials like glass, aluminium and steel for framing and support of panels.

There are no significant constraints on the deployment of PV technology: much less than 1% of the world's land surface is required to achieve 100% renewable energy, and there are minimal security concerns.

Wind and PV energy can support a 100% renewable electricity system in many countries. Required supporting technologies include high-voltage cables to interconnect regions over large areas (hundreds of thousands of square kilometres), in order to smooth out local weather and climate, and storage.

As well as offering the best hope for rapid reduction in global greenhouse gas emissions through reductions in fossil fuel use, PV energy is already produced at vast scale, is already competitive with fossil fuels and is effectively unconstrained by technical, environmental, social and economic factors in most countries.

Extensive renewable electrification of land transport and heating allows PV and wind energy to contribute to large emissions reductions, approaching two-thirds in typical developed countries.

3.1 Introduction

Each year the Earth receives about four orders of magnitude more solar energy than human commercial energy consumption. After accounting for conversion losses (only 10–50% of solar energy incident on a solar collector is successfully collected and converted to a usable form) and inaccessible regions (oceans, the poles, mountains and forests), there is hundreds of times more available solar energy than commercially traded energy consumption.

By far the most widely deployed of solar energy conversion technologies is PV technology. On current trends, PV and wind energy are likely to be the dominant low-emission energy technologies deployed over the next decade. Some of the other low-emission energy technologies may have significant supporting roles in some countries.

3.2 The Solar Resource

The sun provides about 1.3 kilowatts per square metre ($1.3 \, kW/m^2$) to the upper atmosphere of the illuminated half of the Earth. Some is absorbed and some is reflected by the atmosphere. The solar intensity at noon on a sunny day at the Earth's surface is about $1 \, kW/m^2$. Each year the Earth receives about 4×10^{24} joules of solar energy. Most of this energy is in the form of direct-beam radiation; that is, it comes directly from the visible disc of the sun. However, a sizeable fraction appears as diffuse radiation due to scattering from clouds, aerosols and other atmospheric constituents, plus reflected light from the ground.

A relatively small fraction of the solar energy is converted into other usable energy forms and can be harvested as wind energy, hydroelectricity, wave energy, biomass and in other forms.

The sum of the direct and diffuse radiation received by a solar collector is termed global radiation. Some collector systems, such as non-concentrating PV panels, respond to both the direct and diffuse components of sunlight. Other collectors, such as concentrating PV and solar thermal systems, respond only to the direct-beam component – fundamental physical laws limit concentration of scattered light. For this reason, concentrating systems are best suited to dry locations with low levels of cloud and air pollution. The proportion of global radiation that is diffuse is typically 15–50% depending upon location and season, and this solar power will be discarded in concentrating systems. Desert regions have annual diffuse radiation amounting to only 15–20% of the available solar power.

The available solar power depends upon latitude, weather patterns and air clarity. Annual solar energy availability is important, as is its seasonal variation – because it is expensive to store energy harvested in summer for use in winter. In general, low latitudes have much less seasonality in both solar energy availability and energy demand (for heating and cooling).

About two-thirds of the world's population lives in the latitude range +/–35°, where there is generally good solar availability and moderate seasonal variation of both solar energy supply and energy demand compared with higher latitudes. This latitude range is home to most of the populations of Africa, the Middle East, Central and South America, Australasia and Oceania, South East Asia, and India and South Asia. However, most industrialised countries, which are those with the most energy (and greenhouse gas)-intensive economies, lie at higher northern latitudes, including European countries, Russia, Canada and much of the USA, China and Japan.

3.3 Environmental and Social Aspects of Solar Energy

Solar energy collection principally utilises only very common materials. For example, PV systems utilise silicon (for the solar cells); silicon, oxygen and sodium (for the cover glass of the solar module); oxygen, carbon and hydrogen (for the encapsulating plastic); aluminium (for the frame of the solar module); and iron (for the steel support posts), plus some elements in small quantities such as phosphorus, boron, copper and silver. These elements are ubiquitous in the Earth's crust and atmosphere and it is difficult to envisage ever running out of them. The amount of rock that needs to be moved during mining, for a given level of solar energy production, is orders of magnitude smaller than the equivalent for fossil and nuclear energy systems, principally because of the absence of the need for fuel.

Solar energy is available nearly everywhere in vast quantities; it is unlikely that people will ever go to war over access to solar energy, in contrast to the situation with high-energy-density fossil fuels. Utilisation of solar energy entails minimal security and military risks. The highly dispersed nature of millions and billions of individual solar energy collectors entails a robust and resilient energy system with limited utility for warfare and terrorist activity.

About 0.2% of the world's land area would be required to supply all of the world's commercially traded energy requirements from solar energy, assuming that all people consume energy at a similar rate to North Americans. A large segment of the world's energy can be supplied from roof-mounted solar collectors which effectively alienate no land. Another large segment of the world's energy can be supplied from solar collectors in arid regions, in conjunction with high-voltage long-distance transmission of electricity. Relatively little alienation of productive farmland, forests and biosystems is required to achieve a world economy where most of the commercial energy is derived from solar energy.

Solar energy systems do not emit greenhouse gases during operation. However, greenhouse gases, principally carbon dioxide (CO_2), are emitted during the manufacturing phase. The time required to generate enough electricity to displace the CO_2 emissions equivalent to that invested in construction of a solar energy system is currently in the range of 1–3 years (Bhandarib 2015), compared with typical system lifetimes of 20–40 years. Manufacturing intensity, CO_2 and price are directly linked (via material consumption and efficiency), and so CO_2 payback times continue to fall as prices fall. Carbon dioxide payback times are also falling as the proportion of low-emission generators in electricity systems increases; they will eventually fall to a small fraction of 1 year.

Solar energy system manufacturing and operation entails minimal pollution and no noise. Social acceptance is very high. Both the risk and consequences of accidents are very low, in contrast to the cases of fossil fuel and nuclear energy systems.

3.4 Silicon Photovoltaics

Photovoltaic technology produces electricity directly from sunlight, usually without moving parts. Most of the world's PV market (95%) is serviced by crystalline silicon solar cells (Reinders et al. 2015), and this will remain true for the foreseeable future. Sunlight is absorbed by the solar cell and the solar power is converted to electrical power with a conversion efficiency of around 20%. The remaining solar power (80%) becomes waste heat. This process of conversion is called photovoltaics (photo = light, voltaics = voltage).

In a silicon solar cell, sunlight causes electrons to become detached from their host silicon atoms. Near the upper surface is a 'one-way membrane' called a p–n junction. When an electron crosses this junction it cannot easily return, causing a negative voltage to appear on the sunward surface (and a positive voltage on the rear surface). The sunward and rear surfaces are connected together via an external circuit containing a battery or a load in order to extract current, voltage and power from the solar cell (Figure 3.1).

Silicon is the second most abundant element in the Earth's crust after oxygen (it is normally found in the form of silicon dioxide and silicate minerals). Silicon is a light, grey, hard and brittle metallic-like material. Silicon is obtained from quartz (SiO_2) or pure sand. The quartz is mixed with coke or charcoal in a furnace, resulting in the production of metallurgical-grade silicon (98% pure) and CO_2. Millions of tonnes of silicon are produced each year for the aluminium, steel, silicon and other industries, at a price of around USD 2/kg. A few hundred thousand tonnes per year is purified for the PV and electronics industries, resulting in silicon with a purity of 99.99999% and a price of around USD 20/kg.

Silicon has a number of important advantages over alternative PV materials, including elemental abundance, moderate cost, non-toxicity, high efficiency, device performance stability, simplicity (it is a mono-elemental semiconductor), physical toughness, the highly advanced state of knowledge of silicon material and technology and the advantages of incumbency. The latter comprises extensive and sophisticated supply chains, large-scale investment in mass production facilities, deep understanding of silicon PV technology and

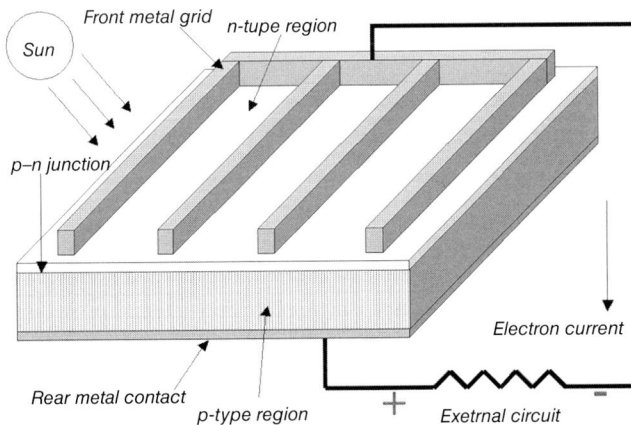

Figure 3.1 Schematic of a typical solar cell.
Source: Courtesy of the author.

markets, and the presence of thousands of highly trained silicon specialists – scientists, engineers and technicians.

In order to produce solar cells, silicon crystalline ingots are grown from a silicon melt at 1400 °C under a non-oxidising ambient atmosphere of argon. In the Czochralski (Cz) process, a narrow seed crystal is lowered into a quartz pot containing liquid silicon, and slowly withdrawn. A cylindrical silicon ingot with a length of several metres, and comprising a single crystal of silicon, grows from the seed over the space of about 10 hours. An alternative process is to slowly cool molten silicon in a pot and to rely on natural crystallisation. The result is a large block of multicrystalline silicon from which square wafers can be cut. Czochralski solar cells have a slight performance advantage but cost slightly more, and the world solar cell market is primarily balanced between Cz and multicrystalline silicon cells.

Solar cells are made on silicon wafers. The wafers have a typical diameter of 156 mm and a thickness of 0.12 mm. The wafers are cut from silicon ingots using a diamond-impregnated wire saw. After ingot growth and wafer slicing, the wafers are cleaned, and the saw damage removed in a silicon etch.

The next step is to diffuse a tiny amount of phosphorus into the sunward surface of the wafer to a depth of about 0.001 mm to create the p–n junction. Then follows deposition of a thin sheet of metal on the rear surface and a grid of metal on the sunward surface to allow extraction of electricity. The design of the front grid is a compromise between having plenty of metal to minimise electrical resistance losses and reducing the amount of metal to minimise reflection of sunlight.

Groups of 50–100 solar cells are electrically connected and encapsulated in thin layers of plastic (ethylene vinyl acetate and polyvinyl fluoride) and laminated behind a tough 3-mm-thick glass cover to form solar modules, each with a power of 300–500 W. An aluminium frame is attached for strength and to assist mounting to frames. A junction box is added to house the electrical terminals. Dozens to millions of solar modules are mounted together and electrically interconnected to form a solar power system.

Most PV systems are mounted on fixed support structures that are tilted up to face the equator, with a tilt equal to the angle of latitude. This maximises annual electricity production. Large ground-mounted PV systems are often mounted on sun-tracking systems to maximise annual output, which can be increased through tracking by 15–20%, depending upon location. Electricity produced by PV modules is conducted to a power conditioning unit that optimises voltage, converts the direct current produced by solar cells to the alternating current used in electrical grids, transforms the voltage to match that of the local grid, and manages interfacing with the local grid.

Photovoltaic systems have unmatched reliability and low maintenance costs due to the lack of moving parts, which explains why they are widely used to provide power in remote regions and on satellites. Manufacturers typically warrant PV modules for 25 years, and in dry locations they may continue to operate for 50 or more years. Specimen modules are exposed to severe accelerated failure testing in order to elucidate and prevent failure mechanisms. Degradation modes of PV modules include physical destruction caused by human action or violent hailstorms; slow chemical changes leading to yellowing of transparent encapsulation materials; and slow ingress of moisture to the solar cells or electrical components causing corrosion of metallic components.

3.5 PV Technology

The best laboratory cells have efficiency around 27%, compared with the theoretical maximum efficiency of 29%. The leading commercial crystalline silicon PV solar cell technology is PERC (passivated emitter and rear cell) technology (Blakers 2019), with solar cell efficiencies of 20–23%. Interdigitated back contact (IBC) and heterostructure with intrinsic thin layer (HIT) silicon solar cells have commercial efficiencies of 23–25%, albeit at a premium price, and are typically deployed where space is limited.

Currently, cadmium telluride (CdTe) and copper indium gallium selenium (CIGS) are the leading non-silicon PV technologies, with about 3% market share each (Fraunhofer 2015). Solar cells based on many other materials are under development but not yet in significant commercial production, notably perovskites. The latter has created substantial interest due to rapid improvement of efficiencies to above 20% for small area laboratory cells, and the possibility of creating highly efficient tandem solar cells in conjunction with crystalline silicon. In a tandem cell, two separate solar cells contribute to the output, each optimised to harvest a different portion of the solar spectrum.

Another branch of photovoltaics is concentrating PV (CPV) systems. Tracking of the sun is required for concentrator systems. Mirrors or Fresnel lenses are used to concentrate light by 10–1000 times onto a small number of highly efficient solar cells. Typically, active cooling of the solar cells is required to remove excessive heat. The best concentrator solar cells have conversion efficiencies of 50% and comprise three or more layers of different semiconductor materials drawn mainly from the group 3 and group 5 columns of the periodic table. Such cells are very expensive per square centimetre compared with conventional silicon solar cells. However, concentration greatly reduces the effective cost per square metre of collector area – essentially most of the solar cells are replaced by much cheaper lenses and mirrors. Concentrating PV technology has only a small market share and will be restricted to areas with plenty of sunshine and low pollution. Concentrating PV technology has much in common with concentrating solar thermal power, since much of the system infrastructure (such as trackers, controllers, lenses, mirrors, waste heat rejection and electrical connection) could be used for either.

3.6 PV Markets

Photovoltaic systems are unusual in that the unit cost of energy is similar for large (megawatt, MW) and small (kilowatt, kW) systems – large systems have lower capital costs but higher regulatory and financing costs, and vice versa. Virtually all other energy sources have strong economies of scale. This confers a major advantage on PV technology

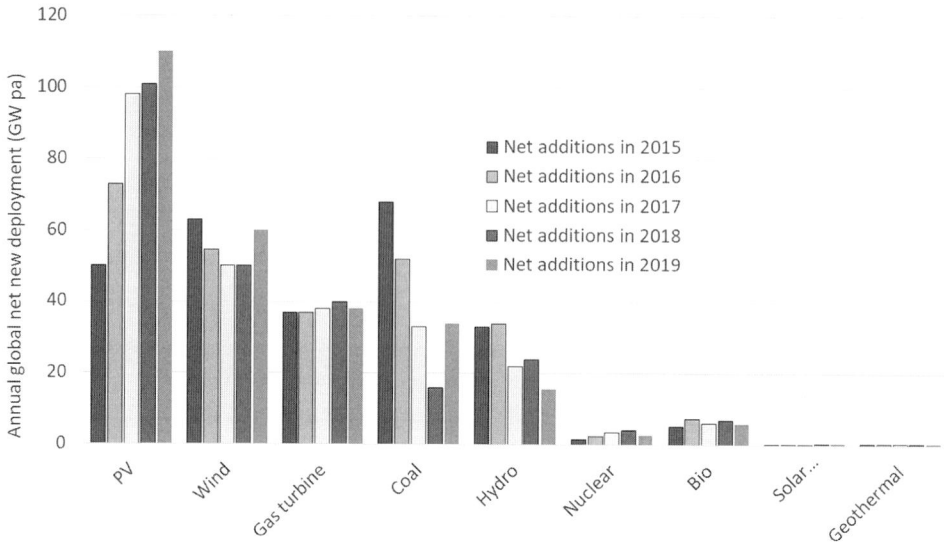

Figure 3.2 New generation capacity added in the year 2015 to 2019 by technology type. Most net new generation capacity is from wind and PV power.

Source: Courtesy of the author. For a colour version of this figure, please see the colour plate section.

since it has markets at every scale from watts to gigawatts for the same basic product – the silicon solar cell.

In earlier decades, PV technology found widespread use in niche markets such as consumer electronics, remote area power supplies and satellites. Throughout the world, remote area energy solutions are based upon various combinations of PV systems, wind, diesel and batteries. Active load management is an important additional feature to minimise the requirement for diesel and batteries.

The PV industry has achieved rapid growth for five decades and costs have declined continuously as production scale increases. Photovoltaic systems are now installed in large ground-mounted power stations and also on tens of millions of house roofs in cities.

Solar PV electricity is now less expensive than domestic and commercial retail electricity from the grid throughout much of the world and has reached cost-competitiveness with wholesale conventional electricity (IRENA 2018). This fact is the reason for the rapid growth in deployment of PV technology worldwide (Blakers et al. 2019). Direct competitiveness with fossil fuels for wholesale energy supply is assisted by carbon pricing and the removal or equalisation of hidden support for fossil fuels. The cost of PV systems can be confidently expected to continue to decline for at least the next decade due to 'learning curve' cost reductions as production continues to increase.

Wind and PV power are now fully competitive with new-build fossil fuel power stations in many countries, leading to very rapid uptake around the world. Around three-quarters of annual global net new generation capacity additions come from wind and PV power (Figure 3.2). Virtually all new generation capacity in some countries such as Australia is from wind and PV power.

3.7 Solar PV Energy in Developing Countries

Solar energy has much to offer developing countries since nearly all are in low latitudes, with good solar resources and low seasonal variability of both energy demand and solar availability. Small amounts of PV electricity can make dramatic differences to living standards, through the enabling of electric services such as lighting, computing, telecommunications, refrigeration, grain grinding and water pumping.

Developing countries generally lack a widespread and robust electricity distribution grid. There are good prospects for organic development of thousands and millions of small solar- and wind-powered systems, which gradually merge to become a national grid. These countries may bypass the centralised electricity distribution systems of high-income countries. An analogy is telephony, for which low- and middle-income countries will rely heavily upon distributed mobile telecommunications rather than fixed lines.

3.8 Solar Energy Systems in Cities

Tens of millions of roof-mounted PV systems have been deployed around the world. These systems have a typical power capacity of 2–20 kW for an individual dwelling, and tens of thousands of kilowatts for commercial buildings. The roof area required amounts to 7–10 m^2/kW. Low-density suburban housing generally has ample roof area to yield enough energy over the course of the year to equal the annual end-use energy of the dwelling. Low-rise commercial and light-industrial building roofs can yield substantial excess quantities of PV electricity for export to the electricity grid. However, in high-density regions of cities there is insufficient unshaded roof space for high penetration of PV electricity into the building energy requirement.

Roof-mounted PV systems generally compete with the retail price of electricity, which is typically two–four times larger than the wholesale price. The levelised cost of obtaining PV electricity from building roofs in moderate latitudes ($<40°$) is well below the retail electricity tariff in many cities. This is driving strong growth in deployment of roof-mounted PV systems; for example, about one in four Australian houses had a roof-mounted PV system in 2019.

The deployment of millions of roof-mounted PV systems causes a large shift in the demand profiles of the distribution networks. Large increases in deployment of thermal and electrical storage allows high penetration of PV electricity in urban areas, which causes a dramatic shift in the economics of the electricity distribution industry. Storage is required because energy demand is often out of phase with solar energy availability. When teamed with effective energy management and control strategies, combined electrical and thermal storage is attractive both for the building owner and the electricity grid operator.

Battery storage can be used to increase self-consumption of roof-mounted PV generation, and effectively manage the electrical network and power system, by taking advantage of the fast response, power-on-demand nature of batteries. However, the high (but declining) costs of batteries means that using battery storage alone to match PV generation to building demand is relatively costly.

Storage of energy in the form of hot water in an insulated tank is very widely deployed. Conventional solar, gas and electric hot water systems are under commercial pressure from the rapidly declining price of electricity from PV systems on building roofs. Advances in highly efficient air-to-water heat pumps allow PV-driven heat pump hot water to become a cost-effective hot water supply option. Heat pumps use electrical energy to move heat from one place (outside the building) to another at a higher temperature (hot water tank). Several units of thermal energy can be delivered per unit of electricity. Heat pump hot water storage can be easily controlled alongside other storage elements in conjunction with roof-mounted PV generation rates and household energy loads.

Thermal storage to provide space heating and cooling in buildings can be accomplished by raising or lowering the temperature of a building when low-cost energy is available, relying on thermal mass to store heat and good insulation to reduce thermal losses. Reverse-cycle air conditioners (which are also heat pumps) are cost-effective in this application and can be powered using roof-mounted PV panels. Heat banks rely on ceramic bricks for high-temperature heat storage, and fans to circulate heat as required. They can be charged using PV electricity during the day for use at night. Cold storage can be accomplished using PV-driven heat pumps to produce cold water and ice during the day. Thermal energy storage via building space heating and cooling can be controlled, along with hot water and battery storage, to optimise the use of PV generation and to manage net household demand.

Reduction in the retail use of gas for low-temperature heating and its replacement with PV-driven delivery of thermal energy in conjunction with heat pumps is an emerging trend in urban buildings. Gas burning in domestic dwellings for delivery of hot water and space heating is relatively expensive and inefficient compared with PV systems. Hotplate cooking, a minor but highly valued use of gas, can be replaced by induction cooktops, which provide the equivalent level of thermal control without flame or excessive heat risk.

Generation of medium-temperature heat (>100 °C) in cities for industrial purposes is usually achieved with gas. The falling price of PV and wind power means that electric furnaces are becoming competitive for high-temperature heating.

3.9 Transport Systems and Solar Fuels

Fossil fuels used for land transport typically account for 12–15% of a developed country's greenhouse gas emissions (IPCC 2014). Emissions can be avoided by moving to electric vehicles and electrically powered public transport, provided that electricity comes predominantly from renewable energy. Electric vehicle sales are rising rapidly, due to cost reductions in vehicles and improvements in automotive systems and batteries. For a driving distance of 8000 km per year, a 1-kW PV panel is required to provide the annual electricity requirements. The fully installed cost of the PV panels will be USD 1000–1500, and typically they will last 25 years – twice the typical lifetime of the car. Thus, an outlay of a few thousand dollars provides the electricity requirements of an electric car for its whole lifetime.

Conversion of most land transport to electricity derived from renewable energy sources appears to be feasible. However, some transport functions, such as ships, aircraft and heavy machinery, cannot easily be met from electrical sources because of the impracticable size, weight and cost of the required storage. Some industrial processes may also be difficult to service with (renewable) electricity.

Synthesis of chemical fuels, utilising solar energy to drive the chemical reactions, allows solar and other renewable electricity sources to substitute for fossil fuels for both transport and as an industrial fuel. There are a limited number of suitable chemical fuels, taking account of material abundance, toxicity, storability and other factors. Notable candidates are carbon-based compounds (methane CH_4, diesel $C_{12}H_{23}$, kerosene $C_{12}H_{26}$), hydrogen (H_2) and ammonia (NH_3). The likelihood is that most synthetic fuels will be 'drop-in' replacements for existing fuels to avoid the need to redesign existing engines; that is, based upon carbon. Synthesis of a low-greenhouse fuel will therefore require an energy source derived from a low-emission technology and a renewable source of carbon, for example by extraction of CO_2 from the air. Extraction of hydrogen by splitting water is the majority of the energy required for fully sustainable synthesis of jet fuel (kerosene), with CO_2 extraction from the air and fuel synthesis being a minority of the energy required.

Chemical fuel synthesis utilising solar energy can be driven either by heat or electricity. It is not feasible to transfer heat over long distances – it must be generated and used locally. Additionally, land for solar collectors is expensive in industrial areas. Thus, direct (local) utilisation of high-temperature solar heat in industry requires locations that have high direct-beam irradiation and low land cost. This is problematic for nearly all current manufacturing localities around the world, including much of China and India (severe air pollution), much of South East Asia (tropical cloudiness), and much of Europe and North America (substantial cloudiness and substantial seasonality of solar insolation). Loss of 50% or more of the global radiation because it is diffuse makes solar concentrators considerably less economic. Electric-driven fuel synthesis can take advantage of rapid reductions in the price of electricity from renewable wind and PV sources. Wind collectors and PV systems can be located remotely in windy/sunny locations to transmit electricity to an industrial centre. Thus, renewable electricity has a significant advantage over solar concentrator heat for heavy industry.

3.10 Long-Distance Transmission of Electricity

Transmission grids within industrialised countries are currently based around a relatively small number of large fossil fuel nuclear and hydropower stations. Increasing the scale of interconnection confers robustness of supply, allows smoothing of total demand by increasing the variety and timing of loads (Blakers et al. 2017), and allows the incorporation of more varied energy storage including pumped hydroelectric storage (Blakers et al. 2021). Additionally, a wide geographical spread of generators connected with high-voltage cables reduces the chance of the simultaneous absence of sufficient sun and wind. Continent-wide transmission grids are emerging and are being strengthened. Long-distance transmission increases competition within markets, as well as allowing 'time shifting' through several time zones from one side of a continent to the other.

Transmission of electricity over long distances generally utilises high-voltage direct current (HVDC) technology. Transmission losses associated with an 800-kV DC power line with 5-GW (gigawatt) capacity is quoted by Siemens as 3% per 1000 km (Siemens 2012). In addition, there is a few per cent conversion loss at the two ends of the cable. Costs of large HVDC cables can be expected to decline substantially as many more are constructed in coming decades. Costs below USD 300/MW-km are likely (Blakers et al. 2012).

These advances in long-distance transmission of electricity, solar PV and wind technology, combined with current levels of hydro and biomass, with investment in storage, can now realistically enable renewables to surpass fossil fuels for worldwide electricity production.

3.11 Conclusion

Solar and wind energy is available on a vast scale for billions of years. Solar PV technology uses only very common materials; has minimal need for mining; has minimal security and military risks; cannot have significant accidents; and has minimal environmental impact over unlimited timescales. It is now cheaper than alternatives for new-build energy systems. It is an ideal long-term energy solution.

References

Bhandarib, K. P., Colliera, J. M., Ellingson, R. J. and Apula, D. S. (2015). Energy payback time (EPBT) and energy return on energy invested (EROI) of solar photovoltaic systems: A systematic review and meta-analysis. *Renewable and Sustainable Energy Reviews*, 47, 133–141.

Blakers, A. (2019). Development of the PERC solar cell. *IEEE Journal of Photovoltaics*, 9, 629–635. Available at: https://ieeexplore.ieee.org/document/8653319.

Blakers, A., Luther, J. and Nadolny, A. (2012). Asia Pacific super grid: Solar electricity generation, storage and distribution. *GREEN: The International Journal of Sustainable Energy Conversion and Storage*, 2, 189–202.

Blakers, A., Lu, B. and Stocks, M. (2017). 100% renewable electricity in Australia. *Energy*, 133, 471–482. Available at: www.sciencedirect.com/science/article/pii/S0360544217309568.

Blakers, A., Stocks, M., Lu, B., Cheng, C. and Stocks, R. (2019). Pathway to 100% renewable electricity. *IEEE Journal of Photovoltaics*, 9, 1828–1833. Available at: https://ieeexplore.ieee.org/document/8836526.

Blakers, A., Stocks, M., Lu, B. and Cheng, C. (2021). A review of pumped hydro energy storage. *Progress in Energy*, 3. Available at: http://iopscience.iop.org/article/10.1088/2516-1083/abeb5b.

Fraunhofer (2015). *Photovoltaics Report*. Freiburg: Fraunhofer Institute for Solar Energy Systems, ISE. Available at: www.ise.fraunhofer.de/content/dam/ise/de/documents/publications/studies/Photovoltaics-Report.pdf.

IPCC (Intergovernmental Panel on Climate Change) (2014). *Climate Change 2014: Synthesis Report. Contribution of Working Groups I, II and III to the Fifth*

Assessment Report of the Intergovernmental Panel on Climate Change. Edited by R. K. Pachauri and L. A. Meyer. Geneva: IPCC. Available at: www.ipcc.ch/report/ar5/syr/.

IRENA (International Renewable Energy Agency) (2018). *Renewable Energy Statistics 2018*. Abu Dhabi: International Renewable Energy Agency. Available at: www.irena.org/publications/2018/Jul/Renewable-Energy-Statistics-2018.

Reinders, A., Freundlich, A., van Sark, W. and Verlinden, P. (2015). *Photovoltaic Solar Energy: From Fundamentals to Applications*. London: Wiley & Sons.

Siemens (2012). *Factsheet Energy Sector*. Abu Dhabi: Siemens.

4

Solar Thermal Energy

JOHN PYE, KEITH LOVEGROVE, PAUL GAUCHÉ AND MARK MEHOS

Executive Summary

Technology Overview and Features

Concentrating solar power (CSP) systems use focusing mirrors to capture and store solar energy in the form of heat. The stored heat, at temperatures approaching 600 °C, can be released as needed and used to power a steam turbine for the dispatchable supply of renewable energy. As such, CSP complements inflexible renewables such as wind and photovoltaics, and offers an alternative to electricity storage systems such as batteries or pumped storage hydro.

The levelised cost of commercial CSP systems built in 2018 was USD 0.00185/kWhe, all of which had at least 4 hours and some had more than 8 hours of thermal energy storage. Costs are trending downward with a learning rate around 20%. Total global deployed capacity in 2018 was 5.48 GWe.

Optimal locations for CSP include Chile, southern Africa, north-western Australia, northern Africa, the Middle East, south-western North America and inland China.

Studies of 100% renewable energy scenarios indicate that it may be optimal to source 10–15% of electricity generation from CSP, with CSP playing the role of providing night-time electricity and adding robustness when a large fraction of generation is inflexible.

Important non-electricity applications of CSP are being developed, including water desalination and greenhouse horticulture. In the future, industrial production of fuels, reduced metals, lime, cement and ammonia may employ CSP as a source of high-temperature process heat.

Low-temperature solar thermal systems are used for small solar domestic hot water systems as well as in agriculture, mining, textiles and district heating.

Sustainable Development and Climate Adaptation Co-benefits

In future decarbonised electricity networks and in some regions, CSP may become the primary source of night-time electricity, due to its low-cost high-capacity energy storage.

Compared to wet-cooled coal power stations, CSP plants today use only 10% of the water on a per-electricity-generated basis. Technologies are in development to drive water use further towards zero.

In arid regions, CSP can be used for the provision of desalinated seawater.

Process heat from CSP may be important in global efforts to decarbonise high-temperature and fossil-fuel-intensive industrial processes.

Concentrating solar power provides a long-term electricity supply with very low ongoing costs, bringing benefits in relation to energy autonomy and regional stability. Concentrating solar power systems can be built with relatively high fractions of local content, and may be beneficial in national industrialisation and development efforts.

Low-temperature solar thermal collectors offer a relatively low-tech pathway to the reduction of emissions in developing areas.

4.1 Introduction

Solar thermal technology provides a wide range of opportunities for climate-resilient global development. High-temperature concentrating solar power (CSP) systems are used to generate flexible, dispatchable renewable electricity in large-scale grid-connected systems and could also soon be used as a heat source for industrial processes such as for desalinated water, fuels, chemical products and refined ores. Most CSP electricity systems include thermal energy storage units, allowing output to continue for hours after sunset. Solar thermal systems, which rely on heating up a working medium to operate, are distinct from solar photovoltaic (PV) technologies that directly convert solar photons into electric current. In addition to CSP, low-temperature solar thermal systems, used for domestic hot water and other applications, are briefly reviewed.

4.2 Concentrated Solar Thermal Energy

4.2.1 High-Level Description of CSP Systems

Concentrating solar thermal power systems work by focusing direct-beam solar radiation (sunlight) onto a receiver surface, where the heat is absorbed. The absorbed heat is transferred to a heat transfer medium (HTM), such as mineral oil, molten salt, water or ceramic particles. The heat is transported via the HTM to the next part of the system, typically a thermal energy storage tank, and then typically used to drive a steam turbine power cycle where the heat is converted into electricity.

The light-focusing part of a CSP system, known as the collector or concentrator, is normally composed of curved or parabolic mirrors. These collectors are referred to as heliostats, troughs, dishes or linear Fresnel reflectors (LFRs) depending on the system type (described in Section 4.2.2). The mirrors need to be of high optical quality and also oriented with sufficient precision as the sun moves, so that the solar radiation can be focused onto a small receiver surface without excessive losses. Tracking systems are needed to ensure that light is continuously reflected onto the receiver as the sun moves. Solar radiation cannot be concentrated past certain optical and thermodynamic limits, dictated primarily by the angular size of the sun in the sky (Lovegrove and Pye 2012).

At the absorbing surface of the receiver, some of the energy will be lost by reflection. Then, as the absorbing surface becomes hot, further energy loss occurs due to re-radiation from the hot surface, and also due to convective air flows on the exterior surfaces. All three of these losses – reflection, re-radiation and convection – are proportional to the area of the receiver opening (the 'aperture'), and hence the efficiency of the system overall can be increased if the concentration of light is as high as possible. Energy not lost by any of these mechanisms is transferred as heat to the HTM, which is then circulated out of the receiver and onward, typically to a thermal energy storage system.

Concentrating solar power systems usually include thermal energy storage capacity, in order to allow the supply of electricity to be matched to the demand. This storage typically involves two tanks containing an energy storage medium (ESM), usually a molten salt material. To 'charge' the storage, the ESM from the 'cold' tank is heated up inside a heat exchanger using the hot HTM from the receiver, and then pumped into the 'hot' tank. The typical molten salt ESM temperature range is from 290 °C ('cold') to 565 °C ('hot'). To discharge the storage, the hot tank ESM is transferred through another heat exchanger in which water is boiled under pressure. The resulting steam is then passed to the steam turbine, and the 'cold' ESM is returned to the cold storage tank.

Power cycles in general, as well as steam turbine power cycles in particular, are more efficient when they have a higher temperature heat input. A simple steam power cycle includes a turbine, condenser, pump and boiler. In the CSP case, the boiler is provided by a heat exchanger connected to the thermal energy storage unit. Higher-efficiency power cycles are more complex, and typically include additional heat exchangers and turbine stages.

The above descriptions are reasonably general, but there are many variations in the types of CSP systems, both constructed and in development. For example, in many systems, the ESM and the HTM are the same medium, molten salt, which means that no heat exchanger is required to charge the storage tank. Not all systems provide storage, and in some cases the HTM in the receiver is water, which is directly transferred into the steam turbine. Other systems use a power cycle based on a gaseous working fluid (such as the Brayton cycle or Stirling cycle of a steam turbine), and there is no water involved. Yet other systems make use of melting/freezing or chemical reactions in the HTM or ESM, instead of simple sensible[1] heating and cooling, as a way to transfer and store energy efficiently. Finally, CSP systems are sometimes hybridised with other conventional power systems, such as coal or gas power stations, allowing continuous or near-continuous supply of power.

The CSP systems described above are all used to produce electricity. However, a growing area of interest for CSP is in the provision of heat for other chemical and industrial processes (see Section 4.2.8).

Energy-intensive products that can be produced using CSP include liquid fuels, hydrogen, ammonia, zinc, iron/steel and cement, with processes for these and others under ongoing development.

---

[1] 'Sensible' (as opposed to 'latent') heating is that which causes a detectable temperature rise, with no phase change.

4.2.1.1 Factors Determining CSP System Performance

Performance of a CSP system depends on direct normal irradiance (DNI), since only a direct beam (as opposed to diffuse radiation) can be focused using mirrors. A global map of the annual DNI is given in Figure 4.1, highlighting the excellent solar resource in many mid-latitude regions (red–purple). Around the equator, higher cloudiness reduces the annual DNI.

As with other power plants, the size of CSP systems is typically reported in terms of *nominal capacity* (also referred to as nameplate or rated capacity). This is the electrical power output when the plant is operating at full-speed design point conditions. The nominal capacity of a CSP plant is mostly constrained by the scale of the power block (turbine and generator). A representative range for the nominal capacity of CSP plants would be 10–250 MWe.[2]

The *capacity factor* of a CSP system is the ratio of the total energy generated by the power block over the course of the year, divided by the amount of power it could have produced had it been running at its nominal capacity for the whole time. A CSP system without any storage, similar to a non-tracking PV system, has a capacity factor of 20–25% (Branker et al. 2011; Romero and González-Aguilar 2014), which reflects the fact that there

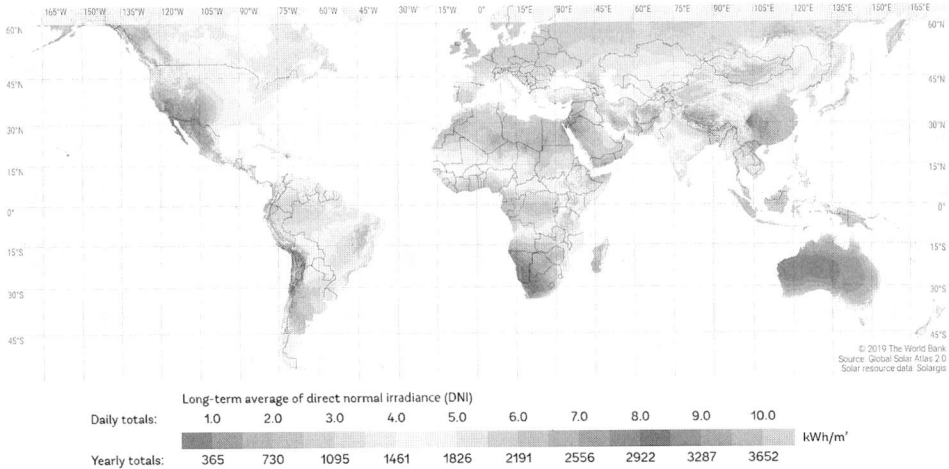

Figure 4.1 Direct normal irradiance (DNI, kWh/m² per year) at sites around the world. Optimal locations for CSP are Chile/Bolivia/Argentina, Australia, South Africa/Namibia and USA/Mexico. Other excellent locations include Spain, India, China, Morocco and Israel.

Source: DNI Solar Map obtained from the Global Solar Atlas 2.0, a free, web-based application is developed and operated by the company Solargis s.r.o. on behalf of the World Bank Group, utilising Solargis data, with funding provided by the Energy Sector Management Assistance Program (ESMAP). For additional information: https://globalsolaratlas.info. For a colour version of this figure, please see the colour plate section.

---

[2] Nominal capacity is often written with units MWe – megawatts electrical. This is to differentiate from megawatts thermal (MWth), which would be used to refer to the rated capacity of, for example, a receiver or a boiler.

are only an average of 6 hours of sunlight each day, once seasonal variations in day length as well as cloudy weather are accounted for. However, higher capacity factors for CSP up to 74% have been achieved through the addition of thermal energy storage (Romero and González-Aguilar 2014). A high overall capacity factor indicates a more cost-effective use of the power block, but this can only be achieved with the addition of storage and a larger collector area, which must be paid for in order to increase the capacity factor of the power block. The capacity factor can also be increased by oversizing the solar collector (see below), or by hybridising the CSP system so that the power block can also be operating using an alternative heat source, such as a gas-fired boiler. The annual output of the plant (expressed as megawatt-hours or gigawatt-hours per year: MWh/y or GWh/y) can then be obtained as the product of the nominal capacity (MW), the capacity factor, and the number of hours in a year.

The size of the thermal energy storage in the CSP plant is normally communicated in terms of the number of hours that the energy contained in the storage can run the power block at its nominal capacity. Storage systems as small as 30 minutes can be useful in improving system performance, since they allow the power block to continue operating through periods of patchy cloud. Storage of 3–6 hours allows a CSP system to preferentially target high-value evening demand, after sunset. Storage of approximately 15 hours allows the power block to be run continuously in summer months (Romero and González-Aguilar 2014) and delivers the lowest levelised cost of energy (see Section 4.2.1.2). Storage of more than 15 hours is technically easy to achieve, however does not yet appear to be valued sufficiently highly for deployment in commercial CSP projects.

The *field efficiency* of CSP systems is the ratio of the thermal output divided by the product of the DNI and the total mirror area. A major factor within the field efficiency is the *cosine efficiency*, which is the ratio of the effective mirror area of the solar field to the total mirror area, as shown in Figure 4.2. The field efficiency varies greatly as the sun moves daily and seasonally, and the main cause is cosine efficiency, which averages ~85% for most commercial system types. Other factors that affect the optical efficiency of a CSP system are mirror reflectivity losses, blocking, shading, atmospheric scattering, spillage and receiver reflection losses. Thermal losses that contribute to reducing the field efficiency include convection and re-radiation from the hot receiver surface, and heat loss in the pipework that conveys the HTM to the storage or power block.

The *solar multiple* is the ratio of the thermal energy rate delivered by the solar field at its design point (typically at solar noon in mid-summer or at equinox, in clear-sky conditions) to the thermal energy rate required in order to supply the power block at its nominal capacity. A solar multiple greater than one is typical, and (for a non-storage system) means that there will often be more power collected than can immediately be used in the power block. It is usually cost-effective to have a solar multiple of 1.2 or 1.3 for non-storage systems, even though that implies some 'dumping' of heat during peak conditions. For systems with storage, the solar multiple can be as high as 4.0, so that enough energy is gathered during the 6 hours of direct sunlight to run the power block for up to the full 24 hours of each day.

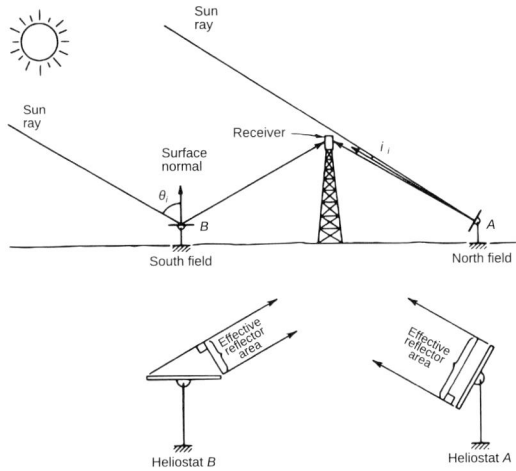

Figure 4.2 Cosine efficiency of a heliostat describes whether mirrors face directly to the sun, presenting a large effective surface area, or whether they are oblique to the sun, presenting a relatively small effective area. Systems with high average cosine efficiency across their entire collector field throughout the year will require less total mirror area.
Source: Stine and Geyer (2001).

As mentioned above, the capacity factor can be increased not only by adding storage but also by hybridising the plant with other heat sources such as natural gas burners. The *solar fraction* is the ratio of solar thermal heat input at the absorber to the total of all heat inputs to the plant. It typically is measured as an average over a day or year. An example of a system with a solar fraction of 100% is the Crescent Dunes system in Nevada, USA. The Ain Beni Mathar plant in Morocco, on the other hand, is a large 472-MWe natural gas combined-cycle power plant with a relatively small solar field offsetting some gas consumption, with a solar fraction of 1.2% (World Bank 2014). The numerous Spanish CSP plants are mostly operated under rules that specify a minimum solar fraction of 85%.

The *power block efficiency* is the efficiency with which the thermal energy supplied to the power block is converted into electrical energy output. Values of 30–40% are typical. Systems with a higher inlet temperature will be more efficient, due to the second law of thermodynamics. Systems with a larger nominal capacity tend to be more efficient, because larger systems can cost-effectively incorporate more energy-saving features such as recuperators and feed-water heaters.

Because of seasonal variations in weather and the path of the sun through the sky, the performance of a CSP plant changes considerably throughout the year. For this reason, simulation of at least one full year of operation of the plant is essential in order to develop (or prove) an optimal system design. Annual performance calculations are made using weather data for the location in question, as well as simplified numerical models for each of the components in the system. Using such models, the combination of storage, power block, receiver and collector can be determined which gives the greatest return on investment.

4.2.1.2 The Levelised Cost of Energy

The levelised cost of energy (LCOE), for CSP as with other energy technologies, is the fixed price at which the generated energy must be sold over the lifetime of the plant in order for all capital and operational costs to be recouped, and for the project to thereby achieve a net present value of zero (Richert et al. 2012; Hernández-Moro and Martínez-Duart 2013). A key parameter in determining the levelised cost of energy is the weighted average cost of capital (WACC), also referred to as the discount rate, an aggregate of the financing costs (loan interest rate, investor dividends) for the project. Care must be taken when comparing LCOE values to ensure that similar assumptions for taxes, profit allowance, government subsidies and other incentives, and other costs and benefits are used in the different analyses. Levelised cost of energy is often criticised for being a poor measure of viability, especially for CSP technology, because, being based on a fixed constant sale price for all the generated electricity, it fails to recognise the value-adding opportunity for a CSP system to sell energy strategically when demand is high. A system optimised naively to give the lowest LCOE may not always result in the most profitable configuration in real market conditions. This question is discussed further in Section 4.2.7. To lower the project risk it is common for CSP plants to sign a power purchase agreement (PPA) with a duration of 20 years or more, to ensure a reliable income for the operational plant. Some discussion of the project benefits of signing a PPA are outlined by Jacobowitz (Jacobowitz and Google 2013). More recently, some project developers have proposed to operate with only a partial PPA, selling a fraction of their electricity into the much more variable but potentially more lucrative spot market.

4.2.2 Types of CSP Systems

Four basic types of CSP systems exist, based on two different ways of concentrating solar radiation (Figure 4.3). Trough and linear Fresnel reflector (LFR) systems are *line-focusing* concentrators, while dish and tower systems are *point-focusing* concentrators. Line-focusing concentrators typically only use single-axis tracking systems, while point-focus systems require two-axis tracking systems. Point-focus systems concentrate their solar radiation onto much smaller spots, allowing these systems to achieve higher temperatures and greater efficiency compared to line-focus systems. However, point-focus systems are more complex in overall design, due to the additional tracking axis required for each reflector component.[3]

The receiver configuration varies for the different system types. Dish and trough systems have a moving receiver which is part of a single integrated tracking structure that also supports the mirrors. Tower and LFR systems have a fixed receiver, which remains stationary, while the mirrors move independently. In tower systems, the mirrors are referred to as heliostats. In both cases, the moving-receiver system is more optically efficient (higher cosine efficiency) than the corresponding fixed-receiver system.

---

[3] One category of point concentrator not covered here is the Fresnel lens. Such concentrators have been used successfully with concentrating photovoltaic (CPV) systems, but have not been commercially adopted for CSP systems.

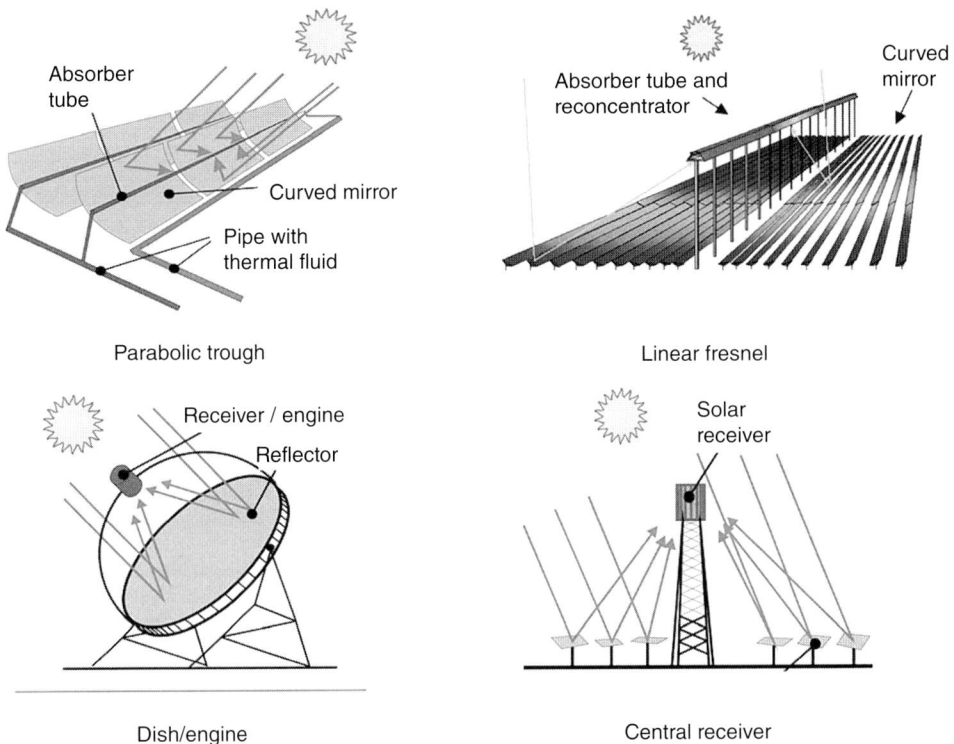

Figure 4.3 Different CSP system configurations. Top: line-focusing systems, (left) trough and (right) linear Fresnel reflector. Bottom: point-focusing systems, (left) dish and (right) tower system with heliostats.

Source: Reproduced from Romero and Steinfeld (2012) with permission from The Royal Society of Chemistry.

There are numerous trade-offs between the different CSP system types. Optical and thermal efficiency, operating temperature, structural cost, wind loads, cost of installation and cost of maintenance are all major factors. Trough systems have been most successful in the market to date, but towers are becoming increasingly commercially competitive. There is no clear winner yet, though, and probably the best system type will be found to vary with location. For example, in hazy areas such as the Middle East, systems with a short distance between the mirror and the receiver are likely to perform better, since there will be lower light-scattering losses (atmospheric attenuation). For dish and LFR systems, commercial activity has recently been relatively limited.

4.2.3 Commercial Deployment of CSP Systems

Early development of CSP systems occurred in the 1980s and early 1990s, culminating in nine Solar Energy Generating System (SEGS) trough systems built in California, with a total capacity of 354 MWe, as well as several prototype-scale tower systems and a wide range of dish systems with a high-efficiency Stirling engine mounted at the focus. From the early 1990s, there was a hiatus of 16 years for large system development, associated with (among other things) the gradually increased security of fossil fuel supplies that occurred after the 1979 oil crisis and the 1990 oil price spike.

However, in the mid-2000s, increased concern about anthropogenic climate change and national energy security led to renewed investment in CSP by governments and industry in several countries, especially Spain. The development in Spain was motivated by a Royal Decree mandating a strong feed-in tariff to support the construction of CSP systems up to a 50-MWe size limit. As a result, from 2002, the Spanish have led the world in the development of a new CSP industry, and the plants now operating there have been a net positive contribution to the Spanish economy despite the considerable cost (Ortega et al. 2013). Since the global financial crisis, progress in Spain has unfortunately slowed. In the USA, a policy of investment tax credits resulted in several much larger-scale projects being built there from 2008. Most recently, CSP development has been concentrated in South Africa, Morocco, the UAE, India, Chile and China, where a number of large systems are currently under construction.

As a result of these and numerous other global efforts to develop CSP, the deployment of CSP has grown rapidly, with a compound annual growth rate of 27% from 2008 to 2018, similar to that of PV (41%) and wind (17%) over the same period. A historical plot of the growth in deployed CSP capacity, including this recent growth, is shown in Figure 4.4. A good body of data on operational and in-development CSP systems is maintained in a database by the US National Renewable Energy Laboratory (NREL) and the International Energy Agency (IEA) Solar Power and Chemical Energy Systems (SolarPACES) network, and is the best publicly available resource for tracking CSP projects (NREL n.d.). A selection of some significant recent CSP plants is given in Table 4.1.

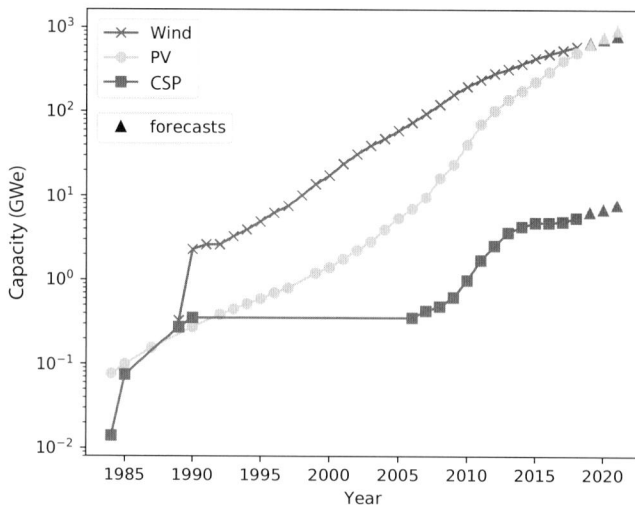

Figure 4.4 Growth in the CSP sector (log-scale plot), compared to that of PV and wind. Triangles indicate systems known to be under construction/development.
Source: Courtesy of the author, based on data from REN21, GWEC, CSP World, IEA.

Table 4.1. *A selection of significant CSP projects from around the world*

Name (Type) Developer	Location	Commissioned	Nominal Capacity	Storage	Significance
SEGS I–IX (trough) Luz Industries	California, USA	1984–1990	354 MWe (total)	Mineral oil (SEGS I only)	The first large-scale commercial CSP systems (nine units), now with over 30 years of operational and maintenance experience.
Andasol 1 (oil trough) ACS Cobra	Andalucía, Spain	2008	50 MWe	7.5 hours molten salt	An early beneficiary of the Spanish feed-in tariff, and the first commercial system to use molten salt storage. Now part of a cluster of three co-located systems. See Figure 4.5.
Gemasolar (salt tower) SENER/Torresol Energy	Andalucía, Spain	2011	19.9 MWe	15 hours molten salt	First commercial molten salt tower system, and first solar power plant of any type to achieve 24-hour operation. See Figure 4.6.
Archimede (salt trough) ENEL	Sicily, Italy	2010	5 MWe equivalent	8 hours molten salt	The first direct salt-heating trough and first integrated solar combined-cycle system (ISCCS), where solar steam is injected into a fossil-fired combined-cycle power plant.
Ivanpah (steam tower) BrightSource	California, USA	2013	392 MWe	—	Largest tower system in the world, made up of three separate towers and heliostat fields in a single location. No storage.
Solana (trough) *Abengoa*	Arizona, USA	2013	250 MWe	6 hours molten salt	A very large trough system with storage. 2 × 140-MWe (gross) steam turbines.
Khi Solar One (steam tower) Abengoa	Northern Cape, South Africa	2016	50 MWe	2.7 hours saturated steam	Technologically ambitious plant with novel sandwich panel heliostats, dry updraft cooling tower and triple cavity receiver designs. Particularly high value-add in this region due to poor electricity infrastructure. See Figure 4.7.

Table 4.1. (*cont.*)

Name (Type) Developer	Location	Commissioned	Nominal Capacity	Storage	Significance
Noor I, II, III (2 × trough, tower) ACWA/Aries/TSK	Ouarzazate, Morocco	2015–2018	143 MWe, 185 MWe, 150 MWe	3, 7 and 7.5 hours molten salt	Together form Ouarzazate Solar Power Station, the world's largest CSP plant at 510 MWe. Financed with loans from World Bank, German bank KfW, European Commission and European Investment Bank. See Figure 4.9.
Jemalong (5 × sodium tower) Vast Solar	NSW, Australia	2018	1.1 MWe	liquid sodium	A demo plant based on modular ~1.25-MWth tower + field units interconnected with pipework carrying a liquid sodium HTM. See Figure 4.8.
Port Augusta (steam tower) Aalborg/eSolar/ Sundrop Farms	Australia	2016	37 MWth	—	A first example of CSP hybridised with greenhouse-based food (tomato) production, providing power for water desalinisation and ventilation. See Figure 4.9.
Miraah (steam trough) Glasspoint	Oman	2017	100 MWth	—	Application of CSP for enhanced oil recovery (EOR) with reduced cost and 80% lower emissions. Troughs are encapsulated in a huge glasshouse to reduce wind loads and soiling. Full project is to be 1021 MWth in scale.
Dunhuang II (salt tower) Beijing Shouhang	Gansu, China	2018	100 MWe	11 hours molten salt	The first >100-MWe tower system in China, and a leader in the new wave of 20+ projects currently under way there since 2016.
Dhursar (linear Fresnel) Areva	Rajasthan, India	2014	125 MWe	—	Largest operational linear Fresnel CSP system to date.

Figure 4.5 Andasol 1, a Spanish 50-MWe trough system with molten salt thermal energy storage. Left: aerial view; centre: molten salt storage tanks; right: one of the parabolic trough collectors.
Source: Courtesy of SENER. For a colour version of this figure, please see the colour plate section.

Figure 4.6 Gemasolar, at Fuentes de Andalucía in Spain. This 19.9-MWe system has 15 hours of thermal energy storage, allowing continuous 24-hour operation in summer months.
Source: Courtesy of SENER. For a colour version of this figure, please see the colour plate section.

4.2.4 Current Costs for CSP

It is challenging to obtain a full picture of the current costs of CSP technology. This is because CSP projects are very large commercial projects, and each one is a major investment for the companies involved. As a result, the solar resource, performance, loan terms,

Figure 4.7 Khi Solar One. Left: 140-m² heliostats; centre: view of solar field, taken from the receiver; right: the tower, incorporating three cavity receivers.
Source: Photo courtesy of John Pye. For a colour version of this figure, please see the colour plate section.

Figure 4.8 Jemalong Solar Thermal Station (left), developed by Vast Solar in central New South Wales, Australia. This modular system has five 1.25-MWth towers, 27-m high (right), with sodium as the HTM, connected to a single 1.1-MWe steam turbine.
Source: Photos from vastsolar.com. For a colour version of this figure, please see the colour plate section.

capital costs and profit margins for these projects are sensitive commercial information which, if leaked, will affect the market value of those companies, hence companies tend to keep these data secret. This situation is very different especially in the case of PV systems, where strong competition on price for domestic-scale PV installations have ensured that there is a very good understanding of cost reductions achieved. This section aims to review some of the limited sources of publicly available cost information relating to CSP systems, but it must be emphasised that a serious evaluation of CSP technology in any given location or market requires detailed analysis of all the local conditions.

Figure 4.9 Left: the Noor III plant near Ouarzazate, Morocco, has a 250-m-high tower and 7400 heliostats, and is co-located with two large parabolic trough plants. Right: Sundrop Farms, Port Augusta, a CSP plant installed to provide combined power and desalinated water for a commercial tomato greenhouse in South Australia.
Source: Photo of Noor III courtesy of SENER. Photo of Sundrop Farms courtesy of sundropfarms.com. For a colour version of this figure, please see the colour plate section.

A previous study which attempts to review all available primary sources on CSP system costs was Hinkley et al. (2013). The primary sources available include a study by Fichtner for the World Bank (Konstantin and Kretschmann 2010; Kulichenko and Wirth 2011), with emphasis on India, Morocco and South Africa, as well a comparative study focused on Australia by Hinkley et al. at CSIRO (the Commonwealth Scientific and Industrial Research Organisation: Hinkley et al. 2011), the tower cost reduction study by Kolb et al. at Sandia (Kolb et al. 2011) and the trough cost reduction study by Turchi et al. (Turchi 2010; Turchi et al. 2010), which incorporates a detailed review by the engineering consulting firm Worley Parsons.

4.2.4.1 Capital Costs

An indicative relative breakdown of capital costs (total installed costs) for two 100-MWe CSP plants with storage in the range 4.5–15 hours, located in South Africa, is given in Figure 4.10.

A recent study of renewable energy project costs by the International Renewable Energy Agency (IRENA) reports that the global average of the total installed cost for CSP plants completed in 2018 was USD 5200/kWe; IRENA 2019). Such costs vary greatly based on the amount of storage (in 2018, they all had at least 4 hours, and two had more than 8 hours), and are also sensitive to fluctuations in steel and glass prices, local wages, transport and the local solar resource (since the size of the solar field will need to be bigger in areas of lower annual DNI).

Concentrating solar power plants can be developed using a large share of local materials and components. A study by Deloitte commissioned by Protermosolar, the Spanish industry association for CSP, determined that for the plants built in Spain in 2010, over 70% of the total investment costs were domestic purchases (Deloitte 2011). In Africa, the recently

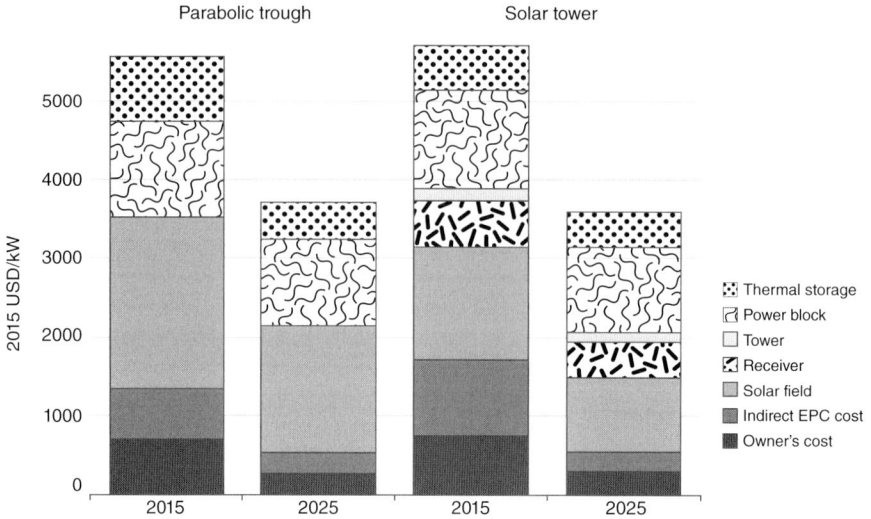

Figure 4.10 Relative breakdown of total installed costs for 100-MWe plants in South Africa. Left: parabolic trough with 13.4 hours of storage; right: a tower system with 15 hours of storage.

Source: IRENA (2016).

operational Noor III plant in Morocco was planned with 35% local content, as a result of a strategy to help advance local industrial development (Moore 2018).

4.2.4.2 Operations and Maintenance Costs

For a 100-Mwe trough system in South Africa, Fichtner estimated that the annual operations and maintenance (O&M) costs would vary between USD 15 and USD 18 per year, depending on the size of the storage system (Konstantin and Kretschmann 2010). This is approximately 2.0% of the system total installed costs each year. For the 100-MWe tower systems in South Africa, the estimate was USD 14–18 per year, again depending on the amount of storage (Konstantin and Kretschmann 2010). This is approximately 1.9% of the system total installed costs each year.

The fractional breakdown of these O&M costs is fairly consistent between these different 100-MWe systems as analysed by Fichtner. The averaged fractional breakdown in O&M costs, calculated from the Fichtner data, is shown in Figure 4.11.

4.2.4.3 Financing Costs

Financing of CSP systems typically requires equity investment as well as a significant loan. The interest rate payable on the loan will depend greatly on the assessed risk of the project, and may be of the order of 6–8% per annum. The return required by equity investors will be higher, of the order of 12% per annum. In many locations, it is necessary for governments to provide various incentives such as investment tax credits, depreciation bonuses, carbon credits or others. Loan guarantees can also be a way to reduce the

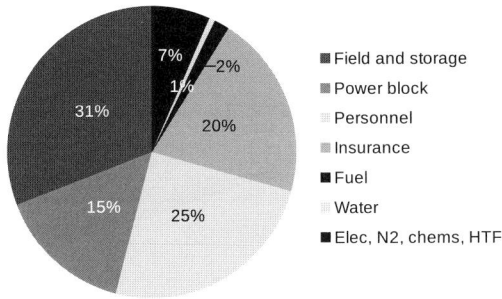

Figure 4.11 An average breakdown of O&M costs for 100-MWe trough and tower systems in South Africa. Fixed costs (field and storage, power block, personnel and insurance) dominate greatly over the other, variable, costs.
Source: Author's analysis.

risk to lenders and investors, hence improving the economic feasibility of a project (Mendelsohn et al. 2012). For the most recent CSP projects, it has been suggested that favourable financing with a WACC as low as 3% must have been obtained (Lilliestam and Pitz-Paal 2018).

In the Fichtner analysis discussed above, variations in carbon credit (USD 0–14 per tonne of carbon dioxide equivalent, or tCO_2e) and overall discount rate (6–8%) have the impact of varying the calculated LCOE in the range USD 0.17–0.22/kWh; these factors strongly affect the overall project feasibility calculation. The more recent report from IRENA gives the latest LCOE achieved in fully completed commercial projects in 2018 as USD 0.185/kWh (IRENA 2019). The LCOE can be broken down into the component used to repay investors and lenders and the component used to cover O&M costs. As much as 80% of the LCOE derives from amortisation of capital costs, with only 20% attributable to operations and maintenance costs (Pitz-Paal et al. 2005; Hinkley et al. 2013; Slaughter 2014).

4.2.4.4 Further Observations on Cost

Direct Normal Irradiance (DNI) (see Figure 4.1). DNI strongly affects the LCOE of a CSP system. The annual DNI affects the amount of electricity generated by a CSP system without altering the capital or operating cost. For example, high DNI in Chile causes the LCOE to be as much as 30% lower for a plant installed there compared to an identical system in Spain (Figure 4.12). Many of the recent plants have been constructed in locations of higher DNI.

Scale: Small CSP systems generally are unable to incorporate a highly efficient power block.[4] This results in the need for more collector area per unit of electrical

---

[4] The main driver here is the efficiency of steam Rankine cycle power blocks at larger scale. Larger power blocks gain large economies of scale due to the cost-effective addition of boiler feed-water heaters and reheat stages. A survey of coal-fired Rankine cycle power blocks in the 1970s showed the cost of power from systems with an average ~2-GWe output being as much as 10% cheaper than that from systems with an average ~200 MWe in size (Christensen and Greene 1976).

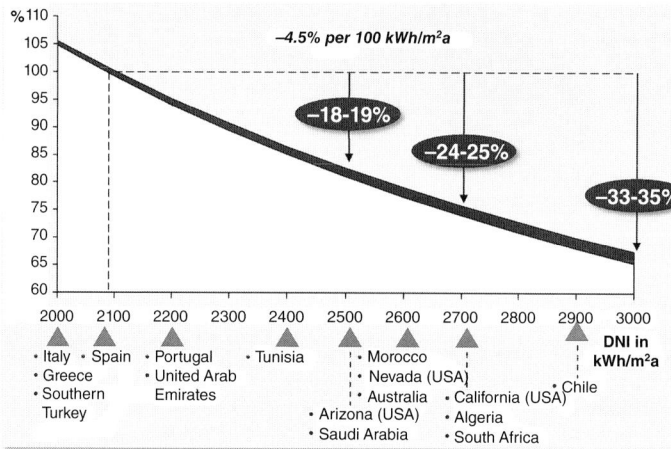

Figure 4.12 Variation of the LCOE for CSP systems as a function of the annual solar DNI. Estimates of DNI here are only approximate; newer data (see Figure 4.1) show sites with 3600 kWh/m² per year in Chile, and 2900 kWh/m² per year in the USA, South Africa and Australia.
Source: A.T. Kearney (2010).

energy delivered, and increases the LCOE. There is, however, an optimal size for CSP systems. For towers, atmospheric attenuation becomes a significant limitation (Ballestrín and Marzo 2012) and appears to have limited fields to a total collector area of the order of 1.30 km^2 (as in the 150-MWe Noor III system), since heliostats in that system extend as far as ~1500 m from the receiver. For trough systems, a larger system incurs greater pumping and thermal losses through the larger collector field, and the largest fields to date have had a total collector area of 1.59 km^2 (250-MWe Solana, Abengoa).

Storage Capacity. The impact of storage capacity on the LCOE of a CSP system is not simple, because while storage has a major capital cost, it can also increase the capacity factor of the turbine, and reduce the amount of energy lost in turbine ramp-up and ramp-down cycling. When whole systems are optimised for LCOE, quite large storage systems of 12 hours or more are observed (Figure 4.13), with a correspondingly larger solar multiple then required to ensure that the storage capacity is efficiently utilised. This is a very important point: CSP systems with a specified power output typically show a lower LCOE with storage than without.

Local Wages and Transport Costs. The LCOE is affected by variations in wages and transport costs from country to country. In some locations, it is necessary to pay a premium for workers in order for them to travel to the remote arid locations commonly associated with CSP systems.

Risk. The financial risk associated with CSP systems is also a major factor in the cost of developing them. As CSP systems become more accepted by banks and investors, the cost

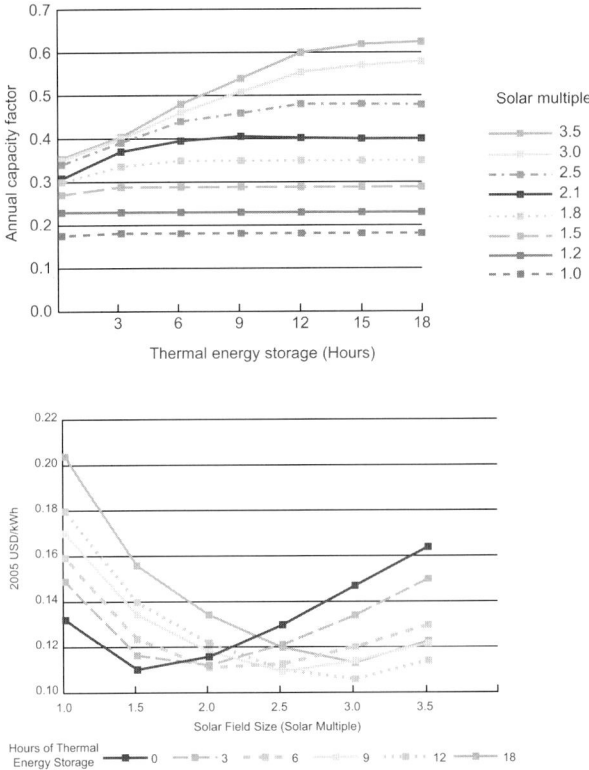

Figure 4.13 For a CSP system in any given location, the capacity factor (top) and the LCOE (bottom) are affected by the storage capacity and the solar multiple (see Section 4.2.1.1). In this example, the lowest LCOE is achieved with 12 hours storage and a solar multiple of 3.0. *Source*: IRENA (2012).

of financing will greatly reduce. This is one of the strongest factors affecting the energy cost from CSP systems.

Government Incentives. Governments can do a great deal to encourage the development of CSP systems in their region. Some options include loan de-risking (Mendelsohn et al. 2012), grant schemes, use of carbon credit revenues and funds for technical assistance (CSP Alliance 2014), or through feed-in tariffs and similar mechanisms that increase project profitability in a way somehow proportional to the higher cost of capital.

The Golden Sunset. PPAs for CSP systems are typically agreed for a period of 20–25 years, corresponding to the expected lifetime of the plant. However, the SEGS systems in California, built during the 1980s, have now been operating for over 30 years in some cases. With all the loans now paid off, it has been profitable for the SEGS systems to continue to operate while selling power to the network for as little as USD 0.057/kWhe

(Shahan 2013). As CSP technology matures, increasing commercial confidence about this end-of-life revenue should help to reduce the cost of CSP to consumers. Richert et al. (2012) give a good analysis of the impact of scale, location, storage capacity, government incentives and time-of-day pricing on the PPA price for CSP plants.

4.2.4.5 Learning Rates

The costs of many new technologies have been shown to decrease in a power-law relationship with the cumulative deployment of the technology. For most industries, learning rates are in the range 10–30% (Dutton and Thomas 1984). Lovegrove et al. (2012) reviewed other studies and found that PV and wind learning ratios were in the range 15–20%.

For CSP systems, establishing accurate learning rate estimates is challenging because few data from commercial plants are publicly released, and because there are so many variations due to location, incentives, etc., as noted above. In a recent study by Lilliestam et al. (2017), it was concluded that CSP learning rates are '20% or higher', significantly higher than the 5–15% assumed in 'most policy analyses'. Some well-known studies (Sargent and Lundy LLC Consulting Group 2003; Pitz-Paal et al. 2005; Kolb et al. 2011) give breakdowns in cost reduction potential in CSP, and help to explain the changes that could be behind these recently observed savings.

4.2.5 Environmental Impacts of CSP

4.2.5.1 Greenhouse Gas Emissions

Concentrating solar power technology offers a 96–98% reduction in greenhouse gas emissions compared to conventional coal generation, and a 92–96% reduction relative to natural gas. Emissions from CSP, PV and wind are all of similar magnitude, and further emissions reductions are expected to be realised through to 2050 (Viebahn et al. 2008, 2010, 2011; Burkhardt et al. 2011, 2012).

4.2.5.2 Water Usage

Water use in CSP systems is an issue due the arid conditions typical of high-DNI locations. Conventional 'wet-cooled' CSP systems make use of recirculating evaporative cooling towers to cool the condenser in the steam Rankine power block, as with typical coal power stations, and consume large amounts of water, ~3 L/kWhe (Bracken et al. 2015). Increasingly, however, CSP systems are migrating to the use of 'dry cooling', where steam is condensed in large air-cooled condensers, either with natural-draft air flow as in Khi Solar One (South Africa), or with fan-forced air flow, as for example in Noor III (Morocco). These systems are estimated to have a total water use of below 300 mL/kWhe (Macknick et al. 2011). Dry cooling reduces water consumption greatly, but comes at the cost of 2–5% reduced electrical output, and a consequent 2.5–8% increase in the LCOE, according to one study (Bracken et al. 2015).

Figure 4.14 A cleaning truck at the Khi Solar One power plant.
Source: Photo courtesy of John Pye.

A third option for cooling of CSP plants is hybrid cooling, employed in the recently operational Crescent Dunes tower system in the USA. In hybrid cooling, water is used only during the hottest weather periods, and the system runs using dry cooling at other times. These systems are estimated to use a total of ~1 L/kWhe of water (Macknick et al. 2011).

Aside from condenser cooling, water is also consumed by the power cycle itself for 'blowdown' of steam, to limit corrosion and scale within the various power cycle components (Cohen et al. 1999). Water consumption for power cycle blowdown is of the order of 100–200 mL/kWhe (Bracken et al. 2015). Finally, a very small amount of water is used to clean mirrors (Figure 4.14), to offset system performance losses due to gradually accumulated dust. Mirror cleaning water use is estimated to be 75–150 mL/kWhe (Vivar et al. 2010; Bracken et al. 2015). Future supercritical CO_2 power cycles for CSP would eliminate the steam blowdown water consumption, while next-generation ultrasonic mirror cleaning may also bring major savings.

4.2.5.3 Other Impacts

Concerns have been raised about the optical glare from CSP plants, and potential risk to aircraft and motorists. These issues can be managed through careful design and site selection, as well as by controlling where the heliostats are aimed when in 'standby' mode (Ho et al. 2014). Another community concern has been the risk to birds and bats. Following detailed studies, in particular in relation to the Ivanpah plant in California, it has been concluded that this risk is minor (Ho 2016; Ho et al. 2016), although future plant operators will certainly be expected to monitor this risk. Finally, arid landscapes can be home for certain animals which in some cases can be endangered, and require care or protection.

4.2.6 Resilience of CSP to Climate Change

As a potential solution to climate problems, it is critical that CSP systems themselves are able to withstand the effects of climate change while maintaining acceptable levels of output. Patt et al. (2013) reviewed CSP and PV technologies with regard to this issue, and found that areas of concern for CSP technology include: (1) increased downtime due to high wind; (2) reduced dry cooling effectiveness due to hotter ambient temperatures; (3) soiling of mirrors due to sand and dust and resulting increased cleaning costs; and (4) potentially prolonged periods of cloud, interrupting system output. The need for CSP systems to survive hail damage has been considered in several projects, but was not noted as an area of concern by Patt et al. (2013). The analysis concluded that the risks to CSP posed by climate change were surmountable, and not of major concern to the technology as a whole.

4.2.7 Concentrated Solar Power in the Electricity Grid

A serious response to the risks of climate change requires the eventual elimination of electricity generation technologies that are net carbon emitters. Wind and PV, having achieved large cost reductions in recent years, will certainly contribute very significantly to this, but the limitation of these technologies is that they are 'inflexible' – they cannot adjust their supply to match a varying demand. This limitation of inflexible generation can be addressed through several strategies, discussed elsewhere in this book (see Chapter 7, which discusses energy storage, and Chapter 16, which discusses applications of solar thermal energy in industry and manufacturing), including centralised and distributed battery storage, centralised pumped hydroelectric energy storage, compressed air storage and demand response.[5] Concentrated solar power, with its integrated energy storage, offers an additional solution to the inflexible generation problem, and offers flexibility that helps to enable matching of time-varying demand and supply profiles. This flexibility in CSP systems has a value, which has been the subject of numerous studies in recent years.

4.2.7.1 The Role for CSP in a Future 100% Renewable Electricity Network

Recently, studies have sought to identify what would be required to transition to a completely renewable electricity network. Generally, these studies take projected costs for different technologies, based on learning rates, and seek to identify the lowest-cost balance of technologies that will meet anticipated demand. Jacobson et al. (2017) developed a 139-country 100% renewable energy scenario for the year 2050, and predicted 9.7% of annual power production would come from CSP, and that CSP would furthermore play a major role in providing peaking power capacity.

Elliston et al. (2013, 2014) completed an extensive analysis of the Australian National Electricity Market (NEM), which supplies electricity to the south-east part of Australia where the majority of the Australian population resides. After optimising the portfolio of PV, CSP, pumped hydro, hydro and biomass-fired gas turbines for lowest-cost generation while meeting the entire NEM demand, they found CSP would contribute significantly to the Australian generation mix, in the range of 7–13% of installed nominal capacity, and in

---

[5] 'Dumping' (throwing away PV- or wind-generated power without sending it to the grid) and 'load shedding' (disconnecting customers from the grid in a controlled way, in order to avoid blackouts) are also options, albeit far less palatable.

the range 13–23% of generated energy. Subject to the cost assumptions of the Australian Energy Technology Assessments (AETA), the conclusions of Elliston et al. are that the cheapest electricity network that Australia could build in 2030, assuming a carbon tax of AUD 20/tCO$_2$, would be one without carbon capture and storage, in which at least 13% of the energy is provided from CSP.

These and other future 100% renewable scenarios (Cochran et al. 2014) are highly dependent on projections of cost reduction, and such projections are notoriously difficult. But on the basis of even conservative cost reduction predictions, it seems likely that CSP will have quite a significant role in meeting the needs of consumers as we seek to deploy lower-emission generation.

4.2.7.2 The Value of CSP in Existing Grids

The analysis of the benefits offered by CSP in existing grids tends to follow a different approach, where the value of CSP is quantified as an incremental modification of an existing network.

Denholm and Mehos and co-workers (Denholm and Mehos 2011; Denholm et al. 2013, 2014; Jorgensen et al. 2013, 2014) define the value to the grid from two aspects. First, *operational value* relates to the reduction of operating costs from the pre-existing part of the electricity network before and after the addition of new plants. When a new CSP plant is added, it reduces the fuel use in existing fossil-fired plants, it reduces the O&M costs on the existing network and it reduces the costs associated with start-ups during periods of varying demand. It also reduces the carbon taxes for the plant, if applicable, corresponding to those periods of operation.

Second, *capacity value* relates to the fact that a new CSP plant eliminates the need for at least a part of a pre-existing power plant. This is determined by considering the hours of peak demand on the whole network, and considering whether the CSP plant is likely to be operational during those hours in the year. It is found that CSP plants typically are able to operate at a close to 100% capacity factor at peak times for the electricity network considered. Hence the capacity value for CSP is high, and the annualised cost of an existing plant on the network can be included in the value that the new CSP plant provides.

Denholm et al. (2014) considered the case of the desert in south-west USA, the region where most of the CSP plants built in that country have been located up to now. Calculating the operational and capacity value for several options, they concluded that a CSP plant has a value of USD 0.08–0.12/kWh (Figure 4.15). Jorgensen et al. (2013) considered the best way to configure a CSP plant having a fixed collector field size. They considered variations in solar multiple and storage capacity and found that the configuration giving the greatest value is that with a small solar multiple (hence a large power block, if the collector field is fixed) and only 6 hours of storage. This is because 6 hours of storage is enough for the peak demand times to be met, and the larger power block size maximises capacity and operational value. Solar Dynamics LLC recently proposed that an even higher value in the grid (perhaps up to USD 0.17/kWh) could be gained from 'CSP peaker' systems with 3 hours of storage and a solar multiple of just 0.5 (Price 2017).

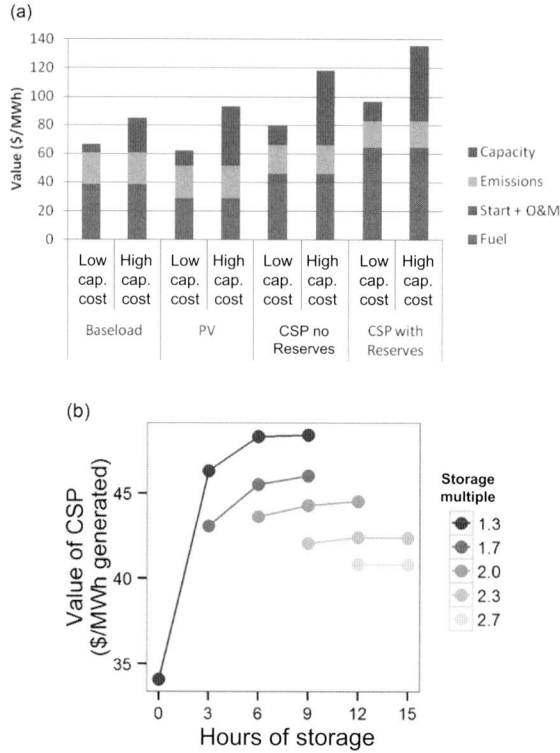

Figure 4.15 (a) Capacity value (displaced amortised capital of installed plants) and operational value (displaced emissions, start + O&M, fuel) for new baseload, PV and CSP plants installed in south-west USA. As there is a range of estimates for the amortised capital cost of installed plants, the plot shows both high and low capital costs as separate bars. (b) Relationship between total plant value and storage size and solar multiple. *Source*: Denholm et al. (2014).

From these analyses it is clear that CSP plants cannot be designed on the basis of the LCOE alone. Project developers should consider that a low-LCOE plant will not necessarily be the most valuable choice for the local electricity network, and that in fact the local electricity network may be prepared to pay a premium price for a different type of system that provides greater capacity or operational value to the network as a whole. Policy-makers can help to ensure that project developers are given incentives to provide the greatest value to the grid, but must ensure that this is not done with too detrimental an impact on the cost of plants being proposed.

The situation analysed for the USA could likely be quite different in developing countries with less extensive electricity networks. In places where there is growing demand and insufficient generation, there will be a high capacity value derived from the installation of CSP systems with substantial amounts of storage. Photovoltaic systems without storage, though cheaper, will not help greatly in the urgent need to supply reliable power at the times it is needed. For these locations, combined CSP and PV plants are being developed, offering a degree of firm capacity at a more competitive cost (Green et al. 2015; Platzer 2015).

4.2.8 Other Markets for CSP

Up to now, commercial development of CSP technology has been mainly focused on electricity generation. However, there is a growing effort to find ways to apply CSP technology to other energy needs in society. These include the production of transport fuels and the supply of industrial process heat for a wide range of industries. This section provides a brief overview of the activities and opportunities in these areas, and the current research efforts.

4.2.8.1 Water Desalination

A major environmental problem today is the unsustainable extraction of groundwater for use in agriculture and cities (see Chapter 13). Desalination already provides 1% of the world's drinking water, and as demand grows and natural sources are depleted, this fraction will continue to grow. As of 2015, 18 000 desalination units were installed worldwide, with a total capacity of 87 million cubic metres per day; 44% of that capacity is in the high-DNI region of the Middle East (Voutchkov 2016). Outside the Middle East, desalination is mostly via the high-pressure process of reverse osmosis, whereas in the Middle East, low fuel costs have led to greater use of distillation (thermal) systems (Voutchkov 2016). The cost of fossil-fuel-powered desalinated water is claimed to be USD $0.8–1.2/m^3$ as of 2016 (Voutchkov 2016).

Concentrating solar power shows excellent potential as a replacement for fossil fuels in powering large-scale desalination systems and was recently estimated to be capable of providing clean water for below EUR $1/m^3$ (= USD $0.78/m^3$ as of 2012) (Lienhard et al. 2012). There is strong opportunity for hybridisation here, since the CSP plant can first produce electricity, and then make use of waste heat from the power block to drive the thermal desalination system.

Recently, a large-scale hybrid desalination system of this type has been built as part of an integrated greenhouse agriculture project in arid land. Sundrop Farms, in Port Augusta, South Australia, has installed a 20-ha greenhouse facility alongside a 39-MWth CSP tower system and a water treatment system, to grow tomatoes using seawater (see Table 4.1 and Figure 4.9) (Staight 2016). The system provides power and water for the plant, and the developers of this system are planning to build several further systems in coming years.

4.2.8.2 Solar Fuels

A significant and growing area of research relates to the use of CSP in the production of transport fuels (Steinfeld 2005; Romero and Steinfeld 2012). Fluids that could play the role of future transport fuels include hydrogen, methane, methanol, dimethyl ether, ammonia and synthesised 'drop-in' fuels.

Hydrogen produced using solar energy needs only water as a feedstock. Two leading approaches are water electrolysis using PV or CSP electricity, and thermochemical water splitting. Electrolysis is already well established commercially but costly, in part due to relatively low efficiency. Thermochemical water splitting, meanwhile, is being heavily researched because of its potential to achieve up to 45% solar-to-fuel conversion efficiency, higher than the ~30% possible for PV- or CSP-driven electrolysis (Wang et al. 2012). Thermochemical water splitting makes use of metal oxides, which are reduced using solar thermal heat input, releasing oxygen. The reduced oxides are then re-oxidised with the addition of water at high temperature. This second reaction releases hydrogen, and the entire two-step process occurs at temperatures of 900 °C or more, depending on the metal oxide material chosen. There are numerous challenges to be overcome in order for thermochemical water splitting to be successful commercially, including heat transfer, chemical conversion, material stability and others.

Hydrogen itself is not an ideal fuel because, in order to be transported with reasonable volumetric energy density, it requires either high pressure (>200 bar) or cryogenic cooling, or else more complex storage techniques such as chemical storage in metal hydrides. Instead of transporting hydrogen directly, one option is to convert the hydrogen into ammonia (NH_3) using nitrogen from the air. Ammonia can be transported at much lower pressure (~10 bar), and still has very good energy density, and can be either burnt as a fuel or processed into other useful chemicals (Zamfirescu and Dincer 2008). Ammonia is still a dangerous material, however, and safety systems would need to be developed or adapted if it were to be adopted as a mass-market fuel.

'Drop-in' solar fuels are synthesised to have a chemical composition close to conventional fossil fuels such as kerosene, gasoline and diesel, all of which are liquid hydrocarbons. Most commonly, this is proposed via the commercially mature Fischer–Tropsch (FT) process, which requires syngas (hydrogen mixed with carbon monoxide) as its input. This syngas can be produced in a number of ways, including via high-temperature solar-driven gasification of biomass (Hertwich and Zhang 2009) or, as recently demonstrated, via CSP-assisted direct air capture of CO_2 (ETH Zurich 2019). One challenge with such processes is that conventional FT synthesis reactors are big and heavy and not suitable for daily ramp-up and ramp-down, as would be preferable when powered from syngas that is produced only during sunlight hours (Hinkley, Naughton, Pye, Saw et al. 2015). Smaller and lighter FT reactors with potential for integration with solar fuels systems are under development (Cao et al. 2009). Other challenges exist with gasification, such as char and tar formation, and availability of high-temperature materials for the gasification reactor. 'Drop-in' synthetic fuels are attractive because they can immediately be integrated with existing infrastructure, and have a market value far higher than other easier-to-produce fuels such as methane. Apart from biomass gasification, solar syngas can also be produced from fossil fuels through steam methane reforming (reacting methane and steam at high temperature, using CSP heat input) and coal gasification (again using CSP as the heat input) (Zedtwitz and Steinfeld 2003; Agrafiotis et al. 2014). These processes are closer to maturity, but result in a fuel with only a modest solar fraction, ~25% (Zedtwitz and Steinfeld 2003). Such processes could be helpful in a transition to a lower-emissions economy, but, on their own, they will not be sufficient for us to achieve our longer-term global emissions reductions targets. On the other hand, large-scale adoption of carbon-neutral solar biofuels will require very large quantities of biomass, production of which will strain our precious water and land resources.

As Lovegrove (2013) points out, the cost of heat from a CSP plant (~AUD 7.50/GJ, 2013 estimates) plus the cost of carbonaceous material (~AUD 2.50/GJ for brown coal, including an AUD 23/tCO$_2$e carbon tax; or ~AUD 0.80/GJ for bagasse) is much less than the cost of diesel (AUD 26.03/GJ, excise free) or oil (AUD 17.88/GJ). There appears, then, to be strong potential to produce solar fuels from carbonaceous materials at a cost that is competitive with conventional fuels. An extensive road-mapping exercise by the IEA SolarPACES organisation is currently nearing completion, which aims to identify the most promising concepts and required research for development of solar fuels processes at commercial scales (Hinkley, Naugthon, Pye, Lipiński et al. 2015).

4.2.8.3 Industrial Process Heat

As efforts are made to achieve reductions in global greenhouse gas emissions, there will need to be major changes to the current practices of providing heat to industrial processes using fossil fuels, currently mostly natural gas and coal. Many of these processes operate at high temperature and replacing them with renewable alternatives is challenging.

Eglinton et al. (2013) reviewed a wide range of high-temperature industrial processes, including in mineral processing and metallurgy. Lab-scale and demonstration-scale experiments have been conducted for several of these processes, including solar zinc reduction and calcination of lime for cement production. Currently, these CSP-driven processes are not commercially viable, but these remain active areas of research. A more recent review of the application of CSP to powering chemical processes is given by Bader and Lipiński (2017).

Considering the broader use of natural gas in industrial processes in the Australian context, Lovegrove et al. (2015) observed a wide range of sectors requiring process heat at temperatures in the range 250–800 °C where CSP systems could potentially provide input. This range of temperatures is highly compatible with current commercial CSP collector technology. Significant sectors include pulp and paper manufacturing, food manufacturing and chemical production. Lovegrove observes, however, that the current relative costs of CSP and natural gas make adoption of CSP in these areas difficult, and that low-temperature solar thermal technologies are closer to viability.

4.3 Low-Temperature Solar Thermal Systems

Although the emphasis of this chapter has been on high-temperature CSP systems, there is a large number and a wide range of systems that provide solar thermal heat for lower-temperature applications.

As of 2018, the IEA estimates that there were 480 GWth of non-concentrating solar thermal collectors installed globally (Weiss and Spörk-Dür 2019). Systems as large as 27 MWth have been installed in recent years for applications in agriculture, mining, textiles and district heating. In addition to these large systems, this sector remains dominated by solar domestic hot water systems, which are either flat plate or evacuated tube systems, and represent ~80% of the total installed capacity globally. Large non-concentrating solar thermal systems continue to be installed and numerous applications in the area of industrial process heat and in agriculture/crop drying have been highlighted by the IEA and IRENA (Kempener et al. 2015). Strong potential for low-temperature solar thermal collectors to cost-effectively replace natural-gas-fired industrial process heat was also highlighted by Lovegrove et al. (2015).

Solar cookers are a low-cost technology with relevance to developing regions. An insulated box receives solar radiation, usually slightly concentrated using one or more manually adjusted reflectors. Food to be cooked is placed inside the box and cooked at temperatures of the order of 90–100 °C. Such cookers tend not to completely displace firewood and other fuels, but field trials from South Africa suggest that owners use them regularly and reduce their fuel use by ~40% (Wentzel and Pouris 2007).

Finally, various passive and active techniques use solar thermal energy to lower or eliminate building heating requirements. Active solar hot air collectors can be placed on roofs, Trombe walls can be incorporated into building facades and well-insulated buildings with good solar orientation ('passive houses') can greatly reduce energy demand from other sources.

4.4 Conclusion

To conclude, solar thermal technologies are growing strongly and have an important role to play in the provision of dispatchable renewable electricity, hot water for domestic and industrial use, and heat for industrial processes, and in the production of solar fuels. Large-scale CSP systems are being developed and constructed currently in Morocco, China, the UAE, Chile, Australia and South Africa, and appear to indicate that CSP is now either at – or very close to – commercial competitiveness in these specific markets where the DNI solar resource is high, and where electricity network constraints, fuel costs or trade barriers limit the use of alternative energy sources. Learning rates in this industry are strong. Next-generation systems currently in development are seeking to reduce costs by incorporating higher-efficiency supercritical CO_2 turbines, cheaper particle-based energy storage, new salt formulations for higher-temperature storage, phase-change energy storage materials, liquid metal heat transfer fluids and other innovations. With its cost-effective energy storage technology, CSP is projected to play a significant and stabilising role in the future electricity grid alongside other renewables, and it has been shown to carry a higher value than other non-flexible renewables as a result.

Acknowledgements

The authors would like to thank Joe Coventry, Michael Smith, Ken Baldwin and Bruce Godfrey for their kind reviews of this text, and their many helpful suggestions.

References

Agrafiotis, C., von Storch, H., Roeb, M. and Sattler, C. (2014). Solar thermal reforming of methane feedstocks for hydrogen and syngas production: A review. *Renewable and Sustainable Energy Reviews*, 29I, 656–682.

A.T. Kearney (2010). *Solar Thermal Electricity 2025*. Dusseldorf: A.T. Kearney GmbH and ESTELA (European Solar Thermal Electricity Association). Available at: www.promes.cnrs.fr/uploads/pdfs/documentation/2010-Solar%20thermal%20electricity%202025%20ESTELA.pdf.

Bader, R. and Lipiński, W. (2017). Solar thermal processing. In M. Blanco and L. R. Santigosa, eds., *Advances in Concentrating Solar Thermal Research and Technology*. Cambridge: Woodhead Publishing.

Ballestrín, J. and Marzo, A. (2012). Solar radiation attenuation in solar tower plants. *Solar Energy*, 86(1), 388–392.

Bracken, N., Macknick, J., Tovar-Hastings, A., Komor, P., Gerritsen, M. and Mehta, S. (2015). *Concentrating Solar Power and Water Issues in the US Southwest*. Technical report NREL/TP-6A50–61376. Golden, CO: Joint Institute for Strategic Energy Analysis. Available at: www.nrel.gov/docs/fy15osti/61376.pdf.

Branker, K., Pathak, M. and Pearce, J. (2011). A review of solar photovoltaic levelized cost of electricity. *Renewable and Sustainable Energy Reviews*, 15(9), 4470–4482.

Burkhardt, J. J., Heath, G. A. and Turchi, C. S. (2011). Life cycle assessment of a parabolic trough concentrating solar power plant and the impacts of key design alternatives. *Environmental Science & Technology*, 45, 2457–2464.

Burkhardt, J. J., Heath, G. and Cohen, E. (2012). Life cycle greenhouse gas emissions of trough and tower concentrating solar power electricity generation: Systematic review and harmonization. *Journal of Industrial Ecology*, 16(1), S93–S109.

Cao, C., Hu, J., Li, S., Wilcox, W. and Wang, Y. (2009). Intensified Fischer–Tropsch synthesis process with microchannel catalytic reactors. *Catalysis Today*, 140, 149–156.

Christensen, L. R. and Greene, W. H. (1976). Economies of scale in US electric power generation. *Journal of Political Economy*, 84, 655–676.

Cochran, J., Mai, T. and Bazilian, M. (2014). Meta-analysis of high penetration renewable energy scenarios. *Renewable and Sustainable Energy Reviews*, 29I, 246–253.

Cohen, G. E., Kearney, D. W. and Kolb, G. J. (1999). *Final Report on the Operation and Maintenance Improvement Program for Concentrating Solar Power Plants*. Technical report SAND99–1290. Albuquerque, NM: Sandia National Laboratories. Available at: www.osti.gov/servlets/purl/8378.

CSP (Concentrating Solar Power) Alliance (2014). *The Economic and Reliability Benefits of CSP with Thermal Energy Storage: Literature Review and Research Needs*. Technical report. CSP Alliance. Available at: www.inship.eu/docs/TES%204% 20the_economic_and_reliability_benefits_of_csp_with_thermal_storage_2014_09_ 09_final.pdf.

Deloitte (2011). *Macroeconomic Impact of the Solar Thermal Electricity Industry in Spain.* Consultants' report. Seville: Protermosolar. Available at: www.solarthermalworld .org/sites/default/files/Macroeconomic_impact_of_the_Solar_Thermal_Electricity_ Industry_in_Spain_Protermo_Solar_Deloitte_21x21.pdf.

Denholm, P. and Mehos, M. (2011). *Enabling Greater Penetration of Solar Power via the Use of CSP with Thermal Energy Storage.* Technical report NREL/TP-6A20–52978. Golden, CO: National Renewable Energy Laboratory. Available at: www.nrel.gov/ docs/fy12osti/52978.pdf.

Denholm, P., Wan, Y.-H., Hummon, M. and Mehos, M. (2013). *An Analysis of Concentrating Solar Power with Thermal Energy Storage in a California 33% Renewable Scenario.* Technical report NREL/TP-6A20–58186. Golden, CO: National Renewable Energy Laboratory. Available at: www.nrel.gov/docs/fy13osti/ 58186.pdf.

Denholm, P., Wan, Y.-H., Hummon, M. and Mehos, M. (2014). The value of CSP with thermal energy storage in the western United States. *Energy Procedia*, 49, 1622–1631.

Dutton, J. M. and Thomas, A. (1984). Treating progress functions as a managerial opportunity. *Academy of Management Review*, 9, 235–247.

Eglinton, T., Hinkley, J., Beath, A. and Dell'Amico, M. (2013). Potential applications of concentrated solar thermal technologies in the Australian minerals processing and extractive metallurgical industry. *Journal of the Minerals, Metals and Materials Society*, 65, 1710–1720.

Elliston, B., MacGill, I. and Diesendorf, M. (2013). Least cost 100% renewable electricity scenarios in the Australian National Electricity Market. *Energy Policy*, 59I, 270–282.

Elliston, B., MacGill, I. and Diesendorf, M. (2014). Comparing least cost scenarios for 100% renewable electricity with low emission fossil fuel scenarios in the Australian National Electricity Market. *Renewable Energy*, 66, 196–204. Available at: https:// arena.gov.au/assets/2017/06/Elliston_2014_5.pdf.

ETH Zurich (2019). Carbon-neutral fuel made from sunlight and air. *ETH Zürich.* 13 June. Available at: https://ethz.ch/en/news-and-events/eth-news/news/2019/06/pr-solar-mini-refinery.html.

Green, A., Diep, C., Dunn, R. and Dent, J. (2015). High capacity factor CSP–PV hybrid systems. *Energy Procedia*, 69, 2049–2059. Available at: https://core.ac.uk/download/ pdf/81170195.pdf.

Hernández-Moro, J. and Martínez-Duart, J. (2013). Analytical model for solar PV and CSP electricity costs: Present LCOE values and their future evolution. *Renewable and Sustainable Energy Reviews*, 20I, 119–132.

Hertwich, E. G. and Zhang, X. (2009). Concentrating-solar biomass gasification process for a 3rd generation biofuel. *Environmental Science & Technology*, 43, 4207–4212.

Hinkley, J., Curtin, B., Hayward, J. et al. (2011). *Concentrating Solar Power: Drivers and Opportunities for Cost-Competitive Electricity.* Commissioned report for the Commonwealth Government of Australia. Canberra: CSIRO. Available at: https:// publications.csiro.au/rpr/download?pid=csiro:EP111647&dsid=DS3.

Hinkley, J. T., Hayward, J. A., Curtin, B. et al. (2013). An analysis of the costs and opportunities for concentrating solar power in Australia. *Renewable Energy*, 57, 653–661.

Hinkley, J. T., McNaughton, R. K., Pye, J. D., Lipiński, W. and Lovegrove, K. M. (2015). Current and future status of solar fuel technologies in Australia. *Journal of the Japan Institute of Energy*, 94, 182.

Hinkley, J. T., McNaughton, R. K., Pye, J., Saw, W. and Stechel, E. B. (2015). The challenges and opportunities for integration of solar syngas production with liquid fuel synthesis. Paper presented at 21st SolarPACES Conference, Cape Town, South Africa, 13–16 October.

Ho, C. K. (2016). Review of avian mortality studies at concentrating solar power plants. *AIP Conference Proceedings*, 1734, 070017.

Ho, C. K., Sims, C. A. and Christian, J. M. (2014). *Evaluation of Glare at the Ivanpah Solar Electric Generating System.* Technical report SAND2014–15847. Albuquerque, NM: Sandia National Laboratories. Available at: https://prod-ng .sandia.gov/techlib-noauth/access-control.cgi/2014/1415847.pdf.

Ho, C. K., Wendelin, T., Horstman, L. and Yellowhair, J. (2016). A method to assess flux hazards at CSP plants to reduce avian mortality. Paper presented at 22nd SolarPACES Conference, Abu Dhabi, UAE, 11–14 October. Available at: www .researchgate.net/publication/317984020_A_method_to_assess_flux_hazards_at_ CSP_plants_to_reduce_avian_mortality.

IRENA (International Renewable Energy Agency) (2012). *Concentrating Solar Power.* IRENA Working Paper. Renewable energy technologies: Cost analysis series Vol. 1: Power Sector. Issue 2/5. Abu Dhabi: International Renewable Energy Agency. Available at: https://irena.org/publications/2012/Jun/Renewable-Energy-Cost-Analysis – Concentrating-Solar-Power.

IRENA (2016). *Power to Change: Solar and Wind Cost Reduction Potential to 2025.* Abu Dhabi: International Renewable Energy Agency. Available at: www.irena.org/-/ media/Files/IRENA/Agency/Publication/2016/IRENA_Power_to_Change_2016.pdf.

IRENA (2019). *Renewable Power Generation Costs in 2018.* Abu Dhabi: International Renewable Energy Agency. Available at: www.irena.org/-/media/Files/IRENA/ Agency/Publication/2019/May/IRENA_Renewable-Power-Generations-Costs-in-2018.pdf.

Jacobowitz, D. and Google (2013). Google's Green PPAs: What, How, and Why. Technical report. *Google.* Available at: https://static.googleusercontent.com/media/ www.google.com/en//green/pdfs/renewable-energy.pdf.

Jacobson, M. Z., Delucchi, M. A., Bauer, Z. A. et al. (2017). 100% clean and renewable wind, water, and sunlight all-sector energy roadmaps for 139 countries of the world. *Joule*, 1, 108–121. Available at: https://web.stanford.edu/group/efmh/219amsar219n/ Articles/I/CountriesWWS.pdf.

Jorgensen, J., Denholm, P., Mehos, M. and Turchi, C. (2013). *Estimating the Performance and Economic Value of Multiple Concentrating Solar Power Technologies in a Production Cost Model.* Technical report NREL/TP-6A20–58645. Golden, CO: National Renewable Energy Laboratory. Available at: www.nrel.gov/docs/fy14osti/58645.pdf.

Jorgenson, J., Denholm, P. and Mehos, M. (2014). Quantifying the value of concentrating solar power in a production cost model. In *Proceedings of ASME 2014 8th International Conference on Energy Sustainability collocated with the ASME 2014 12th International Conference on Fuel Cell Science, Engineering and Technology*, Vol. 1. ASME (American Society of Mechanical Engineers). Available at: https:// asmedigitalcollection.asme.org/ES/ES2014/volume/45868.

Kempener, R., Burch, J., Brunner, C., Navntoft, C. and Mugnier, D. (2015). *Solar Heat for Industrial Processes.* Technology Brief E21. IEA (International Energy Agency)-ETSAP (Energy Technology Systems Analysis Programme) and IRENA (International Renewable Energy Agency). Available at: www.irena.org/publica tions/2015/Jan/Solar-Heat-for-Industrial-Processes.

Kolb, G. J., Ho, C. K., Mancini, T. R. and Gary, J. A. (2011). *Power Tower Technology Roadmap and Cost Reduction Plan*. Sandia report SAND2011–2419. Albuquerque, NM: Sandia National Laboratories. Available at: https://prod-ng.sandia.gov/techlib-noauth/access-control.cgi/2011/112419.pdf.

Konstantin, P. and Kretschmann, J. (2010). *Assessment of Technology Options for Development of Concentrating Solar Power in South Africa for The World Bank*. Stuttgart: Fichtner. Available at: www.climateinvestmentfunds.org/sites/default/files/Presentation%20-%20WB%20(Eskom)%20Project%20-%202010_12_07%20.pdf.

Kulichenko, N. and Wirth, J. (2011). *Regulatory and Financial Incentives for Scaling Up Concentrating Solar Power in Developing Countries*. Energy and mining sector board Discussion Paper 24. Washington, DC: World Bank Group.

Lienhard, J., Antar, M. A., Bilton, A., Blanco, J. and Zaragoza, G. (2012). Solar desalination. *Annual Review of Heat Transfer*, 15, 277–347.

Lilliestam, J. and Pitz-Paal, R. (2018). Concentrating solar power for less than USD 0.07 per kWh: Finally the breakthrough? *Renewable Energy Focus*, 26, 17–21.

Lilliestam, J., Labordena, M., Patt, A. and Pfenninger, S. (2017). Empirically observed learning rates for concentrating solar power and their responses to regime change. *Nature Energy*, 2, 17094.

Lovegrove, K. (2013). The potential for solar fuels in Australia. Paper presented at the Solar Thermal Chemical and Industrial Processes Workshop, University of Adelaide, Australia, 7–8 February.

Lovegrove, K. and Pye, J. (2012). Fundamental principles of concentrating solar power (CSP) systems. In K. Lovegrove and W. Stein, eds., *Fundamental Principles of Concentrating Solar Power (CSP) Systems*. Cambridge: Woodhead Publishing.

Lovegrove, K., Watt, M., Passey, R., Pollock, G., Wyder, J. and Dowse, J. (2012). *Realising the Potential of Concentrating Solar Power in Australia*. Commissioned report for the Australian Solar Institute. Canberra: IT Power (Australia) Pty Ltd and the Australian Solar Institute.

Lovegrove, K., Edwards, S., Jacobson, N. et al. (2015). *Renewable Energy Options for Australian Industrial Gas Users: Background Technical Report*. Background technical report ITP/A0142 rev. 2.0. Canberra: IT Power (Australia) Pty Ltd and ARENA (Australian Renewable Energy Agency). Available at: https://itpau.com.au/wp-content/uploads/2018/08/ITP_REOptionsForIndustrialGas_TechReport.compressed.pdf.

Macknick, J., Newmark, R., Heath, G. and Hallett, K. (2011). *A Review of Operational Water Consumption and Withdrawal Factors for Electricity Generating Technologies*. Technical report NREL/TP-6A20–50900. Golden, CO: National Renewable Energy Laboratory. Available at: www.nrel.gov/docs/fy11osti/50900.pdf.

Mendelsohn, M., Kreycik, C., Bird, L., Schwabe, P. and Cory, K. (2012). *The Impact of Financial Structure on the Cost of Solar Energy*. Technical report NREL/TP-6A20–53086. Golden, CO: National Renewable Energy Laboratory. Available at: www.nrel.gov/docs/fy12osti/53086.pdf.

Moore, S. (2018). *Sustainable Energy Transformations, Power and Politics: Morocco and the Mediterranean*. London: Routledge.

NREL (National Renewable Energy Laboratory) (n.d.). Concentrating solar power projects [data resource]. *SolarPACES*. Available at: www.nrel.gov/csp/solarpaces/.

Ortega, M., del Río, P. and Montero, E. A. (2013). Assessing the benefits and costs of renewable electricity: The Spanish case. *Renewable and Sustainable Energy Reviews*, 27, 294–304.

Patt, A., Pfenninger, S. and Lilliestam, J. (2013). Vulnerability of solar energy infrastructure and output to climate change. *Climatic Change*, 121, 93–102.

Pitz-Paal, R., Dersch, J., Milow, B. et al. (2005). *ECOSTAR: European Concentrated Solar Thermal Road-Mapping*. Roadmap document SES6-CT-2003-502578. DLR (German Aerospace Centre). Available at: www.promes.cnrs.fr/uploads/pdfs/ecos tar/ECOSTAR.Roadmap.pdf.

Platzer, W. (2015). Combined solar thermal and photovoltaic power plants: An approach to 24h solar electricity? Paper presented at 21st SolarPACES Conference, Cape Town, South Africa, 13–16 October.

Price, H. (2017). Dispatchable solar power: Adapting CSP to modern grid needs. Paper presented at 23rd SolarPACES Conference, Santiago, Chile, 26–29 September. Available at: www.solarpaces.org/wp-content/uploads/Hank-Price-Presentation.pdf.

Richert, T., Riffelmann, K.-J. and Nava, P. (2012). LCOE versus LCOE versus PPA bid price: How different financing parameters influence their values. Paper presented at 18th SolarPACES Conference, Marrakech, Morocco, 11–14 September.

Romero, M. and González-Aguilar, J. (2014). Solar thermal CSP technology. *WIREs Energy and Environment*, 3, 42–59.

Romero, M. and Steinfeld, A. (2012). Concentrating solar thermal power and thermochemical fuels. *Energy & Environmental Science*, 5, 9234–9245.

Sargent and Lundy LLC Consulting Group (2003). *Assessment of Parabolic Trough and Power Tower Solar Technology Cost and Performance Forecasts*. Subcontractor report NREL/SR-550-34440. Golden, CO: National Renewable Energy Laboratory. Available at: www.nrel.gov/docs/fy04osti/34440.pdf.

Shahan, Z. (2013). CSP for 5.57 cents/kWh. *CleanTechnica*. 14 June. Available at: cleantechnica.com/2013/06/14/csp-for-5-57-centskwh/.

Slaughter, R. (2014). *Port Augusta Solar Thermal Generation Feasibility Study: Final Balance of Study*. Milestone 4 Report 105-RPT-006. Sydney: Alinta Energy. Available at: https://alintaenergy.com.au/Alinta/media/Documents/Alinta-Energy-Port-Augusta-Solar-Thermal-Generation-Feasibility-Study-Milestone-4-Summary-Report.pdf.

Staight, K. (2016). Sundrop Farms pioneering solar-powered greenhouse to grow food without fresh water. *ABC News*. 2 October. Available at: www.abc.net.au/news/2016-10-01/sundrop-farms-opens-solar-greenhouse-using-no-fresh-water/7892866.

Steinfeld, A. (2005). Solar thermochemical production of hydrogen: A review. *Solar Energy* 78, 603–615.

Stine, W. B. and Geyer, M. (2001). *Power from the Sun*. Available at: www .powerfromthesun.net/index.html.

Turchi, C. (2010). *Parabolic Trough Reference Plant for Cost Modeling with the Solar Advisor Model (SAM)*. Technical report NREL/TP-550e47605. Golden, CO: (NREL) National Renewable Energy Laboratory.

Turchi, C., Mehos, M., Ho, C. K. and Kolb, G. J. (2010). Current and future costs for parabolic trough and power tower systems in the US market. Paper presented at 16th SolarPACES Conference, Perpignan, France, 21–24 September.

Viebahn, P., Kronshage, S., Trieb, F. and Lechon, Y. (2008). *Final Report on Technical Data, Costs, and Life Cycle Inventories of Solar Thermal Power Plants*. DLR (German Aerospace Centre) and CIEMAT (Spanish Centre for Energy, Environment and Technology). Available at: www.solarthermalworld.org/sites/gstec/files/concentrating solar thermal power plants.pdf.

Viebahn, P., Esken, A., Höller, S., Luhmann, H., Pietzner, K. and Vallentin, D. (2010). *RECCS Plus: Comparison of Renewable Energy Technologies (RE) with Carbon Dioxide Capture and Storage (CCS). Update and Expansion of the RECCS study. Final report of Wuppertal Institute*. Berlin: German BMU (Federal Ministry for the

Environment, Nature Conservation and Nuclear Safety). Available at: https://epub
.wupperinst.org/frontdoor/deliver/index/docId/5001/file/5001_RECCSplus_en.pdf.

Viebahn, P., Lechon Y. and Trieb, F. (2011). The potential role of concentrated solar power
(CSP) in Africa and Europe: A dynamic assessment of technology development, cost
development and life cycle inventories until 2050. *Energy Policy*, 39, 4420–4430.

Vivar, M., Herrero, R., Antón, I. et al. (2010). Effect of soiling in CPV systems. *Solar
Energy*, 84, 1327–1335.

Voutchkov, N. (2016). Desalination: Past, present and future. *International Water
Association*. 17 August. Available at: www.iwa-network.org/desalination-past-pre
sent-future.

Wang, Z., Roberts, R. R., Naterer, G. F. and Gabriel, K. S. (2012). Comparison of
thermochemical, electrolytic, photoelectrolytic and photochemical solar-to-hydrogen
production technologies. *International Journal of Hydrogen Energy*, 37,
16287–16301.

Weiss, W. and Spörk-Dür, M. (2019). *Solar Heat Worldwide*. Technical report 2019
edition. Gleisdorf, Austria: SHC (Solar Heating and Cooling) Programme,
International Energy Agency. Available at: www.iea-shc.org/Data/Sites/1/publica
tions/Solar-Heat-Worldwide-2019.pdf.

Wentzel, M. and Pouris, A. (2007). The development impact of solar cookers: A review of
solar cooking impact research in South Africa. *Energy Policy*, 35, 1909–1919.

World Bank (2014). Demonstrating the viability of solar thermal power in Morocco. *The
World Bank*. 15 April. Available at: www.worldbank.org/en/results/2014/04/15/dem
onstrating-the-viability-of-solar-thermal-power-in-morocco.

Zamfirescu, C. and Dincer, I. (2008). Using ammonia as a sustainable fuel. *Journal of
Power Sources*, 185, 459–465.

Zedtwitz, P. and Steinfeld, A. (2003). The solar thermal gasification of coal: Energy
conversion efficiency and CO_2 mitigation potential. *Energy*, 28, 441–456.

5

Nuclear Energy

ANDREW STUCHBERY AND TONY IRWIN

Executive Summary

Nuclear power in 31 countries provides about 11% of the world's electricity with essentially no operating greenhouse gas emissions. Countries with a larger nuclear electricity component consequently have lower per capita greenhouse gas emissions. Nuclear energy is a direct substitute for fossil fuels and can provide electricity straight into existing power grids, and the electricity is delivered with a high capacity factor (often around 90%).

Nuclear fuel has an extremely high energy density, which enables a very high level of energy security. Typically, refuelling times for power reactors are 12–18 months. Most power reactors today require enriched uranium as fuel; however, a country embarking on a nuclear power programme can import prefabricated fuel.

Small modular reactors, suitable for small countries or remote sites, are under development to make nuclear power available with more modest initial investment.

In view of community concerns about possible contamination and proliferation of nuclear weapons, governments must negotiate a 'social licence' for nuclear facilities. A nuclear power programme requires government, legal, regulatory, managerial, technological, human and industry support. It requires compliance with international legal instruments, internationally accepted safety standards, security guidelines and safeguards requirements. A national independent nuclear regulatory organisation is essential.

Reactors typically give many decades of service after which they must be decommissioned. The decommissioning cost is a small fraction of the cost of electricity produced and can be saved over the life of the reactor. A range of satisfactory technical strategies is available to manage nuclear waste. These costs are also a small fraction of the life-cycle cost of electricity produced.

Successive generations of nuclear reactors have become more efficient, more reliable and safer. Revolutionary Generation IV reactors, which aim for excellence in safety, sustainability, cost-effectiveness and proliferation resistance, will be deployed beyond 2030.

Thorium reactors, which use uranium-233 as fuel, have attractive features for future development but are not a short-term solution to mitigate greenhouse gas emissions. Technical impediments still make thorium impractical and uneconomic compared with uranium-fuelled reactors.

Fusion power, which aims to harness the energy that powers the sun with no greenhouse gas emissions, almost no radioactive waste and almost unlimited fuel, is still under development in an international programme. However, there is no prospect that it will deliver a solution to address climate change in the time frame required, although it could provide a solution to the world's energy needs in the latter part of this century.

Nuclear power has advantages for energy security and for energy reliability: that is, generating electricity with high-capacity factors. Nuclear energy already contributes to climate change mitigation as a greenhouse gas emissions-free technology, and can complement wind, solar photovoltaic and hydro to meet the need for firm electricity supply.

5.1 Introduction and World Status

About 11% of the world's electricity is produced with essentially no greenhouse gas emissions by splitting heavy nuclei to release energy: that is, by nuclear fission. This nuclear electricity production avoids the emission of about 2 billion tonnes of carbon dioxide (CO_2) per year; in general, nations with extensive hydro or geothermal resources or a significant nuclear energy component have a lower level of emissions intensity (IEA 2016).

The World Nuclear Association Information Library (WNA n.d.) keeps up-to-date statistics on the generation of nuclear electricity, the number of operable reactors country by country, reactors under construction, and reactors planned and proposed. (*Planned* reactors have approvals, funding or commitment in place for operation within a decade; *proposed* reactors have specific programme or site proposals but uncertain timing for the start of operation.)

Worldwide, in 2016 nuclear reactors generated almost 250 billion kWh (kilowatt-hours) of energy. As of May 2017, the USA remains the largest producer of nuclear electricity, with 99.7 GWe (gigawatts electrical) of operable reactor capacity and 805 billion kWh generated in 2016. The USA is followed by France with 63.1 GWe of operable capacity and 384 billion kWh generated in 2016. China, with 32.6 GWe, is currently fourth in terms of operable reactor capacity, behind Japan with 39.95 GWe, but actually third in nuclear electricity production in 2016, with 211 billion kWh generated in 2016. Moreover, China has the most reactors under construction of any nation. In late April 2017, it had 36 power reactors in operation, 21 under construction and more about to start construction. The goal is to double nuclear power capacity to at least 58 GWe by 2020–2021. While Japan has 42 operable reactors (as of April 2017), it paused nuclear power generation following the Fukushima accident in 2011; 24 reactors are in the process of restart approvals as of April 2017.

A number of nations with established nuclear power capacity, including the USA, France, UK and Russia, are moving ahead with plans to replace retired capacity. Russia in particular has plans to bring online one large reactor per year over the decade to 2028.

One nation withdrawing from nuclear power is Germany. Until the Fukushima accident in March 2011, about 25% of Germany's electricity was generated by 17 reactors. In April 2017, it was generating 14% of its electricity from 8 reactors, with 43% coming from coal.

5.2 Nuclear Technology

5.2.1 Overview of the Nuclear Fuel Cycle

Figure 5.1 shows the nuclear fuel cycle: that is, the activities and processes involved in the production of nuclear electricity. The 'front end' of the nuclear fuel cycle includes the mining and industrial processes that occur before the fuel is loaded into the reactor. Radioactivity levels, mainly from the natural radioactivity of the uranium, are low in the front end of the nuclear fuel cycle. The 'back end' of the nuclear fuel cycle refers to the activities and processes after the fuel has been in the reactor. High levels of radioactivity, which persist for some time, are associated with the back end of the fuel cycle.

Specifically, the front-end activities are the mining of uranium ore and milling to extract uranium from the ore as uranium oxide (mainly U_3O_8), conversion of the oxide to the hexafluoride, enrichment to increase the ^{235}U (uranium-235) isotope fraction from the naturally occurring level of 0.7% to typically 3–5%, and fuel fabrication, which typically involves making ceramic pellets of uranium oxide (UO_2), which are then encased in metal tubes to form fuel rods. The fuel rods are then combined into a fuel assembly ready for insertion into a reactor.

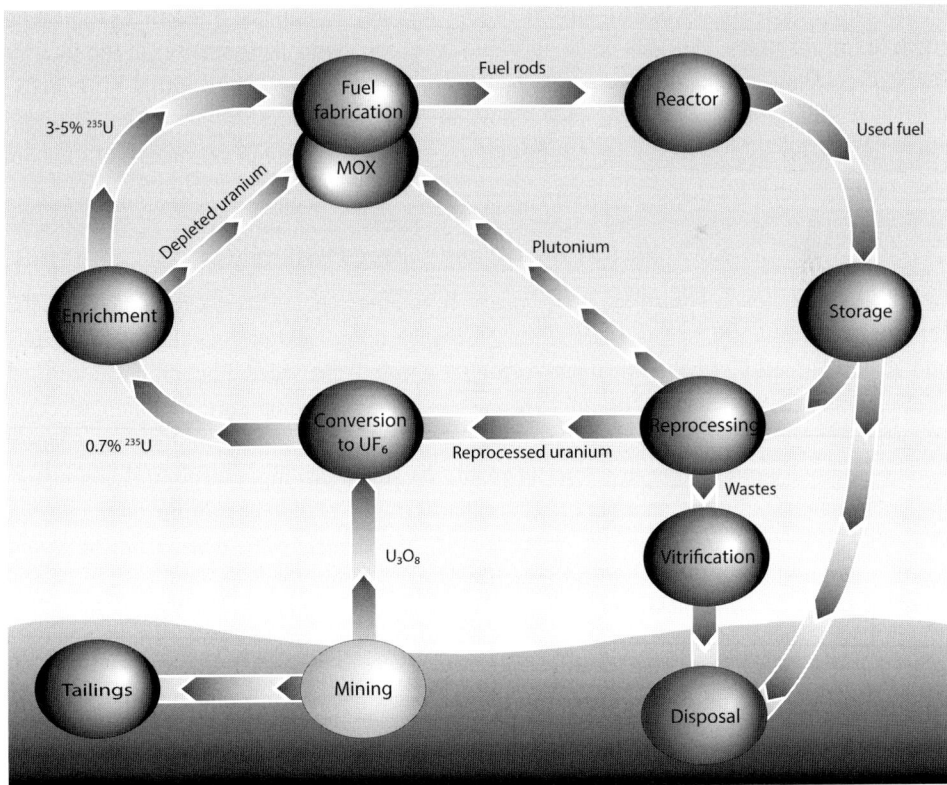

Figure 5.1 Overview of the nuclear fuel cycle.
Source: WNA (2017) (redrawn). For a colour version of this figure, please see the colour plate section.

After the fuel has been used in the reactor for energy production, it is highly radioactive and this radioactivity also produces heat. The spent fuel is typically stored in a cooling pond on-site with the reactor to keep it cool and allow the radioactivity level to decrease. There are then two possible onward paths for the spent fuel. It may be put into long-term storage for final disposal – the so-called 'once-through' cycle – or it may be reprocessed to recover its uranium and plutonium content (about 97%), which 'closes' the nuclear fuel cycle and leaves a much smaller fraction of the spent fuel for disposal.

The front end of the nuclear fuel cycle is an international effort; it can be said that no nation at present (2017) is entirely self-sufficient. The back end of the fuel cycle is less of an international effort to date, largely because each country is ethically and legally responsible for its own waste. However, there have been several proposals for regional and international repositories for high-level nuclear waste. Such facilities would address the problem that not all nations have access to geologically suitable locations for waste disposal, and importantly, multinational repositories are recognised to have non-proliferation advantages; the risk of nuclear material being diverted to weapons production would be reduced within an international collaboration under the auspices of the International Atomic Energy Agency (IAEA).

Opportunities for a country to be involved in the nuclear fuel cycle require careful examination. In 2015, the South Australian Government established a Royal Commission (NFCRC) to examine the risks and opportunities of increased participation in the nuclear fuel cycle. Using evidence-based assessment methodology the NFCRC found some activities were viable and some were not viable in South Australia (Government of South Australia 2016).

5.2.2 Nuclear Fission and Energy Production

In a nuclear reactor, a controlled chain reaction generates heat: fission is induced by the capture of neutrons by the nuclear fuel in the reactor core. However, as the nucleus splits and releases energy, several neutrons are also emitted. Provided one of these emitted neutrons induces the fission of another nucleus in the reactor core, a chain reaction is sustained.

The electricity production in a nuclear plant, via steam driving a turbine, is essentially the same as in a coal plant. In most existing nuclear power plants the thermal efficiency for electricity production is about 33%. Thus, a 3000-MWt (megawatts thermal) plant would produce about 1000 MWe (megawatts electrical).

Because the energy released per fission is about 40 million times higher than that released in burning a carbon atom, the energy density of nuclear fuel is proportionately higher than conventional fuels. For example, a 1000-MWe nuclear reactor requires about 27 tonnes of fresh fuel per year whereas a coal power station producing the same electricity requires more than two and a half million tonnes of coal. This small volume of nuclear fuel enables a very high level of energy security, which has been a primary motivation for the uptake of nuclear power in several countries.

5.2.3 Nuclear Fuel and Enrichment

The nuclear chain reaction requires a fuel that undergoes fission when it captures a neutron. Such materials are said to be *fissile*. The only naturally occurring fissile material is the isotope of uranium, ^{235}U, which constitutes only 0.7% of natural uranium.

Other energy-relevant fissile materials must be converted from *fertile* material by suitable nuclear reactions. Two important fertile isotopes are abundant in nature: ^{238}U (99.3% of natural uranium) and ^{232}Th (thorium-232: 100% isotopic abundance). The fissile isotopes ^{239}Pu (plutonium-239) and ^{233}U are produced from fertile ^{238}U and ^{232}Th, respectively, by neutron irradiation in a reactor.

Although some reactors were designed to operate using natural uranium (usually by nations that did not have access to enriched material at the time), most power reactors today require enriched uranium as fuel. Enriched uranium has the ^{235}U proportion increased from 0.7% to typically 3–5%. In general, the additional cost for enrichment pays off in terms of reactor performance and increased times between refuelling.

Because the chemical properties of ^{235}U and ^{238}U are the same and the masses of the two isotopes differ by only a little over 1%, separating them to produce ^{235}U-enriched uranium is technically challenging. Nevertheless, a number of methods have been devised. Electromagnetic separation in a mass spectrograph produced enough enriched uranium for one atomic bomb during the Second World War; however, that process has never been economically viable. The first economical process was gaseous diffusion, which makes use of the fact that ^{235}UF$_6$ gas molecules have a higher average velocity than ^{238}UF$_6$ molecules and therefore have a slightly higher probability to diffuse through a porous membrane. After many decades of production in the USA, this method has now been phased out and all commercial gaseous diffusion plants are now retired. Today, most isotope enrichment uses gas centrifuge technology. Uranium hexafluoride gas is fed into a container that is rotated at high speed such that the heavier ^{238}UF$_6$ molecules tend to move to the outside. Although this method was known in the 1940s it only became viable with the development of new materials and improved bearing technology. Gas centrifuge separation is expected to remain the predominant enrichment process for some decades.

Laser technology holds promise for improved isotope separation in the future. Differences in the nuclear properties of ^{235}U and ^{238}U cause small differences in atomic structure and hence in the ways that their atoms absorb light. A laser can thus be tuned to ionise ^{235}U but leave ^{238}U unchanged, and the ionised ^{235}U is then extracted by an electric field. Although this process is highly selective, laser enrichment of uranium in either atomic or molecular form has not yet proved commercially viable. This assessment may change, however, with the proprietary and classified SILEX (separation of isotopes by laser excitation) process. A commercial-scale SILEX plant to enrich up to 8% ^{235}U has been licensed (September 2012) in the USA but as there is at present (2017) a worldwide oversupply of enrichment capacity, there may be little incentive to proceed to the commercial scale in the near future.

Efforts such as the 'Megatons to Megawatts' programme, which was completed in December 2013, have made ^{235}U and ^{239}Pu from weapons available for energy production.

Weapons-grade uranium is down-blended with depleted uranium to make fuel for power and research reactors. Weapons-grade plutonium can be blended with natural or depleted uranium to form a mixed-oxide, or MOX, fuel for power reactors. This source of nuclear fuel for energy production is projected to diminish over the next few years; however, MOX fuel will continue to be manufactured using plutonium recovered from PUREX (plutonium uranium redox extraction) reprocessing of spent fuel.

5.2.4 *Thermal and Fast Reactors*

The neutrons emitted during fission have an average energy of the order of 1 MeV. These *fast* neutrons have a relatively small probability of inducing the fission of another fuel nucleus; however, if they are slowed to *thermal* energies (0.025 eV) the probability of inducing fission increases by a factor of about 600. The process of slowing the neutrons, called moderation, is based on scattering the neutrons off the nuclei in a moderator. Depending on the reactor type, the moderator may be ordinary light water (H_2O), heavy water (D_2O) or carbon, often in the form of graphite.

Although the first nuclear electricity was produced by a fast reactor (Experimental Breeder Reactor-1, or EBR-I, in December 1951), and there have been a few fast reactors that have produced electricity into the grid, the vast majority of nuclear electricity production is accomplished with thermal reactors; that is, reactors that slow the neutrons from fission to thermal energies by scattering in a moderator.

Thermal reactors have the advantages that they can operate with relatively low (or no) fuel enrichment; that they have accumulated many years of operational experience; and that the most common types are not, in themselves, a proliferation risk.

In contrast with thermal reactors, fast reactors have no moderator. The coolant is typically a high-mass element such as sodium or lead. The probability that a fast neutron will induce fission is much smaller than for a thermal neutron, so higher fuel enrichment is necessary; however, the number of neutrons emitted when fission does occur is higher, which compensates in part. The additional neutrons per fission also allow for the possibility to breed fissile material from fertile material in the reactor core. With a closed fuel cycle and breeding, nuclear fuel would be available for many thousands of years (WNA 2018).

The waste produced by fast reactors does not contain the long-lived transuranic nuclei that are problematic in spent thermal reactor fuel. In fact, fast reactors burn these transuranic isotopes to shorter-lived fission fragments and can potentially become part of a strategy to manage nuclear waste. By producing energy from reused thermal reactor fuel, it is possible to reduce the time required for the waste to reach background levels from over a hundred thousand years to a few hundred years (Till and Chang 2011: ch. 11).

Despite their advantages, fast reactors have not been widely used for electricity production. In part this is because of technical, and hence economic, challenges. It is also partly due to the increased potential for proliferation with a closed fuel cycle that includes reprocessing. Nevertheless, fast reactors may become more common in the future. More than half of the designs chosen for development as Generation IV reactors (see Section

5.3.2) are fast reactors; these new reactor designs seek to address the technical and proliferation concerns while benefiting from the advantages of a closed fuel cycle, or at least to produce shorter-lived waste which is easier to manage (Gen IV 2002).

5.2.5 Breeding and Conversion

As described previously, conversion is the process whereby neutron capture in the reactor core produces fissile material from fertile material. All thermal reactors are converters – neutron captures on ^{238}U, which is typically more than 95% of the uranium present, build up an equilibrium level of ^{239}Pu. This fissile isotope then contributes to the energy produced by the reactor. For Canada Deuterium Uranium (CANDU) reactors fuelled with natural uranium, the ^{239}Pu produced by conversion produces about the same amount of energy as ^{235}U. For a typical pressurised water reactor operating with 3–5% enriched uranium, about 30% of the energy production is from the ^{239}Pu produced by conversion (Bodansky 2005: ch, 7).

Conversion becomes *breeding* when a reactor makes more fissile material than it consumes. Breeding requires that a sufficient number of neutrons be emitted per fission, which depends both on the fissile isotope and whether fission is induced by thermal or fast neutrons. Thermal reactors with ^{235}U as fuel cannot become breeders; however, fast reactors with ^{239}Pu as fuel can. In principle, breeding is possible with both thermal and fast reactors based on a ^{232}Th–^{233}U cycle. This property is one of the incentives for developing thorium reactor technology.

5.3 Technology Alternatives

5.3.1 Existing Technology

Table 5.1 lists the nuclear power plants in operation as of the end of December 2015. The most common reactor type is the pressurised water reactor (PWR), followed by the boiling water reactor (BWR). Both are thermal reactors moderated and cooled by ordinary (light) water. With a typical fuel enrichment of 3–5%, off-load refuelling (replacing one-third of the core) is done every 12–18 months. The long time the fuel spends in the reactor produces a mix of plutonium isotopes that renders the plutonium in PWR and BWR spent fuel unsuitable for nuclear weapons. Thus, spent fuel from these reactors is not considered a proliferation risk.

For the PWR (see Figure 5.2), the primary cooling circuit through the reactor core is kept under pressure so that the water does not boil and steam is produced in a secondary circuit. BWRs were designed to be somewhat simpler, with one water circuit through the reactor core and the steam turbine.

The pressurised heavy water reactor (PHWR) and gas-cooled graphite-moderated reactor (Magnox), which can operate on natural uranium, were developed in Canada (CANDU) and the UK, respectively, when these nations did not have access to enriched uranium (PHWRs are now sometimes operated with slightly enriched uranium). Both

Table 5.1. *Nuclear power plants in operation at 31 December 2015*

Reactor Type	Main countries	Number	Electrical net output (GWe)	Fuel	Coolant	Moderator
Pressurised water reactor (PWR)	USA, France, Japan, Russia, China	282	264	Enriched UO_2	H_2O	H_2O
Boiling water reactor (BWR)	USA, Japan, Sweden	78	75	Enriched UO_2	H_2O	H_2O
Pressurised heavy water reactor (PHWR) 'CANDU'	Canada, India	49	25	Natural UO_2	D_2O	D_2O
Advanced gas-cooled reactor (AGR)	UK	14	8	Natural U, enriched UO_2	CO_2	Graphite
Light water graphite reactor (RBMK and EGP)	Russia	15	10	Enriched UO_2	H_2O	Graphite
Fast neutron reactor (FNR)	Russia	3	1.4	PuO_2 and UO_2	Liquid sodium	None
	TOTAL:	**441**	**383.4**			

Source: IAEA (2016).

reactor types have on-load refuelling. The UK developed a higher-temperature gas-cooled graphite-moderated reactor, the advanced gas-cooled reactor (AGR) and 14 of these are in operation, but it is unlikely that any more will be built. The construction of new PHWRs continues in India.

The Russian RBMK is a graphite-moderated water-cooled reactor. It has a smaller version called EGP. After the Chernobyl accident in 1987 a number of design changes were introduced to the RBMK fleet to improve safety. Even so, the RBMK is a poor design from the point of view of both safety and safeguards (they were designed for plutonium production) and no more RBMK reactors will be built.

5.3.2 Future Technology Beyond 2030: Generation IV

To enable an understanding of how nuclear power reactors have developed, they are commonly identified in four generations:

Generation I were the early prototypes; most of these reactors are now shut down.

Generation II reactors are the majority of the current operating reactors: PWRs, BWRs, CANDUs and AGRs.

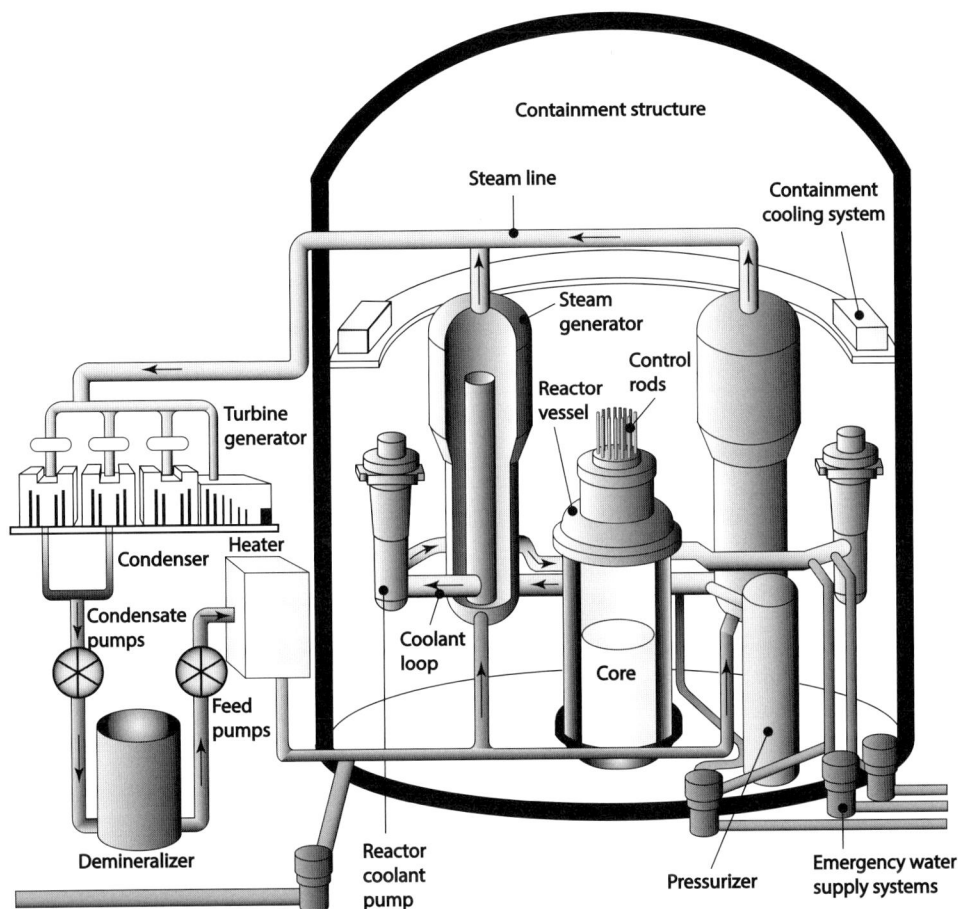

Figure 5.2 Pressurised water reactor (PWR).
Source: US NRC (n.d.) (redrawn). For a colour version of this figure, please see the colour plate section.

Generation III reactors are developments of the Generation II types, but with improved safety features and reduced construction time. Further developments with particularly passive safe features are identified as Generation III+. Examples of these reactors under construction are the advanced boiling water reactor (ABWR), the AP1000 (an advanced passive 1000-MWe PWR), the economic simplified BWR (ESBWR), and the European pressurised reactor (EPR).

Generation IV reactors are more revolutionary designs. The aim of Generation IV systems is to excel in safety, sustainability, cost-effectiveness and proliferation resistance. Although the main application is still electricity generation, many designs can also broaden the opportunities for the uses of nuclear energy by supplying process heat for use in industrial applications like hydrogen production, or burning actinides in spent fuel from current light water reactors (LWRs).

The two main projects driving Generation IV development are:

- The IAEA-led International Project on Innovative Nuclear Reactors (INPRO) (IAEA n.d.a); and
- The USA-led Gen IV International Forum (GIF), which had 13 member countries, all with significant nuclear power programmes. In 2016, Australia became the 14th member of GIF, bringing particular expertise in materials for high-temperature reactors. In the first phase of the project, GIF members selected six reactor systems for further developments. Many of these technologies were examined back in the early days of nuclear power, but were rejected typically because the supporting technology, particularly advanced materials, was not available then, but such technology is becoming available now.

Table 5.2 lists the six Generation IV reactor types.

A sodium-cooled fast reactor (SFR) has been combined with electrometallurgical pyroprocessing and fuel fabrication on-site for a closed-loop fuel cycle in the Integral Fast Reactor (IFR) project. This system can recycle actinides from LWR spent fuel, resulting in the efficient use of uranium. Only fission products remain as radioactive waste, reducing the storage time of the waste fuel by thousands of years. The process was demonstrated at Experimental Breeder Reactor-2, or EBR II, in the USA between 1963 and 1994, and is now being revisited by GE-Hitachi as the PRISM (power reactor innovative small module) reactor (Hitachi n.d.).

Generation IV reactors are expected to become widely deployed after 2030.

5.4 Small Modular Reactors

Most power reactors under construction worldwide are units of over 1000 MWe. These are too big for countries with small grid systems or for remote areas. Thus, there is a developing interest in small modular reactors (SMRs). 'Small' by IAEA definition is less than 300 MWe, but typically they are around 100 MWe. 'Modular' means assembled at a factory off-site, with the economy and high quality assurance of factory mass production and a simple standard design (IAEA n.d.b). The complete reactor vessel is transported as one unit to the site, reducing site construction time/costs and reducing the probability of project delays. The initial investment is much less than that of a big reactor and modules can be easily added as extra capacity is required, with operating units generating cash flow to support the additional modules.

Many SMRs are designed to be multipurpose – in addition to supplying electricity they can also be used for desalination or to supply process heat.

A major advantage of SMRs is their natural (passive) safety. No electrical supplies or pumps are required to cool the reactor as this is achieved by natural convection and gravity coolant feed. This feature ensures the reactor will remain safe under severe accident conditions. These features also make an SMR simpler to operate and maintenance costs are reduced. The reactor containment can be installed below ground, providing protection against external hazards and unauthorised interference.

Table 5.2. *Generation IV reactors*

Reactor type	Fast/thermal	Coolant	Fuel	Prior technology	Fuel cycle	Size(s) (MWe)	Uses	Projected prototype date
Gas-cooled fast (GFR)	Fast	Helium	^{238}U + ^{235}U or ^{239}Pu	None	Closed on-site	1100	Electricity and actinide recycling	After 2021
Lead-cooled fast (LFR)	Fast	Lead or Pb–Bi	^{238}U + ^{235}U or ^{239}Pu	Russian submarines	Closed regional	20–600	Electricity, actinide recycling and hydrogen	2021 construction licence issued for BREST demonstration plant in Russia
Molten salt fast (MSFR)	Fast	Fluoride salts	UF in salt	MSFR UK Atomic Energy Research Establishment	Closed	1000	Electricity and hydrogen	After 2025
Molten salt thermal (MSR)	Thermal	Fluoride salts	UF/Th in salt or UO$_2$ pebbles	Molten Salt Research Experiment (MSRE) Oak Ridge National Laboratory, USA	Open	1000–1500	Electricity and hydrogen	After 2021
Sodium-cooled fast (SFR)	Fast	Sodium	^{238}U and MOX	Fast neutron reactors since 1951	Closed	<50, 300–600, 600–1500	Electricity and actinide recycling	Commercial operation in Russia since 1980
Supercritical water (SCWR)	Thermal or fast	Water	UO$_2$	BWR & coal-fired supercritical	Open (thermal), Closed (fast)	300–700, 1000–1500	Electricity	By 2025
Very high temperature gas (VHTR)	Thermal	Helium	UO$_2$ TRISO prism or pebbles	Dragon (UK), AVR (Germany)	Open	100-300	Electricity, process heat, hydrogen	2021 commissioning of HTR-PM in progress

Note: TRISO = tristructural-isotropic fuel. *Source*: Gen IV (2002).

Table 5.3. *SMRs and their status (September 2020)*

Country	Reactor	Module size	Status
USA	Generation mPower	180 MWe	Basic design completed
USA	NuScale	77 MWe	Design certification application approved August 2020
South Korea	KAERI SMART	100 MWe	Design approval 2012, memorandum of understanding 2015 with Saudi Arabia for deployment
Argentina	CNEA/ INVAP CAREM	27 MWe	Under construction, operation scheduled for 2023
Russia	KLT-40S	35 MWe	Floating plant completed, in commercial operation May 2020 at Pevek
China	CNNC/ NPIC ACP-100	100 MWe	Preliminary safety analysis report approved, site for first deployment identified

There is extensive experience of much of the technology employed by SMRs. For many years they have been the power supply for submarines and icebreakers where totally reliable power is essential.

Examples of SMRs based on this proven PWR technology are listed in Table 5.3.

Also under development are SMRs using Generation IV technology:

Fast neutron SMRs are very compact due to the high-conductivity liquid metal coolant. They operate at higher efficiencies due to their higher operating temperatures. An example is the Toshiba 4S, which is a 10-MWe SMR designed for remote locations that currently rely on expensive diesel generators. The 4S can operate for 30 years before refuelling is required.

Very high temperature gas reactors use helium as coolant, with outlet temperatures up to 900 °C and tristructural-isotropic (TRISO) fuel. Following the experience of operating the 10-MWth experimental VHTR at the Institute of Nuclear and New Energy Technology in China since 2000, two 105-MWe demonstration units are now under construction in Shandong Province.

5.5 Thorium Fuel Cycle

Thorium reactors are often suggested to be superior to uranium reactors in several respects. This assessment requires qualification. Fundamentally, it must be remembered that natural thorium consists entirely of the fertile isotope ^{232}Th. Thus, thorium can be used with fissile ^{235}U or ^{239}Pu to convert ^{232}Th to fissile ^{233}U, but it is not a substitute for fissile ^{235}U; rather it is an alternative to fertile ^{238}U.

As thorium is three to four times more abundant than uranium, it could provide a source of nuclear fuel well into the future. The advantages of thorium-based nuclear fuel cycles are several, and include nuclear properties that make conversion of ^{232}Th to ^{233}U favourable.

Fission of ^{233}U yields a high number of neutrons per neutron absorbed over a wide range of neutron energies, enabling breeding in thermal as well as fast reactors. There is an intrinsic proliferation resistance because ^{232}U is unavoidably produced with ^{233}U and the daughters of ^{232}U emit strong gamma radiation. The waste from a thorium cycle is expected to be easier to manage in the long term as fewer long-lived minor actinides (isotopes of neptunium, americium and curium) are produced.

However, along with the advantages, there are disadvantages. The pathway for conversion of ^{232}Th to ^{233}U is via ^{233}Pa (protactinium-233), which has a 27-day half-life. This relatively long lifetime means that about a year is required for all of the irradiated ^{232}Th to become ^{233}U. In comparison, the intermediate isotope in ^{238}U–^{239}Pu conversion is ^{239}Np (neptunium-239), with a 2.35-day half-life, so the equivalent cooling time is about a month. The ^{232}U that hinders proliferation also makes handling the irradiated thorium fuels difficult in the industrial processes needed for the Th–U cycle. The high melting point of ThO_2, which might be an advantage in an accident scenario, also makes fuel preparation more difficult, and ThO_2 being chemically inert hinders reprocessing, which is important for a cycle that inherently aims at breeding.

The IAEA concludes that despite a long history of research from the 1950s, the database and experience with thorium fuels and thorium fuel cycles is not yet adequate for widespread commercial use. Technical impediments still make thorium impractical and uneconomic compared with uranium (IAEA 2005).

Certain reactor types such as molten salt reactors (MSRs) and accelerator-driven systems are often discussed in the context of thorium reactors. It is important to note that these reactor types are not unique to the thorium cycle and their inherent safety features are not tied to thorium fuel.

To sum up, a full thorium cycle is a longer-term prospect: at present it is not a technically or economically viable option for immediate implementation (IAEA 2005).

Box 5.1 **Fusion Energy**

KENNETH G. H. BALDWIN

Fusion energy – the energy that powers the sun – is created by the fusing of light nuclei: principally the isotopes of hydrogen (deuterium and tritium). This requires the creation and magnetic confinement of the extremely high-temperature – 100 000 000 °C – plasmas (ionised gases) needed for the fusion process to burn. However, fusion power offers the prospect of unlimited, high-density energy, given that the fuel required is in enormous abundance in seawater or can be manufactured as a by-product of the fusion process itself (Chen 2011).

Fusion can act as a direct replacement for other large thermal power stations whether these be coal, gas or conventional nuclear, but without any greenhouse gas emissions and with almost no radioactive waste products (although some of the containment system can potentially require safe storage for up to a century or so). A further advantage is that there is no prospect of nuclear proliferation arising from the fusion process, and the fusion reactor cannot 'melt down' because once the plasma is no longer contained the fusion reaction stops.

Indeed, it is entirely conceivable that all zero-carbon energy sources (including renewables, hydro and conventional nuclear) could potentially be regarded as decarbonisation stopgaps on the road to a completely fusion future, although clearly other zero-carbon sources will continue to have niche applications.

The key issue is the time frame by which fusion power will be realised. Fusion research under way since the 1950s has thus far been able to demonstrate a 63% energy return (Wesson 2004) for a brief period on the Joint European Torus (JET) based at Culham in the UK. The realisation of net fusion energy gain is the goal of the USD 20 billion international ITER fusion programme funded by 30 nations and based in Cardarache, France, which aims to generate 500 MW of electrical power for a 50-MW input over ~400 seconds. The goal is to achieve this by 2030, following which a further fusion reactor is planned – DEMO – which aims to demonstrate continuous net fusion power generation as a prelude to the commercial introduction of fusion reactors. Initial indications are that the cost of fusion power plants will be similar to that of nuclear fission (Cook et al. 2002).

Why should these predictions be any more believable than historical claims of the imminent prospects for fusion breakeven? The answer lies in the inexorable progress towards the physical scaling laws that predict the achievement of fusion breakeven – the so-called Lawson criterion (see Figure 5.3), which is a triple product of plasma density (n), temperature (T) and containment

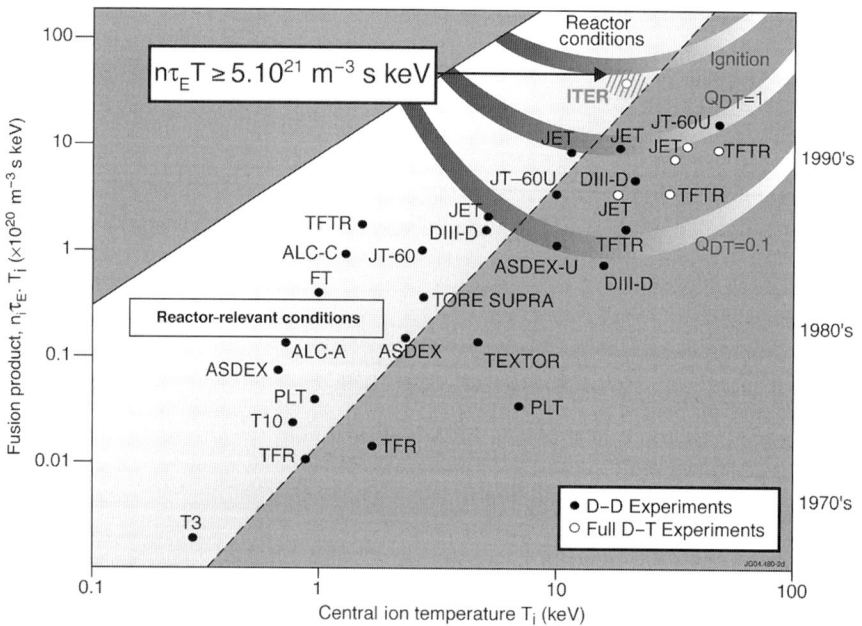

Figure 5.3 The triple product for fusion breakeven (left vertical axis) as a function of plasma temperature (horizontal axis in kiloelectron volts: 10 keV ~ 100 000 000 °C). The progress of fusion experiments around the world over time (right-hand axis) is shown leading to projected breakeven (Q_{DT}) by the ITER programme.

Source: euro-fusion.org. For a colour version of this figure, please see the colour plate section.

time (τ_E). Thus far, all the experiments being undertaken in fusion reactors around the world have been tracking the physical models that predict breakeven by ITER according to the Lawson criterion.

So, assuming that physical laws don't show up any new surprises, then the modelling indicates that it is simply a matter of scaling up the size of the reactor to the dimensions of the ITER project in order to demonstrate a net energy gain. Even if this is achieved by the 2030 goal, commercial fusion power stations are unlikely to be introduced until the second part of this century, meaning that intermediate zero-carbon electricity sources will still play a crucial role in the decarbonisation of world economies.

5.6 Reprocessing and Waste Storage

Nuclear power plants produce radioactive waste ranging from very low-level waste to high-level waste, of which the high-level waste (HLW) associated with spent fuel is of primary concern, as it must be managed and eventually disposed of responsibly and safely. High-level waste contains such high levels of radionuclides that it must be isolated from the accessible environment. It also generates significant heat from radioactive decay. Thus, deep geological disposal facilities with engineered barriers are indicated for HLW disposal. It is important to recognise, however, that the volume of radioactive waste produced by the nuclear power industry is very small compared with the volumes of toxic waste produced in the industrialised world. Moreover, HLW, which contains most (about 95%) of the radioactive inventory, is a small fraction of the total volume of nuclear waste.

A typical 1000-MWe LWR discharges about $20 \, \text{m}^3$ or 27 tonnes of used fuel per year, which corresponds to a disposal volume of $75 \, \text{m}^3$ following encapsulation if it is treated as waste. Such spent fuel, however, still consists of about 95% uranium, and contains ^{235}U at a higher enrichment than natural uranium. It also contains traces of plutonium, with ^{239}Pu the dominant isotope. One option is to reprocess the spent fuel to recover the fissile material for reuse, and also reduce the volume of HLW, which then consists of highly radioactive (but short-lived) fission fragments and some long-lived transuranic radionuclides. With or without reprocessing, the initial radioactivity and heat is generated predominantly by short-lived fission products and so the radioactivity and heat from spent fuel falls by a factor of 1000 over a period of 40–50 years. Permanent underground disposal as a synthetic rock or vitrified glass would normally follow such a cooling period.

Reprocessing therefore makes best use of uranium resources and helps manage nuclear waste. But whether reprocessing of spent fuel is economically advantageous is an open question. The decision to reprocess or not is therefore made on other grounds. Although the plutonium from power reactors contains a high level of ^{240}Pu, which renders it unsuitable for weapons use, the perception often remains that reprocessing is associated with an increased risk of proliferation of nuclear weapons. Another consideration is that used fuel

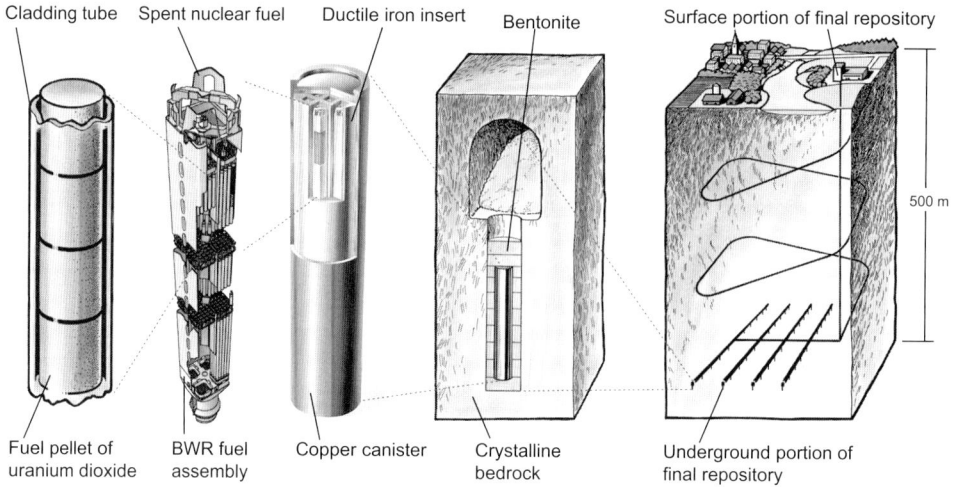

Figure 5.4 Swedish KBS-3 system for deep geological disposal of used fuel.
Source: Figure supplied courtesy of SKB (Swedish Nuclear Fuel and Waste Management Company). Illustrator: Jan Rojmar. For a colour version of this figure, please see the colour plate section.

is subject to international safeguards because of its uranium and plutonium content, but separated and vitrified HLW is not. On balance, there is no economic incentive at present to engage in reprocessing, so, given also the proliferation concerns, a country newly engaging in nuclear power production would best avoid it.

The IAEA safety guide (IAEA 2011) for HLW recommends deep geological disposal. The most advanced HLW repository is at Forsmark in Sweden. Sweden has 10 operating nuclear power plants and the Swedish Nuclear Fuel and Waste Management Company (SKB) was established in 1977. As illustrated in Figure 5.4, complete spent fuel assemblies are encapsulated in copper canisters 4.8 m high, weighing 25 tonnes. Each canister, holding 12 fuel assemblies, is stored in holes at a depth of 500 m in 1.9-billion-year-old granite. The void between the bedrock and the canisters is filled with bentonite clay to absorb any leakage.

A licence application for the volunteer site at Forsmark was lodged in 2011 and the Land and Environment Court and the Swedish Nuclear Regulator presented their conclusion to the government in 2017. Further information was submitted in 2019. In October 2020, the Osthammar Municipal Council voted yes to hosting the facility at Forsmark. The Swedish Government is expected to announce its decision in 2021.

This KBS-3 system has also been adopted for the deep geological disposal site at Okiluoto, Finland. The Finnish Government granted approval for the construction of this facility in November 2015.

The final report of the South Australian NFCRC identified an opportunity for South Australia to establish waste storage and disposal facilities for used nuclear fuel and intermediate-level waste from overseas. The NFCRC found that there was a global need and an economic opportunity for a commercial waste facility in South Australia.

5.7 Decommissioning

About 110 commercial power reactors have been retired from operation. The facility is decommissioned and demolished leaving the site available for use. At all sites the first action is to remove the fuel and coolant, which removes most of the radioactivity. There are then two main options for decommissioning: immediate dismantling and deferred dismantling.

5.7.1 Immediate Dismantling

This option has the following advantages: personnel with site knowledge are still available; waste costs are known; no ongoing maintenance or surveillance is required; current experience of dismantling similar types may be available and the site can be immediately reused.

However, the main disadvantage of this option is that the material is more radioactive. As a consequence, there is an increased potential for higher doses to workers, and costs for dismantling and waste storage are higher. Also, the funds for dismantling must be immediately available.

5.7.2 Deferred Dismantling ('Safstor')

The facility is placed in a safe storage regime for typically 30–60 years before full decommissioning. This approach has the following advantages: lower radioactivity levels; lower radiation doses to workers; less demand for remote-handling equipment; and the possibility of better technology for dismantling, which could be developed in the interim. The disadvantages include the ongoing maintenance and surveillance required, the loss of personnel with intimate site knowledge and the fact that the site cannot be reused immediately.

One situation when deferred dismantling could be the best option is where there are several reactors on one site and it would be better to wait until they are all shut down and then progressively dismantle them.

In the USA, utilities collect 0.1 to 0.2 cents/kWh to fund decommissioning (WNA 2020).

Nine large US power reactors have been completely decommissioned. Decommissioning BWR costs are higher than those for PWR decommissioning, due to more material becoming radioactive in the single-loop BWR. Graphite-moderated reactors (the main reactor type in the UK) are more expensive to decommission as the graphite accumulates radionuclide contamination while in service, particularly carbon-14, and then has to be treated as intermediate-level waste.

5.8 Regulatory and Governance: Safety, Security, Safeguards

Safeguards are measures applied by the IAEA to verify that non-proliferation commitments made by states under safeguards agreements with the IAEA are fulfilled. 'Safeguards' refers to the total system for accounting of nuclear materials. In practice, each facility has its own safeguards department, which maintains records of all nuclear material on the site. A team of IAEA inspectors will regularly visit the site and verify the quantities and types of nuclear materials. A state-level regulatory authority is responsible for the state's obligations and will perform independent checks.

Modern safeguards implementation is based more on safeguards considerations for the state as a whole, rather than solely on the quantity and type of nuclear materials.

Security is the physical protection of nuclear materials against removal and sabotage. States are party to the *Convention on the Physical Protection of Nuclear Materials,*[1] which has a graded approach for security measures.

Safety is the protection of people and the environment from ionising radiation, the minimisation of accidents and the mitigation of any events as described in the IAEA guide, *Fundamental Safety Principles* (IAEA 2006). An important safety principle is 'defence in depth', which includes the use of access controls, physical barriers, redundant and diverse key safety functions, and emergency response measures to prevent accident progression and mitigate the consequences of severe accidents.

The IAEA defines five independent levels of defence in depth to compensate for potential human and mechanical failures:

(1) prevent deviations from normal operation;
(2) detect and control deviations;
(3) incorporate safety features, safety systems and procedures to prevent core damage;
(4) mitigate the consequences of accidents;
(5) mitigate radiological consequences of significant releases of radioactive materials.

The other key safety principle is safety culture: 'that assembly of characteristics and attitudes in organisations and individuals which establishes that, as an overriding priority, nuclear plant safety issues receive the attention warranted by their significance' (IAEA 1991: 1).

5.9 Development of a Nuclear Power Programme

A nuclear power programme is a major undertaking for any country, requiring government, legal, regulatory, managerial, technological, human and industry support throughout its lifetime. It requires compliance with international legal instruments, internationally accepted safety standards, security guidelines and safeguards requirements. Success depends on a long-term energy plan and the competence and credibility of the organisations and individuals responsible, in particular the nuclear regulatory organisation, which must be independent. The IAEA guide *Milestones in the Development of a National Infrastructure for Nuclear Power* (IAEA 2007) lists all the issues that must be considered.

[1] *Convention on the Physical Protection of Nuclear Materials,* opened for signature 3 March 1980, INFIRC274/Rev1 (entered into force 8 February 1987). Available at: www.iaea.org/publications/documents/conventions/convention-physical-protection-nuclear-material.

A prerequisite for developing nuclear power in many countries is wide community support and support from both sides of politics. Generally, nuclear power has high upfront construction costs, which are balanced later by lower fuel costs. However, investing in nuclear construction is not commercially viable in a situation where a policy change could prevent operation and remove the ability to recover construction costs.

5.10 Conclusion

Advanced PWR technology is at present the dominant type of reactor technology for electricity production. It is likely to remain dominant in the immediate future, particularly for larger reactors being installed in Asian and the Middle Eastern countries with larger economies. Examples include four AP1000 reactors in China. These are large Generation III+ plants with passive safety.

Small modular reactors may impact the take-up of nuclear power, especially in smaller economies and remote locations. However, SMRs will probably not be a big player in climate change mitigation unless staged installation builds up significant capacity.

Fast reactors are an attractive option for the management of nuclear waste, by burning spent LWR fuel; they also offer long-term energy security through breeding, should this prove necessary. Generation IV reactors, to be deployed beyond 2030, aim for high temperatures to enable desalination along with electricity and hydrogen production, and to deliver process heat.

Uranium–plutonium cycles will dominate nuclear power generation in the immediate term, with thorium cycles possibly becoming economically viable towards the end of the century. Accelerator-driven systems, which are often associated with thorium as a nuclear fuel, will be used to produce neutrons and serve as research facilities in the first half of the twenty-first century. They may later play a combined role in energy production and waste transmutation.

References

Bodansky, D. (2005). *Nuclear Energy: Principles, Practices, and Prospects*, 2nd ed. New York: Springer-Verlag.

Chen, F. F. (2011). *An Indispensable Truth: How Fusion Power Can Save the Planet*. New York: Springer-Verlag.

Cook, I., Miller, R. L. and Ward, D. J. (2002). Prospects for economic fusion electricity. *Fusion Engineering and Design*, 63–64, 25–33.

Gen IV (Generation IV International Forum) (2002). *A Technology Roadmap for Generation IV Nuclear Energy Systems*. Issued by the US DOE Nuclear Energy Research Advisory Committee and the Generation IV International Forum. GIF-002-00. Available at: www.gen-4.org/gif/upload/docs/application/pdf/2013-09/gif_rd_outlook_for_generation_iv_nuclear_energy_systems.pdf.

Government of South Australia (2016). *Nuclear Fuel Cycle Royal Commission Report*. Adelaide: Government of South Australia. Available at: http://nuclearrc.sa.gov.au/media-centre/nuclear-fuel-cycle-royal-commission-report-delivered/.

Hitachi (n.d.). PRISM. *Nuclear Power Plants GE Hitachi*. Available at: https://nuclear
.gepower.com/build-a-plant/products/nuclear-power-plants-overview/prism1.

IAEA (International Atomic Energy Agency) (n.d.a) International project on innovative
nuclear reactors and fuel cycles (INPRO). *IAEA.org*. Available at: www.iaea.org/
services/key-programmes/international-project-on-innovative-nuclear-reactors-and-
fuel-cycles-inpro.

IAEA (n.d.b). Small modular reactors. *IAEA.org*. Available at: www.iaea.org/topics/small-
modular-reactors.

IAEA (1991). *Safety Culture: A Report by the International Nuclear Safety Advisory
Group*. INSAG Series No. 4. International Atomic Energy Agency. Available at:
www.iaea.org/publications/3753/safety-culture.

IAEA (2005). *Thorium Fuel Cycle: Potential Benefits and Challenges*. IAEA Tecdoc 1450.
International Atomic Energy Agency. Available at: www.iaea.org/publications/7192/
thorium-fuel-cycle-potential-benefits-and-challenges.

IAEA (2006). *Fundamental Safety Principles*. IAEA Safety Standards Series No. SF-1.
International Atomic Energy Agency. Available at: www.iaea.org/publications/7592/
fundamental-safety-principles.

IAEA (2007). *Milestones in the Development of a National Infrastructure for Nuclear
Power*. IAEA Nuclear Energy Series NG-G-3.1. International Atomic Energy
Agency. Available at: www.iaea.org/publications/7812/milestones-in-the-develop
ment-of-a-national-infrastructure-for-nuclear-power.

IAEA (2011). *Geological Disposal Facilities for Radioactive Waste: Specific Safety Guide*.
IAEA Safety Standards Series No. SSG-14. International Atomic Energy Agency.
Available at: www.iaea.org/publications/8535/geological-disposal-facilities-for-radio
active-waste.

IAEA (2016). *Nuclear Power Reactors in the World: 2016 Edition*. Reference Data Series
No. 2. International Atomic Energy Agency. Available at: www.iaea.org/publica
tions/11079/nuclear-power-reactors-in-the-world.

IEA (International Energy Agency) (2016). *CO_2 Emissions from Fuel Combustion:
Highlights*. Paris: International Energy Agency. Available at: https://emis.vito.be/
sites/emis.vito.be/files/articles/3331/2016/CO2EmissionsfromFuelCombustion_
Highlights_2016.pdf.

Till, C. E. and Chang, Y. I. (2011). *Plentiful Energy: The Story of the Integral Fast
Reactor*. Charles E. Till and Yoon Il Chang.

US NRC (Nuclear Regulatory Commission) (n.d.). Pressurized water reactors. *U.S. NRC*.
Available at: www.nrc.gov/reactors/pwrs.html.

Wesson, J. (2004). *Tokamaks*, 3rd ed. Oxford: Clarendon Press.

WNA (World Nuclear Association) (n.d.). Information library. *World Nuclear Association*.
Available at: www.world-nuclear.org/information-library.aspx.

WNA (2017). Nuclear fuel cycle overview. *World Nuclear Association*. Available at:
www.world-nuclear.org/info/Nuclear-Fuel-Cycle/Introduction/Nuclear-Fuel-Cycle-
Overview/.

WNA (2018). Processing of used nuclear fuel. *World Nuclear Association*. Available
at: www.world-nuclear.org/information-library/nuclear-fuel-cycle/fuel-recycling/pro
cessing-of-used-nuclear-fuel.aspx.

WNA (2020). Decommissioning nuclear facilities. *World Nuclear Association*. Available
at: www.world-nuclear.org/information-library/nuclear-fuel-cycle/nuclear-wastes/
decommissioning-nuclear-facilities.aspx.

6

Hydropower

JAMIE PITTOCK

Executive Summary

Hydropower is the largest source of renewable energy in the world (85%) and provides 16.3% of the world's electricity in 159 countries; it is expected at least to double by 2050.

The scope for expansion of hydropower is considerable, but adverse environmental and social impacts need to be managed; hundreds of millions of people have been adversely affected by hydropower development. Poor communities often rely on fish for nutritious food and hydro projects can significantly reduce this resource.

There are a variety of environmental impacts of small versus large hydro dams, and a number of remedial measures are available. Freshwater biodiversity needs to be conserved with a changing climate; different types of freshwater habitats may become higher priorities for conservation.

Climate change is impacting hydro generation through changed snow melts and river flows, greater evaporation and more frequent extreme events, such as flooding and droughts. Hydropower infrastructure needs to have margins to cope with extreme events and adapt to changing conditions, for example, by redesigning spillways and reinforcing dam walls.

Evaporation from reservoirs can be reduced by floating solar arrays that also generate power.

Relicensing at specified intervals can provide a framework for renovation, removal or changes to minimise impacts and maximise benefits of dams.

Planning of dams needs to be undertaken on a whole-of-river-basin scale to minimise impacts and maximise benefits.

The World Commission on Dams (2000) recommended strategic priorities for more sustainable dam development, emphasising public consultation, careful review of options, identification of impacts and opportunities, sustaining ecosystems and livelihoods, recognising entitlements and sharing benefits with impacted people, complying with laws and sharing transboundary rivers for peace, development and security.

The International Hydropower Association's Hydropower Sustainability Assessment Protocol (2010) is one codification of good hydropower development practices, spanning the process from early planning to operations across 20 topics.

Hydropower has an important role in decarbonising energy systems by providing firming capacity (generating electricity to match demand) to complement intermittent generation from solar and wind generators. Pumped storage hydropower can store excess electricity from intermittent generators for later use (see Chapter 7).

6.1 Introduction and Context

Hydropower is currently the world's largest source of renewable electricity and generation is forecast to at least double by 2050. Lower carbon emissions from hydropower reservoirs compared to other (e.g. fossil-fuelled) generators and the ability to facilitate greater deployment of wind and solar power are driving support for greater development of hydropower in national policies. However, instream hydropower plants are having severe negative environmental and social impacts. Emerging hydropower technologies offer some potential for more electricity with lower impacts. This section explores how society may enjoy the benefits while reducing the environmental and social impacts of existing and proposed hydropower developments, through better planning, design and reoperation (rejuvenation of old reservoirs).

6.2 Types of Hydropower

Hydropower has been a widely used mechanical energy source for thousands of years: its deployment increased during the industrial revolution of Western nations and has been used to generate electricity since the nineteenth century (IEA 2012). Conventional hydropower involves damming a river to store water in order to generate electricity on demand. Often this involves catching higher wet season flows and storing the water to generate electricity in dry seasons, but these kinds of changes to river flows are ecologically disastrous, as elaborated in Section 6.4. The ability for storage dams to store this potential energy is vital for complementing baseload power stations, for meeting peak energy requirements, and for 'black start' capacity to resume services after a grid failure (IEA 2012). Run-of-river hydro schemes generate baseload electricity in proportion to river inflows, and they lack a storage capacity.

Pumped storage hydropower, where excess electricity is used to pump water from a lower to an upper reservoir, allows later generation of electricity on demand at around 80% efficiency (IEA 2012). The benefits of pumped storage systems are discussed more fully in Chapter 7. Pumped storage reservoirs can be sited off rivers, considerably reducing their environmental impact (Pittock 2019). The ability of storage and pumped storage hydropower reservoirs to store energy makes them important to support greater deployment of intermittent solar and wind generators.

An often-neglected mitigation option is the opportunity to retrofit older dams to produce more electricity from the same structures (IEA 2012). For instance, an assessment of four old dams in Brazil concluded that repowering could increase energy production at each dam by 46 to 205% (WWF 2004). Further, many existing storages could be re-engineered as pumped storage projects (IEA 2012). Reoperated old dams can also provide environmental flows to restore downstream ecosystems and enhance livelihoods (Postel and Richter 2003; Krchnak et al. 2009).

Increasingly, pipelines originally for purposes such as water supply and sewerage are being fitted with micro hydro generators. Further technological changes can be expected as marine and hydrokinetic electricity generators are being developed to capture 'the energy from waves, tides, ocean currents, the natural flow of water in rivers, and marine thermal gradients' (US DoE 2015: 1). These technologies will raise new questions on how to maximise the benefits while minimising adverse social and environmental impacts. This illustrates the importance of the preparation and adoption of sustainability standards for hydropower developments (discussed in Section 6.6) that are sufficiently broad and flexible to manage changing hydropower technologies. Hydropower technologies are related to the emerging tidal and wave power marine generators.

Proponents also argue that hydropower enhances resilience in water and food security. For example, the International Energy Agency (IEA) states: 'Hydropower reservoirs can also regulate water flows for freshwater supply, flood control, irrigation, navigation services and recreation. Regulation of water flow may be important to climate change adaptation' (IEA 2012: 5). In practice, these claims have little substance, since storing water for energy generation on demand usually conflicts with keeping a reservoir partly empty to catch flood peaks, or for seasonal release of water to grow crops (Bates et al. 2008). Hydropower proponents have been criticised for exaggerating the non-hydropower benefits of their reservoirs in order to seek approval for their projects (WCD 2000).

6.3 State of Hydropower Development

Hydropower, operating in 159 countries, currently generates 16.3% of the world's electricity, which is 85% of renewable power globally. In theory, there is great potential to expand hydropower, especially in Africa, Asia and Latin America. For instance, the IEA *Hydropower Technology Roadmap* foresees, by 2050, a doubling of global capacity up to almost 2000 GW (gigawatts) and of global electricity generation over 7000 TWh (terawatt-hours) (IEA 2012). The IEA also predicts expansion of pumped storage hydropower by a factor of three to five in that time period.

Middle-income countries with increasing energy demands are rapidly expanding their hydropower production or importing hydropower from neighbouring states, as many of these nations have further economically feasible hydropower potential (IEA 2012). However, severe environmental impacts from such developments are likely as globally the greatest fish biodiversity coincides with the rivers in the countries with the greatest hydroelectric power development potential (Opperman, Hartmann et al. 2015).

Decisions are required on trading off development of low-carbon electricity generation from hydropower to mitigate climate change against the negative environmental and social impacts of this technology. As an example, the World Wildlife Fund (WWF), a non-government environmental organisation, published a comprehensive energy policy proposal that included limited expansion of hydropower (WWF 2007). They propose hydropower development by nation or river basin of up to 30% of economically feasible potential, to preserve ecologically significant river reaches in a free-flowing state while enabling electricity generation in developing nations. Globally, of 2270 GW of economically feasible large hydropower potential, 740 GW (33%) had been developed by 2007, 445 GW were planned (20%) and 120 GW were under construction (5%). In 2007, the WWF considered that 250 GW from large hydropower sites could be developed with relatively low impacts, plus an additional 20 GW from medium hydropower sites, 100 GW from small and 30 GW from reoperating existing dams (WWF 2007).

Hydropower has been categorised as an energy source with low greenhouse gas emissions under the Kyoto Protocol's Clean Development Mechanism (CDM) (CDM Executive Board 2009). This low greenhouse gas emissions status is contested given those from the production of construction materials like concrete; from development, including from deforestation of the reservoir area; or from operation, from the release of methane and carbon dioxide from decomposition of inundated organic matter (Fearnside 2004; Harvey 2006; Rosa et al. 2006; Pacca 2007; Weisser 2007; IHA 2010b; IEA 2012).

6.4 Environmental and Social Impacts of Hydropower Development

The often severe environmental and social impacts of hydropower dams are now outlined, along with discussion of how these costs may be minimised. Socially, somewhere between 40 and 80 million people have been displaced by dam developments (WCD 2000) and a further half a billion people down river from major dams have been negatively affected by such impacts as changes in flooding regimes, declines in crop production and fish catch, and loss of lands in places like river deltas (Richter et al. 2010). Hydropower developments have been extensively criticised for failing to obtain prior informed consent from people impacted by dam projects or adequately compensating those affected (WCD 2000). The recommendations of the World Commission on Dams (WCD) and the Hydropower Sustainability Assessment Protocol (described in Section 6.6) are two attempts to establish processes to minimise the impacts of dam developments on people.

The environmental impacts of hydropower dams differ little from those of other types of barrages and this assessment draws on information from all types of dams (WCD 2000). Much of the environmental debate over the impacts of dams is mistakenly focused on the area of terrestrial ecosystems inundated by impoundments, overlooking the significant impacts that any dam can have in degrading freshwater ecosystems (Garcia de Leaniz 2008). The conservation of fish and other aquatic biodiversity deserves special attention. Globally, freshwater biota (living organisms) has declined more than that from any other biome (biological community), and is least well protected by conservation measures (MEA 2005). Dam developments are one of the major reasons for the loss of freshwater wildlife, due to the resulting changes to water flows and the barriers they form to species migration (WCD 2000; MEA 2005; Poff et al. 2007).

A number of organisations are promoting small hydropower development as a less ecologically and socially damaging form of hydropower development. The ecological impacts of dams are outlined in Table 6.1 together with a list of possible remedial measures and the relative impacts of larger versus smaller hydropower dams. While the impacts of hydropower dams on freshwater ecosystems are severe, a number of these problems are reduced with smaller dams. Importantly, there are structural measures such as the addition of fish ladders that can partly reduce environmental damage caused by dams (Cowx and Welcomme 1998; Krchnak et al. 2009; Pittock and Hartmann 2011) (Table 6.1). These are most effective and cheaper when built into new dams rather than retrofitted at a later time. However, these engineering measures never fully eliminate environmental damage. Considering fish ladders, for example: they are unlikely to work for all aquatic wildlife species; they do not work adequately in rivers with the largest fish migrations; and, in a cascade of dams, each fish ladder takes a toll on the passing fish, reducing the population to some degree (Larnier and Marmulla 2004; Agostinho et al. 2007; ICEM 2010).

Maintaining the ecosystem services provided by healthy freshwater ecosystems for people is not a trivial question or a matter of 'balance' between the environment and development. Freshwater ecosystems provide a diverse range of benefits to people: more per unit area than any other biome (Costanza et al. 1997; MEA 2005). An ecosystem services framework has been developed to more systematically evaluate the costs and benefits of intact rivers versus developments such as hydropower projects to better inform decision makers. This framework has been adopted by national governments through international institutions such as the UN Environment Programme and Intergovernmental Panel on Biodiversity and Ecosystem Services (Brink 2011; Díaz et al. 2015).

One ecosystem service provided by rivers is fish. Often, poorer, rural residents rely on fish as a key source of protein and micronutrients (Richter et al. 2010; Ziv et al. 2012; Béné et al. 2015). The value of these fisheries in the subsistence and informal economy is poorly accounted for in development decisions in places like the Mekong River basin (Baran and Myschowoda 2009; Arthur and Friend 2011). An assessment of proposed hydropower development in the Mekong basin suggests that planned hydropower developments may supply 8.3% of the electricity for these nations but destroy 26–42% of the fisheries by 2030, exacerbating poverty (ICEM 2010). Great care is required in developing hydropower projects to avoid perverse impacts on people who depend on riverine resources. Further, replacement of fish by other food sources requires considerably more land and water that may result in further greenhouse gas emissions and socio-economic impacts (Orr et al. 2012).

Under the Convention on Biological Diversity and the Ramsar Convention on Wetlands, national governments have committed to conserving the full range of biodiversity through: more sustainable use, provision of environmental flows in rivers, better catchment management and designation of representative nature reserves (Ramsar 2008; CBD 2010). To fulfil these obligations, among other measures, some rivers should remain undeveloped and in many countries free-flowing rivers are beginning to receive legal protection (WWF 2006).

Table 6.1. *Relative impacts on freshwater ecosystems of large compared to small hydropower dams on the environment*

Ecological impacts	Remedial measures	Impact of large hydropower dams	Impact of small hydropower dams
Block wildlife passage, e.g. migratory fish	Add wildlife passage devices, e.g. fish ladders	Providing effective passage is not possible on the largest dams, resulting in the loss of species and reduction in fisheries.	Wildlife passage devices are more practical on smaller dams but there are cumulative impacts of each dam on migratory fish.
Turbines kill fish fry and other wildlife moving down river	Add fish-friendly or aerating turbines	A major problem that reduces wildlife populations. It has rarely been addressed. Better turbines may partly reduce the death of wildlife.	A problem that is exacerbated by the lesser attention paid to wildlife-friendly turbines on small dams but it is partly ameliorated where there are wildlife passage devices.
Thermal pollution	Add thermal pollution control devices, e.g. multilevel offtake tower	Thermal pollution is a major impact that curtails wildlife breeding for hundreds of kilometres downriver of problem dams. It is readily fixed by control devices but these have rarely been fitted to dams.	The small depth and run-of-river operations of many small hydropower dams mean that thermal pollution is not a widespread problem.
Changes to water flows	Add large-water-release structures; reregulating dams; coordinated reservoir operations; release of environmental flows	The major impact: one that has rarely been addressed. Large-release structures and provision of environmental flows may partly reduce the ecological impacts.	The run-of-river operations of smaller hydropower dams mean that altered flows are not a widespread problem.
Sediment capture in reservoirs	Add bypass tunnels and bottom outlets that can be used to flush sediments through impoundments	A substantial impact resulting in major changes to the geomorphology and ecology of rivers and deltas. Flushing sediment through large impoundments is generally not possible.	The small depth and run-of-river operations of many small hydropower dams mean that sediment flushing is more feasible.

Source: Modified from Pittock and Hartmann (2011).

6.5 Climate Change and Hydropower Development

The advent of climate change is an opportunity and threat to hydropower production, and may negatively impact both freshwater ecosystems and hydropower infrastructure. Climate change has five key implications for hydropower development, as follows.

First, climate change policies are increasing support for low-carbon energy generation, especially from hydropower (Pittock 2010). In particular, middle-income countries with increasing energy demands are rapidly expanding their hydropower production or are importing hydropower from neighbouring states, as many of these nations have further economically feasible hydropower potential (Government of China 2007; Government of Brazil 2008; PMCCC 2008; IEA 2012).

Second, hydropower producers are particularly vulnerable to climate change. Hydropower is being negatively affected through changes in snow melt and river flow, greater evaporation from reservoirs, more frequent and severe extreme events like floods and droughts, and increased sediment loads (WCD 2000; Bates et al. 2008; Opperman, Hartmann et al. 2015). While at a continental scale the impact of increased and decreased climate-induced changes in run-off on hydropower generation may balance out (Hamududu and Killingtveit 2012), in key regions, decreases in production are projected or being experienced, reducing power generation (Opperman, Hartmann et al. 2015), harming the economies of some developing nations that are overdependent on hydropower, such as Brazil (Bates et al. 2008).

The third implication is that climate change will transform the ways in which hydropower is developed and deployed. Climate change policies are increasing funding for low-carbon energy generation (Pittock 2010). Fine-resolution climate data sufficient to reliably engineer dams may not be available, requiring service providers to adopt robust risk management approaches (Hallegatte 2009). These include: first upgrading existing infrastructure; building infrastructure that is reversible or can be used under a range of conditions; building in larger safety margins to cope with extreme events; and enabling more rapid and incremental responses, such as the addition of smaller and decentralised infrastructure (WCD 2000; World Bank 2004; Hallegatte 2009). It is being proposed that reservoirs with flood retention capacity are reoperated to use that storage space instead for additional power generation (Opperman, Galloway et al. 2009; Opperman, Hartmann et al. 2015). Some of the extra revenue will be invested in restoring floodplains to give rivers room to flood safely, providing additional environmental and social benefits. Technological change is also likely, including greater deployment of pumped storage hydropower to back up intermittent wind and solar power generators, as well as hydro-kinetic hydropower generators (Pittock 2010; IEA 2012; US DoE 2015). As a result, hydropower operators will need to manage much greater uncertainty in infrastructure planning, development and operation.

Fourth is the question of dam safety and operations. Dams are designed to operate under a specific range of hydrological conditions; however, climate change will change water availability, flows and the frequency of extreme events. Hence, dams need to be designed or rebuilt to work in a non-stationary environment (WCD 2000; IPCC 2007; Milly et al.

2008). The main reasons for dam failure are internal erosion, insufficient shear strength of foundations and overtopping (WCD 2000). Dam spillways are designed to release water in floods up to a certain size (e.g. safely release the floods expected up to the 1 in 1000 year extreme event); however, the return interval of large floods is expected to increase with climate change (WCD 2000; IPCC 2007; Bates et al. 2008). Even before the impacts of climate change are felt, countries with considerable resources like Australia, China and the USA are struggling to maintain the safety of existing dams (Pittock and Hartmann 2011). To maintain or reduce the level of risk on existing dams, spillways need to be redesigned and often dam walls will need reinforcement (WCD 2000). Changes in river discharge are expected in every populated river basin globally, making changes to infrastructure essential, including removal of unsafe or redundant infrastructure (Palmer et al. 2008). Further, climate change-induced erosion is anticipated to increase sedimentation in dams, reducing storage capacity and amplifying the flood risk (Palmer et al. 2008). More resources need to be invested in the regulation, safety and renovation of existing hydropower infrastructure.

Fifth are the implications of hydropower dams for adaptation of freshwater biodiversity to climate change. The freshwater biota lives in a narrow range of water temperature and flow conditions (Olden and Naiman 2009). One implication of climate change is that water at the temperatures that these species live in may, in future, be found at locations closer to the poles or at higher altitude (Palmer et al. 2008). There is already evidence of movement of the distribution of fish species (Daufresne and Boet 2007). Under these circumstances it will be more important to reduce the barrier effect of dams to enable further changes in fish distribution (Olden and Naiman 2009), and hydropower may need to be reoperated to provide better environmental flows (Null et al. 2013). Further, different types of freshwater habitats may become higher priorities for conservation, including for reservation from development or restoration. Examples include free-flowing rivers and river reaches that gain water from aquifers, and those with a north–south alignment (Pittock and Finlayson 2011; Lukasiewicz et al. 2013). Consequently, different river reaches may need to be reserved from hydropower development.

6.6 Better Planning for Hydropower Development

Many of the social and environmental impacts of hydropower development could be reduced with earlier and more comprehensive assessment and better planning. Environmental assessment of individual projects usually occurs at a time in the development cycle when so much is invested in decisions like particular dam sites that it is too late to significantly reduce impacts. Instead, earlier strategic environmental assessment is required, which considers hydropower development at a whole-of-river-basin scale to enable alternative dam locations to be considered (Opperman, Hartmann et al. 2015). The benefits are illustrated retrospectively for the Penobscot River in the USA, where relicensing of a cascade of dams resulted in enhancement of some dams to maintain hydropower generation and removal or bypassing of others, which will reopen an additional 60% of the river basin to fish migration (Opperman, Apse et al. 2011).

Globally, the WCD recommended seven strategic priorities and associated policy principles for better dam development and management to guide decision-making, and 26 advisory guidelines based on good practice from around the world. The seven priorities are: gaining public acceptance through prior, informed consent; undertaking comprehensive options assessment on options for best delivering energy or water services; addressing the impacts and opportunities to better use existing dams; sustaining river ecosystems and the livelihoods of people who depend on them; recognising entitlements and sharing benefits with impacted people; ensuring compliance with the law; and sharing transboundary rivers for peace, development and security.

While establishing a principled basis for better dam-related developments, these WCD priorities and guidelines have proven challenging to put into practice (Pittock 2010). Subsequently, the International Hydropower Association (IHA), with government and non-government stakeholders, has developed the global Hydropower Sustainability Assessment Protocol, a publicly available and practical framework for developers and others to assess the performance of planned and existing hydropower projects (IHA 2010a). The protocol applies to the four main stages of hydropower development (early stage, preparation, implementation and operation) and enables assessments using objective evidence against some 20 topics (Foran 2010; IHA 2010a). Application of this protocol should become standard practice for financiers, developers and regulators of hydropower projects to maximise their benefits and minimise their impacts.

There is a range of other frameworks, planning and regulatory tools available to better design and manage hydropower schemes in a changing climate while minimising environmental and social impacts (Opperman, Hartmann et al. 2015).

6.7 Managing the Life Cycle of Hydropower Developments

Dams are commonly perceived as permanent, yet they have a finite life, eventually silting up, failing or requiring decommissioning (WCD 2000). Globally, it is estimated that 0.5–1% of the world's reservoir volume is lost annually (Howard 2000). Dams require regular maintenance and periodic refurbishment to remain safe and deliver services (Howard 2000). Their purposes can be adjusted and their performance improved by changing operating rules, or upgrading or retrofitting new equipment, and unsafe or redundant dams can be removed (Bermann 2007; Krchnak et al. 2009; Pittock and Hartmann 2011; Opperman, Hartmann et al. 2015). The benefits of an iterative review of dams to enhance their performance or remove them can only be achieved with better governance. While a number of cyclical dam review systems exist, the most effective for achieving change is periodic dam relicensing, as practised by the US Federal Energy Regulatory Commission (Russo 2000; Pittock and Hartmann 2011). Key elements of an ideal regulatory system to optimise hydropower infrastructure performance would include periodic (time-limited) relicensing of all infrastructure. Such relicensing would be overseen by an independent regulatory agency that would take decisions in the public interest through a transparent process, involving public participation (Pittock and Hartmann 2011). Each dam would have an identified owner who would be required to apply the best available technologies to maximise safety, socio-economic and environmental performance. Dam renovation could minimise current non-climate impacts, improve migration of aquatic wildlife and even attenuate some climate impacts on the freshwater biota (Viers 2011).

6.8 Conclusion

Hydropower has a growing role to play in providing electricity, especially in developing countries, and also as a facilitator of greater use of solar and wind energy in electricity grids. Technological innovation may see new types of hydrokinetic generation emerge whose environmental and social implications for management of rivers, estuaries and seas are yet to be discerned.

However, hydropower developments are having significant environmental and social impacts. Further, the feasibility and impacts of such projects will become more complex with climate-induced changes to hydrology. Governments and hydropower developers should minimise environmental and social impacts and maximise benefits through: better planning of projects, deploying infrastructure that can adjust to changes in hydrology and incorporating environmental mitigation devices into structures. The non-stationary impacts of climate change, changing societal needs and new knowledge require more iterative and professional governance. Periodic relicensing of hydropower projects is needed to enhance safety and to ensure that the socio-economic benefits are maximised while the social and environmental costs are minimised.

References

Agostinho, C. S., Agostinho, A. A., Pelicice, F., Almeida, D. A. d. and Marques, E. E. (2007). Selectivity of fish ladders: A bottleneck in neotropical fish movement. *Neotropical Ichthyology*, 5, 205–213.

Arthur, R. I. and Friend, R. M. (2011). Inland capture fisheries in the Mekong and their place and potential within food-led regional development. *Global Environmental Change*, 21, 219–226.

Baran, E. and Myschowoda, C. (2009). Dams and fisheries in the Mekong basin. *Aquatic Ecosystem Health & Management*, 12, 227–234.

Bates, B. C., Kundzewicz, Z. W., Wu, S. and Palutikof, J. P., eds. (2008). *Climate Change and Water: Technical Paper of the Intergovernmental Panel on Climate Change*. Geneva: IPCC (Intergovernmental Panel on Climate Change) Secretariat.

Béné, C., Barange, M., Subasinghe, R. et al. (2015). Feeding 9 billion by 2050: Putting fish back on the menu. *Food Security*, 7, 261–274.

Bermann, C. (2007). Impasses e controvérsias da hidreletricidade. *Estudos Avançados*, 21, 139–153.

Brink, P. T., ed. (2011). *The Economics of Ecosystems and Biodiversity in National and International Policy Making*. Oxford: Earthscan.

CBD (Convention on Biological Diversity) (2010). Decision X/28. Inland waters biodiversity. UNEP/CBD/COP/DEC/X/28. Montreal, Canada: *Convention on Biological Diversity*. Available at: www.cbd.int/decisions/cop/10/28/25.a.

CDM (Clean Development Mechanism) Executive Board (2009*). Approved Consolidated Baseline and Monitoring Methodology ACM0002: Consolidated Baseline Methodology for Grid-Connected Electricity Generation from Renewable Sources.* ACM0002/Version 10. Sectoral Scope: 01. EB 47. United Nations Framework Convention on Climate Change. Bonn: United Nations Framework Convention on Climate Change.

Costanza, R., d'Arge, R., de Groot, R. et al. (1997). The value of the world's ecosystem services and natural capital. *Nature*, 387, 253–260.

Cowx, I. G. and Welcomme, R. L. (1998). *Rehabilitation of Rivers for Fish*. Oxford: FAO and Fishing News Books.

Daufresne, M. and Boet, P. (2007). Climate change impacts on structure and diversity of fish communities in rivers. *Global Change Biology*, 13, 2467–2478.

Díaz, S., Demissew, S., Carabias, J. et al. (2015). The IPBES Conceptual Framework: Connecting nature and people. *Current Opinion in Environmental Sustainability*, 14, 1–16.

Fearnside, P. M. (2004). Greenhouse gas emissions from hydroelectric dams: Controversies provide a springboard for rethinking a supposedly 'clean' energy source. An editorial comment. *Climatic Change*, 66, 1–8.

Foran, T. (2010). *Making Hydropower More Sustainable? A Sustainability Measurement Approach Led by the Hydropower Sustainability Assessment Forum*. Chiang Mai: Mekong Program on Water, Environment and Resilience.

Garcia de Leaniz, C. (2008). Weir removal in salmonid streams: Implications, challenges and practicalities. *Hydrobiologia*, 609, 83–96.

Government of Brazil (2008). *Executive Summary: National Plan on Climate Change*, English version. Brasília: Interministerial Committee on Climate Change, Government of Brazil. Available at: www.mma.gov.br/estruturas/imprensa/_arqui vos/96_11122008040728.pdf.

Government of China (2007). *China's National Climate Change Program*, English version. Beijing: National Reform and Development Commission, Government of China. Available at: https://en.ndrc.gov.cn/newsrelease_8232/200706/ P020191101481828642711.pdf.

Hallegatte, S. (2009). Strategies to adapt to an uncertain climate change. *Global Environmental Change*, 19, 240–247.

Hamududu, B. and Killingtveit, A. (2012). Assessing climate change impacts on global hydropower. *Energies*, 5, 305–322.

Harvey, L. D. D. (2006). The exchanges between Fearnside and Rosa concerning the greenhouse gas emissions from hydro-electric power dams. *Climatic Change*, 75, 87–90.

Howard, C. D. D. (2000). Operations, monitoring and decommissioning of dams. Contributing paper prepared as input to the World Commission on Dams Thematic Review IV.5: Operation, Monitoring and Decommissioning of Dams. Cape Town, South Africa: World Commission on Dams.

ICEM (International Centre for Environmental management) (2010). *MRC Strategic Environmental Assessment of Hydropower on the Mekong Mainstream*. Final report. Hanoi: International Centre for Environmental Management. Available at: www .mrcmekong.org/assets/Publications/Consultations/SEA-Hydropower/SEA-FR-sum mary-13oct.pdf.

IEA (International Energy Agency) (2012). *Technology Roadmap: Hydropower*. Technology report. Paris: International Energy Agency. Available at: www.iea.org/ reports/technology-roadmap-hydropower.

IHA (International Hydropower Association) (2010a). *Hydropower Sustainability Assessment Protocol*. London: International Hydropower Association.

IHA (2010b). *Hydropower and the Clean Development Mechanism*. Policy statement. London: International Hydropower Association.

IPCC (Intergovernmental Panel on Climate Change) (2007). *Climate Change 2007: Impacts, Adaptation and Vulnerability. Contribution of Working Group II to the Fourth Assessment Report of the Intergovernmental Panel on Climate Change.* Edited by M. L. Parry, O. F. Canziani, J. P. Palutikof, P. J. van der Linden and C. E. Hanson. Cambridge: Cambridge University Press.

Krchnak, K., Richter, B. and Thomas, G. (2009). *Integrating Environmental Flows into Hydropower Dam Planning, Design, and Operations*. Washington, DC: World Bank Group.

Larnier, M. and Marmulla, G. (2004). Fish passes: Types, principles and geographical distribution – an overview. In R. Welcomme and T. Petr, eds., *Proceedings of the Second International Symposium on the Management of Large Rivers for Fisheries, Vol. II*. Bangkok: Food and Agriculture Organization of the United Nations. Available at: www.fao.org/3/ad526e/ad526e0g.htm#bm16.

Lukasiewicz, A., Finlayson, C. M. and Pittock, J. (2013). *Identifying Low Risk Climate Change Adaptation in Catchment Management While Avoiding Unintended Consequences*. Synthesis and integrative research final report. Gold Coast, Australia: National Climate Change Adaptation Research Facility. Available at: https://nccarf.edu.au/identifying-low-risk-climate-change-adaptation-catchment-man agement-while-avoiding/.

MEA (Millennium Ecosystem Assessment) (2005). *Ecosystems and Human Well-being: Wetlands and Water*. Synthesis. Washington, DC: World Resources Institute. Available at: www.millenniumassessment.org/documents/document.358.aspx.pdf.

Milly, P. C. D., Betancourt, J., Falkenmark, M. et al. (2008). Stationarity is dead: Whither water management? *Science*, 319, 573–574.

Null, S. E., Ligare, S. T. and Viers, J. H. (2013). A method to consider whether dams mitigate climate change effects on stream temperatures. *Journal of the American Water Resources Association*, 49, 1456–1472.

Olden, J. D. and Naiman, R. J. (2009). Incorporating thermal regimes into environmental flows assessments: Modifying dam operations to restore freshwater ecosystem integrity. *Freshwater Biology*, 55, 86–107.

Opperman, J. J., Apse, C., Banks, J., Day L. R. and Royte, J. (2011). The Penobscot River (Maine, USA): A basin-scale approach to balancing power generation and ecosystem restoration. *Ecology and Society*, 16, 7. Available at: www.ecologyandsociety.org/vol16/iss3/art7/.

Opperman, J. J., Galloway, G. E., Fargione, J., Mount, J. F., Richter, B. D. and Secchi, S. (2009). Sustainable floodplains through large-scale reconnection to rivers. *Science*, 326, 1487–1488.

Opperman, J. J., Hartmann, J. and Harrison, D. (2015). Hydropower within the climate, energy and water nexus. In J. Pittock, K. Hussey and S. Dovers., eds., *Climate, Energy and Water: Managing Trade-offs, Seizing Opportunities*. Cambridge: Cambridge University Press.

Orr, S., Pittock, J., Chapagain, A. and Dumaresq, D. (2012). Dams on the Mekong River: Lost fish protein and the implications for land and water resources. *Global Environmental Change*, 22, 925–932.

Pacca, S. (2007). Impacts from decommissioning of hydroelectric dams: A life cycle perspective. *Climatic Change*, 84, 281–294.

Palmer, M. A., Liermann, R., Nilsson, C. et al. (2008). Climate change and the world's river basins: Anticipating management options. *Frontiers in Ecology and the Environment*, 6, 81–89.

Pittock, J. (2010). Better management of hydropower in an era of climate change. *Water Alternatives*, 3, 444–452.

Pittock, J. (2019). Pumped-storage hydropower: Trading off environmental values? *Australian Environment Review*, 33, 195–200.

Pittock, J. and Finlayson, C. M. (2011). Australia's Murray-Darling Basin: Freshwater ecosystem conservation options in an era of climate change. *Marine and Freshwater Research*, 62, 232–243.

Pittock, J. and Hartmann, J. (2011). Taking a second look: Climate change, periodic re-licensing and better management of old dams. *Marine and Freshwater Research*, 62, 312–320.

PMCCC (Prime Minister's Council on Climate Change) (2008). *National Action Plan on Climate Change*. New Delhi: Government of India.

Poff, N. L., Olden, J. D., Merritt, D. M. and Pepin, D. M. (2007). Homogenization of regional river dynamics by dams and global biodiversity implications. *Proceedings of the National Academy of Sciences*, 104, 5732–5737.

Postel, S. and Richter, B. (2003). *Rivers for Life: Managing Water for People and Nature*. Washington, DC: Island Press.

Ramsar (Ramsar Convention) (2008). *The Ramsar Strategic Plan 2009–2015*. Gland: Ramsar Convention Secretariat. Available at: www.ramsar.org/document/the-ramsar-strategic-plan-2009-2015.

Richter, B. D., Postel, S., Revenga, C. et al. (2010). Lost in development's shadow: The downstream human consequences of dams. *Water Alternatives*, 3, 14–42.

Rosa, L. P., dos Santos, M. A., Matvienko, B., dos Santos, E. O. and Sikar, E. (2006). Scientific errors in the Fearnside comments on greenhouse gas emissions (GHG) from hydroelectric dams and response to his political claiming. *Climatic Change*, 75, 91–102.

Russo, T. N. (2000). US Federal Energy Regulatory Commission. Contributing paper prepared as input to the World Commission on Dams Thematic Review IV.5: Operation, Monitoring and Decommissioning of Dams. Cape Town, South Africa: World Commission on Dams.

US DoE (Department of Energy) (2015). *Marine and Hydrokinetic Energy Projects: Fiscal Years 2008–2014*. US Department of Energy Wind and Water Power Technologies Office Funding in the United States. Washington, DC: US Department of Energy. Available at: www.energy.gov/sites/prod/files/2015/04/f22/MHK-Project-Report-4-14-15.pdf.

Viers, J. H. (2011). Hydropower relicensing and climate change. *Journal of the American Water Resources Association*, 47, 655–661.

WCD (World Commission on Dams) (2000). *Dams and Development: A New Framework for Decision-Making*. The report of the World Commission on Dams. London: Earthscan. Available at: www.internationalrivers.org/resources/dams-and-development-a-new-framework-for-decision-making-3939.

Weisser, D. (2007). A guide to life-cycle greenhouse gas (GHG) emissions from electric supply technologies. *Energy*, 32, 1543–1559.

World Bank (2004). *Water Resources Sector Strategy: Strategic Directions for World Bank Engagement*. Washington, DC: World Bank. Available at: http://documents.worldbank.org/curated/en/941051468765560268/Water-resources-sector-strategy-strategic-directions-for-World-Bank-engagement.

WWF (World Wildlife Fund) (2004). *Repowering Hydroelectric Utility Plants as an Environmentally Sustainable Alternative to Increasing Energy Supply in Brazil.* Brasília: WWF Brazil. Available at: http://assets.panda.org/downloads/brazilupgradinghydropowerreport.pdf.

WWF (2006). *Free-Flowing Rivers: Economic Luxury or Ecological Necessity?* Gland: WWF International. Available at: https://wwf.panda.org/?63020/Free-flowing-rivers-Economic-luxury-or-ecological-necessity.

WWF (2007). *Climate Solutions: WWF's Vision for 2050.* Gland: WWF International. Available at: https://wwf.panda.org/?122201/Climate-Solutions-WWFs-Vision-for-2050.

Ziv, G., Baran, E., Nam, S., Rodríguez-Iturbe, I. and Levin, S. A. (2012). Trading-off fish biodiversity, food security, and hydropower in the Mekong River basin. *Proceedings of the National Academy of Sciences*, 109, 5609–5614.

7

Energy Storage

LACHLAN BLACKHALL, EVAN FRANKLIN, BJORN STURMBERG,
ALEXEY M. GLUSHENKOV AND HEDDA RANSAN-COOPER

Executive Summary

Energy storage can take many forms but this chapter addresses energy storage mechanisms that are capable of absorbing and storing energy via a reversible process, prior to then converting the majority of that stored energy into electrical energy.

The role and importance of energy storage is changing with the introduction of renewable energy generation such as wind and solar photovoltaics, whose output is inherently variable. This increasing generation variability has created a need for energy storage to provide energy balancing, something that will become increasingly vital as the percentage of renewable generation in power systems globally increases over the decades ahead.

This chapter discusses the different requirements for energy balancing within renewable-based power systems over various timescales, including long- (weeks to seasonal), medium- (hours to days) and short-term (seconds to minutes) ones. The requirements for balancing services across these differing timescales will be met by different forms of energy storage, highlighting the need for a portfolio of energy storage mechanisms and technologies.

Energy storage also provides other benefits for operating modern power systems. For example, energy storage can provide network services for maintaining networks within voltage and thermal limits, underpin resilience to contingency events occurring within power systems (for example when a generator or transmission line unexpectedly fails) and provide black start capabilities when generation needs to be restarted from rest.

This chapter discusses a multitude of energy storage mechanisms that include pumped storage hydro (PSH) systems and various forms of battery storage, as well as other forms of energy storage with varying levels of technical and commercial maturity.

An important issue explored within this chapter is the integration of energy storage into electricity grids. It is noted that there is significant experience to support the integration of PSH systems, although work remains to be done to determine the best ways to integrate battery storage, which can offer a range of benefits.

This chapter also reviews social research related to energy storage uptake, noting that equity and sustainability issues are important in the transition to new energy systems.

Finally, the chapter presents an outlook for energy storage where it is noted that the uptake of energy storage globally is on track to meet the International Energy Agency's

Sustainable Development Scenario. The outlook also emphasises the importance of better understanding how to robustly calculate the levelised cost of storage in order to allow comparisons with levelised cost of energy generation metrics.

7.1 Introduction

Energy storage is widely regarded as one of the key elements, or perhaps *the* key element, required to enable successful operation of secure, reliable and resilient electrical power systems in a future with increasingly high and very high levels of variable renewable and distributed energy generation. Highly controllable energy storage technologies are capable of providing various power system balancing and management functions. Such storage technologies can provide balancing between available and required power on timescales ranging from milliseconds to days. These have hitherto been largely provided by high-capacity-factor synchronous generators with their inherent operating characteristics. Without these balancing and management functions, electrical power systems are generally unable to maintain a reliable supply of energy when and where it is needed, while also maintaining dynamic stability or system security. A suite of energy storage technologies can be deployed to meet these functional power system needs.

In the context of electrical power systems, we define an energy storage system as one that is capable of absorbing and storing energy over some period of time via a reversible process, prior to then converting the majority of that stored energy into electrical energy that can, in a controllable manner, be injected into an electrical power system. We consider energy storage systems those that convert either electrical energy or some other form of input energy into an energy storage medium. However, we exclude energy storage systems that are incapable of converting stored energy back into electrical energy. We make this definition based on the utility of the storage technology in providing services necessary for the operation of electrical power systems. Thus, we do, for example, consider thermal storage supplied directly by a heat source that is used primarily to generate electrical energy, but we exclude from our consideration any thermal storage supplied directly via consumption of electrical energy but which is subsequently only used to service non-electrical loads. We recognise the potential benefits of these latter technologies (for example, hot water systems and building space heating/cooling) in helping to balance and manage electricity networks and the power system. Nonetheless, we argue that these are best considered alongside other flexible and controllable loads which also provide demand response capabilities. For similar reasons, we also exclude from our consideration various other forms of energy storage that may be encountered in everyday life (storage of energy-rich liquid fuels or gas for example) but which are not, generally speaking, readily available via a bidirectional energy storage system.

In this chapter, we first outline the current state of energy storage deployed in power systems today, highlighting key global energy storage projects. We go on to describe the various balancing and management functions that are required by the power system, before describing individual energy storage technologies and capabilities in greater detail. Subsequently, we provide perspectives on the integration of energy storage into the grid,

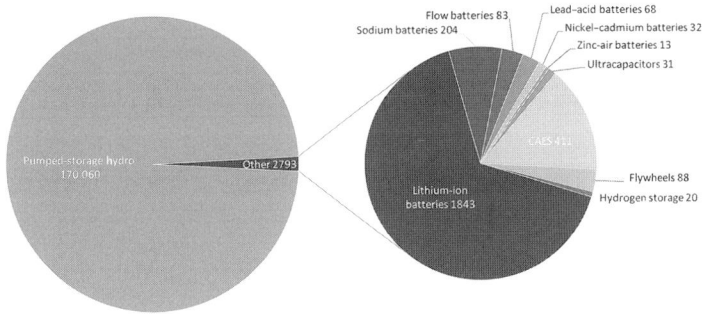

Figure 7.1 Overwhelmingly, energy storage installed in global electricity grids is provided by PSH. Total global installed grid-scale energy storage capacity (power capacity, mega-watts (MW)) by technology type.[1]

Source: US DOE (n.d.). For a colour version of this figure, please see the colour plate section.

including the social implications of integrating energy storage. We conclude this chapter with a brief summary of the future outlook for energy storage.

In our assessment, and based on the status and availability of storage technologies today, we expect a mix of distributed and central battery storage, pumped hydro and possibly some other currently less-developed storage technologies, to contribute significantly to the balancing and management of electrical power systems in the long term as we make the transition to systems based largely on variable renewable generation. In particular:

- Pumped hydro technology will be used for the provision of inertia, primary frequency response and secondary spinning reserve, medium-term energy balancing, voltage stability and black start capabilities.
- Battery storage will provide very fast dynamic primary frequency response, secondary response (or spinning reserve) services, short-term and medium-term energy balancing, as well as local demand and generator smoothing, network voltage management, as well as facilitating grid forming for islanded or microgrid operations.

Overwhelmingly the most significant form of energy storage installed in global power systems today is pumped storage hydro (PSH). After the first decade of the twenty-first century, PSH represented over 99% of all installed energy storage globally (IEA 2014), a dominance which continues today. The prominence of PSH in electricity systems to date has arisen due to its long history of development (Barbour et al. 2016), and because of its ability to manage diurnal load variations in power systems comprising generators with limited flexibility (Rogner and Troja 2018). In the decade to 2020, installed energy storage capacity has continued to grow; PSH has still dominated new capacity additions, but is accompanied by a growing share of alternative and emerging energy storage technologies (see Figure 7.1).

At the start of 2020, the global power capacity of grid-scale reversible electricity storage is estimated to be 173 GW (gigawatts). A definitive source of data on global energy storage

---

[1] Flywheel storage capacity does not include grid-connected installations that store energy via flywheels for the dedicated purpose of generating high-energy pulses in fusion research facilities.

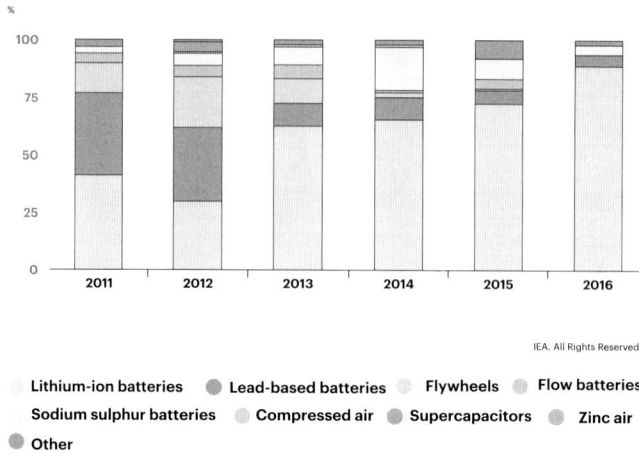

- Lithium-ion batteries ● Lead-based batteries ○ Flywheels ● Flow batteries
- Sodium sulphur batteries ● Compressed air ● Supercapacitors ● Zinc air
- ● Other

Figure 7.2 For non-PSH storage, battery storage will dominate the growth of energy storage capacity over the years ahead.
Source: IEA (2019b). For a colour version of this figure, please see the colour plate section.

capacity does not readily exist, particularly given the proliferation of small-scale systems. Our figures are based on the US Department of Energy Global Energy Storage Database (US DOE n.d.), which is widely accepted as a reliable source of up-to-date information on grid-scale operational storage projects, but which does not reflect behind-the-meter storage and likely also underrepresents small battery storage systems. This total storage capacity is equivalent to approximately 2.5% of the world's total electricity generation capacity (IRENA 2019). The share of grid-scale reversible storage capacity held by PSH has declined to around 98%, or to less than approximately 95% of global energy storage capacity if behind-the-meter storage and solar thermal storage are also included.

The decline in PSH's dominance is set to continue, with the rise of non-PSH storage technologies, most notably lithium-ion batteries. The reported 2018 annual deployment of non-PSH storage, for example, nearly doubled from 2017, with over half of this newly installed capacity being located behind the meter (IEA 2019b). The vast majority of behind-the-meter storage installed to date, which is generally accompanied by on-site solar photovoltaics, has also been based on lithium-ion batteries (Zinaman et al. 2020). This trend towards lithium-ion technology being the dominant or almost sole non-PSH storage technology being deployed, illustrated in Figure 7.2, is being driven primarily by the declining costs of lithium battery technology, owing to the continuing scale-up in manufacturing of batteries for electric vehicles (EVs) (IEA 2019b).

While the global install base for battery storage is growing, it is important to realise that battery storage, particularly lithium battery storage, is starting from a relatively small install base. By way of example, in 2017 the Hornsdale Power Reserve (HPR) was installed in Australia (Neoen 2017) and at the time of commissioning was the world's largest lithium battery, with a capacity of 100 MW (megawatts)/129 MWh (megawatt-hours). This single installation represented approximately 10% of all lithium battery storage installed globally by 2017 (Robson and Bonomi 2018; US DOE n.d.).

Battery storage in the grid will also increase through global uptake of EVs. In 2018, another 1.98 million EVs were sold, increasing the global EV footprint to 5.12 million (IEA 2019a). While much of this energy storage is not accessible for power system participation or support, this is likely to change from 2019 as vehicle-to-grid (V2G) capable EVs start being manufactured globally.

7.2 Key Global Energy Storage Projects

Given the increasing importance of energy storage in supporting energy security, reliability and resilience, it is instructive to understand some of the key energy storage projects already operating and under development globally. In recent years, the addition of new storage capacity has included major innovative projects using both PSH and battery storage technologies.

7.2.1 Australia

Perhaps the most prominent example of a utility-scale battery system was the 2017 installation of the Hornsdale Power Reserve (HPR) – or, colloquially, the 'Tesla Big Battery' – in South Australia. At the time of installation, this was the world's largest lithium battery installation (Neoen 2017).

The HPR provides a total of 129 MWh of energy storage, and is rated for 100 MW discharge and 80 MW charge (Australian Energy Market Operator 2018a). The HPR development has been notable for several reasons, but perhaps most importantly it has demonstrated the opportunities for battery storage to participate in markets for energy and ancillary services, contributing directly to supporting energy reliability and energy security in the Australian National Electricity Market (NEM) in:

- energy arbitrage;
- reserve energy capacity;
- network loading control ancillary services; and
- frequency control ancillary services.

Crucially, the participation of the HPR in the NEM has been profitable for its operators (Parkinson 2019), demonstrating the commercial opportunities for deploying battery storage and supporting the case for the wider deployment of these battery storage capabilities around Australia, and in similar jurisdictions globally.

Alongside the HPR, Australia is notable for its uptake and demonstration of virtual power plant capabilities through several projects and initiatives. Virtual power plants are widely considered to be an important capability for harnessing the uptake of residential battery storage systems for participation in markets for energy, ancillary services and network services (Australian Energy Market Operator 2018b).

In addition to battery storage projects, several major PSH developments are also well under way in Australia. Most notably, the pumped hydro Snowy Scheme expansion

('Snowy 2.0') will see 2 GW of new generation and pumping capacity added to the existing generation capacity, designed primarily to store energy generated from the solar- and wind-resource-rich regions in south-eastern Australia and then supply this energy to major load centres on the east coast as required (Snowy Hydro 2019). A closely related project is under development by Hydro Tasmania, this time designed to exploit the wind-resource-rich coastal regions of Tasmania and to utilise the existing and potential high-voltage DC interconnects to the major south-eastern Australian load centres. This project, or group of projects, is commonly being referred to as the 'Battery of the Nation Project', and includes proposals for projects amounting to multi-gigawatts of new capacity (Hydro Tasmania 2018).

7.2.2 USA

While the HPR is colloquially referred to as the 'Tesla Big Battery', it will shortly no longer be the biggest Tesla or Lithium battery in the world, after the announcement of several projects with the Pacific Gas and Electric utility in California (California Public Utilities Commission 2018). These projects, which will replace existing gas generation power plants, will add over 2 GWh of energy storage across four individual installations. Installations under way in California continue to demonstrate the leading role that this US state has taken in supporting the uptake of renewable generation and energy storage.

In California, the California Public Utilities Commission (CPUC) is also a global leader in driving the development and uptake of standards for the connection and integration of distributed and residential energy and battery storage capabilities through their Rule 21 activities (California Public Utilities Commission n.d.). Through driving the adoption of international standards and guiding implementation (Sunspec Alliance 2016), it is hoped that manufacturers of energy storage, as well as their customers, will be better placed to take advantage of the benefits of the interconnection and integration of energy and battery storage systems in global electricity systems.

7.2.3 Asia

Korea continues to be a leading light in the deployment of grid-scale and behind-the-meter installations, contributing to one-third of installed global storage capacity in 2018 (IEA 2019b). This leading role is largely attributed to favourable policy measures and is no doubt assisted by the leading global role of Korean industry in driving the development and manufacturing of battery storage capabilities.

China is also contributing to the uptake of energy storage in Asia, providing an important global demonstration of alternatives to lithium-based batteries, with a significant deployment of vanadium flow batteries in Dalian, China (Weaver 2017). Deployment of vanadium storage technology is an important demonstration of this technology's maturation since it was first developed at the University of New South Wales, Australia, in 1985. China is also home to the recently commissioned Liyang Pumped Storage Power Station

(US DOE n.d.). The power station has a generation capacity of 1500 MW, provided by six 250-MW reversible Francis pump-turbines.

In recent years, China has added significant new PSH capacity into their power system, in an effort to add the required flexibility to the country's power system. China is currently responsible for more than half of all PSH capacity additions globally, with this trend expected to continue for some years to come (IEA 2019d). Recent PSH projects include the Shenzhen Pumped Storage Power Station, a 1200-MW facility perhaps most notable for being constructed only 20 km from the centre of a major metropolitan city (Guangdong Hydropower Planning and Design Institute 2013).

7.2.4 Europe

From the European perspective, Germany is one of the global leaders in the uptake and integration of residential battery systems into power systems. Uptake is being driven by incentives, tariffs and technology availability (Colthorpe 2018). However, Europe is also interesting due to the diversity of energy storage capabilities that have been, and are being, deployed. In addition to having one of only two compressed air energy storage systems (CAESs) (Crotogino et al. 2001) globally, Europe is also home to the Andasol solar power station – a concentrated solar power station with capacity of 150 MW (NREL n.d.).

Europe has arguably also led the way in the deployment of advanced technology pumped hydro plants with the recently completed Linth–Limmern hydro storage plant, consisting of variable-speed pump/turbine sets with a total capacity of 1450 MW (Keller 2016), along with a number of other new facilities being developed elsewhere in Europe.

The variability in uptake of storage technologies in different countries provides insights into the importance of taking a socio-techno-economic perspective when analysing storage uptake and deployment globally.

7.3 The Role of Energy Storage Systems in Renewables-Based Power Systems

Electricity systems and markets have been designed to solve the fundamental problem of ensuring that the supply and demand of energy is in equilibrium. To achieve this outcome, the electricity system must ensure that supply meets demand at all times throughout the geographic area that is covered by the electricity system.

To satisfy this requirement, power systems generally operate on a generating unit dispatch basis. A power system operator will forecast electricity demand over a forward horizon of dispatch intervals for each major region of the system and, based on the available generating units, determine the optimal dispatch required to meet that demand at least cost (which in many cases is determined by the related operation of an energy market). Such a system works well provided that there are, at each dispatch interval, a sufficient number of dispatchable generating units available. Conventional generators, such as thermal, gas and hydro, may inherently be operated to provide at least some degree of schedulability, and are thus well suited to dispatch in this manner. In power systems dominated by conventional generators there is generally no need, at power systems level at least, for the participation of energy storage assets.

Owing to an improved understanding of the impacts of climate change and the required response to reduce fossil fuel use, communities and governments globally are increasingly looking to reduce the carbon intensity of electricity production. This desire, coupled with the improving economics of renewable generation (Graham et al. 2018), sees the long-term trend for energy generation as being towards large amounts of renewable generation. These renewable energy generation sources are covered in detail in other chapters in this book.

The characteristics of renewable generation units are quite different to the fossil-fuel-fired generating units described above. Chief among these differences is the fact that they are not dispatchable or schedulable, instead providing an output which is highly dependent upon the raw power resource (the sun or the wind), which is itself variable. While solar and wind resource forecasting is improving considerably (Blaga et al. 2018; Sobri et al. 2018; Liu et al. 2019), there will always be some inherent variability of output across the timescales of power system operation.

Hence, with increasing levels of variable renewable generation in power systems, there will be an increasing need to manage increased generation variability. Energy storage that is capable of being operated across the timescales of power system operation will therefore ensure both energy reliability and security.

In order to frame the various energy storage technologies in terms of their roles in providing energy system balancing, it is important to consider the energy versus power characteristics of particular storage capabilities, as well as considering the various time-scales on which energy storage is needed in the power system.

7.3.1 Storage Capacity: Energy Versus Power

When considering energy storage capabilities, it is always necessary to carefully consider two key dimensions: energy and power. Energy capacity is a measure of the total amount of energy (MWh) that can be stored or can be delivered in one complete cycle by an energy storage system. Power capacity is the maximum rate of energy delivery (MW) to/from an energy storage system. Energy storage is often also referred to in terms of number of 'hours' of storage. This is simply the energy capacity divided by the power capacity, and hence 'hours' refers to the period of time that the system is able to operate at nominal or maximum power. Power capacity and energy capacity requirements are usually determined by the particular application for which a system is needed, but may also be limited by design constraints.

7.3.2 Long-Term (Weeks to Seasonal) Energy Balancing

Long-term storage may be required in a power system, or a region of a power system, when either electricity demand or generation (supply) has significant dependence upon seasonal climate patterns or strong weather variations within the year, and where the other modes of generation are unable to compensate for such variations.

Power systems have generally been designed to handle this long-term and seasonal variability in demand by ensuring an over-capacity in generation that is able to respond on long timescales and is backed by appropriate energy storage. This type of long-term energy storage is typically achieved by maintaining large water storage reservoirs in conventional hydropower facilities (as is the case in countries such as Norway and regions such as Tasmania, Australia) and/or by stockpiling fuel (or otherwise guaranteeing fuel supply) used by thermal power generation facilities.

However, as power systems become increasingly dominated by solar and wind generation, and hence contain relatively less conventional generation, variations in supply will occur seasonally or on a timescale of weeks, owing primarily to seasonal variations in solar and wind resources. In such cases, new long-term storage capacity will be required. The quantity of long-term storage required in such cases will depend upon factors such as technology and geographical diversity of the generation fleet, so a probabilistic approach will generally be needed. Storage designed to cover long-term energy balancing needs will require a very large energy capacity to power capacity ratio.

7.3.3 Medium-Term (Hours to Days) Energy Balancing

Perhaps the most immediate and obvious need for energy storage in the power system is to balance energy supply and demand on a medium-term timescale of hours to days. Electricity demand exhibits a quite strong diurnal pattern, with typically low demand overnight, higher demand during daytime hours and, for many regions, a particularly high demand in the morning and evening shoulder periods (see Figure 7.3). Furthermore, generation from solar is inherently diurnal in nature, while wind generation exhibits similar diurnal variation in many locations (Mulder 2014).

The net result of such diurnal variability in solar and wind generation can be that rapid and considerable variations in power generation are required from remaining generators in the system. In a system with a high proportion of solar and wind generation capacity, the net demand variations and associated ramping requirements can exceed the technical capabilities of other generating plants. In such cases, it becomes critical to incorporate storage that is capable of reducing net demand variations and ramping requirements over these medium-term timescales.

The level of medium-term storage capacity required in a system is largely dictated by the scarcity or surplus of supply of energy from the generation portfolio across the day. A large proportion of new energy storage being deployed presently is essentially targeting this need, and is being installed in power systems with high levels of variable renewable generation. Market drivers are generally supporting these deployments: diurnal wholesale energy price variations incentivising utility-scale energy storage projects, and retail tariff arrangements (e.g. time-of-use pricing and import/export tariff differentials) incentivising behind-the-meter small-scale storage deployment.

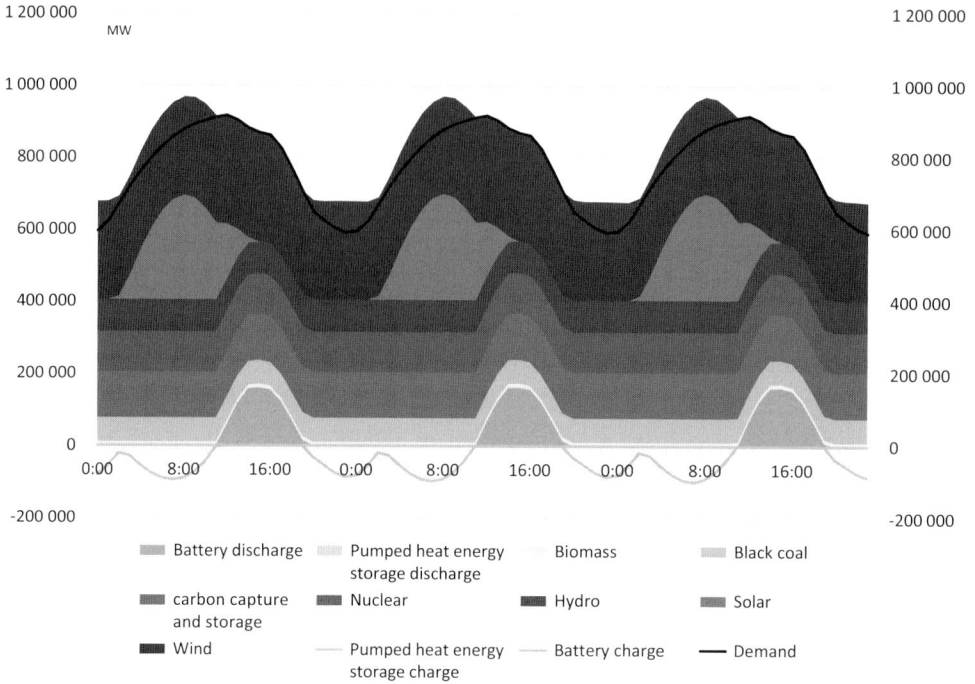

Figure 7.3 An example energy demand profile and generation mix from the North American region during spring, showing the diurnal energy storage cycle that provides medium-term energy balancing.

Source: Graham et al. (2018). For a colour version of this figure, please see the colour plate section.

7.3.4 Short-Term (Seconds to Minutes) Energy Balancing

The balance between supply and demand in a power system must be carefully maintained on all timescales, but it is arguably the short (seconds to minutes) timescale that can be the most challenging to manage, particularly as conventional generating units exit the system and are replaced by variable renewable generators. As with other timescales, demand fluctuation constantly occurs on short timescales. However, the most significant imbalance between supply and demand is when a so-called contingency event occurs, when either a large load or large generating unit is unexpectedly disconnected from the power system. After such an event, the supply and demand balance must be restored within a matter of seconds, to avoid partial or full system blackout. With insufficient time to bring new generating units online, existing generators (at least some of them) must immediately increase or decrease their power output until balance is restored.

In power systems with decreasing levels of generation sourced from synchronous generators and increasing levels of renewable generation, there is both less system inertia and also fewer generators capable of providing conventional primary frequency response. New energy storage technologies will play a critical role in providing the necessary system security functionality in future power systems by operating on this short timescale. Battery storage in particular has already demonstrated the ability of energy storage to provide near-to-instantaneous response to contingency events in the Australian grid (Australian Energy Market Operator 2018a).

7.3.5 Network Services Provision

From the previous discussion, it is clear that energy storage is able to provide energy balancing needs at the whole-of-power-system level. However, with distributed energy and battery storage becoming increasingly prevalent in distribution networks, energy storage will also be available for use by network operators to manage constraints (power flow and voltage limits) in distribution networks. Through technical regulation or economic incentives, network operators will be able to use energy storage to reduce peak reverse power flow and voltage rise on networks during periods of high coincident solar generation, and limit peak demand and voltage drop on networks when there is high demand.

7.4 Pumped Storage Hydro

As detailed previously, PSH is currently the dominant form of worldwide energy storage. Pumped storage hydro relies upon established technology and is able to provide a broad range of support services for the electricity grid. Conceptually, it is very simple to understand: water is pumped up a height difference to an upper reservoir when there is excess energy generating capacity available (i.e. when energy costs are low) and is subsequently released to a lower reservoir, via turbine and generator to produce power, when demand (and hence cost) is high. Typically, in the order of 80% of the electrical energy required to provide pumping is returned to the electrical power system during generation, although the power loss is highly dependent upon the specific generator/pump configuration and the hydraulic design. Either the upper or lower reservoir may form part of a conventional hydropower scheme, or alternatively, they may both be purpose-built for a dedicated pumped hydro system. The basic physics of hydro storage dictates that energy storage capacity increases linearly with height difference (head) between the top surface of the upper reservoir and the turbine/pump, and thus sites with a large height difference are generally favoured and are cheaper on a stored energy capacity basis. A schematic of a PSH system is shown in Figure 7.4, and Figure 7.5 shows the 1.77-GW La Muela plant in Spain'.

7.4.1 A Changing Role for PSH

Pumped storage hydro systems were first built in the early twentieth century, with the majority of schemes built in the 1960s and 1970s (Rogner and Troja 2018). The main driver for establishing these plants was to accommodate diurnal demand variations when the supply side was dominated by large, inflexible thermal generating units designed for operation at a relatively constant output. In particular, the units are unable to operate below a lower limit and hence there emerged a need to create a 'baseload'. A large number of such

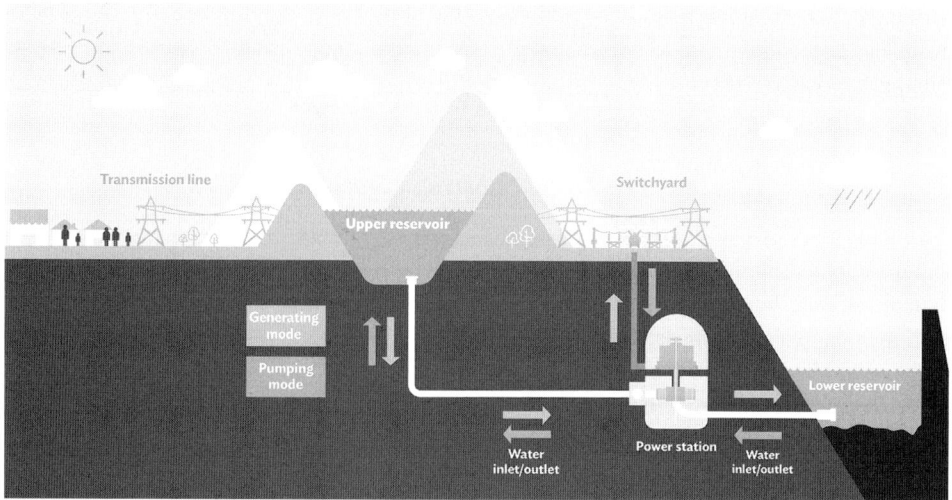

Figure 7.4 Pumped storage hydro schematic.
Source: Hydro Tasmania. For a colour version of this figure, please see the colour plate section.

Figure 7.5 The 1.77-GW La Muela plant constructed in Spain in 2013.
Source: Patel (2013); reproduced with permission of Iberdrola.

plants were built to coincide with the development of large nuclear power stations. Pumping generally occurred overnight when demand was low, with generation occurring during peak demand periods during the day.

The recent growth in variable renewable generation has led to a resurgence of interest in PSH. However, the key drivers for deployment have changed significantly since the majority of PSH was deployed over the previous several decades. Newly deployed PSH plants are built to store energy at various times of the day when supply from variable renewable generators is plentiful, and discharge at periods where demand is high and/or renewable generation output is low. The use case for PSH is changing significantly, with the need for more frequent cycling, highly variable power consumption or production during pumping and generating modes, and energy balancing on a wider range of time-scales. Importantly, despite PSH being well established and demonstrated over many decades, the specific technologies deployed and the way they need to be operated is changing markedly to suit these evolving needs.

7.4.2 PSH System Components

Pumped hydro plants can be constructed to be configured in a number of different ways, the details of which significantly alter the flexibility they offer the power system and variety of other system services they can provide. Pumped storage systems consist of four basic components:

- upper reservoir;
- lower reservoir;
- water conveyance system; and
- powerhouse, including turbine/pump and electric machine set.

The design of the upper reservoir and/or lower reservoir, and the height difference between them, dictates the energy storage capacity of pumped hydro. The size and capacity of the water conveyance system (tunnels and penstocks) and of the pump/turbine and machine dictate the power capacity.

For PSH plants built specifically for bidirectional energy flow, one or both reservoirs will typically be a relatively small 'turkey's nest'-type dam, with water depth in the order of 10–20 m. An example of such a reservoir is at Taum Sauk pumped hydro facility (Rogers et al. n.d.), shown in Figure 7.6. The second reservoir may be similarly constructed, or alternatively may be a much larger reservoir created by natural topology and conventional dam construction. Pumped storage hydro systems may be closed-loop, with no inflows at either reservoir, or open-loop, having natural inflows at one reservoir. Although new conventional hydro developments may not be possible now in many locations in the world, numerous sites for off-river PSH have been identified around the world (Blakers et al. n.d.). In many cases, existing conventional hydropower facilities can be retrofitted with pumped storage capabilities, by building either a new upper or lower reservoir (dependent upon existing topology). However, an entirely new powerhouse is usually required in either scenario, since pump/turbine sets in pumped hydro facilities are required to be situated at a sufficient depth below the intake/outlet of both reservoirs, which is not typically the case for conventional hydro plants. The powerhouse is normally situated below ground for this reason, with its specific contents (combination of pump/turbine and electric motor/generator) determining the type of operation and energy balancing functionality it can provide.

Figure 7.6 Construction during 2009 of the upper reservoir at Taum Sauk pumped storage facility, Missouri, USA.
Source: KTrimble (2009).

7.4.3 Pump/Turbine Configurations and Implications

The functionality and flexibility that a PSH facility can provide depends upon the hydraulic design and the configuration of the pump/turbine and electric machines. Pumped storage hydro generating units can be categorised into four types:

- reversible pump/turbine with fixed-speed electric machine;
- reversible pump/turbine with variable-speed electric machine;
- separate pump and turbine, each with fixed- (or variable-) speed electric machine;
- combined pump and turbine (ternary set) with single fixed-speed electric machine.

Reversible Francis-type pump/turbines are able to be operated in generating mode at a fixed speed, and produce a wide range of power outputs. This is the normal mode of operation of a conventional hydropower generating unit, utilising a synchronous machine for power generation. In this mode of operation, a fixed-speed pumped hydro unit provides inertia to the power system and can (within the limitations of the hydraulic design, distance between reservoirs and penstock length) rapidly vary its output (in the order of 1% per second). Thus, it can provide short-term energy balancing functionality, including primary frequency response. In pumping mode, the plant similarly provides system inertia but is essentially restricted, owing to pump efficiency limitations, to operation at near to full power. Hence, this type of unit would not have the ability to ramp up and down for energy balancing purposes. To change operation from generating to pumping mode, this type of unit must first be desynchronised from the power system, brought to a halt, have its direction reversed and then be brought up to speed before synchronising with the power system again. A similar process is required to change the other way: from pumping to turbining mode. This changeover may take in the order of 2–6 minutes (Kruger 2018), depending upon specific system design, with start-up from standstill occupying a similar duration. This reversible, fixed-speed configuration is the cheapest option available (West et al. 2018) and is quite well suited to medium-term energy balancing, but does not provide all of the flexibility needed to manage under a highly fluctuating supply/demand balance on short timescales. Indeed, at times when there is high instantaneous penetration of wind and solar generation in a power system, when pumping operation is most likely, this configuration does not inherently provide the desired flexibility of operation unless operated concurrently with other energy balancing systems. The overwhelming majority of existing pumped storage plants are based on this reversible, fixed-speed configuration.

One way to overcome the major limitation of fixed-speed pumping systems just outlined is to replace fixed-speed synchronous machines with asynchronous machines coupled to the power system via a power electronics converter. A reversible pump/turbine with variable-speed machine is able to provide full flexibility in both generating and pumping modes. The required variable-speed functionality can be achieved by employing either a synchronous machine with full-power electronics converter interface or a doubly-fed induction machine with partial power electronics convertor. This solution is estimated to increase total project cost by 10% compared to using fixed-speed machines (West et al. 2018). Despite this additional cost, it is almost certainly preferable in situations where a high degree of short-term energy balancing is required. Despite their advantages, variable-speed pumped hydro units provide no inertia to the power system, whether in generating or pumping mode, thus exposing the system to larger and faster system frequency deviations in the absence of other mitigating technologies. Changeover time (from generator to pumping, or vice versa) is of similar order in this case to that for fixed-speed machines. Many of the recently constructed pumped storage plants, deployed in systems with high variable renewable generation content, are based on this variable-speed technology.

The limitations, in pumping mode, of using a single reversible turbine with synchronous/fixed-speed electric machine can be overcome by having more operating machines, thereby creating a system with a separate pump and turbine. Flexibility in pumping mode, the ability to quickly ramp power consumption up or down, is achieved by fixed power pumping from one set and simultaneous variable generation from another, as illustrated in Figure 7.7. With appropriate sizing of pumping and generating units, changeover from generation to consumption or vice versa can be quick and seamless since there is no reversal of machine direction required. In addition, net generation is controlled by varying the generating unit output in the same way as is done for conventional hydro generating units. This mode of operation can be quite easily achieved in a large pumped hydro facility containing multiple units. However, losses will increase significantly for this mode of operation, and the total available power capacity (generation or load) of the facility will also be reduced significantly during periods where such flexibility is being provided.

The final configuration being considered by pumped storage proponents, though not yet commonly deployed, combines the turbine and pump into one integrated set that is driven by a single electric machine. This is known as a ternary set, and is usually accompanied by a hydraulic short circuit, as shown in Figure 7.8. The pump and turbine are both coupled to

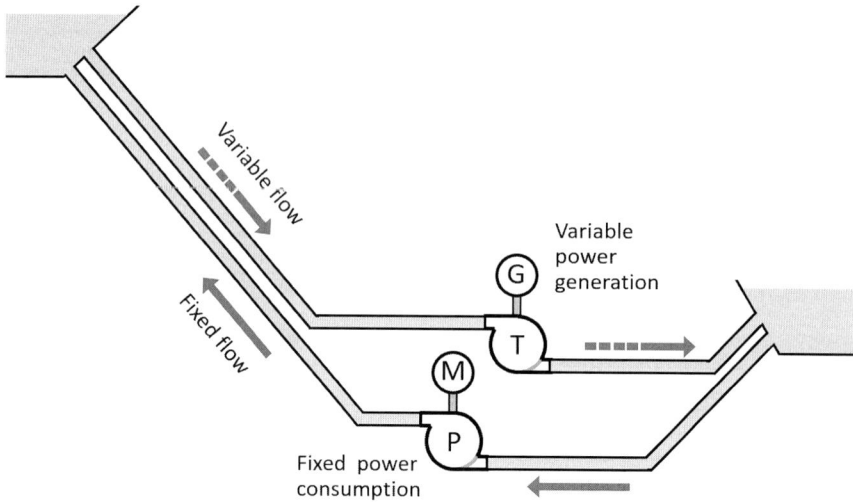

Figure 7.7 Flexibility during pumping mode can be created in PSH plants using synchronous machines, by running some units in pumping mode and others in flexible, generating mode.

Source: Courtesy of Evan Franklin.

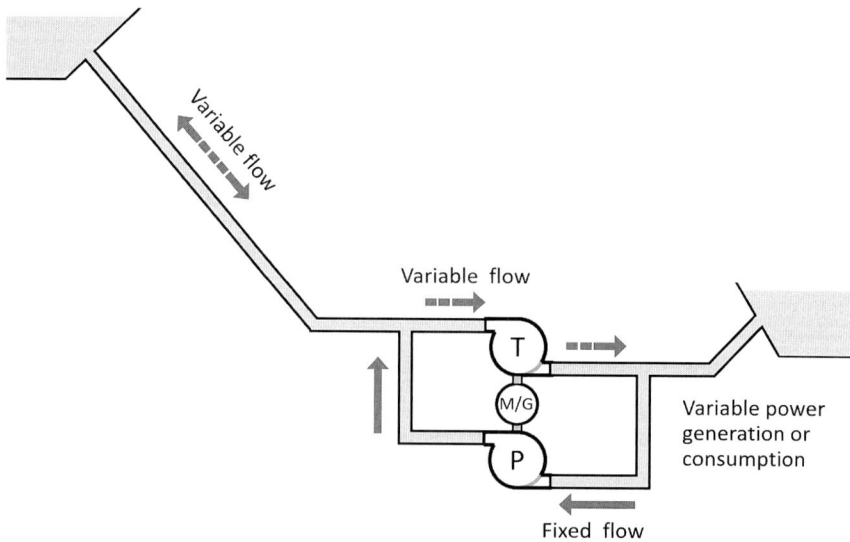

Figure 7.8 Ternary sets, utilising a single synchronous machine with integrated pump and turbine and a hydraulic short circuit, afford maximum flexibility and seamless changeover from pumping to turbining mode.

Source: Courtesy of Evan Franklin. For a colour version of this figure, please see the colour plate section.

Table 7.1. *Key features of different PSH plant pump/turbine configurations*

Pump/turbine	Electric machine	Key features
Reversible pump/turbine	Fixed-speed (synchronous)	Lowest cost Provides system inertia No flexibility in pumping mode Long changeover time Suited for medium-term balancing
Reversible pump/turbine	Variable-speed and power electronics	Increased total cost Provides no system inertia Full flexibility in pumping mode Long changeover time Suited for short-term and medium-term balancing
Separate pump and turbine	Two fixed-speed machines (synchronous)	Provides system inertia Flexibility in pumping mode Short changeover time Significantly higher losses Suited to existing facilities, but total plant capacity reduction Suited for short-term and medium-term balancing
Ternary set (combined pump/turbine)	Single fixed-speed machine (synchronous)	Moderate cost increase Provides system inertia Flexibility in pumping mode Some higher losses Short changeover time Suited for short-term and medium-term balancing

the (usually synchronous speed) machine, and either can be run individually or both can be run simultaneously. As is the case for completely separate sets, the turbine can be run as needed to provide flexibility during pumping mode. Fast changeover between modes of operation is facilitated by this arrangement. A ternary set allows for flexible operation over the almost full power range in both pumping and generating modes. Employing ternary sets is estimated to add 15% to total project costs, compared to fixed-speed reversible pump/turbine design (West et al. 2018).

A summary of these key pumped storage pump/turbine-machine set configurations and characteristics are provided in Table 7.1.

7.5 Battery Storage

Battery storage is based upon the storage of energy in chemical bonds that can be released via a chemical reaction and directly converted to electrical energy in a reversible process.

The process is enabled by electrochemical 'cells', which facilitate the reversible electro-chemical reactions to take place at their electrodes.

7.5.1 Battery Technologies and Chemistries

There are a variety of battery technologies that include conventional enclosed aqueous battery systems (such as lead–acid (Pb–acid) and nickel–cadmium (Ni–Cd) batteries), high-temperature batteries (sodium–sulphur (Na–S) and the so-called ZEBRA batteries), flow batteries (with vanadium and zinc–bromine (Zn–Br) chemistries as the most representative examples) and lithium-ion (Li-ion) batteries with two dominant positive electrode chemistries.

Traditionally, Pb–acid batteries have had broad industrial uses for standby applications in which secure power is critically important, such as data centres, national security and telecommunications (May et al. 2018). Grid-related energy storage is perhaps a natural extension of these uses, and lead-based batteries can be deployed due to their reliability, long cycle life (under optimised charge and discharge limits) and applicability to utility and smaller-scale domestic and commercial energy storage. An individual Pb–acid cell consists of a Pb negative plate (grid) and a PbO_2 positive plate (grid) with an aqueous sulphuric acid electrolyte, which allows for an attractive voltage of 2.05 V for such a cell. There are also now newer carbon-enhanced Pb batteries (containing a capacitor-type carbon component in their negative electrode in addition to Pb). While lead-containing materials in these batteries are toxic, efficient recycling processes are available in some countries and the recycling rate is close to 100%.

Nickel–cadmium batteries are relatively expensive with respect to Pb–acid batteries, but are attractive due to their energy density (40–60 Wh/kg versus 35–40 Wh/kg for Pb–acid batteries) (Breeze 2018). Another clear advantage is their relative robustness under electrical and mechanical abuse conditions. A major drawback of Ni–Cd batteries is the toxicity of their components, with cadmium being a particularly toxic heavy metal. Another concern is the memory effect, requiring periodic full discharge of these batteries. The transition from cadmium in this type of batteries is possible. In terms of energy density, a superior performance (up to 40% higher) is achieved in a related type of nickel–metal hydride (Ni–MH) batteries but their cost is at a substantial premium compared to that of Ni–Cd batteries.

High-temperature batteries are alternatives for grid applications (Gür 2018). Sodium–sulphur batteries incorporate molten sodium (in the negative electrode), sulphur (in the positive electrode) and a solid ceramic electrolyte, and operate at typical temperatures of 300–360 °C, which requires independent heaters to complement these battery systems. Most of the installed high-temperature Na–S batteries are located in Japan and the USA. These batteries have attractive energy density (>200 Wh/kg) and long lifetime of 15–20 years. A Na–NiCl$_2$ (ZEBRA) battery is another variation of a high-temperature battery operating at 270–350 °C. Molten sodium acts as the negative electrode in this cell, and $NiCl_2$ (Ni in the discharged state) as the positive electrode, with molten sodium tetrachloroaluminate ($NaAlCl_4$) used as an electrolyte (Dustmann 2004).

Another class of technologies suitable for grid integration applications is flow batteries (Weber et al. 2011; Ke et al. 2018). Flow batteries are able to possess very large capacities due to active materials stored in external tanks, utilised by pumping these liquids via independent anode and cathode loops through a chamber containing two terminals and a separator. In vanadium redox batteries, utilising arguably the most prominent flow battery chemistry on the market today, aqueous vanadium salts are pumped through, and the oxidation state of vanadium species changes from V^{5+} to V^{4+} on the positive carbon electrode, and from V^{2+} to V^{3+} on the negative carbon electrode in discharge. The capacity of this system is regulated by the size of the external tanks, and infinite capacities can be, theoretically, achievable. Other notable flow battery chemistries include the zinc–bromine (Zn–Br_2) system, which offers higher energy density but which also comes with practical challenges and is currently being pursued by only a small number of proponents. Environmental release of vanadium and bromine in both of the above types of batteries presents potential risks.

7.5.2 Lithium-Ion Batteries: Current Industry Status, Resource Requirements, Battery Systems in Deployment

Among the various types of batteries, Li-ion batteries are becoming dominant in grid and grid-related applications. These batteries utilise organic (non-aqueous) electrolytes, which allows an individual cell to possess a high nominal voltage of 3.2–3.85 V (depending on the electrode configuration) and a high energy density (100–265 Wh/kg). A Li-ion battery cell operates via the so-called 'rocking chair' mechanism in which lithium ions sequentially insert and de-insert into/from the positive and negative electrodes; each of these processes is accompanied by a redox reaction in a corresponding electrode. In particular, during the battery discharge, lithium leaves the negative electrode and inserts from the electrolyte into the positive electrode, causing electrons to flow in the external circuit.

Most of the current grid and home installations are dominated by two Li-ion battery configurations, and commercial batteries are available from Tesla, Enphase Energy and numerous other market participants. While graphite is typically used as the negative electrode material, two dominant chemistries of positive electrodes are employed in energy storage for grid applications. The first type of Li-ion batteries uses lithium iron phosphate ($LiFePO_4$) as the positive electrode material. While this type of battery has a lower energy density, and a larger number of cells in the module are needed to reach the required capacity and voltage, the high reliability and safety makes this type of Li-ion battery an attractive candidate for home and office energy storage. A common alternative Li-ion battery format involves lithium nickel manganese cobalt oxide (NMC) as the positive electrode material. These batteries have a higher voltage and capacity, leading to their inherent higher energy density.

As discussed, Li-ion batteries are emerging as a dominant battery type for energy storage in grids with high penetration of renewables. However, the potential limitations of this battery technology are the availability of the required raw materials. The geographic distribution of lithium in particular is uneven (predominantly located in Chile, Bolivia, Australia and China), and its total content in Earth's crust is low. Lithium-ion batteries with NMC cathodes also have additional drawbacks due to the scarcity of cobalt and nickel. Potentially, resource limitations may pose long-term challenges for the use of Li-ion batteries at a large scale. In this context, there is a growing need for recycling of spent Li-ion batteries (Li et al. 2018; Lv et al. 2018). The techniques for recovering materials from used batteries are not well developed, requiring substantial improvements before matching older battery technologies such as Pb–acid batteries, where most of the materials can be successfully recovered.

Driven by the raw material limitations of Li-ion batteries, there has been considerable attention in recent years devoted to the development of room-temperature Na-ion batteries. Instead of using lithium ions in their operation, sodium ions act as ionic shuttles in the same 'rocking chair' mechanism. Due to the similar chemical properties of sodium and its natural abundance (global reserves exceeding those of lithium by about 1000 times), Na-ion batteries may represent a sustainable alternative to Li-ion batteries (Hwang et al. 2017). Commercial prototypes are currently being tested in trial applications (such as e-bicycles) but the expectation for this emerging technology (should it be proven operational) is for it to be deployed in large stationary energy storage such as grid-related applications.

7.5.3 Interface with the Power System

Battery storage systems, regardless of size and technology, are inherently DC power sources. These sources are connected to the AC power system via a power electronics interface that converts the DC power source into an AC power source. The power electronics and associated control software determines the direction and amount of power flow at each instant in time, ensuring operation within the limits of the battery hardware itself.

These power electronics capabilities also underpin the provision of both real and reactive AC power and are capable of ramping power up and down or reversing power flow, with a typical response time of milliseconds.

7.5.4 Electric Vehicles as Energy Storage

Battery electric vehicles (BEVs), also known as all-electric vehicles or EVs, use battery technology that is very similar to stationary battery storage. Some vehicle manufacturers, such as Tesla, even use the same battery cells in both vehicles and stationary storage (Brown 2017). Similarly to stationary battery storage, BEVs can connect to the grid via AC coupling, where the vehicle contains an AC to DC inverter, or DC coupling where the inverter is located outside of the vehicle in the charging infrastructure. Therefore, BEVs can provide all the same benefits to the energy system as stationary storage, with the crucial differentiation that BEVs are a large net load, with energy eventually expended on mechanical propulsion.

The coincidence in technological specifications, particularly around Li-ion battery chemistries, has provided efficiencies of scale to both stationary and mobile storage. While BEVs have historically been the main driving force behind increasing production volumes and decreasing costs, this is beginning to change as both markets mature, with utility-scale stationary storage projects in particular doubling their capacity in 2018 to 33 GW (Hering 2019). The divergence in battery technologies is occurring due to both desire for higher power density in BEVs and greater cycling capacity in stationary storage, as well as the high prices of raw materials such as cobalt (Maloney 2018).

Battery electric vehicles and typical residential battery storage have different energy storage capacities. While distributed energy resource (DER) storage systems typically have energy capacities of up to 5–15 kWh (kilowatt-hours), BEV energy capacities begin at 40–62 kWh for a Nissan LEAF (Kane 2019) and are 60–100 kWh in Tesla models S and X (Lambert 2019a). There is also likely to be considerable deployment of battery electric freight vehicles, whose battery capacities range from 40–300 kWh for urban distribution trucks such as those manufactured by StreetScooter (StreetScooter n.d.) and Volvo (Lambert 2019b), to an advertised 800 kWh for the Tesla Semi (Lambert 2018).

The large net load of BEVs may become a significant challenge for distribution networks, which will host all but the largest BEV chargers, while motivating an increase in distributed generation and storage. These stresses will be exacerbated by the deployment of DC fast-chargers with power ratings of between 45 kW to 350 kW (Dow 2018; EV SafeCharge 2019).

At the same time, the advent of newly available BEVs, such as the Nissan LEAF, that support V2G functionalities (Kane 2018) creates opportunities for BEVs to enhance grid resilience and energy reliability and security. Managing the integration of BEVs into the power system, and unlocking the benefits of V2G BEVs, will depend critically on innovations in planning, control and coordination. While these innovations may build on learnings from stationary storage, there are unique elements to BEVs, such as their non-deterministic availability: for example, when they are being used as a means of transport and thus not connected to the grid. Such non-deterministic availability is an important consideration for the delivery of network and grid services where the provision or absorption of power is highly time sensitive.

7.6 Other Energy Storage Technologies

7.6.1 Thermal Storage

Thermal storage media represents one of the largest energy storage capacities globally. In the built environment, energy is stored primarily in hot water systems designed for direct use or space heating, but is also stored in the fabric of buildings for and via space heating systems. With a typical household hot water storage system, for example, containing in the order of 20 kWh of energy, one billion homes represents around 20 000 GWh of energy storage capacity. This is greater than twice the total estimated energy storage capacity from all PSH and battery systems globally. Much of this stored energy is sourced via the electrical power system, meaning that thermal energy storage could play an important role in managing electricity systems with large amounts of variable renewable generation. However, we do not consider this in any detail in this chapter since it cannot be considered as a reversible or bidirectional storage technology. Instead, thermal storage of this type is best considered elsewhere as part of the narrative on flexible loads, demand response and energy efficiency.

In this section we limit our discussion to those thermal storage technologies which provide generation flexibility or injection of power into the energy system.

7.6.1.1 Concentrating Solar Power with Thermal Storage

Concentrating solar thermal power (CSP or CST) plants with thermal storage involve the use of a large array of mirrors to concentrate sunlight onto a 'receiver' where the energy is collected by heating a fluid. The fluid can be stored and then used later, when needed, to generate steam and run a turbine/generator to produce electricity. Concentrating solar power's particular benefit is that its configuration allows energy storage as an easily and cost-effectively integrated part of the system. Systems with as much as 15 hours of storage capacity have been installed (e.g. Gemasolar, Spain, and Crescent Dunes, USA), achieving a commercial supply of 24-hour solar energy (Fitzpatrick 2013). Storage is achieved, in most practical implementations, via molten salts.

Energy storage in CSP plants is not currently bidirectional in nature, being designed to store energy from the solar energy source and subsequently inject that energy into the electricity system as needed. However, there is no clear technological impediment to augmenting plants to convert electrical energy into heat for storage in the same medium, thus becoming bidirectional storage systems.

Concentrating solar thermal power plants primarily use synchronous generators to interface directly with the electrical power system. They operate a thermal cycle in much the same way that a modern, conventional thermal power plant does. Consequently, a CSP plant's output operating characteristics (minimum output, start time and ramp rates) are also quite similar and so are able to provide energy balancing across short-term and medium-term timescales. The use of a synchronous generator means that CSP plants automatically provide power system inertia, and the primary frequency response can be an integral part of operation; furthermore, the integrated thermal storage facilitates secondary 'spinning reserve' functionality in the same manner as conventional thermal power plants.

A more complete treatment of CSP technology is provided in Chapter 4.

7.6.1.2 Direct, Reversible Thermal-Electric Storage Technologies

There is an emerging family of thermal-electric storage technologies which have been proposed, but have not been implemented at any significant scale. These technologies allow direct bidirectional storage of energy in high-temperature thermal storage mediums. One proposed technology at development stage uses molten salts, similar to the technology now prevalent in CSP projects, but converts electrical energy to the molten salt storage and subsequently generates electrical energy via a closed-cycle Brayton engine (Laughlin 2017). Other emerging direct thermal storage technologies in the early stages of development include technologies based on molten silicon as the energy storage medium (1414 Degrees n.d.; Chu 2019).

7.6.2 Compressed Air Energy Storage

Compressed air energy storage (CAES) involves storing energy as pressurised air, either through compression of air into a fixed volume (isochoric CAES) or by inflating a volume by injecting air at a (relatively) constant pressure (isobaric CAES). This stored potential energy (more precisely the exergy) is converted into electricity by expansion of the pressurised air through a turbine to drive a generator.

Compressed air energy storage systems have similar energy and power properties to PSH systems, which typically makes them best suited to load shifting energy across hours or days and discharging over a few hours (Garvey 2018). Similarly to hydro turbines, they provide the power system with black start backup (Cavanagh et al. 2015) and inertia (which, depending on the turbine, may exceed the inertia constant of hydro turbines) (Banks et al. 2017). Like pumped hydro, CAES systems have a longer life but lower cycle efficiencies than batteries (CAES having lower efficiencies than pumped hydro) (Energy Storage Association n.d.a), and have to date required bespoke engineering for each installation.

There are three thermodynamic approaches that can be employed in driving CAES: following diabatic, adiabatic and isothermal pressure–volume (P–V) curves in compression and expansion. The closer the P–V curve is to an isotherm, the more efficient the process, the larger the energy storage density and the shorter the start-up time (Budt et al. 2016).

Diabatic CAES (DCAES) is the least efficient process because the heat generated through compression is ejected as waste. The system therefore requires an external heat source (typically fossil-fuel-powered) to inject heat prior to expansion at the time of power generation.

Adiabatic CAES (ACAES) systems capture this heat, store it and reintroduce it at the expansion step. The heat can be extracted in one – high-temperature (>500 °C) – step, or in multiple – lower-temperature, 200–400 °C –steps. Each additional step adds complexity but improves start-up time from cold conditions, due to reduced thermal stress, from around 15 minutes to 5 minutes, and enables more off-the-shelf components to be used.

Isothermal CAES (ICAES) systems are yet to be commercially demonstrated. In this concept, compression and expansion occur at a slower rate to minimise temperature increases that deviate the P–V curve away from the isotherm and produce efficiency losses. These concepts have focused primarily on piston machinery.

To date, most CAES developments have been isochoric DCAES and ACAES systems. Two large isochoric DCAES plants have been operating for over 20 years in Huntorf, Germany, and McIntosh, USA. These both use solution-mined salt caverns as large impermeable storage spaces and have power ratings of 290 MW (Crotogino et al. 2001) and 110 MW (Seltzer 2017) respectively. Recently there has been renewed interest in CAES, with companies such as Hydrostor innovating with isobaric ACAES designs underwater and in water-filled mines (ARENA 2019).

7.6.3 Flywheel Storage

Flywheels convert electrical energy into the kinetic energy of a rotating mass, typically made of steel or fibre composite (Chen et al. 2009). Other key components are vacuum housing and bearings, either mechanical or magnetic, to suspend the rotor mass and reduce friction; a motor-generator that both provides the power to accelerate the mass and decelerates it to convert energy back into electricity; and power electronics connected with the motor-generator. The inherently non-hazardous nature of these components is one of the advantages of flywheel storage.

The energy and power capacities of flywheels can be tuned independently. The size and speed of the rotor determines the energy capacity, and the power rating of the motor-generator determines the power capacity (Hebner et al. 2002). As a mechanical system, the frictional losses in flywheels scale inversely with the flywheel size, so flywheels typically have power and energy capacities exceeding a kilowatt. The main applications of flywheel storage have been frequency regulation, voltage support and smoothing of power profiles from wind farms, all of which have suited high power ratios (Ding and Zhi 2016). Large systems include a 20-MW facility in the USA (Beacon Power n.d.) and a 2-MW facility in Canada (Spears 2014).

Advantages of flywheels as a storage technology include the negligible degradation of cycling, leading to long lives estimated at over 100 000 cycles (Energy Storage Association n.d.b), fast response times of under a second (Amiryar and Pullen 2017), and cycling efficiencies of over 90% (Pena-Alzola et al. 2011). In contrast to batteries, the power output of flywheels degrades less across their discharge cycle. The major disadvantage of flywheels is their high self-discharge rate, which limits the duration for which they can store energy to less than 15 minutes. Recent developments claim to address this shortcoming and extend storage times to more than 4 hours (Amber Kinetics n.d.).

7.7 Integrating Energy Storage into the Grid

While ensuring that we have sufficient uptake of energy storage is of critical importance, there is also considerable work to be done to understand how best to integrate all of these storage technologies into global electricity systems. Due to its strong global uptake and ability to provide inertia, global market and system operators already understand how to integrate and operate PSH (Rogner and Troja 2018). However, further work is needed to enable emerging energy storage capabilities, particularly those with a power-electronics-based interface to the power system, to be integrated and operated effectively.

From a technical perspective, there are important questions about how power electronics should behave. While energy storage systems with a power electronics interface provide no natural inertia, they can be designed to behave like synchronous machines, delivering virtual inertia (Fang et al. 2018) to the grid. In microgrids, energy storage systems with a power electronics interface can already provide grid-forming capabilities (Singh et al. 2015).

In this context, there are important questions about whether power-electronics-interfaced storage should provide virtual inertia, provide grid-forming capabilities or operate in new ways to provide a new class of synchronisation services. Better understanding these challenges and opportunities for providing stability services is the subject of a significant European project, MIGRATE (European Commission 2016).

In addition to technical interface questions, it is also important to better understand the implications for operating the grid with a significant amount of distributed and distribution-connected energy storage. Within the transmission network, it is possible to decouple the management of voltage and frequency due to the low resistance compared to reactance of the conducting equipment in the transmission network. Consequently, in the transmission network, real power is typically used to maintain frequency and reactive power is used to manage voltage.

In the distribution network, resistance and reactance can be of similar order, so both real and reactive power have an impact on system frequency and voltage. It will therefore be particularly important to understand the implications of this on the value, performance and operation of service delivery from DER assets into markets for energy, ancillary and network services.

In Australia (Energy Networks Australia 2018) and the UK (Energy Networks Association n.d.) issues of coordination are primarily focused on the need for a distributed/distribution system operator (DSO) or a distributed/distribution market operator (DMO), with considerable activity around the development and trialling of appropriate orchestration technology for consumer-owned storage devices (Scott et al. 2019). In addition to questions of coordination there are also important considerations about the behind-the-meter optimisation and control strategies being utilised to operate energy storage assets and their impact on energy networks (Ratnam et al. 2016) and the broader electricity system and markets. These considerations become particularly important given the high level of decentralisation that many global grids will have over the coming decades, as seen in Figure 7.9.

7.8 Social Implications for the Deployment of Energy Storage

There is a dearth of research on the social dimensions of the transition to a grid with significantly higher rates of storage. Implications for institutional and regulatory conditions, social equity and energy consumption behaviours and patterns, not to mention public acceptance, have yet to be fully explored (Devine-Wright et al. 2017). New actors such as aggregators and 'prosumers' are entering the system, with key questions about their associated roles and responsibilities and interactions with incumbents and regulatory bodies yet to be worked through. As with many technological transitions, research on the implications of these changes is playing catch-up.

Decentralization ratio

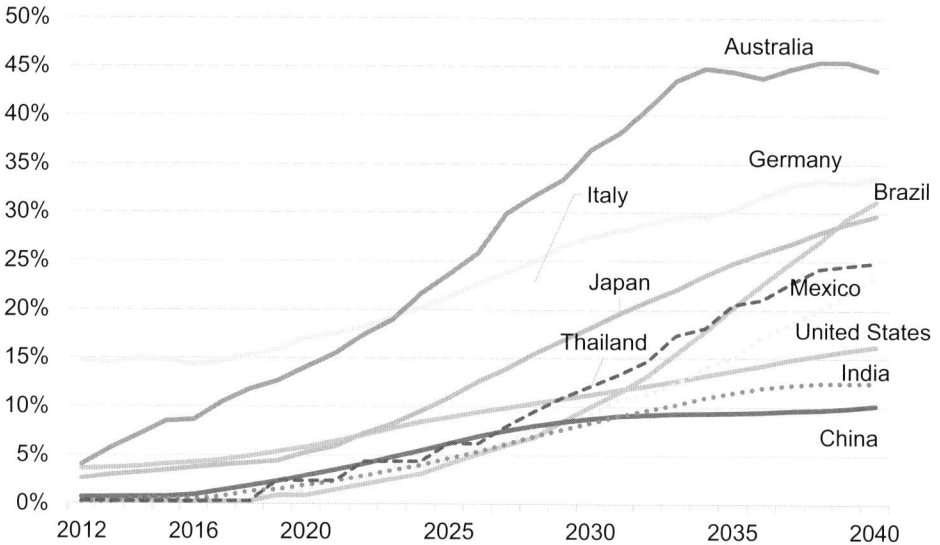

Figure 7.9 Many grids globally will demonstrate increasingly high levels of decentralisation over the decades ahead.

Note: Decentralisation ratio is the ratio of non-grid-scale capacity to total installed capacity. For a colour version of this figure, please see the colour plate section.

Source: Bloomberg NEF (2017).

More generally, while there are cogent arguments for integrating social sciences in energy research, social science concepts and methodologies remain underutilised in battery and storage research (Sovacool 2014). Researchers and practitioners interested in how storage will be integrated into a changing grid have an opportunity to build on several decades of energy social science to ensure that the transition can not only reduce costs, but also improve well-being and sustainability goals.

There is already a well-developed literature in the energy social sciences, across several subdisciplines in the social sciences (e.g. human geography, sociology, science and technology studies, behavioural economics, policy studies, etc.), that can contribute to emerging questions about storage grid integration. Topics with active research agendas include experiences and impacts on demand management (Hui and Walker 2018), energy efficiency (Lutzenhiser 2014), solar photovoltaic installation (Bulkeley et al. 2016), community energy (Hicks and Ison 2018) and smart meters (Lovell 2017), as well as studies that have examined how technology innovation can be scaled up (Kemp 1994). There is a significant research gap on experiences and perceptions of battery storage. Devine-Wright et al. (2017) have developed a research agenda for the social science of storage, which is informed by public acceptance research and research on governance and innovation. This research agenda provides a multiscale, multi-method series of questions to explore the various dimensions of acceptance across regulation, markets and innovation, and sociocultural acceptances. This is an important call to social researchers who have neglected battery storage to date, but must also be extended to researchers in the technical fields whose research can be broadened out to consider social dimensions alongside technical require-ments and considerations.

To date, there are only a handful of studies that examine perceptions of prospective storage integration into the grid (Romanach et al. 2013; Ginninderry Energy Research Team 2017; Jones et al. 2018), only one of which uses an in-depth method (focus groups) to capture detail and nuance (Ambrosio-Albalá et al. 2019). The latter study, conducted in the UK, found that perceptions of community-level battery storage were complex, influ-enced by specific local energy cultures including forms of energy consumption, costs, expectations of family members, previous experiences, perceptions of government and the municipal authority, and expectations about the technologies. All these factors were explored as likely to shape acceptance and adoption of battery storage at the household and community level.

There are even fewer studies that examine the experiences of households (or businesses) that own a battery, whether providing support to the network or not. We now turn to a brief summary of one of these studies – the CONSORT project on Bruny Island, Tasmania, Australia – in order to highlight the importance of considering the social dimensions of storage integration. CONSORT was a multidisciplinary research project and industry collaboration involving three universities, a network service provider and a technology start-up with a battery management system.

The technological innovation that underpinned the demonstration in CONSORT was a software platform that allowed the distribution network to trade energy with households, via control of the residential battery system, in near real time, so that energy from household batteries could be used to support the network at peak times, at overall least cost (for the network and for the households). Several dozen households were provided a subsidy to select a package of technology which included solar photovoltaics, a battery, an inverter and an energy management system. The social research activity covered focus groups, interviews, household observation and energy diaries with all of the participating households (for a summary of this research, see Watson et al. 2019).

The in-depth longitudinal research carried out through this study revealed a number of important insights for the roll-out of storage at a household scale. While the majority of households felt generally comfortable with the technology in their homes, all but one experienced some degree of confusion, frustration and stress associated with the selection, installation and ongoing use of the technology. While the study was small in scale, its findings challenged a common assumption among industry and government reports on DER: that householders are likely to be willing and unproblematic participants in DER sharing with distribution networks at early stages of technological development. Research carried out on new models should integrate the social dimension of technology roll-out so that learnings about technology deployment and user needs can be integrated early on.

7.9 Outlook for Energy Storage

In their 2017 report, IRENA provide a forecast breakdown, by storage type, of the growth of energy storage capability out to 2030:

Total electricity storage capacity in energy terms may grow from an estimated 4.67 TWh in 2017 to between 6.62 TWh and 7.82 TWh in the Remap Reference case in 2030, which is 42–68% higher than in 2017. In the Remap Doubling case, where the share of renewable energy in the global energy system is doubled from 2014 levels, electricity storage capacity could increase to between 11.89 TWh and 15.27 TWh in 2030, or 155–227% higher than in 2017. *(IRENA 2017: 103)*

Globally, the uptake of energy storage is on track to meet the International Energy Agency's Sustainable Development Scenario (IEA 2019c); however, there are clearly several factors that will underpin the forecast growth in energy storage broadly and battery storage specifically.

Importantly, it is clear that the price of energy storage will be key to its long-term uptake, particularly in the contribution of energy storage to the levelised cost of energy (LCOE) supply across regional or national electricity systems. Unlike traditional calculations for the LCOE, the levelised cost of storage is heavily dependent on its use case. Recent analyses taking this into account (Lazard 2018) do highlight significant cost declines across most use cases and energy storage technologies.

Given that storage is required in the power system on multiple timescales, and that the cost of energy storage capacity for different technologies can vary markedly (increasing design energy capacity for a PSH facility comes at relatively low marginal cost, compared with doing so for a battery system) it is apparent that more than one storage technology will be required. On current cost expectations, it appears likely that PSH technologies will secure a dominant position in providing storage at medium to long timescales, while battery technologies will gain traction for short to medium timescale applications.

For battery storage generally and lithium batteries in particular, future costs will depend heavily on global commodity prices and supply chains. The dependence of many battery storage technologies on rare earth metals has implications for price, based on international geopolitics as well as human rights and environmental concerns for global supply chains (Amnesty International 2017).

Ultimately, as we discussed in this chapter, energy storage is necessary to underpin the global adoption of renewable generation. As a consequence, future growth in the adoption and deployment of energy storage will also heavily depend on the global ambition for adopting renewable generation sources, the rate at which those renewable technologies are taken up and the subsequent demand for the firming capacity required to maintain secure and reliable power systems.

References

1414 Degrees (n.d.). How. *1414 Degrees*. Available at: https://1414degrees.com.au/how/.

Amber Kinetics (n.d.). Flywheel energy storage. Available at: www.amberkinetics.com.

Ambrosio-Albalá, P., Upham, P. and Bale, C. S. (2019). Purely ornamental? Public perceptions of distributed energy storage in the United Kingdom. *Energy Research and Social Science*, 48, 139–150.

Amiryar, M. E. and Pullen, K. R. (2017). A review of flywheel energy storage system technologies and their applications. *Applied Sciences*, 7, 286.

Amnesty International (2017). Industry giants fail to tackle child labour allegations in cobalt battery supply chains. *Amnesty International.* 15 November. Available at: www.amnesty.org/en/latest/news/2017/11/industry-giants-fail-to-tackle-child-labour-allegations-in-cobalt-battery-supply-chains/.

ARENA (Australian Renewable Energy Agency) (2019). South Australian zinc mine to be converted into Australia's first compressed air facility for renewable energy storage. *ARENA.* 8 February. Available at: https://arena.gov.au/news/south-australian-zinc-mine-to-be-converted-into-australias-first-compressed-air-facility-for-renewable-energy-storage/.

Australian Energy Market Operator (2018a). *Initial Operation of the Hornsdale Power Reserve Battery Energy Storage System.* Australian Energy Market Operator. Available at: www.aemo.com.au/-/media/Files/Media_Centre/2018/Initial-oper ation-of-the-Hornsdale-Power-Reserve.pdf.

Australian Energy Market Operator (2018b). *NEM Virtual Power Plant (VPP) Demonstrations Program.* Consultations paper. Australian Energy Market Operator. Available at: www.aemo.com.au/-/media/Files/Electricity/NEM/DER/2018/NEM-VPP-Demonstrations-program.pdf.

Banks, J., Bruce, A. and MacGill, I. (2017). Fast frequency response markets for high renewable energy penetrations in the future Australian NEM. Paper presented at the Asia-Pacific Solar Research Conference 2017, Melbourne, 5–7 December. Available at: www.ceem.unsw.edu.au/sites/default/files/documents/024_J-Banks_DI_Peer-reviewed.pdf.

Barbour, E., Wilson, I. G., Radcliffe, J., Ding, Y. and Li, Y. (2016). A review of pumped hydro energy storage development in significant international electricity markets. *Renewable and Sustainable Energy Reviews*, 61, 421–432.

Beacon Power (n.d.). Stephentown, New York. *Beacon Power.* Available at: https://beaconpower.com/stephentown-new-york/.

Blaga, R., Sabadus, A., Stefu, N., Dughir, C., Paulesu, M. and Badescu, V. (2018). A current perspective on the accuracy of incoming solar energy forecasting. *Progress in Energy and Combustion Science*, 70, 119–144.

Blakers, A., Stocks, M., Lu, B., Cheng, C. and Nadolny, A. (n.d.). Global Pumped Hydro Atlas. The Australian National University. Available at: http://re100.eng.anu.edu.au/global/index.php.

Bloomberg NEF (New Energy Finance) (2017). *New Energy Outlook 2017.* London: Bloomberg New Energy Finance.

Breeze, P. (2018). *Power System Energy Storage Technologies.* London: Academic Press.

Brown, A. (2017). Tesla's Gigafactory to build Model 3 motors in addition to batteries. *The Drive.* 18 January. Available at: www.thedrive.com/news/7008/teslas-gigafactory-to-build-model-3-motors-in-addition-to-batteries.

Budt, M., Wolf, D., Span, R. and Yan, J. (2016). A review on compressed air energy storage: Basic principles, past milestones and recent developments. *Applied Energy*, 170, 250–268.

Bulkeley, H., Powells, G. and Bell, S. (2016). Smart grids and the constitution of solar electricity conduct. *Environment and Planning A: Economy and Space*, 48, 7–23.

California Public Utilities Commission (n.d.). Rule 21 interconnection. *California Public Utilities Commission.* Available at: www.cpuc.ca.gov/Rule21/.

California Public Utilities Commission (2018). Resolution E-4949. Available at: http://docs.cpuc.ca.gov/PublishedDocs/Published/G000/M229/K550/229550723.PDF.

Cavanagh, K., Ward, J., Behrens, S., Bhatt, A., Ratnam, E., Oliver, E. and Hayward, J. (2015). *Electrical Energy Storage: Technology Overview and Applications.*

Canberra: Commonwealth Scientific and Industrial Research Organisation. Available at: www.aemc.gov.au/sites/default/files/content/7ff2f36d-f56d-4ee4-a27b-b53e01ee322c/CSIRO-Energy-Storage-Technology-Overview.pdf.

Chen, H., Cong, T. N., Yang, W., Tan, C., Li, Y. and Ding, Y. (2009). Progress in electrical energy storage system: A critical review. *Progress in Natural Science*, 19, 291–312.

Chu, J. (2019). Sun in a box. *MIT Technology* Review. 27 February. Available at: www .technologyreview.com/s/612798/sun-in-a-box/.

Colthorpe, A. (2018). Germany reaches 100k home battery storage installations. *Energy Storage News*. 28 August. Available at: www.energy-storage.news/news/germany-reaches-100k-home-battery-storage-installations.

Crotogino, F., Mohmeyer, K.-U. and Scharf, R. (2001). Huntorf CAES: More than 20 years of successful operation. Available at: www.fze.uni-saarland.de/AKE_Archiv/ AKE2003H/AKE2003H_Vortraege/AKE2003H03c_Crotogino_ea_HuntorfCAES_ CompressedAirEnergyStorage.pdf.

Devine-Wright, P., Batel, S., Aas, O., Sovacool, B., Carnegie Labelle, M. and Ruud, A. (2017). A conceptual framework for understanding the social acceptance of energy infrastructure: Insights from energy storage. *Energy Policy*, 107, 27–31.

Ding, K. and Zhi, J. (2016). Wind power peak–valley regulation and frequency control technology. In N. Wang, C. Kang and D. Ren, eds., *Large-Scale Wind Power Grid Integration*. Academic Press, pp. 211–232.

Dow, J. (2018). VW's Electrify America opens California's first 350kW ultra-fast charger, before cars can actually use it. *electrek*. 6 December. Available at: https://electrek.co/ 2018/12/06/electrify-america-first-350kw-charger-california/.

Dustmann, C. (2004). Advances in ZEBRA batteries. *Journal of Power Sources*, 127, 85–92.

Energy Networks Association (n.d.). Future worlds: Consultation. *Energy Networks Association*. Available at: www.energynetworks.org/electricity/futures/open-net works-project/future-worlds-consultation.html.

Energy Networks Australia (2018). A joint Energy Networks Australia and Australian Energy Market Operator (AEMO) project. *Energy Networks Australia*. Available at: www.energynetworks.com.au/joint-energy-networks-australia-and-australian-energy-market-operator-aemo-project.

Energy Storage Association (n.d.a). Mechanical energy storage: CAES. *Energy Storage Association*. Available at: http://energystorage.org/compressed-air-energy-storage-caes.

Energy Storage Association (n.d.b). Mechanical energy storage: Flywheels. *Energy Storage Association*. Available at: http://energystorage.org/energy-storage/technolo gies/flywheels.

European Commission (2016). Massive InteGRATion of power Electronic devices. *Cordis*. Available at: https://cordis.europa.eu/project/rcn/199590/factsheet/en.

EV SafeCharge (2019). DC fast charging explained. *EV SafeCharge*. Available at: https:// evsafecharge.com/dc-fast-charging-explained/.

Fang, J., Li, H., Tang, Y. and Blaabjerg, F. (2018). Distributed power system virtual inertia implemented by grid-connected power converters. *IEEE Transactions on Power Electronics*, 33, 8488–8499.

Fitzpatrick, E. (2013). Solar storage plant Gemasolar sets 36-day record for 24/7 output. *Renew Economy*. 8 October. Available at: https://reneweconomy.com.au/solar-stor age-plant-gemasolar-sets-36-day-record-247-output-12586/.

Garvey, S. (2018). Let's store solar and wind energy: By using compressed air. *The Conversation*. 24 October. Available at: https://theconversation.com/lets-store-solar-and-wind-energy-by-using-compressed-air-103183.

Ginninderry Energy Research Team (2017). *Householder Attitudes to Residential Renewable Energy Futures*. Canberra: Riverview Projects. Available at: https://ginninderry.com/wp-content/uploads/2016/09/Ginninderry-2017-Householder-Attitudes-to-Residential-Renewable-Energy-Futures.pdf.

Graham, P. W., Hayward, J., Foster, J., Story, O. and Havas, L. (2018). *GenCost 2018: Updated Projections of Electricity Generation Technology Costs*. Canberra: Commonwealth Scientific and Industrial Research Organisation. Available at: www.csiro.au/~/media/News-releases/2018/renewables-cheapest-new-power/GenCost2018.pdf.

Guangdong Hydropower Planning and Design Institute (2013). Shenzhen Pumped Storage Power Station. *GPDI*. 19 December. Available at: www.gpdiwe.com/en/webview/?artid=40815.

Gür, T. (2018). Review of electrical energy storage technologies, materials and systems: Challenges and prospects for large-scale grid storage. *Energy & Environmental Science*, 10, 2696–2767.

Hebner, R., Beno, J. and Walls, A. (2002). Flywheel batteries come around again. *IEEE Spectrum*, 39, 46–51.

Hering, G. (2019). Amid global battery boom, 2019 marks new era for energy storage. *S&P Global Market Intelligence*. 11 January. Available at: www.spglobal.com/marketintelligence/en/news-insights/trending/9GIYsd7qF8tNpiopwH7KSg2.

Hicks, J. and Ison, N. (2018). An exploration of the boundaries of 'community' in community renewable energy projects: Navigating between motivations and context. *Energy Policy*, 113, 523–534.

Hui, A. and Walker, G. (2018). Concepts and methodologies for a new relational geography of energy demand: Social practices, doing-places and settings. *Energy Research & Social Science*, 36, 21–29.

Hwang, J., Myung, S. and Sun, Y. (2017). Sodium-ion batteries: Present and future. *Chemical Society Reviews*, 12, 3529–3614.

Hydro Tasmania (2018). *Battery of the Nation: Analysis of the Future National Energy Market*. Hobart: Hydro Tasmania. Available at: www.hydro.com.au/docs/default-source/clean-energy/battery-of-the-nation/future-state-nem-analysis-full-report.pdf.

IEA (International Energy Agency) (2014). *Technology Roadmap: Energy Storage*. Technical report. Paris: International Energy Agency. Available at: www.iea.org/reports/technology-roadmap-energy-storage.

IEA (2019a). Electric vehicles. *Tracking Transport*. Paris: International Energy Agency. Available at: www.iea.org/topics/tracking-clean-energy-progress (these data accessed 2019).

IEA (2019b). Energy storage. *Tracking Energy Integration*. Paris: International Energy Agency. Available at: www.iea.org/topics/tracking-clean-energy-progress (these data accessed 2019).

IEA (2019c). *Sustainable Development Scenario*. Paris: International Energy Agency. Available at: www.iea.org/reports/world-energy-model/sustainable-development-scenario.

IEA (2019d). Will pumped storage hydropower expand more quickly than stationary battery storage? Analysis from Renewables 2018. *IEA.org*. 4 March. Available at: www.iea.org/articles/will-pumped-storage-hydropower-expand-more-quickly-than-stationary-battery-storage.

IRENA (International Renewable Energy Agency) (2017). *Electricity Storage and Renewables: Costs and Markets to 2030*. Abu Dhabi: International Renewable Energy Agency. Available at: www.irena.org/-/media/Files/IRENA/Agency/Publication/2017/Oct/IRENA_Electricity_Storage_Costs_2017.pdf.

IRENA (2019). Renewable energy now accounts for a third of global power capacity. *IRENA.org*. 2 April. Available at: www.irena.org/newsroom/pressreleases/2019/Apr/ Renewable-Energy-Now-Accounts-for-a-Third-of-Global-Power-Capacity.

Jones, C. R., Gaede, J., Ganowski, S. and Rowlands, I. H. (2018). Understanding lay-public perceptions of energy storage technologies: Results of a questionnaire conducted in the UK. *Energy Procedia*, 151, 135–143.

Kane, M. (2018). Nissan LEAF is Germany's first V2G-approved electric car. *Inside EVs*. 23 October. Available at: https://insideevs.com/news/340585/350ought-leaf-is-ger manys-first-v2g-approved-electric-car/.

Kane, M. (2019). Here is the Nissan LEAF e+ 62 kWh battery: Video. *Inside EVs*. 9 January. Available at: https://insideevs.com/news/342009/here-is-the-nissan-leaf-e-62-kwh-battery-video/.

Ke, X., Prahl, J., Alexander, J., Wainright, J., Zawodzinski, T. and Savinell, R. (2018). Rechargeable redox flow batteries: Flow fields, stacks and design considerations. *Chemical Society Reviews*, 23, 8721–8743.

Keller, T. (2016). Could this be the most extreme power plant in the world? *GE Reports: Hydropower*. 7 June. Available at: www.ge.com/reports/how-the-swiss-turned-an-alpine-peak-into-a-battery-the-size-of-a-nuclear-plant/.

Kemp, R. (1994). Technology and the transition to environmental sustainability: The problem of technological regime shifts. *Futures*, 26, 1023–1046.

Kruger, K. (2018). Li-ion battery versus pumped storage for bulk energy storage: A comparison of raw material, investment costs and CO_2 footprints. Paper presented at HydroVision Conference, Charlotte, USA, 27 June. Available at: www.voith.com/ corp-de/VH_Paper_Battery-versus-Pumped-Storage-_2018_HydroVision_en.pdf.

KTrimble (2009). Aerial photo of Taum Sauk reservoir under construction [image]. *Wikimedia Commons*. 22 November. Available at: https://commons.wikimedia.org/ wiki/File:TaumSaukReservoir_underconstruction.jpg.

Lambert, F. (2018). Tesla Semi production version will have closer to 600 miles of range, says Elon Musk. *electrek*. 2 May. Available at: https://electrek.co/2018/05/02/tesla-semi-production-version-range-increase-elon-musk/.

Lambert, F. (2019a). Tesla releases new Model S battery pack, makes massive price drop, kills base Model X pack. *electrek*. 1 March. Available at: https://electrek.co/2019/03/ 01/tesla-model-s-model-x-prices-options/.

Lambert, F. (2019b). Volvo delivers its first electric trucks. *electrek*. 21 February. Available at: https://electrek.co/2019/02/21/volvo-delivers-first-electric-trucks/.

Laughlin, R. (2017). Pumped thermal grid storage with heat exchange. *Renewable and Sustainable Energy*, 9. Available at: https://doi.org/10.1063/1.4994054.

Lazard (2018). *Levelized Cost of Energy Analysis: Version 4.0*. Available at: https://www .lazard.com/perspective/levelized-cost-of-energy-and-levelized-cost-of-storage-2018/.

Li, L., Zhang, X., Li, M., Chen, R., Wu, F., Amine, K. and Liu, J. (2018). The recycling of spent lithium-ion batteries: A review of current processes and technologies. *Electrochemical Energy Reviews*, 1, 461–482.

Liu, H., Chen, C., Lv, X., Wu, X. and Liuy, M. (2019). Deterministic wind energy forecasting: A review of intelligent predictors and auxiliary methods. *Energy Conversion and Management*, 195, 328–345.

Lovell, H. (2017). Are policy failures mobile? An investigation of the Advanced Metering Infrastructure Program in the State of Victoria, Australia. *Environment and Planning A: Economy and Space*, 49, 314–331.

Lutzenhiser, L. (2014). Through the energy efficiency looking glass. *Energy Research & Social Science*, 1, 141–151.

Lv, W., Wang, Z., Cao, H., Sun, O., Zhang, Y. and Sun, Z. (2018). A critical review and analysis on the recycling of spent lithium-ion batteries. *ACS Sustainable Chemistry & Engineering*, 6, 1504–1521.

Maloney, P. (2018). Electric vehicle and stationary storage batteries begin to diverge as performance priorities evolve. *Utility Dive*. 1 August. Available at: www.utilitydive.com/news/batteries-for-electric-vehicles-and-stationary-storage-are-showing-signs-of/528848/.

May, G., Davidson, A. and Monahov, B. (2018). Lead batteries for utility energy storage: A review. *Journal of Energy Storage*, 15, 145–157.

Mulder, F. (2014). Implications of diurnal and seasonal variations in renewable energy generation for large scale energy storage. *Journal of Renewable and Sustainable Energy*, 6. Available at: https://doi.org/10.1063/1.4874845.

Neoen (2017). *Hornsdale Power Reserve*. Available at: https://hornsdalepowerreserve.com.au/.

NREL (National Renewable Energy Laboratory) (n.d.). Concentrating solar power projects by project name. *Concentrating Solar Power Projects*. Available at: https://solarpaces.nrel.gov/projects.

Parkinson, G. (2019). Tesla Big Battery delivered a $22 million profit in 2018. *Renew Economy*. 14 May. Available at: https://reneweconomy.com.au/tesla-big-battery-delivered-a-22-million-profit-in-2018-2018/.

Patel, S. (2013). Spain inaugurates 2-GW pumped storage facility. *Power*. 30 November. Available at: www.powermag.com/spain-inaugurates-2-gw-pumped-storage-facility/.

Pena-Alzola, R., Sebastián, R., Quesada, J. and Colmenar, A. (2011). Review of flywheel based energy storage systems. In *2011 International Conference on Power Engineering, Energy and Electrical Drives, PowerEng2011*. Piscataway, NJ: IEEE, pp. 1–6.

Ratnam, E., Weller, S. and Kellett, C. (2016). Central versus localized optimization-based approaches to power management in distribution networks with residential battery storage. *International Journal of Electrical Power & Energy Systems*, 80, 396–406.

Robson, P. and Bonomi, D. (2018). *Growing the Battery Storage Market 2018: Exploring Four Key Issues*. White paper. Dufresne Research. Available at: https://energystorageforum.com/files/ESWF_Whitepaper_-_Growing_the_battery_storage_market.pdf.

Rogers, J., Watkins, C. and Hoffman, D. (n.d.). *Overview and History of the Taum Sauk Pumped Storage Project*. Rolla, MO: Missouri University of Science and Technology. Available at: https://web.mst.edu/~rogersda/dams/2_43_Rogers.pdf.

Rogner, M. and Troja, N. (2018). *The World's Water Battery: Pumped Hydropower Storage and the Clean Energy Transition*. IHA Working Paper. London: International Hydropower Association. Available at: www.hydropower.org/publications/the-world-e2-80-99s-water-battery-pumped-hydropower-storage-and-the-clean-energy-transition.

Romanach, L., Contreras, Z. and Ashworth, P. (2013). *Australian Householders' Interest in the Distributed Energy Market* [survey]. Report No. EP133598. Canberra: Commonwealth Scientific and Industrial Research Organisation (CSIRO). Available at: http://apvi.org.au/wp-content/uploads/2013/11/CSIRO-Survey-Report.pdf.

Scott, P., Gordon, D., Franklin, E., Jones, L. and Thiebaux, S. (2019). Network-aware coordination of residential distributed energy resources. *IEEE Transactions on Smart Grid*, 10, 6528–6537.

Seltzer, M. A. (2017). Why salt is this power plant's most valuable asset. *Smithsonian Magazine*. 4 August. Available at: www.smithsonianmag.com/innovation/salt-power-plant-most-valuable-180964307/.

Singh, M., Lopes, L. and Ninad, N. (2015). Grid forming battery energy storage system (BESS) for a highly unbalanced hybrid mini-grid. *Electric Power Systems Research*, 127, 126–133.

Snowy Hydro (2019). *Snowy 2.0 Project Update*. Snowy Hydro Limited. Available at: www.snowyhydro.com.au/our-scheme/snowy20/.

Sobri, S., Koohi-Kamali, S. and Abdul Rahim, N. (2018). Solar photovoltaic generation forecasting methods: A review. *Energy Conversion and Management*, 156, 459–497.

Sovacool, B. K. (2014). What are we doing here? Analyzing fifteen years of energy scholarship and proposing a social science research agenda. *Energy Research & Social Science*, 1, 1–29.

Spears, J. (2014). Ontario electricity gets taken for a spin. *The Star*. 7 November.

StreetScooter (n.d.). *StreetScooter*. Available at: www.streetscooter.com/de.

Sunspec Alliance (2016). *IEEE 2030.5 Common California IOU Rule 21 Implementation Guide for Smart Inverters*. Common Smart Inverter Profile Working Group. Available at: www.pge.com/includes/docs/pdfs/shared/customerservice/nonpgeuti lity/electrictransmission/handbook/rule21-implementation-guide.pdf.

US DOE (Department of Energy) (n.d.). DOE OE Global Energy Storage Database. Available at: www.sandia.gov/ess/global-energy-storage-database/.

Watson, P., Lovell, H., Ransan-Cooper, H., Hann, V. and Harwood, A. (2019). CONSORT Bruny Island battery trial. Project final report. Available at: http://brunybatterytrial .org/wp-content/uploads/2019/05/consort_social_science.pdf.

Weaver, J. F. (2017). World's largest battery: 200MW/800MWh vanadium flow battery: Site work ongoing. *electrek*. 21 December. Available at: https://electrek.co/2017/12/ 21/worlds-largest-battery-200mw-800mwh-vanadium-flow-battery-rongke-power/.

Weber, A., Mench, M. M., Meyers, J., Ross, P., Gostick, J. and Liu, Q. (2011). Redox flow batteries: A review. *Applied Electrochemistry*, 41, 1137–1164.

West, N., Watson, P. and Potter, C. (2018). *Pumped Hydro Cost Modelling*. ENTURA-10686B. Cambridge, Tasmania: Entura. Available at: www.aemo.com.au/-/media/ Files/Electricity/NEM/Planning_and_Forecasting/Inputs-Assumptions-Methodologies/2019/Report-Pumped-Hydro-Cost-Modelling.pdf.

Zinaman, O., Bowen, T. and Aznur, A. (2020). *An Overview of Behind-the-Meter Solar-Plus-Storage Regulatory Design*. Report/Contract No. IAG-17-2050. USA: National Renewable Energy Lab. Available at: www.nrel.gov/docs/fy20osti/75283.pdf.

8

The Hydrogen Economy

FIONA J. BECK, DAVID GOURLAY, MICHELLE LYONS AND
MAHESH B. VENKATARAMAN

Executive Summary

The global energy system will have to undergo a significant transformation over the next few decades to reduce greenhouse gas emissions and avoid dangerous climate change. There is an urgent need to find carbon-free forms of energy that can be produced cheaply, used across multiple emissions sectors and stored to balance energy grid intermittency. Hydrogen is a particularly interesting candidate for a zero-carbon energy vector, as it has a high energy density and can be produced and used in a variety of ways.

Hydrogen is already used extensively by industry, and techniques for production and handling are well established. However, the current supply chain is relatively simple and relies heavily on fossil fuels, resulting in large amounts of carbon dioxide emissions.

The value chain for a future hydrogen economy will be more complex and require different technologies to be further developed and scaled up. These include:

- hydrogen production processes with low or no carbon emissions;
- methods for large-scale storage and transport to allow hydrogen to be widely and safely traded across the world; and
- technologies to enable hydrogen to be used in a wide range of new applications in industry, transport, heating and cooking, and the electricity sector.

The emergence of the hydrogen economy faces a number of economic, social acceptance and regulatory challenges. Governments around the world have a role to play in providing policies to address potential market failures, socialising the widespread use of hydrogen and establishing international regulations and standards.

Hydrogen produced with zero-carbon emissions has the potential to be a major new globally traded commodity which could enable countries with high energy needs and limited renewable energy potential to decarbonise their economies. As hydrogen can be produced and exported by a broad range of countries, it could diversify the global energy supply and increase energy security.

International cooperation and governance will be needed to facilitate the development of the hydrogen economy. A wide variety of public multilateral and national institutions are working in this space to provide information and analysis for decision makers, build technical capacity, coordinate research, development and demonstration funding, provide international financing and set international standards.

8.1 Introduction: Why Hydrogen, and Why Now?

The hydrogen economy was first described in a visionary paper by Bockris and Appleby in 1972 (Bockris and Appleby 1972). Their concept is simple: electricity from renewable sources like wind and solar should become the primary source of global energy, and where direct electrification is not feasible, fossil fuels should be replaced with hydrogen generated from renewable energy.

Global momentum to develop a hydrogen economy has never been stronger. All countries have committed to the Paris Agreement, which seeks to limit global temperature increases to reduce the risks and impacts of climate change.[1] In 2018, the Intergovernmental Panel on Climate Change (IPCC) issued a stark warning: global green-house gas (GHG) emissions must reach net zero by 2050 to limit global temperature levels to 1.5 °C above pre-industrial levels and avoid the worst effects of climate change (IPCC 2018). This will require a rapid transformation of the current global energy system, which is highly dependent on fossil fuels.

Investments in renewable electricity generation are already driving the energy trans-formation. However, current renewable energy sources are mostly intermittent (e.g. solar photovoltaic (PV) and wind energy) and not easily stored or transmitted over long distances. As the proportion of renewable energy on electricity grids increases, so does the need for grid-balancing services, including electricity storage. Additionally, not all emissions sectors can directly replace fossil fuel use with renewable electricity. These include heavy freight, aviation and industries like iron and steel, cement, chemicals and aluminium. A third of global energy emissions currently have no commercially viable alternative to fossil fuels (IRENA 2018). To truly decarbonise, we need to find carbon-free energy vectors and fuels that can replace fossil fuels across the spectrum of emissions sectors.

Hydrogen is a particularly promising candidate for a carbon-free energy vector. It has a high specific energy (i.e. energy per unit mass) and is very versatile in terms of how it can be produced and used. In particular, it can be generated from the electrolysis of water using renewable energy, resulting in a gaseous fuel similar to fossil fuel sources like natural gas, but without the same carbon dioxide (CO_2) emissions or supply limitations. Hydrogen produced in this way could allow renewable energy to be stored on a very large scale, mitigating the intermittency of resources like wind and solar, and opening up the possibility of exporting local renewable resources around the world. Hydrogen could play a role in decarbonising a wide range of previously hard-to-abate emissions sectors, including domestic heating and cooking, transportation and heavy industrial processes, where it

---

[1] *Paris Agreement Under the United Nations Framework Convention on Climate Change*, opened for signature 16 February 2016. Available at: https://unfccc.int/process-and-meetings/the-paris-agreement/the-paris-agreement.

can be used as both a fuel and a feedstock. This versatility has led IRENA to identify hydrogen as the possible 'missing link' in the energy transition (IRENA 2018: 45).

This chapter provides an overview of pathways and implications of a future hydrogen economy. It explores current and future options for hydrogen production, storage, transport and use, including future supply and demand scenarios. It analyses economic, social and regulatory barriers to hydrogen uptake and examines potential policy responses and international governance mechanisms to assist governments and international organisations to overcome these barriers. It also considers the positive impacts a renewable hydrogen industry would have on global energy security and international trade.

8.2 The Hydrogen Value Chain

In 2018, nearly 70 million tonnes of pure hydrogen were consumed, mostly for ammonia production and petrochemical refining (IEA 2019d). Hydrogen for these industrial applications is generally produced on-site from fossil fuels. The value chain in a future hydrogen economy will be much more complicated, requiring production methods with low or no carbon emissions, and for hydrogen to be stored and transported across the world for use in a range of sectors.

This section outlines a variety of technologies for production, transport and utilisation that will need to be developed and integrated at scale to enable the hydrogen economy, with a focus on the relative challenges and opportunities.

8.2.1 Production

8.2.1.1 Current Production Techniques

Global hydrogen production is currently dominated by thermochemical processes requiring carbon-based feedstock. Fossil fuels contain hydrocarbons (organic compounds made up of carbon and hydrogen), which are reacted with water at high temperatures to produce hydrogen and CO_2.

The most common technology, responsible for almost half of total dedicated hydrogen production, is steam methane reforming (SMR), which uses natural gas as the feedstock. Steam methane reforming is a high-temperature process (700–1000 °C) in which steam reacts with methane in natural gas in the presence of a catalyst to produce hydrogen, carbon monoxide and a relatively small amount of CO_2. A further reaction between the carbon monoxide and steam produces more CO_2 and pure hydrogen. This is an endothermic process, and the heat is provided by burning a part of the fossil fuel feedstock, resulting in additional CO_2 emissions. Steam methane reforming is up to 75% efficient[2] and is currently the cheapest hydrogen production technology; however, the cost is very sensitive to gas prices.

---

[2] Throughout the chapter, efficiency is defined as the usable energy output compared to all energy inputs, assuming the lower heating value (LHV) of hydrogen.

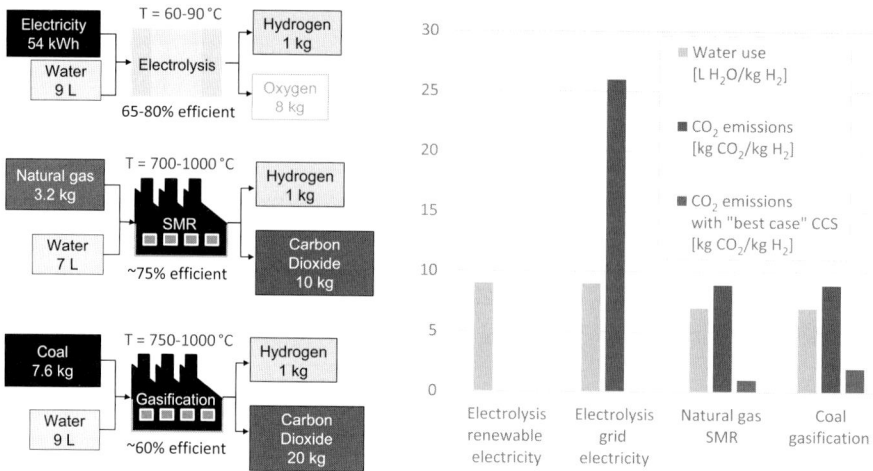

Figure 8.1 Comparison of SMR, coal gasification and electrolysis hydrogen production technologies, showing process inputs, temperatures and outputs. Bar chart compares water usage and CO_2 emission intensities for the different processes with and without 'best-case' CCS, equivalent to 90% CO_2 capture and retention rate.

Source: Authors' analysis; emissions and water use for different production techniques is taken from IEA (2019d); fossil fuel inputs taken from Milbrandt and Mann (2009). For a colour version of this figure, please see the colour plate section.

Another roughly 30% of hydrogen is generated by partial oxidation of hydrocarbons (POX). In POX, methane (e.g. from natural gas) or other heavy hydrocarbons are partially combusted in a low-air atmosphere to produce 'syngas', which is further reformed to produce pure hydrogen. A further 18% of global hydrogen is produced by coal gasification. During gasification, coal is reacted with oxygen and steam, at high temperature (and in some cases high pressures), to form syngas, which again is further refined to form pure hydrogen. All of these methods produce CO_2 as a by-product of the process.

The GHG emissions from current hydrogen production are significant (see Figure 8.1). Producing 1 kg of hydrogen with SMR generates 9–12 kg of CO_2 (Cetinkaya et al. 2012; Committee on Climate Change 2018), not including the fugitive emissions from the extraction of the natural gas. Coal gasification is even more emission-intensive, at 19–23 kg CO_2/kg H_2 (per kilogram of hydrogen) (Muradov 2017; Committee on Climate Change 2018). The IEA calculated the total annual CO_2 emissions from hydrogen production to be 830 MT (megatonnes) of CO_2 per year (IEA 2019c), which is comparable to the total annual CO_2 equivalent emissions from Germany in 2018.

8.2.1.2 Low-Carbon Hydrogen

In order to be a low-carbon process, existing fossil-fuel-based hydrogen production technologies can be coupled with carbon capture and storage (CCS). The use of CCS, an integrated suite of technologies, can prevent large quantities of CO_2 from being released into the atmosphere: CO_2 is separated from the waste stream of a polluting source, compressed and transported to the site of suitable underground rock formations, and then injected underground, usually at a depth of more than one kilometre, for long-term storage (Global CCS Institute 2018). The production of hydrogen through CCS integration with SMR is also called the 'blue hydrogen' pathway.

Carbon capture and storage technology is not 100% effective: some CO_2 will escape into the atmosphere, and additional CO_2 is emitted due to the extra power needed to compress, transport and store CO_2. Capture efficiencies are expected to reach 80–90% (Muradov 2017), but current projects have reported much lower capture rates, as low as 31% (IRENA 2019). Additionally, there are currently no international standards for compulsory monitoring to ensure that the captured CO_2 is not released at a later date. Currently, most CCS plants use the captured CO_2 in enhanced oil recovery, which can result in significant amounts of CO_2 escaping back to the atmosphere (IRENA 2019): one study demonstrated that different enhanced oil recovery projects had very different CO_2 leakage rates, varying between 4% and 72% (Olea 2015).

Hydrogen production via SMR and coal gasification are good candidates for CCS, as the CO_2 is released in a concentrated stream, making it easy to capture. However, as of 2019, there are only two dedicated SMR-based hydrogen production plants that have demonstrated successful CCS integration, namely, Air Products' SMR in Port Arthur, Texas, and Quest in Alberta, Canada. Several more CCS-integrated low-carbon hydrogen production projects are at the planning and feasibility stages. These include the Hydrogen 2 Magnum (H2M) in the Netherlands, H21 North of England, Hynet North West, Ervia Cork CCS and HyDeploy in the UK and the Hydrogen Energy Supply Chain (HESC) in Australia (Global CCS Institute 2018).

Integrating CCS to SMR increases the cost of hydrogen production. The International Energy Agency (IEA) estimates roughly a 50% increase in capex (capital expenditure), an additional 10% for fuel and doubling of opex (operational expenditure) due to CO_2 storage and transport costs (IEA 2019d).

Methane cracking (MC) could be a low-emission alternative to the SMR–CCS technology. In the MC process, methane (CH_4) from natural gas is subjected to a thermal treatment at high temperatures in the absence of air, producing solid carbon and pure hydrogen. If the required heat is provided by burning the hydrogen produced in the process, or through a renewable source, then the process can (theoretically) be made emissions-free. Methane cracking for production of hydrogen has been demonstrated at a commercial scale; however, the process has a relatively low efficiency and high cost compared to SMR. New reactor/process designs have demonstrated higher energy conversion efficiency on a laboratory scale (Geißler et al. 2016) but scalability of this technology is yet to be demonstrated (Weger et al. 2017).

Low-emissions technologies relying on fossil fuels for hydrogen production have been proposed as an important stepping stone towards a fully renewable hydrogen economy, supporting the development of global hydrogen supply chains. However, it is not yet clear if these technologies can scale up sufficiently rapidly. There is a risk that fossil-fuel-based production will ramp up much faster than the required CCS facilities, resulting in significant CO_2 emissions. The International Renewable Energy Agency (IRENA) has also expressed some concern that investment in CCS could divert limited capital away from the renewable energy technologies towards fossil fuels, ultimately slowing the transition to a decarbonised energy system (IRENA 2019). Additionally, these technologies do not mitigate the fugitive GHG emissions that occur during the extraction of fossil fuels, which can be significant.

8.2.1.3 Zero-Carbon Hydrogen

Electrolysers driven by renewable electricity are likely to be the main production technology for zero-carbon hydrogen in the near term. Electrolysis is the process of splitting water into hydrogen and oxygen using electricity. Although only a few per cent of hydrogen is currently produced this way, it is a mature technology. Hydrogen generation with electrolysis does not produce any CO_2 emissions, and has the potential to be a truly 'zero-carbon' technology if the electricity used in the process is itself generated from renewable sources. Hydrogen produced this way is referred to interchangeably as 'green' or 'renewable' hydrogen. It should be noted that significant emissions are released if the electrolysers are powered using the electricity generated with fossil fuels; the IEA calculated average emissions at 26 kg CO_2/kg H_2 for grid-connected electrolysers, assuming a 'world average electricity mix' (IEA 2019d: 53); while the Australian National Hydrogen Strategy have estimated emissions to be 40.5 kg CO_2/kg H_2 for electrolysers connected to the Australian electricity grid (COAG Energy Council Hydrogen Working Group 2019).

The most common commercial electrolysers employ two electrodes separated by a thin membrane in an alkaline electrolyte. The membrane allows the ions in the electrolyte to conduct electricity between the electrodes, but keeps the product gases separate so that pure hydrogen and oxygen can be collected independently. Current electrolyser technologies run at 60–80 °C and are around 65% efficient, with incremental efficiency enhancements up to roughly 80% expected over the next decade (Schmidt et al. 2017).

Alkaline electrolysers currently dominate the market and have the lowest capital costs; however, proton exchange or polymer electrolyte membrane (PEM) technologies are becoming competitive (Buttler and Spliethoff 2018). The term 'PEM' refers to technologies that employ a solid electrolyte to conduct ions and separate product gases. Such technologies have the advantage of running at higher current densities and can react more quickly to changes in the electricity supply, which may be particularly relevant for systems powered by variable renewable electricity (Schmidt et al. 2017). Another alternative is high-temperature electrolysers, called solid-oxide electrolyser cells (SOECs). These are still in demonstration phase, but could offer significant electricity-to-hydrogen efficiency enhancements over standard electrolysers by running at 650–1100 °C (Hauch et al. 2008). They require a heat source to generate steam, which could be provided from renewable sources such as solar thermal, or from waste heat from industrial process.

In general, standard electrolysers require roughly 54 kWh (kilowatt-hours) of electrical energy and 9 litres of water to produce 1 kg of hydrogen. They also need a range of supporting systems, known as the balance of plant, including pumps, compressors, heat exchangers and gas lines. Additionally, most electrolysers include a water treatment system to remove impurities found in potable water that could otherwise affect the reaction and contaminate the evolved gases. Electrolysers are highly modular and are generally sold in standard-sized units rated by their maximum power input (typically in the 100 kW–1 MW range).

Producing hydrogen with renewable energy and electrolysis is still significantly more expensive than hydrogen produced with SMR. However, recent analysis from IRENA has suggested that 'lowest-cost wind and solar projects can provide hydrogen at a cost comparable to that of hydrogen produced from fossil fuels' with integrated CCS (IRENA 2019). Predictably, costs for renewable hydrogen are sensitively dependent on electricity prices and the cost of electrolyser systems. However, running electrolyser systems at high capacity factors[3] is also critical to keep costs low (Bruce et al. 2018). It is likely that the cost of electrolysers will fall as the industry scales up, with estimated experience rates of 18% (Schmidt et al. 2017).

8.2.1.4 Water Use

Water is used as a feedstock for almost all hydrogen production technologies, with the notable exception of MC. Producing 1 kg of hydrogen requires 7–9 litres of water, depending on the production pathway (see Figure 8.1) (IEA 2019d). To put this into perspective: the fuel cycle of conventional natural gas consumes less than a third of a litre per kilogram of gas produced, and roughly 1 L/kg for shale gas.[4]

Large-scale hydrogen production could result in an undesirable burden on freshwater supplies in the future. For example, the Australian National Hydrogen Strategy estimated that 'Under strong hydrogen growth settings, water consumption in Australia may be the equivalent of about one-third of the water now used by the mining industry' (COAG Energy Council Hydrogen Working Group 2019). To avoid this, desalination plants could be co-located with hydrogen production in coastal areas. Desalination would necessarily add to overall system costs, but the additional energy requirements are negligible. The energy required for desalination with reverse osmosis is 4 Wh/L (Caldera et al. 2017) or 36 Wh/kg H_2 – less than a thousandth of the energy needed to generate 1 kg of hydrogen with electrolysis.

8.2.1.5 Emerging Technologies

There are several hydrogen production technologies under active development that have the potential to disrupt the hydrogen value chain in the future.

A range of different technologies aim to split water into hydrogen and oxygen directly using the power of the sun. Concentrated solar thermal hydrogen production uses the heat of concentrated sunlight (at 500–2000 °C) to run a thermochemical reaction to split water. This technology has the potential to have higher solar-to-hydrogen efficiencies than solar-power-driven electrolysers. Photoelectrochemical (PEC) systems use solar-cell-like

---

[3] The capacity factor is defined as the actual hydrogen output divided by the maximum possible hydrogen output of a given system over a period of time.
[4] Estimated from data in Meldrum et al. (2013).

components made of semiconductors to provide the current and voltage needed for water electrolysis using only sunlight. These systems could potentially be cheaper than renewable-energy-driven electrolysis; however, questions remain over how much the cost of hydrogen can be reduced, as they could require more complex balance of plant (Shaner et al. 2016). Photocatalytic hydrogen production takes advantage of materials that can absorb sunlight and generate high-energy electrons, which can initiate the water splitting reaction and release oxygen and hydrogen without any additional power, external wiring, membranes or additional components. Photocatalyst systems have very low solar-to-hydrogen conversion efficiencies but are attractive as the materials have the potential to be very cheap and scalable. However, it is not yet clear if this low materials cost will translate into a lower production cost, as additional gas separation systems will be needed to collect and store pure hydrogen.

Another class of technologies aim to leverage microbial processes for hydrogen generation. These include photobiological processes, where waste water is combined with microorganisms and sunlight; and microbial electrolysis, where an electric current is used to drive the reaction instead of sunlight (Feng et al. 2015). Alternatively, microbes can be used in a fermentation process called microbial biomass conversion, which breaks down organic matter to produce hydrogen without any additional energy input. These systems are interesting as they do not use potable water sources.

8.2.2 Transport and Storage

The low volumetric density of hydrogen makes it challenging to economically store and transport large quantities of hydrogen fuel. For comparison, one litre of hydrogen gas at atmospheric pressures and ambient temperatures contains roughly one-third of the energy of the same volume of natural gas. Developing safe, low-cost and scalable techniques to increase the amount of hydrogen that can be stored in a given volume is a key challenge to realising a future global hydrogen economy.

8.2.2.1 Mature Storage and Transport Technologies

As hydrogen has been used as an industrial feedstock for many decades, there are well-established processes for safe handling and storage. However, the current hydrogen supply chain is much simpler and on a smaller scale than that envisioned for the hydrogen economy. The majority of hydrogen is currently consumed at the point of production and only about 15% is distributed off-site (IEA 2019d).

Small volumes of hydrogen gas can be mechanically compressed and stored in pressurised vessels. Under high pressure, the energy density is improved but specialised high-pressure tanks are needed, and a large amount of energy is required to compress the gas, making it costly for large volumes. To further improve energy densities, hydrogen can be converted to a cryogenic liquid. This is also a mature technology but is highly energy intensive and requires complex storage vessels. Compressed or liquefied hydrogen can be transported in tanks by rail or truck.

Very large volumes of hydrogen gas can be pumped underground and stored in suitable geological features, like salt caverns and depleted natural gas reservoirs. Similar structures are already used for long-term storage of natural gas. Geological storage is very promising, as it has a very high efficiency (i.e. most of the gas pumped in can be recovered successfully), good economies of scale and low operational costs, although it is limited to areas with storage capacity. Several successful pilot projects have already been demonstrated, most notably hydrogen storage in salt caverns at Teeside in England by Imperial Chemical Industries (Foh et al. 1979).

Compressed or liquefied hydrogen can also be stored and distributed in pipelines. While 100% hydrogen pipelines are already used in industry, the extent of these networks is very limited (IEA 2019d). Unfortunately, there are several key differences between natural gas and hydrogen which mean that existing natural gas pipelines and distribution networks cannot be directly used for pure hydrogen (Melaina et al. 2013). Hydrogen is a very small molecule that can diffuse easily, requiring specialised components such as valves and seals to contain it securely. Diffusion also causes the phenomenon of *metal embrittlement*, which occurs when steel and certain other alloys commonly used in industry are exposed to hydrogen. Absorption of hydrogen by the metal can lead to sudden and unpredictable cracking and fracturing of components, and the risks increase if the gas is stored at high pressures. Scaling up hydrogen distribution networks would be costly due to the specialised components required.

8.2.2.2 Large-Scale Transport of Hydrogen

Technologies and supply chains for the large-scale and long-distance transport of hydrogen will need to be developed. It is instructive to consider the natural gas industry, which is already adept at trading large quantities of gaseous fuel across large distances. The global supply chain for liquefied natural gas (LNG) includes liquefaction, pipelines, large-scale storage, port infrastructure and shipping tankers. The hydrogen economy will require a similar supply chain; however, there are specific challenges related to handling liquid hydrogen (LH2) that will make it unlikely that existing infrastructure can be used as-is. As well as the dangers of hydrogen embrittlement, and the need for specialised plant components, hydrogen has a much lower boiling point (–253 °C compared to –163 °C for LNG) meaning that more energy is needed to liquefy the gas – between 25% and 40% of the energy embedded in the hydrogen (assuming the low heating value of hydrogen) (IEA 2019d). Maintaining this low temperature also requires much more complex and expensive storage tanks and insulation. In addition, LH2 has only 40% of the energy density of LNG – meaning that bigger tanks are needed to carry the equivalent amount of fuel. While the technology for handling LH2 already exists, the stringent requirements for hydrogen and the lower density will mean that the costs are likely to be significantly higher than for LNG.

Chemical storage offers another route for increasing the density of hydrogen fuel, while simultaneously making it easier to store and transport. Ammonia is the most mature hydrogen carrier and is produced by reacting hydrogen with nitrogen in the Haber–Bosch process. Aqueous ammonia is 1.5 times denser than LH2 and can be stored and transported at ambient conditions, although it is usually pressurised and transported in large tanks. Once again, there is a significant energy penalty in converting hydrogen to ammonia – between 7% and 18% of the total energy contained in the hydrogen (IEA 2019d). Reconverting ammonia back to hydrogen is a less mature process and incurs an additional energy penalty of less than 20%. Ammonia is also highly toxic, flammable and corrosive, and careful handling is needed to ensure safety. While ammonia production is an extensive industry in its own right, with applications in fertilisers, refrigerants, pharmaceuticals and textiles, very large-scale and widespread distribution as needed in the hydrogen economy may introduce challenges.

Liquid organic hydrogen carriers (LOHCs) are another promising option for liquid-phase hydrogen storage and transport. These material systems consist of organic compounds that reversibly react with hydrogen by catalytic hydrogenation and dehydrogenation. The result is a hydrogen-rich liquid that can be stored at ambient temperatures and pressures, and is compatible with existing liquid fuel infrastructure such as oil pipelines. Unlike ammonia, the materials themselves are not used up in the process of releasing the hydrogen, and need to be returned to place of origin for reuse. The most advanced LOHC is methylcyclohexane (MCH), and is currently being trialled by several organisations developing hydrogen supply chains (IEA 2019c). Methylcyclohexane uses toluene as a carrier molecule, which is a low-cost chemical already widely used as a solvent in paints, lacquers and leather processing, among other applications. Unfortunately, toluene is toxic, and while safe handling procedures have been developed for its current commercial use, it may cause difficulties in scaling up MCH for large-scale hydrogen storage. A range of non-toxic LOHCs are under development. The main challenges are to reduce material costs and the amount of energy needed to bind and release the hydrogen, while maintaining high-capacity hydrogen storage (Preuster et al. 2017).

Other potential carrier material systems are under active research and development. Solid-phase metal hydrides can chemically bond with hydrogen (chemisorption), releasing heat (He et al. 2016). The hydrogen can be recovered by a heating step. Hydrogen can also be physically adsorbed (physisorption) onto the surface of porous media like carbon-based nanomaterials, metal–organic frameworks and polymers (Dalebrook et al. 2013).

8.2.3 Applications of Hydrogen

8.2.3.1 Existing Uses of Hydrogen in Industry

The vast majority of the hydrogen produced today is used in the industrial sector. One-third of existing capacity is used by the petroleum industry where it serves two purposes – first, catalytic cracking and hydrogenation (hydrocracking) of heavy hydrocarbons to produce gasoline and diesel and, second, to upgrade low-quality crude oil by removing impurities. Nearly one-third of the remaining hydrogen is used for the production of ammonia, which in turn finds application in the manufacture of fertilisers, a range of nitrogenous compounds and, directly, as a cleaning agent, refrigerant and anti-microbial agent.

Hydrogen is also used to produce methanol, an important organic solvent, fuel additive and additive for polymer/resin production. Additionally, hydrogen is used in the iron and steel industry for annealing, as a blanketing gas and as forming gas (a mixture of nitrogen and hydrogen). Hydrogen also finds some general industrial uses, albeit in smaller quantities, such as propellant fuel, semiconductors, glass production, hydrogenation of fats and as a cooling agent.

8.2.3.2 New Applications of Hydrogen in Industry

As well as decarbonising existing applications, zero- or low-carbon embedded hydrogen could be used to replace fossil fuels in a range of industry applications that cannot be easily electrified, including metallurgical extraction and the production of high-temperature heat.

The steel industry accounts for roughly 7–8% of the total global GHG emissions from fossil fuels use. Producing steel from iron ore is energy intensive and has a significant carbon footprint due to the use of coke as both a fuel and a reductant. New steelmaking routes using hydrogen as a reducing agent (H2-DRI) have the potential to completely eliminate the GHG emissions from steelmaking, if renewable hydrogen is used (Vogl et al. 2018).

Industrial high-temperature heat (>400 °C) is responsible for 3% of global energy sector emissions and is required in a range of industrial applications such as calcination, annealing, forging and rolling. Most of these processes burn fossil fuels to provide the heat, although some specific applications employ electric resistance furnaces or microwave technology. Electrifying these processes is likely to be challenging and, instead, hydrogen, or derivatives of hydrogen, could be combusted to provide high-temperature heat with no carbon emissions. Unfortunately, converting from fossil fuel combustion to hydrogen is also not straightforward and will require changes in equipment, as well as fuel handling procedures.

Renewable hydrogen can also be combined with CO_2 to produce syngas, which can be further processed to produce a variety of synthetic fuels, including gasoline, diesel and methanol, via a range of commercial synthesis processes. This is particularly interesting as it offers a way of converting renewable electricity to 'drop-in' fuels that can be used in place of standard fossil fuels. Synthetic fuels contain carbon and will release CO_2 again when combusted. However, if the CO_2 used for the process is sourced from a waste stream the process can be considered to have low life-cycle emissions. It should be noted that biomass-derived synthetic liquid fuels[5] is a competing technology for 'drop-in' fuels, with a lower emissions profile (National Academies of Sciences Engineering and Medicine 2016).

8.2.3.3 Decarbonising Transport

Hydrogen could play a role in decarbonising a range of transportation systems in two ways: by direct use of hydrogen in fuel cells, and by using hydrogen-based fuels in modified internal combustion engines.

---

[5] Biomass-based synthetic fuels also emit CO_2, but this carbon is biogenic, having been extracted from the atmosphere by plants. It can be considered carbon neutral if the associated land-use change impacts are negligible.

Fuel cell electric vehicles (FCEVs) are a type of electric drivetrain vehicle that uses hydrogen as on-board energy storage rather than relying on batteries. Fuel cells are essentially reverse electrolysers: hydrogen is reacted with oxygen from the air to produce electricity in an electrochemical cell. The most mature technology is PEM, which uses a solid polymer between the electrodes in place of a liquid electrolyte.

Fuel cell electric vehicles have two main advantages over battery electric vehicles (BEVs), due to the fact that energy is stored chemically (in the bonds of hydrogen), instead of electrochemically (in batteries). Unlike batteries that require long charging times (from 30 minutes, for super charging, to hours), refuelling with hydrogen can be done in a matter of minutes. Additionally, the specific energy of hydrogen is much greater than that of batteries, meaning that FCEVs can store more fuel on board and have a longer range than BEVs.

However, it is important to note that FCEVs fuelled by renewable hydrogen will always be less efficient than BEVs run directly on renewable energy. In an FCEV, electricity is first converted to hydrogen in an electrolyser, then converted back into electricity in the on-board fuel cell, with a significant energy loss associated with each conversion process. In comparison, a BEV is only limited by the efficiency of the battery. By one analysis, FCEVs powered by renewable energy have overall electricity-to-wheel efficiencies of 41%, compared to 86% for BEVs (Committee on Climate Change 2018). FCEVs also require specialised infrastructure and refuelling stations, and are a less mature technology. For this reason, FCEVs should be considered a complementary technology to BEVs, for applications not well suited to battery technologies. Fuel cell electric vehicles are most promising for vehicles that need long ranges or high energy intensity per kilometre (i.e. heavy vehicles), and for vehicles that are heavily used – making the long charging times required for batteries unsuitable. In particular, vehicles such as forklifts, buses, trains, ferries and trucks that have predesignated routes and predictable refuelling needs are particularly suited to hydrogen fuel cells.

Hydrogen-based fuels could enable the decarbonisation of the maritime and aviation sectors, which are both very energy intensive and require very long ranges and are not well suited to electrification. Synthetic 'drop-in' fuels could be used directly, or ammonia can be burnt in modified internal combustion engines (National Academies of Sciences Engineering and Medicine 2016).

8.2.3.4 Replacing the Use of Natural Gas for Domestic Use and Commercial Buildings

Hydrogen gas can be combusted for space heating and cooking, and could replace the use of gas in urban settings. However, it is much more energy efficient to use the electricity directly, using modern technologies such as heat pumps and induction cookers, rather than using the electricity to generate, store and transport hydrogen for these applications (Committee on Climate Change 2018).

Coupling hydrogen production with the natural gas networks could, however, prove useful during the transition to the hydrogen economy. Hydrogen gas can be blended into existing natural gas networks with concentrations of up to 10–20% by volume without having to modify the network infrastructure or appliances at the point of use (Melaina et al. 2013). This could provide a relatively large market for the nascent renewable hydrogen industry, as well as providing built-in distribution and storage. However, due to the low volumetric energy density of hydrogen compared to methane, blending up to 20% hydrogen into the gas network would only reduce carbon emissions from domestic gas use by up to 7%.[6]

Due to the material properties of hydrogen, it is unlikely that existing natural gas networks could be used for the distribution of higher concentrations of hydrogen without substantial upgrades (Melaina et al. 2013). In addition to embrittlement issues and the need for specialised components, differences in gas density, ignition temperature and flame velocity mean that gas meters and end-use appliances may also need modification. Widespread use of pure hydrogen in domestic and urban settings would also pose specific safety challenges: for example, hydrogen is odourless and burns with a clear flame, making it difficult to detect leaks and fires.

Box 8.1 **System Integration**

One of the advantages of hydrogen as an energy carrier is that it is very versatile: it can be generated, converted and used in a range of ways in the electricity, industry and transport sectors.

Figure 8.2 demonstrates some of the ways that the technologies in the hydrogen value chain could be integrated into the energy network during the transition to a zero-carbon economy.

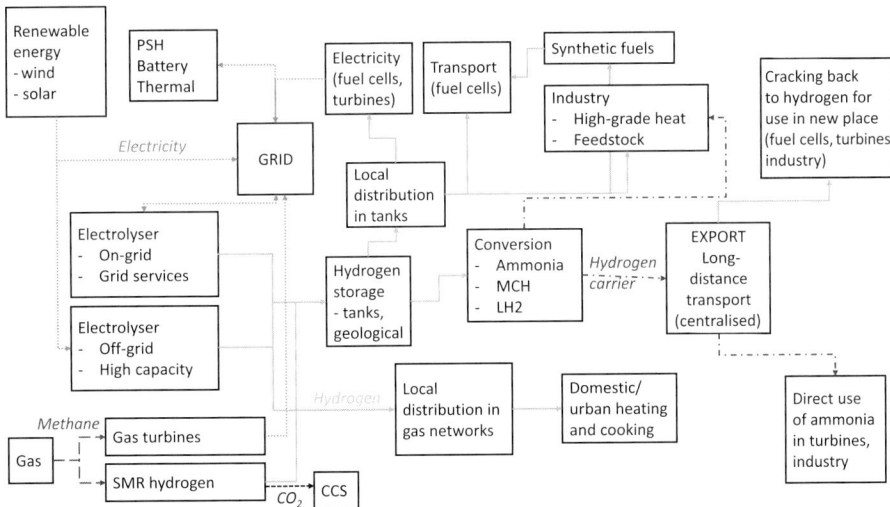

Figure 8.2 Schematic showing the possible ways that hydrogen technologies could be integrated into the larger energy system.

Source: Authors' summary. For a colour version of this figure, please see the colour plate section.

---

[6] Calculation by Associate Professor Matthew Stocks at The Australian National University.

It is important to consider that each conversion, transport and storage process will incur an energy and concomitant cost penalty. This means that although the flexibility of hydrogen as an energy carrier allows multiple pathways from production to use, not all of these routes will be cost-effective.

8.2.3.5 Renewable Electricity Generation and Storage

Hydrogen can be used to generate electricity in two ways: in fuel cells or by combustion in a gas turbine. Like electrolysers, fuel cells for stationary electricity applications are highly modular, ranging from kilowatt to megawatt scales. For very large applications (>100 MW), gas turbines can be used, analogous to standard natural gas electricity production; however, pure hydrogen gas and ammonia turbines are still at the demonstration phase. Hydrogen can also be blended into natural gas up to concentrations of 20% and co-fired in existing gas turbines. This could be used during the transition to a hydrogen economy as another source of hydrogen demand and to reduce emissions from gas-fired electricity generation, although as mentioned above, this is likely to provide only modest CO_2 emissions reductions.

Converting renewable electricity into hydrogen and back into electricity incurs a significant energy penalty, and the round-trip efficiency for storing electricity as hydrogen is very low, at only 35–40%. It is unlikely that hydrogen electrolysers and fuel cells will be able to compete with more established storage technologies – like batteries or pumped storage hydro (PSH) – for small- to medium-capacity electricity storage, as the fuel cells, hydrogen storage and electrolysers needed are still too expensive. However, hydrogen has the potential to be used for high-capacity electricity storage, employing underground geological sites to store large quantities of hydrogen in order to mitigate the effects of seasonal variation in renewable resources. Seasonal hydrogen storage would be beneficial for high penetration of renewables on the grid, and could be economically viable in Northern Europe even with a relatively low round-trip efficiency of 45% (DNV-GL 2018; IRENA 2019). The potential for international trade is discussed in Section 8.4.1.

8.3 Challenges and Policy Responses

The IEA (IEA 2019d) lists three major challenges to overcome if a hydrogen economy is to emerge. These are: the complexity of supply chains and infrastructure; uncertainty regarding policy and technology; and the need to establish regulations, standards and acceptance. These are discussed below. To establish a hydrogen economy, governments will have an important role in overcoming these challenges, which could be addressed through policy and engagement multilaterally, bilaterally, nationally or subnationally.

The hydrogen opportunity is likely to be different from country to country. Some countries will have comparative advantages in particular areas, such as access to a surplus of low-cost renewable electricity that can be used for electrolysis or existing manufacturing demand, that will shape their hydrogen industry in a particular direction. An important role for governments is to identify these areas of comparative advantage, and their most viable hydrogen pathways, and use these as the basis for scaling up the industry. One way governments can do this is by developing a hydrogen strategy or plan. As of mid-2019, nine of the G20 countries and the EU had a hydrogen plan in place (IEA 2019b). Determining and leveraging countries' areas of competitive advantage is a major focus of these plans (Kosturjak et al. 2019). Establishing and adhering to a national plan can reduce policy uncertainty for businesses looking to invest in hydrogen infrastructure or applications.

8.3.1 Economic Challenges

Market failures could prevent a hydrogen economy from emerging or scaling up efficiently. Most pertinent is that hydrogen will have to compete against fuels that produce carbon emissions, such as oil and natural gas. As of 2018, 87% of global GHG emissions were not subject to any form of carbon pricing, while less than 1% are subject to a carbon price equivalent to the social cost of carbon (Jenkins 2019). All else being equal, this will result in a continued overuse of fossil fuels, and an underuse of green hydrogen.

As discussed above, while there are established technologies to produce hydrogen, the industry needs to take risks and innovate to learn how to produce, transport and deploy hydrogen at scale. Knowledge is often thought of as a public good as it is available for all to benefit from. Companies can undervalue the activities required to do this, as they take on all the associated risk while the benefits are available to all companies in the industry (Gillingham and Sweeney 2011; Carbon Pricing Leadership Coalition 2017). This could result in the rate of innovation in hydrogen being below what is required to scale up the industry efficiently. This is a problem that exists in all new industries and is often used as a justification for government research and development (R&D) support.

A hydrogen economy will require investment in complementary infrastructure across the supply chain. These investments will need to be synchronised in time, scale and technology pathway. A failure to coordinate can lead to a suboptimal rate or type of investment (Rodrik 2004; Bento 2008; Gillingham and Sweeney 2011). For example, some jurisdictions may adopt an ammonia-based supply chain, while others prefer LOHCs, severely limiting the efficiency of hydrogen trade. For firms investing in hydrogen production, demand uncertainty is a major risk that will affect decisions about the scale of their investments, while potential hydrogen users face similar risks from the supply-side uncertainty about whether sufficient hydrogen will be available. At the extreme end, coordination failures can prevent investment altogether, although they can be overcome with inter-industry collaboration.

Investment in hydrogen infrastructure is likely to have high capital costs. Innovative projects can struggle to secure the necessary capital (or secure it at too high a cost) as lenders do not have access to sufficient information to adequately assess the risk of the project (Carbon Pricing Leadership Coalition 2017). This will be a barrier to market entry for new projects, although it should be noted that this challenge is not specific to hydrogen and is common across new industries, particularly in energy systems.

Governments can implement a range of policy measures to address these market failures. Carbon pricing can help to ensure hydrogen's emissions reduction potential is adequately valued in the market. However, the carbon price required to make hydrogen competitive with alternative fuels is likely to be unfeasibly high, at least in the short term. At the same time, carbon pricing alone is unlikely to be enough to establish a hydrogen industry due to the presence of multiple market failures (Bataille et al. 2018). Government support for R&D, knowledge sharing and, particularly, early pilot and demonstration projects can improve the local stock of knowledge, addressing an undersupply of innovation. If coordination failures are not resolved within the industry, governments can overcome these by making sure the necessary investments are made across the supply chain. This could be done while addressing credit constraints, such as through providing grants or guarantees to hydrogen projects to 'de-risk' them, or by providing finance with a higher risk tolerance or at concessional rates. State-owned investment banks, such as Germany's KfW or Australia's Clean Energy Finance Corporation (CEFC), would be well placed to provide this. Governments could also reduce the risk of coordination failures by establishing dedicated hydrogen 'clusters' with shared infrastructure (IEA 2019d), or by facilitating cooperation between businesses.

Ensuring there is reliable demand for hydrogen will require more than just putting the infrastructure in place. Many new hydrogen applications are not yet cost-competitive with alternative technologies. Deploying targeted subsidies or tax breaks in the short term could bring down the costs of these applications (IRENA 2018). During the transition to a hydrogen economy, governments could also use mandated targets to set a base level of demand to provide certainty to investors in hydrogen production or supply chains. This could include mandating hydrogen be blended in domestic natural gas supply (up to local technical limits) or targeting a percentage of existing industrial hydrogen use to come from low emissions sources.

8.3.2 Safety, Social License and Acceptance Challenges

Like all chemical fuels, hydrogen has specific handling requirements and safety considerations. Hydrogen is non-toxic, but it is highly flammable, prone to leakage and can cause the embrittlement of metals. It therefore requires specialised infrastructure for safe distribution and storage. Both ammonia and toluene used in hydrogen carriers are toxic. As discussed above, hydrogen and synthetic fuels based on hydrogen have been in use by industry and the energy sector for decades. Safe handling procedures and some standards already exist. However, international trade, and widespread use of hydrogen in urban settings, will bring with them new health and safety challenges. These challenges posed by hydrogen, while real and unique, are not more severe than the risks associated with other fuels that have broad use and acceptance today, such as oil and natural gas (IEA 2019d).

Nevertheless, the public will have to accept these risks for hydrogen use to become widespread in people's vehicles or homes, or for it to be produced, stored or transported near population centres. Incidents such as the explosion of one of Norway's three hydrogen refuelling stations in June 2019, caused by escaped hydrogen which ignited (Huang 2019), highlights the reality of these risks.

Studies conducted to date suggest that most people do not have predisposed attitudes towards hydrogen, either positive or negative. The vast majority of respondents to an Australian study conducted in 2018 (81%) gave a neutral response when asked about their perceptions of hydrogen. At the same time, only 3% of respondents had positive perceptions of hydrogen, while 13% made negative associations, such as to the hydrogen bomb or the Hindenburg airship (Lambert and Ashworth 2018). This is consistent with other studies, conducted in Europe and Asia, which have also found mostly neutral associations with hydrogen (Hickson et al. 2007; Zimmer and Welke 2012; Schmidt and Donsbach 2016; Itaoka et al. 2017). The Australian study also found that respondents had limited overall awareness of hydrogen's properties or its uses (Lambert and Ashworth 2018).

A hydrogen industry can earn a social licence to operate by building a strong safety record. Governments will also be essential in establishing public trust that these risks are well managed. Through public awareness campaigns that assume no prior knowledge, governments can build on the public's limited existing awareness and lack of preconceived notions to establish a positive public perception of hydrogen. Governments should expect industry to contribute to building awareness and acceptance for hydrogen, and a coordination approach that brings together governments, industry and academics is likely to increase the chance of success (Lambert and Ashworth 2018).

The public also expects governments to ensure the safety of hydrogen by implementing the right standards and safety regulations (Lambert and Ashworth 2018). This can also increase acceptance, as the belief that domestic hydrogen applications conform to international standards can itself be an important factor in building public trust (Chen et al. 2016; Lambert and Ashworth 2018).

8.3.3 Regulatory Challenges

At present, countries' regulatory regimes could be a barrier to hydrogen uptake in new applications. This is because hydrogen was not considered when regulations were drafted for sectors such as gas infrastructure, transportation and power generation. This means that in most jurisdictions there are likely to be gaps where required regulations or standards do not exist, as well as regulations that could actively restrict the use of hydrogen in certain applications. The types of regulation needed to support a hydrogen industry fall under three categories: the functional use of hydrogen, the safety of hydrogen use and commercial frameworks for hydrogen business activities (Bruce et al. 2018).

In Europe, the HyLAW project has reviewed the legislation and regulations relevant to hydrogen production and applications across 23 European countries to determine the legal barriers to its uptake. For functional use and safety, the project found major structural barriers preventing the injection of hydrogen into existing gas networks, stemming from the fact the regulations governing gas networks were written to account for the physical properties of natural gas (Floristean et al. 2018). The project also found commercial barriers of varying severity that could affect hydrogen use for electricity generation or in buildings, while FCEVs were relatively unaffected (Floristean et al. 2018).

To ensure the right standards are in place, governments should first review their own regulatory settings to identify any relevant gaps or barriers. Following this, governments should look to global standards as the basis for hydrogen-specific regulations before considering country-specific requirements. This is important to ensure aspects such as safety and technical compatibility are the same across jurisdictions. The current state of international standards, and further work required on this, are discussed in the following section.

8.4 International Trade and Governance

8.4.1 International Trade

Hydrogen has the potential to be a major new globally traded commodity; however, there are significant variations in estimates of its share of future global energy use. At the high end, the Hydrogen Council envisions hydrogen accounting for 18% of final global energy demand in 2050 (Hydrogen Council 2017). Modelling exercises have predicted that hydrogen has the potential to account for 3% by 2050 (Chapman et al. 2019).

In the short term, government policy and especially targets are likely to be the strongest factors in determining the size of the global hydrogen market. As of late 2019, several jurisdictions have announced hydrogen targets for the decade from 2020 to 2030 and beyond, the majority of which are for deployment numbers of FCEVs or the associated infrastructure (i.e. number of fuel stations). The hydrogen roadmaps of Japan and South Korea are major drivers behind hydrogen's global momentum (IEA 2019d). Both countries rely on imported fossil fuels for the majority of their energy needs, have poor renewable energy potential and large manufacturing bases. As a result, both countries view hydrogen as a promising route to decarbonise their economies. Both countries have targets for hydrogen use in their roadmaps and both acknowledge that a large proportion of this will have to be imported.

Although ambitious, these targets are well below the scale of the hydrogen economy envisioned by the Hydrogen Council and are likely to be insufficient to draw in significant investment. For example, Japan envisions a hydrogen supply chain of 300 000 tonnes by 2030, with a long-term target of up to 10 million tonnes (METI 2017). This, however, is a fraction of existing industrial demand for hydrogen (70 million tonnes in 2018) (IEA 2019d). In the long term, the size of a global hydrogen trade will depend on a number of factors, including the cost-competitiveness of production and storage with existing fuels, whether global interest translates into infrastructure investment and geopolitical concerns such as energy security.

Countries with high renewable energy potential could establish hydrogen production capability to take advantage of this new export opportunity. This includes countries in regions such as North Africa and the Middle East, as well as Australia, Chile, New Zealand and Norway (IEA 2019d). The ability to produce large quantities of hydrogen at low cost will be one factor in determining which countries are able to become major exporters. Other important factors will include proximity to markets, existing trading relationships and infrastructure.

8.4.2 Energy Security

The IEA defines energy security as the uninterrupted availability of energy sources at an affordable price (IEA 2019a). Other conceptions of energy security take a broader view to encompass factors such as environmental sustainability and energy efficiency (Ang et al. 2015). Energy security has different implications over different time frames. Short-term energy security concerns are focused on managing sudden changes in the supply–demand balance, while long-term concerns deal with ensuring timely investments to supply energy in line with economic developments and environmental needs (IEA 2019a).

The development of the hydrogen economy has important implications for global energy security, particularly over the long term. These implications are largely positive. For example, hydrogen provides an opportunity to diversify global energy supply chains. Traditionally, fossil fuel industries have relied on a limited number of actors, and continuity of energy supply depended on a range of volatile political, economic and ecological factors (Sheffield and Sheffield 2007). Comparatively, hydrogen is more flexible as it can be produced in a range of ways from different fuel sources by a much wider variety of country actors. Hydrogen could also enhance global energy security through its wide application across emissions sectors. As an energy vector, renewable hydrogen provides a form of energy storage similar to fossil fuel sources like LNG, but without supply limitations or an emissions footprint. If renewable and low-carbon hydrogen can overcome the high production and storage costs which exist at present, they have a range of energy security advantages over fossil fuels.

Renewable hydrogen has energy security advantages over other vectors for international renewable energy trade, such as high-voltage direct current (HVDC) cables. For example, renewable hydrogen offers more flexibility in terms of its transport and storage than HVDC transmission of renewable energy. It circumvents many of the geopolitical complexities with regulating energy trade across multiple national electricity markets that have resulted in slow development of transnational HVDC networks (IEA 2019d). It can be stored and shipped anywhere in the world, particularly when converted to ammonia (European Commission 2003). While the transmission efficiency of HVDC cables transporting renewable energy is higher than renewable hydrogen production at present, there is little analysis directly comparing costs between these technologies.

8.4.3 International Governance to Facilitate Trade of Green Hydrogen

A complex array of international organisations is engaged in the development of the hydrogen economy. This is symptomatic of international energy governance more broadly, but also a reflection of the diverse range of end uses for renewable hydrogen. Existing international governance actors perform a variety of functions. These include information-sharing and capacity-building, the coordination of research, development and demonstration funding, international financing and setting international standards for hydrogen production, storage and transport. The following sections provide an overview of existing international governance relevant to the development of a hydrogen economy. This list is not exhaustive; it does not detail, for example, energy subgroups within multilateral organisations like the G20 or ASEAN (the Association of Southeast Asian Nations). However, it provides an overview of the key public multilateral institutions facilitating the development of the hydrogen economy at present.

8.4.3.1 Information-Sharing and Capacity-Building

A number of international organisations perform knowledge-sharing and capacity-building functions related to the development of the hydrogen economy. These include the International Energy Agency (IEA), the International Renewable Energy Agency (IRENA) and the International Partnership for Hydrogen and Fuel Cells in the Economy (IPHE).

The IEA is an autonomous body within the Organisation for Economic Co-operation and Development (OECD) framework, founded in 1974 in the wake of the 1973 oil crisis. It undertakes analysis across all energy sectors (although historically it has focused more on fossil fuel sectors), develops global scenarios and advocates policies to enhance the reliability, affordability and sustainability of energy for its 30 member countries and beyond. Since 1977, the IEA has coordinated a Hydrogen Technology Collaboration Program (TCP) to accelerate hydrogen implementation and utilisation.

Recently, the IEA has increased its analytical focus on the development of hydrogen technologies and their potential to create a more sustainable and secure global energy supply (IEA 2019a). For example, in 2019 the G20, through its Japan Presidency, commissioned the IEA to undertake analysis on the current state of the hydrogen industry as well as recommendations for its future development (IEA 2019b). The report carried a broad analysis across all hydrogen production routes and uses, and made a range of recommendations seeking to assist in future growth of a green hydrogen economy (IEA 2019d). In May 2019, the IEA also began coordinating a new hydrogen partnership though the Clean Energy Ministerial to drive international collaboration on policies, programmes and projects to accelerate the commercial deployment of hydrogen and fuel cell technologies across all sectors of the economy (Clean Energy Ministerial 2019).

The IPHE was established in 2003 to facilitate the development of the hydrogen economy through information-sharing and capacity-building among its members. It consists of 19 member countries and the European Commission. It currently has two working groups: one focused on education and outreach and another on regulations, codes, standards and safety (IPHE 2019).

Lastly, as the only multilateral institution focused solely on renewable energy, IRENA also has an interest in undertaking analysis and building capacity to facilitate the development of a hydrogen economy. Founded in 2009, IRENA is a relatively new organisation with 160 member countries and another 23 states in accession (IRENA 2019). IRENA is seeking to provide policy-makers with analysis around the technology outlook for renewable hydrogen (IRENA 2018). For example, a recent report focused on the role of hydrogen in the energy transition, hydrogen supply economics and the existing challenges that restricted hydrogen production (IRENA 2019).

8.4.3.2 Research and Development Funding

Although funding for research, development and demonstration (R&D&D) has traditionally been the domain of national governments, in 2015 a multilateral R&D&D initiative, Mission Innovation, was launched on the eve of the Paris Agreement. It is a global initiative of 24 countries and the EU, which seeks to accelerate the pace of clean energy innovation to achieve performance breakthroughs and cost reductions, to provide widely affordable and reliable clean energy solutions (Mission Innovation 2019b). Together, Mission Innovation members account for approximately 80% of global government funding for R&D&D research. Mission Innovation has a number of innovation challenges designed to facilitate global collaborations and innovation in key technology areas. In 2018, it announced the creation of a new innovation challenge focused on renewable and clean hydrogen. Its objective is to accelerate the development of a global hydrogen market by identifying and overcoming key technology barriers to the production, distribution, storage and use of hydrogen at gigawatt scale (Mission Innovation 2019a).

8.4.3.3 International Financing

Like most emerging energy sectors, the development of the hydrogen economy will require large capital investments and the establishment of new global value chains that span numerous countries (IEA 2019d). A range of international financing tools will be needed to assist in the development of the hydrogen economy. For example, multilateral development banks can facilitate finance flows to assist in commercialising renewable hydrogen projects and assisting in the development of the hydrogen economy. Many countries also have export finance agencies to facilitate international trade and mitigate export risks.

8.4.3.4 International Standards

Widely adopted international codes and standards are vital to establishing the hydrogen economy and for lowering regulatory barriers to trade (IEA 2019d). International standards are mainly developed through the International Organization for Standardization (ISO); an independent, non-governmental international organisation with a membership of 164 national standards bodies. The ISO develops voluntary consensus-based, international standards, including a number related to hydrogen production, storage and transportation through the ISO Technical Committee 197 (International Standards Organization 2019). At present, there are 21 hydrogen standards that have either been published or are under development through the ISO. These include standards on safety of hydrogen systems (ISO/TR 15916:2015), gaseous hydrogen fuelling station standards (ISO 19880-3:2018,

ISO 19880-5:2019, ISO/FDIS 19880-1) and industrial, commercial and residential application of hydrogen generators using water electrolysis (ISO 22734:2019).

The International Electrotechnical Commission (IEC), a sister organisation of the ISO, has also developed some standards relating to hydrogen fuel cell technologies. However, there are some existing gaps in international standards on hydrogen, particularly around utilisation of gas networks. As mentioned above, the IPHE is also seeking to coordinate global standards for hydrogen internationally, with the support of member countries, particularly Japan.

8.4.3.5 Sectoral Organisations

One of the benefits of renewable hydrogen is its potential to be used in sectors where other renewable fuels are not suitable, such as transport and heavy industry. The International Maritime Organization (IMO) and the International Civil Aviation Authority (ICAO) are both engaged in the development of the hydrogen economy in their respective sectors to meet emissions reductions commitments. For example, the IMO has committed to reduce emissions by 50% from 2008 levels by 2050. To achieve this target, alternative fuels for shipping will need to be developed. Using hydrogen with marine fuel cells is one promising technology being explored, although capital and fuel costs remain prohibitively high (Balcombe et al. 2019). Like the IMO, the ICAO recognises the potential of renewable hydrogen as an alternative fuel which provides abatement opportunities to replace bunker fuel use in the aviation sector with a sustainable alternative.

8.4.3.6 International Certification and Other Measures for Renewable and Low-Carbon Hydrogen

Using hydrogen does not produce GHG emissions; however, only the use of green hydrogen will ensure the supply chain is zero emissions. The international certification of green and low-carbon hydrogen can assist in the development of the hydrogen economy by providing assurance to consumers on how and where a product was produced. By guaranteeing the provenance of renewable hydrogen, international certification will create a market where there is demand for zero- or low-emissions hydrogen. Research has shown that consumers care about the emissions intensity of hydrogen and are far more supportive of renewable hydrogen production as opposed to hydrogen using CCS (Lambert and Ashworth 2018).

International certification could take the form of a 'guarantee of origin' scheme such as that provided by the EU's CertifHy scheme (CertifHy 2015). The CertifHy scheme is managed and operated from a central registry, which issues, transfers and cancels guarantees of origin for green and low-carbon hydrogen. This is similar to other national and private registries for emissions units. There is a need for a global guarantee of origin scheme for green and low-carbon hydrogen, as the CertifHy scheme has only recently completed its pilot phase and only operates in the EU. A global scheme could potentially be linked to international standards for hydrogen production developed through the ISO and other international standards organisations. It could also possibly be utilised through

Article 6 of the Paris Agreement, depending on how rules for carbon markets are implemented through this agreement.

In the absence of the development of a guarantee of origin scheme, there is scope to facilitate trade of renewable hydrogen bilaterally, plurilaterally or multilaterally through development or modification of free trade agreements.

8.5 Conclusion

While the hydrogen economy is not a new concept, there are promising signs that the development of renewable hydrogen energy production may be about to scale up substantially. Globally, there is an urgent need to find carbon-free forms of energy that can be produced cheaply, used across multiple emissions sectors and stored to balance energy grid intermittency. While renewable and low-carbon hydrogen is still significantly more expensive than other zero or low-carbon energy sources, hydrogen's versatility in terms of end use, and capacity to be converted into other chemicals like ammonia for transportation and storage, mean that it has unique advantages as an energy fuel and vector. Renewable and low-carbon hydrogen also provides the opportunity to improve global energy security through diversification of energy supply and other co-benefits such as improved air quality. This is not to say it is a silver bullet. Producing, transporting and storing hydrogen has significant safety considerations and its conversion to chemical form results in significant efficiency losses. However, in a global energy landscape searching for opportunities to decarbonise rapidly, development of the hydrogen economy offers a significant opportunity.

References

Ang, B. W., Choong, W. L. and Ng, T. S. (2015). Energy security: Definitions, dimensions and indexes. *Renewable and Sustainable Energy Reviews*, 42, 1077–1093.

Balcombe, P., Brierly, J., Lewis, C. et al. (2019). How to decarbonise international shipping: Options for fuels, technologies and policies. *Energy Conversion and Management*, 182(January), 72–88.

Bataille, C., Guivarch, C., Hallegatte, S. et al. (2018). Carbon prices across countries. *Nature Climate Change*, 8, 648–650.

Bento, N. (2008). Building and interconnecting hydrogen networks: Insights from the electricity and gas experience in Europe. *Energy Policy*, 36, 3019–3028.

Bockris, J. O. and Appleby, A. J. (1972). The hydrogen economy: An ultimate economy? *The Environment This Month*, 1, 29–35.

Bruce, S., Temminghoff, M., Hayward, J. et al. (2018). *National Hydrogen Roadmap: Pathways to an Economically Sustainable Hydrogen Industry in Australia*. Canberra: Commonwealth Scientific and Industrial Research Organisation (CSIRO). Available at: www.csiro.au/en/Do-business/Futures/Reports/Hydrogen-Roadmap.

Buttler, A. and Spliethoff, H. (2018). Current status of water electrolysis for energy storage, grid balancing and sector coupling via power-to-gas and power-to-liquids: A review. *Renewable and Sustainable Energy Reviews*, 82, 2440–2454.

Caldera, U., Bogdanov, D., Afanasyeva, S. et al. (2017). Role of seawater desalination in the management of an integrated water and 100% renewable energy based power sector in Saudi Arabia. *Water*, 10. DOI: 10.3390/w10010003.

Carbon Pricing Leadership Coalition (2017). *Report of the High-Level Commission on Carbon Prices*. Washington, DC: World Bank. Available at: www .carbonpricingleadership.org/report-of-the-highlevel-commission-on-carbon-prices.

CertifHy (2015). Overview of the market segmentation for hydrogen across potential customer groups, based on key application areas. *CertifyHy*. Available at: www .certifhy.eu/images/D1_2_Overview_of_the_market_segmentation_Final_22_June_ low-res.pdf.

Cetinkaya, E., Dincer, I. and Naterer, G. F. (2012). Life cycle assessment of various hydrogen production methods. *International Journal of Hydrogen Energy*, 37, 2071–2080.

Chapman, A., Itaoka, K., Hirose, K. et al. (2019). A review of four case studies assessing the potential for hydrogen penetration of the future energy system. *International Journal of Hydrogen Energy*, 44, 6371–6382.

Chen, T.-Y., Huang, D.-R. and Huang, A. Y.-J. (2016). An empirical study on the public perception and acceptance of hydrogen energy in Taiwan. *International Journal of Green Energy*, 13, 1579–1584.

Clean Energy Ministerial (2019). Hydrogen initiative: An initiative of the clean energy ministerial. *Clean Energy Ministerial*. Available at: www.cleanenergyministerial.org/ initiative-clean-energy-ministerial/hydrogen-initiative.

COAG Energy Council Hydrogen Working Group (2019). *Australia's National Hydrogen Strategy*. Canberra: COAG Energy Council. Available at: www.industry.gov.au/data-and-publications/australias-national-hydrogen-strategy.

Committee on Climate Change (2018). *Hydrogen in a Low-Carbon Economy*. London: Committee on Climate Change. Available at: www.theccc.org.uk/publication/hydro gen-in-a-low-carbon-economy.

Dalebrook, A. F., Gan, W., Grasemann, M., Moret, S. and Laurenczy, G. (2013). Hydrogen storage: Beyond conventional methods. *Chemical Communications*, 49, 8735–8751.

DNV-GL (2018). Hydrogen: Decarbonising heat. *DNVGL.com*. Available at: www.dnvgl .com/oilgas/natural-gas/hydrogen-decarbonizing-the-heat.html.

European Commission (2003). *Hydrogen Energy and Fuel Cells: A Vision of Our Future*. EUR Community Research 20719. Luxembourg: Office for Official Publications of the European Communities. Available at: www.fch.europa.eu/sites/default/files/docu ments/hlg_vision_report_en.pdf.

Feng, Y., Liu, Y. and Zhang, Y. (2015). Enhancement of sludge decomposition and hydrogen production from waste activated sludge in a microbial electrolysis cell with cheap electrodes. *Environmental Science: Water Research and Technology*, 1, 761–768.

Floristean, A., Brahy, N. and Kraus, N. (2018). *HyLAW: List of Legal Barriers*. Available at: www.hylaw.eu/sites/default/files/2019-01/D4.2 - List of legal barriers.pdf.

Foh, S., Novil, M., Rockar, E. and Randolph, P. (1979). *Underground Hydrogen Storage: Final Report [Salt Caverns, Excavated Caverns, Aquifers and Depleted Fields]*. Chicago, IL: US Department of Energy and Environment. Available at: www.osti .gov/biblio/6536941.

Geißler, T., Abánades, A., Heinzel, A. et al. (2016). Hydrogen production via methane pyrolysis in a liquid metal bubble column reactor with a packed bed. *Chemical Engineering Journal*, 299, 192–200.

Gillingham, K. and Sweeney, J. (2011). Market failure and the structure of externalities. In B. Moselle, J. Padilla and R. Schmalensee, eds., *Harnessing Renewable Energy in Electric Power Systems: Theory, Practice, Policy*. Routledge, pp. 69–92.

Global CCS Institute (2018). *The Global Status of CCS 2018*. Global CCS Institute. Available at: www.globalccsinstitute.com/resources/global-status-report/previous-reports/.

Hauch, A., Ebbesen, S. D., Jensen, S. H. and Mogensen, M. (2008). Highly efficient high temperature electrolysis. *Journal of Materials Chemistry*, 20, 2331–2340.

He, T., Pachfule, P., Wu, H., Xu, Q. and Chen, P. (2016). Hydrogen carriers. *Nature Reviews Materials*, 1, 1–17.

Hickson, A., Phillips, A. and Morales, G. (2007). Public perception related to a hydrogen hybrid internal combustion engine transit bus demonstration and hydrogen fuel. *Energy Policy*, 35, 2249–2255.

Huang, E. (2019). A hydrogen fueling station fire in Norway has left fuel-cell cars nowhere to charge. *Quartz*. 12 June. Available at: https://qz.com/1641276/a-hydrogen-fueling-station-explodes-in-norways-baerum/.

Hydrogen Council (2017). *Hydrogen Scaling Up: A Sustainable Pathway for the Global Energy Transition*. Hydrogen Council. Available at: https://hydrogencouncil.com/wp-content/uploads/2017/11/Hydrogen-scaling-up-Hydrogen-Council.pdf.

IEA (International Energy Agency) (2019a). Energy security. *IEA.org*. Available at: www.iea.org/topics/energysecurity.

IEA (2019b). IEA contribution to G20 energy in 2019. *IEA.org*. 28 June. Available at: www.iea.org/articles/iea-contribution-to-g20-energy-in-2019/.

IEA (2019c). *IEA Hydrogen Technology Collaboration Program: Renewable Hydrogen Production*. Paris: International Energy Agency.

IEA (2019d). *The Future of Hydrogen*. Paris: International Energy Agency. Available at: www.iea.org/reports/the-future-of-hydrogen.

International Standards Organization (2019). ISO/TC 197: Hydrogen technologies. *ISO.org*. Available at: www.iso.org/committee/54560.html.

IPCC (2018). *Global Warming of 1.5 °C: An IPCC Special Report on the Impacts of Global Warming of 1.5 °C Above Pre-Industrial Levels and Related Global Greenhouse Gas Emission Pathways, in the Context of Strengthening the Global Response to the Threat of Climate Change, Sustainable Development, and Efforts to Eradicate Poverty*. Edited by V. Masson-Delmotte, P. Zhai, H.-O. Pörtner et al. Cambridge: Cambridge University Press. Available at: www.ipcc.ch/sr15/.

IPHE (International Partnership for Hydrogen and Fuel Cells in the Economy) (2019). *International Partnership for Hydrogen and Fuel Cells in the Economy*. Available at: www.iphe.net.

IRENA (International Renewable Energy Agency) (2018). *Hydrogen from Renewable Power: Technology Outlook for the Energy Transition*. Abu Dhabi: International Renewable Energy Agency. Available at: www.irena.org/publications/2018/Sep/Hydrogen-from-renewable-power.

IRENA (2019). *Hydrogen: A Renewable Energy Perspective*. Abu Dhabi: International Renewable Energy Agency. Available at: www.irena.org/-/media/Files/IRENA/Agency/Publication/2019/Sep/IRENA_Hydrogen_2019.pdf.

Itaoka, K., Saito, A. and Sasaki, K. (2017). Public perception on hydrogen infrastructure in Japan: Influence of rollout of commercial fuel cell vehicles. *International Journal of Hydrogen Energy*, 42, 7290–7296.

Jenkins, J. D. (2019). Why carbon pricing falls short and what we can do about it. *Kleinman Center for Energy Policy*. 24 April. Available at: https://kleinmanenergy.upenn.edu/policy-digests/why-carbon-pricing-falls-short.

Kosturjak, A., Dey, T., Young, M. D. and Whetton, S. (2019). *Advancing Hydrogen: Learning from 19 Plans to Advance Hydrogen from Across the Globe*. Future Fuels CRC. Available at: www.energynetworks.com.au/resources/reports/advancing-hydro gen-learning-from-19-plans-to-advance-hydrogen-from-across-the-globe-ffcrc/.

Lambert, V. and Ashworth, P. (2018). *The Australian Public's Perception of Hydrogen for Energy*. Australian Renewable Energy Agency. Available at: https://arena.gov.au/ assets/2018/12/the-australian-publics-perception-of-hydrogen-for-energy.pdf.

Melaina, M., Antonia, O. and Penev, M. (2013). Blending hydrogen into natural gas pipeline networks: A review of key issues. *Contract*, 303(March), 275–300.

Meldrum, J., Nettles-Anderson, S., Heath, G. and Macknick, J. (2013). Life cycle water use for electricity generation: A review and harmonization of literature estimates. *Environmental Research Letters*, 8, 015031.

METI (Japanese Ministry of Economy, Trade and Industry) (2017). *Basic Hydrogen Strategy*. Ministerial Council on Renewable Energy, Hydrogen and Related Issues. Available at: www.meti.go.jp/english/press/2017/pdf/1226_003b.pdf.

Milbrandt, A. and Mann, M. (2009). *Hydrogen Resource Assessment: Hydrogen Potential from Coal, Natural Gas, Nuclear, and Hydro Power*. Technical report NREL/TP-560-42773. Golden, CO: National Renewable Energy Laboratory. Available at: www .nrel.gov/docs/fy09osti/42773.pdf.

Mission Innovation (2019a). IC8: Renewable and clean hydrogen. *Mission Innovation*. Available at: http://mission-innovation.net/our-work/innovation-challenges/renew able-and-clean-hydrogen/.

Mission Innovation (2019b). Overview. *Mission Innovation*. Available at: http://mission-innovation.net/about-mi/overview/.

Muradov, N. (2017). Low to near-zero CO_2 production of hydrogen from fossil fuels: Status and perspectives. *International Journal of Hydrogen Energy*, 42, 14058–14088.

National Academies of Sciences Engineering and Medicine (2016). Sustainable alternative jet fuels. In *Commercial Aircraft Propulsion and Energy Systems Research: Reducing Global Carbon Emissions*. Washington, DC: The National Academies Press.

Olea, R. A. (2015). CO_2 retention values in enhanced oil recovery. *Journal of Petroleum Science and Engineering*, 129, 23–28.

Preuster, P., Papp, C. and Wasserscheid, P. (2017). Liquid organic hydrogen carriers (LOHCs): Toward a hydrogen-free hydrogen economy. *Accounts of Chemical Research*, 50, 74–85.

Rodrik, D. (2004). *Industrial Policy for the Twenty-First Century*. CEPR Discussion Papers No. 4767. London: Centre for Economic Policy Research.

Schmidt, A. and Donsbach, W. (2016). Acceptance factors of hydrogen and their use by relevant stakeholders and the media. *International Journal of Hydrogen Energy*, 41, 4509–4520.

Schmidt, O., Gambhir, A., Staffell, I., Hawkes, A., Nelson, J. and Few, S. (2017). Future cost and performance of water electrolysis: An expert elicitation study. *International Journal of Hydrogen Energy*, 42, 30470–30492.

Shaner, M. R., Atwater, H. A., Lewis, N. S. and McFarland, E. W. (2016). A comparative technoeconomic analysis of renewable hydrogen production using solar energy. *Energy & Environmental Science*, 9, 2354–2371.

Sheffield, J. W. and Sheffield, Ç., eds. (2007). *Assessment of Hydrogen Energy for Sustainable Development*. NATO Science for Peace and Security Series C: Environmental Security. Dordrecht: Springer Netherlands.

Vogl, V., Åhman, M. and Nilsson, L. J. (2018). Assessment of hydrogen direct reduction for fossil-free steelmaking. *Journal of Cleaner Production*, 203, 736–745.

Weger, L., Abánades, A. and Butler, T. (2017). Methane cracking as a bridge technology to the hydrogen economy. *International Journal of Hydrogen Energy*, 42, 720–731.

Zimmer, R. and Welke, J. (2012). Let's go green with hydrogen! The general public's perspective. *International Journal of Hydrogen Energy*, 37, 17502–17508.

Example Economies

9

Decarbonisation Strategies and Economic Opportunities in Australia

AMANDINE DENIS-RYAN, FRANK JOTZO, PAUL GRAHAM, STEVE HATFIELD-DODDS,
PHILIP ADAMS, ROB KELLY, SCOTT FERRARO, ANDY JONES, ANNA SKARBEK, JOHN
THWAITES, SARAH LEVY AND NIINA KAUTO

Executive Summary

The chapter synthesises the collaborative Deep Decarbonisation Pathways Project (DDPP) involving 16 countries, representing around three-quarters of global greenhouse gas (GHG) emissions, and aimed at showing how individual countries can achieve very low-carbon economies consistent with limiting global warming to 2 °C over pre-industrial levels. While the project was completed in 2014, its general findings are still relevant today.[1]

The chapter's authors are the authors of the Australian Deep Decarbonisation study. The four 'pillars' of decarbonisation are identified as:

- strong energy efficiency improvements, halving energy intensity by 2050;
- make electricity supply low- or zero-carbon – through renewables, nuclear or fossil fuels with carbon capture and storage (CCS);
- electrification and fuel switching – in transport, industry and buildings; and
- reducing non-energy emissions.

Decarbonisation of electricity generation and electrification with zero- or low-carbon electricity are the keys to emissions reductions. Depending on national circumstances, which differ greatly between the 16 countries studied, these can be based on increased renewables and/or combinations of renewables with nuclear and fossil fuels with CCS. Hydrogen is expected to be significant as an energy carrier by 2040.

A particular focus in the chapter is Australia. Australia's emissions from electricity generation accounted for 40% of national emissions. The shift from coal to renewable energy has begun, and with it emissions from electricity have started to decline. However, stationary energy emissions and emissions from industry and transport have continued to rise. Australia has unencumbered prospects for a zero-carbon renewables-based electricity supply that is much larger than today's total power system, as well as widespread electrification.

---

[1] An update of the technology review and scenarios was conducted in 2020 (CWA 2020).

Emissions from buildings can be reduced by 97% by 2050 compared to 2012, through decarbonised electricity replacing gas and energy efficiency improvements. In the industry sector, modelling shows that Australia can double value added while reducing emissions by 50%. Electrification is the key, including electricity for heating, and electric arc reduction in steelmaking. Transport emissions can be reduced by energy efficiency improvement, electrification and the introduction of biofuels and, later, hydrogen.

The remaining emissions, mostly from industry and agriculture and in part related to exports, could be fully compensated for by carbon sequestration through afforestation, for a zero net emissions outcome by 2050.

Decarbonisation would not affect economic prosperity: very low net emissions outcomes in Australia could be achieved while GDP grows at 2.4% average, leading to a 150% larger economy in 2050. Trade would keep growing, with exports increasing by 3.5% per annum. The structure of the economy does not need to change much.

Australia is rich in renewable energy potential, particularly solar and wind, which could establish a new comparative advantage in low-cost zero-carbon energy replacing that based on fossil fuels. This could re-enable energy-intensive manufacturing industries, including for export, such as aluminium smelting and the production of biogas and hydrogen, providing clear prospects for new industries to emerge that will thrive in a low-carbon world economy.

9.1 Introduction

To avoid unacceptable risks of dangerous climate change, the increase in global mean surface temperature must be limited. Limiting temperature rise to 2 °C has been agreed by 141 countries since 2009 and is the focal point in the international climate negotiations (UNFCCC n.d.). The scale of emissions reductions required to achieve the 2 °C limit means that energy and industrial emissions must more than halve by 2050, and net greenhouse gas (GHG) emissions must then approach zero during the second half of this century (Edenhofer et al. 2014).

Global carbon dioxide (CO_2) emissions from today's energy systems and industry are around 34 billion tonnes per year (Edenhofer et al. 2014). This will need to decrease to around 15 billion tonnes by 2050 to have a 50% chance of achieving the 2 °C limit, and to around 11 billion tonnes to have a greater than two-thirds chance (Edenhofer et al. 2014). If the higher figure is apportioned equally across the global population, it equates to 1.6 tonnes per capita by 2050 (assuming a global population of 9.6 billion in 2050, in line with the medium fertility projection of the UN Population Division (UN DESA 2013)).

While countries, cities and corporations have made commitments to reducing emissions, the world is not on track to stay within the 2 °C limit. If stronger pledges are not made, the world is on a trajectory to an increase in global mean temperature of 3.7 °C to 4.8 °C compared to pre-industrial levels. When accounting for full climate uncertainty, this range extends from 2.5 °C to 7.8 °C by the end of the century (IPCC 2014a). These trajectories would lead to catastrophic climate change resulting in grave and irreversible harm to human well-being and development prospects in all countries (IPCC 2014a).

9.1.1 The Global Deep Decarbonisation Pathways Modelling Exercise

The Deep Decarbonisation Pathways Project (DDPP) was a collaborative initiative, convened by the Sustainable Development Solutions Network (SDSN) and the Institute for Sustainable Development and International Relations (IDDRI). Its objective was to understand and show how individual countries could transition to a very low-carbon economy, a process referred to as 'deep decarbonisation', in order to illustrate how the world could meet the internationally agreed target of limiting the increase in global mean surface temperature to well below 2 °C.

The DDPP comprised 16 Country Research Teams, composed of leading researchers and research institutions from countries representing over 74% of global GHG emissions and different stages of development: Australia, Brazil, Canada, China, France, Germany, India, Indonesia, Italy, Japan, Mexico, Russia, South Africa, South Korea, the UK and the USA. (Note: Italy committed to the DDPP after the 2014 UN SDSN report was published, and so is included in the 2015 updated figures.) Working within a common global framework, Country Research Teams developed deep decarbonisation pathways (DDPs) consistent with the objective of limiting the increase in global mean surface temperature to 2 °C. These pathways take into account national socio-economic conditions, development aspirations, infrastructure stocks, resource endowments and other relevant factors. (Note: economic and social costs and benefits have not been included at this stage.) Dividing the global carbon budget between countries is a contentious issue. While a 2050 global average per capita emissions level was used as a benchmark, the pathways focus rather on common, bold actions that will be eventually needed within nearly all countries (IEA 2013).

Keeping below the 2 °C limit is challenging but feasible (IEA 2013). The pathways outline areas that require international support and increased international cooperation, such as research, development, demonstration and deployment (RDD&D) of low-carbon technologies. Countries need to act quickly and in a determined and coordinated manner to keep the 2 °C limit within reach. (Note: it is assumed here that the world invests massively, through global international collaborative public–private partnerships, in the development and early deployment of the necessary technologies.) It requires steep declines in carbon emissions from all sectors of the economy through a profound transformation of energy and production systems, industry, agriculture, land use and other dimensions of human development.

In this chapter, we – the principal authors and modellers of the Australian DDPP study – lay out key results and insights from the Australian deep decarbonisation research.

9.1.2 The Pillars of Decarbonisation

Stabilising GHG concentrations requires large-scale transformation of current systems, especially in energy supply and distribution. The energy supply sector is the largest energy user, accounting for 49% of all energy-related GHG emissions and 35% of anthropogenic GHG emissions (IEA 2012), up 13% from 22% in 1970 (JRC and PBL

2013). The transformation pathway (Edenhofer et al. 2014) required to reduce GHG emissions includes: switching from fossil-fuel-based technologies to low-carbon alternatives such as renewable energy, nuclear power and carbon capture and storage (CCS); implementing energy efficiency improvements; and fugitive emissions reductions in fuel extraction as well as in energy conversion, transmission and distribution systems (IPCC 2014b).

For the DDPP, transformation pathways have been categorised into three 'pillars' (outlined below), which all countries need to act on to successfully decarbonise their energy systems. Within these pillars, each country's DDP shows a wide variety of different approaches based on national circumstances including: socio-economic conditions; the availability of renewable energy resources; and national political imperatives regarding the development of renewable energy, nuclear power, CCS and other technologies. In Australia there is a fourth pillar of non-energy emissions reduction, as non-energy emissions account for over one-third of total emissions.

1. **Energy efficiency**: Greatly improved energy efficiency in all energy end-use sectors including: passenger and goods transportation, through improved vehicle technologies, smart urban design and optimised value chains; residential and commercial buildings, through improved end-use equipment, architectural design, building practices and construction materials; and industry, through improved equipment, material efficiency and production processes and reuse of waste heat.
2. **Low-carbon electricity**: Decarbonisation of electricity generation through the replacement of existing fossil-fuel-based generation with renewable energy (e.g. hydro, wind, solar and geothermal), nuclear power, and/or fossil fuels (coal, gas) with CCS.
3. **Electrification and fuel switching**: Switching end-use energy supplies from highly carbon-intensive fossil fuels in transportation, buildings and industry to lower-carbon fuels, including: low-carbon electricity, other low-carbon energy carriers synthesised from electricity generation (such as hydrogen), sustainable biomass or lower-carbon fossil fuels.
4. **Non-energy emissions**: These emissions can be reduced through process improvements, material substitution, best-practice farming and implementation of CCS. In addition, carbon can be stored in the soil and vegetation, in particular through reforestation, and offset some of the emissions created by other sectors.

9.1.3 Country Results: Similarities and Differences

The DDPP shows that deep decarbonisation with continued economic growth can be achieved in all countries involved. In 2014, results showed that in aggregate, CO_2 energy emissions could fall to 12.3 gigatonnes of CO_2 ($GtCO_2$) by 2050, a 45% reduction from the 22.3 $GtCO_2$ that the initial 15 countries emitted in 2010 (SDSN and IDDRI 2014). The 2015 update puts the aggregate figures for the current 16 countries at 9.8–11.9 $GtCO_2$, a 48–57% reduction below 2010 levels.

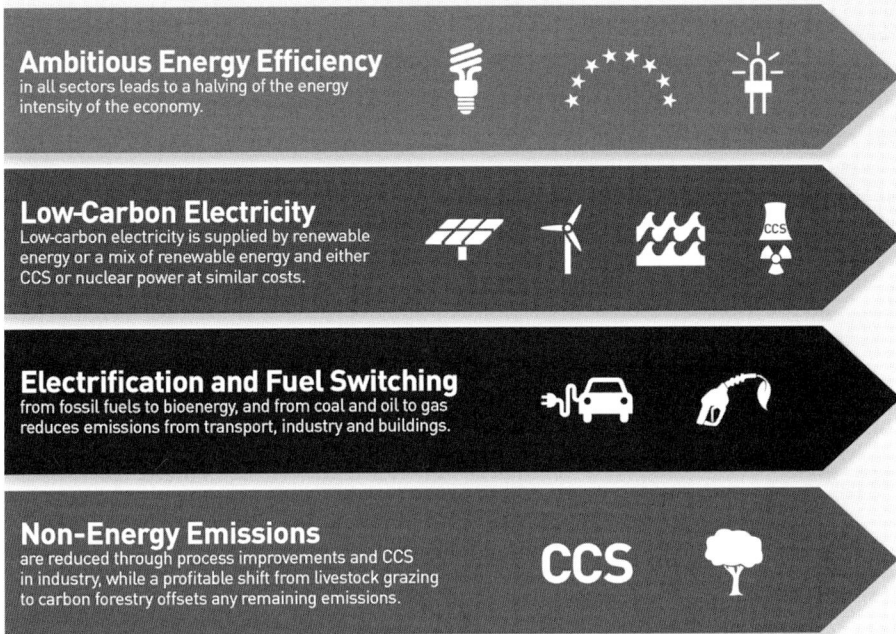

Figure 9.1 Pillars of deep decarbonisation.
Source: Courtesy of the authors.

All of the 16 DDPP countries could achieve an 80–96% reduction in average energy-related emissions per GDP between 2010 and 2050 (Table 9.1). (Note that most countries in the DDPP other than Australia have only reported CO_2 emissions from energy.) This is while population is expected to increase by an aggregate average of 17%, and GDP growth is expected to be around 250% (an average rate of 3.1% per year).

This is led by strong reductions in energy emissions from developed economies and a reversal of growth trajectories in emissions from developing economies. It is important to note that these results represent an initial analysis of the technical feasibility of DDPs within each country: no definitive judgements based on the details of the country DDPs or their aggregate results should therefore be drawn at this stage. All DDPs assume either continued or rapid economic growth to 2050 and the average energy consumption per capita converges to 2 metric tonnes of oil equivalent (toe) by 2050.

9.1.3.1 Sector Contributions

Across the 15 DDPs, different sectors contribute to different levels of CO_2 emissions reductions, as per Table 9.2 (SDSN and IDDRI 2014).

Electrification and the decarbonisation of electricity is a crucial element to decarbonisation. By 2050, power generation is almost completely decarbonised in all countries using varying mixes of renewable energy, nuclear power and/or fossil fuels with CCS. On average, emissions per kilowatt-hour (kWh) are reduced by a factor of 15 below the

Table 9.1. *Estimated change in population, GDP and energy emissions between 2010 and 2050 (%)*

	Population	GDP	Energy emissions
Australia	+57	+148	−71
Brazil	+16	+267	+8
Canada	+41	+203	−100
China	+2	+535	−34
France	+11	+104	−81
Germany	−10	+30	−92
India	+25	+1547	+41
Indonesia	+31	+752	−9
Japan	−24	+64	−84
Mexico	+34	+227	−37
Russia	−15	+163	−87
South Africa	+40	+353	−39
South Korea	−4	+153	−85
UK	+22	+160	−83
USA	+42	+166	−86

Note: Energy consumption is believed to decline in absolute terms in high-income countries and increase in middle-income countries. This is due to energy efficiency improvements outweighing population and GDP growth in high-income countries and improved energy access and rapid GDP growth in middle-income countries.

Table 9.2. *Sectoral shares of CO_2 energy emissions, 2050 compared to 2010*

	Sectoral share 2010 (21.8 $GtCO_2$)	Sectoral share 2050 (11.5 $GtCO_2$)	Emissions in 2050 compared to 2010
Electricity	40%	11%	85% reduction
Buildings	10%	11%	56% reduction
Passenger transport	12%	10%	56% reduction
Freight transport	8%	17%	13% increase
Industry	30%	51%	14% reduction

2010 value. Liquid and gas fuel supplies could be decarbonised using biomass fuels with low embedded carbon emissions and synthetic fuels such as hydrogen (DDPP 2015). The energy supply mix depends on a country's potential for renewable energy, geological storage capacity for CCS and social preferences and degrees of public support for nuclear power and CCS. In the UK's DDP, generation system decarbonisation to 2030 could be largely achieved through the use of wind, nuclear and CCS. Bioenergy is increasingly used in electricity and hydrogen production. By 2040, hydrogen starts to play an important role in the transport, industry and electricity sectors (DDPP 2015).

All 15 DDPs show a decrease in residential energy use, primarily due to electrification (i.e. energy efficiency improvements combined with increased solar photovoltaic (PV) take-up) and an increased use of solar thermal energy and combined heat and power. Electrification is also a crucial element to decarbonising the energy used in passenger transport: this is supported by other strategies such as using biofuels and fuel cell vehicles powered by renewable hydrogen (SDSN and IDDRI 2014). In the US DDP, electricity balancing is done primarily through liquid hydrogen production. Hydrogen and biofuels are used in tandem to decarbonise the transportation (liquid) fuel supply. This allows higher levels of natural gas to remain (Williams et al. 2014).

In industry, the sectoral share of emissions increases but the absolute number decreases: by 2050, industrial emissions account for 51% of total emissions, up from 30% in 2010. In all 15 DDPs, energy consumption in industry is reduced through extensive energy efficiency measures. Most countries can achieve a significant reduction in the CO_2 intensity of energy used in industry through a combination of electrification, fuel switching and CCS. In the China DDP, the application of CCS technologies in power generation and industry sectors is key to achieving deep emissions reductions after 2030. In 2050, CCS is projected to help achieve an annual net removal of around 2737 $MtCO_2$ (870 $MtCO_2$ in the industry sector and 1867 $MtCO_2$ in the power generation sector), this is 32% less than that in the absence of CCS (Bataille et al. 2015).

Freight transport differs from other sectors as there is an increase in total CO_2 emissions. Freight transport is more difficult to decarbonise than passenger transport as low-carbon technology options are less mature. Options such as electrification or sustainable biofuels are challenging to deploy at the scale needed to achieve significant CO_2 reductions. These results highlight the importance of finding additional and innovative ways to reduce emissions in these two sectors (SDSN and IDDRI 2014). Despite its challenges, Canada's DDP outlines the extensive use of third-generation biofuels for various transport modes, especially long-distance and heavy transport (Bataille et al. 2015).[2]

9.2 Building Decarbonisation Pathways for Australia

The aim of the Australian report is to help focus the national climate change debate on the importance of taking a long-term perspective when designing short- and medium-term climate strategies. Backcasting analysis from a 2050 time horizon can help identify technological pathways and the actions needed to achieve them. Government and industry can help focus the national climate debate on this 2050 time horizon by participating in the development and ongoing review of DPPs, and by preparing for their implementation.

---

[2] Since this research was completed, significant progress has been achieved in electrification technologies, especially for short-haul application, and renewable ammonia and hydrogen look promising for several long-haul applications.

The analysis presented demonstrates that, in a low-carbon world, Australia can decarbonise via a range of pathways while maintaining prosperity. It was developed using a combination of well-established modelling tools, with a prominent role for least-cost economic modelling methodology.

9.2.1 Analytical Framework and Modelling Ensemble

The illustrative pathway explores the types of technology transitions that could occur in each sector of the Australian economy as it decarbonises, and the potential associated economic impacts, based on technologies known today. It does not assume major technological breakthroughs, major structural changes in the economy or substantial lifestyle changes. Like all such modelling exercises, it does not attempt to predict the future or claim that the results represent the most likely outcome. This study does not advocate a particular scenario as the most desirable course of action.

The model used for the Australian DDP is grounded in economic modelling, supported by a sectoral analysis of technical emissions reduction potential. The analysis uses a combination of bottom-up sectoral models brought together in a national economic model. The models are well established and have been used in similar exercises before. Figure 9.2 shows a schematic diagram of the main models, processes and data.

Modelling of the Australian economy was carried out by the Centre of Policy Studies using the Monash Multi-Regional Forecasting (MMRF) multi-sector general equilibrium model. The MMRF model is a single-country multiregional model of Australia and its six states and two territories. The current version of the model distinguishes 58 industries,

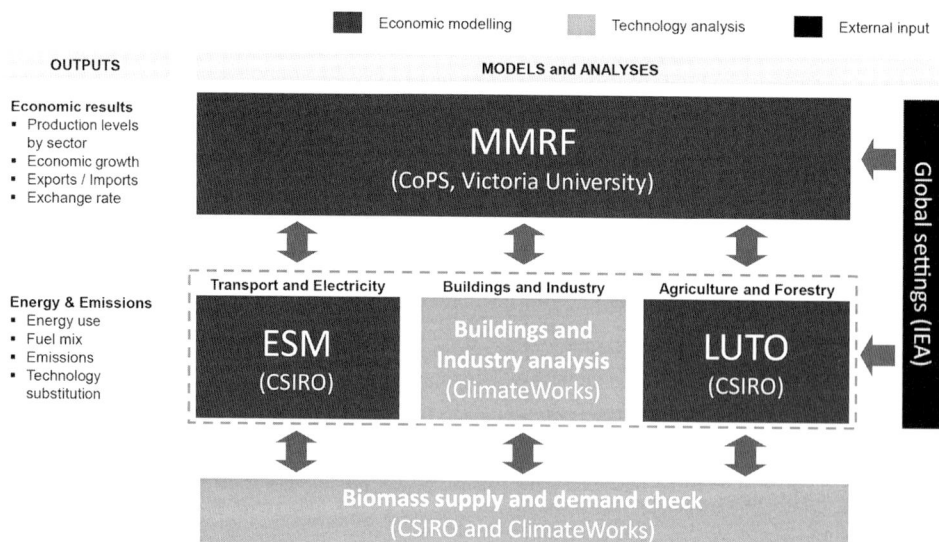

Figure 9.2 The modelling framework.
Source: Courtesy of the authors.

63 products produced by the 58 industries and eight states/territories. At the state/territory level, it is a fully specified bottom-up system of interacting regional economies. More detail is provided in Adams and Parmenter (2013) and Adams et al. (2014). In line with best practice, this general equilibrium model was run in conjunction with detailed sectoral analysis of the technical and economic potential for emissions reduction. This is widely viewed as the benchmark approach in Australia and internationally, and has been the norm in recent Australian climate policy analysis (Garnaut 2008).

The sectoral analyses and modelling includes:

- **Economic modelling of the electricity and transport sectors**. The CSIRO (Commonwealth Scientific and Industrial Research Organisation) Energy Sector Model (ESM) provides least-cost solutions for meeting electricity and transport demand trajectories under given abatement incentives. It builds upon an assessment of the resources and technologies available, as well as the physical constraints applying to those technologies. The results of the modelling were used to inform the MMRF model about fuel demand, technology mix and the activity growth in electricity generation and transport subsectors. Projected emissions trajectories for those sectors were directly taken from the ESM results. The ESM has been extensively applied in Australia, including in every major analysis of climate policy in Australia since 2006 (Reedman and Graham 2011), in most cases working with the MMRF model.

- **Detailed analysis of the emissions reduction opportunities in the buildings and industry sector**. ClimateWorks conducted a detailed bottom-up analysis of the potential for energy efficiency, fuel switching (e.g. from coal/oil to gas and gas to biogas/biofuel or electricity), direct emissions reduction opportunities and deployment of CCS in buildings and industry. The findings from this analysis were used as inputs to the MMRF analysis, and to calibrate the energy and emissions results from the model. Detailed assumptions and sources considered are provided in CWA et al. (2014).

- **Economic modelling of the carbon forestry potential**. The CSIRO Land Use Trade-Offs (LUTO) model was used to develop the potential for land sector sequestration of carbon from non-harvest carbon plantings (including single species eucalypt plantations and mixed species plantings providing carbon and biodiversity benefits), where this would be more profitable than traditional agricultural activities (crops, livestock) under projected future input and output prices, carbon abatement incentives and associated impacts on agricultural production. These results were used to inform the MMRF model about changes in land use and forestry activity, and the supply of land sector offsets. Details of the modelling framework and sensitivity analysis are provided in Bryan et al. (2014).

- **Check on biomass supply and use**. Finally, ClimateWorks and CSIRO collaborated to ensure that the volume of biomass use across the Australian economy was consistent with available resources. The ESM is able to resolve competition for biomass resources between the electricity and transport sectors. Existing CSIRO modelling was also called upon to determine whether biomass volumes were consistent with projected agricultural and carbon forestry activities (Bryan et al. 2014).

The UN SDSN and IDDRI provided a number of global settings for the modelling such as global demand for energy, population and economic growth, technology costs and fuel prices, drawn primarily from the work of their project partner, the International Energy Agency (IEA). A number of features of the modelling framework are new compared to previous domestic modelling exercises. In particular, it is the first time that:

- the updated carbon forestry LUTO model has been used for a whole-of-economy analysis;
- a much more thorough investigation of mitigation potential from industrial production, buildings and transport was used to calibrate the results of the MMRF model in terms of energy use and emissions associated with those sectors;
- a strong shift to electrification in industry and buildings was modelled in Australia;
- modelling of an Australian emissions reduction pathway occurred in the context of a harmonised international modelling exercise.

In addition, many assumptions have been updated since similar modelling was conducted by the Garnaut Review and Commonwealth Treasury Department of Australia. For example:

- the cost of many renewable generation technologies has decreased significantly, for example solar PV technology already costs almost half of what previous studies estimated it would cost in 2030 (SKM MMA 2011 cited in Australian Treasury 2011; IEA–PVPS 2013);
- the cost of technologies used to manage variable electricity supply has decreased significantly, in particular the cost of batteries; and
- biofuels for aviation were thought infeasible in the original modelling by the Garnaut Review (2008) while the first fully biofuel-powered commercial international flight was completed in 2014 (Amyris 2014).

This approach provides a flexible and robust framework, drawing on demonstrated and well-documented models where possible. All modelling frameworks have strengths and weaknesses, however, and the following limitations of the analysis should be noted:

- The consistency of global settings provided by UN SDSN and IDDRI, and the response of the Australian economy and other countries participating in the project, was not tested. It is likely that the sum of individual country responses will not reflect global settings. However, this was a deliberate compromise in the project to focus at a country level rather than integrated global-level modelling. Of particular interest for Australia is the extent to which the demand for Australia's energy-intensive exports changes as other countries implement their decarbonisation pathways.
- This study has checked that biomass consumption does not exceed expected limits, which is an improvement on previous modelling of the same scale and scope; however, it has not checked whether the distribution of biomass resources is economically optimal. The ESM does economically optimise biomass distribution between the electricity and transport sectors, but the remaining distribution of biomass for other energy purposes within different parts of the direct energy use sector was imposed rather than modelled.

- Except where imposed by a scenario assumption, the major driving human behavioural assumption is cost minimisation/profit maximisation, based on economic theory. This is a fairly safe assumption for projecting large asset purchases. However, it may be less reliable for projecting consumer goods purchases, such as electric vehicles and solar panels, where other factors can play a significant role in decision-making.
- The UN SDSN and IDDRI provided a number of global settings for the modelling, such as global demand for energy, population and economic growth, and technology costs and fuel prices, drawn primarily from the work of project partner, the IEA.

9.2.1.1 High-Level Descriptions of the Modelled Scenarios

The modelled scenarios prioritise continued economic growth and focus on technological solutions, with less emphasis on change in economic structure or consumption patterns beyond current projections. Indeed, modelled consumption pattern changes are limited to continuation of recent trends (e.g. continued move towards smaller cars) and modelled changes to the economic structure are limited to a reflection of assumed changes in global trends (e.g. reduced global demand for coal). In terms of technology options, as broad a range of technology options are reflected as possible, rather than focusing on one particular technology, as well as using technologies that exist today or are at some stage of development, rather than anticipating future technological breakthroughs.

By design, the modelled scenarios are ambitious, given that they achieve deeper emissions reductions domestically than previous equivalent exercises, and that they are anchored in the context of a world embarking on a DDP. The analysis makes relatively conservative assumptions in terms of potential future energy efficiency improvements and future technology costs in that context, as well as in terms of future industrial production. For example, technologies, design and behaviour change strategies already exist today to improve the energy efficiency of residential and commercial buildings more than the 50% assumed in the model: since undertaking the modelling, it appears that battery storage costs for households and commercial buildings will fall more quickly (i.e. the Tesla announcement; see BNEF 2000). For example, it has not been assumed that Australia's heavy industry moves offshore, but instead looks at how it could be decarbonised if it remains in Australia. In a similar way, potential step changes in technology have not been included in the example pathway. Qualitatively alternative options have also been explored should some technologies included in the example pathway not be available, or end up being more costly than estimated in the modelling.

9.2.2 Australia's Emissions Profile and Reduction Options

Australia's per capita GHG emissions are amongst the highest in the world. The high emissions intensity of our economy is primarily due to:

- the predominance of coal-fired generation in Australia's electricity supply;
- the relatively large role of energy- and emissions-intensive industrial activity in Australia's economy;

Table 9.3. *Australia's GHG emissions in 2012, (%) total emissions 555 MtCO$_2$ equivalent*

Source	Emissions (%)
Agriculture and forestry	19
Residential	19
Manufacturing	20
Mining	15
Services	14
Electricity, water and gas	5
Transport	5
Construction	2

- the relatively low cost of energy historically and resultant relatively slow progress in improving energy efficiency in many parts of the economy;
- the relatively large role of agriculture in the economy, including beef production; and
- the long-distance transport requirements resulting from the large distances between urban centres.

Between 1990 and 2010, Australia's GHG emissions, excluding land use and forestry, have grown by approximately 30%. Electricity sector emissions grew by nearly 60% between 1990 and 2009 but fell 18% between 2009 and 2013. Emissions from stationary energy (other than electricity), transport, industrial processes and fugitive emissions have all risen substantially since 1990. Emissions from agriculture have risen slightly and emissions from waste have declined. Reductions in forestry emissions are the main reason why overall emissions including land use did not greatly increase. Increasing emissions from energy use between 1990 and 2010 were roughly offset by reduced deforestation and increased plantation forestry (CWA et al. 2014; DOE 2014).[3]

Since 1990, the overall emissions intensity of Australia's economy has almost halved and emissions per capita have decreased by approximately 25% (DOE 2014). Near-term business-as-usual projections see emissions rising over the rest of the decade, as large-scale resources projects (in particular natural gas extraction and liquefaction) come online and the outlook for further reductions in deforestation is limited (CWA 2010). However, falling electricity demand may temper increases in emissions from the electricity sector.

The Australian economy can achieve net zero emissions by 2050 via a range of options. The illustrative pathway for Australia applies four pillars of energy system decarbonisation across the various sectors of Australia's economy: ambitious energy efficiency; low-carbon electricity; fuel switching and electrification; and non-energy emissions. The following sections detail how each pillar contributes to the illustrative deep decarbonisation pathway for Australia. (The data presented for 2012 (Table 9.3) are

---

[3] For more recent analysis of emissions trends, refer to CWA (2018).

directly extracted from the model and may in some instances differ slightly from official energy and emissions statistics.)

9.2.3 Overall Decarbonisation Pathway for Australia

The illustrative pathway shows that Australia could reach net zero emissions by 2050 (see Table 9.4) and achieve its share of the global carbon budget. (Note: in this pathway, the cumulative emissions to 2050 are compatible with Australia's carbon budget, as recommended by Australia's Climate Change Authority (2014), an independent body established under the *Climate Change Act 2011*. This would require strong mitigation action in all sectors of the economy, in the context of a strong global decarbonisation effort.)

For Australia to contribute commensurately to the objective of limiting global temperature increase to less than 2 °C, our energy-related emissions would need to decrease by an order of magnitude by 2050. Australia's DDP presents a pathway by which these emissions are reduced by over 80% on 2012 levels (16.6 tCO_2 per capita) to 3.0 tCO_2 per capita in 2050 (note: this result corresponds to the 100% renewable grid electricity scenario; with the CCS or nuclear scenario, energy emissions would be reduced to 3.2 tCO_2 per capita) and further reduced to 1.6 tCO_2 per capita if emissions directly attributable to the production of exports are excluded (Table 9.5).

Table 9.4. *GHG emissions per capita by source, tCO_2 equivalent per capita, 2012 and 2050*

	2012	2050
Fuel combustion	16.6	3.0
Agriculture	3.9	2.9
Fugitive, process and waste	3.7	1.1
Forestry	0.5	−7.0
Total	24.6	0.0

Table 9.5. *Energy emissions per capita by sector, tCO_2 per capita, 2012 and 2050*

	2012	2050 total	2050 net of exports
Transport	4.2	1.1	0.8
Industry	5.9	1.7	0.6
Buildings	5.1	0.1	0.1
Other	1.5	0.2	0.1
Total	16.6[*]	3.0[*]	1.6

[*] 82% reduction from 2012 to 2050 total.

Table 9.6. *Renewable energy potential and status in Australia*

Renewable energy source	Potential in Australia
Solar	• 58 million PJ (petajoules) solar radiation per year • 10 000 × today's energy use
Wind	• 10 000 PJ per year • >1.7 × today's energy use
Biomass	• ~1000 PJ per year by 2050 • One-sixth of today's energy use
Hydro	• 216 PJ per year • Limited additional potential
Geothermal	• 441 000 PJ of recoverable heat per year • >70 × today's energy use
Ocean	• Supplying <10% of electricity demand by 2050 may require as little as 150 km of coastline

Note: 'Potential' refers to technical potential except for solar (theoretical) and biomass (technical–environmental). 'Today's energy use' refers to total net energy consumption in 2012–2013 (5884 PJ). *Source*: BREE (2014).

Deep decarbonisation can be achieved in Australia while real GDP grows at 2.4% per year on average, resulting in an economy nearly 150% larger than today in 2050 (CWA et al. 2014). (Note: the economic results presented in this section correspond to the illustrative pathway, based on the 100% renewables grid electricity scenario. It is not expected that economic results would change significantly if other electricity scenarios were used instead. See the power sector results for more detail around electricity scenarios.)

Trade would also keep growing, with exports growing at 3.5% per annum. This result is consistent with the findings of many other reports that show that decoupling GDP growth from CO_2 emissions growth is achievable (Edenhofer et al. 2014).

9.2.3.1 Renewable Energy Opportunities

Australia is rich in renewable energy opportunities and has substantial potential for geological sequestration. The potential for generating energy from renewable resources in Australia is far greater than Australia's total energy use today. The challenge for Australia is not the availability of renewable resources, but harnessing their potential (see Table 9.6) (CWA et al. 2014).

9.2.3.2 Land-Use Sector Biosequestration Opportunities

Australia has vast land available for carbon forestry to mitigate residual emissions. Australia has more arable land per capita than any other G20 country (Figure 9.3). This represents a significant opportunity for a range of carbon forestry plantings, which could offset residual emissions from electricity generation, industrial processes and agriculture. Recent modelling by CSIRO has found that carbon plantings could profitably deliver

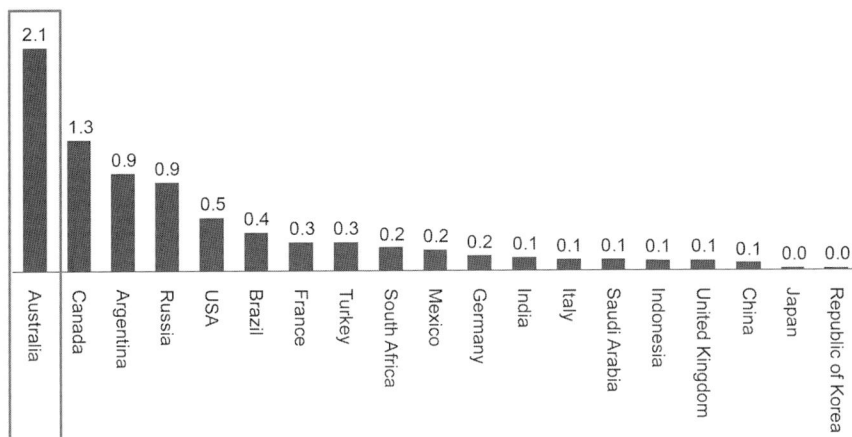

Figure 9.3 Arable land in G20 countries, 2011, hectares per person.

Note: Arable land (hectares per person) includes land defined by the UN Food and Agriculture Organization as land under temporary crops (double-cropped areas are counted once), temporary meadows for mowing or for pasture, land under market or kitchen gardens and land temporarily fallow. Land abandoned as a result of shifting cultivation is excluded.

Source: Akbar et al. (2014).

significant carbon abatement between today and 2050, given the value of carbon emissions reductions implicit in our scenarios. Realising this potential would increase land sector incomes. But it would also require many challenges to be overcome, such as establishing supply chains, as well as supporting services and managing impacts on water availability and food production (Bryan et al. 2014).

If the challenges were overcome, and if there is strong demand for emissions offset credits from other countries, this could potentially enable Australia to become a net exporter of carbon offsets by 2050, when the cost of reducing residual emissions in other countries is higher than the cost of planting carbon forests in Australia (CWA et al. 2014).

9.2.3.3 Opportunities for Australia in a Decarbonised World

Australia's abundant renewable energy resources as well as its geological storage potential could form the basis of a new comparative advantage in low-carbon electricity generation, replacing the existing comparative advantage derived from fossil fuels. The realisation of this comparative advantage could eventually result in a revival of energy-intensive manufacturing industries such as aluminium smelting, and the potential to develop renewable energy carriers for export markets, such as biogas or hydrogen.

The prerequisite for these opportunities is that all major producing economies face strong carbon constraints, either through their domestic frameworks or through import demand favouring products from zero- or low-carbon sources. Australia also has the opportunity to be a global leader in CCS expertise and technology development thanks to its great potential for CCS. The DDPs of a number of countries show very large volumes of CCS, implying a large demand for research and engineering services.

Prospects for the extraction, refining and export of minerals such as non-ferrous metals and ores, uranium, lithium and other precious metals are also good. With ore grades declining globally and energy input costs rising, Australia's world-leading eco-efficient metal ore comminution technologies and services sector can help the mineral industry globally reduce its carbon intensity compared to business as usual.

Australia's substantial potential for carbon forestry, bioenergy generation and biose-questration, and harnessing renewable energy resources located in regional Australia, could also contribute to the economic revitalisation of regional and rural communities, biodiversity protection, improved water quality and energy security (Eady et al. 2009).

Australia's agribusiness sector continues to grow, and there are significant opportunities to reduce emissions significantly from paddock to plate, to further enhance Australia's clean, green reputation for food exports.

Australia's leadership since the Sydney Olympics in low-carbon buildings and precincts, and innovations in everything from low embodied carbon building materials to precinct software-modelling tools, evidences the potential of the Australian green building services sector to expand and increase service exports into Asia. There are also opportunities to improve the productivity of the construction sector through the use of low-carbon prefabricated building materials to speed up the rate of construction.

There is potential also for Sydney to become a centre for carbon trading, this century, further enhancing Australia's financial services sector.

9.3 Decarbonising Australia's Electricity Supply

Decarbonising the electricity sector is critical for decarbonising the Australian economy, as it enables other sectors to decarbonise their activities as well, switching from fossil fuels to electricity where feasible (e.g. buildings, passenger transport).

Low-carbon electricity could be supplied by renewable energy (RE) or a mix of renewable energy and either CCS or nuclear power at similar costs. These three technology types form the basis of the scenarios used to understand possible pathways to deep decarbonisation for Australia, as well as the majority of the generation mix in all DDPP models by 2050 (see Figure 9.4).[4]

While all three scenarios lead to similar emissions intensities by 2050, the 100% renewables grid results in the lowest emissions by 2050. Solar PV systems are projected to play a particularly important role across various scenarios, largely due to the rapid reduction in PV module costs, which have dropped over 90% between 1982 and 1992 and more than 50% between 2010 and 2012 (CWA et al. 2014). Large-scale solar was already at USD 3400/kW in 2012 (BREE 2014) and commercial (>10 kW) rooftop systems are between USD 1500/kW and USD 2100/kW (IEA 2013). For these reasons, solar PV was selected for the whole-of-economy modelling (CWA et al. 2014).

---

[4] Please note that more recent analyses suggest 100% renewables is the most cost-effective way to decarbonise the electricity sector, with greater shares of solar, wind and hydro generation.

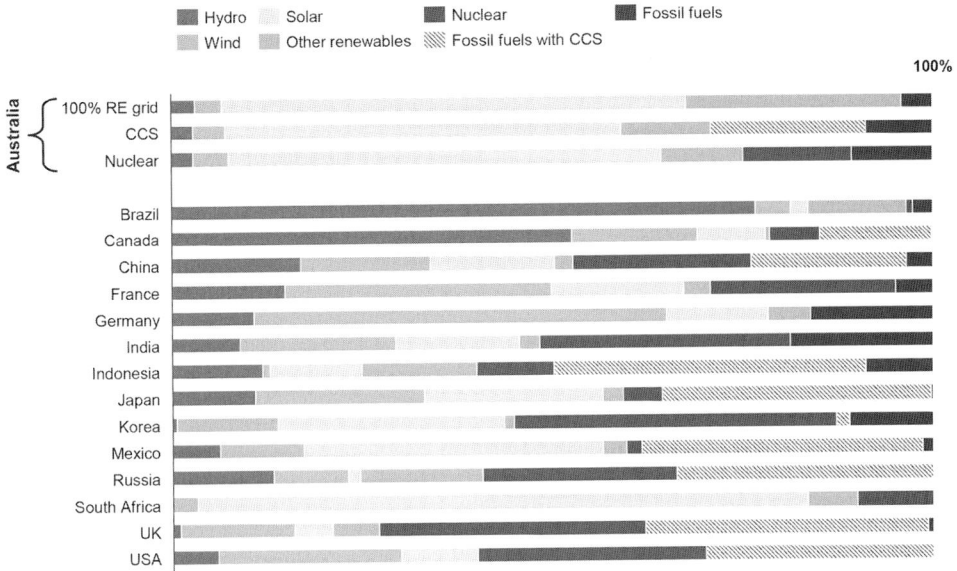

Figure 9.4 Electricity generation mix in 2050 for the 15 DDPs.
Note: To achieve deep decarbonisation, a variety of scenarios beyond these three are possible. A wider mix of technologies may be less likely, given that any of those technologies requires significant investment to deploy and support them (e.g. storage facilities for CCS, intermittency management for renewables, and radioactive waste management infrastructure for nuclear), so a focused strategy is likely to result in lower system costs.
Source: Denis et al. (2014: 18).

The nuclear scenario could achieve similar emissions levels if constraints were applied to the share of gas generation allowed in the generation mix, which was not done in this modelling exercise. Electrification across all sectors drives a 2.5-fold increase in electricity demand by 2050.

The modelled scenarios highlight a number of key findings (CWA et al. 2014):

- All scenarios include a dominant share of renewables, driven by the decrease in cost of renewable technologies such as solar and wind over recent years with a minimum penetration of 48% by 2030 and 71% by 2050 (Trancik 2014). They are expected to be the lowest-cost technologies to achieve decarbonisation until their penetration requires significant additional costs for the management of variability.
- The major difference between scenarios is how the variability of wind and solar is managed. In the CCS and nuclear scenarios, backup for variability is met by these technologies[5] combined with peaking gas, while in the 100% renewables grid scenario

---

[5] For the nuclear scenario, this is the case for the east coast. In the main scenario, nuclear generation has been excluded for Western Australia given the high level of uncertainty on the future structure of the grid and the potential transmission and distribution costs associated in creating a large enough grid in the state to accommodate large-scale nuclear plants. If nuclear generation was included in Western Australia, then nuclear generation would amount to 27% of total generation by 2050.

it is met by combining storage with renewables and use of non-variable renewable technologies such as geothermal.[6]

- Solar becomes the dominant technology by 2050. The high share of solar power (either PV or solar thermal) in the electricity generation mix is a reflection both of its cost advantages and also that a third of electricity consumption occurs in Western Australia, due to increases in mining activity and electrification of mining processes, where conditions for solar power are particularly favourable. Taking into account the need to invest in backup capacity to cover variable supply, solar becomes more profitable than wind power towards 2050.

Projected wholesale electricity prices for the three scenarios differ very little over the projection period, reflecting that the technology costs of the higher-cost renewable technologies, nuclear and CCS, are in a similar range. An analysis of retail prices was conducted for the 100% renewable grid scenario. Including an increase in transmission and distribution costs, it is expected that retail prices would increase at an average rate of 0.9% per year or around 40% to 2050, but average household electricity use (excluding for electric vehicles) would fall by half, so that average household power bills could be reduced by 30%. Taking into account a projected 56% increase in average per capita incomes to 2050, the share of electricity expenditure in household income is halved, on average, for households that are using only electricity today. (Note: if there were any costs involved in reducing electricity use per capita via energy efficiency measures, the cost of these measures would need to be included in a full analysis of electricity supply and end-use costs.)

9.4 Energy Efficiency and Fuel Switching in Buildings, Industry and Transport

Ambitious energy efficiency in all sectors (see Table 9.7) could lead to a halving of the energy intensity of the economy. In the illustrative pathway, the final energy use associated with each dollar of GDP halves by 2050. This is driven by strong improvements in energy efficiency in all sectors of the economy (CWA et al. 2014).

Technological and fuel-related solutions are primarily used to reduce emissions as these solutions are broadly applicable at a national scale and largely result in maintaining our current lifestyles (CWA et al. 2014). Electrification and fuel switching from fossil fuels to bioenergy, and from coal and oil to gas, reduces emissions from transport, industry and buildings (see Table 9.8).

9.4.1 Buildings

Emissions from residential and commercial buildings could reduce by 97% between 2012 and 2050, even with a substantial increase in the number of households and size and output of the commercial sector. This emissions reduction is achieved through energy

---

[6] Please note that more recent analyses suggest storage (battery, pumped hydro) is likely to be more cost-effective than non-variable renewable technologies.

Table 9.7. *Energy efficiency improvements in the productive sectors 2010–2050*

	Improvement by 2050	Activity	Example
Buildings sector	Reduction in energy use by 2050: over 50% in residential; just under 50% in commercial.	Efficient new builds and replacing equipment by best-practice models at the end of its useful life.	LEDs (light-emitting diodes) can reduce energy use by almost 80% compared to halogen globes, and can even provide 25% savings compared to efficient compact fluorescent lamps (CWA 2013).
Manufacturing sector	40% decrease in energy intensity of production.	Process improvements and equipment upgrades for existing plants. Implementing best-practice technologies at the time of construction.	Reducing thermal losses from heating processes (furnaces, kilns and boiler systems). Capturing waste heat to preheat materials and reducing the fuel inputs required to perform other industrial processes (CWA 2013).
Mining sector	40% decrease in energy intensity of production.[*]	Operational improvements.	Changing the gradient of the slope upon which vehicles travel, reducing the amount of time vehicles stop and start and improving load management (CWA et al. 2014).
Transport sector	Improvements in energy efficiency: 70% cars and light vehicles; 30% aviation; 22% marine; 17% rail; and 15% freight (Cosgrove et al. 2012).	Electrification and fuel efficiency improvements. Continuation of trend towards smaller vehicles. New technology development.	Hybrid vehicles commercially available today achieve up to 65% improvement in fuel efficiency compared to an average car (Allianz Australia 2014). Today, the A380 is already 18% more efficient per passenger seat than the previous generation of large aircraft (Cosgrove et al. 2012).

[*] Mining energy efficiency improvements are counterbalanced by a structural increase in energy intensity. Past energy intensity trends show that every year, around 3% more energy is needed to extract a similar volume of minerals as the year before, due in particular to a degradation in ore quality and increasingly difficult access to good resources. As a result, mining energy intensity doubles between today and 2050. *Source*: See citations throughout table.

Table 9.8. *Fuel switching improvements in the productive sectors 2010–2050*

	Improvement by 2050	Activity	Example
Buildings sector	Near elimination of emissions from buildings.	Move from gas to electricity.	Electricity from a decarbonised energy system used for all heating, hot water and cooking equipment.
Industry sector	Approximately 60% reduction in energy emissions.	Shift from coal and oil use towards electricity, bioenergy and gas.	Shift to electricity for heating processes, a shift in mining from trucks to electricity-based technologies such as conveyors for materials handling.
Transport sector	Oil use lowered 85%; CO_2 emissions reduced by two-thirds.	Shift from oil to natural gas. Shift from internal combustion engines to electric and hybrid drivetrains, and to a lesser extent hydrogen fuel cells.	Air travel can be replaced by electric fast rail, especially between east coast cities. Biofuels can replace oil use in aviation. Marine and rail sectors can switch from oil to natural gas and biofuels.

efficiency and a switch to a decarbonised energy supply. In the modelled pathway, energy use per household reduces by over 50%, while heating and cooling and other equipment use continues to increase. In the commercial sector, energy use per square metre reduces by just under 50% and energy use per dollar of value-add in the commercial sector reduces by almost 70% (CWA et al. 2014).

Converting direct fuel combustion such as gas stoves, gas hot water and boilers to electrical systems powered by decarbonised electricity supply results in substantial emissions reductions. Achieving these emissions reductions requires a near full conversion of the energy system to low-emissions generation, such as renewables, fossil fuels with CCS, or nuclear. In order to achieve this substantial decarbonisation, rates of improvement in energy efficiency need to increase from current low levels. This will not require a substantial technological leap from today's best available technology and the cost of energy saved is likely to offset the costs required for investment in more efficient buildings and equipment (CWA et al. 2014).

9.4.2 Industry

In industry, modelling shows that industrial value added could more than double in real terms between 2012 and 2050. In parallel, emissions could decrease by more than 50%. Most of the reduction in energy emissions intensity would result from the reduction in emissions intensity of the fuel mix, in particular through electrification. Electrification is a key component of deep decarbonisation. It is likely to be driven by three major technology groups: increase in iron and steel production from electric arc furnace technology; shift to electricity for heating processes; and shift from trucks to conveyors for materials handling in mining.

It is assumed that most coal and oil use in manufacturing could be switched to gas, with half of the remaining oil use able to be shifted to biofuels in mining, and that 15% of remaining gas use could be shifted to biomass/biogas in manufacturing. In most sectors, it is assumed that process emissions intensity could be reduced by over two-thirds, through a combination of sector-specific measures and implementation of CCS. Similarly, it is assumed that fugitive emissions intensity could be reduced by around half to three-quarters in resource extraction sectors, mostly through the use of methane flaring/oxidation and CCS. Non-energy emissions are well suited for the use of CCS, given the relatively high purity of CO_2 outflows (CWA et al. 2014).

9.4.3 Transport

As electricity generation switches to low-carbon energy sources, such as renewable energy technologies, nuclear power or CCS, electricity becomes the least emissions-intensive energy source. This drives widespread electrification across transport, buildings and industry and results in substantial decreases in emissions from these sectors. As a result, electricity's share in final energy use increases from 22% today to 46% in 2050. Fuel switching from fossil fuel to bioenergy and from coal to gas drives further emissions reductions (CWA et al. 2014).

Bioenergy utilised in the modelled pathway is exclusively sourced from second- and third-generation feedstocks, meaning there is no significant impact on agricultural production. These feedstocks include agricultural and forest residues (e.g. bagasse, stubble and sawmill residues), wastes and energy crops such as pongamia (a type of oilseed tree), grasses, algae and coppice eucalyptus (mallee), and account for approximately 1000 PJ of potential. Some of these feedstocks could be replaced and/or complemented by wood waste from the newly planted carbon forests, or other feedstock depending on the relative costs of production (CWA et al. 2014). Overall, road transport GHG emissions decrease by 69% to 23 $MtCO_2$ equivalent ($MtCO_2e$) by 2050. Non-road transport emissions increase to 5% above their current level by 2050, reflecting the greater challenges in abating emissions from those transport modes (CWA et al. 2014).

Table 9.9 summarises the status of the major technologies involved in the deep decarbonisation scenario for the transport sector. As can be seen, the technologies required for decarbonisation are already being developed; integration and commercialisation is where most progress is needed (CWA et al. 2014).

Table 9.9. *Summary of technology status in the transport sector*

Current technology status

Legend:
- ✓ Yes
- ✔ Yes, in some cases
- Empty No

Element of pathway	Modelling assumptions	Mature technology	Cost effective	Equal performance	Widespread implementation	Examples of current progress	Improvement required
Change in transport activity levels / mode shift	• 15% mode shift from air to rail; • 5% shift from air to teleconferencing	✓	✓	✓	✓	• Some companies are implementing initiatives to reduce air travel • Fast train replaces air travel in many countries (e.g. Japan, France)	• Development of infrastructure for mode shift
Energy efficiency (for oil based vehicles)	Decrease in energy intensity 2012–2050 • Card: ~75% • LCVs: ~80% • Buses: ~50% • Trucks: ~75% • Rail: ~20-25% • Air: ~33% • Water: ~30%	✓	✓		✓	• Australian new light vehicle efficiency improved by 20% from 2002 to 2012 • A380 is 14% more efficient than the average, and new aircraft 80% more efficient than they were in the 1960s • Efficiency in Australian rail transport has increased of 33% since 1990	• Uptake of best available technology and processes, often cost-effectively over the life of the vehicles • Continuous improvement and development
Electric vehicles (EVs) and plug-in hybirds (PIHs)	• EVs: 34% of cars/LCVs by 2050 • Plug-in hybirds: 38% cars/LCVs	✓	✓	✓		• 5% of 2013 Dutch sales are EV. PIH • Range for EVs now up to 425km • Tesla Model S EV voted best overall vehicle by consumer reports	• Reduction in costs through large scale commercialisation and supply chain development • Development of infrastructure of recharging

Technology	Projection					Current status	Development priorities
Fuel cell cars	• 22% cars/ LCVs in 2050	✓				• Demonstration cars exist • Commercial vehicles expected to be released by Toyota in 2015	• Reduction in costs through large scale commercialisation and supply chain development • Development of hydrogen infrastructure and supply chain
Shift to gas and biofuels	• 70% gas for road frieght • 50% biofuels for air transport	✓	✓	✓		• Gas used for 3.9% of road transport in 2012 • First 100% biofuel-powered international commercial flisht 2014	• Development of gas supply infrastructure • Adaptation for Australian applications • Large-scale supply of biofuels
Biomass availability/ production	• Additional 1050 PJ (~7×current bioenergy use) • 70% from energy crops including coppice eucalupt, pongamia, grasses and algae • 30% forestry and agricultural residues (bagasse and stubble)	✓	✓	✓		• Bioenergy contributed 4% of total energy use in 2012–13, mainly from bagasse, wood and wood wastes • Cultivation of grasses is highly developed • Australia has two of the world's largest commercial algae plants producing high-value (non-energy) products	• Development of supply chain for energy crops & stubble that are not yet used for energy. • Reduction in costs through large scale commercialosation & supply chain development • ongoing management of land use impacts and competition for resources
Bioenergy production/ use	• Liquid biofuels represent the clear majority (85%) to be used in mining and transport • 12% biocoke as reductant in iron and steel production • 3% biogas/ biomas in manufacturing	✓	✓	✓		• Significant use of bioenergy where feedstock is cheap, readily available and easily converted into energy, e.g. heat and power from bagasse and landfill biogas, and first-generation biofuels from food industry wastes/co-products • Some progress in producing bioenergy from feedstocks that are less readily converted into liquid biofuels, e.g. Mackay pilot plant (QLD) producing ehtanol from various feedstocks	• Further development and commercialisatino of advanced biofuels and biocoke • Significant supply chain development, integration and deployment to • Management of lifecycle emissions through entire supply chain

Source: CWA et al. (2014), where further details and references may be found.

9.5 Reduction of Non-Energy Emissions and Carbon Forestry

Non-energy emissions from industry are reduced through CCS and process improvements while a profitable shift from livestock grazing to carbon forestry offsets any remaining emissions; see Table 9.10 (CWA et al. 2014).

Australia has substantial potential to offset emissions via land sector sequestration. The illustrative pathway includes a shift in land use towards carbon forestry, where profitable for landholders under carbon abatement incentives, but it does not include the sale of permits into overseas markets, nor the purchase of permits from other countries. The analysis found that there is more than enough economic potential to shift land use to carbon forestry to offset all residual emissions to 2050, allowing Australia to reach net zero emissions by 2050 (CWA et al. 2014).

Modelling was used to estimate the volume of carbon that would be profitable to supply, where delivery of carbon credits would provide higher economic return than competing agricultural land uses. The illustrative carbon forestry scenarios are calibrated to provide 4.3 $GtCO_2e$ and 4.8 $GtCO_2e$ over the period to 2050, exactly offsetting the difference between the cumulative emissions budget assumed for the project and the projected emissions from all sources for the 100% renewables energy scenario and the CCS scenario. The projected supply of profitable sequestration is up to 16 GtCO2e over the period and three to four times the volume of credits required to achieve the project's cumulative emissions budget, with zero net emissions by 2050 (CWA et al. 2014).

Table 9.10. *Emissions reduction activities in forestry and agriculture*

	Improvement by 2050	Activity	Example
Industry	Process and fugitive emissions reduced.	Process optimisation and equipment upgrades.	Partial use of bio-coke in iron and steel production, CCS and increased combustion/catalysing of gases with high global warming potential.
Agriculture	Emissions reduced from farming and livestock.	Best practice farming techniques.	For beef production this includes intensification of breeding, improvement in feeding and pasture practices, as well as enhanced breeding and herd selection for lower livestock methane emissions (Herrero et al. 2013).
Carbon forestry	Residual emissions offset to enable deep decarbonisation by 2050.	Forestry biosequestration.	Large shifts in land use from agricultural land (in particular livestock grazing) to carbon forestry.

9.6 Economic Implications

Under the deep decarbonisation pathway, the overall structure of Australia's economy does not change significantly (see Table 9.11). The commercial sector's contribution to the economy continues to grow at a similar rate as over the past four decades (ABS 2014), while the share of manufacturing continues on a gradual decrease, although more slowly than to date (ABS 2014). The agricultural and forestry sector maintains a similar share of GDP. Traditional mining and manufacturing industries continue to grow in terms of real value added (but more slowly than other sectors, so they shrink as a percentage of the total), including iron ore (138%), metal ore (150%), other mining (329%), other chemical production (113%) and aluminium, iron and steel production (37%), with a decrease in coal, oil and petroleum (CWA et al. 2014).

Although the overall structure of the economy remains largely the same, significant changes occur within some sectors. Modelling by the International Energy Agency (IEA 2013) estimates that under a deep decarbonisation scenario, global demand for coal decreases by 40% between today and 2050, resulting in a decrease in the unit price for coal, while global demand for oil decreases by 30% over the same period. At the same time, coal demand from Australia's key export markets (namely Japan, South Korea and China) decreases by more than 50%, strongly impacting fossil fuel extraction in Australia, and lowering coal prices (IEA 2013). In all three of the electricity generation scenarios in this analysis, electricity generated from coal decreases. For the 100% renewables scenario and nuclear scenarios the decrease is 100%, whereas under the CCS scenario it is 70% (CWA et al. 2014). The decline in the contribution of these sectors to the Australian economy is offset by the increase in renewable electricity generation (excluding hydro) and gas extraction. Furthermore, the analysis shows job creation in the renewable electricity generation sector is double the job losses from the coal-fired electricity generation sector. (Note: any changes in job numbers in a particular industry will be compensated by changes in employment in other parts of the economy over the long time periods considered in this modelling.)

The DDPP results show that global demand for gas increases under a deep decarbonisation scenario, largely due to its use in industry and road freight. This is supported by the

Table 9.11. *Sectoral contribution to GDP (%)*

Sector	2012	2050
Commercial	64	72
Mining	10	8
Manufacturing	9	7
Energy and construction services	8	7
Transport	4	4
Agriculture and forestry	3	2
Power	1	1
Residential	0.4	0.2

Note: Numbers do not add to 100% due to rounding.

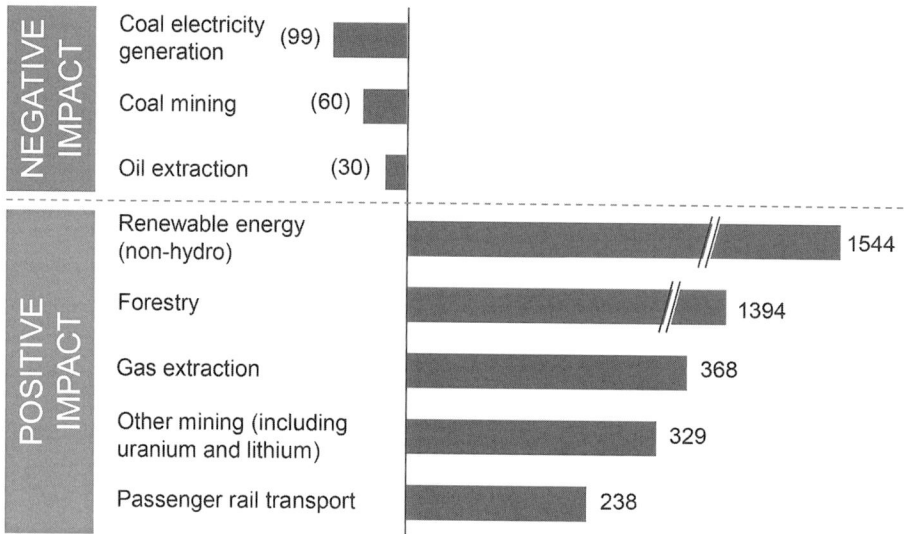

Figure 9.5 Key sectors impacted by decarbonisation, growth in value added between 2012 and 2050 in %.
Source: Courtesy of the authors.

IEA (2013) results, which suggest that global demand for gas increases by around 15% between today and 2050. This increase in global demand contributes to an increase in gas extraction in Australia.

Carbon forestry expands strongly due to its role in offsetting non-energy GHG emissions. Also, global demand for minerals such as uranium and lithium increases (IEA 2013), leading to increased mining production in Australia. Results from the DDPP show that nuclear power quadruples across the 12 DDPP countries, due to its role in decarbonising the electricity systems of many countries (in particular China and the USA), while global demand for lithium is driven by the widespread uptake of batteries in the electricity and transport sectors. Figure 9.5 shows an extract of the economic results for the illustrative pathway. It shows the sectors most impacted, either positively or negatively, by deep decarbonisation.

9.6.1 Deep Decarbonisation Can Be Achieved While Maintaining Prosperity

The modelling results show that deep decarbonisation can be achieved while real GDP grows at 2.4% per year on average, resulting in an economy nearly 150% larger than today in 2050. (Note: the economic results presented in this section correspond to the illustrative pathway, based on the 100% renewables grid electricity scenario. It is not expected that economic results would change significantly if other electricity scenarios were used instead.) Trade would also keep growing, with exports growing at 3.5% per annum. This result is consistent with the findings of many other reports that show that decoupling GDP growth from CO_2 emissions growth is achievable (Edenhofer et al. 2014).

This study explores the feasibility of Australia achieving deep decarbonisation, in the context of international action to limit global warming to 2 °C. Although cost comparisons with a no-action scenario are not the purpose of this report, many stakeholders will be interested in the projected cost of achieving this decarbonisation pathway. Devising an appropriate reference scenario with which to compare the decarbonisation scenario is inherently difficult. Australia taking no abatement action is inconsistent with both current Government and Opposition policy, and would appear risky from a geopolitical perspective, given that most major countries are already undertaking measures to cut emissions. A realistic scenario of 'no action' by Australia in the context of global deep decarbonisation would thus need to consider potential responses by our trading partners, including possible trade sanctions and other adverse ramifications in Australia's external political and commercial relationships.

Furthermore, there are well-known fundamental shortcomings in estimating the economic cost of emissions reductions using the type of economic models employed in the present analysis, and similar exercises:

- Standard computable general equilibrium models of an economy, in this case the MMRF model, do not include the physical and economic impacts of climate change, and therefore ignore the economic benefits arising from a reduction in climate change impacts over time. These omitted benefits include direct economic effects, benefits from reduced risk of extreme climate impacts or crossing global tipping points, and non-market values, for instance the existence of iconic ecosystems such as coral reefs (Garnaut 2008).
- The models ignore immediate co-benefits from climate change mitigation. For example, there are health and labour productivity benefits from reduced urban air pollution from combustion of fossil fuels (including transport), and benefits for energy security from more stable energy system costs and reduced demand for oil imports (Fleurbaey et al. 2014).
- Standard economic models are not well suited to represent the potential for growth-enhancing effects of mitigation policies. For example, increased energy efficiency could lead to productivity gains throughout the economy, and increased global innovation could lift trend global growth rates. Research by the ClimateWorks Foundation for the World Bank suggests that improving energy efficiency performance could boost global GDP by USD 1.6–2.6 trillion per annum above business as usual by 2030 (Akbar et al. 2014).

Comparisons between estimated economic activity in the policy and reference case is therefore not a useful guide to whether deep decarbonisation is beneficial. Recent research demonstrates that the world pursuing strong mitigation action, compatible with a 2 °C target, is desirable and could yield economic benefits (Global Commission on the Economy and Climate 2014). A significantly warmer world would pose impacts and risks that are generally seen as economically, socially and environmentally unacceptable (IPCC 2014a).

Nevertheless, it is useful to compare estimates of economic costs (see Table 9.12). The main insight is that the estimated costs in the DDPP modelling study are in the same range as previous modelling exercises. This is despite emissions levels in the present study's

Table 9.12. *Comparison with cost estimates in previous studies* (%)

	Annual average growth		Annual average growth relative to reference case		Economic parameters at 2050, compared to reference case			Emissions level relative to reference, 2050		Emissions reductions from 2000 to 2050	
	GDP	GNI per person	GDP	GNI per person	GDP	GNI per person	PFC[*]	DE[***]	NE[***]	DE[***]	NE[***]
DDPP scenario	2.4	1.1	–0.19	–0.12	–6.6	–4.6	–4.9	–100	–100	–100	–100
2011 Treasury 'high price' scenario	2.5	1.0	–0.21	–0.19	–4.7	–7.1	–8.3	–66	–	–42	–80
2008 'Garnaut 25' scenario	2.2	1.1	–0.14	–0.17	–5.8	–6.7	–6.5	–83	–	–69	–90

[*] Real private consumption; [**] Domestic emissions; [***] Net emissions. GDP, gross domestic product, GNI, gross national income.
Source: DDPP (2015). Australian Treasury (2011) and Garnaut (2008).

analysis coming down to much lower levels than in previous exercises, and the technical reference case overstating the amount of action required. Like all previous studies, this study finds that Australia can achieve substantial reductions in emissions while maintaining robust economic growth (Australian Treasury 2011).

The primary reasons why this study finds greater emissions reductions could be achieved at similar macroeconomic effects are that:

- the observed and projected costs for many zero- and low-emissions technologies have fallen significantly over recent years;
- this analysis includes a much more thorough investigation of mitigation potential from industrial production, buildings and transport; and
- the analysis conducted for this study suggests a larger supply of profitable carbon sequestration from land-use change and forestry.

9.6.2 Decarbonisation of Australia's Economy Can Be Achieved Without Significant Structural Shifts or Lifestyle Changes

Deep decarbonisation in the illustrative pathways is mostly characterised by technological transitions. Our everyday needs continue to be met in a manner similar to what we experience today. Similarly, since deep decarbonisation does not drive significant structural shifts in the economy, employment is predominantly in the services sector, as it is today, as well as in the industrial sector.

Due to widespread electrification, the most noticeable change is that the majority of household and commercial building energy comes from electricity, supplied via a centralised grid, as it is today, or via localised distributed generation, augmented by battery storage.

Buildings are heated and cooled more efficiently, they contain more efficient, smarter appliances and the services they offer are mostly electrified. For example, highly efficient (possibly induction) electric cooktops replace gas stoves and electric heaters replace gas heaters.

A wide variety of vehicles are available, mostly fuelled by low-carbon fuels, including electricity and hydrogen, while plug-in hybrid vehicles are available for longer trips, or where vehicle range is a constraint. Air travel is available, with biofuels gradually taken up in the aviation sector, and rail provides a viable alternative to air travel on some routes. There are a number of lifestyle changes that could further reduce emissions, which have not been included in the analysis. These changes may also be driven by other social, economic and environmental factors, and could include:

- Smaller houses, greater range of tolerance in heating/cooling requirements (where feasible), less travel, more widely available public transport, less emissions-intensive consumer products and decreased beef consumption.
- Substitution of business travel with teleconferencing, and preferential sourcing of less emissions-intensive products and services.

9.7 Conclusion

Deep decarbonisation creates opportunities for Australia. In a decarbonised world, Australia's abundant renewable energy resources, as well as its geological storage potential, could form the basis of a new comparative advantage in low-carbon electricity generation, replacing the existing comparative advantage derived from fossil fuels. The realisation of this comparative advantage could eventually result in a revival of energy-intensive manufacturing industries such as aluminium smelting, and the potential to develop renewable energy carriers for export markets, such as biogas or hydrogen. Australia's substantial potential for carbon forestry, bioenergy generation and biosequestration could also contribute to the economic revitalisation of regional and rural communities, biodiversity protection and improved water quality (Eady et al. 2009).

The technologies required for decarbonisation are available or under development. However, further efforts in commercialisation, enhancement and integration will improve cost-competitiveness and performance. The rate of development of low-emissions technologies has progressed rapidly in recent years, with many technologies now mature in Australia or other similar economies. For some, such as energy efficiency, cost savings can be achieved without any policy measures. Some of the technologies in the illustrative pathway would require further development to improve performance or reduce costs. Where technologies are not yet mature, pilot projects will demonstrate the potential of the technology to be deployed at a large scale. Deployment allows continuous improvement through 'learning by doing' in the manufacturing, supply and operation of the technology. For example, the development of solar PV technology has seen prices reduce by approximately 10% per year with production increasing by 30% per year over the last 30 years (Trancik 2014). There has been rapid improvement in battery storage for renewables integration and electric vehicles. Electric vehicles can already offer superior performance to conventional internal combustion engines.

While decarbonisation is a significant transition for Australia, it is achievable. Deep decarbonisation requires a transition in how we both use and produce energy, and also how we manage our land to sequester remaining emissions. Shifts of this nature have been made in the past: Australia's economy has demonstrated that it is flexible, adaptable and resilient, and has a long history of benefiting from new trends in the global economy. In the past, Australia's prosperity has been built on gold mining and wool production; today the main export drivers include tourism, education, manufactured exports (including food and beverage products), coal and iron ore production and other minerals extraction. The transitions were made without damage to Australia's overall economic fortune. In fact, adapting to new circumstances historically has benefited Australia. Many factors will drive change in Australia's economy over the next decades. Likely global growth areas that Australia can benefit from are agribusiness, renewable hydrogen and green metals, tourism and international education. Decarbonisation would be just one of many influences.

References

ABS (Australian Bureau of Statistics) (2014). 5206.0 – Australian national accounts: National income, expenditure and product, Mar 2014: Table 6. Gross value added by industry, chain volume measures. *Australian Bureau of Statistics.* Available at: www.abs.gov.au/AUSSTATS/abs@.nsf/DetailsPage/5206.0Mar%202014? OpenDocument.

Adams, P. D. and Parmenter, B. R. (2013). Computable general equilibrium modelling of environmental issues in Australia: Economic impacts of an emissions trading scheme. In P. B. Dixon and D. Jorgenson, eds., *Handbook of CGE Modelling, Vol. 1A.* Oxford: North-Holland, pp. 553–657.

Adams, P. D., Parmenter, B. R. and Verikios, G. (2014). An emissions trading scheme for Australia: National and regional impacts. *Economic Record,* 90, 316–344.

Akbar, S., Kleiman, G., Menon, S. and Segafredo, L. (2014). *Climate-Smart Development: Adding Up the Benefits of Actions That Help Build Prosperity, End Poverty and Combat Climate Change.* Main report, Vol. 1. Working Paper 88908. Washington. DC: International Bank for Reconstruction and Development/The World Bank and ClimateWorks Foundation. Available at: http://documents.worldbank.org/curated/en/ 794281468155721244/pdf/889080WP0v10RE0Smart0Development0Ma.pdf.

Allianz Australia (2014). Top fuel efficient cars. *Allianz Car Insurance News.* Available at: www.allianz.com.au/car-insurance/news/top-fuel-efficient-cars (site discontinued).

Amyris (2014). First international commercial flight completed with newly approved Amyris-Total aviation biofuel [press release]. *Amyris.com.* 31 July. Available at: https://investors.amyris.com/2014-07-31-First-International-Commercial-Flight-Completed-With-Newly-Approved-Amyris-Total-Aviation-Biofuel.

Australian Treasury (2011). *Strong Growth, Low Pollution: Modelling a Carbon Price.* Canberra: Commonwealth Government of Australia.

Bataille, C., Sawyer, D. and Melton, N. (2015). *Pathways to Deep Decarbonization in Canada.* Sustainable Development Solutions Network and the Institute for Sustainable Development and International Relations. DOI: 10.13140/ RG.2.2.30696.70401.

BNEF (Bloomberg New Energy Finance) (2000). Battery pack prices cited below $100/ kWh for the first time in 2020, while market average sits at $137/kWh. *BloombergNEF.* 16 December. Available at: https://about.bnef.com/blog/battery-pack-prices-cited-below-100-kwh-for-the-first-time-in-2020-while-market-average-sits-at-137-kwh/.

BREE (Australian Bureau of Resources and Energy Economics) (2014). *2014 Australian Energy Update.* Canberra: Commonwealth Government of Australia.

Bryan, B, Nolan, M, Harwood, T. et al. (2014). Supply of carbon sequestration and biodiversity services from Australia's agricultural land under global change. *Global Environmental Change,* 28, 166–181.

CCA (Climate Change Authority) (2014). *Reducing Australia's Greenhouse Gas Emissions: Targets and Progress Review: Final Report.* Melbourne: Climate Change Authority. Available at: www.climatechangeauthority.gov.au/reviews/ targets-and-progress-review-3.

Cosgrove, D., Gargett, D., Evans, C., Graham, P. and Ritzinger, A. (2012). *Greenhouse Gas Abatement Potential of the Australian Transport Sector.* Technical report from the Australian Low Carbon Transport Forum. Australia: Commonwealth Scientific and Industrial Research Organisation (CSIRO). Available at: https://publications .csiro.au/rpr/pub?pid=csiro:EP117670.

CWA (ClimateWorks Australia) (2010). *Low Carbon Growth Plan for Australia.* Melbourne: ClimateWorks Australia. Available at: www.climateworksaustralia .org/wp-content/uploads/2019/10/climateworks_lcgp_australia_full_report_mar2010-compressed.pdf.

CWA (2013). *Tracking Progress Towards a Low Carbon Economy: Overview.* Melbourne: ClimateWorks Australia. Available at: www.climateworksaustralia.org/wp-content/ uploads/2019/10/climateworks_trackingprogress_overview_july2013.pdf.

CWA (2018). *Tracking Progress to Net Zero Emissions.* Melbourne: ClimateWorks Australia. Available at: www.climateworksaustralia.org/resource/tracking-progress-to-net-zero-emissions/.

CWA (2020). *Decarbonisation Futures.* Melbourne: ClimateWorks Australia. Available at: www.climateworksaustralia.org/resource/decarbonisation-futures-solutions-actions-and-benchmarks-for-a-net-zero-emissions-australia/.

CWA, ANU (Australian National University), CSIRO (Commonwealth Scientific and Industrial Research Organisation) and CoPS (Centre for Policy Studies) (2014). *Pathways to Deep Decarbonisation in 2050: How Australia Can Prosper in a Low Carbon World.* Technical report. Melbourne: ClimateWorks Australia. Available at: www.climateworksaustralia.org/wp-content/uploads/2014/09/climateworks_pdd2050_ technicalreport_20140923-1.pdf.

DDPP (Deep Decarbonization Pathways Project) (2015). *Pathways to Deep Decarbonization 2015 Report.* Sustainable Development Solutions Network (SDSN) and Institute for Sustainable Development and International Relations (IDDRI). Available at: www.iddri.org/en/publications-and-events/report/pathways-deep-decarbonization-2015-synthesis-report.

Denis, A., Jotzo, F., Ferraro, S. et al. (2014). *Pathways to Deep Decarbonization in 2050: How Australia Can Prosper in a Low Carbon World.* Sustainable Development Solutions Network (SDSN) and Institute for Sustainable Development and International Relations (IDDRI). Available at: www.iddri.org/sites/default/files/old/ Publications/AU_DDPP_report.pdf.

DOE (Australian Department of the Environment) (2014). National greenhouse gas inventory: Kyoto Protocol classifications. *Australian Greenhouse Emissions Information System.* Available at: http://ageis.climatechange.gov.au/NGGI.aspx.

Eady, S., Grundy, M., Battaglia, M. and Keating, B. (2009). *An Analysis of Greenhouse Gas Mitigation and Carbon Sequestration Opportunities from Rural Land Use.* St Lucia, Queensland: Commonwealth Scientific and Industrial Research Organisation (CSIRO). Available at: https://publications.csiro.au/rpr/pub?pid= changeme:822.

Edenhofer, O., Pichs-Madruga, R., Sokona, Y. et al. (2014). Technical summary. In O. Edenhofer, R. Pichs-Madruga, Y. Sokona et al., eds., *Climate Change 2014: Mitigation of Climate Change. Contribution of Working Group III to the Fifth Assessment Report of the Intergovernmental Panel on Climate Change.* Cambridge: Cambridge University Press, pp. 33–107. Available at: www.ipcc.ch/site/assets/ uploads/2018/02/ipcc_wg3_ar5_technical-summary.pdf.

Fleurbaey, M., Kartha, S., Bolwig, S. et al. (2014). Sustainable development and equity. In O. Edenhofer, R. Pichs-Madruga, Y. Sokona et al., eds., *Climate Change 2014: Mitigation of Climate Change. Contribution of Working Group III to the Fifth Assessment Report of the Intergovernmental Panel on Climate Change.* Cambridge: Cambridge University Press, pp. 283–350. Available at: www.ipcc.ch/site/assets/ uploads/2018/02/ipcc_wg3_ar5_chapter4.pdf.

Garnaut, R. (2008). *The Garnaut Climate Change Review: Final Report*. Melbourne: Cambridge University Press.

Global Commission on the Economy and Climate (2014). *Better Growth, Better Climate: The New Climate Economy Report*. Synthesis Report. Washington, DC: The Global Commission on the Economy and Climate. Available at: https://newclimateeconomy .report/2016/wp-content/uploads/sites/2/2014/08/BetterGrowth-BetterClimate_ NCE_Synthesis-Report_web.pdf.

Herrero, M., Havlík, P., Valin, H. et al. (2013) Biomass use, production, feed efficiencies, and greenhouse gas emissions from global livestock systems. *Proceedings of the National Academy of Sciences*, 110, 20888–20893.

IEA (International Energy Agency) (2012). *CO_2 Emissions from Fuel Combustion*. International Energy Agency. Paris: OECD Publishing. Available at: www.oecd-ilibrary.org/energy/co2-emissions-from-fuel-combustion-2012_co2_fuel-2012-en.

IEA (2013). *World Energy Outlook 2013*. Paris: IEA. Available at: www.iea.org/reports/ world-energy-outlook-2013.

IEA–PVPS (IEA Photovoltaic Power Systems Programme) (2013). *Trends 2013 in Photovoltaic Applications: Survey Report of Selected IEA Countries Between 1992 and 2012*. Report No. IEA-PVPS T1–23:2013. Paris: IEA.

IPCC (Intergovernmental Panel on Climate Change) (2014a). *Climate Change 2014: Impacts, Adaptation, and Vulnerability. Part A: Global and Sectoral Aspects. Contribution of Working Group II to the Fifth Assessment Report of the Intergovernmental Panel on Climate Change*. Edited by C. B. Field, V. R. Barros, K. J. Mach et al. Cambridge: Cambridge University Press. Available at: www.ipcc .ch/site/assets/uploads/2018/02/WGIIAR5-PartA_FINAL.pdf.

IPCC (2014b). *Climate Change 2014: Mitigation of Climate Change. Contribution of Working Group III to the Fifth Assessment Report of the Intergovernmental Panel on Climate Change*. Edited by O. Edenhofer, R. Pichs-Madruga, Y. Sokona et al. Cambridge: Cambridge University Press. Available at: www.ipcc.ch/report/ar5/wg3/.

JRC (Joint Research Centre) and PBL (Netherlands Environmental Assessment Agency) (2013). Emission Database for Global Atmospheric Research (EDGAR). Release Version 4.2 FT2010 [database]. European Commission. Available at: https://edgar .jrc.ec.europa.eu/archived_datasets.php.

Reedman, L. and Graham, P. (2011). *Road Transport Sector Modelling: Supplementary Report on Clean Energy Future and Government Policy Scenarios*. A report for the Department of Treasury. Australia: Commonwealth Scientific and Industrial Research Organisation (CSIRO). Available at: https://treasury.gov.au/sites/default/ files/2019-03/c2011-sglp-supplementary-CSIRO_Supp.pdf.

SDSN (Sustainable Development Solutions Network) and IDDRI (Institute for Sustainable Development and International Relations) (2014). *Pathways to Deep Decarbonization: 2014 Report*. Sustainable Development Solutions Network and the Institute for Sustainable Development and International Relations. Available at: www.globalccsinstitute.com/archive/hub/publications/184548/pathways-deep-decar bonization-2014-report.pdf.

Trancik, J. (2014). Renewable energy: Back the renewables boom. *Nature News & Comment*. 19 March. Available at: www.nature.com/news/renewable-energy-back-the-renewables-boom-1.14873.

UN DESA (UN Department of Economic and Social Affairs, Population Division) (2013). *World Population Prospects: The 2012 Revision. Volume I: Comprehensive Tables*. New York: United Nations.

UNFCCC (United Nations Framework Convention on Climate Change) (n.d.). Information provided by Parties to the Convention relating to the Copenhagen Accord. *United Nations Climate Change.* Available at: https://unfccc.int/process/confer ences/pastconferences/copenhagen-climate-change-conference-december-2009/state ments-and-resources/information-provided-by-parties-to-the-convention-relating-to-the-copenhagen-accord.

Williams, J. H., Haley, B., Kahrl, F. et al. (2014). *Pathways to Deep Decarbonization in the United States.* Sustainable Development Solutions Network and the Institute for Sustainable Development and International Relations. Available at: https://usddpp .org/downloads/2014-technical-report.pdf.

10

Decarbonisation Strategies and Economic Opportunities in Indonesia

UTJOK W. R. SIAGIAN AND RETNO GUMILANG DEWI

Executive Summary

Indonesia is one of the largest emitters of greenhouse gases (GHGs) in the world, and one of the globe's biodiversity hotspots. It is highly exposed to the effects of climate change but also has wide-ranging options for emissions reduction, including the development of low-carbon and renewable energy.

Indonesia's main sources of GHGs are LULUCF (land use, land-use change and forestry), at 48%; and energy, at 37%. In the energy sector, the main GHG emitter is coal combustion followed by oil. Under the baseline (business as usual, or BAU) scenario, the use of coal will keep increasing, causing an increase of national carbon intensity.

Considering that the major sources of GHG emissions are LULUCF and energy, to achieve the climate change targets in Indonesia's intended Nationally Determined Contribution (NDC) to the UN Framework Convention on Climate Change, Indonesia will need to focus initially on reducing emissions from the land-use and energy sectors. Other areas are to be addressed progressively.

This chapter focuses on Indonesia's sustainable energy transition challenge: to meet rising energy demand due to a growing economy and population and simultaneously improve energy efficiency and decarbonise the energy supply. It also shows how this can provide a platform to help decarbonise the buildings, industry and transport sectors.

Modelling undertaken as part of the UN Deep Decarbonisation Pathways Project, featured in this chapter, identifies that the main strategies to achieve decarbonisation are: energy efficiency measures, electrification of energy end use and deep decarbonisation of the power generation sector. Fuel switching, from coal to gas and oil, to biofuels, and eventually to renewables, will reduce emissions.

A renewable energy target of at least 31% by 2050 is a key part of the decarbonising policy. The Deep Decarbonisation Pathways Project (DDPP) is expected to achieve a 50% reduction in GHGs compared to BAU by 2050.

Emissions from the industrial sector will continue to rise as will those from energy demand, in part to supply electrification of end use. Energy demand by 2050 will grow by 300% compared to 2010 but at a lower rate than the economy – 2.8% compared with 5.4–5.8%. Indonesia's NDC forecasts a 29% reduction in GHGs below the baseline emission by 2030. Further reduction up to 41% could be achieved if international assistances were made available.

In the transport sector, Indonesia plans to increase low-emission mass transportation such as rail and promote: fuel switching to gas and biofuels, electric vehicles (eventually powered from renewable sources), and structural change towards a more service-oriented economy. Liquid fuel demand is expected to grow by 50% by 2050 but carbon intensity will be reduced by substitution of biofuels.

To achieve decarbonisation in Indonesia, energy sector major investment will be required, amounting to 1.2% of GDP in 2020. However, decarbonisation investment as a fraction of GDP will fall to 0.7% by 2050. Investment in low-carbon options is expected to amount to 5% of total investment and will be offset to some extent by reduced investment in fossil fuels. Decarbonisation will require significant expenditure on infrastructure, including transport systems and transmission networks including subsea cables. It is noted that nuclear generation poses an issue of social licence and significant community education would be required to gain acceptance of this technology.

In implementing mitigation actions in all sectors, an enabling environment is required, including appropriate policies and regulations. Currently, national and regional development plans include climate change in the development agenda. Nevertheless, the implementation still requires more specific policy/regulatory supports such as: incentives for decarbonisation, a carbon tax, renewable energy made mandatory, an energy consumption cap, energy labelling of appliances and building codes. Increased budgetary resources and capacity-building will also be required as part of the strategy to ensure that the full potential of cost-effective emissions reduction measures is realised.

This chapter also shows that the economic costs of inaction for Indonesia are significant, because Indonesia is exposed to risks of unmitigated climate change through extreme temperatures, sea-level rises, more extreme weather events, drought or floods, spread of vector-borne diseases and loss of agricultural productivity, biodiversity and ecological resilience.

Conversely, investments in a low-carbon economy reduce exposure to these risks and help build a more resilient and prosperous future. For instance, distributed low-carbon renewable energy, microgrids using smart energy storage options (i.e. batteries and pumped hydro) and decarbonising transport have the potential to significantly reduce Indonesia's dependence on fossil fuels and biomass for energy, and help to improve energy security creating greater prosperity. In addition, large increases in the deployment of solar photovoltaic (PV) panels are expected to occur due to recent significant decreases in the cost of solar PV technology. Despite recent wind power development in Sulawesi, the contribution of wind is not expected to be deployed at a large scale because Indonesia's wind resource is considered not large.

10.1 Introduction: Why is Action on Climate Change a Priority for Indonesia?

Indonesia, like all nations, has a responsibility to contribute to the global endeavour to achieve a 2 °C limit on global warming over pre-industrial levels. This is strongly in

Indonesia's interests as Indonesia is among the countries that will be most negatively affected by unmitigated climate change. While climate variability and trends differ vastly across the region and between seasons, according to the *Fifth Assessment Report* of the Intergovernmental Panel on Climate Change (IPCC), the observed and projected evidences of climate change in South East Asia, including Indonesia, are:

- annual total wet-day rainfall has increased by 22 mm per decade, while rainfall from extreme rain days has increased by 10 mm per decade;
- between 1955 and 2005, the ratio of rainfall in the wet compared to the dry seasons increased;
- an increased frequency of extreme events has been reported in northern South East Asia;[1] and
- rainfall intensity increased in much of the region.[2] Future increases in precipitation extremes related to the monsoon are very likely in East, South and South East Asia.

The projected detrimental impacts of climate change include droughts, forest fires, smoke aerosols and the vulnerability of livelihoods in agrarian communities.[3] Sea-level rise in South East Asia, including Indonesia, is projected to reach 40–60 cm at the end of the century (Kompas Daily 2016). Indonesia's population and economic resources will be impacted greatly by sea-level rise, because many of Indonesia's large cities such as Jakarta, Surabaya, Semarang and Makassar are in low-lying coastal areas. Kalimantan will lose many of its low-lying forests. The country's 80 000 km of coastline and hundreds, if not thousands, of islands will be negatively affected by sea-level rise. Sea-level rise will also negatively affect the fate of Indonesia's marine resources such as coral reefs, fisheries, mangroves, etc.

Meanwhile, Indonesia is among the largest greenhouse gas (GHG) emitters. Therefore, it is reasonable and appropriate to expect that Indonesia will actively contribute to the global effort to prevent a 2 °C increase in global temperatures by 2050 through the decarbonisation of its economy.

Decarbonisation is not only needed for preventing the 2 °C increase but also is expected to generate significant co-benefits including:

- reduced local pollution due to changes in transport, industry, power plants and residential buildings;
- economic stimulation as a consequence of investment in new, lower-carbon-emitting technologies, and energy efficiency increasing the economy's productivity;
- improved energy security through realising the potential of domestic renewable energy sources, which are among the largest in the world; and

---

[1] A decrease in such events has been reported in Myanmar.
[2] In Peninsular Malaya during the south-west monsoon season, total rainfall and the frequency of wet days decreased, but during the north-east monsoon, total rainfall, the frequency of extreme rainfall events and rainfall intensity all increased over the peninsula.
[3] Vulnerability of agrarian communities also arises from their geographic settings, demographic trends, socio-economic factors, access to resources and markets, water consumption, farming practices and lack of adaptive capacity.

- jobs creation from energy efficiency and renewable energy development, such as plantation work for producing biofuel feedstock.

10.2 The Study Scope and Methodology

Energy is one of the most important sectors that contribute to GHG emissions in Indonesia. According to the Indonesia Second Biennial Update Report, in 2016 the share of energy activities in total GHG emissions was 37%, the second highest after land use, land-use change and forestry (LULUCF (43%)). However, between 2000 and 2016 emissions from energy grew at a faster rate (3.6% per annum) than LULUCF (1.3% p.a.) (Republic of Indonesia 2018). This indicates the growing importance of energy in Indonesian GHG emissions. Therefore, mitigation of the energy sector will play a crucial role in Indonesian efforts to reduce GHG emissions.

This Indonesian case study explores a development trajectory of Indonesian energy and energy-using sectors, which could contribute to the global endeavour to stabilise average global surface temperature rises at 2 °C (above pre-industrial levels) between now and 2050, while still maintaining at least 5% economic growth per annum. As a comparison, Indonesia's energy sector's trajectory under a business as usual (BAU) or unmitigated/inaction case is also analysed.

The chapter addresses two main questions: (i) is it technically feasible for Indonesia to decarbonise its energy system? and (ii) what investment would be required to achieve such a decarbonisation? It shows how a transition to low-carbon energy and energy-using sectors is feasible using existing technologies.

Finally, the chapter considers policy levers available to underpin and better incentivise investment in a transition to a low-carbon, resilient and prosperous future.

The exploration of the technical potential for decarbonisation was carried out using the deep decarbonisation modelling dashboard developed by the Deep Decarbonisation Pathways Project (DDPP), organised by the Sustainable Development Solutions Network and l'Institut du Développement Durable et des Relations Internationales (SDSN and IDDRI). This study analyses Indonesia's energy development trajectory for 2010 to 2050. In exploring the potential for decarbonisation, the country's socio-economic conditions, development aspirations, infrastructure stock, resource endowments and other relevant factors are taken into account. The economic analysis of this Indonesian decarbonisation development pathway is limited to the cost required for investment in infrastructure/facilities that will be deployed for decarbonisation. The future costs of decarbonisation technology are evaluated using the SDSN cost estimates that have taken into account the future learning curve resulting from global deployment of the technology.

10.3 National Circumstances

10.3.1 Demography

The Republic of Indonesia is the largest archipelago in the world, located between the Pacific and Indian Oceans, bridging two continents: Asia and Oceania. It consists of

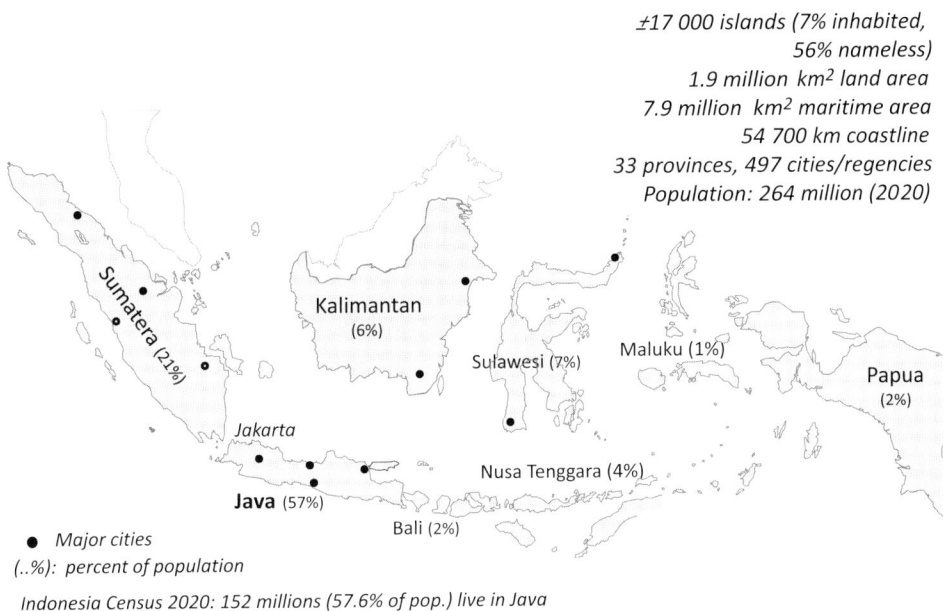

±17 000 islands (7% inhabited,
56% nameless)
1.9 million km² land area
7.9 million km² maritime area
54 700 km coastline
33 provinces, 497 cities/regencies
Population: 264 million (2020)

● *Major cities*
(..%): percent of population

Indonesia Census 2020: 152 millions (57.6% of pop.) live in Java

Figure 10.1 Map of Indonesia with basic statistics.
Source: Courtesy of Utjok W. R. Siagian, based on data from BPS (2020).

approximately 17 000 islands with a population of 264 million. The majority (almost 80%) of Indonesians live in the western part of Indonesia, on the islands of Java and Sumatera (see Figure 10.1). Administratively, the Republic of Indonesia is divided into 34 provinces.

Of the 200 million hectares (ha) of land territory, about 50 million ha are devoted to agricultural activities. There are nearly 20 million ha of arable land, of which about 40% is wetland (e.g. rice fields), 40% is dry land and 15% is shifting cultivation.

During the past five decades, Indonesia's population has continuously increased from 119 million in 1971 to 264 million in 2020 (BPS 2020). However, its annual population growth rate is decreasing, from 1.98% (1980–1990) to 1.4% (2000–2016). It is projected that the population will exceed 300 million by 2050. Unemployment and underemployment are still relatively high; hence poverty remains a challenge. The employment rate has improved in the past 10 years. The unemployment decreased from about 10.3% in 2006 to around 5.6% in 2016 (BPS 2016).

The distribution of the population follows the distribution of the country's economic activity, which is concentrated in the western part of Indonesia: on the islands of Java and Sumatera. In 2010, more than 136 million people (57% of the population) lived on the Island of Java, and around half inhabited urban areas. Provinces with more than 50% of their inhabitants living in urban areas are Jakarta (100%), Riau (83%), Banten, Yogyakarta and West Java (more than 60%).

Indonesia's economy has grown rapidly in the last decade. Its GDP has increased from 2300 trillion Indonesian rupiah (IDR) (USD 248 billion) in 2004 to IDR 12 400 trillion (USD 932 billion) in 2016. This corresponds to a per capita GDP increase from IDR

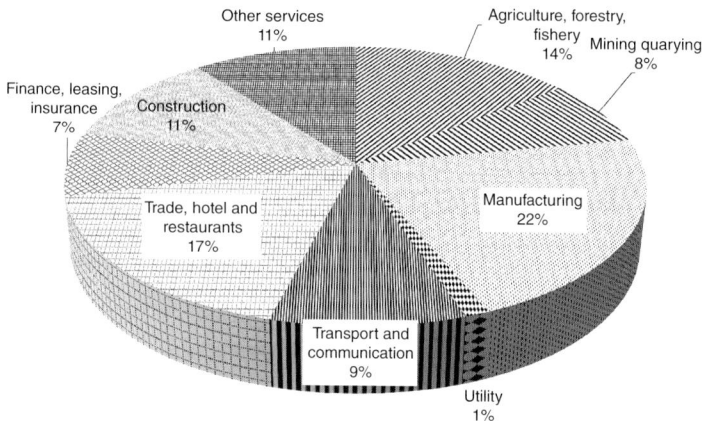

Figure 10.2 Distribution of Indonesian GDP in 2016.
Source: BPS (2017).

10.5 million (USD 1132) in 2004 to IDR 47.8 million (USD 3605) in 2016. During this period, Indonesian economic growth averaged 5.7% per annum.

Over the last 50 years, Indonesia's economy has experienced a structural transformation, from a largely agricultural-based economy to a largely industrial and services-based economy. Figure 10.2 shows the aggregate economic structure (GDP) in 2012. It also shows that the share of industry and services accounted for 86% of the economy. Major contributors in the industrial sectors are manufacturing, mining and extraction, and construction; while trading, hotels, restaurants, finance, real estate, transport and telecommunication are the major contributors in the commerce and service sectors.

Indonesian life expectancy at birth has improved substantially in the past four decades, from only 47.9 years in 1970 to 70.9 years in 2016. In the education sector, as the result of sustained efforts, Indonesian adult literacy in 2011 was 95%, which is significantly higher than it was in 1970, when it was only 79%.

Prior to 1999, Indonesia had been successful in alleviating poverty. In 1970, 60% of the population (70 million people) lived in absolute poverty. By 1990, the number had dropped to 27 million, or 15% of the population, and continued to improve up to 1997, when the figure decreased to 20 million. However, in 1999, for the first time in years, Indonesia experienced a severe 18-month drop in the country's social and economic condition, resulting in over 100 million people living below the poverty line. Despite successful recovery following the country's economic and political reforms since 2000, in 2016 the Indonesian poor totalled 28 million people (11% of the population). According to the Medium-Term Development Plan (Kementarian PPN/Bappenas 2014), the Government of Indonesia planned to implement various development and welfare programmes to reduce the poverty rate to between 6.5% and 8.0% of the population by 2019 but data for that year are not available at the time of writing.

10.3.2 Energy Resources and Trends

Indonesia is well endowed with a wide range of energy resources (Table 10.1). Oil, gas and coal have been exploited intensively for supplying domestic demand as well as for export. Prior to 2004, Indonesia was a net oil exporter and a member of OPEC. Increasing demand and diminishing reserves has made Indonesia a net oil importer since 2004. The table also shows that Indonesia has significant renewable resources (averages are shown over the whole country). Currently, the utilisation of these resources is still limited because, in the past, oil fuels and electricity were heavily subsidised: renewable energy prices could not compete with subsidised fossil fuel energy.

The energy sector is an important driver of economic activity and also generates government revenues from sales of natural resources to domestic and export markets, royalties and taxes. The schematic of Indonesian energy flows is shown in Figure 10.3.

Energy consumption has been growing in line with economic and population growth. Between 2000 and 2016, total final energy demand grew on average 2.3% annually, from 113 million tonnes oil equivalent (TOE) to 163 million TOE. The development of Indonesian energy supply and demand during 2010–2016 (MEMR 2017) is summarised in Figures 10.4 to 10.8.

The industrial, residential and transport sectors dominated the final energy consumption rates (Figure 10.4). High energy consumption growth occurred in the transport (6.9% per year) and commercial (4.6% per year) sectors. The growth was much higher than that of the industrial (1.8% per year) and residential (0.9% per year) sectors. By fuel type (Figure 10.5), the energy system is still dominated by oil, which accounts for around 37.0% of the total consumption, followed by biomass (26.1%), gas (15.6%), coal (11.4%)

Table 10.1. *Indonesia's energy resources*

Energy resources		
Fossil fuels	Reserve	Resource
Oil, billion barrels	7.4	–
Natural gas, TSCF	150	–
Coal, billion tonnes	29	119
Coal bed methane, TSCF	–	453
Shale gas, TSCF	–	574
Renewable energy	**Potential power**	
Hydro	75 000 MW	
Geothermal	29 000 MW	
Micro hydro	750 MW	
Biomass	14 000 Mwe	
Solar	4.80 kWh/m^2 per day	
Wind	17–133 W/m^2 swept area	

Note: TSCF = trillion standard cubic feet; MW = megawatts; Mwe = megawatts equivalent; kWh = kilowatt-hours. *Source*: Indonesia DEN (2015).

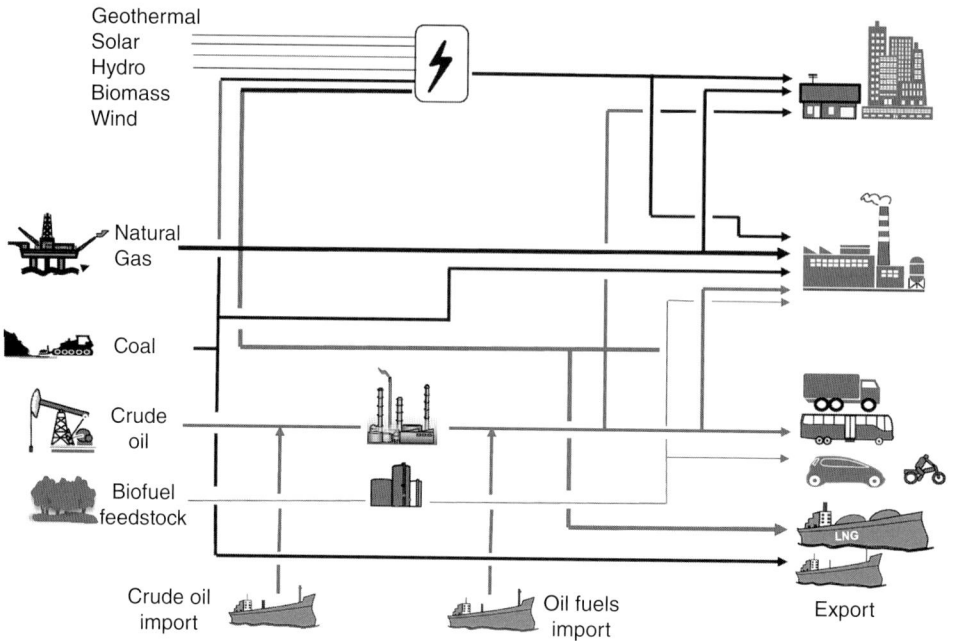

Figure 10.3 Indonesian energy flows.
Source: Courtesy of Utjok W. R. Siagian.

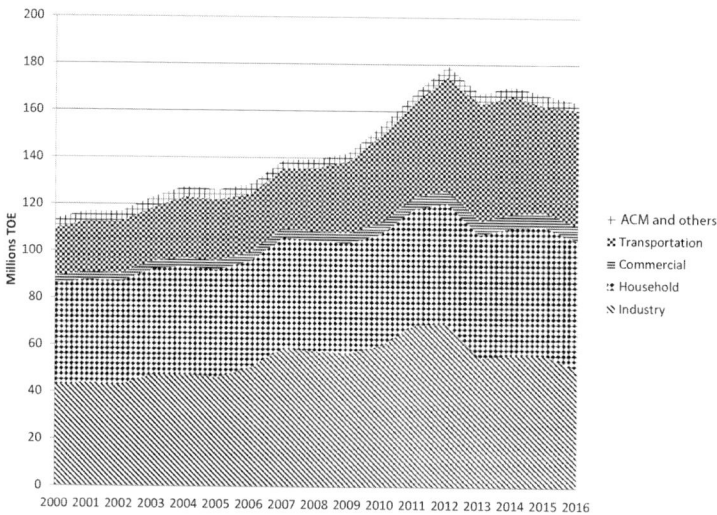

Figure 10.4 Development of final energy demand by sector.
Note: ACM = agriculture, construction, and mining.
Source: Data and Information Center, MEMR of Indonesia (MEMR 2017).

and electricity (9.9%). High demand growth occurred in coal (10.8% per year), electricity (6.8% per year) and gas (4.8% per year). The gas consists of natural gas (pipe gas or compressed natural gas, CNG) and liquid petroleum gas (LPG). High growth in gas demand was in the residential sector, due to a government policy that switched the kerosene

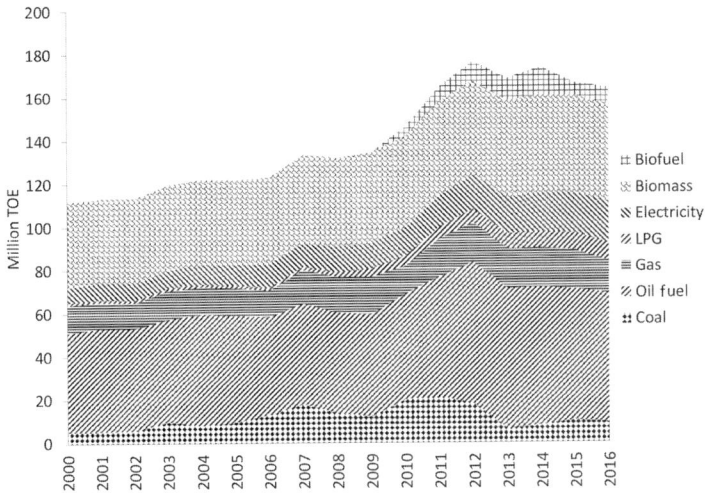

Figure 10.5 Development of final energy demand by fuel type.
Source: MEMR (2017).

subsidy to an LPG subsidy. Coal consumption as final energy occurred solely in the industrial sector. The high growth in coal consumption was due to the removal of an industrial diesel oil subsidy and cheap coal, resulting in industries switching from diesel oil to coal. The high growth of coal also resulted from the easy access to coal and the fact that domestic coal supply is abundant. Switching from diesel oil to gas in industry is more difficult than switching from diesel to coal, not only due to the higher price of gas when compared to coal but also due to limited gas accessibility (i.e. limited infrastructure).

Between 2000 and 2016, primary energy supply grew at a rate of 2.9% per year, from 145 million to 231 million TOE. As shown in Figure 10.6, primary energy supply has been dominated by oil, followed by coal and natural gas. Ever since the decline of domestic oil production, the government has, for energy security reasons, been attempting to move away from oil by promoting energy that is abundantly available in the country – coal, natural gas and renewables. These attempts have resulted in the high growth in coal supply (11.5% per year), which is significantly higher than the growth in oil (2.8% per year) and natural gas (3.9% per year). These rates of growth have resulted in a decreased oil percentage share in the supply mix, from 41% in 2000 to 34% in 2016, and an increase in the coal percentage share, from 9% in 2000 to 24% in 2016.

Around 10% of final energy consumption was in the form of electricity. The electricity demand was met by different types of power plants: coal, gas, hydropower, geothermal and oil fuels (Figure 10.7). In the past 16 years, the plants that experienced high annual growths were geothermal (10%) and coal (9%). The high growth rate of coal plants has increased the share of coal in the power generation mix from 37% in 2000 to 55% in 2016. Despite the high growth rates for geothermal, the relative percentage share of geothermal in the power generation mix remains small; 4.4% in 2016. The rate of oil used for power generation is largely attributable to the distributed diesel plants that were installed to boost

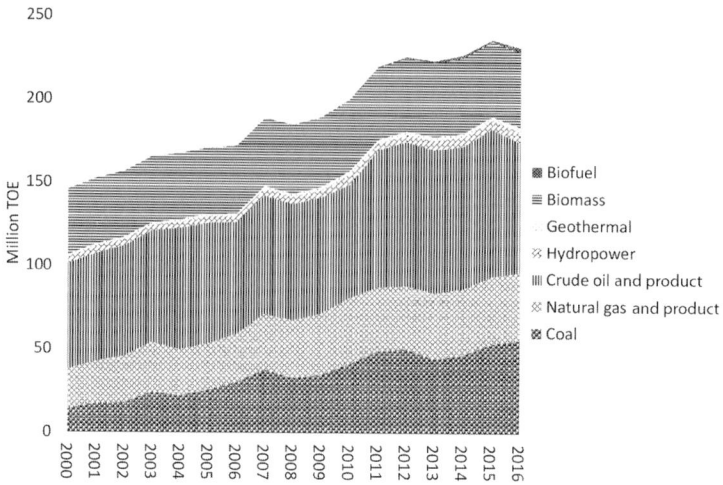

Figure 10.6 Development of primary energy supply.
Source: MEMR (2017).

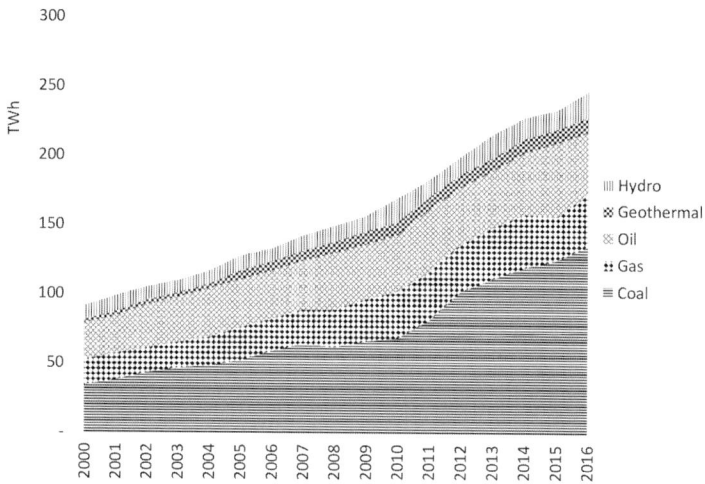

Figure 10.7 Development of power generation mix.
Source: MEMR (2017).

rural electrification in the 1980s and 1990s. Attempts to substitute these diesel plants with other types of power plants have been made but with limited success, and consequently the share of electricity from oil fuels remains high: 18% in 2016.

The future challenge in the energy sector is to meet rising energy demand due to a growing economy and population growth by: (i) improving energy efficiency to reduce energy demand growth; and (ii) decarbonising the energy supply. To guide the development of Indonesian energy, the National Energy Council (DEN) endorsed the National Energy Policy in 2014. The main features of the energy policy are as follows:

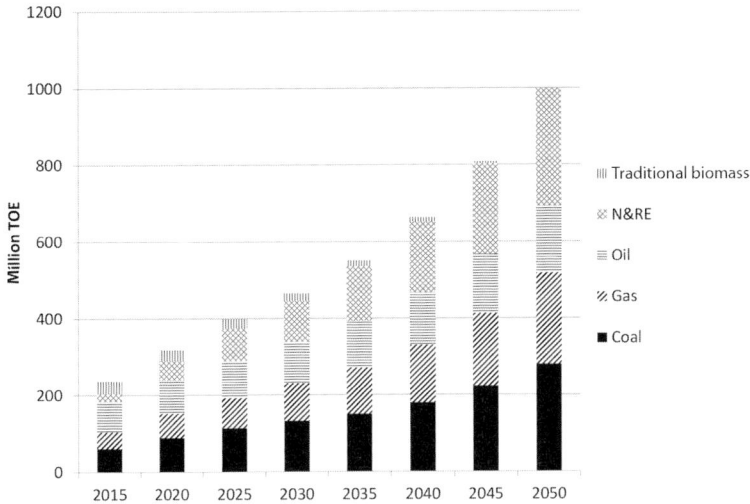

Figure 10.8 National Energy Council's energy supply scenario.
Source: Indonesia DEN (2015).

- to strive for energy security (move away from oil – reduced to 19.5% of supply in 2050; promote more abundantly available resources such as natural gas and coal and prioritise the use of domestic resources for satisfying the domestic market);
- to increase energy efficiency (with an energy elasticity target of less than 1 in 2025); and
- to promote the development of renewable energy (targeting 31% of supply mix in 2050).

The existing energy policies heavily emphasise the achievement of energy security and independence as the ultimate objectives. Therefore, Indonesian energy demand is to be satisfied as much as possible using domestic resources. With the current Indonesian policy framework, the Indonesian National Energy Council has developed a long-term energy supply scenario (Figure 10.8). Under current Indonesian government policy settings, they find that, despite high growth in new and renewable energy (N&RE), Indonesia's primary energy supply will remain dominated by fossil energy for the coming three decades. The high share of fossil energy consumption in Indonesia and the coupling of energy and economic growth will make the energy sector the largest contributor of GHG emissions, assuming the mitigation actions in LULUCF are successful.

10.3.3 Current GHG Emissions

According to the 2018 Indonesian GHG inventory (Republic of Indonesia 2018), in 2016 Indonesia emitted around 1478 MtCO$_2$e (megatonnes of carbon dioxide equivalent). This represents an increase of 452 MtCO$_2$e in emissions when compared to 2000. The major contributions of emissions are from LULUCF and peat fire, which together account for 43% of the total emissions. Emissions from energy activities contribute 37% of the total emissions (Figure 10.9; see also Table 10.2).

Table 10.2. *Comparison of emissions growth among sectors of activity*

Average annual growth, 2000–2012	
Energy	3.6%
IPPU	1.7%
Agriculture	1.1%
LULUCF	1.3%
Waste	3.8%
Total	**2.2%**

Note: IPPU = industrial processes and product use; LULUCF includes peat fire.
Source: Republic of Indonesia (2018).

1026 MtCO$_2$

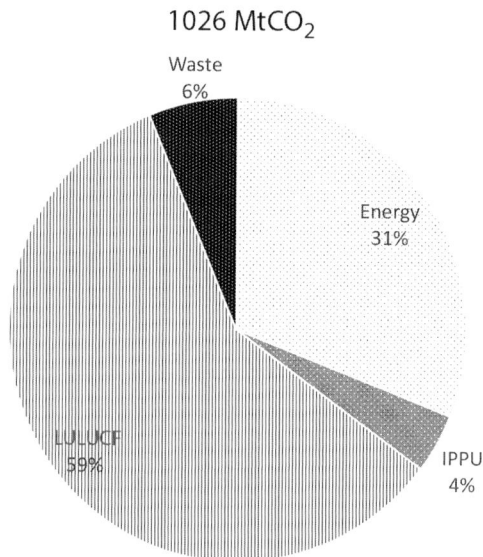

Figure 10.9 Distribution of Indonesia GHG emissions in 2000.
Source: Republic of Indonesia (2018).

In 2016, energy activities emitted around 538 MtCO$_2$e. This figure is accounted for primarily by power generation, transportation and industrial activities (Figure 10.10). In the fuel combustion emission category, coal is the major emission source. Coal is the main fuel in power generation. It is also a major energy source in industrial activities. The second major emission source is oil combustion, in the transport sector and building sector (which is the aggregate of commerce, offices and residential buildings). In the end-use sector, one-half of the direct combustion emissions is from fuel burning in industrial activities. Around 68% of emission from electricity generation is accounted by residential, commerce and office buildings (aggregated as the building sector) and the remaining 38% is accounted by

Energy 2016
538 MtCO$_2$e

Others 2%
Fugitive 4%
Commercial 1%
Residential 6%
Transport 25%
Electricity generation and petrochemical refinement 46%
Industry 16%

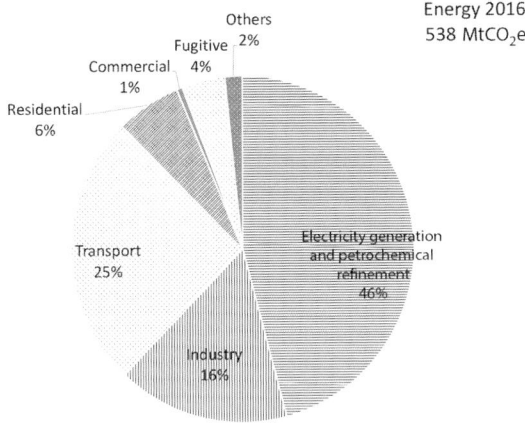

Figure 10.10 Distribution of GHG emission sources within energy sector.
Source: Republic of Indonesia (2018).

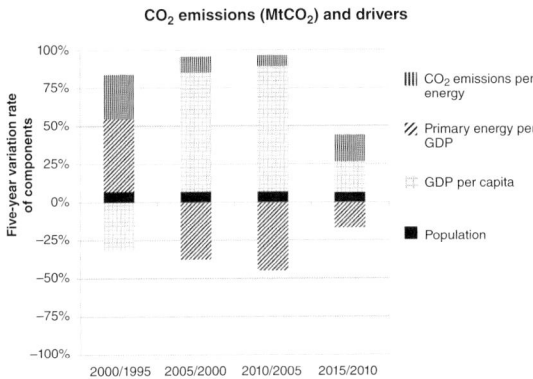

CO$_2$ emissions (MtCO$_2$) and drivers

Five-year variation rate of components

100%
75%
50%
25%
0%
−25%
−50%
−75%
−100%

2000/1995   2005/2000   2010/2005   2015/2010

‖‖ CO$_2$ emissions per energy

⁒ Primary energy per GDP

▱ GDP per capita

■ Population

Figure 10.11 Breakdown of historical energy-related CO$_2$ emissions.
Source: Processed from MEMR (2013) and Republic of Indonesia (2018).

the industry sector. Electricity use in transport (trains) is very small and therefore the indirect emissions of electric trains are negligible.

As shown in the breakdown of energy-related emissions (Figure 10.11), the main driver of GHG emissions over the past decade has been economic activity, which grew at a rate of 5% to 6% per year. Due to the growing importance of coal, a small increase in the carbon intensity of energy can also be observed.[4] On the other hand, the Indonesian economy has experienced important improvements in energy efficiency, as captured by the steady decrease in energy use per GDP.

---

[4] Coal replaced oil, which is less carbon intensive.

10.3.4 Indonesia's Nationally Determined Contribution

To contribute to global effort in preventing a greater than 2 °C increase of global average temperature in the middle of this century, Indonesia in November 2016 submitted its first Nationally Determined Contribution (NDC) to the UN Framework Convention on Climate Change (UNFCCC) (Republic of Indonesia 2016). Indonesia's NDC outlines the country's climate plan, covering its mitigation and adaptation strategies, planning process and strategic approach to combating climate change. It is indicated in the NDC that Indonesia intends to reduce its emissions through mitigation actions, targeting an emissions reduction of 29% below the country's baseline emission (BAU scenario) by 2030. This reduction target (the unconditional reduction target) is to be achieved by using national resources. Through international assistance and cooperation, Indonesia is committed to further reduce its emissions up to 41% below its BAU scenario by 2030 (the conditional emissions reduction target).

Indonesia's emission level, under a BAU scenario, is projected to reach 2.869 $GtCO_2e$ in 2030. Referring to this BAU emission level, Indonesia's unconditional and conditional emissions reduction targets in 2030 are, respectively, 0.83 $GtCO_2e$ and 1.08 $GtCO_2e$. The unconditional emissions reduction target is to be achieved through implementing effective land use and spatial planning, sustainable forest management, improved agriculture and fishery productivity, energy conservation, promotion of clean and renewable energy resources and improved waste management. The specificity of Indonesia's NDC is still limited to these targets, the BAU level and the narration of mitigation action programmes. The delineation of the emissions reduction targets and the actions through which the targets are to be realised in specific sectors is not yet indicated in the NDC document. Currently, the Ministry of Environment and Forestry is the focal point for NDC in engaging relevant line ministries and other stakeholders to delineate the NDC into specific mitigation actions and reduction targets in each sector (agriculture, forestry and land use, energy, waste and IPPU).

10.4 Energy Sector Decarbonisation Scenario and Strategy

10.4.1 Decarbonisation Scenario

In this study, one decarbonisation scenario is envisaged that employs decarbonisation through energy efficiency measures, electrification of end use and deep decarbonisation of power generation. For comparison, the BAU scenario is also evaluated. The BAU scenario is in accordance with the energy mix target stated in Indonesia's National Energy Policy (NEP),[5] which was established with national energy security as the main consideration.

The decarbonisation scenario assumes that Indonesia undertakes various mitigation actions in the energy sector to a degree that will significantly contribute to global efforts in preventing global temperature increase of more than 2 °C by 2050. According to the DDPP, to prevent the (2 °C) global temperature increase, countries have to reduce their

---

[5] National Energy Policy, Indonesia. Available at: www.bphn.go.id/data/documents/14pp079.pdf.

Table 10.3. *Development indicators and energy service demand drivers*

	2010	2020	2030	2040	2050
Population (millions)	234	252	271	289	307
GDP per capita (USD/capita)	2306	3655	5823	9319	14 974
Electrification rate	70%	85%	99%	99%	99%
Poverty indicator	12%	8%	3%	3%	2%

carbon footprint to 1.5 tCO_2 per capita per year by 2050. In this Indonesian study, the per capita carbon intensity target is 1.7 tCO_2 per capita, slightly lower than the country's current energy sector emission level (around 1.9 tCO_2 per capita). Indonesia is a developing country, which aspires to continue its economic development for increasing the wealth of its people. Economic development, together with population increase, will drive energy consumption and, if not mitigated, CO_2 emissions. A very deep decarbonisation target, say to 1 tCO_2 per capita, would be very difficult, if not unrealistic, to achieve. Besides, Indonesia has a large forest sector, which has the potential to be a net emission sink in the future to offset the extra emissions from the energy and other sectors (i.e. process emissions, waste and agriculture).

The energy development drivers, such as GDP growth and population growth, for the decarbonisation scenario and for the BAU scenario are identical. Being a developing country, the Indonesian economy and population are projected to grow significantly in the next four decades. Until 2030, the economy is expected to grow between 5.4% and 5.8% per year. Afterwards, the growth is expected to remain steady at 5.4% per year. Indonesia's population is expected to grow steadily around 1.1% per year until 2020 and then slow to an average rate of around 0.6% per year. The projection of these energy service demand drivers and other relevant development indicators for 2010–2050, applied for the BAU and decarbonisation scenarios, are shown in Table 10.3.

The factors that differentiate the BAU scenario and the decarbonisation scenario are those related to energy demand and supply assumed for each scenario, among others: technical performance of the energy system (intensity of energy consumption, energy consumption per unit of building floor area, etc.), the type of energy/fuel to be used and the type of industry to be developed in the future. The different assumptions made for the BAU and decarbonisation scenarios are shown in Table 10.4.

In each scenario, the constraint that is imposed in deploying renewable energy systems is the availability of domestic renewable energy resources that can reasonably be deployed. The types of technologies deployed are limited to those that have reached commercial stage or are expected to be commercial in the near future.

10.4.2 Decarbonisation Strategy (How to Decarbonise)

This section explores the decarbonisation process for the energy sector in Indonesia from 2010 to 2050, aiming to assist economic development while simultaneously decreasing

Table 10.4. *Assumptions in the BAU and decarbonisation scenarios*

Subsector	Technology/fuel type and size	Unit	2010	BAU 2050	DEC 2050
Residential	Unit energy consumption	MJ/m^2	145	200	170
	Final electricity	%	43	50	70
Commercial	Commercial floor space	Billion m^2	0.4	1.0	0.8
	Unit energy consumption	MJ/m^2	460	800	600
Passenger transport	Passenger-kilometres (PKM)	Trillion PKM	1.1	2.50	2.10
	Bus	%	30	30	35
Car (personal and taxi)	Electric vehicle share, by vehicle kilometres travelled (VKMT)	%	0	5	20
	Ethanol	%	0	10	20
	Biodiesel	%	0	15	20
Bus	Electric share, VKMT	%	0	3	5
	Biodiesel	%	0	25	30
	Compressed pipeline gas	%	0	15	30
Air	Biofuel	%	0	5	20
Freight trucks	Share of tonne kilometres (TKM): Diesel	%	100	50	40
	Share of TKM: CNG	%	0	20	30
Freight rail	Biodiesel	%	0	20	30
Iron and steel manufacturing	Energy intensity	GJ/tonne	6.00	5.00	4.00
	Electricity final	%	30	30	40
Cement manufacturing	Energy intensity	GJ/tonne	4.00	3.50	3.00
Small/medium manufacturing	Energy intensity	MJ/USD	12.5	10.0	8.0
	Share of industry value added (VA)	%	90	85	85
	Electricity final	%	8	25	35
	Direct biomass combustion	%	5	10	15
	Biodiesel	%	2	10	25
Large industry	Energy intensity	MJ/USD	23.0	20.0	18.0
	Electricity final	%	6	10	40
	Direct biomass combustion	%	0	5	15
	Biodiesel	%	0	5	25
Power sector	Coal	%	49	27	6
	Fuel oil	%	12	2	1
	Natural gas	%	30	16	20
	Nuclear	%	0	5	16
	Hydropower	%	6	16	16
	Wind	%	0	1	1
	Solar photovoltaics	%	0	5	10
	Biomass	%	0.1	8	12
	Geothermal	%	3	15	14
	Biofuel	%	0	5	4

Note: BAU = business-as-usual scenario; DEC = decarbonisation scenario.

GHG emissions. It is envisaged that decarbonisation can be implemented via three main avenues: energy efficiency measures; switching from high carbon intensity fossil energy systems to energy systems with lower carbon intensity or zero GHG emissions; and changes to the country's economic structure from an energy-intensive economy to a less energy-intensive economy.

Energy efficiency measures reduce energy consumption while maintaining delivery of the energy services needed for economic development. They should be applied to the energy supply side (power plants) as well to the demand side (vehicles, transport, residential appliances, etc.).

Fuel switching in power plants from coal to natural gas is one example of switching from a high carbon intensity to a lower carbon intensity fuel. Substituting diesel oil consumption in vehicles with biodiesel is another example of fuel switching. Replacing gas stoves with electric stoves, where the electricity is generated by renewable power plants, is also considered fuel switching.

Decarbonisation through structural economic change will happen when a country with significant industrial GDP share (an energy-intensive economy) is transformed into a service-oriented economy (less energy-intensive economy) while maintaining the same level of economic development.

Indonesia's decarbonisation pathway is composed of a combination of these measures:

- energy efficiency measures in the building, transport, industry and power generation sectors;
- fuel switching from high-carbon fossil fuels to lower-carbon fossil fuels on the demand side (from coal to gas, from oil to gas or biofuel) in the building, transport and other industries; electrification of end-use devices (i.e. substitution of on-site combustion energy systems for off-site electric energy systems) in the building, transport and industry sectors;
- decarbonisation of the electric power sector through large-scale deployment of zero- or lower-carbon-emitting power generation systems (i.e. solar PV panels, geothermal, hydropower, biodiesel, nuclear, natural gas and wind);
- wind power development – already established in South Sulawesi but with limited prospects due to Indonesia's relatively weak wind resource;
- widespread deployment of solar PV systems, given that the solar resource quality is of a similarly high level throughout the country;
- geothermal and hydropower deployment, constrained by the location of the resource relative to the location of demand;
- assuming long-distance subsea cables are economically feasible, development of large hydro resources in Papua (25 gigawatts (GW)) to supply power to the western part of Indonesia; and
- structural changes in the economy (i.e. decrease in the role of industry in the formation of national GDP through service-oriented sector substitution) are expected to result in less demand for energy.

10.5 Results of Decarbonisation Strategy

This section summarises the high-level results of the analysis for the two scenarios. The summary includes comparison of values in 2010 and 2050 for final energy consumption, primary energy supply, emissions and drivers of decarbonisation. By implementing decarbonisation strategies, Indonesia's energy-related CO_2 emissions will continue to grow but in a sustainable manner. This will be achieved by realising deep decarbonisation later towards 2050, as shown in Figure 10.12. Indonesian CO_2 emissions will first increase (due to economic development) and then decrease at later stage (as a result of decarbonisation measures). Compared to the BAU scenario, the decarbonisation scenario results in much lower emissions intensity: in 2050, emissions for the decarbonisation scenario are half the emissions for the BAU scenario. Both scenarios result in decreased carbon energy intensity. However, the decarbonisation scenario has a larger intensity reduction: from $4\,tCO_2/TOE$ in 2010 to $1.7\,tCO_2/TOE$ in 2050 (in BAU scenario it decreases to $2.8\,tCO_2/TOE$).

Despite decarbonisation efforts, emissions from the industrial sector will continue to increase, from $152\,MtCO_2$ in 2010 to $241\,MtCO_2$ in 2050. Emissions from electricity generation will also increase, from $144\,MtCO_2$ in 2010 to $184\,MtCO_2$ in 2050. Although the carbon intensity of power generation is expected to decrease significantly, economic and population growth coupled with electrification of end use will continue to increase emissions from the electricity sector. Decarbonisation will occur in the transportation sector, from $111\,MtCO_2$ in 2010 to $88\,MtCO_2$ in 2050, due to a combination of effects from mode shift to public transport, electrification of transport and fuel substitution to biofuels.

10.6 Final Energy Demand and Emissions

As a developing country, Indonesia expects its economy to continue to grow. This, combined with increasing population, will lead to an increase of energy demand. By

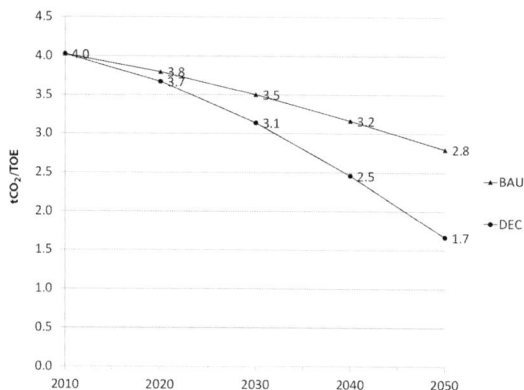

Figure 10.12 Pathways for Indonesia's sectoral energy-related emissions reductions.
Note: BAU = business-as-usual scenario, DEC = decarbonisation scenario.
Source: Adapted from authors' modelling.

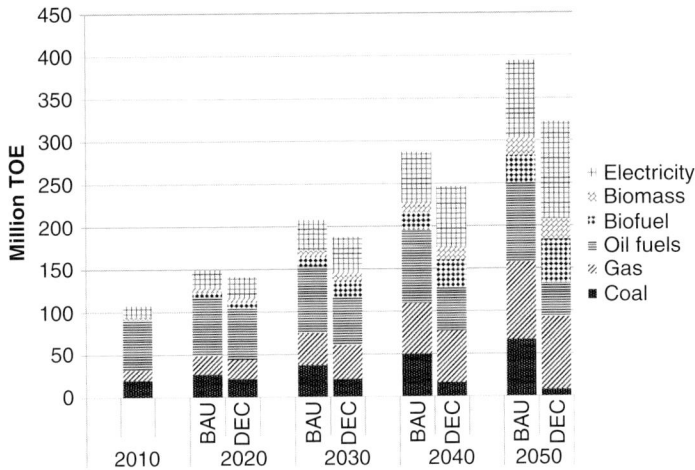

Figure 10.13 Final energy demand evolution.
Note: BAU = business-as-usual scenario, DEC = decarbonisation scenario.
Source: Adapted from authors' modelling.

2050, the final energy demand under a decarbonisation scenario is estimated to be 300% of the 2010 level (Figure 10.13). This corresponds to an average growth rate of 2.8% per annum, much lower than economic growth, which is in the range of 5.4–5.8% per year, indicating that the energy elasticity is less than 1. The moderate growth in demand is attributed to the results of energy efficiency measures. Important changes from 2010 to 2050 include: in terms of sectoral consumption, there will be a significant increase of the share of industrial energy consumption (from 49% in 2010 to 74% in 2050); in terms of energy type, there will be significantly increased share of electricity, gas and biofuels and decrease of oil fuels. Compared to the BAU scenario, the decarbonisation scenario has 20% lower energy demand in 2050.

10.6.1 Industry Sector

The industry sector is one of the major energy consumers in Indonesia. Currently, this sector accounts for around 49% of total energy consumption. Electrification of end use, fuel switching to lower-carbon fuels (from coal to gas) and bioenergy (solid biomass wastes and biofuels) are the dominant strategies for decarbonisation in industry. In addition, CO_2 emissions reductions are also realised through industrial efficiency improvement (decreasing energy intensity). Between 2010 and 2050, energy sector demand is expected to increase by 350%, from 53 MTOE in 2010 to 238 MTOE in 2050. This corresponds to an average growth rate of 3.8% per year. The BAU scenario will have higher energy demand in 2050 (288 MTOE). Decarbonisation measures would reduce the emission intensity of fuels in the industry sector from 3.7 tCO_2/TOE in 2010 to 1.6 tCO_2/TOE in 2050, lower than that would be obtained without decarbonisation (2.8 tCO_2/TOE) (Figure 10.14).

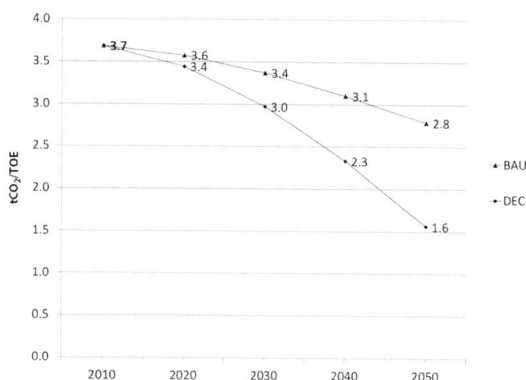

Figure 10.14 Final energy demand and emission intensity in the industry sector.
Note: BAU = business-as-usual scenario, DEC = decarbonisation scenario.
Source: Adapted from authors' modelling.

10.6.2 Building Sector

The building sector (i.e. residential and commercial buildings) will expand along with population, per capita income and commercial sector development. For the residential sector, increasing per capita income will increase energy consumption, but this will be balanced by more efficient appliances and the expectation that homes will remain relatively small. For commercial buildings, increases in the size of the service economy and modernisation of building equipment will result in increased energy consumption. From 2010 to 2050 energy demand from the building sector is projected to increase 112% from 17 MTOE to 36 MTOE, which corresponds to an average growth of 1.7% per year. Compared to the BAU scenario, decarbonisation results in 25% lower energy demand in 2050.

Decarbonisation in the building sector would result primarily from fuel switching, from oil to gas/LPG and from fuels to electricity, along with deployment of more energy-efficient electric appliances. Switching from on-site fuel combustion to electricity would reduce direct emissions from buildings, and a decarbonised electricity generation sector would also lead to emissions reductions. Decarbonisation measures would reduce the energy-related emission intensity of the building sector from 6.3 tCO$_2$/TOE in 2010 to 1.7 tCO$_2$/TOE in 2050 (lower than the carbon intensity under the BAU scenario, 3.2 tCO$_2$/TOE) (Figure 10.15).

10.6.3 Transport Sector

The energy demand in the transport sector is expected to increase significantly with economic development and population growth. In the passenger transport sector, the decarbonisation strategy includes modal shifts to mass transport, electrification of vehicles, fuel switching to less carbon-emitting fuels (oil to gas), use of more energy-efficient vehicles and extensive use of biofuels. Similar strategies are also applied to freight transport. A shift of freight transport from road to (electrified) railway is expected to

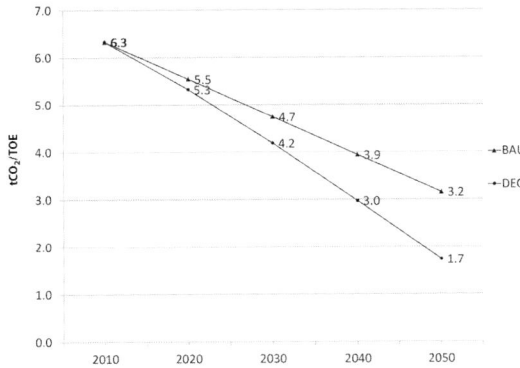

Figure 10.15 Final energy demand and emission intensity in the building sector.
Note: BAU = business-as-usual scenario, DEC = decarbonisation scenario.
Source: Adapted from authors' modelling.

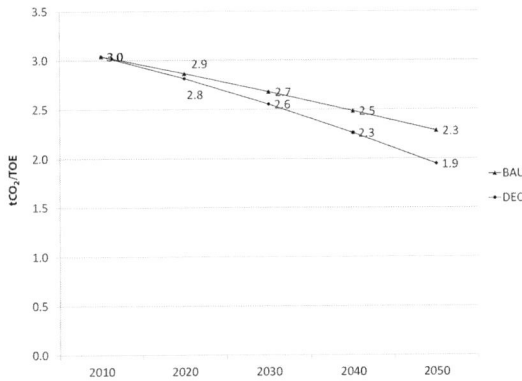

Figure 10.16 Energy demand and emission intensity in the transport sector.
Note: BAU = business-as-usual scenario, DEC = decarbonisation scenario.
Source: Adapted from authors' modelling.

decrease CO_2 emissions. As a result of modal shifts, it is expected that the share of passenger transport using personal vehicles in total passenger transport demand decreases from 60% in 2010 to 40% in 2050. In 2050, it is expected that 30% of personal cars will be electric vehicles. In the period between 2010 and 2050, transport energy demand is expected to increase 130% from 37 MTOE in 2010 to 47 MTOE in 2050, which corresponds to an average growth rate of 0.6% per year. Compared to BAU, the decarbonisation scenario has 19% lower energy demand in 2050. Decarbonisation measures could reduce the emission intensity from 3.0 tCO$_2$/TOE in 2010 to 1.9 tCO$_2$/TOE in 2050 (lower than the BAU scenario, 2.3 tCO$_2$/TOE) (Figure 10.16).

10.6.4 Primary Energy Supply

Under the decarbonisation scenario, primary energy demand in 2050 is projected to be 288% higher than that of 2010 (Figure 10.17). This corresponds to an average growth rate

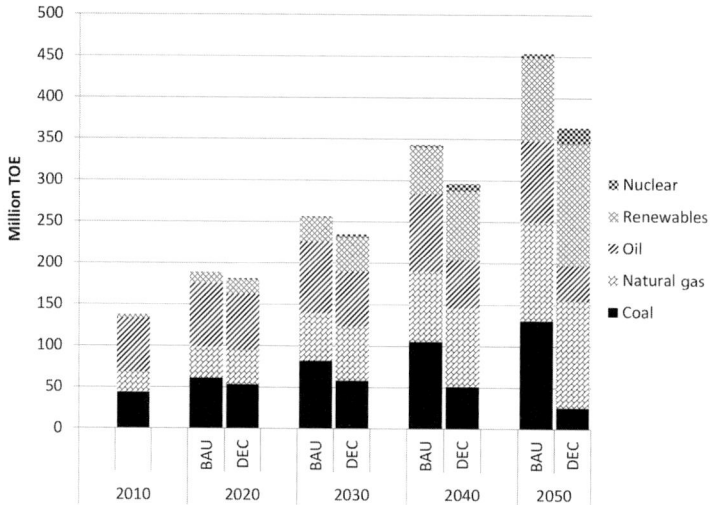

Figure 10.17 Pathway of primary energy, by source.
Note: BAU = business-as-usual scenario, DEC = decarbonisation scenario.
Source: Adapted from authors' modelling.

of 2.5% per year. This indicates that there will be improved energy conversion efficiency (note that final energy demand from 2010 to 2050 increases by 300% or an average growth of 2.8% per year). By fuel type, there will be a significant change of shares: new and renewable energy will increase from a mere 3% in 2010 to 45% in 2050; the coal share will decrease from 32% in 2010 to only 7% in 2050. Compared to BAU, decarbonisation will make possible a 20% lower energy supply in 2050.

10.6.5 Electricity Generation

As the result of a significant increase in electrification of end use, electricity generation is expected to increase sharply from 165 TWh (terawatt-hours) in 2010 to around 1390 TWh in 2050, which corresponds to an average growth rate of 5.5% per year. Compared to BAU, the decarbonisation scenario has a 24% higher electricity demand in 2050 (due to more extensive electrification of end use). Deployment of renewables (solar PV, geothermal, hydropower, wind), nuclear and gas to substitute coal power plants is expected to result in a drastic decrease of carbon intensity of electricity generation, from 871 gCO_2/kWh in 2010 to 133 gCO_2/kWh in 2050 (lower than BAU carbon intensity, 305 gCO_2/kWh) (Figure 10.18).

10.6.6 Liquid Fuel Production

In total, liquid fuel demand will continue to increase with economic development and increase of population. Between 2010 and 2050, liquid fuel demand will increase from 60 MTOE in 2010 to 91 MTOE in 2050, which corresponds to an average growth of 1% per year. Due to more extensive energy efficiency measures, liquid fuel demand in the

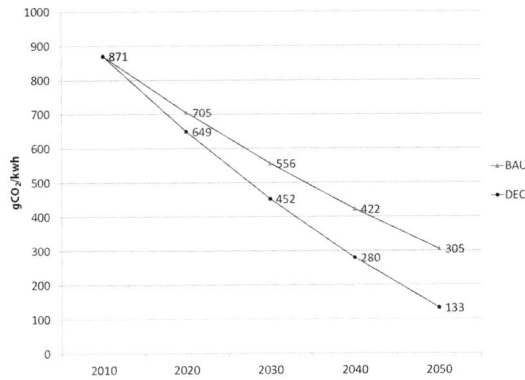

Figure 10.18 Pathway of electricity generation and carbon intensity.
Note: BAU = business-as-usual scenario, DEC = decarbonisation scenario.
Source: Adapted from authors' modelling.

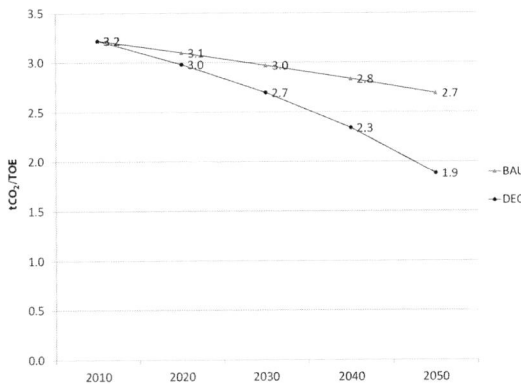

Figure 10.19 Demand and carbon intensity of liquid fuels.
Note: BAU = business-as-usual scenario, DEC = decarbonisation scenario.
Source: Adapted from authors' modelling.

decarbonisation scenario is expected to be 24% lower than that of the BAU scenario. Due to substitution of oil fuel with biofuel, the carbon intensity of liquid fuels is projected to decrease from $3.2\,tCO_2/TOE$ in 2010 to $1.9\,tCO_2/TOE$ in 2050 (lower than the BAU scenario, which decreases to $2.7\,tCO_2/TOE$) (Figure 10.19).

10.6.7 Decarbonisation Pillars

Indonesian decarbonisation is composed of three pillars: energy efficiency measures, electrification of end use and decarbonisation of the power sector.

Energy efficiency measures in the building sector include efficient building design, more efficient appliances and energy-efficient lifestyle changes. Efficient transportation is expected to be achieved through reduced transport demand (transport- and energy-conscious urban design) and extensive development of mass transport systems.

Efficiency measures are also to be taken in electricity generation systems: deployment of more efficient thermal power plants and improvement of electricity transmission and distribution systems (thus reducing transmission and distribution losses). Use of energy-efficient industrial equipment and development of less energy-intensive industry constitute efficiency measures in the industrial sector. These efficiency measures, combined with structural changes in the country's economy (i.e. decreasing the role of industry in the formation of national GDP through service sector substitution), are expected to contribute to the significant decrease of overall energy input per USD GDP formation, from 199 TOE per million USD in 2010 to 70 TOE per million USD in 2050 (Figure 10.20).

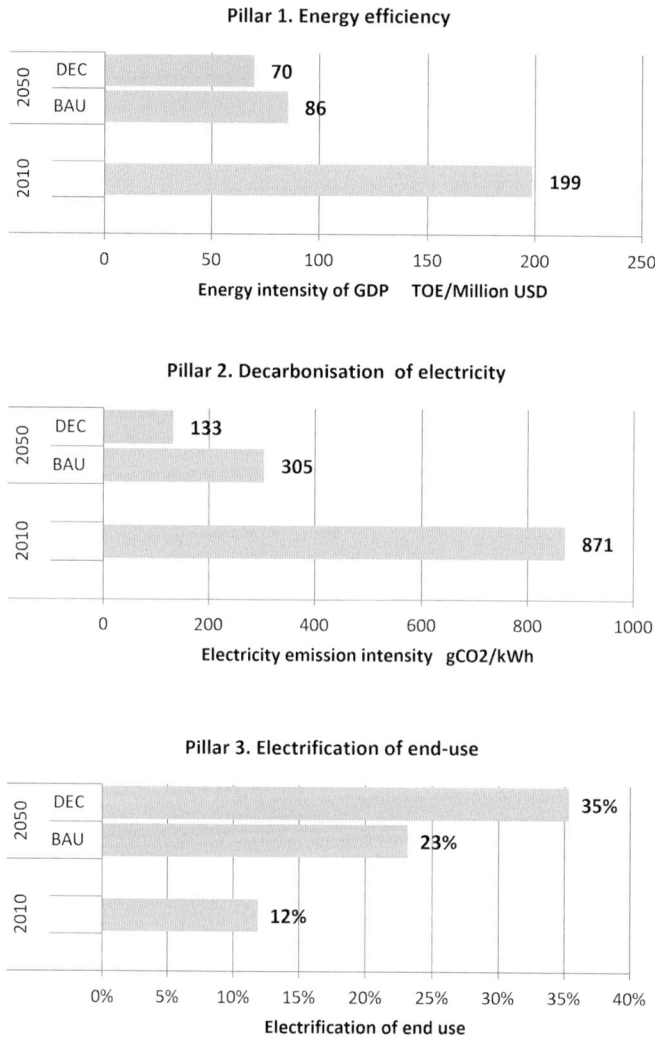

Pillar 1. Energy efficiency

Pillar 2. Decarbonisation of electricity

Pillar 3. Electrification of end-use

Figure 10.20 Pillars of decarbonisation.
Note: BAU = business-as-usual scenario, DEC = decarbonisation scenario.
Source: Adapted from authors' modelling.

Decarbonisation of the power sector is to be achieved through deployment of renewables (geothermal, hydropower, biomass, solar and wind), natural gas and nuclear power. The decarbonisation is projected to sharply decrease the carbon emission intensity of power generation from 871 gCO_2/kWh to 133 gCO_2/kWh.

The replacement of in situ combustion energy systems with electrically operated devices (end-use electrification) will increase the share of electricity in final energy demand and, combined with a decarbonised electricity generation system, will result in an overall decarbonised energy system. Electrification of end use is projected to take place in all sectors: fossil-fuelled heating systems (stoves/heaters/boilers) in buildings and industry would be replaced by electric heating systems, as well as deployment of electric vehicles to replace internal combustion engine vehicles. Between 2010 and 2050, the share of electricity in final energy demand is expected to triple, from 12% in 2010 to 35% in 2050.

10.7 Investment Needs

Decarbonisation requires investing specifically in low-carbon options, notably low- or zero-carbon-emitting power plants, low- or zero-carbon fuel production units and the procurement of low- or zero-carbon-emitting vehicles. Investment is needed for Indonesia's decarbonisation in these three key sectors. Indeed, Indonesia's energy sector decarbonisation strategy features significant, steady increases in investments (Figure 10.21).[6] These decarbonisation investment needs are calculated using the DDPP investment calculator developed by the SDSN/IDDRI project team. The investment calculator considers the learning curve of each technology.

To put the investment numbers in perspective, we assess them as a share of GDP. Given the GDP trends, investments in low-carbon options for these three key activities correspond to a maximum of 1.2% of GDP in 2020, before decreasing gradually towards 0.7% in 2050. These investments, although not negligible, are likely to be manageable for the Indonesian economy. Another way of interpreting these investment needs is to assess them relative to total macroeconomic investment of Indonesia, which has been rising from 22% of GDP in the early 2000s to about 35% of GDP in the early 2010s (Economy Watch n.d.). If one assumes that gross capital formation reverts to the lower end of the range, investment in low-carbon options would still represent less than 5% of total investments in the Indonesian economy. It should be noted that investments in low-carbon technologies under decarbonisation occur in parallel with a significant reduction in fossil fuel production (for both domestic uses and exports), triggering, in turn, a reduction of investments in the fossil fuel sectors.

---

[6] Note: the investment needs presented in this section do not include the additional infrastructure needed to support the operation of plants such as construction for gas pipelines or regasification plant (imported LNG) and to support the operation of electric vehicles or compressed natural gas (CNG) such as recharging/refuelling stations; the costs associated with energy efficiency measures in buildings and industry have not been included.

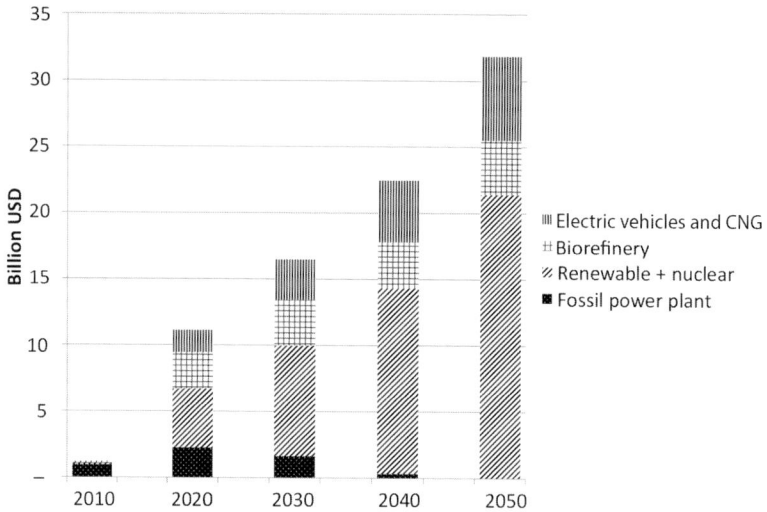

Figure 10.21 Investment needed for capacity addition of power plant and refineries and for vehicle procurement.

Source: Adapted from authors' modelling.

10.8 Challenges, Opportunities and Enabling Conditions

Technological changes on a wide range of low-carbon technologies – electric vehicles, high-efficiency power plants and others – are the keys of Indonesia's decarbonisation scenario. Some of these are still in the demonstration phase, or require important technical progress if their cost is to decline, which is a condition for their mass deployment. Decarbonisation realisation thus depends on the development, and maturing in the coming years, of these crucial technologies. There are also some proven decarbonising technologies (including solar power, wind, biofuel and geothermal). Solar power and wind power are now generally competitive with fossil fuels (petroleum diesel, coal power plants) while some technologies require further technical development and fast learning rates so that they become available at competitive, affordable prices. This also requires the right energy pricing policy.

Most of the technologies envisaged to be used in the pathway are imported. It is imperative, therefore, that Indonesia begins developing these technologies domestically. To speed up the process, efficient and large-scale international cooperation is a crucial enabling condition.

In addition, the deep decarbonisation transformation is characterised by a need for massive infrastructure development, such as infrastructure to enable mass public transport, new railways, gas transmission and subsea electrical transmission. Therefore, one of the main challenges of the pathway is to finance the infrastructure investment, and most notably, how to redirect investment flows towards these low-carbon options.

Indonesia's economy is expected to continue to offer significant investment opportunities. Given its fast growth and size, the Indonesian economy is expected to be able to absorb the investment needs for the realisation of decarbonisation transformation. The main challenge lies in the country's capacity to reorient investment decisions towards low-carbon solutions, in a drastic change compared to past and current decisions that are largely targeted on the development of fossil energy sources.

The deployment of nuclear power could play a role in decarbonisation but should be considered in the context of the need to decarbonise the entire Indonesian energy system. Nuclear poses a special challenge of social acceptability, which would require a social campaign and public debate. In addition, nuclear also poses safety risks due to the fact that Indonesia has a large probability of earthquakes.

In November 2016, Indonesia submitted its first NDC to the UNFCCC, which commits to emissions reduction of 29% below unmitigated (BAU) emissions in 2030. However, Indonesia is a non-Annex I country and climate concern has not yet been fully internalised into the Indonesian development agenda. To embrace the deep decarbonisation pathway, the government has first to internalise climate change, making it an integral part of the national development agenda.

To realise decarbonisation, Indonesia has to deal with enormous and unprecedented challenges: internalising climate change into the national agenda, attracting financing for infrastructure investment and technology development, technology transfer, a social campaign to facilitate the social acceptance of nuclear energy and the right energy pricing policy for renewable sources. To overcome some of these challenges, international cooperation is needed, especially when it comes to infrastructure financing and technology transfer. The government should begin to seek international cooperation, and find assistance for infrastructure development. In addition, the government must seek international partners for the transfer of the technologies necessary for deep decarbonisation.

This study also envisages that decarbonisation in Indonesia, despite all these challenges, in fact holds economic opportunities, and therefore goes hand in hand with the country's national development objectives. Developing renewable energy sources, and other less-carbon-emitting technologies, could stimulate economic development and create jobs. Energy efficiency measures could improve economic productivity. A deep decarbonisation scenario for Indonesia also assumes that the country's economic structure would shift towards a more service-oriented economy and that, in a decarbonised world, Indonesia's economy would be less dependent upon unstable revenues from fossil fuel exports. Deep decarbonisation also holds two additional benefits. First, reduced local pollution from transport, industry, power plants and residential energy. Second, decarbonisation can lead to improved energy security through developing the potential of domestic renewables.

The Indonesian decarbonisation scenarios have been built upon assumptions that the country maintains steady economic growth, has built sufficient infrastructure to enable electricity access for almost all households and that the poverty rate will be reduced. It can be said, therefore, that realisation of Indonesian decarbonisation would be compatible with the country's socio-economic development objective and priorities.

The materialisation of the described potential low-carbon path requires a significant, if not radical, change from the current BAU approach. First, the public must be made fully aware about the country's need for climate change mitigation. This can be achieved through a continuous and sustained public campaign through various media, and embedding climate change in the education curriculum at all levels. Second, it is necessary to internalise and mainstream climate change into Indonesia's development agenda. Development planning should include climate change in its selection criteria. Third, the government has to establish policies that will enable the materialisation of the described decarbonisation potential. These polices include: (i) incentives (such as renewable energy targets, feed-in tariffs), (ii) regulatory standards (such as emissions control, minimum energy performance standards for electric devices, minimum energy efficiency standards for new buildings) and (iii) a carbon tax. In addition, the government should allocate more budget for promoting research and development and the demonstration of new low-carbon technologies and capacity-building activities (education and training) in climate-related fields.

10.9 Conclusion

This study finds that Indonesia has the technical potential to very significantly reduce its energy-related CO_2 emissions, while at the same time developing its economy and achieving economic growth. The Indonesian decarbonisation scenario used in this study will achieve 536 Mt in 2050, which, in per capita terms, translates to $1.7 \, tCO_2$ per capita. At this emission level, Indonesia will significantly contribute to the global efforts to prevent temperature increases above 2 °C in 2050.

Indonesia decarbonisation could be technically achieved through the implementation of three decarbonisation pillars: energy efficiency measures, electrification of end use and decarbonisation of the electricity sector (massive deployment of renewable and zero-carbon power generation).

Energy-related CO_2 emissions will continue to increase in the 2010–2030 period (due to economic development at a rate of 5.4–5.8% per year) and then decrease afterwards. This emission evolution is associated with the assumed trajectory: that is, gradual development of low-carbon options in the short term, followed by intensive decarbonisation.

The decarbonisation scenario assumes the total deployment of renewable energy (solar PV, hydro, geothermal and wind) to replace most, if not all, coal and oil power plants. In addition to renewables, some fraction of the power plants could be nuclear-powered. This scenario further assumes that large-scale solar PV technology is deployable, and that large hydro resources in Papua (eastern Indonesia) are utilisable, to cater for demand in the western part of Indonesia through long-distance subsea cables.

Enormous investment in energy infrastructure is a major challenge for Indonesia's decarbonisation but the investment challenge is likely manageable, considering that it represents only a small fraction of total investments throughout the economy, especially in the context of the country's fast economic growth, which is assumed in our scenarios. The main challenge, therefore, is to develop adequate policy incentives and schemes to reorient investments towards low-carbon options.

Finally, it should be noted that, in all these projections, the Indonesian economy is able to develop and grow (at around 5%) while simultaneously realising deep decarbonisation.

Box 10.1   **India: Enhancing Renewables through Policy Innovation**

KENNETH G. H. BALDWIN

1. An Emerging Economic Giant

India is a major developing country, with the second largest population in the world. Its economy is ranked seventh in terms of GDP (World Bank n.d.), and is projected to move significantly higher in these rankings in the coming decades (PricewaterhouseCoopers 2017).

India also has ambitious emissions reduction goals, driven largely by national government policy. This is reflected in and underpinned by their NDC to the Paris emissions reduction goal: India has committed to reduce emission intensity of GDP by 33–35% of 2005 levels by 2030.[7]

In terms of electricity, India has now achieved a relatively high access rate (IEA 2019), reaching over 95% of the population, and is on a rapidly rising trajectory for electricity consumption per capita (IEA 2018). However, while electricity generation in India is currently dominated by coal (76%) (BP 2019), India has a target of 450 GW renewable generating capacity by 2030 (Climate Action Tracker n.d.). India's conditional NDCs further commit to providing 40% of its electricity-generating capacity from non-fossil-fuel sources (Climate Action Tracker n.d.).

2. Renewable Installation

India is currently deemed to be on track with its Paris targets (Climate Action Tracker n.d.), in large part due to its focus on increasing renewable electricity generation. Figure 10.22 shows the annual installed generation capacity for wind and solar (Burke et al. 2019), plus other renewables comprising mostly hydro (currently 9% of total generation (BP 2019)), as well as biomass. In particular, India has significant solar resource potential, and, to a lesser extent, wind.

3. The Role of Policy

Examination of the reasons behind the rapid uptake of wind and solar in India reveals a number of key institutional and policy factors (Burke et al. 2019). Institutionally, India has developed a strong platform for renewables, including the creation of a Ministry of Non-Conventional Energy Sources back in 1992.

In terms of policy, there are a number of key factors at play:

- renewable energy investment has been prioritised at prime ministerial level;
- reverse auctions for solar and wind have for some time been used to pave the way for significant price reductions through competition – see Figure 10.23 (Burke et al. 2019);
- renewable purchase obligations have been used to meet renewable energy targets;
- renewable energy certificates have enabled interstate renewable energy purchases to be counted towards the meeting of renewable targets.

---

[7] *Paris Agreement Under the United Nations Framework Convention on Climate Change*, opened for signature 16 February 2016. Available at: https://unfccc.int/process-and-meetings/the-paris-agreement/the-paris-agreement.

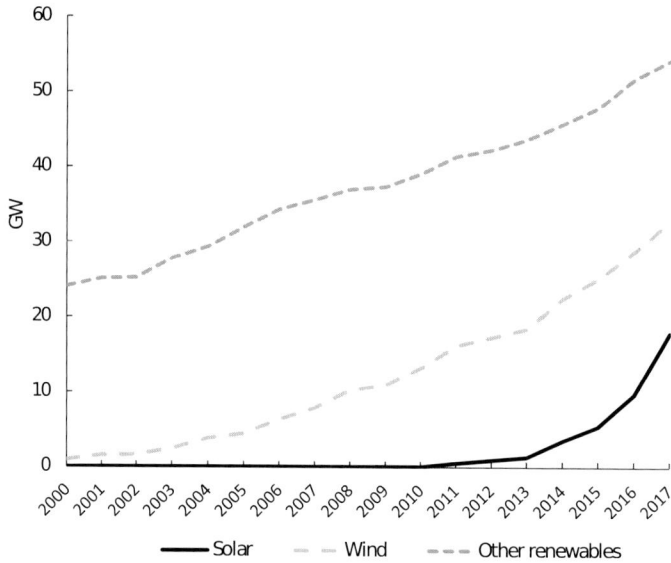

Figure 10.22 Annual installed capacity for solar, wind and other renewables.
Source: Burke et al. (2019).

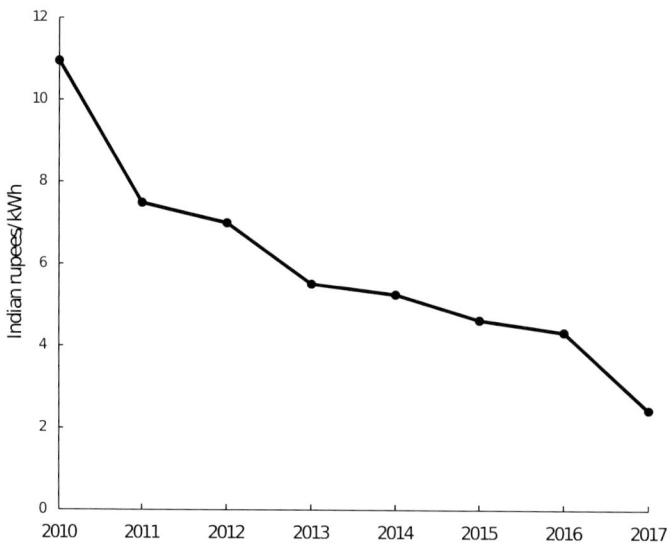

Figure 10.23 India's reverse auction minimum price for utility-scale solar PV installations.
Source: Burke et al. (2019).

However, there are also countervailing factors that challenge the rapid deployment of renewables:

- entrenched fossil fuel incumbency exists; for example, through the adoption in the past of power purchase agreements with more than 20-year time frames;

- uptake of rooftop solar panels has been limited, in part, by delays in disbursing subsidies;
- delays in transmission infrastructure discriminate against utility wind and solar;
- slow roll-out of smart meters and other smart grid infrastructure inhibits rooftop solar panels.

4.   The Way Forward

Despite such barriers to realising rapid deployment of renewable energy, prospects exist for even faster development of India's significant solar and wind potential. Key to this will be:

- continuing high-level political commitment and ambitious targets;
- fostering an attractive investment environment by reducing regulatory uncertainty and removing investment barriers;
- moving from fossil fuel subsidies to emissions pricing;
- encouraging international trade and investment, including renewable energy trading;
- continuing innovation in policy development and the planning of renewable energy zones, with a focus on ensuring full utilisation of renewable energy from the highest-resource locations in the country.

In this way, the benefits of the rapid global improvement in the cost-effectiveness of renewable energy can be captured to enhance the development of such a rapidly growing economy.

References

BP (2019). *BP Statistical Review of World Energy 2018*. London: BP. Available at: www .bp.com/content/dam/bp/business-sites/en/global/corporate/pdfs/news-and-insights/ speeches/bp-statistical-review-of-world-energy-2018.pdf.

BPS (Budan Pusat Statistik) (2016). Available at: www.bps.go.id/linkTabelStatis/view/id/ 1267.

BPS (2017). *Produk Domestik Bruto Indonesia Menurut Pengeluaran, 2012–2016*. Available at: www.bps.go.id/publication/2017/06/05/684ad7c8506e2765d964522a/ produk-domestik-bruto-indonesia-menurut-pengeluaran–2012-2016.html.

BPS (2020). Sensus Penduduk 2020 [data resource]. Available at: www.bps.go.id/sp2020/.

Burke, P. J., Widnyana, J., Anjum, Z., Aisbett, E., Resosudarmo, B. and Baldwin, K. G. H. (2019). Overcoming barriers to solar and wind energy adoption in two Asian giants: India and Indonesia. *Energy Policy*, 132, 1216–1228.

Climate Action Tracker (n.d.). India [data resource]. *Climate Action Tracker*. Available at: https://climateactiontracker.org/countries/india/.

Economy Watch (n.d.). Indonesia investment (% of GDP) statistics. *Economy Watch*. Available at: www.economywatch.com/economic-statistics/Indonesia/Investment_ Percentage_of_GDP/.

IEA (International Energy Agency) (2018). *World Energy Statistics and Balances 2018*. Paris: International Energy Agency. Available at: https://webstore.iea.org/world-energy-statistics-and-balances-2018.

IEA (2019). SDG7: Data and projections. *IEA.org*. Available at: www.iea.org/reports/ sdg7-data-and-projections/access-to-electricity.

Indonesia DEN (National Energy Council) (2015). *Outlook Energi Energy Indonesia 2015*. Jakarta: Dewan Energi Nasional. Available at: www.esdm.go.id/assets/media/con tent/content-indonesia-energy-outlook-2015-1vgcv6t.pdf.

Kementarian PPN/Bappenas (2014). *Rencana Pembangunan Jangka Menengah Nasional (RPJMN) 2015–2019* [*National Medium-Term Development Plan*]. Jakarta: Kementarian PPN/Bappenas.

Kompas Daily (2016). *Kompass Daily*. 1 April 2016.

MEMR (Indonesia Ministry of Energy and Mineral Resources) (2013). *2013 Handbook of Energy & Economics Statistics of Indonesia*. Jakarta: Pusdatin–ESDM (Ministry of Energy and Mineral Resources). Available at: www.esdm.go.id/assets/media/content/content-handbook-of-energy-economic-statistics-of-indonesia-2013-997ndnz.pdf.

MEMR (2017). *2017 Handbook of Energy & Economics Statistics of Indonesia*. Jakarta: Pusdatin–ESDM. Available at: www.esdm.go.id/assets/media/content/content-hand book-of-energy-economic-statistics-of-indonesia-2017–1.pdf.

PricewaterhouseCoopers (2017). *The Long View: How Will the Global Economic Order Change by 2050?* London: PricewaterhouseCoopers. Available at: www.pwc.com/gx/en/issues/economy/the-world-in-2050.html.

Republic of Indonesia (2016). *First Nationally Determined Contribution*. Submitted to United Nations Framework Convention on Climate Change. Available at: www4.unfccc.int/sites/ndcstaging/PublishedDocuments/Indonesia%20First/First%20NDC%20Indonesia_submitted%20to%20UNFCCC%20Set_November%20%202016.pdf.

Republic of Indonesia (2018). *Indonesia: Second Biennial Update Report Under the United Nations Framework Convention on Climate Change*. Jakarta: Kementerian Lingkungan Hidup dan Kehutanan.

World Bank (n.d.). GDP. In World development indicators [data resource]. *The World Bank Data Catalog*. Available at: https://datacatalog.worldbank.org/dataset/world-development- indicators.

Cities and Industry

11

Cities

XUEMEI BAI, TIMOTHY M. BAYNES, ROBERT WEBB, CHRIS RYAN
AND MICHAEL H. SMITH

Executive Summary

Cities are currently responsible for over 70% of global carbon dioxide emissions from energy use. Building and upgrading infrastructure for cities in developing countries could release up to 226 gigatonnes of carbon dioxide by 2050, if these cities are to enjoy the same level of per capita infrastructure as cities in developed countries today.

The drivers of greenhouse gas (GHG) emissions from cities vary across different urban economies, geography, wealth, consumption behaviour and urban form, although the largest direct and indirect sources of GHG emissions come from buildings, industry and transport.

Cities are subject to a range of climate change conditions including rising temperatures, sea-level rise, extreme weather events, drought and water scarcity. Climate hazards compound with stressors such as poverty, and climate change will hit cities in developing countries and the poor population harder.

Mitigation and adaptation measures interact in a complex way, and sometimes climate actions interact with each other to generate unintended consequences. A systems approach such as integrated planning, and a strategy that recognises synergies and possible conflicts, are crucial in order to achieve optimal outcomes.

Cities and precincts are at a good scale for innovation and experimentation to address climate change, but alignment with national levels of governance is essential to enable effective planning and action.

Many cities across the world are taking actions to combat climate change. There are an increasing number of cities committing to 100% renewable energy and net zero emissions by 2030, with some already achieving it.

Key enablers of change include: a shared longer-term vision for a city and its region and effective stakeholder engagement across government, private and community sectors; relevant, credible and accessible knowledge to support evidence-based decision-making; and aligned institutional arrangements.

11.1 Introduction: Scope and Purpose of This Chapter

Cities are the main contributor of climate change through greenhouse gas (GHG) emissions – up to 75% of GHG emissions from final energy use are from cities (Grubler et al. 2012; Seto et al. 2014). At the same time, cities are expected to be significantly affected by climate change, such as through higher average temperatures, longer heatwaves, more flood events, changing rainfall patterns and associated droughts, higher risks of bushfire, and coastal inundation with higher storm surges combined with sea-level rises (Corburn 2009; Grimmond 2011; McDonald et al. 2011; Willems et al. 2012; IPCC 2013; Bai et al. 2018; Rosenzweig et al. 2018). Such impacts will hit developing countries' cities harder, as these cities are often facing more compound social, economic and environmental issues with limited capacity to respond to the impacts (World Bank 2010a, 2010b; Nagendra et al. 2018; Pelling et al. 2018). The share of world population living in cities is estimated to approach 70% by 2050. Most of this future population growth is projected to be added to developing countries' cities, where climate change will compound with many other existing challenges. Therefore, building resilient and low-carbon cities will be crucial for the future of humanity. Rapid transformations in cities are required to achieve the target of limiting global warming to 1.5 °C while adapting to the impacts of climate change (IPCC 2018; Solecki et al. 2018).

This chapter begins with the rationale of addressing climate change at the city level, followed by an overview of the state-of-the-art knowledge of urban mitigation and adaptation. Finally, we present a solution-oriented perspective, with particular emphasis on how to bring about transformative change towards building resilient and low-carbon cities.

Throughout the chapter, a strong emphasis is placed on a systemic understanding and approach (Bai et al. 2016a), considering the synergies and trade-offs between mitigation and adaptation across different sectors and the challenges and opportunities such interlinkages bring about. This is because cities are complex systems, and hence designing climate solutions is best done through an integrated lens to optimise synergies and reduce conflicts where possible.

The city-level discussion in this chapter is complemented by subsequent chapters investigating specific sectors: buildings and precincts, industry, urban water, transport, again with particular emphasis on the synergies and conflicts between mitigation, adaptation and sustainable development. Together, these chapters are designed to:

(1) inform a more integrated approach by local and national governments in addressing climate change mitigation, climate change adaptation and sustainable development strategies;

(2) help urban practitioners, investors and business leaders actively seek out the most effective leverage points or new sweet spots – namely, actions and investments that simultaneously achieve adaptation, mitigation and sustainable development; and

(3) shed some light on the key role of people's behaviour and consumption choices, and design programmes, incentives and educational schemes appropriately.

11.2 Why Is Action on Climate Change a Priority for Cities?

In addition to the above-mentioned direct impacts and risks from climate change, there are a number of reasons why action on climate change in cities is critical and a priority. We distil these into six issues: cascading impacts; lock-in; co-benefits; consumer behaviour; capacity for cities to take action; and the city or region level of governance as an appropriate point of intervention.

First, cities will face various cascading negative impacts under climate change. Although not exhaustive, the following are some key considerations.

Heat. Increased temperatures and longer heatwaves, combined with the urban heat island (UHI) effect, will result in higher peak electricity demand, which means demand for more infrastructure and associated higher costs to cities. The energy demand for air conditioning in summer is projected to increase more than 30-fold by 2100 due, significantly, to unmitigated climate change: from nearly 300 terawatt-hours (TWh) in 2000 to about 4000 TWh in 2050 and more than 10 000 TWh in 2100 (Isaac and Van Vuuren 2009). Higher temperatures and longer heatwaves, combined with extreme drought, will heighten the risks of bushfire, as manifested in Australia in 2019–2020, with compounding negative health impacts (Yu et al. 2020).

Water. Shortages of water are already experienced by approximately 150 million people living in cities. Climate change modelling scenarios suggest that this could rise to as many as 1 billion by 2050 due to predicted more extreme drought (McDonald et al. 2011) (see Chapter 13). Many cities are dependent on hydropower for electricity, the capacity of which can be rapidly reduced by drought.

Exposed Assets. These include rising infrastructure and capital asset exposure in large coastal cities. The amount of urban assets exposed to hazards from unmitigated climate change in cities such as Dhaka (Bangladesh) could rise 60-fold (Hanson et al. 2011). More extreme and intense cyclones, storms and flooding have already caused significant damage to many cities in southern USA (up to New York), the Caribbean, Central America, Asia and the Pacific in the last decade. The literature suggests that there are limits to some coastal cities' capacity to adapt cost-effectively to sea-level rises of greater than 3–5 metres. In low-lying coastal cities, even small sea-level rises can seriously disrupt storm-water drainage systems.

Supply Chain Interruption. Cities are highly dependent and reliant on just-in-time supply chains for food, transport fuels, many raw materials and goods and services. This means cities are more vulnerable to extreme-weather-related disruption of them.

Essential Infrastructure. Also vulnerable are urban infrastructures providing essential 'systems of provision' (energy, water, food, waste, transport, shelter) (Ryan 2005), which have complex interdependencies and are exposed to indirect or flow-on impacts from climate change (Bai et al. 2018; Ryan et al. 2019). Extreme weather events negatively impact on electricity grids: for example, heatwaves causing demand spikes from air conditioning or storms bringing down powerlines. This can then cause power cuts, posing risks to critical services such as sewerage treatment plants. Rainfall beyond historical peaks can overwhelm existing storm-water systems, with flow-on impacts on transport and other infrastructure.

Second, the investment and design decisions made in cities in the next two decades will largely determine whether the world can successfully chart a course to a low-carbon economy and avoid dangerous climate change. Much of the world's future urban areas and their infrastructure is yet to be built. This will happen over the next two to three decades. The very long design life of the urban built environment and infrastructure (energy, water, transport and telecommunications) means that the decisions taken will be critical to dealing with climate change.

There are critical differences in approach between mature and rapidly growing cities. In the former case, mitigation is limited by the historical lock-in of infrastructure systems that were developed through the era of low-cost fossil fuels. Such cities often have embedded dependencies on high flows of fossil fuels for maintaining urban structure and functions. Much of the remaining carbon budget will be used in embodied energy if humanity continues to build new buildings for over 2 billion people by 2050 using 'traditional' materials (Müller et al. 2013). As shown in Figure 11.1, it is estimated that building and upgrading infrastructure for cities in developing countries

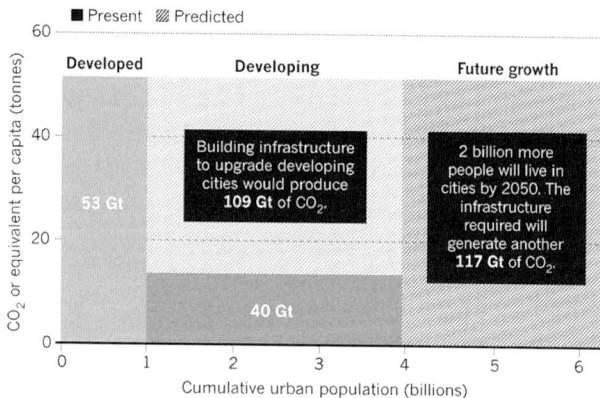

Figure 11.1 Estimated carbon emissions from future urban development.
Note: Assumptions made in generating this figure: (a) CO_2 or equivalent per capita is based on Müller et al. (2013); (b) per capita infrastructure in developing country cities will catch up with that of the developed cities today; (c) per capita urban infrastructure in developing cities is calculated as twice that of rural residents, an assumption based on the current income disparity between urban and rural areas. Developed countries are as listed in Annex I to the Kyoto Protocol; developing countries are those not listed in Annex 1. For a colour version of this figure, please see the colour plate section.
Source: Bai et al. (2018). © Nature.

could release 226 Gt (gigatonnes) of carbon dioxide (CO_2) by 2050, which is more than four times the amount used to build existing developed-world infrastructure (Bai et al. 2018).

From a global perspective, it is imperative that, regardless of the city context, *all* urban development (buildings, infrastructure and urban planning) in the next two decades will be low- or zero-carbon. The *Deep Decarbonization Pathways Project* reports (SDSN and IDDRI 2015) show that cities that adopt integrated approaches to climate change mitigation can become close to net climate neutral by 2050 (Chapter 9 discusses the Australian DDPP). For each urban sector that is a large source of GHG emissions – buildings, industry, transport, urban water utilities – this chapter through to Chapter 16 provide detailed analyses of how each of these sectors can reduce emissions in ways that maximise adaptation synergies.

Third, many of the largest and most profitable climate change mitigation opportunities lie in cities, which can unlock significant productivity, environmental and social co-benefits such as reduced urban air pollution. For instance, the New Climate Economy modelling result shows that urban climate change mitigation pathways could add as much as USD 17 trillion to global GDP by 2050 (above a business-as-usual approach) (Gouldson et al. 2015). As we show in this chapter, this is a significant underestimate as it fails to include (a) many mitigation measures and (b) air pollution reduction and other co-benefits in its analysis.

Fourth, the consumption power of cities and their citizens means the choices and behaviours of consumers in cities can make significant differences to the amount and pace of GHG emissions increase or reduction. In addition, cities often are the loci of cultural innovation, which means a significant opportunity to bring about new low-carbon consumer behaviour.

Fifth, many of the key actors needed to achieve the transition to a low-carbon resilient future – the private sector (finance, corporations, insurance), the public sector (government, innovation/research and development, educational) and civil society institutions (non-governmental organisations (NGOs) and community organisations) – are based in cities. Such close proximity means higher capacity for cities to take coordinated action, and enables ease of partnership development and also the development of innovation clusters within cities, both of which are critical to enabling the transition to a low-carbon resilient economy this century.

Finally, and importantly, cities have the right level of governance where actions on climate change mitigation and adaptation can take place. There is growing recognition of the key role of city-scale governments in addressing both mitigation of, and adaptation to, climate change in partnership with national governments. Leading cities already understand this and a coalition of the Global Covenant of Mayors, the C40 Cities Climate Leadership Group, United Cities and Local Government (UCLG), Local Governments for Sustainability (ICLEI Global n.d.) and UN Habitat have committed to a common action and reporting programme to drive decarbonisation and adaptation initiatives from the local level (GcoM 2019).

11.3 Understanding Cities' Contribution to Global Warming and Mitigation Potentials

11.3.1 Understanding Cities' Contribution to Climate Change

Estimates of the contribution of cities to climate change vary greatly due to disparities in the accounting methods, urban boundary definitions and data availability, within and between countries (UN Habitat 2011). According to UN Habitat (2011: 51), there is no globally accepted definition of an urban area or city, and there are no globally accepted standards for recording emissions from subnational areas. In addition, there is little clarity on the relative allocation of responsibility from production- or consumption-based approaches (UN Habitat 2011). This is a continuing knowledge gap identified in the 2014 Intergovernmental Panel on Climate Change (IPCC) reports (Seto et al. 2014), and expanding observation and establishing comparable cross-city data is identified as one of the six research priorities in cities and climate change (Bai et al. 2018).

The most significant, and most certain, contribution of cities to GHG emissions now and most likely into the future is through end-use energy consumption from fossil fuel sources. The World Energy Outlook (IEA 2008) estimated that cities are responsible for 71% of global CO_2 emissions from energy use. In the Global Energy Assessment's urban chapter (Grubler et al. 2012), the world's urban energy consumption lies between 180 and 250 EJ (exajoules) per year, depending on the definition of administrative, spatial or functional boundaries. This is between 56% and 78% of global final energy use and, given the corresponding global emissions, Seto et al. (2014) estimate that between 53% and 87% of CO_2 emissions from final energy use are from cities. This figure changes significantly when overall CO_2 emissions are considered. In Africa, for example, the urban share of energy-related CO_2 emissions is high (64% to 74%) but the urban share across all emissions and all sectors is low (21% to 30%) (Marcotullio et al. 2013), reflecting a high urban share in energy use but a lower share in other sources of CO_2 emissions (e.g. from agriculture and land-use change) in the region.

The extent of the responsibility of cities for GHG emissions depends on the scope of assessment and how we attribute responsibility for emissions to sectors of society (Baynes and Wiedmann 2012). If we say that it is in production that choices are made about technologies used and their energy or emissions intensity, then we have a direct or production-based attribution of responsibility for emissions. This is often appropriate in terms of seeking interventions within a specific territory where production activity occurs.

Such an approach, however, ignores the large immediate and remote hinterlands of modern cities that support an economy, service-based or otherwise. It also encourages an 'out of sight, out of mind' mentality, where a consumer benefits from goods produced elsewhere without being held responsible for the emissions. Steininger et al. (2014) show that a consumption-based approach, which is an indirect attribution, can improve both cost-effectiveness and justice at global level, if unilateral climate policies of industrialised countries are consumption-based. Up to 70% of urban household consumption-based emissions in China are from other regions, and this number seems to be consistent across a household-survey-based study (Lin et al. 2013) and an input–output-model-based study

(Mi et al. 2016). This consumption-based assessment is relevant for broadly applied mitigation policies such as carbon pricing, and is more comprehensive in that it includes *all* upstream production activity, wherever it takes place (Minx et al. 2009; Lenzen and Peters 2010). In a developed nation like the UK, the impact of consumption is relatively homogeneous over different scales and urban situations, while the territorial emissions vary with settlement type and local economic structure (Minx et al. 2013). The consumption-based approach is certainly comprehensive, but placing responsibility for emissions on the consumer is not without issues: a producer also derives benefit in the form of income while causing emissions that a consumption-based assessment will attribute to a consumer. There are hybrid approaches that seek to represent the territorial activity and include essential flows across the boundary of the defined urban area (Ramaswami et al. 2008, 2011; Hillman and Ramaswami 2010).

As noted in the latest IPCC report (Seto et al. 2014), the literature on emissions from cities suffers from being mostly about energy-related CO_2 with less consideration of other emissions, few data on small to medium cities and a bias towards reporting on developed regions. There are few attempts at global and comprehensive assessments of urban emissions. Accounting for CO_2, methane (CH_4), nitrous oxide (N_2O) and sulphur hexafluoride (SF_6) emissions produced only directly within the city, Marcotullio and colleagues estimated that the urban areas' contribution was approximately 37% of global emissions for the year 2000. Including upstream emissions from electricity consumption, this rose to 49% of global emissions (Marcotullio et al. 2013). While it is likely that urban areas dominate global CO_2 emissions from the perspective of final energy consumption, there are still significant CO_2, methane and nitrous oxide emissions (e.g. due to land-use change and agriculture) that occur outside urban areas but can be linked back to urban activities through production and consumption chains. Clearly then, an integrated approach is needed to manage those urban activities that release GHG within the city, as well as those activities that require GHG emissions beyond the city boundaries.

11.3.2 Understanding Drivers of Urban Emissions

There is a large disparity in urban carbon profiles depending on demographics, local geography and climate, urban form and structure, affluence, attitudes and social and economic conditions. The 2011 report *Cities and Climate Change* (UN Habitat 2011) and other literature (Doherty et al. 2009; Bai et al. 2012; Minx et al. 2013) have identified major factors that influence and drive GHG emissions from urban areas, namely:

Demographics: The size and rate of increase of the urban population are important determinants of urban GHG emissions. According to the UN, by 2050, nearly all of the world's net population growth, estimated at 2.6 billion, is expected to occur in urban areas, and more than 80% of urban dwellers in 2050 will be in the developing world. More people will be living in cities of one million people or less, and it is at this level where we observe the greatest increase in energy demand with increase in population (Grubler et al. 2012).

The urban economy: This relates to the city's economic structure, as well as its types of economic activities and how carbon-intensive they are. For instance, Shanghai's industrial sector consumed 90% of the city's energy between 1990 and 2005. Cities like London, Paris and New York have become dominantly service-based economies over the last 50 or more years, but these activities are not impact-free. In Japan, for example, household expenditure on services induced 14% of total domestic CO_2 emissions in 2000, exceeding the 11% induced by household expenditure on electricity and public transportation combined (Nansai et al. 2009).

The wealth of urban residents: Cities tend to generate more income per capita as they increase in size (Bettencourt et al. 2007) and this income growth has been linked to energy use and, importantly, indirect emissions (Lenzen et al. 2004). The OECD anticipates that developing countries' national GDP will increase fourfold between 2015 and 2050 (OECD 2016). A report by the McKinsey Global Institute (Dobbs et al. 2011) expected that the 600 largest cities will expand 1.6 times as fast as the global population and produce 60% of global GDP by 2025. This translates into 735 million households with an average income of USD 32 000. There is a large urban–rural income disparity in developing countries, which means generally the urban emissions per capita in these countries are greater than their national average. But the reverse is true in developed countries, and there is a global trend of convergence in urban emissions per capita.

Consumer behaviour and choices: Most of the world's goods and services are consumed in cities, so consumer choices and behaviours within cities influence urban GHG emission. In concert with infrastructure choices, there needs to be a change in behaviour choices: for example, building higher-density residences needs to come with tolerance of higher-density living, which is difficult in the context of a general long-term decline in urban density (Angel et al. 2010); service-oriented consumption will only work if, for example, people are willing to forgo the convenience of individual ownership of (service-providing) products, such as personal vehicles, instead choosing car-share or public transport arrangements. Occupants' behaviour has the potential to reduce building energy consumption by up to 20% (Zhang et al. 2018). Because of the urbanisation of the global population, consumption choices increasingly reflect the influence of 'urban lifestyles'. These patterns of urban consumption are, on the one hand, 'structural' (or 'obligatory'): broadly determined by the physical structuring of urban form and the economic and energy structure; and on the other hand, choices and behaviours or practices that are more sociocultural, reflecting experiences and expectations learned within the daily life of an urban community.

Geography and climate: A city's geographic location and climate influences the amount of energy required for heating, cooling and lighting (Singh and Kennedy 2015), but the combination of population and affluence at a given location matters. Further, climate change is likely to change the current pattern of links between geographic location and energy consumption. Isaac and Van Vuuren (2009) simulated changes in residential space conditioning (heating and cooling) in response to climate change. The coincidence of a greater population being more able to afford air

Atlanta and barcelona have similar populations but very different carbon emissions

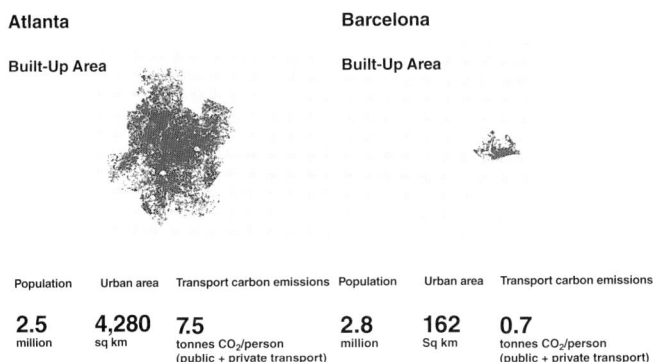

Atlanta

Built-Up Area

Barcelona

Built-Up Area

Population	Urban area	Transport carbon emissions	Population	Urban area	Transport carbon emissions
2.5 million	**4,280** sq km	**7.5** tonnes CO_2/person (public + private transport)	**2.8** million	**162** Sq km	**0.7** tonnes CO_2/person (public + private transport)

Figure 11.2 Comparison between the relationship between urban area and per capita transport emissions for Atlanta City, USA, and Barcelona, Spain.
Source: Bertaud and Richardson (2004), with additional data on transport energy from Kenworthy et al. (1999).

conditioning, in locations where there would be a greater number of hot days, meant that air-conditioning energy demand increased by 72%. At the regional scale, considerable impacts can be seen, particularly in South Asia, where energy demand for residential air conditioning could increase by around 50% due to climate change.

Urban form, design and planning choices: Urban structure and form have significant effects on direct and indirect (embodied) emissions. Sprawling cities tend to have higher per capita transport emissions than more compact ones as shown by comparisons between, for instance, Atlanta and Barcelona (see Figure 11.2).[1] Compact cities likely have less total traffic needs, and make mass public transportation systems more feasible. Other dimensions of urban design that have an impact on emissions are: land-use mix, connectivity, accessibility, and the combination of these measures can be more than the sum of the parts.

As mentioned above, the largest direct and indirect (through electricity and embodied energy in other goods and services) sources of GHG emissions come from buildings, industry and transport. Clearly, the infrastructure choices of cities and their hinterlands can significantly influence their impact. A study of Chinese cities shows that up to 31% of CO_2 emissions reduction can be achieved by updating a small fraction of existing infrastructure (Shan et al. 2018).

Embodied energy in urban infrastructure/building stock: Many of the materials used for urban infrastructure and buildings are highly carbon-intensive. For example, the production of steel and cement contributed 9% and 7%, respectively, of GHG emissions in 2006 (Allwood et al. 2010). The embodied carbon in infrastructure as

---

[1] It is important to note that this does not mean that overall per capita emissions in more compact cities is necessarily lower than in more spread-out cities. That depends on various factors such as embodied and operational energy-related emissions in high-density, high-rise buildings/precincts. Once buildings are built over four to six storeys in height, the embodied energy of steel and cement structures combined with additional heating, ventilating and air conditioning and elevator operational energy demand can outweigh the overall transport emissions reductions from a more compact city.

of 2008 was roughly 122 ($-20/+15$) GtCO$_2$ (gigatonnes of CO$_2$) (Müller et al. 2013). Reducing embodied energy in construction materials will be essential in reducing overall GHG emissions from cities.

Energy structure: Assuming all other factors are equal, a nation's/region's energy structure also directly affects how much carbon is released from cities, especially under the consumption-based accounting method. If the electricity supplied to the city is predominantly produced via coal power plants, the city will have a much larger carbon footprint than those cities using electricity from hydropower or nuclear sources. Conversely, a 100% renewable energy supply would mean a much lower carbon footprint from the city.

11.3.3 Reducing Greenhouse Gas Emissions from Cities

Given the multidimensional and complex drivers underlying urban GHG emissions, an integrated approach is needed to reduce these emissions. The technical ability and financing options are largely there already, as illustrated in depth in other chapters in this book. In addition to these sector-specific technical and financial solutions, there are a variety of levers that are particularly relevant at the city scale, as listed below:

- Urban and transport planning levers to encourage a transition to low-carbon resilient sustainable cities:
  - enable shorter trips and reduce congestion through compact and mixed-use planning;
  - ensure equitable access to rapid public transport and affordable car/bike-sharing schemes to reduce the need for car ownership;
  - provide electric transport infrastructure, such as charging stations to enable e-buses, e-scooters, e-bikes and e-vehicles;
  - plan for other rapidly evolving transport technologies (e.g. hydrogen fuel cells, autonomous vehicles);
  - establish new regulations and standards for new commercial and residential estate developments to enable and encourage the uptake of existing solutions such as solar passive housing;
  - involve communities to co-design and co-produce planning solutions (Ryan et al. 2016; Gaziulusoy and Ryan 2017).
- Other levers include the following:
  - increasing investment in green and blue infrastructure with multiple social, environmental and resilience co-benefits.
  - decarbonisation of electricity and upper-stream supply chains by developing renewable energy, and aim for 100% renewable electricity supply in the city. Renewable energy investment in cities is rising in many countries (Newman 2017), and some cities can already meet their electricity needs largely through hydroelectric power and other renewables. Barcelona's low transport emissions are connected to urban form, but it also has low emissions from stationary energy uses because its electricity supply is dominated by nuclear power (Baldasano et al. 1999). The Australian Capital

Territory has achieved 100% renewable electricity at the end of 2019, which will reduce its carbon emissions by about 40% from its 1990 level;

o reducing the embodied carbon in urban building and infrastructure by encouraging the use of lightweight structures, substituting carbon-intensive construction materials such as cement and steel with low-carbon materials such as timber or bamboo, promoting reuse and recycling in the construction sector;

o separating the need for physical infrastructure from the need for *services* from infrastructure: for example, reducing the need for new vehicles and use of highways through promoting collaborative consumption of mobility (car-share and ride-share) rather than providing a vehicle (Baynes and Müller 2016);

o encouraging behaviour change on multiple fronts, including reducing energy and water usage in the home and office (e.g. switching off lights, taps and appliances when not needed, utilising natural ventilation to warm or cool homes), turning aesthetic gardens into productive permaculture gardens with food produce and laying chickens;

o learning from front-runner cities and transferring/contextualising innovative solutions proven effective elsewhere (Gouldson et al. 2015; Peng et al. 2019).

11.4 Managing Risks of, and Adapting to, Climate Change

11.4.1 Urban Climate Risks and Adaptation Overview

Climate risks, and their negative impacts on health, livelihoods and other aspects of well-being, and on the world's underpinning built and natural assets, are increasing (Revi et al. 2014; Güneralp et al. 2015). They fall disproportionately on urban areas, reflecting the concentration of people, built assets, economic activities and output in cities, as well as the fact that many cities are located in areas highly affected by climate change. The vulnerabilities and adaptation deficits (Fankhauser and McDermott 2014) are greater again for low- and middle-income countries, and especially for the informal settlements typically found in these countries – often found along with inadequate infrastructure and services, and more limited economic, technical and institutional capacities. Rapidly growing small and medium-sized cities are at higher risks as they often have insufficient financial, technical and knowledge capacities, and so need urgent boosts of resilience (Birkmann et al. 2016). At an individual level, the impact of hazards such as sea-level rise and flooding compound with existing stressors such as poverty, and the poor living in cities in the Global South often have much less means to withstand these impacts (Pelling and Garschagen 2019).

Contemporary cities are complex interdependent systems that are vulnerable to climate-related hazards from both gradual change (e.g. increasing temperatures and sea-level rise) and extreme events (e.g. Hurricanes Irene (2011), Sandy (2012) and Harvey and Maria (2017) in north-east America, the 2009 south-east Australian heatwave and the European heatwave in 2019). Human casualties, economic losses in the trillions of dollars, sustained power outage and disrupted water and other supplies are just some examples of the negative impacts of these events (Godschalk 2003; Hallegatte 2009).

This section provides a brief overview of climate adaptation for key urban sectors, with some implications for integrated cross-sector adaptation strategies. The detailed impacts and potential responses for many of the individual sectors have been covered elsewhere in this book, including the analysis for several infrastructure sectors (energy, transport, water, buildings) and industry sectors (manufacturing, mining, agriculture and food), so the detail will not be repeated here. A challenge for cities, however, is to understand and develop those strategies that operate across all urban issues (like urban strategy and planning, or stakeholder engagement), and to manage or facilitate the many complex systems interactions between individual sectors as they affect broader urban goals and outcomes.

11.4.1.1 Urban Food Systems and Agriculture

Food availability and prices can be impacted in both the short term (via climate extremes) and longer term (via increasing and persistent drought conditions, and competition for use of scarce agricultural land and water), and this can aggravate food and nutrition insecurity, especially for urban dwellers in lower-income countries (Cohen and Garrett 2010; Crush and Frayne 2011; FAO 2019). Responses to this include transformation of food production throughout the production and supply chain (Godfray et al. 2010; FAO 2019), including modifying consumer behaviours and diets, and supporting opportunities for urban and peri-urban agriculture (Cohen and Garrett 2010; Lee-Smith 2010).

11.4.1.2 Urban Ecosystems and Biodiversity

Urban ecosystems are increasingly threatened by both urbanisation and climate change. Responses include preservation and protection of selected natural urban ecosystems, but also enhancements through the development of green and blue infrastructure in conjunction with urban development (Foster et al. 2011; La Greca et al. 2011). Maintenance and development of urban ecosystems services have economic, social and environmental benefits (Elmqvist et al. 2015). Urban resilience to natural hazards is enhanced through ecosystems-based adaptation, again with significant potential co-benefits when an integrated systems view is taken (Munang et al. 2013; Brink et al. 2016; Geneletti et al. 2016).

11.4.1.3 Urban Resource, Waste and Pollution

Climate change can aggravate air pollution as well as natural resource shortages, especially water. Urban industries and infrastructure have the opportunity to reduce waste and pollution (in addition to GHGs), through circular economy approaches (Ellen Macarthur Foundation 2013). There is also the opportunity to modify consumer behaviours significantly to reduce waste (e.g. Hammer et al. 2011; UN Habitat 2011).

11.4.1.4 Disaster Risk and Emergency Management

There is increasing evidence of more frequent and extreme weather events and damage as a result of climate change (UNISDR 2015). National and local governments have key roles in addressing these risks, and in supporting the broader development of resilient cities (Rockefeller Foundation n.d.; Godschalk 2003). Best practice includes the development of 'Prevent, Prepare, Respond, Recover' strategies. These, especially the Prevent (e.g. land-use planning, building regulations, flood protection) and Recover (e.g. 'build back better') strategies, should be closely integrated with climate adaptation responses at all scales (local, subnational, national and international) (IPCC 2014). The Preparation and Response strategies require major engagement with and preparation by communities, as well as effective emergency services.

11.4.1.5 Health Systems Management

Human health and well-being in urban areas can be impacted by climate change in multiple direct and indirect ways (The Lancet Commission 2015), and the most vulnerable in society are also often the most exposed to such health impacts. Adaptation responses in and across other sectors can materially reduce exposure and health impacts, especially the indirect impacts (Bowen and Ebi 2015).

The largely public health and emergency management sectors themselves also have key planning and service roles to play in helping reduce risks and manage the actual impacts, direct or indirect. Thus health services can respond through (e.g. McMichael et al. 2008; Huang et al. 2011; Cheng and Berry 2013):

- public health risk and impact assessments;
- public health education, awareness and warning programmes;
- flexible health services capacity-building;
- coordination with emergency services and other responses to extreme events.

11.4.1.6 Economic Issues and Management

Extreme weather events damage infrastructure and other assets that are crucial to local (and often distant) economies, and slower-onset climate changes can change the medium- and longer-term comparative economic advantage of cities and their hinterlands. Cities in low-elevation coastal zones are especially vulnerable (McGranahan et al. 2007). Businesses can be indirectly as well as directly impacted (Hallegatte et al. 2011; da Silva 2012) and the low-income urban livelihoods often dependent on the informal sector can be especially vulnerable (Moser and Satterthwaite 2009). It is often cheaper to prevent the impacts than to retrofit existing assets (McGranahan et al. 2007) and anticipatory adaptation can be much cheaper than the costs and losses of unmanaged risks (Hallegatte et al. 2013). Persistent and poorly managed climate risks increasingly deter external industry and financial investors (Hallegatte et al. 2007), which in turn undermines economic performance.

11.4.1.7 Urban Strategy and Planning

Local government has functional and leadership roles in urban climate adaptation, both as an actor in their own right as infrastructure and service providers, and as an influencer of private sector, household and citizen decisions and behaviours (Wilbanks et al. 2007; Revi et al. 2014). This includes providing, regulating and influencing housing choices, which are heavily impacted by natural disasters (Jacobs and Williams 2011) and are especially vulnerable in developing country informal settlements. Effective spatial, master and precinct planning for both new and renewed development provides key opportunities for integrating climate adaptation and resilience, including through aligned development policies, regulations and codes. Innovative examples of precinct design are emerging, which incorporate solar panels with battery storage, while addressing other sustainability goals such as water-sensitive urban design, energy efficiency, landscape conservation, social housing and community participation (Wiktorowicz et al. 2018; Newton et al. 2019).

11.4.1.8 Urban Governance

The role of local government in managing risks and adapting to climate impacts is prominent, but it needs to be supported by aligned policies at regional, province/state and national levels. This includes not only supporting potentially contentious local decisions, but also encouraging local government to overcome the inherent spatial, temporal and institutional-scale mismatches between climate change issues and urban governance, and planning effectively at broader spatial scales across local government jurisdictions, over longer timescales and with effective collaboration across institutional scales (Bai 2007).

Implementing the above approaches requires effective multilevel governance (e.g. Sanchez-Rodriguez 2009), as well as engagement with, and initiatives from, grassroots communities and the private sector. Often this requires overcoming that fact that adaptation has received less attention (Bulkeley 2010) and is less understood by the public and by policy-makers than climate change mitigation; and that national adaptation responses have focused more on natural-resource-based sectors like ecosystems and agriculture rather than the urban domain (Roberts 2010). It also requires that local and other levels of government increasingly take an adaptive whole-of-systems approach to decision-making. At this stage, and in practice, well-intentioned regional, city-wide and local direction and guidance are often not translated effectively into individual decisions (Storbjörk and Uggla 2014).

However, while such integrated governance approaches are challenging, and still emerging, the number and extent of urban adaptation responses has grown significantly in the last decade (e.g. Hunt and Watkiss 2011; Revi et al. 2014; Eames et al. 2017) and is supported by broader city networks and development agencies (Bulkeley and Castán Broto 2013; Bulkeley and Newell 2015; Fünfgeld 2015).

Many adaptation measures can also have co-benefits within and across sectors which often translate into economic value and savings. The catalogue of direct and indirect economic impacts reinforces the fact that an effective response to urban climate risks is as much an economic as a social and environmental issue. In fact, adaptation action could often be justified economically in its own right, while also reinforcing social and environmental outcomes (Allen and Clouth 2012).

11.4.2 Linking Adaptation with Mitigation and Development Strategies

It is increasingly recognised that mitigation and adaptation measures interact in a complex way (McEvoy et al. 2006), and sometimes even well-intended climate actions compound each other to generate unintended consequences (Ürge-Vorsatz et al. 2018). Figure 11.3

Adaptation strategies

Mitigation strategies ↓	Emergency risk reduction	Insurance	Urban planning and zoning regulations	Design guidelines	Neighbourhood watch programmes and safety nets	Education and capacity building	Health and livelihoods	Resilient energy installations	Water and wastewater adaptive management	Inland and coastal flood protection	Climate-proof infrastructure (for example, transportation)	Wetland restoration	Green roofs and walls	Green space and bioswales
Urban design and form	+		+	[a]	++		+	+[b]	++	+	[c]	---	+	--[d]
Modal shift, mobility services, traffic optimisation	+		++	[e]			+[f]			-	++			--
High-efficiency, low-emissions, smaller vehicles	+	-		-				--[g]		+				
Low-energy demanding, heat-resistant architecture	+	+		++ [h]	+	+	++[i]	+	+[j]		+		[k]	++
High-efficiency appliances and equipment									+					
Energy-efficient and low-carbon urban industries							+	+	+	+				++
High-performance operation of buildings				+			+	+[l]	+[m]				+[n]	
Reducing urban heat islands (such as white and green surfaces, green infrastructure)		+	++[o]	++			+	+	--	+		++	++	++[p]
Infrastructure-integrated renewable energy systems generation	++			+				++[q]	--[r]					
Fuel switch to low(er) carbon generation	+		+					++	-[s]	-			+	
Affordable low-carbon, durable construction materials; timber infrastructure	[t]		+	+				++[u]		+		+	[v]	
Carbon capture and utilisation in construction materials														
Lifestyle, behaviour, choices, sustainable consumption and production, sharing economy, circular economy	+		+[w]			+	++		+	+				+[x]

[a] Urban design for optimised adaptation and mitigation may coincide or compromise each other. [b] Building orientation, height, and spacing can help reduce need for cooling units[4]. [c] Flood protection may compromise urban design best serving mitigation purposes. [d] Maximising compact urban design can reduce green space areas. [e] Urban designs best serving disaster risk reduction or adaptation needs may compromise the energy efficiency of the transport system. [f] Traffic optimisation results in improved air quality; modal shift typically results in more activity, that is, health gains. [g] Increased vehicular air conditioning will increase transport emissions. [h] In heat-prone regions design guidelines may prioritise the availability of mechanical cooling to reduce health risks, exacerbating emissions. [i] Very high-efficiency buildings with heat recovery ventilation have major health and welfare benefits. [j] High-efficiency buildings often also manage water resources efficiently. [k] In heat-prone regions design guidelines may prioritise the availability of mechanical cooling to reduce health risks, exacerbating emissions, but otherwise the synergies are dominant. [l] High-performance operation of buildings will increase the efficiency of mechanical cooling. [m] High-performance operation typically also extends to better water management. [n] Green roofs will improve energy efficiency and operation of building. [o] Green space will reduce urban heat islands and reduce risk of flooding. [p] Renewable energy reduces risk of power loss during extreme events. [q] Energy dependency on pumping water from flooding. [r] Some small-scale energy generation technologies require water resources. [s] Timber infrastructure may be less resilient to disasters than conventional ones. [t] Utilising lightweight construction and phase-change materials (PCMs), solar heat can be absorbed by PCMs, in turn improving thermal regulation of buildings while also reducing energy, heating and cooling[4]. [u] Climate-proof infrastructure could utilise timber; in other cases it needs to rely on concrete. [v] Incorporating institutions and stakeholders into planning can improve lifestyle choices of city as a whole. [w] Integrated approaches encourage more stakeholders to engage in the project, as multiple sectors and institutions are impacted by the adaptation and mitigation efforts. [x] Experiencing biodiversity has been proven to improve life quality and environmental consciousness.

Figure 11.3 Key interactions between urban mitigation and adaptation strategies.
Source: Ürge-Vorsatz et al. (2018). For a colour version of this figure, please see the colour plate section.

illustrates such interactions. For example, nature-based adaptation measures such as wetland restoration and green roofs and walls that help reduce flooding also contribute to mitigation goals of reducing an urban heat island. On the other hand, adaptation-oriented design guidelines in heat-prone regions may prioritise the availability of mechanical cooling to reduce health risks, which may exacerbate emissions.

Pathways for adaptation and mitigation must be planned coherently to gain the advantages of synergies, and avoid investments in infrastructure, processes or policies that subsequently become barriers to transformation. This is often not the case in current practice – many of the cities that have adaptation responses have not taken a holistic systems view, nor captured the many interdependencies (synergies and trade-offs) between potential adaptation responses, climate mitigation and other broader sustainable development goals.

It is important to recognise the need for a systems approach to assessing adaptation, mitigation and broader sustainable development options. The implementation of this approach can be greatly aided by governance agencies that can act coherently across scale and sectors.

It is also important to identify the ultimate desired outcomes with a scope wide enough to recognise connections between sectors of society or the economy. These sectors may be required to support the outcome or be affected indirectly. By mapping out the connections beyond the proximal point of intervention, we may follow and discover chains of influence and possibly feedbacks (see, for example, Seto et al. 2014: 945). Where positive feedbacks reinforce the desired outcome, this suggests the initial intervention will flourish (Meadows 1999).

Other sectors of society or the economy (e.g. food and waste disposal) may have their own policies that have competing, opposing or symbiotic relationships with a given climate change intervention. The adaptation or mitigation interventions for an initiating sector need to be assessed alongside the positive and negative implications for these other actions in the policy 'ecosystem', including objectives outside climate change such as sustainable development goals. The relevance to cities of the UN's Sustainable Development Goals and targets has been well recognised (UCLG 2016), as has been the need to consider interdependencies between these goals and targets (Breuer et al. 2019). The trade-offs or synergies may not appear immediately; there may be a delay or the effect may be cumulative over time.

11.5 Towards Resilient Low-Carbon Cities: Challenges, Opportunities, Enablers

Cities will increasingly be a focus for global action on mitigation and adaptation in response to climate change. It has also been apparent in the international efforts to develop a climate agreement (at least from the United Nations Conference of the Parties (COP 16) in Copenhagen in 2010) that 'subnational' agencies – principally cities – are becoming powerful actors in setting targets and implementing policies (Acuto 2016). While governments at all levels have a role to play, cities generally are well placed to respond to global issues like climate change: they have types of governance more responsive to local concerns; they are able to rapidly and effectively communicate to their citizens; they have an appropriate scale for experimentation with changes to urban conditions and exploring associated benefits of mitigation and adaptation (Peng and Bai 2018). Cities and metropolitan areas can build a sense of identity and belonging and develop cultures that are action-oriented and that reward innovation and creativity. Such innovations and experimentations conducted in cities can have significant importance for sustainability and low-carbon transition (Bai et al. 2010; Ryan et al. 2016; Peng and Bai 2018; Peng et al. 2019), and can be broadened and upscaled through effective intercity networks.

While the centrality of cities is increasingly evident, there are clear challenges in delivering the necessary changes, along with a need to recognise more clearly the emerging opportunities and to focus on the enablers for transformational change. This final section summarises the challenges, opportunities and enablers for future resilient low-carbon cities.

11.5.1 The Challenges

Cities face multiple challenges as they seek to respond to climate change.

11.5.1.1 Meaningful Measures and Targets

The challenges here are quite different for climate mitigation and climate adaptation or resilience.

In principle, the measure for mitigation is more straightforward, as the principle objective is reducing GHG emissions. The drivers of GHG emissions are reasonably understood and need to be taken into account particularly for fast-developing cities where the profiles will change rapidly. However, as noted earlier, for a particular city it is crucial to understand the distinction between direct (or production, territorial) versus indirect (or consumption, embedded) measures of GHG emission sources – the distinction in attribution of sources can better inform appropriate targets and the value of various interventions. For example, we have seen that decarbonisation will require combinations of changes in energy *production* (most importantly in a switch to non-carbonaceous energy supply) along with changes in (end-use) *consumption* of energy from technical efficiency improvements. This will affect all the core 'systems of provision' (Ryan et al. 2019) for the city (energy, water, food, transport, waste disposal, shelter, information).

All of the above will require some degree of change to established urban form. For example, 'passive' design to deliver resilience to sudden flooding now widely includes the provision of permeable surfaces for roads and footpaths (etc.) and small vegetation areas that can absorb storm water (often called 'rain gardens') and even larger sacrificial spaces in parks and city squares that are able to (infrequently) accommodate temporary flooding.

Transformative changes are needed to increase resilience in the face of climate shifts and weather extremes, but improvements in resilience are not as easily measured, partly due to the lack of rigorous and practical indicators for resilience. Programmes tend to rely on mapping and reporting factors that are hard to quantify, so while the principles can be defined, measuring progress is difficult to address (see for example the Rockefeller Resilient Cities framework at Rockefeller Foundation (n.d.)).

11.5.1.2 Dealing with Interconnectedness and Physical, Economic and Social/Cultural Lock-in

Systems of provision in cities develop complex interconnections as they evolve, making it difficult to address transformation one system at a time, without running the risk of unintended or countervailing impacts (e.g. using desalination to solve water issues, but increasing energy demand).

Systems of provision are structured (over time) through investments in technology and physical infrastructure; they become embedded in the physical and structural form of the city. Transformation then becomes – in part – a process of 'dis-embedding' (Ryan 2013; Ryan et al. 2019) or overcoming physical and technological lock-in, with associated costs and disruption. This interconnectedness of city form extends to culture and social identity. As Glaeser (2011) argues: '[t]he enduring strength of cities reflects the profoundly social nature of humanity'. Cities are a social invention and as a city develops over time, the social/cultural becomes intermingled with the physical in ways that retain and reproduce meaning (Harvey 2012; Ryan 2013; Sorkin 2013). In other words, the physical city is not just a background against which the play of life takes place; approaches to transformation have to deal with our deeply human connection to our constructed environment.

11.5.1.3 Responding to External Technology, Economic and Social Changes

At the same time as local factors need to be addressed, the city is inevitably influenced by externally driven technology, economic and social changes. To address climate change, solutions need to be sustainable, and that includes being viable under externally as well internally driven change.

11.5.1.4 Managing across Spatial and Temporal Scales

Cities are connected to their hinterlands through supply chains and environmental impacts, and through integrated infrastructure planning that meets rural, regional and urban needs. Many of these complex issues are first played out in peri-urban areas as development encroaches on traditional land uses and settlements. In most cities there are also different urban issues and solutions for city centres, middle suburbs and outer suburbs. And increasingly, it is recognised that a city's decisions have remote national and international consequences, and the city is part of a tele-connected network of cities.

Most urban transformational strategies also require an assessment across multiple timescales, from the immediate and often incremental step-outs, to the 50-or-more-year timescales associated with major infrastructure decisions.

11.5.1.5 Recognising the Diversity of Cities

Cities are diverse and care needs to be taken in assessing their specific adaptation and development needs as well as their unique carbon status so that low-carbon initiatives can be tailored and measured appropriately. Local context has to be taken into account: for example, in Saudi Arabia, desalination is an absolutely viable adaptation for climate-induced water shortages, whereas the trade-offs with energy usage will be quite different elsewhere.

11.5.1.6 Dealing with Uncertainties

Each of the above challenges introduces elements of uncertainty in terms of predicting the effects of climate-related impacts and the effectiveness measures (Hulme 2009). Such uncertainties challenge established ideas about 'science' and pose new dilemmas for established processes of policy formation. The broad predictions of risk available from climate modelling, which decrease in certainty as they are narrowed in time and space, require new approaches to decision-making. For democratic governance, the task of gaining citizen support for actions is difficult and easily disrupted by interest groups (Oreskes and Conway 2010). These issues also underpin the difficulty of moving to a resilience perspective in policy and action programmes to deal with climate shifts and extreme events, where the challenge is often shifting from a disaster management approach to disaster prevention (Kreimer et al. 2003), which is also contingent on uncertain projections.

11.5.2 The Opportunities

While the challenges are indeed significant, there are many positive opportunities for effective intervention.

11.5.2.1 Many Practical Solutions Have Already Emerged

The complexity and vulnerability of critical infrastructure in cities, as well as their potential for economic, political and cultural action, has brought increasing attention to the question of what cities can and should do to mitigate emissions and increase adaptive capacity. What is possibly most significant is that this attention to the 'cities agenda', in mitigation and building resilience to changing climate, is not restricted to *theoretical* analysis of future scenarios, and there has been a rapid growth of programmes of practical action *in and by* cities (Newton 2013). For example, 19 global cities committed to make new buildings net zero carbon by 2030 and all buildings by 2050 (C40 Cities 2018), and over 50 organisations have committed to net zero-carbon buildings by 2030 (World Green Building Council 2019). The Australian Capital Territory and City of Sydney have pledged to source 100% electricity from renewable sources in 2020 and have achieved it already.

11.5.2.2 Leveraging Synergies between Mitigation, Adaptation and Sustainable Development

When interventions take account of the complex interdependencies across urban areas and between sectors, they can achieve larger outcomes. As noted above, mitigation and adaptation strategies can be complementary and often have broader co-benefits in health and well-being, and economic and environmental outcomes, which is made possible by diverse interactions and linkages among social, economic and environmental processes within the city. Thus, developing effective policies that maximise synergies and co-benefits while minimising trade-offs will be key priority areas for cities. This would require a systems approach (Bai et al. 2016a), and the fact that this is rarely the case in practice means there is a large potential to tap into.

11.5.2.3 Reconceptualisation of Design Principles

There are also new 'city programmes' that can be characterised as aiming for a *re-conceptualisation* of the design principles, or the underlying systems architecture of cities, for achieving resilient, healthy and productive futures. The clear intention is to influence the planning and design of new city developments by defining approaches or a framework that deliver sustainable outcomes. For example, the '*20 minute city*' (active city concept) proposes to frame urban design around mobility objectives that would force a radical change in urban form, structure and life, where citizens are able to meet most of their daily needs with a public transport, short walk or bicycle ride of less than 20 minutes. As another example, a radical repositioning of the relationship between the city and nature is the underlying objective of the '*biophilic cities*' concept (Biophilic Cities n.d.) with the reintroduction of nature in the city subverting an historical idea of the city as a refuge from nature. These high-level design principles are not silver bullets, but used in conjunction with traditional planning and design approaches, they can guide a move towards better urban outcomes.

11.5.2.4 Cumulative Learning, Networking and Confidence

Some of the above approaches have arisen as a response to national GHG emissions agendas and targets, but there has also been a growth in regional and global city programmes involving networking and cooperation between cities that transcend national borders. Many of those programmes have energy transformation to reduce GHG emissions as their core focus (setting CO_2 targets, improving energy efficiency/productivity, increasing renewable energy supply, etc.); others also include identification and commitment to adaptation planning and action; and some have sought to emphasise broader resilience, and/or reduce the city's ecological footprint.

Many of the initiatives are experimental in nature, and the diversity of city and local contexts means that learning from other cities always needs to be translated to local relevance and needs. However, this translation is possible and, at the very least, the sharing of experience demonstrates potential for change and builds mutual confidence in the power of city-based initiatives. While it is too early to assess the impact of these city-led action programmes on the international effort for a global agreement on climate mitigation and adaptation, such smaller-scale local experimentation and shared learning is a key response to the challenge of uncertainty mentioned above (Bulkeley and Castán Broto 2013). In addition, city networks can play a key role in facilitating cross-city learning and broadening and upscaling of successful initiatives in low-carbon and resilient cities (Bai et al. 2019; Frantzeskaki et al. 2019).

11.5.3 The Enablers

There are some critical enablers, or prerequisites, to meeting the challenges and realising the opportunities discussed above. Webb et al. (2018) identify the following as important in this:

- a shared longer-term vision for a city and its region, accompanied by more immediate step-out innovations and experiments, the latter with built-in flexible pathways wherever possible;
- effective stakeholder engagement across government, private and community sectors;
- relevant, credible and accessible knowledge to support evidence-based decision-making;
- aligned institutional arrangements, with increasingly collaborative governance across multiple goals, silos and sectors, and evidence-based rather than political decision-making.

11.5.3.1 Overarching Visions Guiding the Learning from Local Innovation

As mentioned previously a key approach to managing uncertainty is to foster local innovation, experimentation and learning. However, this is most likely to be effective if carried out within the context of collective (cross-sector, cross-community) visioning for the city and its region (McPhearson et al. 2017). In developing future visions and scenarios, there is a natural tendency for citizens to underestimate the potential rate of change (Biggs et al. 2014), and the role of disruptive changes (technological, social, organisational) need to be taken into account in a way that encourages speculative exploration of possibilities for transformation (Ryan 2013; Ryan et al. 2015; Ryan et al. 2019). Universities and researchers can have important roles in this process: for example, introducing scenarios for future climate impacts and changes in other socio-economic drivers, communicating the potentials for unintended consequences and non-linear changes and articulating the outcomes of actions and alternative and desirable pathways. In fact, many programmes make design researchers and professional designers available for the community and there is an increasing attention to the visualisation of transformed urban conditions (infrastructure and life) as a way of negotiating the exploration of options for transformation (Ryan et al. 2016; Gaziulusoy and Ryan 2017; Ryan et al. 2019).

11.5.3.2 Cross-Sector and Community Engagement, Values and Behaviours

The broader visions, investments and local innovations, and urban decision-making more broadly, need to be co-developed, supported and underpinned by even more fundamental societal values and behaviours. Achieving a shared vision on plausible and desirable futures needs to combine both top-down and bottom-up approaches (Bai et al. 2016b). Broader visioning and aligned 'top-down' policies and programmes can 'give licence' to bottom-up approaches, and challenging objectives for change – strong CO_2 targets, clear resilience and social objectives and so on, taking into account broader sustainability goals. 'Bottom-up' initiatives often aim at widening involvement in shaping future transformation, not only building support for transformative change but also encompassing more deliberatively democratic approaches to 'co-designing' futures, engaging stakeholders at the community/precinct/suburb level. This means a shift from traditional approaches of communication and consultation to collaboration and empowerment. The right type of engagement depends on the nature of the issue – ranging from the broadest spatial and temporal scales to local immediate issues.

11.5.3.3 Relevant, Credible and Accessible Knowledge

Many of the challenges in practice mirror key knowledge gaps and science challenges. Bai et al. (2018) identified six research priorities in cities and climate change, including:

(i) expanding observation and establishing comparable data across cities;
(ii) understanding interactions between climate and other processes in cities;
(iii) studying climate change mitigation and adaptation in informal settlements;
(iv) understanding the impacts and harnessing disruptive technologies;
(v) supporting innovative solutions to achieve broader societal transformations; and
(vi) linking climate change mitigation and adaptation to the global sustainability context.

The authors call for stronger partnerships between researchers, policy-makers and other city stakeholders in knowledge co-production. Improved approaches to sharing and translation of knowledge, including membership of the growing urban networks – nationally and internationally – also enhances accessibility and relevance as well as growing confidence in the adoption and/or translation of new knowledge and solutions.

11.5.3.4 Aligned Institutional and Governance Arrangements

The final enabler of (and too often the main barrier to) urban transformation is aligned institutional arrangements, including increasingly collaborative governance across multiple goals, silos, scales and sectors, and evidence-based rather than mostly political decision-making. Institutional changes are called for to facilitate collaboration within and between sectors across scales, incubation of local innovations, development of a system-based coherent policy and financing of the effective implementation of the policy. A transparent and participatory approach to urban governance is critical for the transition towards low-carbon and resilient cities to be just and equitable (Newell and Mulvaney 2013; Leach et al. 2018).

Institutional and governance challenges are universal but with different aspects. In developed countries, the challenge is to bring change and greater flexibility to often entrenched and long-lived institutional arrangements. In developing countries with rapidly growing cities, the opportunity exists to shape new and more flexible institutional arrangements at a much earlier stage of development, but the available resource and capacity challenge is greater (Nagendra et al. 2018). In both cases, better-aligned institutional and governance arrangements, together with associated leadership, arguably provides the single greatest enabler for fulfilling the transformational potential of cities that is now increasingly evident, and for developing the speed and urgency required to address climate change and other sustainable development goals.

References

Acuto, M. (2016). Give cities a seat at the top table. *Nature News*, 537, 611.
Allen, C. and Clouth, S. (2012). *A Guidebook to the Green Economy. Issue 1: Green Economy, Green Growth, and Low-Carbon Development: History, Definitions and a Guide to Recent Publications*. Division for Sustainable Development, UN

Department of Economic and Social Affairs (UN DESA). Available at: https://sustainabledevelopment.un.org/content/documents/GE Guidebook.pdf.

Allwood, J., Cullen, J. and Milford, R. (2010). Options for achieving a 50% cut in industrial carbon emissions by 2050. *Environmental Science & Technology*, 44, 1888–1894.

Angel, S., Parent, J., Civco, D., Blei, A. and Potere, D. (2010). *A Planet of Cities: Urban Land Cover Estimates and Projections for All Countries, 2000–2050*. Cambridge, MA: Lincoln Institute of Land Policy. Available at: www.lincolninst.edu/publications/working-papers/planet-cities.

Bai, X. (2007). Integrating global environmental concerns into urban management: The scale and readiness arguments. *Journal of Industrial Ecology*, 11, 15–29.

Bai, X., Roberts, B. and Chen, J. (2010). Urban sustainability experiments in Asia: Patterns and pathways. *Environmental Science & Policy*, 13, 312–325.

Bai, X., Dhakal, S., Steinberger, J. and Weisz. H. (2012). Drivers of urban energy use and main policy leverages. In A. Grubler and D. Fisk, eds., *Energizing Sustainable Cities: Assessing Urban Energy*. London: Earthscan, ch. 12.

Bai, X., Surveyer A., Elmqvist, T. et al. (2016a). Defining and advancing systems approach for sustainable cities. *Current Opinion in Environmental Sustainability*, 23, 69–78.

Bai, X., van der Leeuw, S., O'Brien, K. et al. (2016b). Plausible and desirable future in the Anthropocene: A new research agenda. *Global Environmental Change*, 39, 351–362.

Bai, X., Dawson, R. J., Ürge-Vorsatz, D. et al. (2018). Six research priorities for cities and climate change. *Nature*, 555, 23–25.

Bai, X., Colbert, M., McPhearson, T., et al. (2019). Networking urban science, policy and practice. *Current Opinion in Environmental Sustainability*, 39, 114–122.

Baldasano, J., Soriano, C. and Boada, L. (1999). Emission inventory for greenhouse gases in the City of Barcelona, 1987–1996. *Atmospheric Environment*, 33(23), 3765–3775.

Baynes, T. and Müller, D. (2016). A socio-economic metabolism approach to sustainable development and climate change mitigation. In R. Clift and A. Druckman, eds., *Taking Stock of Industrial Ecology*. Springer International Publishing, pp. 117–135. Available at: http://link.springer.com/chapter/10.1007%2F978-3-319-20571-7_6.

Baynes, T. M. and Wiedmann, T. (2012). General approaches for assessing urban environmental sustainability. *Current Opinion in Environmental Sustainability*, 4, 458–464.

Bertaud, A. and Richardson, H. W. (2004). Transit and density: Atlanta, the United States and Western Europe. In H. W. Richardson and C.-H. C. Bae, eds., *Urban Sprawl in Western Europe and the United States*. Taylor and Francis.

Bettencourt, L., Lobo, J., Helbing, D., Kuhnert, C. and West, G. (2007). Growth, innovation, scaling, and the pace of life in cities. *Proceedings of the National Academy of Science*, 104, 7301–7306. Available at: www.pnas.org/content/104/17/7301.full.pdf.

Biggs, C., Ryan, C., Bird, J., Trudgeon, M. and Roggema, R. (2014). *Visions of Resilience: Design-led Transformation for Climate Extremes*. Melbourne: Victorian Eco-Innovation Lab, The University of Melbourne. Available at: http://hdl.handle.net/11343/165204.

Biophilic Cities (n.d.). *BiophilicCities*. Available at: www.biophiliccities.org/.

Birkmann, J., Welle, T., Solecki, W., Lwasa, S. and Garschagen, M. (2016). Boost resilience of small and mid-sized cities. *Nature*, 537, 605.

Bowen, K. J. and Ebi, K. L. (2015). Governing the health risks of climate change: Towards multi-sector responses. *Current Opinion in Environmental Sustainability*, 12, 80–85.

Breuer, A., Janetschek, H. and Malerba, D. (2019). Translating sustainable development goal (SDG) interdependencies into policy advice. *Sustainability*, 11, 2092.

Brink, E., Aalders, T., Ádám, D. et al. (2016). Cascades of green: A review of ecosystem-based adaptation in urban areas. *Global Environmental Change*, 36, 111–123.

Bulkeley, H. (2010). Cities and the governing of climate change. *Annual Review of Environment and Resources*, 35, 229–253. Available at: www.annualreviews.org/doi/abs/10.1146/annurev-environ-072809-101747.

Bulkeley, H. and Castán Broto, V. (2013). Government by experiment? Global cities and the governing of climate change. *Transactions of the Institute of British Geographers*, 38, 361–375.

Bulkeley, H. A. and Newell, P. (2015). *Governing Climate Change*. London: Routledge.

C40 Cities (2018). 19 global cities commit to make new buildings 'net-zero carbon' by 2030. *C40 Cities: Media*. 23 August. Available at: www.c40.org/press_releases/global-cities-commit-to-make-new-buildings-net-zero-carbon-by-2030.

Cheng, J. J. and Berry, P. (2013). Health co-benefits and risks of public health adaptation strategies to climate change: A review of current literature. *International Journal of Public Health*, 58, 305–311.

Cohen, M. J. and Garrett, J. L. (2010). The food price crisis and urban food (in) security. *Environment and Urbanization*, 22, 467–482.

Corburn, J. (2009). Cities, climate change and urban heat island mitigation: Localising global environmental science. *Urban Studies*, 46, 413–427.

Crush, J. S. and Frayne, G. B. (2011). Urban food insecurity and the new international food security agenda. *Development Southern Africa*, 28, 527–544.

da Silva, J., Kernaghan, S. and Luque, A. (2012). A systems approach to meeting the challenges of urban climate change. *International Journal of Urban Sustainable Development*, 4, 125–145.

Dobbs, R., Smit, S., Remes, J., Manyika, J., Roxburgh, C. and Restrepo, A. (2011). *Urban World: Mapping the Economic Power of Cities*. McKinsey Global Institute. Available at: www.mckinsey.com/~/media/McKinsey/Featured%20Insights/Urbanization/Urban%20world/MGI_urban_world_mapping_economic_power_of_cities_full_report.pdf.

Doherty, M., Nakanishi, H., Bai, X. and Meyers, J. (2009). *Relationships between Form, Morphology, Density and Energy in Urban Environments*. GEA Background Paper. Canberra: CSIRO Sustainable Ecosystems. Available at: www.iiasa.ac.at/web/home/research/Flagship-Projects/Global-Energy-Assessment/GEA_Energy_Density_Working_Paper_031009.pdf.

Eames, M., Dixon, T., Hunt, M. and Lannon, S., eds. (2017). *Retrofitting Cities for Tomorrow's World*. Oxford: Wiley-Blackwell.

Ellen MacArthur Foundation (2013). *Towards the Circular Economy: Economic and Business Rationale for an Accelerated Transition*. Cowes: Ellen MacArthur Foundation. Available at: www.ellenmacarthurfoundation.org/assets/downloads/publications/Ellen-MacArthur-Foundation-Towards-the-Circular-Economy-vol.1.pdf.

Elmqvist, T., Setälä, H., Handel, S. N. et al. (2015). Benefits of restoring ecosystem services in urban areas. *Current Opinion in Environmental Sustainability*, 14, 101–108.

Fankhauser, S. and McDermott, T. K. J. (2014). Understanding the adaptation deficit: Why are poor countries more vulnerable to climate events than rich countries? *Global Environmental Change*, 27, 9–18.

FAO (Food and Agriculture Organization) (2019). *The State of Food Security and Nutrition in the World 2019: Safeguarding Against Economic Slowdowns and Downturns*. Rome: Food and Agriculture Organization. Available at: www.fao.org/3/ca5162en/ca5162en.pdf.

Foster, J., Lowe, A. and Winkelman, S. (2011). *The Value of Green Infrastructure for Urban Climate Adaptation*. Washington, DC: The Center for Clean Air Policy.

Available at: https://ccap.org/resource/the-value-of-green-infrastructure-for-urban-climate-adaptation/.

Frantzeskaki, N., Buchel, S., Spork, C., Ludwig, K. and Kok, M. T. (2019). The multiple roles of ICLEI: Intermediating to innovate urban biodiversity governance. *Ecological Economics*, 164, 106350.

Fünfgeld, H. (2015). Facilitating local climate change adaptation through transnational municipal networks. *Current Opinion in Environmental Sustainability*, 12, 67–73.

Gaziulusoy, I. and Ryan, C. (2017). Shifting conversations for sustainability transitions using participatory design visioning. *The Design Journal*, 20, suppl. 1, S1916–S1926.

GcoM (Global Covenant of Mayors) (2019). The founding partners. *Global Covenant of Mayors*. Available at: www.globalcovenantofmayors.org/about/.

Geneletti, D. and Zardo, L. (2016). Ecosystem-based adaptation in cities. An analysis of European urban climate adaptation plans. *Land Use Policy*, 50, 38–47.

Glaeser, E. (2011). *Triumph of the City: How Our Greatest Invention Makes Us Richer, Smarter, Greener, Healthier and Happier*. New York: Penguin Press.

Godfray, H. C. J., Beddington, J. R., Crute, I. R. et al. (2010). Food security: The challenge of feeding 9 billion people. *Science*, 327, 812–818.

Godschalk, D. (2003). Urban hazard mitigation: Creating resilient cities. *Natural Hazards Review*, 4, 136–143.

Gouldson, A., Colenbrander, S., Sudmant, A. et al. (2015). *Accelerating Low-Carbon Development in the World's Cities*. Working Paper. The New Climate Economy. Washington, DC: The Global Commission on the Economy and Climate. Available at: http://newclimateeconomy.report/workingpapers/.

Grimmond, C. S. B. (2011). Climate of cities. In I. Douglas, D. Goode, M. Houck and R. Wang, eds., *Routledge Handbook of Urban Ecology*. Abingdon: Routledge, pp. 103–119.

Grübler, A. and Fisk, D. (2013). *Energizing Sustainable Cities*. London: Earthscan.

Grubler, A., Bai, X., Buettner, T. et al. (2012). Urban energy systems. In T. B. Johansson, A. Patwardhan, N. Nakicenovic and L. Gomez-Echeverri, eds., *Global Energy Assessment: Toward a Sustainable Future*. Cambridge: Cambridge University Press, pp. 1307–1400. Available at: www.iiasa.ac.at/web/home/research/Flagship-Projects/Global-Energy-Assessment/Chapte18.en.html.

Güneralp, B., Güneralp, İ. and Liu, Y. (2015). Changing global patterns of urban exposure to flood and drought hazards. *Global Environmental Change*, 31, 217–225.

Hallegatte, S. (2009). Strategies to adapt to an uncertain climate change. *Global Environmental Change*, 19, 240–247.

Hallegatte, S., Hourcade, J. and Ambrosi, P. (2007). Using climate analogues for assessing climate change economic impacts in urban areas. *Climatic Change*, 82, 47–60.

Hallegatte, S., Henriet, F. and Corfee-Morlot, J. (2011). The economics of climate change impacts and policy benefits at city scale: A conceptual framework. *Climatic Change*, 104, 51–87.

Hallegatte, S., Green, C., Nicholls, R. and Corfee-Morlot, J. (2013). Future flood losses in major coastal cities. *Nature Climate Change*, 3, 802–806.

Hammer, S., Kamal-Chaoui, L., Robert, A. and Plouin, M. (2011). *Cities and Green Growth: A Conceptual Framework*. OECD Regional Development Working Papers 2011/08. OECD Publishing. Available at: https://dx.doi.org/10.1787/5kg0tflmzx34-en.

Hanson, S., Nicholls, R., Ranger, N. et al. (2011). A global ranking of port cities with high exposure to climate extremes. *Climatic Change*, 104, 89–111.

Harvey, D. (2012). *Rebel Cities: From the Right to the City to the Urban Revolution*. London: Verso.

Hillman, T. and Ramaswami, A. (2010). Greenhouse gas emission footprints and energy use benchmarks for eight U.S. cities. *Environmental Science & Technology*, 44, 1902–1910.

Huang, C., Barnett A., Wang, X., Vaneckova, P., FitzGerald, G. and Tong, S. (2011). Projecting future heat-related mortality under climate change scenarios: A systematic review. *Environmental Health Perspectives*, 119, 1681–1690.

Hulme, M. (2009). *Why We Disagree about Climate Change: Understanding Controversy, Inaction and Opportunity*. Cambridge: Cambridge University Press.

Hunt, A. and Watkiss, P. (2011). Climate change impacts and adaptation in cities: A review of the literature. *Climatic Change*, 104, 13–49.

ICLEI Global (n.d.). *ICLEI Global*. Available at: www.iclei.org/.

IEA (International Energy Agency) (2008). *World Energy Outlook 2008*. Paris: International Energy Agency. Available at: www.iea.org/reports/world-energy-outlook-2008.

IPCC (Intergovernmental Panel on Climate Change) (2013). *Climate Change 2013: The Physical Science Basis. Contribution of Working Group I to the Fifth Assessment Report of the Intergovernmental Panel on Climate Change*. Edited by T. F. Stocker, D. Qin, G.-K. Pattner et al. Cambridge: Cambridge University Press. Available at: www.ipcc.ch/report/ar5/wg1/.

IPCC (2014). Summary for Policymakers. In O. Edenhofer, R. Pichs-Madruga, Y. Sokona et al., eds., *Climate Change 2014: Mitigation of Climate Change. Contribution of Working Group III to the Fifth Assessment Report of the Intergovernmental Panel on Climate Change*. Cambridge: Cambridge University Press. Available at: www.ipcc .ch/report/ar5/wg3/.

IPCC (2018). *Global Warming of 1.5 °C: An IPCC Special Report on the Impacts of Global Warming of 1.5 °C Above Pre-Industrial Levels and Related Global Greenhouse Gas Emission Pathways, in the Context of Strengthening the Global Response to the Threat of Climate Change, Sustainable Development, and Efforts to Eradicate Poverty*. Edited by V. Masson-Delmotte, P. Zhai, H.-O. Pörtner et al. Cambridge: Cambridge University Press. Available at: www.ipcc.ch/sr15/.

Isaac, M. and van Vuuren, D. (2009). Modeling global residential sector energy demand for heating and air conditioning in the context of climate change. *Energy Policy*, 37, 507–521.

Jacobs, K. and Williams, S. (2011). What to do now? Tensions and dilemmas in responding to natural disasters: A study of three Australian state housing authorities. *International Journal of Housing Policy*, 11, 175–193.

Kenworthy, J. R., Laube, F. B. and Newman, P. (1999). *An International Sourcebook of Automobile Dependence in Cities, 1960–1990*. Boulder, CO: University Press of Colorado.

Kreimer, A., Arnold, M. and Carlin, A. (2003). *Building Safer Cities: The Future of Disaster Risk*. Washington, DC: World Bank.

La Greca, P., La Rosa, D., Martinico, F. and Privitera, R. (2011). Agricultural and green infrastructures: The role of non-urbanised areas for eco-sustainable planning in a metropolitan region. *Environmental Pollution*, 159, 2193–2202.

Leach, M., Reyers, B., Bai, X. et al. (2018). Equity and sustainability in the Anthropocene: A social–ecological systems perspective on their intertwined futures. *Global Sustainability*, 1. DOI: 10.1017/sus.2018.12.

Lee-Smith, D. (2010). Cities feeding people: An update on urban agriculture in equatorial Africa. *Environment and Urbanization*, 22, 483–499.

Lenzen, M. and Peters, G. (2010). How city dwellers affect their resource hinterland. *Journal of Industrial Ecology*, 14, 73–90.

Lenzen, M., Dey, C. and Foran, B. (2004). Energy requirements of Sydney households. *Ecological Economics*, 49, 375–399.

Lin, T., Yu, Y., Bai, X., Feng, L. and Wang, J. (2013). Greenhouse gas emissions accounting of urban residential consumption: A household survey based approach. *PLoS ONE*, 8, e55642.

Marcotullio, P., Sarzynski, A., Albrecht, J., Schulz, N. and Garcia, J. (2013). The geography of global urban greenhouse gas emissions: An exploratory analysis. *Climatic Change*, 121, 621–634.

McDonald, R., Green, P., Balk, D. et al. (2011). Urban growth, climate change, and freshwater availability. *Proceedings of the National Academy of Sciences*, 108, 6312–6317.

McEvoy, D., Lindley, S. and Handley, J. (2006). Adaptation and mitigation in urban areas: Synergies and conflicts. *Proceedings of the Institution of Civil Engineers-Municipal Engineer*, 159, 185–191.

McGranahan, G., Bulk, D. and Anderson, B. (2007). The rising tide: Assessing the risks of climate change and human settlements in low elevation coastal zones. *Environment and Urbanization*, 19, 17–37.

McMichael, A. J., Friel, S., Nyong, A. and Corvalan, C. (2008). Global environmental change and health: Impacts, inequalities, and the health sector. *BMJ*, 336, 191–194.

McPhearson, T., Iwaniec, D. M. and Bai, X. (2017). Positive visions for guiding urban transformations toward sustainable futures. *Current Opinion in Environmental Sustainability*, 22, 1–8.

Meadows, D. (1999). *Leverage Points: Places to Intervene in a System*. Hartland, VT: The Sustainability Institute. Available at: http://donellameadows.org/archives/leverage-points-places-to-intervene-in-a-system.

Mi, Z., Zhang, Y., Guan, D. et al. (2016). Consumption-based emission accounting for Chinese cities. *Applied Energy*, 184, 1073–1081.

Minx, J., Wiedmann, T., Wood, R. et al. (2009). Input–output analysis and carbon footprinting: An overview of applications. *Economic Systems Research*, 21, 187–216.

Minx, J., Baiocchi, G., Wiedmann, T. et al. (2013). Carbon footprints of cities and other human settlements in the UK. *Environmental Research Letters*, 8, 035039.

Moser, C. and Satterthwaite, D. (2009). Toward pro-poor adaptation to climate change in the urban centers of low- and middle-income countries. In R. Mearns and A. Norton, eds., *The Social Dimensions of Climate Change: Equity and Vulnerability in a Warming World*. Washington, DC: The World Bank, pp. 231–258. Available at: https://openknowledge.worldbank.org/bitstream/handle/10986/2689/520970PUB0EPI11C010disclosed0Dec091.pdf?sequence=1&isAllowed=y.

Müller, D., Liu, G., Løvik, A. et al. (2013). Carbon emissions of infrastructure development. *Environmental Science & Technology*, 47, 11739–11746.

Munang, R., Thiaw, I., Alverson, K., Liu, J. and Han, Z. (2013). The role of ecosystem services in climate change adaptation and disaster risk reduction. *Current Opinion in Environmental Sustainability*, 5, 47–52.

Nagendra, H., Bai, X., Brondizio, E. S. and Lwasa, S. (2018). The urban south and the predicament of global sustainability. *Nature Sustainability*, 1, 341.

Nansai, K., Kagawa, S., Suh, S., Fujii, M., Inaba, R. and Hashimoto, S. (2009). Material and energy dependence of services and its implications for climate change. *Environmental Science & Technology*, 43, 4241–4246.

Newell, P. and Mulvaney, D. (2013). The political economy of the 'just transition'. *The Geographical Journal*, 179, 132–140.

Newman, P. (2017). The rise and rise of renewable cities. *Renewable Energy and Environmental Sustainability*, 2, 1–5.

Newton, P. W. (2013). Regenerating cities: Technological and design innovation for Australian suburbs. *Building Research & Information*, 41, 575–588.

Newton, P., Prasad, D., Sproul, A. and White, S., eds. (2019). *Decarbonising the Built Environment: Charting the Transition*. London: Palgrave Macmillan.

OECD (2016). GDP long-term forecast [indicator]. *OECD.org*. Available at: https://data .oecd.org/gdp/gdp-long-term-forecast.htm#indicator-chart: 10.1787/d927bc18-en.

Oreskes, N. and Conway, E. M. (2010). *Merchants of Doubt: How a Handful of Scientists Obscured the Truth on Issues from Tobacco Smoke to Global Warming*. London: Bloomsbury Press.

Pelling, M. and Garschagen, M. (2019). Put equity first in climate adaptation. *Nature,* 569, 327–329.

Pelling, M., Leck, H., Pasquini, L. et al. (2018). Africa's urban adaptation transition under a 1.5° climate. *Current Opinion in Environmental Sustainability*, 31, 10–15.

Peng, Y. and Bai, X. (2018). Experimenting towards a low-carbon city: Policy evolution and nested structure of innovation. *Journal of Cleaner Production*, 174, 201–212.

Peng, Y., Wei, Y. and Bai, X. (2019). Scaling urban sustainability experiments: Contextualization as an innovation. *Journal of Cleaner Production*, 227, 302–312.

Ramaswami, A., Hillman, T., Janson, B., Reiner, M. and Thomas, G. (2008). A demand-centered, hybrid life-cycle methodology for city-scale greenhouse gas inventories. *Environmental Science & Technology*, 42, 6455–6461.

Ramaswami, A., Chavez, A., Ewing-Thiel, J. and Reeve, K. (2011). Two approaches to greenhouse gas emissions foot-printing at the city scale. *Environmental Science & Technology*, 45, 4205–4206.

Revi, A., Satterthwaite, D. E., Aragón-Durand, F. et al. (2014). Urban areas. In C. B. Field, V. R. Barros, D. J. Dokken et al., eds., *Climate Change 2014: Impacts, Adaptation, and Vulnerability. Part A: Global and Sectoral Aspects. Contribution of Working Group II to the Fifth Assessment Report of the Intergovernmental Panel on Climate Change*. Cambridge: Cambridge University Press, pp. 535–612. Available at: www .ipcc.ch/report/ar5/wg2/.

Roberts, D. (2010) Prioritizing climate change adaptation and local level resilience in Durban, South Africa. *Environment and Urbanization*, 22, 397–413.

Rockefeller Foundation (n.d.). *The City Resilience Framework. 100 Resilient Cities*. The Rockerfellar Foundation. Available at: www.rockefellerfoundation.org/wp-content/ uploads/100RC-City-Resilience-Framework.pdf.

Rosenzweig, C., Solecki, W., Romero-Langao, P., Mehrotra, S., Dhakal, S. and Ali Ibrahim, S., eds. (2018). *Climate Change and Cities: Second Assessment Report of the Urban Climate Change Research Network*. Cambridge: Cambridge University Press.

Ryan. C. (2005). Sustainable production and consumption systems. In K. Hargroves and M. Smith, eds., *The Natural Advantage of Nations: Business Opportunities, Innovation and Governance in the 21st Century*. London: The Natural Edge Project, Earthscan.

Ryan, C. (2013). Eco-acupuncture: Designing and facilitating pathways for urban transformation, for a resilient low-carbon future. *Journal of Cleaner Production*, 50, 189–199.

Ryan, C., Twomey, P., Gaziulusoy, A. I. and McGrail, S. (2015). *Visions 2040: Results from the First Year of Visions and Pathways 2040: Glimpses of the Future and Critical Uncertainties*. University of Melbourne, Melbourne: CRC for Low Carbon Living/Victorian Eco-Innovation Lab (VEIL).

Ryan, C., Gaziulusoy, I., McCormick, K. and Trudgeon, M. (2016). Virtual city experimentation: A critical role for design visioning. In J. Evans, A. Karvonen and R. Raven, eds., *The Experimental City*. London: Taylor & Francis, ch. 5.

Ryan, C., Twomey P., Gaziulusoy I. et al. (2019). Visions, scenarios and pathways for rapid decarbonisation of Australian Cities by 2040. In P. Newton, D. Presard, A. Sproul and S. White, eds., *Decarbonising the Built Environment: Charting the Transition*. London: Palgrave Macmillan.

Sanchez-Rodriguez, R. (2009). Learning to adapt to climate change in urban areas: A review of recent contributions. *Current Opinion in Environmental Sustainability*, 1, 201–206.

SDSN (Sustainable Development Solutions Network) and IDDRI (Institute for Sustainable Development and International Relations). *Pathways to Deep Decarbonization: 2014 Report*. Sustainable Development Solutions Network and the Institute for Sustainable Development and International Relations. Available at: www .globalccsinstitute.com/archive/hub/publications/184548/pathways-deep-decarboniza tion-2014-report.pdf.

Seto, K., Dhakal, S., Bigio, A. et al. (2014). Human settlements, infrastructure, and spatial planning. In O. Edenhofer, R. Pichs-Madruga and Y. Sokona, eds., *Climate Change 2014: Mitigation of Climate Change. Contribution of Working Group III to the Fifth Assessment Report of the Intergovernmental Panel on Climate Change*. Cambridge: Cambridge University Press, pp. 923–1000. Available at: www.ipcc.ch/site/assets/uploads/2018/02/ipcc_wg3_ar5_chapter12.pdf.

Shan, Y., Guan, D., Hubacek, K. et al. (2018). City-level climate change mitigation in China. *Science Advances*, 4, eaaq0390.

Singh, S. and Kennedy, C. (2015). Estimating future energy use and CO_2 emissions of the world's cities. *Environmental Pollution*, 203, 271–278.

Solecki, W., Rosenzweig, C., Dhakal, S. et al. (2018). City transformations in a 1.5 °C warmer world. *Nature Climate Change*, 8, 177.

Sorkin, M. (2013). *Twenty Minutes in Manhattan*. New York: North Point Press.

Steininger, K., Lininger, C., Droege, S., Roser, D., Tomlinson, L. and Meyer, L. (2014). Justice and cost effectiveness of consumption-based versus production-based approaches in the case of unilateral climate policies. *Global Environmental Change*, 24, 75–87.

Storbjörk, S. and Uggla, Y. (2014). The practice of settling and enacting strategic guidelines for climate adaptation in spatial planning: Lessons from ten Swedish municipalities. *Regional Environmental Change*, 15, 1133–1143.

The Lancet Commission (2015). Health and climate change: Policy responses to protect public health. *The Lancet*, 386, 1861–1914.

UCLG (United Cities and Local Governments) (2016). *The Sustainable Development Goals: What Local Governments Need to Know*. United Cities and Local Governments. Available at: www.uclg.org/sites/default/files/the_sdgs_what_local gov_need_to_know_0.pdf.

UN Habitat (2011). *Global Report on Human Settlements 2011: Cities and Climate Change*. New York: UN Habitat. Available at: http://unhabitat.org/books/cities-and-climate-change-global-report-on-human-settlements-2011/#.

UNISDR (UN International Strategy for Disaster Reduction) (2015). *The Human Cost of Weather Related Disasters 1995–2015*. United Nations Office for Disaster Risk Reduction. Available at: www.undrr.org/publication/human-cost-weather-related-dis asters-1995-2015.

Ürge-Vorsatz, D., Rosenzweig, C., Dawson, R. et al. (2018). Locking in positive climate responses in cities. *Nature Climate Change*, 8, 174–177.

Webb, R., Bai, X., Stafford Smith, M. et al. (2018). Sustainable urban systems: Co-design and framing for transformation. *Ambio*, 47, 57–77.

Wiktorowicz, J., Babaeff, T., Breadsell, J., Byrne, J., Eggleston, J. and Newman, P. (2018). WGV: An Australian urban precinct case study to demonstrate the 1.5 °C agenda including multiple SDGs. *Urban Planning*, 3, 64–81.

Wilbanks, T., Romero Lankao, P., Bao, M. et al. (2007). Industry, settlement and society. In M. Parry, O. Canziani, J. Palutikof et al., eds., *Climate Change 2007: Impacts, Adaptation and Vulnerability. Contribution of Working Group II to the Fourth Assessment Report of the Intergovernmental Panel on Climate Change*. Cambridge: Cambridge University Press, pp. 357–390. Available at: www.ipcc.ch/site/assets/uploads/2018/02/ar4-wg2-chapter7-1.pdf.

Willems, P., Olsson, J. and Arnbjerg-Nielsen, K. (2012). *Impacts of Climate Change on Rainfall Extremes and Urban Drainage Systems*. London: IWA Publishing.

World Bank (2010a). *World Development Report 2010: Development and Climate Change*. Washington, DC: World Bank. Available at: https://openknowledge.worldbank.org/handle/10986/4387.

World Bank, ed. (2010b). *Cities and Climate Change: An Urgent Agenda*. Urban Development Series Knowledge Papers 63704. Washington, DC: The International Bank for Reconstruction and Development/The World Bank. Available at: http://documents.worldbank.org/curated/en/194831468325262572/Cities-and-climate-change-an-urgent-agenda.

World Green Building Council (2019). Net Zero Carbon Buildings Commitment surpasses 50 signatories in latest status report. *World Green Building Council*. 28 May. Available at: www.worldgbc.org/news-media/net-zero-carbon-buildings-commitment-surpasses-50-signatories-latest-status-report.

Yu, P., Xu, R., Abramson, M. J., Li, S. and Guo, Y. (2020). Bushfires in Australia: A serious health emergency under climate change. *The Lancet Planetary Health*, 4, PE7–E8. Available at: www.thelancet.com/journals/lanplh/article/PIIS2542-5196(19)30267-0/fulltext.

Zhang, Y., Bai, X., Mills, F. and Pezzey, J. (2018). Rethinking the role of occupant behavior in building energy performance: A review. *Building and Energy*, 172, 279–294.

12

Buildings and Precincts

MICHAEL H. SMITH, PETER NEWTON, ALAN PEARS, AMANDINE DENIS-RYAN
AND ESHAN AHUJA

Executive Summary

The building and precincts sector contributes 25% of global greenhouse gas emissions and up to 40% of national electricity demand. Climate change mitigation via improved integrated building design, energy efficiency, on-site solar photovoltaics and energy storage, and low embodied carbon building materials, combined with behaviour change, can make a significant contribution to nations' efforts to meet their Paris Climate Change Agreement commitments.

Decarbonising national electricity grids, combined with these building sector approaches to climate mitigation, can result in the building sector becoming net carbon neutral by 2050 at the latest.

Climate mitigation opportunities in the building sector represent overall the largest and most profitable climate change mitigation potentialities of any sector. This is partly because of significant technical improvements and price reductions in enabling technologies. As a result, most mitigation opportunities for new and existing buildings offer returns on investment with under 7–8-year paybacks via energy and water bill savings, as well as other significant co-benefits.

Conversely, lack of action of climate mitigation and adaptation risks locked-in emissions, lack of resilience to climate change impacts and stranded assets in the buildings sector, given the long design lives of buildings, up to 100 years. Inaction in this sector also makes achievement of the Paris Climate Change Agreement target impossible. The chapter shows that most climate change mitigation strategies for this sector also simultaneously help the sector to adapt to climate change.

Hence, the chapter evidences the case for the global green building movement to embrace integrated 'climate-smart' green building design, construction and operation, which optimises new and existing green buildings to achieve both mitigation and adaptation goals synergistically and cost-effectively.

The climate-smart building agenda is a high priority for this sector because it can help improve the well-being, productivity and health of occupants, and provide other social equity benefits. Investing in the overall energy and water efficiency of buildings also improves asset value, attracts premium tenants and results in lower tenant turnover.

This chapter overviews the range of key stakeholders and decision makers, and how each can best play their part to enable the needed changes in this sector to achieve a net zero-carbon resilient future. It also looks at the role of governments to address major market and informational failures, and what policies are needed to underpin efforts by all these key actors.

Finally, there are many climate change mitigation tools for design and urban planning, including life-cycle-analysis databases and software packages, as well as green building rating tools and cost curve studies, to help reduce greenhouse gas emissions in the buildings and precincts sector. However, there are relatively few tools for climate change adaptation in the building and precincts sector and even fewer to assist designers, planners and other key decision makers best create an integrated approach to climate change mitigation and adaptation and sustainable development. This chapter addresses this major knowledge gap. It shows how an integrated approach to sustainable development, mitigation and adaptation in the buildings and precincts sector can be achieved.

12.1 Introduction

Buildings have always been built to help adapt to local climates and provide shelter and safe places to meet, live, communicate, trade, celebrate, and commemorate. Residential and commercial buildings therefore ideally should have been built to protect their occupants from extreme temperatures and weather events, as well as other disruptions, damage and displacement. But this is often not the case because many buildings and precincts were built to building standards and urban planning approaches developed before the risks of climate change were being incorporated into building codes. A significant percentage of existing buildings have been built to old building standards that did not consider the implications of climate change and had not been built to withstand level 4–6 hurricanes, increasingly intense bushfires and extreme 1-in-100-year flooding events. So, a significant percentage of existing building stock is currently exposed to risks of damage from more intense extreme weather events such as cyclones, hailstorms and flooding. Buildings and cities are also forecast to be increasingly vulnerable to water shortage over the next century. At the same time, a warming climate can store more water vapour in the air, resulting in a forecast for extreme flooding events, by 2100, having 10% to 60% greater intensity compared to average flooding between 1960 and 1990. Many buildings and precincts also will be increasingly vulnerable to flooding from the combination of storm surges and sea-level rises this century due to climate change.

While this sector is vulnerable to extreme weather events, and in the longer term to sea-level rises, this chapter will show that there is much that can be done to reduce risks and negative impacts from climate change through adaptation strategies which simultaneously also significantly reduce greenhouse gas (GHG) emissions and vice versa. According to the Intergovernmental Panel on Climate Change (IPCC) (Lucon et al. 2014), among energy end users, the buildings and precincts sector represents the largest and most profitable climate change mitigation opportunities. This is because the buildings sector contributes over 25% of global GHG emissions through operational energy-usage-related emissions and the embodied-energy-related emissions from the extraction and production of building materials. This combined with the fact that there have been significant technical improvements and price reductions in enabling technologies – energy-efficient lighting, appliances, heating and cooling; solar photovoltaic (PV) systems; next-generation battery storage, combined with smart meters and smart ICT building technologies – means that there are many mitigation opportunities with reasonable to good return on investment.

Combining integrated approaches to energy efficiency, on-site low-carbon energy production (i.e. solar PV panels), low embodied energy building materials and behaviour change can result in the sector becoming close to net climate neutral by 2035–2050. Conversely, if urgent action is not taken in this sector, it makes it impossible to achieve net zero emissions by 2050 and achieve the Paris Climate Change Agreement targets and adaptation goals. Buildings' long design lives make it imperative that new buildings are designed and built to be zero net emission and resilient to climate change impacts to ensure that national economies can achieve mitigation and adaptation goals. This is because it is far cheaper to design and construct buildings and precincts to be net zero emission and resilient to extreme weather events than to retrofit buildings later.

To date, most of the literature has focused on how to either adapt buildings and precincts to climate change (see ASBEC 2012; ABCB 2014) or reduce net emissions from this sector (Lucon et al. 2014), and not on what potential there is holistically to design or retrofit buildings and precincts simultaneously to be both zero emission and resilient to climate change.

12.2 Mapping Potential for Climate Change Mitigation and Adaptation Synergies

Clearly, the ideal would be to identify and implement synergistic mitigation and adaptation opportunities which reduce GHG emissions and energy and water costs, while improving the resilience of new and existing buildings and precincts (Table 12.1). How to do this is outlined next. As Table 12.1 shows, most well-designed climate change mitigation strategies have significant climate change adaptation co-benefits, and vice versa. Hence, it is vital that these synergies are better understood as outlined below.

12.3 Achieving Zero Net Carbon by 2050: Are There Any Significant Adaptation Trade-offs?

As mentioned earlier, historically there has been a common perception that adaptation and mitigation measures inevitably are trade-offs. The list of mitigation and adaptation synergies outlined in Table 12.1 clearly debunks this mistaken assumption. Still, some may wonder whether focusing on identifying and implementing adaptation/mitigation nexus measures would divert funds from achieving a rapid transition to a net zero-carbon building sector. Pursuing the climate-resilient measures listed in Table 12.1, which simultaneously

Table 12.1. *Cost-effective climate-smart measures for the building sector: mitigation and adaptation synergies and benefits*

	Climate-smart residential and commercial building/precinct sector	
Climate-smart buildings	Climate change mitigation benefit	Climate change adaptation benefit
Integrated approaches to energy efficiency and renewable energy: mitigation and adaptation benefits		
1. Energy efficiency retrofit of existing (and design of new) buildings (i.e. integrating all energy efficiency measures listed below (items 3–29) plus solar PV systems)	Energy savings: (i) Residential: 50–75% (total energy use); (ii) Commercial buildings: 25–50% (total heating, ventilation and air conditioning (HVAC)) and 30–60% (lighting retrofits) (Lucon et al. 2014). Global net reductions in end-use energy by 2050 of 60–70% (Laitner et al. 2012) below business-as-usual (BAU) baseline (Lucon et al. 2014). Combining energy efficiency improvements with fuel switching (i.e. natural gas to efficient electric appliances) and on-site solar PV panels plus storage has the potential to achieve close to net zero emissions by 2050.	Produces buildings resilient to risks of damage from extreme weather events. Makes buildings habitable despite power cuts as they are self-sufficient in energy and water (as outlined below). Reduces heat island effect, particularly due to solar PV panels (Masson et al. 2014). Reduces peak electricity demand and risk of blackouts.
Specific building envelope energy efficiency strategies: mitigation and adaptation benefits		
2. Integrated approach to improving building envelope energy efficiency	Under BAU, air-conditioning loads are forecast to increase exponentially over 20-fold by 2050 (IPCC 2013). A systematic approach to improving the building envelopes of existing and new buildings can achieve a reduction of 25–100% in mechanical heating and cooling loads and an overall 40–50% reduction in emissions below BAU baseline by 2050 (IEA 2010). Global carbon dioxide (CO_2) one-time offset potentials from cool roofs and pavements	Improves maintaining thermal comfort of buildings and health of occupants, despite a warming climate. Reduces air-conditioning loads. Reduces air-conditioning-related peak electricity demand and, if applied across cities and nations, reduces the risks of blackouts. Maintains thermal comfort level, reduces peak electricity demand and allows buildings to be

		occupied and used with adequate comfort even if there are power outages.
	amount to 78 GtCO$_2$ (gigatonnes of CO$_2$) (Menon et al. 2010).	
3. Passive solar design, white roofs and green walls, cool pavements, permeable landscapes and shade trees	40–60% air-conditioning savings in a series of buildings where these strategies were used in warm to hot climates (Rosenfeld et al. 1995).	
4. Energy-efficient window treatments: blinds, awnings, sun shades, shutters, storm panels (US DoE n.d.)	Can reduce solar heat gain through windows in summer – 40–90% – significantly reducing heat gain and thus air-conditioning loads and energy consumption (US DoE n.d.); can also cut heat losses in cold climates, thus reducing heating loads and energy consumption.	Reduces risks from damage to windows from extreme weather events (cyclones, hailstorms, flooding) and wind-blown debris. Reduces risks of windows being the weak point that enables bushfire embers, hailstorms and (the wind gusts from) hurricanes to get inside buildings and further damage them significantly.
5. Energy-efficient resilient glass windows	Can reduce solar heat gain and losses, reducing heating and cooling loads by 30–40%.	'The quality of glass plus multiple panes used in energy-efficient windows reduce the likelihood that fire will cause breakage and embers will get into buildings. Efficient multiple-pane windows are also more resistant to breakage by windstorms.' (Mills 2003)
6. Designing buildings to utilise natural ventilation	Reduces the need for mechanical HVAC systems up to 100%.	Maintains comfort even without mechanical ventilation. For example, the natural and passive control system of traditional housing in Kerala, India, has been shown to maintain bedroom temperatures of 23–29 °C even as outdoor temperatures vary from 17 °C to 36 °C (Dili et al. 2010).
7. Adjustable windows	Uses internal and external temperature differentials to cool or heat buildings without using any energy. Adjustable windows also allow occupants to open windows when there are cool breezes. The literature shows that in office buildings, where occupants can open windows, it had a significant cooling effect (Rijal et al. 2008).	Gives occupants of buildings additional options to cool or heat buildings without using any electrical or mechanical energy.

Table 12.1. (*cont.*)

Climate-smart buildings	Climate-smart residential and commercial building/precinct sector	
	Climate change mitigation benefit	Climate change adaptation benefit
8. Effective insulation and sealed building envelope	Air-sealing retrofits alone can save an average of 15–20% of annual heating and air-conditioning energy use in US houses (Francisco et al. 1998; Mills 2003). A study by the International Energy Agency (IEA) indicated that air sealing alone can reduce the need for heating by 20–30% (IEA 2010).	Reduces vulnerability of buildings to extreme temperatures. Reduces air-conditioning and heating loads. Reduces risks of flooding or cyclonic wind gusts damaging buildings.
9. Energy-efficient air conditioning	Air conditioners: 50–75% improvement possible in the energy efficiency of current air-conditioning stock (Lucon et al. 2014). Coefficient of performance (COP) 5.0–6.5 in Japan: this means up to 50% energy savings since current COP in most systems is 3.0 or less (Waide et al. 2011).	Enables business and households to adapt to more extreme hot weather and risks of heatwaves. Reduces summer peak electricity demand.
10. Ceiling fans	50–57% energy savings potential improvement possible compared to current average market performance (Letschert et al. 2012; Sathaye et al. 2013). For instance, the Aerotron E503 ceiling fan: at 5–18 W (watts, low to high speed), it is much more efficient than typical fans which use 45–100 W.	Reduces risks of exposure to high temperatures. Ensures air movement when sealing up building on extremely hot days, helping to keep people cooler.
Specific lighting and appliance energy efficiency: mitigation and adaptation benefits		
11. Energy-efficient lighting	Replacing inefficient lighting such as standard incandescent 15 lm/W (lumens per watts, a higher number is more efficient) bulbs with more efficient bulbs such as:	More energy-efficient lighting significantly reduces unwanted heat gains compared with inefficient halogen or incandescent lighting. This reduces air-conditioning loads and improves building occupant comfort to help

		adapt to risks of heatwaves and higher temperatures.
	• compact fluorescent lamps (60 lm/W); • best currently available white-light LEDs (100 lm/W); or • current laboratory LEDs (250 lm/W); can result in a 50% reduction below BAU by 2030 (Pantong et al. 2011). This would cut global lighting costs from current USD 250 billion per year (2018 figures) to USD 125 billion per year by 2020–2030.	
12. Daylighting	Savings in lighting energy use of 40–80% in the daylit perimeter zones of office buildings (Levine et al. 2007). Solar tube technologies now allow many residential and commercial buildings to be significantly lit during the day by natural light. Daylighting is also effective for warehouse lighting.	Allows continued facility occupancy and operation during power outages. Reduces peak electricity demand and thus risks of blackouts on extremely hot days. (Note: poorly designed daylighting can increase electricity demand if it allows in too much heat. It is important to check the design of the daylighting technology to ensure it lets light but not excess heat into the building.)
13. Appliances	Energy efficiency savings (and equivalent GHG emissions reductions) from advanced appliances: • ovens: −45%; • microwave ovens: −75%; • dishwashers: up to −45%; • clothes washers: −28% (by 2030, globally); • clothes dryers: factor of 2 reduction; and • office computers: −40% (Lucon et al. 2014).	Energy-efficient appliances produce less unwanted heat, reducing air-conditioning loads in buildings. This helps to adapt buildings to risks of extreme high temperatures and heatwaves.
Specific energy/water efficiency nexus opportunities: mitigation and adaptation benefits		
14. Water efficiency	Using less hot water cuts energy use.	Residential and commercial buildings: 50–80% reduction in mains water use by water-saving fixtures compared to older standard fixtures (see Chapter 13).

Table 12.1. (*cont.*)

Climate-smart buildings	Climate-smart residential and commercial building/precinct sector	
	Climate change mitigation benefit	Climate change adaptation benefit
15. Cutting water leakage	Repairing and preventing water leakage reduces water supply and treatment-related energy.	Reduces water losses by as much as 30%, helping to adapt to risk of drought, water restrictions and water price rises.
16. Energy/water-efficient hot water systems	Typical efficiency factor for gas electric water heaters in the USA 0.67 and 0.8 in the EU, while the latest EcoCute Japanese heat pump water heater system has a COP of 4.5.	Reduces contribution to urban heat island effect. Reduces the amount of water needed to be provided by hot water services.
17. Low-flow showerheads	Roughly 30% of energy use in the home is for heating water; therefore, such an investment in showerheads pays itself back within months (von Weizsacker et al. 1997).	Reduces showering water consumption by up to 75% (von Weizsacker et al. 1997).
18. Low-flow faucet aerators	Reduces the energy costs of heating water by up to 50% (Cohen et al. 2009).	Inexpensively – low-flow faucet aerators cost only around USD 5 – reduces faucet water flow by 30–50% (range is based on aerator type and faucet use).
19. Water-efficient appliances	Front-loading domestic washing machines are 40–75% more water and energy efficient than top-loading (Cohen et al. 2009). New-generation top-loaders are getting better too. High spin effectiveness cuts electricity use and drying time.	Reduces phosphorus and salt loads in waste water, as water-efficient washers use less detergent.
20. Rainwater harvesting/use and grey-water reuse	Reduced energy used to treat and transport water, but can be offset by inefficient pumps and small-scale on-site water treatment if not designed well.	Reduces risks of flooding and storm-water events when integrated with local water management systems. Increases water security during times of drought.

Strategy		
21. Green drought-tolerant landscaping with trees	Reduces water requirements and therefore reduces energy consumption to pump water through drip-irrigation systems.	Significantly reduces water requirements to maintain landscapes. Reduces urban heat island effect and air-conditioning loads on buildings through shading from trees. Cuts heat stress for humans and vegetation. Can increase fire risk, though careful plant selection reduces fire risk.

Specific low-carbon, low embodied energy material strategies: mitigation and adaptation benefits
As buildings continue to become much more energy efficient (and thus their percentage contribution of operational energy falls), the initial embodied energy and global warming potential of the different construction materials will contribute more to the overall carbon footprint of buildings across their life cycle.

22. Building timber products including engineered timbers	In comparing a wooden frame and a reinforced concrete frame, the manufacture of materials for the wooden building uses 28% less primary energy and emits 45–50% less CO_2 than the manufacture of materials for concrete buildings (Sathre and Gustavsson 2009).	Engineered timbers can be made from fast-growing forests that are 10–12 years old. Offers a way of revitalising the forestry industry by creating high-value-added products that store significant amounts of carbon.
23. Low-carbon cements for concrete slabs and pavements, car parks	Low-carbon cement offers an 80% efficiency improvement and 80% reduction in GHG emissions per tonne, compared to Portland cement (Smith et al. 2010).	Uses 20% less water than Portland cement. Thermal mass can be used for cooling. Light-coloured low-carbon cements also have higher albedo (30–50% reflection) than black-coloured bitumen (<10% reflection) for car parks and thus have significant potential to reduce urban heat island effect.
24. Low-carbon bricks and ceramics	Emissions can be reduced by using waste materials as feedstocks, improving kiln efficiencies and using low-carbon energy sources. For instance, bricks made from fly ash waste produce lower emissions. Carbon-neutral bricks fired with sawdust are now commercially available in Australia.	Reduces the amount of clay and quarried materials used in bricks, freeing up more land to remain in food production to help adapt to climate change and contribute to food security.

Table 12.1. (*cont.*)

	Climate-smart residential and commercial building/precinct sector	
Climate-smart buildings	Climate change mitigation benefit	Climate change adaptation benefit
25. New steel roof materials	If steel roofing is required, new steel roofing uses a nanotech coating that reflects and re-radiates more heat. This lowers the temperature of the roof and roof cavity significantly, thus reducing air-conditioning energy requirements. Photovoltaic systems integrated into roofing materials are close to becoming available. Improving durability and weight reduction cuts embodied GHGs and maintenance costs.	Reduces impact of hot weather on internal temperatures of buildings using steel roofing. Reduces air-conditioning loads, so reducing peak electricity demand and thus risks of blackouts. Steel is easily recyclable at relatively low energy cost.
26. Bamboo building materials	Growing bamboo stores carbon at rates comparable to fast-growing trees. China, in particular, has shown that bamboo can be used for many building and furniture products that help to further store carbon for longer, helping to achieve greater mitigation. Engineered bamboo building products are further expanding how bamboo can be used to help reduce the embodied energy carbon footprint of buildings materials.	Bamboo provides food and raw materials (provisioning services) for consumers in developing and developed countries. It regulates water flows, reduces water erosion on slopes and along riverbanks, can be used to treat waste water and can act as windbreaks in shelterbelts, offering protection against storms (regulating services).
Other strategies to mitigate and adapt to climate change		
27. Elevate building structures and walkways	In tropical climates, raised building structures have been used to maximise natural ventilation from sea breezes through the floor as part of the tropical architecture movement. In regions vulnerable to flooding, buildings and precincts can be designed and built above potential flood levels.	Post-Hurricane Katrina, in New Orleans, USA, new buildings are required to be built raised off the ground. Walkways can also be designed between buildings so residents can easily move between buildings in times of flooding. This reduces risks of damage to buildings and contents as a result of flooding, storm surges or sea-level rises.

Additional strategies for low-income homes		
28. Energy-efficient lighting and solar power	Replaces kerosene or biomass energy sources (including forest sources), reducing GHG and black carbon emissions. Compared to kerosene lamps, LED lighting is 100 times more energy efficient.	Reduces deforestation, helping to conserve hydrological services, including reducing risks of flooding and soil erosion (Prasad et al. 2001). Improves adaptive capacity through access to information via radio and mobile phones.
29. Energy-efficient cookers	Compared to open fires, advanced biomass stoves provide fuel savings of 30–60% and reduce indoor air pollution levels by 80–90% for models with chimneys (Ürge-Vorsatz et al. 2012). For example, in the state of Arunachal Pradesh, advanced cookstoves with an efficiency of 60% have been used in place of traditional cookstoves with an efficiency of 6–8% (Rawat et al. 2010).	Reduces demand for wood to the point where local people can grow sustainable woodlots on limited land. Saves time that can be used more productively, with health co-benefits. Reduces deforestation, helping to conserve hydrological services including reducing risks of flooding and soil erosion (Prasad et al. 2001).

Source: Compiled by Smith from sources referenced throughout.

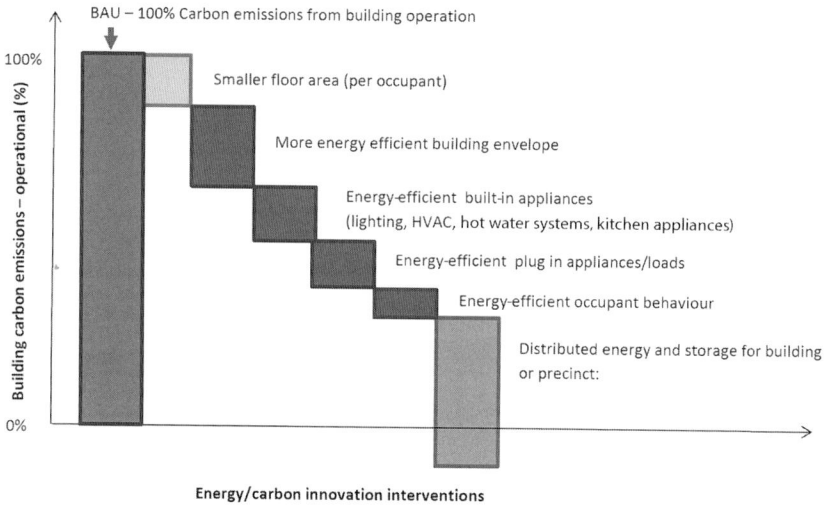

Figure 12.1 Staircase of measures to create carbon-negative buildings.
Source: Deng and Newton (2016).

achieve both mitigation and adaptation benefits, will realise close to 100% of the climate change mitigation potential for this sector. Such is the potential for synergistic overlap.

There are established pathways available for designers (in *as designed* buildings), contractors (in *as built* buildings) and occupants (in *as operated* buildings) to achieve highly energy-efficient buildings – which also have the potential to be carbon negative, when efficient all-electric buildings are combined with on-site solar PV-distributed generation. These measures are illustrated diagrammatically in Figure 12.1.

It is technically cost-effective to achieve significant energy efficiency improvements, such as 50–75% for new residential buildings (von Weizsacker et al. 2009; Lucon et al. 2014); and for new commercial buildings, 25–50% for total HVAC and 30–60% in lighting retrofits (von Weizsacker et al. 2009; ClimateWorks Australia 2010; Lucon et al. 2014). Combined, global net reductions in end-use energy demand compared with BAU could be as high as 60–70% by 2050 through energy efficiency (von Weizsacker et al. 2009; Laitner et al. 2012). The literature finds significant potential to improve building envelope thermal efficiency and appliance efficiency, enabling many buildings to be largely powered by on-site renewable sources such as solar PV systems. This enables the global buildings sector to become net climate neutral by 2050 (Lucon et al. 2014).

As the energy efficiency and use of on-site solar PV systems of buildings increases, the search for further reductions in the carbon profile of buildings will necessarily throw the spotlight onto embodied energy and carbon from construction and building products and materials (Newton et al. 2012). In the USA, the UK, Europe and Australia, progress is being made. In the UK, for instance, the embodied energy of all new building- and construction-related products and materials is put into a central life-cycle analysis database, which is building-information-modelling compatible (see Section 12.6.2). This enables designers, planners and builders to include embodied energy considerations easily in their design decision-making and modelling. This can enable construction and property companies to work with their supply chains to further decarbonise the manufacture of building products and materials. In Australia, for instance, innovations in low-carbon cement (Smith et al. 2010) and engineered timbers and investments in green steel are providing options to reduce the embodied carbon and energy of building materials.

12.4 Sustainable Development Co-benefits

There are significant financial, productivity, cost of living and health co-benefits from climate-smart buildings for all key stakeholders. Studies show that climate-smart buildings improve asset value and cost less to operate and maintain, securing tenants more quickly, enjoying lower tenant turnover and improving the productivity of occupants (Romm and Browning 1994; von Weizsacker et al. 2009; Smith 2013). The World Green Building Council reported in 2018 that new green building investments resulted in a 14% reduction in operating costs, and a 7% increased asset value, with green building investments paying back within 7 years. The World Green Building Council also reported that green building retrofit investments resulted in a 13% decrease in 5-year operating costs and an increased asset value of 5%, with a 6-year payback on average (World Green Building Council 2019).

Low-carbon commercial 'green buildings' (and precincts) consistently outperform traditional buildings in terms of comfort and productivity. Better lighting and fresher air, as well as better control over temperature and lighting, have been shown to directly affect productivity, reduce absenteeism and help businesses retain the best staff (Loftness et al. 2003). For instance, there have been up to 11% gains in labour productivity from improved ventilation (Loftness et al. 2003), and up to 23% gains (Loftness et al. 2003) in labour productivity from improved lighting design. Staff costs, including salaries and benefits, typically account for about 90% of service sector business tenant operating costs: hence, labour productivity improvements can significantly improve business tenants' bottom line.

Smart or green buildings provide numerous social and health co-benefits, especially for low-income households, by reducing exposure to extreme temperatures, improving thermal comfort and reducing indoor air pollution. There is significant potential for job creation from programmes to encourage low-carbon resilient new buildings/precincts and retrofit existing buildings/precincts. According to the US Green Building Council, green construction supported over two million workers in the USA by 2008, up from one million just 6 years previously (US Green Building Council 2009).

More energy-efficient buildings also reduce the need for electricity and gas infrastructure investment. For instance, in Australia, cutting peak demand by just 1 kW (kilowatt), the equivalent power used to run a small oil heater, can save almost AUD 1000 in electricity system infrastructure investment, reducing electricity prices for everyone (ASBEC and CWA 2018).

12.5 Roles and Responsibilities of Key Stakeholders

There are a myriad of stakeholders and decision makers who influence whether or not a transition to low-carbon and resilient buildings and precincts is achieved, as shown in Figure 12.2.

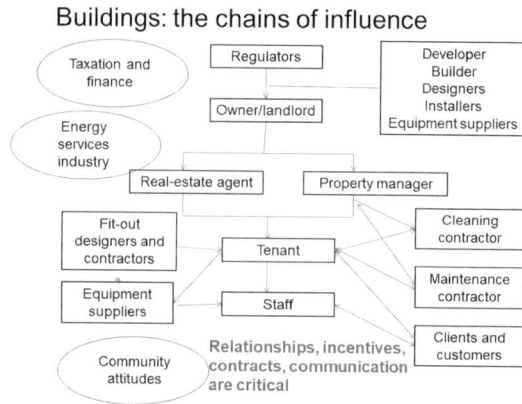

Figure 12.2 Buildings: key actors, relationships and chains of influence.
Source: Courtesy of Alan Pears.

Everyone, from designers, architects and engineers, to construction companies, property managers, tenants and owner-builders, to investors, banks and insurance companies, as well as urban planners, local government planning boards, building code boards and government policy-makers, influences the mitigation and adaptation performance of the building and precincts sector. The IEA has found that less than 20% of the potential for climate change mitigation in buildings has been realised to date. What then are the responsibilities for each of these key actors, and opportunities for them to play their part?

12.5.1 Designers

Traditionally, the design and construction of a building and its systems has been 'largely linear, in which design elements and system components are specified, built and installed without consideration of optimization opportunities in the following design and building phases, thus losing key opportunities for the optimization of whole buildings as systems' (Lucon et al. 2014: 686). Thus, it is at the design phase of built-environment projects that whole-system synergies between mitigation and adaptation can best be considered, and optimised for lowest cost. Using a whole-of-system approach (integrated design process) to identify and implement mitigation and adaptation opportunities helps to enable the full mitigation and adaptation potential to be realised at the same or less cost as conventional buildings. This is the secret to achieving cutting-edge green building and precinct outcomes in practice.

12.5.2 Investors, Property and Construction Companies and Tenants

The construction and property market is a highly capital-intensive market. It is in the investor's economic interest and fiduciary duty to choose to invest in construction and property companies which are proactively addressing the risks and opportunities of climate change by implementing the mitigation/adaptation nexus strategies outlined in Table 12.1 (Smith 2013).

This is recognised by the 392 institutional investors with USD 32 trillion in assets collectively under management, who have committed to *The Investor Agenda* to take effective climate action through shifting investment priorities into low-carbon and climate-resilient projects, policy advocacy in support of the Paris Agreement[1] implementation, and construction and property sector corporate engagement (CDP 2018). It is highlighted by the 650 institutional investors with USD 87 trillion in assets backing the Carbon Disclosure Project's annual environmental disclosure request to thousands of construction and property companies and cities; and by the 296 investors from 29 countries with USD 31 trillion in assets that are signatories to Climate Action 100+. The Carbon Disclosure Project and Climate Action 100+ are non-governmental organisation (NGO) initiatives that engage corporate GHG emitters on behalf of institutional investors and ask them to curb emissions to achieve the goals of the Paris Agreement.

Numerous property and construction companies are now undertaking the detailed climate change risks and opportunities analysis required, because of both the importance of good stakeholder relations and recognition that proactive action on climate change is simply good business practice. It is financially prudent to make sure that, over the next 20–50 years, residential or commercial building assets are not going to have prohibitive annual insurance costs (or, even worse, become relatively worthless stranded assets) due to climate change risks such as sea-level rises, greater flooding risks or loss of access to water sources. Also, increasingly in many cities, it is becoming harder to rent out commercial buildings that are not energy and water efficient, as premium tenants, such as government departments, seek to rent in buildings where operational costs for energy and water are low.

In Australia, the majority of property and construction companies have committed to become net zero emissions by 2025 to 2050 across all their property stock (ClimateWorks Australia 2019). For example:

- Dexus, a real-estate investment trust with AUD 26 billion worth of assets, aims to be net zero emissions by 2030.
- Mirvac, a property group managing over AUD 18 billion worth of assets, is aiming for net negative carbon emissions by 2030.
- Lendlease Australia is aiming for net zero emissions by 2025 (ASBEC and CWA 2018).

Being proactive in this space also enables property owners and tenants to explore ways to share the benefits of such operational cost reductions. Increasingly, property owners and tenants are signing 'green leases', so that landlord and tenant can work together better to reduce energy usage. This offers substantial benefits to property owners; for instance, energy-efficient lighting in shopping centres generates much less heat than inefficient lighting, and reduces the overall air-conditioning load in these centres. It has been shown

---

[1] *Paris Agreement Under the United Nations Framework Convention on Climate Change*, opened for signature 16 February 2016. Available at: https://unfccc.int/process-and-meetings/the-paris-agreement/the-paris-agreement.

that green leases are often critical to attracting and retaining premium tenants who pay premium prices, such as government departments. In addition, as explained above, property owners who enter into green-lease agreements with their tenants realise numerous financial benefits for both parties in addition to the energy efficiency savings. Interestingly, government-funded guides of how to do green leases usually focus on climate change mitigation and not adaptation, and completely ignore potential adaptation/mitigation synergies. There is a need for green-lease agreements and guides to be updated to reflect the win–win adaptation/mitigation opportunities outlined in Table 12.1. A very pragmatic contribution that could be made in this sector is that the Carbon Disclosure Project's standard questionnaire template should add a number of questions to ensure that companies are encouraged to consider and implement mitigation/nexus opportunities.

12.5.3 Insurers

The insurance industry has a key role to play in sending a price signal to the marketplace to drive climate change adaptation planning and implementation in all sectors, including this sector. As discussed, insurance companies are already refusing to provide flood insurance for regions particularly vulnerable to flooding or storm surges and are already factoring in potential risks and liability from climate change into their pricing models. Dr Evan Mills has explained that it is in the insurance industry's interests to examine how it could better incentivise climate change mitigation strategies which reduce risk exposure for this sector through reduced premiums (see Table 12.2). Mills (2003) identified 78 energy efficiency and renewable energy technological measures that can reduce insurance losses and manage risks in the buildings sector, especially those associated with high temperatures and heatwaves, power failures, fire and wind damage and indoor air-quality hazards.

Similarly, it has been recognised since 2003 that it is in the fiduciary interests of the insurance industry to reward customers who are taking action on climate change mitigation measures that reduce risk exposure (Mills 2003). In the US market, for example, insurers are now offering 'green building' products and services especially designed for new green buildings and upgrades, either following a loss or in the course of normal renovations.

12.5.4 Industry Groups

Green building industry councils are also important actors in helping to catalyse cost-effective, low-carbon building/precinct market transformation by increasing the rate of market diffusion of best-practice design and technology practices to achieve climate change mitigation and adaptation in the sector. They contribute to this in many countries, by providing market recognition of best practice, capacity-building and knowledge-sharing (through rating tools) and training; and by building networks of their members who wish to achieve Paris targets.

Table 12.2. *Current and emerging building energy assessment and rating frameworks and tools operating in Australia (similar trends exist in other countries)*

Building tool	Developer organisation	URL/Key reference
As designed		
NatHERS (Nationwide House Energy Rating Scheme)	Commonwealth Department of Industry and Science, CSIRO (Commonwealth Scientific and Industrial Research Organisation)	www.nathers.gov.au
BASIX (Building Sustainability Index)	NSW Department of Planning and Environment	www.planningportal.nsw.gov.au/basix
BESS (Built Environment Sustainability Scorecard)	CASBE (Council Alliance for a Sustainable Built Environment, Victoria)	bess.net.au
As designed and constructed		
Green Star Design and As Built	Green Building Council of Australia	https://new.gbca.org.au/rate/rating-system/design-and-built/
As constructed		
As Built Verification	Building Verification Forum	www.bvc.org.au
As operated		
NABERS (National Australian Built Environment Rating System)	NSW Office of Environment and Heritage	www.nabers.gov.au
Point of sale		
Liveability Property Features Appraisal Form	The Centre for Liveability Real Estate	www.liveability.com.au
ACT (Australian Capital Territory) Mandatory Disclosure	ACT Planning and Land Authority	www.planning.act.gov.au

For instance, the World Green Building Council's *Net Zero Carbon Buildings Commitment* has seen, as of 2019, over 50 large construction and property companies around the world sign up and commit to net zero-carbon operating emissions within their portfolios by 2030, and to be policy advocates for all buildings to be net zero in operation by 2050, in line with the Paris Agreement targets (World Green Building Council 2019). To implement this, national green building councils have set up nine national net zero-carbon certification schemes (in Australia, Brazil, Canada, France, South Africa, Germany, Sweden, India and the USA) and two frameworks (in the Netherlands and the UK), within which, as of 2019, 390 buildings globally have already been certified as net zero carbon. The world and national green building industry councils have historically focused more on climate change mitigation than climate adaptation and resilience issues. Currently 'green building' rating systems mainly provide points for climate change mitigation but they are starting to add adaptation/resilience criteria to existing green building/precinct/community rating systems to ensure that they incentivise efforts by developers to identify more effectively, and implement, smart mitigation/adaptation synergies. Green building councils are to be commended for publishing frameworks on climate change adaptation but these tools tend not yet to emphasise the significant mitigation/adaptation nexus synergies available. So green building industry councils could, as a next practical step, use this chapter to review their green building/precinct/community rating tools to reflect more effectively the key role of adaptation and the opportunities, through good design, to simultaneously achieve mitigation/adaptation nexus opportunities.

12.6 Tools to Optimise Building and Precinct Design, Construction and/or Retrofit

12.6.1 Digital Information Platforms for Low-Carbon Built Environment Design and Assessment

Clearly, a major transformation is required in the way we design, build and experience buildings, precincts and metropolitan areas. Such a transformation will need to be based on an emerging digital platform of building information models, precinct information models and city information models. These will enable integrated modelling, virtual representation and real-time automated multifactor performance assessment at the planning and design stage. This platform is emerging. The sustainability and resilience required of twenty-first-century cities will *also* necessitate a more fundamental and extensive retrofitting and redesign (Newton 2015). Urban retrofitting that is regenerative needs innovation at multiple scales – building, precinct and metropolitan level – all of which needs to be aligned/integrated in order to deliver the necessary step change in performance (see Table 12.3). There are now numerous tools to assist in design assessment and decision-making for buildings, precincts and cities that are adapted to decarbonising the built environment. The extent to which those tools are connected to leading digital information platforms will influence the effectiveness and efficiency (productivity) of the building, precinct and urban design process.

12.6.2 Building Information Modelling

In the residential building sector (unlike commercial buildings), building information modelling (BIM) is yet to gain traction. Many housing design practices still operate predominantly with two-dimensional CAD (computer-aided design) systems and the associated decision support, performance assessment and rating tools that have emerged to date are either scorecards or software, where design information needs to be re-entered from drawings and specifications (Table 12.2). Building information modelling overcomes this inefficiency, with its ability to automate information entry processes and make real-time assessments of performance *during* the design process (see BuildingSmart Australia n.d.), providing direct performance feedback on design options (and costs) as they affect energy rating.

Table 12.3. *Current and emerging precinct assessment and rating frameworks and tools operating in Australia*

Precinct tool	Developer organisation	URL/Key reference
Green Star Communities	Green Building Council of Australia	https://new.gbca.org.au/rate/rating-system/communities/
LEED-ND	United States Green Building Council	www.usgbc.org/leed/rating-systems/neighborhood-development
CASBEE	Japan Sustainable Building Consortium	www.ibec.or.jp/CASBEE/english/
BREEAM for Communities	BRE Group UK	www.breeam.com/discover/technical-standards/communities/
EnviroDevelopment Certification	Urban Development Institute of Australia	www.envirodevelopment.com.au
IS (Infrastructure Sustainability) Rating	Infrastructure Sustainability Council of Australia	www.isca.org.au/is_ratings
PrecinX	NSW Government	www.landcom.com.au
CCAP Precinct	Kinesis	http://kinesis.org/ccap-precinct
Thriving Neighbourhoods	ICLEI (International Council for Local Environmental Initiatives)	www.lgfocus.com.au/editions/2012-10/creating-thriving-neighbourhoo.php
One Planet Communities	Bioregional	www.bioregional.com/one-planet-living
BESS (Built Environment Sustainability Scorecard)	CASBE (Council Alliance for a Sustainable Built Environment)	http://bess.net.au

12.6.3 *Precinct Information Modelling and Decarbonising Precincts*

Precinct-scale design assessment is the least developed of the built environment models and tools, which include product declarations, whole-building modelling and city modelling. Yet precincts constitute the critical operational scale at which a city is assembled (greenfields), is rebuilt (brownfields, greyfields) and is operated (where residents spend large proportions of their day either in domestic or workplace settings). They are the 'building blocks' of our cities (Sharifi and Murayama 2013) and represent the scale at which urban design makes its contribution to city performance. It has been argued, however (Codoban and Kennedy 2007), that the unsustainable nature of today's cities is due in part to poor planning at the precinct level. For example, the high levels of car usage and traffic congestion are a reflection of an absence of: mixed-use development, variety in housing types (especially medium density) and walkability and public transit access. These features have been designed into urban precincts in recent decades. Purely in CO_2 terms, the variability in the housing and transport attributes of different suburbs means that precinct-scale carbon emissions can vary by as much as 50% across Australian cities (Crawford and Fuller 2011; Newton et al. 2012). Precincts constitute a critical focus for

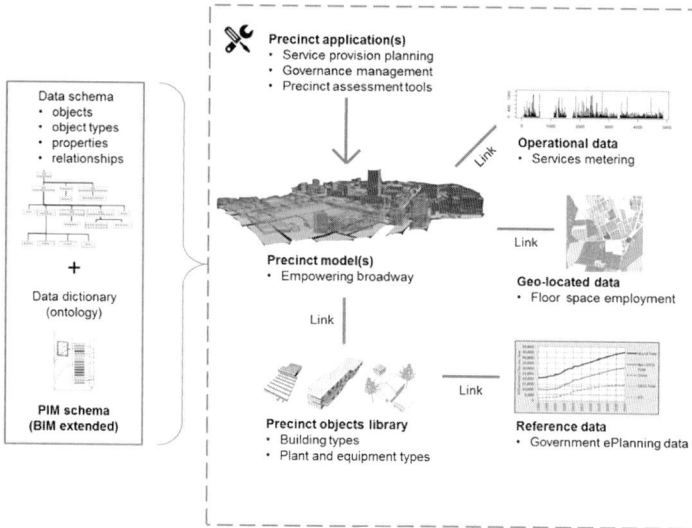

Figure 12.3 Operational structure of the PIM platform.
Source: Plume et al. (2015). For a colour version of this figure, please see the colour plate section.

the achievement of any carbon neutrality target for cities, since this is the scale at which an optimal combination of urban design innovation, urban technology innovation and behaviour change can jointly occur.

It is also at this scale where a convergence of digital technologies is required to support built-environment planning and management. Building information modelling has emerged as a platform to support construction sector industries focused on buildings, while geographic information systems (GIS) are focused on planning and management of land and infrastructure assets. Precincts are at the interface of these two digital technologies, and precinct information modelling (PIM) has emerged as a critical platform for more effective planning, design and management of spatial data relevant at that scale, variously termed neighbourhood, district and community (Newton et al. 2013). The operational structure of a PIM has been outlined by Plume et al. (2015; see Figure 12.3), and has at its centre a data schema and data library that, together, enable the representation of any type and scale of precinct, based on a specification of the multiple 'objects' that together comprise the built environment of that area; for example, types of buildings (residential, commercial, retail), transport objects (roads, paths), greenspace, etc.

The PIM thus facilitates linkage to external data sources capable of informing attributes of those objects (e.g. embodied energy/carbon, operating energy, etc.), enabling it to become a comprehensive repository of all relevant information required for precinct analysis. Since it is held in an accessible, open standard format, any number of software applications can be used to interact with the model to better understand or manage the precinct that is represented. This enables any precinct to be modelled to accommodate the disparate needs of the range of analysis and operational activities that support more sustainable performance throughout its life cycle (Figure 12.4).

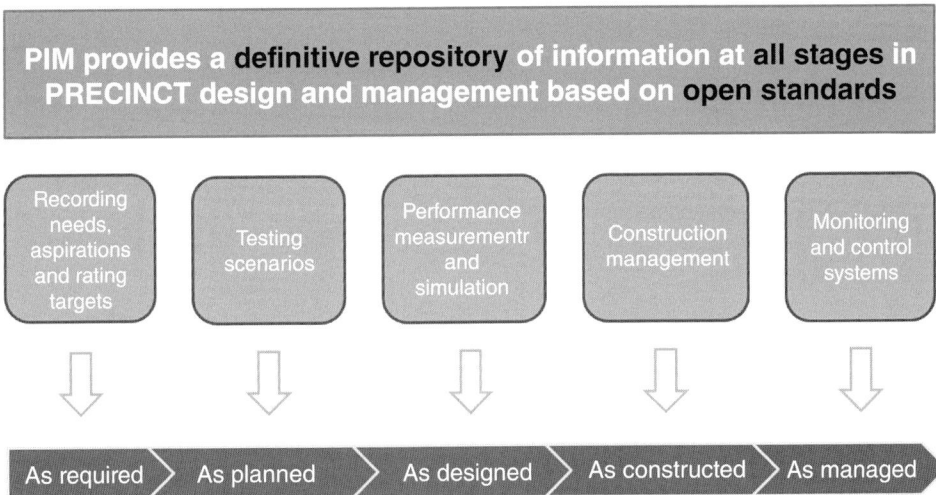

Figure 12.4 Life-cycle management of precincts using PIM.
Source: Plume et al. (2015). For a colour version of this figure, please see the colour plate section.

A review of precinct-scale assessment and rating systems currently operating in Australia reveals an increasing demand for such tools (Table 12.3).

Some assessment and rating systems are more narrowly focused (e.g. CCAP Precinct is focused on carbon), while most enable assessment across core built environment systems such as energy, water, transport and waste, and others have an even broader scope. They can be divided into assessment tools (alternatively termed design decision support tools), which model the performance of the precinct in the specific areas listed above, and rating tools (e.g. Green Star Communities; EnviroDevelopment, IS Rating), which take the outputs from the assessment modelling as a basis for industry 'labelling' or certification. Currently, most precinct assessments and ratings are evaluated against sets of benchmarks established by industry groups (e.g. Green Building Council) or municipalities (e.g. CASBE). To date, precinct assessment tools have been slow to take advantage of PIM and three-dimensional modelling, although research prototypes are now emerging (Newton 2019; Newton and Taylor 2019).

The value of a PIM-enabled precinct assessment and visualisation tool was demonstrated in a brief, one-month exercise focusing on envisioning a low-carbon water-sensitive future for Fishermans Bend (Figure 12.5). Fishermans Bend is a 250-hectare brownfield precinct adjacent to Docklands and the CBD (central business district) in Melbourne, required to accommodate between 80 000 to 120 000 residents and between 40 000 and 60 000 commercial jobs over the next 40 years (CRC for Water Sensitive Cities and CRC for Low Carbon Living 2015).

Fishermans Bend needs an urban design response for its buildings that aspires to be low-carbon, efficient, biophilic (i.e. optimising the exposure of buildings and their occupants to natural elements) and water-sensitive (i.e. minimising the import of potable water into, and export of waste water from, the precinct by maximising the use of rainwater harvesting by buildings, and greywater recycling within buildings).

Figure 12.5 Mutopia representation of Docklands and Fishermans Bend precincts, Melbourne (with the CBD in the background).
Source: CRC for Water Sensitive Cities and CRC for Low Carbon Living (2015). For a colour version of this figure, please see the colour plate section.

For the design assessment modelling of a future Fishermans Bend precinct, a spectrum of building energy ratings, ranging from current practice to international best practice, were employed to provide an estimate of the total amounts of energy and water (and CO_2 emissions) required by residential and commercial buildings in the precinct under varying development scenarios. This enabled a comparative performance assessment to be made as to the scale of environmental benefits to be achieved from adopting current versus best-practice performance targets in design briefs for developers (the latter representing what should be prescribed as a target for those developers wanting to be involved in the creation of Melbourne's largest inner-city precinct). Using world leadership performance (represented as 10 star (NatHERS)/6 star (NABERS)) for all buildings, compared to current practice, delivers a 43% reduction in annual energy use and carbon emissions, 51% reduction in water demand and a 30% reduction in lifetime carbon emissions.

Innovative building and precinct design as represented in the modelling here is a necessary but not sufficient step towards achieving a carbon-neutral built environment. It will also require low- or zero-carbon distributed generation solutions capable of operating at precinct scale, likely in combination with precinct-scale water, waste and mobility technologies and practices. It also requires that the embodied energy of buildings, precincts and cities be taken into account.

12.6.4 City Information Modelling for Low-Carbon Metropolitan Planning

Assessments of performance at the city scale are typically broader than those for buildings and aspire to cover the multiple precincts within city limits. Some of the performance indicators tend to include: productivity, competitivity, liveability, sustainability and social inclusivity. They cover the traditional goals of sustainable urban development. Governance is sometimes added as a fourth dimension. Resilience is also a concept that needs to be added to the pantheon of performance dimensions for cities: the ability to manage and learn from major challenges and to bounce back after some adversity or shock (exogenous or endogenous) to the system (Newton and Doherty 2014). The long-term strategic plans for each of the nation's metropolitan regions attempt to present a blueprint for development capable of delivering improved performance on each of these dimensions. An ability to evaluate city performance at a process level and an outcome level is now seen by government as fundamental. The latter typically exists as studies of city performance involving sets of single indicators. Of greater value are indexes that combine several indicators into a smaller set of lenses on city performance (e.g. the index of vulnerability assessment for mortgage, petrol and inflation risks and expenses, also known as the VAMPIRE index; socio-economic indexes for areas; and ecological and carbon footprints) that often reveal striking variations within and between cities, which call for some public policy response (e.g. the Brookings Institute 2008 study of carbon footprints of American cities; Brown et al. 2008). Also emerging are dual-factor studies that assess the covariation of combinations of leading indicators or indexes revealed by bivariate mapping and graphing; for example, sustainability and liveability (Newton 2012); happiness and GDP (Worldwatch Institute 2008); health and income inequality (Wilkinson and Pickett 2009); and sustainability and equity (UNDP 2011) among others.

Less common are city models that attempt to represent the interaction of multiple elements of an urban system. Attempts to evaluate the impact of alternative forms of land use and transport configuration – the two most fundamental components of a city's structure and performance – has generally been lacking. Calthorpe Associates (2011) has developed a macro-scale model for California capable of examining future development scenarios to 2050 that involve alternative land-use options (primarily relating to rates of infill, levels of density and housing mix) and policy options (BAU versus 'green' scenarios). The scenarios revealed significant variation in outcomes, relating to land consumption, urban travel, energy use, fiscal impact and GHG emissions (over 80% reduction in transport CO_2 emissions were achievable by more compact urban development and progressively stronger vehicle and fuel policies).

In Australia, the 1997 National Inquiry into Urban Air Quality (AATSE 1997; Newton 1997) was the first to examine the nature of any link between urban form (transport networks and associated travel patterns, location of housing, location of jobs, population distribution and density) and environmental performance (energy use, GHG emissions and urban air quality). The modelling assessed several archetypal urban forms (see Figure 12.6) in the context of future scenarios of population growth, employment and housing distribution and transport infrastructure investment (see Newton 1997; Newton and Manins 1999 for detailed descriptions of scenarios and assumptions).

Low-density suburban development (the BAU scenario characteristic of the 1990s – still representative of much greenfield development) was the worst performer across all indicators. The compact city (the bulk of new population and jobs located in the established

Table 12.4. *Percentage improvement in CO_2 emissions compared to the BAU scenario*

Future scenario related to urban form	Reduced CO_2 emissions from transport (%)
Compact city	31.5
Edge city (poly-centred)	21.7
Corridor city	15.5
Fringe city	20.7
Ultra city	16.0

Source: Newton (1997: 114).

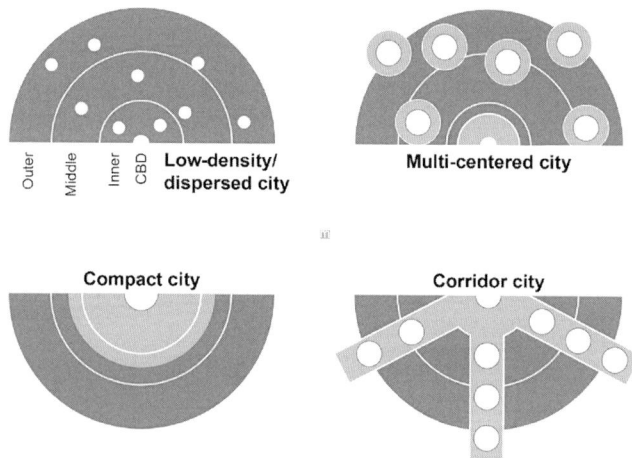

Figure 12.6 Archetypal urban forms.
Source: Newton (1997). For a colour version of this figure, please see the colour plate section.

public-transport-rich inner and middle suburbs, together with associated transport infra-structure investment) performed best in terms of transport-based energy consumption and CO_2 emissions (see Table 12.4), with a reduction of more than 30% in energy use and GHG emissions compared to low-density sprawl. Reductions in total transport CO_2 emissions compared to the BAU scenario were also greater for all other urban forms. Subsequent studies (ASBEC 2010; Newton et al. 2012) have confirmed the need for more intentional and integrated planning of transport, housing and employment – compared to market-led development – if the carbon footprints of cities are to be substantially reduced.

12.7 Role of Government: Coordination and Policy Reform

While there is much that can and is being achieved without policy change, climate-change-related policy measures are needed in this sector to address significant market and non-market barriers to improving energy efficiency and penetration of on-site renewable energy generation and to realise the economic and social benefits outlined above. These include:

- split incentives, especially in the building sectors;
- lack of access to capital: this is particularly relevant for small to medium businesses and low-income households with limited capacity to retrofit existing buildings;
- monopolistic behaviour by large incumbent local property developers or industrial building product manufacturers, which allows them to use their market power to block emerging, more innovative low-carbon solutions, building materials and competitors;
- imperfect and asymmetric information: a large percentage of construction and property developer firms, and businesses which operate in commercial buildings, have reported that a lack of skills and lack of relevant and reliable information are significant barriers to identifying and implementing low-carbon resilient buildings (Australian Industry Group 2012);
- inappropriate financial settings, which allow distortions to arise when high discount rates are applied to future costs and impacts, due to the longevity of impacts of decisions on building design and location; and
- lack of coordination due to the wide range of actors, whose decisions affect the emissions, energy usage and level of environmental friendliness of buildings.

Different levels of government have a clear role to support research, development and deployment, provide information, regulate for minimum standards and provide economic incentives to support the adoption of best practice to address market failures. In addition, government can play a key role by leading by example as a purchaser, owner/manager and user of building services (Lucon et al. 2014). But to date, it is very important to note that climate change policy reform, whether by government or in the form of recommendations from NGOs, focuses either on climate change mitigation or on climate change adaptation. No one, to date, has provided a policy roadmap that integrates both mitigation and adaptation. So, the following builds upon the existing IPCC Fifth Assessment Report policy option summary for the building sector to look at how complementary policy measures can be designed to best achieve mitigation and adaptation outcomes simultaneously. Some of the commonly used government planning and policy options for mitigation and adaptation can be better integrated as follows to encourage more investment in adaptation/mitigation synergies.

12.7.1 Government Planning

All levels of government – municipal, state/province and national – need to contribute to climate change mitigation and adaptation and must do so in a coordinated way. This should start with creating a vision, and urban and regional plans that maximise smart climate change mitigation and adaptation synergistic opportunities through: encouraging smart green urban development and water-sensitive urban design; zoning that enables north–south orientation to maximise solar radiation on roofs for solar PV systems; and requiring residential and commercial blocks to provide natural shade in summer to reduce air-conditioning costs. Issues around good governance in this area are discussed in Chapter 22. Government planning and development approval areas also can be key areas for reform to no longer require new suburbs and developments to provide natural gas supply infrastructure. In the Australian capital, Canberra, a new gas-free suburb is being built that is forecast to save new all-electric solar PV homes AUD 16 000 over 10 years, while costing approximately AUD 5000 extra each upfront to build. The UK Independent Committee on Climate Change has argued that new homes and new suburbs should not be connected to natural gas and instead use efficient electric heat pumps, similarly citing a strong business case for this recommendation.

The Australian Sustainable Built Environment Council, which represents all the major property companies that build new suburbs in Australia, concluded in 2018:

Research shows that newly constructed buildings will be 'all electric' ... [t]his is currently more cost-effective than installing gas connections in new buildings. This is because electric heating and hot water technology is becoming increasingly efficient, so despite gas being cheaper per unit of energy than electricity, electric appliances have lower running costs than their gas-powered counterparts ... An all-electric approach would also reduce the cost of building homes in new suburbs, as it avoids the need for costly new gas infrastructure to be built. *(ASBEC 2018: 27)*

In the USA, a number of jurisdictions have formally banned all new homes connecting to natural gas. A number of construction and property companies in Australia, such as Mirvac and Dexus, have publicly committed to building all-electric, renewably powered buildings (ClimateWorks Australia 2019).

12.7.2 Setting Short- and Long-Term Emission Targets to Provide Industry Certainty

Since the climate science is clear that economies need to decarbonise by 2050 at the latest, and buildings have long design lives, it is logical that, to achieve this decarbonisation goal, nations need to adopt scientifically based long-term targets for the building sector. There is growing consensus of the need to ensure new buildings are designed to be net zero energy, carbon neutral and resilient by 2030, and all buildings net carbon neutral by 2045. This aligns with the World Green Building Council's plan, announced in 2016, for all new buildings to achieve the net zero energy standard by 2030, and all existing buildings to be retrofitted to net zero energy by 2050. To help provide the property and construction industry with certainty, some nations are developing 'zero-carbon ready' or 'net zero energy' national building code trajectories. For instance, in Australia, the Council of Australian Government (COAG) Energy Ministers have agreed to, and publicly launched, a national construction code 'low-energy building' trajectory (COAG 2019). These trajectories provide examples of how national building codes can be utilised between now and 2030 to ratchet up minimum building standards to transition to zero-emission new buildings over the next decade. In some of these trajectories, it is recommended that on-site solar PV systems be mandated for new commercial and residential buildings as part of the national building code updates, to complement their focus on energy efficiency. This can help achieve a rapid transition to nearly net zero emission buildings (ASBEC 2018). California, for instance, has mandated that most new households built from 2020 must have solar PV panels, enabling them to be net zero energy. Massachusetts has enacted a plan for net zero-carbon emissions from all buildings in the city by 2040. Germany, Austria and Switzerland have pioneered a 'Passivhaus' design that, combined with solar PV systems, enables buildings to be net zero energy or nearly net zero energy. As of 2019, there are over 5000 net zero energy buildings around the world.

12.7.3 Climate-Smart National Building and Planning Codes

Governments have key policy levers such as national building construction codes and planning rules (e.g. 'precinct codes') within which it is possible to mandate many aspects of the transition to zero-emission, resilient buildings and precincts.

12.7.3.1 Reviewing and Updating Building Codes to Be 'Climate-Smart'

Governments can use national building codes to specify much more than just energy and water efficiency standards. They can also require new buildings, where appropriate, to have solar PV systems and energy storage, to be 'electric-vehicle-ready' and to have climate adaptation features. To date, governments have typically considered climate mitigation and adaptation issues as separate when looking at ways to update national building codes. Based on the analysis of adaptation/mitigation synergies in this chapter (Table 12.1), building codes and standards review processes need to do a better job of accurately accounting for actual costs/benefits of synergistic climate change mitigation/adaptation measures. For instance, improving the thermal efficiency of the building envelope of buildings not only helps to reduce energy operational costs but also has climate change adaptation co-benefits through reduced impact of extreme heat, reduced vulnerability to embers from bushfires and lower peak electricity demand. Higher-quality, better-insulated thermal efficiency of building envelopes also reduces noise pollution in cities. These multiple benefits of higher thermal efficiency of building envelopes are often not suffi-ciently taken into account when setting minimum building envelope standards. This is an example of how reviews of national construction codes need to take a more integrated approach to embedding climate change mitigation and adaptation goals.

12.7.3.2 Reviewing and Updating Planning Rules, Codes and Development Approval Review Processes to be 'Climate-Smart'

Governments with planning authority can and should also review and update planning rules, codes, development approval review processes and planning design principles and guidelines to be 'climate-smart'. Governments should review planning regulations and identify opportunities to require sustainable, climate-smart built environments as part of precinct developments. Through planning rules, relevant codes and planning precinct codes, it is possible to improve outcomes for optimal solar orientation of precinct blocks, as well as providing solar access, living infrastructure (i.e. mature tree retention), perme-able surfaces, effective water-sensitive urban design and bushfire and flooding risk man-agement for new developments.

Governments with planning authority can also proactively support and educate the market by incorporating climate-smart requirements into their publicly available planning design principles and design guidelines. These tend to include water-sensitive urban design, and flood and bushfire risk reduction guidelines, which developers need to comply with to ensure their development applications are approved.

12.7.4 Minimum Energy and Water Efficiency Appliance, Lighting and Industrial Equipment Standards

Governments should adopt and update mandatory minimum energy and water efficiency performance standards (MEPS) and labels across lighting, appliances and industrial equipment with effective product testing to ensure integrity of the programme. At the same time, governments need to phase out inefficient lighting and other major inefficient energy- and water-using technologies which have been superseded by more efficient technologies. This is one of the most cost-effective policies available to help nations achieve their 2030–2050 climate change mitigation and adaptation targets. Minimum efficiency performance standards schemes can be complemented by other energy efficiency policies to achieve market transformation of key enabling technologies, through programmes such as energy efficiency retailer obligation schemes. Over 50 nations have MEPS schemes.

12.7.5 White Certificate or Retailer Energy Efficiency Obligation Schemes

Energy efficiency retailer obligation and white certificate schemes require electricity retailers to achieve an energy savings contribution target or equivalent GHG emissions reduction target each year. They allow the retailer to meet that target by helping electricity consumers (usually building sector consumers) to save energy, by providing energy efficiency upgrades to (usually) commercial and residential buildings, either for free or at a discounted cost. Over 50 countries have these schemes. They allow rapid, safe roll-out with high compliance of energy efficiency products across cities and nations. They provide GHG abatement at half the cost of renewable energy technology pull policies. Along with minimum energy efficiency building and appliance standards, retailer energy efficiency obligation schemes have proven to be in the top three building sector energy efficiency policies in terms of achieving significant GHG reductions compared to BAU.

12.7.6 Encouraging Use of Low Embodied Energy/Carbon Building Materials and Products

As mentioned earlier in the chapter, the search for further reductions in the carbon profile of buildings will necessarily throw the spotlight onto embodied energy and carbon from construction and building products and materials. Governments can follow the lead of the UK and inform the market about the embodied energy of all new building- and construction-related products and materials, put into a central LCA database, which is building-information-modelling compatible. This enables designers, planners and builders to include embodied energy considerations easily in their design decision-making and modelling. This can enable construction and property companies to work with their supply chains to further decarbonise the manufacture of building products and materials. New analysis tools and databases for embodied energy assessment of building materials have been developed by the Cooperative Research Centre (CRC) for Low Carbon Living to enable Australia's National Carbon Offsets Certification Scheme for Buildings and Precincts to be extended to include Scope 3 emissions (Wiedmann et al. 2019). (See Chapter 16 for more coverage of ways to decarbonise the production of steel and cement and also alternative lower-carbon building materials such as engineered timbers.)

12.7.7 Decarbonisation of the Electricity Grid

Governments have a key role in setting targets for and incentivising decarbonisation of the grid, and also in providing incentives for commercial buildings and residential precincts to better utilise on-site renewable energy and next-generation storage to meet their energy needs. Nations, regions and cities which have achieved net 100% renewable electricity make it much easier for the energy end-use sectors like the buildings sector to subsequently transition to net zero emissions.

12.7.8 Complementary Measures

In addition to the above six main policy levers available to achieve large measurable reductions in GHG emmisions in the buildings and precincts sector as well as adapt to climate change, there is a portfolio of additional complementary policy measures that can be used to achieve this transition still more effectively.

Economic Incentives to Encourage Adoption and Investment in Best Practice in New Buildings. This can be done at the national and/or subnational level, through government taxation, grants and incentives which relate to buildings (e.g. stamp duty, first-home-buyer grants). These can be used to influence the sustainable design of new builds and renovations of buildings. For instance, first-home-buyer grants should only be given if the new home is going to be net zero emissions, by being all-electric with solar PV panels.

Development Incentives That May Be Better Utilised. This involves reviewing potential development incentives for accelerated planning approvals for new buildings/ precinct developments that meet certain climate change mitigation, climate change adaptation and sustainable development criteria. By contrast, new building/precinct development should be prevented in areas of high risk from sea-level rise, flooding and bushfires (Newton et al. 2018). Clear guidelines should be created for a planned retreat over time for coastal developments to enable the property and residential sectors to adjust and plan. Proactively planning for a managed retreat could save the global economy trillions by 2100 in avoided loss and damage (Kahn 2014).

Mandatory Disclosure of Energy and Water Efficiency Performance of Buildings at Point of Sale or Lease. Mandatory disclosure enables potential lessors or buyers to have this information before they make their decision. Studies validate that this policy works, with energy-efficient households commanding a premium both in home sales and with rental tenants (Fuerst and Warren-Myers 2018).

Minimum Energy Efficiency and Insulation Performance Standards for Rental Properties. These standards would reduce energy costs and improve public health. These are best phased in over a clearly defined time period to give landlords time to invest and upgrade their properties. Governments should offer grants or incentives to landlords to undertake energy efficiency upgrades, on condition that they do not pass on the costs of these upgrades in higher rental rates. Boulder, Colorado, is a global leader in successfully implementing minimum rental energy efficiency standards over the last decade (Petersen and Lalit 2018). The UK and parts of Europe have also adopted this policy reform.

Energy Performance Contracting Services Provided by Energy Services Companies (ESCOs). Governments can promote ESCOs, which aim to increase the market and quality of energy efficiency and water efficiency service offers, in which savings are guaranteed and investment is covered from energy and water cost savings. Governments can facilitate uptake of energy efficiency performance contracting by commissioning and promoting education and training in how to implement them. Governments can also host and promote workshops for ESCO providers and large energy users to build trust in ESCO services and financing. The global ESCO market as of 2019 is worth over USD 26 billion (IEA 2019).

Recycling Targets and Waste Reduction. There should be a reduction in GHG-related emissions from construction and demolition waste, and waste produced from the operation of buildings, through developing state- and city-wide waste strategies that aim over time to reduce waste to zero. Recycling targets of 80–90% for construction and demolition waste are increasingly being adopted by governments around the world.

Biodiversity Conservation Policies. There needs to be a reduction in negative impacts on biodiversity by encouraging green urban development, water-sensitive urban design and biodiversity corridors through urban landscapes. Encouragement of enhanced biodiversity outcomes should be part of green urban development for new residential estates and suburbs through urban planning, building policy and biodiversity conservation policy, through government mechanisms. For instance, there should be work to maintain old trees that provide hollows and habitat for native birds; biodiversity programmes and landscaping projects could be run with urban schools to give the next generation practical experience and understanding of the need to improve biodiversity in urban built environment development. Governments can secure investment by requiring major developments to ensure that overall there is no net negative impact on biodiversity and the environment, ensuring that property developers invest in natural capital.

Information and Research and Development Support. Governments can address information gaps by creating a 'one-stop-shop' web portal on climate change mitigation and adaptation for the building sector. This will:

- provide information on national climate change data, such as expected temperature changes, flooding risk and other hazards, to facilitate adaptation decision-making;
- help people keep up to date with the most recent advice and data provided to government;
- allow built-environment professionals and communities to understand the predicted impacts of climate change for their local areas and to take appropriate action to enhance resilience;
- give stakeholders access to information, case studies and tools to help with integration of the mitigation/adaptation nexus;
- produce a national guide on climate change mitigation/adaptation nexus opportunities for the buildings sector; and
- promote tools and information that are freely available.

Governments can also fund relevant research and development (R&D) and demonstration projects; for example:

- fund research into the cost–benefit analysis of taking an integrated climate change mitigation/adaptation approach to building retrofit and new building design, compared to approaches that simply seek to optimise building design in terms of either climate change mitigation or adaptation;
- fund research into how to create 'green lease' frameworks that best enable property asset owners and tenants to work together to realise beneficial climate change mitigation/adaptation nexus opportunities;
- fund R&D into how best to design and build low-carbon, resilient buildings and precincts.

Improved Transparency Related to Insurance Options. It is important to recognise the roles and responsibilities of insurers and government in providing coverage for areas at risk from climate-change-related events. Governments have a role to play to:

- increase transparency around insurance funding and risk assessment processes;
- provide plain English information about risks and the potential to obtain coverage;
- ensure that renters and low-income residents have access to appropriate insurance; and
- examine the appropriateness of a reinsurance pool or other government-backed mechanism to ensure coverage in the event of negative impacts from extreme weather events.

Governments also have a role in encouraging insurers to:

- insure properties in flood-, cyclone-, storm-surge- or bushfire-prone areas and to deal fairly with legacy investments while also managing a planned retreat over time from these areas;
- reward customers with reductions in insurance premiums if they are proactively reducing risk exposure through mitigation/adaptation measures.

Leading by Example. Finally, governments have the potential to lead by example and demonstrate the benefits of low-carbon resilient buildings/precincts to the private sector. Governments can help to build domestic markets for low-carbon resilient buildings: as they are owner-operators of their buildings for over 30+ years, it makes budget sense for them to invest in low-carbon, adaptation and sustainability measures, even if they have a >7–8-year payback. By specifying higher standards for competitive tenders for building contracts for new government buildings, schools, hospitals and public/social housing, governments can stimulate competition and innovation among the building design and construction industry to rapidly bring down costs for higher-standard green buildings. Governments can also utilise energy efficiency and water efficiency performance contracts to help finance upgrades of existing government buildings as well as street-lighting upgrades. Governments can also lead by insisting that government will not rent buildings that are not designed to be low-carbon, energy and water efficient and resilient. It makes sense for government departments to rent energy and water-efficient buildings as these reduce operating costs for government tenants while helping to transform the market. This has proven to be a powerful leverage for change, because in many cities, government department tenants are regarded as premium customers by property companies. Governments can also progress this next-generation adaptation/mitigation opportunity by setting benchmarks for new government buildings and committing to retrofit targets for existing government buildings. Governments can work with private property owners to improve mitigation/adaptation nexus retrofits within properties leased by government.

12.8 Conclusion

Climate mitigation opportunities in the buildings and precinct sector represent overall the largest and most profitable climate change mitigation opportunities of any sector (Lucon et al. 2014). Partly in recognition of this, over 50 globally or nationally significant construction and property companies have formally committed to achieving net zero emissions for their entire portfolio of property. Buildings and precincts also play a critical role to help humanity adapt to risks from extreme weather events. Due to the array of market and informational failures, as well as the range of stakeholders, government leadership, coordination and an effective portfolio of policies, as outlined above, are essential to underpin and support private-sector leadership to achieve a whole-of-sector transition to a net zero emission resilient buildings and precincts sector.

References

AATSE (Australian Academy of Technological Sciences and Engineering) (1997). *Urban Air Pollution in Australia*. An inquiry by the Australian Academy of Technological Sciences and Engineering for the Commonwealth Minister for the Environment.

ABCB (Australian Building Codes Board) (2014). *Resilience of Buildings to Extreme Weather Events*. Final paper. Australian Building Codes Board. Available at: www.abcb.gov.au/Resources/Publications/Consultation/Resilience-of-Buildings-to-Extreme-Weather-Events.

ASBEC (Australian Sustainable Built Environment Council) (2010). *The Second Plank Update: A Review of the Contribution That Energy Efficiency in the Buildings Sector Can Make to Greenhouse Gas Emissions Abatement*. The Allen Consulting Group. Available at: www.asbec.asn.au/research-items/the-second-plank-update-report-2010/.

ASBEC (2012). *Preparing for Change: A Climate Change Adaptation Framework for the Built Environment*. Australian Sustainable Built Environment Council. Available at: www.asbec.asn.au/files/ASBEC%20Preparing%20for%20Change%20Report%20FINAL.pdf.

ASBEC (2018). *The Bottom Line*. Building Code Energy Performance Trajectory Project: Interim report. Australian Sustainable Built Environment Council. Available at: www.asbec.asn.au/wordpress/wp-content/uploads/2018/03/180208-ASBEC-CWA-The-Bottom-Line-household-impacts.pdf.

ASBEC and CWA (ClimateWorks Australia) (2018). *Built to Perform: An Industry Led Pathway to a Zero Carbon Ready Building Code*. Australian Sustainable Built Environment Council. Available at: www.asbec.asn.au/wordpress/wp-content/uploads/2018/10/180703-ASBEC-CWA-Built-to-Perform-Zero-Carbon-Ready-Building-Code-web.pdf.

Australian Industry Group (2012). *Energy Shock: Pressure Mounts for Efficiency Action*. Australian Industry Group.

Brown, M., Sarzynski, A. and Southworth, F. (2008). Shrinking the carbon footprint of metropolitan America. *Brookings Institute*: 29 May. Available at: www.brookings.edu/research/shrinking-the-carbon-footprint-of-metropolitan-america/.

BuildingSmart Australia (n.d.). *BuildingSmart Australia*. Available at: https://buildingsmart.org.au/.

Calthorpe Associates (2011). *Vision California – Charting Our Future: Statewide Scenarios Report*. Berkeley, CA: Farmland Information Center. Available at: https://farmlandinfo.org/publications/vision-california-charting-our-future-statewide-scenerios-report/.

CDP (Carbon Disclosure Project) (2018). Nearly 400 investors with $32 trillion in assets step up action on climate change. *CDP.net*. 13 September. Available at: www.cdp.net/en/articles/investor/nearly-400-investors-with-32-trillion-in-assets-step-up-action-on-climate-change.

ClimateWorks Australia (2010). *Commercial Buildings Opportunities*. ClimateWorks Australia and Carbon Trust Australia. Available at: www.climateworksaustralia.org/resource/commercial-buildings-emissions-reduction-opportunities/.

ClimateWorks Australia (2019). *Net Zero Momentum Tracker: Property Sector Report*. ClimateWorks Australia. Available at: www.climateworksaustralia.org/resource/net-zero-momentum-tracker-property-sector-report/.

COAG (Coalition of Australian Governments) Energy Council (2019). *Trajectory for Low Energy Buildings*. Coalition of Australian Governments. Available at: www.coagenergycouncil.gov.au/publications/trajectory-low-energy-buildings.

Codoban, N. and Kennedy, C. A. (2008). The metabolism of neighbourhoods. *Journal of Urban Planning and Development*, 134, 21–31.

Cohen, R., Ortez, C. and Pinkstaff, C. (2009). *Making Every Drop Work: Increasing Water Efficiency in California's Commercial, Industrial, and Institutional (CII) Sector*. Natural Resources Defense Council. Available at: www.nrdc.org/water/cacii/files/cii.pdf.

Crawford, R. and Fuller, R. (2011). Energy and greenhouse gas emissions implications of alternative housing types for Australia. In *State of Australian Cities National Conference 2011 Proceedings*. Melbourne: State of Australian Cities.

CRC for Water Sensitive Cities and CRC for Low Carbon Living (2015). *Ideas for Fishermans Bend*. Discussion Paper. Melbourne: Cooperative Research Centre for Water Sensitive Cities. Available at: https://watersensitivecities.org.au/content/ideas-for-fishermans-bend/.

Deng, G. and Newton, P. (2016). *Assessing the Impact of Solar PV on Domestic Electricity Consumption in Sydney: Exploring the Prospect of Rebound Effects.* Sydney: CRC for Low Carbon Living.

Dili, A. S., Naseer, M. A. and Varghese, T. Z. (2010). Passive control methods of Kerala traditional architecture for a comfortable indoor environment: Comparative investigation during various periods of rainy season. *Building and Environment*, 45, 2218–2230.

Fuerst, F. and Warren-Myers, G. (2018). Does voluntary disclosure create a green lemon problem? Energy-efficiency ratings and house prices. *Energy Economics*, 74, 1–12.

Francisco, P. W., Palmiter, L. and Davis, B. (1998). Modeling the thermal distribution efficiency of ducts: Comparisons to measured results. *Energy and Buildings*, 28, 287–297.

IEA (International Energy Agency) (2010). *Policy Pathways: Energy Performance Certification of Buildings.* Paris: International Energy Agency. Available at: www.iea.org/reports/policy-pathway-energy-performance-certification-of-buildings.

IEA (2019). *Energy Service Companies.* Paris: International Energy Agency. Available at: www.iea.org/topics/energyefficiency/escos/.

IPCC (Intergovernmental Panel on Climate Change) (2013). Summary for policymakers. In T. F. Stocker, D. Quin, G.-K. Pattner et al., eds., *Climate Change 2013: The Physical Science Basis. Contribution of Working Group I to the Fifth Assessment Report of the Intergovernmental Panel on Climate Change.* Cambridge: Cambridge University Press. Available at: www.ipcc.ch/site/assets/uploads/2018/02/WG1AR5_SPM_FINAL.pdf.

Kahn, B. (2014). Adapting to sea level rise could save trillions by 2100. *Climate Central.* 3 February. Available at: www.climatecentral.org/news/adapting-to-sea-level-rise-could-save-trillions-by-2100-17034.

Laitner, J. A. S., Nadel, S., Elliott, R. N., Sachs, H. and Khan, A. S. (2012). *The Long-Term Energy Efficiency Potential: What the Evidence Suggests.* Report No. E121. Washington, DC: American Council for an Energy-Efficient Economy. Available at: www.garrisoninstitute.org/downloads/ecology/cmb/Laitner_Long-Term_E_E_Potential.pdf.

Letschert, V. E., Desroches, L.-B., Ke, J. and McNeil, M. A. (2012). *Estimate of Technical Potential for Minimum Efficiency Performance Standards in 13 Major World Economies: Energy Savings, Environmental and Financial Impacts.* Berkeley, CA: Lawrence Berkeley National Laboratory. Available at: https://eta-publications.lbl.gov/sites/default/files/lbnl-5723e_pdf.pdf.

Levine, M., Ürge-Vorsatz, D., Blok, K. et al. (2007). Residential and commercial buildings. In B. Metz, O. R. Davidson, P. R. Bosch et al., eds., *Climate Change 2007: Mitigation. Contribution of Working Group III to the Fourth Assessment Report of the Intergovernmental Panel on Climate Change.* Cambridge: Cambridge University Press, pp. 387–446. Available at: www.ipcc.ch/site/assets/uploads/2018/02/ar4-wg3-chapter6-1.pdf.

Loftness, V., Hartkopf, V., Gurtekin, B., Hansen, D. and Hitchcock, R. (2003). Linking energy to health and productivity in the built environment: Evaluating the cost–benefits of high performance building and community design for sustainability, health and productivity. Paper presented at the Greenbuild International Conference and Expo 2003, Pittsburgh, 12–14 November. Available at: http://mail.seedengr.com/documents/LinkingEnergytoHealthandProductivity.pdf.

Lucon, O., Ürge-Vorsatz, D., Zain Ahmed, A. et al. (2014). Buildings. In O. Edenhofer, R. Pichs-Madruga, Y. Sokona et al., eds., *Climate Change 2014: Mitigation of Climate Change. Contribution of Working Group III to the Fifth Assessment Report of the*

Intergovernmental Panel on Climate Change. Cambridge: Cambridge University Press, pp. 671–738. Available at: www.ipcc.ch/site/assets/uploads/2018/02/ipcc_wg3_ar5_chapter9.pdf.

Masson, V., Bonhomme, M., Salagnac, J.-L., Briotett, X. and Lemonsu, A. (2014). Solar panels reduce both global warming and urban heat island. *Frontiers in Environmental Science*, 4 June. Available at: http://journal.frontiersin.org/Journal/10.3389/fenvs.2014.00014/abstract.

Menon, S., Akbari, H., Mahanama, S., Sednev, I. and Levinson, R. (2010). Radiative forcing and temperature response to changes in urban albedos and associated CO_2 offsets. *Environmental Research Letters*, 5, 014005.

Mills, E. (2003). Climate change, insurance and the buildings sector: Technological synergisms between adaptation and mitigation. *Building Research & Information*, 31, 257–277. Available at: www.researchgate.net/publication/228596673_Climate_change_insurance_and_the_buildings_sector_Technological_synergisms_between_adaptation_and_mitigation.

Newton, P., ed. (1997). *Reshaping Cities for a More Sustainable Future: Exploring the Link between Urban Form, Air Quality, Energy and Greenhouse Gas Emissions.* Research Monograph No. 6. Melbourne: Australian Housing and Urban Research Institute.

Newton, P. (2012). Liveable and sustainable? Socio-technical challenges for twenty-first-century cities. *Journal of Urban Technology*, 19, 81–102.

Newton, P. (2015). Framing new retrofit models for regenerating Australia's fast growing cities. In M. Eames, T. Dixon, M. Hunt and S. Lannon, eds., *Retrofitting Cities for Tomorrow's World.* London: Wiley-Blackwell, pp. 183–206.

Newton, P. (2019). The performance of urban precincts: Towards integrated assessment. In P. Newton, D. Prasad, A. Sproul and S. White, eds., *Decarbonising the Built Environment: Charting the Transition.* Singapore: Palgrave Macmillan, pp. 357–386.

Newton, P. and Doherty, P. (2014). The challenges to urban sustainability and resilience. In L. Pearson, P. Newton and P. Roberts, eds., *Resilient Sustainable Cities.* London: Routledge, ch. 2.

Newton, P. and Manins, P. (1999). Cities and air pollution. In J. F. Brotchie, P. W. Newton, P. Hall and J. Dickey, eds., *East-West Perspectives on 21st Century Urban Development.* Aldershot: Ashgate, pp. 277–304.

Newton, P. W. and Taylor, M. A. P., eds. (2019). *Precinct Design Assessment: A Guide to Smart Sustainable Low Carbon Urban Development.* Sydney: CRC for Low Carbon Living.

Newton, P., Pears, A., Whiteman, J. and Astle, R. (2012). The energy and carbon footprints of urban housing and transport: Current trends and future prospects. In R. Tomlinson, ed., *The Unintended City.* Melbourne: Commonwealth Scientific and Industrial Research Organisation (CSIRO) Publishing.

Newton, P., Marchant, D., Mitchell, J., Plume, J., Seo, S. and Roggema, R. (2013). *Performance Assessment of Urban Precinct Design: A Scoping Study.* Sydney: Cooperative Research Centre for Low Carbon Living. Available at: www.lowcarbonlivingcrc.com.au/sites/all/files/publications_file_attachments/rp2001_-_performance_assessment_urban_precinct_design-final_0.pdf.

Newton, P., Bertram, N., Handmer, J., Tapper, N., Thornton, R. and Whetton, P. (2018). Australian cities and the governance of climate change. In R. Tomlinson and M. Spiller, eds., *Australia's Metropolitan Imperative. An Agenda for Governance Reform.* Melbourne: CSIRO Publishing, pp. 193–210.

Pantong, K., Chirarattananon, S. and Chaiwiwatworakul, P. (2011). Development of energy conservation programs for commercial buildings based on assessed energy saving potentials. *Energy Procedia: 9th Eco-Energy and Materials Science and Engineering Symposium*, 9, 70–83. Available at: www.sciencedirect.com/journal/energy-procedia/vol/9/suppl/C.

Petersen, A. and Lalit, R. (2018). *Better Rentals, Better Cities*. Boulder, CO: Rocky Mountain Institute. Available at: www.rmi.org/wp-content/uploads/2018/05/Better-Rentals-Better-City_Final3.pdf.

Plume, J., Marchant, D. and Mitchell, J. (2015). *PIM: An Open Digital Information Standard throughout the Urban Development Lifecycle*. Project Progress Report. Sydney: Cooperative Research Centre for Low Carbon Living.

Prasad, R., Maithal, S. and Mirza, A. (2001). Renewable energy technologies for fuelwood conservation in the Indian Himalayan Region. *Sustainable Development*, 9, 103–108.

Rawat, J. S., Sharma, D., Nimachow, G. and Dai, O. (2010). Energy efficient chulha in rural Arunachal Pradesh. *Current Science,* 98, 1554–1555.

Rijal, H. B., Tuohy, P., Nicol, F., Humphreys, M. A., Samuel, A. and Clarke, J. (2008). Development of an adaptive window-opening algorithm to predict the thermal comfort, energy use and overheating in buildings. *Journal of Building Performance Simulation*, 1, 17–30.

Romm, R. and Browning, W. (1994). *Greening the Building and the Bottom Line: Increasing Productivity through Energy-Efficient Design*. Boulder, CO: Rocky Mountain Institute. Available at: https://rmi.org/insight/greening-the-building-and-the-bottom-line/.

Rosenfeld, A. H., Akbari, H., Bretz, S., Fishman, B. L., Kurn, D. M., Sailor, D. and Taha, H. (1995) Mitigation of urban heat islands: Materials, utility programs, updates. *Energy and Buildings*, 22, 255–265

Sathaye, N., Phadke, A., Shah, N. and Letschert, V. (2013). *Potential Global Benefits of Improved Ceiling Fan Efficiency*. Berkeley, CA: Lawrence Berkeley National Laboratory. Available at: https://eta.lbl.gov/sites/all/files/publications/lbnl.5980e.pdf.

Sathre, R. and Gustavsson, L. (2009). Using wood products to mitigate climate change: External costs and structural change. *Applied Energy*, 86, 251–257.

Sharifi, A. and Murayama, A. (2013) A critical review of seven selected neighbourhood sustainability assessment tools. *Environmental Impact Assessment Review*, 38, 73–87.

Smith, M. (2013). *Assessing Climate Change Risks and Opportunities for Investors: Property and Construction Sector*. Canberra: The Investor Group on Climate Change (IGCC) and The Australian National University (ANU). Available at: https://igcc.org.au/wp-content/uploads/2016/04/Property-and-Construction-1.pdf.

Smith, M., Hargroves, K., Desha, C. and Stasinopoulos, P. (2010). Factor 5 in eco-cement: Zeobond Pty Ltd. *Ecos Magazine*, 21, 149. Available at: www.ecosmagazine.com/?act=view_file&file_id=EC149p21.pdf.

UNDP (UN Development Programme) (2011). *Human Development Report 2011*. New York: UN Development Programme. Available at: http://hdr.undp.org/sites/default/files/reports/271/hdr_2011_en_complete.pdf.

Ürge-Vorsatz, D., Petrichenko, K., Antal, M. et al. (2012). *Best Practice Policies for Low Carbon & Energy Buildings: Based on Scenario Analysis. Research Report*. Center for Climate Change and Sustainable Policy (3CSEP) for the Global Buildings Performance Network. Available at: www.gbpn.org/sites/default/files/08.CEU%20Technical%20Report%20copy_0.pdf.

US DoE (Department of Energy) (n.d.). Energy efficient window attachments. *Energy.gov*. Available at: http://energy.gov/energysaver/energy-efficient-window-treatments.

US Green Building Council (2009). *Green Jobs Study*. US Green Building Council. Available at: https://s3.amazonaws.com/legacy.usgbc.org/usgbc/docs/Archive/General/Docs6435.pdf.

von Weizsacker, E., Lovins, A. B. and Lovins, L. H. (1997). *Factor Four: Doubling Wealth, Halving Resource Use*. London: Earthscan.

von Weizsacker, E., Hargroves, K., Smith, M., Desha, C. and Stasinopoulos, P. (2009). *Factor Five: Transforming the Global Economy through 80% Increase in Resource Productivity*. London: Earthscan.

Waide P., Klinckenberg, F., Harrington, L. and Scholand, J. (2011). Learning from the best: The potential for energy savings from upward alignment of equipment energy efficiency requirements? In G. Trenev and P. Bertoldi, eds., *Proceedings of the 6th International Conference: EEDAL'11 Energy Efficiency in Domestic Appliances and Lighting*. European Commission Joint Research Centre, pp. 485–496. Available at: https://e3p.jrc.ec.europa.eu/publications/proceedings-6th-international-conference-eedal11-energy-efficiency-domestic-appliances.

World Green Building Council (2019). Net Zero Carbon Buildings Commitment surpasses 50 signatories in latest status report. *World Green Building Council*. 28 May. Available at: www.worldgbc.org/news-media/net-zero-carbon-buildings-commitment-surpasses-50-signatories-latest-status-report.

Wiedmann, T., Teh, S. H. and Yu, M. (2019). ICM database: Integrated carbon metrics embodied carbon life cycle inventory database [data resource]. *Research Data Australia*. Available at: https://doi.org/10.26190/5df6aa5d5effd.

Wilkinson, R. and Pickett, K. (2009). *The Spirit Level: Why More Equal Societies Almost Always Do Better*. London: Allen Lane.

Worldwatch Institute (2008). State of the World 2008. Available at: www.slideshare.net/actionforhappiness/wellbeing-and-happiness-an-introduction-8993368/4-But_higher_GDP_doesnt_always mean greater life satisfaction.

13

Urban Water

ANDREA TURNER, MICHAEL H. SMITH AND STUART WHITE

Executive Summary

Without significant progress on mitigation, the costs of adaptation to climate change will become prohibitive. This is especially the case for the water sector as a result of unmitigated climate change risks, making droughts more intense and frequent as well as causing more severe rainstorms, flooding and cyclones, and increasing water scarcity in cities. Climate change also risks melting glaciers and snow, upon which over 2 billion people globally depend for part of their water supply.

Many urban water systems have been built without adequately factoring in the risks of climate change. These risks are already impacting cities in diverse ways: for instance, extreme droughts, or sewer systems becoming overwhelmed by storms, sending raw sewage into streets, rivers and into drinking water. Declining water availability risks higher energy and carbon intensity of water; for example, due to inter-catchment transfers and increased desalination requirements.

It is therefore in the interests of all key actors in this sector, such as investors, insurers, urban water utilities, urban planners, governments, businesses and communities, to work together to mitigate emissions and adapt systems. This chapter documents a number of climate change mitigation strategies that also yield significant climate adaptation co-benefits and vice versa, including:

- improving water efficiency (adaptation) in buildings, industry, energy generation and agriculture sectors in ways that also improve energy efficiency (mitigation);
- reducing urban water leakage, improving water security (adaptation) and also reducing the amount of energy needed to treat the water being delivered to cities (mitigation);
- urban managed aquifer recharge, which diversifies low embodied energy urban water supplies (mitigation) while also improving urban water security (adaptation) and holding back salinisation of coastal aquifers from sea-level rise (adaptation);
- investing in floating solar photovoltaic arrays on dams, on irrigation channels and at water treatment plants, reducing evaporation losses (adaptation) while providing a zero-carbon power source (mitigation);
- using pumped hydro both to provide storage for renewables (mitigation) and also to create greater resilience for the grid to withstand extreme peak demand events due to higher temperatures and urban air-conditioning loads in summer (adaptation); and

- effective catchment management and source water protection to maximise freshwater availability (adaptation) while reducing the amount of water treatment and associated energy inputs required (mitigation).

The chapter also shows how pursuing these mitigation/adaptation nexus strategies can help improve sustainable development goals of improved productivity, public health, new jobs in water/energy efficiency functions and better social equity outcomes.

13.1 Introduction

This chapter focuses on climate change risks and climate change mitigation/adaptation opportunities to reduce exposure to those risks for the urban water sector. Rural and regional water-related issues associated with, especially, agriculture and the resources sector are addressed in Chapters 19 and 20, respectively. While urban water is a smaller proportion of total water usage, compared to regional water usage, which includes agricultural uses, it is very important for a number of reasons.

First, over 50% of people live in cities, and we are heading towards a projected 80% of people living in cities by 2030. Partly due to this, urban water demand and consumption is growing rapidly in emerging and low-income countries. This is being achieved, in many cities, by unsustainable levels of extraction of groundwater sources. Climate change risks, such as more extreme drought, when combined with unsustainable groundwater extraction puts human well-being and economic growth of these cities at risk in the long term.

Second, urban water sectors are significantly exposed to climate-change-related risks. Over 150 million people already live in cities which experience a perennial water shortage (defined as less than 100 litres per person per day of sustainable surface and groundwater flow within their urban setting). Averages across all climate change scenarios suggest this could grow to 1 billion by 2050 (McDonald et al. 2011). There is much that developing and rapidly emerging economies can learn from cities like Singapore and nations like Israel and Australia, where freshwater consumption per capita has peaked and where there has been decoupling from GDP growth over the last decade.

Third, urban water contributes as much as 80% of the embodied energy in water consumption. If nations choose to invest in energy-intensive water supply systems to adapt to climate change, such as desalination plants, it will be harder and more expensive to decarbonise cities.

Finally, there are high operating costs associated with the delivery of urban water, and its subsequent treatment and disposal, as well as the capital works required to meet urban water security expectations.

This chapter first discusses the potential risks and impacts of climate change on the urban water sector and the contribution that the urban water sector makes to greenhouse gas emissions. It shows that, because of the strong climate–water–energy nexus, the range of potential climate change adaptation/mitigation synergies to address these risks is great. It proposes 10 key strategies for urban water sector suppliers and end users that enable sustainable development and climate change mitigation and adaptation strategies to be better optimised and simultaneously achieved. It provides evidence of the value policy-makers can create by providing incentives for investors to invest in these win–win–win solutions. The chapter ends with a proposed path for urban water futures for key actors.

13.2 The Integrated Mitigation/Adaptation Imperative

To understand how best to identify and design adaptation–mitigation nexus strategies for the urban water sector, it is important to first understand the main risks to the sector from climate change. The water sector is exposed to a wide variety of risks of unmitigated climate change. The Intergovernmental Panel on Climate Change (IPCC) Fifth Assessment notes that climate change will lead to 'changes in all components of freshwater systems' (Jiménez Cisneros et al. 2014).

13.2.1 Higher Risks of Extreme Drought

In some regions of the planet, it is likely that long-term climate change will result in a greater frequency of extreme droughts (IPCC 2012). For instance, unmitigated climate change is predicted to double the frequency of drought in Jordan by 2100 (Rajsekhar and Gorelick 2017). Based on an average across all climate change scenarios, this could result in up to one billion people being exposed to ongoing water shortages by 2050 (McDonald et al. 2011). Many parts of the world have experienced significant droughts in the last decade, where climate change is highly likely to be a contributing factor, including: Australia (1997–2012) (Turner et al. 2016), Syria (2006–2011) (Harisson and Gleick 2014) and California, USA (2010–2015) (Williams et al. 2015). These countries are particularly vulnerable to drought. The severe droughts in Syria (Mohtadib et al. 2015) and Sudan have been found to have been a security threat multiplier contributing to the conflicts in those countries.

13.2.2 Risks to Groundwater Supplies

The risks from drought coincide with many countries increasingly using groundwater at levels that are not sustainable, in many cases 'mining' groundwater resources that are not renewable in human lifetimes. The risks of reduced groundwater availability are likely to increase with anthropogenic climate change in dry parts of the world. Groundwater flow through shallow aquifers is an integral part of the hydrological cycle and therefore affected by climate variability and change through recharge processes (Chen et al. 2002) and regional human impacts (Petheram 2001). Groundwater extraction has been increasing and groundwater levels falling around the world due to over-pumping since mid-last century. Any climate-change-related decreases in rainfall and subsequent groundwater recharge are therefore additional burdens to overstretched systems.

13.2.3 Higher Risks of Extreme Flooding Events

At the same time, the risk of more intense rainfall events will rise in some parts of the world. Over the past few decades, flood damage has constituted about a third of the economic losses inflicted by natural hazards worldwide (Munich Re Group 2005). The economic losses associated with floods worldwide have increased by a factor of five between the periods 1950–1980 and 1996–2005 (Kron and Berz 2007). By the end of the twenty-first century, the number of people exposed annually to the equivalent of a twentieth-century 1-in-100-year river flood is projected to be three times greater for very high emissions than for very low emissions (Jiménez Cisneros et al. 2014). Climate change could increase the annual cost of flooding in the UK almost 15-fold by the 2080s under high-emission scenarios. If climate change increased European flood losses by a similar magnitude, annual costs could increase by up to USD 120–150 billion, for the same high-emission scenarios (ABI 2009).

13.2.4 Risks to Water Quality

Climate change is expected to change the quantity and timing of surface water run-off, which can influence water quality. While increased run-off can be expected throughout the tropics and higher latitudes, the dry tropical and mid-latitude regions will experience reduced run-off. Timings will also be impacted: summer flows will be reduced and winter flows increased, with earlier peak flows (Hennessy et al. 2007) and possibly drier conditions, which increase pollutant concentrations. This is a concern especially for groundwater sources that are already of low quality, such as Bangladesh, where arsenic is a problem (Black and King 2009). More frequent and severe forest wildfires could also seriously degrade urban water quality through negatively impacting urban water catchments (Emelko et al. 2011; Smith et al. 2011). Another risk to urban coastal water quality arises from the combination of declining rainfall flows into coastal aquifers, rising extraction rates and rising sea levels, increasing the risk of saline intrusion of coastal aquifers. This has potential to contaminate coastal aquifers and make them too saline to drink without some form of desalination, and to reduce the productivity of agricultural crops if drawn from the same reserves. There is also the long-term risk of sea-level rise that could negatively impact urban storm-water drainage and sewage disposal in coastal and low-lying areas, as well as the quality of the receiving waters.

13.2.5 Risks of Higher Urban Energy and Water Consumption

The water–energy–climate nexus refers to the strong interdependence between the two elements of modern infrastructure (the water and energy systems) and the phenomenon of long-term climate change (Hussey and Pittcock 2012). As previously indicated, the water and energy sectors are highly interdependent, with the urban water sector being a significant energy user, and the energy sector a key user of water. This strong interdependence creates risks that nations could respond to effects of climate change, such as drought, in

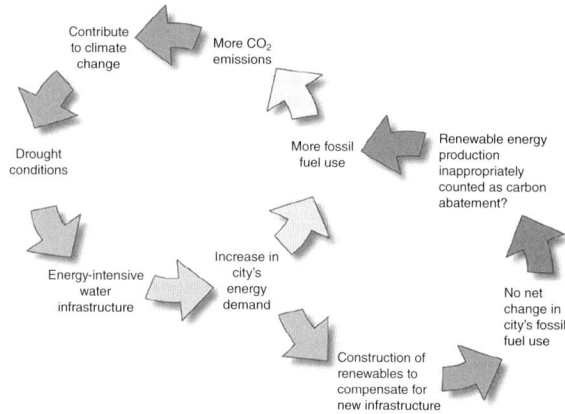

Figure 13.1 Example of a vicious circle.
Source: Retamal et al. (2010).

suboptimal ways that negatively impact on climate change mitigation. For instance, some nations' responses to recent droughts have resulted in increased levels of treatment, increases to the distances involved in inter-catchment transfers, and the construction of major supply infrastructure with a high energy intensity, such as desalination plants and advanced waste water recycling plants. Unabated, this produces a 'vicious circle', as seen in Figure 13.1, in that this increasing energy consumption contributes to global climate change, and the responses to drought increase the overall energy use associated with the water supply system. This is an instance where the actions associated with climate change adaptation, in this case the construction of major water supply infrastructure, moves us in a direction which is opposite to where we want to go in terms of climate change mitigation. In other words, climate change adaptation and climate change mitigation actions are misaligned and *maladaptation* arises.

It is therefore important to understand and map strategies that enable climate change adaptation and mitigation simultaneously. It is also important to better appreciate and understand the economic, social and environmental benefits of investing in such mitigation–adaptation synergies. Finally, it is important then to consider how best key actors and decision makers can incorporate these insights into good decision-making processes for water utilities, urban water planners and urban water end users.

13.3 Climate-Smart Urban Water Systems: Maximising Mitigation and Adaptation Synergies

To date, much consideration has been given to climate change adaptation or mitigation strategies for the water services sector (EPA 2015), but no studies have looked at the potential synergies between adaptation and mitigation. The following shows there are many such synergies that can help to both adapt to climate-change-related risks and achieve mitigation. It is possible to improve resilience of the urban water supply, consumption and treatment system by integrating demand-side and supply-side measures by both water suppliers (i.e. water utilities) and water end users (i.e. residential households and non-residential institutions and businesses). Research suggests there are at least 10 major adaptation–mitigation synergistic opportunities for the urban water sector. These are summarised in Table 13.1 and then further elaborated. The discussion follows the urban water cycle, from urban water sources and supply of urban water, to water usage in cities and towns, and finally to urban water and sewage treatment. Section 13.3.1 explicitly outlines how smart adaptation–mitigation synergistic strategies address the major climate change risks to the urban water sector.

13.3.1 Ten Mitigation/Adaptation Nexus Strategies

This section highlights 10 strategies that can help address climate change risks, through a combination of adaptation and mitigation. These strategies include some on the supply side of the urban water system, some on the demand side, and some that apply to the waste water system.

13.3.1.1 Strategy 1: Adapting to Risks of Loss of Water Quality with Catchment Management and Source Protection

Catchment management and source water protection is a vital first step to protect water quality. The condition of the catchment – the area over which rainwater is caught and drains into a water supply – significantly affects the quality of water. Effective catchment management and source water protection maximises freshwater availability (adaptation) while reducing the amount of water treatment and associated energy inputs required (mitigation). Investing in forest protection, reforestation, stream bank restoration, improved agricultural practices and forest fire management in water catchment areas can reduce the amount of pollutants flowing into supplies of drinking water. The 2014 report, *The Urban Water Blueprint* (McDonald and Shemie 2014), which analysed 534 cities and 2000 watersheds, demonstrated the value of investing in natural living infrastructure to maintain water quality and reduce water treatment costs. It found that water quality for more than 700 million people in the world's largest 100 cities could be significantly improved by adopting conservation practices in watersheds.

13.3.1.2 Strategy 2: Adapting to Risks of Higher Temperatures and Rates of Evaporation from Dams

Evaporation losses from dams in urban water systems can be significant. For instance, in South East Queensland, Australia, which has a semi-tropical climate, water losses from dams due to evaporation are estimated to average 300 GL (gigalitres) per year, which means that two and a half units of water need to be collected and stored for each unit of water supplied (UWSRA 2012). Increasingly, urban water supply or treatment companies are investing in reducing water losses associated with higher temperatures by either shading them with solar arrays or using chemical thin films to reduce losses. Solar PV panels can

Table 13.1. *Urban water demand- and supply-side options: adaptation and mitigation synergies*

Water utility adaptation and mitigation nexus synergies	Government/water utility/end-user demand- and supply-side options: adaptation/mitigation nexus synergies
Reduce water losses and associated embodied energy through: • improving source protection strategies (Strategy 1); • cutting water losses due to evaporation losses (Strategy 2); and • reducing system water leakage (Strategy 3).	Reduce water demand by water end users, which also saves energy. Water utilities and governments can encourage water efficiency with water end-use customers (households and businesses) through: • water restrictions and real-time pricing to encourage behaviour change; • water-efficient showerhead swap programmes; • toilet replacement programmes; • washing machine rebate programmes; • encouraging businesses to use more water-efficient equipment (i.e. waterless woks and pre-rinse valve retrofit programmes); and • requiring business water efficiency plans to be developed.
Improve water utility energy efficiency through: • optimisation of aeration systems; • improved ventilation systems and pumping systems (e.g. variable-speed drives); • fleet fuel efficiency improvements; and • building energy efficiency.	Improve energy and water efficiency of households, and businesses, through: • legally enforceable energy and water efficiency use standards (e.g. for appliances); • improving building code standards for thermal insulation and energy efficiency to reduce air-conditioning loads and associated water usage in buildings; and • improving energy efficiency of domestic hot-water systems.
Diversify water supply through: • rain and storm-water harvesting; and • managed aquifer storage and recovery (Strategy 4).	Diversify water supply for water end users through: • utilising commercial buildings with large roof areas for rainwater harvesting and large rainwater tanks (the high pressure from which can be used to pump water to where it is needed); and • designing new residential and commercial building precincts to be able to utilise recycled water created using low-carbon energy inputs.
Generate energy from water-powered sources through: • hydro energy generation using existing dams; and • installation of mini-turbine engines along high-flow pipes to generate electricity (e.g. in Portland, Oregon, USA, Scotland, Hong Kong).	Water end users: generate energy from waste water streams through: • harvesting of biogas from biosolids from waste water treatment plants; and • recovering energy from high-strength waste from industrial and agricultural customers.

Table 13.1. (*cont.*)

Water utility adaptation and mitigation nexus synergies	Government/water utility/end-user demand- and supply-side options: adaptation/mitigation nexus synergies
Urban water utility: use water to store energy through: • new on-river or existing on-river dammed pumped hydro energy storage. This is responsible globally for 99% of all current energy storage; and • the 'energy island' concept: creates a 30–40-m differential between the height of sea levels and a lake in the middle of the energy island. In times of excess renewable supply, water is pumped from the lake to the sea. When there is insufficient renewable energy, water is allowed to flow from the sea into the lake, generating electricity (DNV KEMA Energy 2013).	Water end users: use potential energy from rainwater harvesting in tall water tanks. Commercial facilities have significant roof space at relatively high elevation. These facilities can therefore capture significant rainwater to fill very large rainwater tanks with significant height. This stores significant potential energy to provide pressure to ensure the water can be piped to where it is needed using gravity. As the Australian *Your Home* guide states for the residential sector: • gravity can be a reliable, silent way to supply rainwater without external power; • the tank can be placed on a stand, or a garden tap fitted near the bottom of a tank to provide sufficient pressure to fill a watering can or a pool, or slowly water a lower garden by hose; and • if the rainwater collection area and tank are more than 15 m higher than the house, gravity pressure can be sufficient for all domestic use (Australian Government n.d.).
Power water supply and waste water treatment with up to 100% renewable energy. Urban hydroelectric dams and treatment plants can produce much of their own electricity to power their processes through: • hydroelectricity; and • solar photovoltaic (PV) arrays based on floats on water utility dams. Water treatment plants can meet their energy needs with biogas cogeneration plants as well as solar PV floating arrays.	Power households and businesses with 100% renewable energy. See Chapters 12 and 16.

cut evaporation rates by 90%. This can reduce water losses significantly because annual loss of water from dam storages through evaporation can potentially exceed 40% of water stored (Craig et al. 2007). For instance, one Indian study concluded that floating solar PV arrays, if used widely on large water reservoirs, could save 16 233 billion litres of water per year and contribute significantly to meeting India's renewable energy targets (Sharma and

Kothari 2016). The multiple benefits include reducing water evaporation and improving efficiency of energy production from solar PV panels through cooling with water. There is also the ability to store excess solar PV energy production in pumped hydro energy storage, and reduced costs from not having to purchase any land for the solar PV farm. Floating solar PV structures have been, or are in the process of being, built in the UK, China, the USA, Australia, India, Brazil, Thailand, Italy, France, Japan, Korea, Singapore, Switzerland and Israel (World Bank Group et al. 2018).

13.3.1.3 Strategy 3: Reducing Vulnerability to Drought through Reducing Urban Water Leakage and Urban Non-Revenue Water

Of all distributed water globally, 25–50% is lost due to illegal connections, inaccurate metering, leakages and deteriorating infrastructure. This is known as non-revenue water. Urban water leakage varies from as little as 3% (Tokyo) or 6% (Singapore) to as much as over 60%. Reducing leakage rates through leakage and pressure management not only stops the waste of water needed for growing populations (adaptation) but also reduces the amount of energy needed to treat and pump the water being delivered to customers (mitigation).

13.3.1.4 Strategy 4: Adapting to Risks of Salinisation of Coastal Urban Aquifers through Managed Coastal Aquifer Recharge and Recovery

Salinisation of urban water aquifers can occur from a combination of sea-level rise, over-extraction of urban aquifer groundwater and declines in rainfall over coastal urban water catchments. Managed aquifer recharge is the purposeful recharge of water to aquifers. It can help act as a barrier to saline water intrusion. There is significant potential to increase managed aquifer recharge in coastal cities. For instance, the potential of managed aquifer recharge has so far been assessed in three Australian cities: Perth (100–250 GL/year) (Scatena and Williamson 1999), Adelaide (20–80 GL/year (Hodgkin 2005), with 60 GL/ year achievable using urban storm water (Wallbridge and Gilbert 2009)) and Melbourne (100 GL/year) (Dudding et al. 2006). These studies suggest that managed aquifer recharge could not only help to maintain a saline intrusion barrier, but also be a major contributor to diversifying and increasing water supplies.

13.3.1.5 Strategy 5: Adapting to Risks of Rising Peak Electricity Input Costs for Urban Water Utilities by Improving Energy Efficiency

The IPCC Fifth Assessment warned that air-conditioning-related energy demand could increase 100-fold by 2100, resulting in higher peak electricity demand and increases in electricity costs (Jiménez Cisneros et al. 2014). Electricity costs are usually between 5% and 30% of total operating costs among urban water utilities. Energy efficiency opportunities generate 10–30% energy savings per measure and typically have 1- to 5-year payback periods. Hence, the business case for improving the energy efficiency of water provision is very strong. The following measures help to improve energy efficiency:

- increasing gravity-fed supply by diverting water production and supplying from pumping to gravity-fed where possible; and
- saving energy in pumps and pumping by upgrading to more efficient pumps, and through better matching pumps to their duty with variable-speed drives.

The energy-saving potential of the sector at its current level of operation is globally in the range of 34 to 168 TWh (terawatt-hours) per year.

13.3.1.6 Strategy 6: Reducing Risks of Not Meeting Peak Urban Electricity Demand by Turning Urban Gravity-Fed Water Supply Pipe Systems into Electricity Generators

Gravity-fed urban water systems can have excess water pressure. Also, piped urban water systems can be used to store excess electricity in low-demand periods by, first, moving urban water further back up the urban water system during low electricity demand periods and then, second, running that water down the urban water piping system to generate electricity to help meet peak electricity demand. Energy recovery turbines (ERTs) are micro hydro turbines installed in urban water systems at locations where throttling of pressure is beneficial or required (Po-An and Karney 2014). Examples of ERTs within water supply systems in the USA, Europe and Australia range from 40 kW (kilowatts) to 5 MW. As the Australian Water Services Association have stated, regarding hydro energy generation and the installation of mini-turbine engines along high-flow pipes to generate electricity, '[t]he technology to implement these opportunities is available now, and is expected to be a future focus of on-site renewable energy generation for water utilities' (WSAA 2012).

13.3.1.7 Strategy 7: Maximising Energy Efficiency Opportunities through New Technologies and Behavioural Changes in Water End Use (Residential and Commercial)

There are a wide range of climate-resilient strategies for water end-user residential and commercial buildings that have mitigation and adaptation co-benefits, including reducing water leakage losses, improving water efficiency in ways that also improve energy efficiency (Table 13.2), reducing the need for evaporative cooling, and landscaping with drought-tolerant plants and shade trees.

13.3.1.8 Strategy 8: Maximising Energy Efficiency Opportunities through New Technologies and Behavioural Changes in Water End Use (Industry, Business)

The industry chapter of this book (Chapter 16) shows clearly that there are many win–win climate change adaptation/mitigation opportunities that reduce both water and energy costs to business, such as:

- **Water-efficient/energy-efficient technologies**: such as waterless conveyor-belt lubricants, water-efficient spray nozzles and spray guns (Sydney Water 2013), clean-in-place technologies (i.e. sensors), steam traps and condensate return systems, water- and energy-efficient cooling-tower technologies (Sustainability Victoria 2006).

Table 13.2. *Technologies and behaviour changes to reduce urban water demand, save energy and cut greenhouse gas emissions (residential and commercial buildings)*

End use/sector	Enabling technologies/behaviours
Residential buildings	• Various water-efficient showerhead designs exist to reduce water consumption by up to 50% along with reducing the requirement for water heating. For example, in Australia, showerheads now commonly use between 7.5 and 9 L/min where once they might have used >15 L/min depending on the pressure in the system (Turner et al. 2010).
	• Water-efficient aerators/flow restrictors also typically reduce faucet/tap water flows by 30–50% and reduce the associated energy costs of heating water.
	• Water-efficient appliances such as front-loading domestic washing machines assist in reducing water demand significantly. While the water usage of top-loading washing machines in Australia has decreased in recent years from around 140 to 100 L/wash, front-loading machines can typically use less than 70 L/wash depending on size (Fyfe et al. 2015).
	• Dual-flush (4.5/3-L) toilets are capable of significantly reducing water usage compared to, for example, the 12-L single-flush toilets typically used in Australia and 20-L single-flush toilets typically used in the USA pre-2000s. In Australia, measured savings from toilet rebate programmes have ranged from 20 to 30 kL per household per year, saving significant levels of water and energy associated with potable water delivery (Turner et al. 2010).
Commercial buildings	• In commercial buildings, leakage/baseflow can typically be in the order of 10–20%. In some audits, losses of 80% have been found. With the installation of submetering systems and associated best-practice management, such leaks and baseflows can be observed, investigated and rectified, saving thousands of litres of water in individual buildings annually.
	• Waterless urinals use liquid-repellent coatings and a lighter-than-urine biodegradable trap liquid to prevent odours. They cost less to install and can save between 150 to 230 kL per conventional unit per year (Hawken et al. 1999). Hundreds of thousands of waterless urinals have been installed around the world, savings millions of litres of water each year (Cohen et al. 2009). Similarly, ultra-efficient urinals have been developed by manufacturers that flush as little as 0.5 L per flush.
	• Hybrid dry air/water cooling systems for large buildings have been optimised to reduce typical consumption of water by as much as 75% (von Weizsacker et al. 2009). The Bond, Australia's first building to receive a five green star 'as built' rating, for example, uses chilled beam technology for cooling on a commercial scale, assisting in the building achieving a 30% to 40% reduction in energy use compared to a typical office building and a 30% reduction in greenhouse gas emissions (GBC n.d.).
	• In the Sydney region alone, there are over 5000 registered cooling towers that can typically use 20 to 33% of commercial building water

Table 13.2. (*cont.*)

End use/sector	Enabling technologies/behaviours
	usage. Using simple best-practice checklists to manage such systems more effectively, installing submetering within the system and increasing the number of cycles of concentration can dramatically reduce the water and energy usage of cooling towers. In addition, alternative water sources such as condensate, rainwater and treated storm water and recycled water can all be used in cooling towers and cooling-tower water can be used for other purposes such as toilet flushing (Sydney Water 2007).
Commercial buildings: restaurants	• Efficient food steamers can use 90% less water and 60% less energy than older-style units, have shorter cook times, higher production rates and reduced heat losses (Sydney Water 2007).
	• Water-efficient commercial dishwashers can save 25% of water usage. Payback periods for installing small efficient commercial dishwashers range between 1 year and 4 years.
	• Waterless woks in Asian restaurants, which use air instead of water for cooling, can save nearly 90% of water consumption with similar short payback periods (Sydney Water 2009).
	• Pre-rinse spray valves account for a significant proportion of water consumption in commercial kitchens. Replacing a traditional pre-rinse spray valve can save between 25% and 80% of this water and associated energy.

- **Generating electricity from biogas from industrial waste water**: For instance, in the food and beverage sector there are significant biomass-based energy opportunities from separating and utilising the biomass in waste water that is commercially viable. This is because it helps to reduce water treatment costs while also providing significant on-site energy generation for food processing plants.
- **Treating and reusing water from industry**: According to the IPCC Fifth Assessment, 'Approximately 47% of wastewater produced in manufacturing sectors globally is still untreated. As the share of recycled or reused material is still low, waste treatment technologies and energy recovery can also result in significant emission reductions from waste disposal' (Fischedick et al. 2014: 744).

Numerous businesses have shown that it is possible to dramatically reduce freshwater intensity of their products and services. This is shown in Chapter 16.

13.3.1.9 Strategy 9: Reducing Risks from Extreme Weather Events Leading to Loss of Electricity to Urban Waste Water Treatment Plants by Increasing Energy Security

Extreme weather events have often resulted historically in loss of electricity to waste water treatment plants. This poses significant health risks to cities. It is technically possible for many waste water treatment plants to meet their energy needs through improving the energy efficiency of waste water treatment plants and then investing in combinations of cogeneration and renewable energy. There are already some examples that are 100% powered by renewable energy such as in the USA, France, Germany and Australia, to name a few. Most of these plants achieve 100% renewable status through a combination of cogeneration and solar PV technology. For many waste water treatment plants, it makes economic sense to invest in biogas cogeneration plants. Also, increasingly, waste water treatment plants are investing in on-site floating solar PV arrays to reduce evaporation and generate electricity.

13.3.1.10 Strategy 10: Reducing Risks of Extreme Flooding and the Urban Heat Island Effect, through Water-Sensitive Urban Design Strategies Combined with Green Urban Development Strategies

When a rainstorm is extreme, the run-off can cause overflows and flooding from outdated sewer systems that combine both raw sewerage and storm water in a single pipe. This tide of pollutants ends up in surrounding waterways, which serve as drinking water sources and recreational areas. Numerous cities have embarked on innovative storm-water run-off fixes that rely not so much on the old 'grey infrastructure' of huge, piped systems and sewerage treatment plants, but rather on new green infrastructure to collect and treat storm water at the local level across cities. These can achieve multiple sustainable development, adaptation and mitigation synergies and co-benefits through (i) reducing flooding risks, (ii) providing alternative urban water sources, (iii) recharging coastal aquifers, (iv) reducing the urban heat island effect, (v) providing public 'green' spaces, and (vi) encouraging active transport (cycling, walking) through creating safe urban corridors with high urban amenity.

There is a significant body of literature showing that water-sensitive urban design combined with green urban development can have a cooling effect on cities to address the urban heat island effect. The literature shows that water-sensitive urban design should be implemented strategically into the urban landscape, targeting areas of high heat exposure, with many distributed water-sensitive urban design features at regular intervals to promote infiltration and evapotranspiration, and to maintain tree health.

13.4 Economic, Productivity and Social Co-benefits

There are significant financial, productivity and health co-benefits from climate-resilient pathways in this sector for investors, water utility companies, businesses, homeowners, tenants and government. There are also significant economic benefits from rapid climate change mitigation, which reduces the costs of adaptation for the water sector significantly, costed at over USD 100 billion per annum. Without climate change mitigation, the costs of adaptation for the water sector are prohibitive. Urgent climate change mitigation will significantly reduce the following adaptation costs.

13.4.1 Improved Global and National Productivity and Economic Growth

Research by McKinsey Global Institute suggests that reducing urban water leakage and improving water efficiency globally could add USD 165 billion and USD 120 billion, respectively, to global GDP per annum above business as usual (BAU) by 2030 (Dobbs et al. 2011). At the national level, for a country like the USA, a USD 10 billion investment in water efficiency programmes would boost GDP by USD 13–15 billion above BAU, creating 150 000 jobs (Mitchell et al. 2008) and saving 35 trillion litres (Mitchell et al. 2008; Wallis 2014). More specifically, for the US economy:

- the economic output benefits of investments in water efficiency range between USD 2.5 and 2.8 million, per million dollars of direct investment;
- GDP benefits range between USD 1.3 and 1.5 million, per million dollars of direct investment; and
- direct investment in the order of USD 10 billion in water/energy efficiency could save between 20 and 40 trillion litres of water, with resulting energy reductions as well (Mitchell et al. 2008).

13.4.2 Water-Efficient Cities

The potential to improve urban water efficiency internationally is significant (see Box 13.1). Take the relatively water-efficient cities in the state of California. Cooley et al. (2010) show that despite the progress already made in improving water efficiency, a further 30% of California's urban water could be saved. This has been borne out during the most recent drought where average savings reached over 25% of pre-drought levels in 2015 alone (SWRCB 2016) and where more potential remains. At a subsector level, Heberger et al. (2014) find that, for California, water efficiency gains of 30–60% are possible for the commercial, institutional and industrial sectors, and 40–60% for households, from widespread adoption of water-saving appliances and fixtures, along with replacement of lawns with water-efficient landscapes.

13.4.3 Creating Jobs

Modelling shows that, for the USA, direct investment in the order of USD 10 billion in water/energy efficiency programmes can boost US GDP by USD 13 to 15 billion, and employment by 150 000 to 220 000 jobs (Mitchell et al. 2008). It could save up to 35 trillion litres of water, with resulting energy reductions as well.

13.4.4 Cost-Effectiveness

The Water Resources Group has shown that using a mix of water efficiency, demand management and supply augmentation and diversification technologies can reduce pressure on freshwater resources significantly and help nations meet supply–demand gaps without

---

Box 13.1 **Best-Practice Decoupling GDP from Urban Water Consumption**

A number of cities around the world have shown what is possible if purposeful policies are implemented. Jerusalem, Israel, Los Angeles and San Diego, California, Austin, Texas, and Sydney and Melbourne, Australia, for instance, have all achieved significant reductions in water per capita intensity of at least 20–30% over the last 15 years (Postel 1997). Seattle, USA, since the 1990s, has reduced per capita water usage by 35% (Cooley and Gleick 2009). In the last 15 years, Cape Town, South Africa, has managed to reduce water use by 30%, despite a population increase of 30% over the same period. Singapore also has achieved significant levels of relative decoupling (UNEP 2016). In places like Australia, this significant reduction in average water demand can be attributed to the increased efficiency of basic appliances such as toilets, showers and washing machines (Turner et al. 2016).

An outstanding example of rapid reduced urban water demand in response to drought comes from Brisbane and the surrounding area of South East Queensland, Australia. The Queensland Government and relevant authorities ran extensive water efficiency, demand management and alternative water source programmes (e.g. rainwater tanks for residential homes). This resulted in the achievement of a 50% per capita reduction – from 300 L per person per day to below 150 L per person per day – in residential potable water usage from 2005 to 2010. These savings were achieved through innovative demand management programmes that broke new ground in terms of both programme reach and speed of implementation, such as 'Target 140', 'Waterwise' and the 'One to One' programme, targeting both structural and behavioural aspects of water usage in the residential sector and employing a sophisticated understanding of water customers (Turner et al. 2016). Subsequent to the drought, demand has risen slightly but has not bounced back to pre-drought levels. The vast majority of water savings achieved per capita were highly cost-effective compared to the relatively high cost of large, centralised water infrastructure projects such as desalination (Turner et al. 2016). For further decoupling case studies see co-authors Professor Urama and Dr Smith's contributions to *Options for Decoupling Economic Growth from Water Use and Water Pollution* (UNEP 2016).

---

significant increases in freshwater usage (Dobbs et al. 2011). For instance, for China, 55 levers utilising over 20 water efficiency and over 25 supply diversification technologies were identified to close the projected base case supply–demand gap of 201 billion cubic metres in the seven basins.

13.4.5 Improving Water Quality

Investment to improve water quality by preventing and effectively removing water pollutants is also economically efficient. The studies reviewed by the OECD (2008) show that national measures to reduce pollution in agricultural run-off and storm water – including introducing targeted measures to reduce a variety of different pollutants such as arsenic and nitrates – results in health benefits costed to be in excess of USD 100 million for large OECD economies. Other benefits from climate-resilient pathways for this sector include:

- **Global security**: The US Intelligence Community (IC) *Global Water Security* report in 2012 (US Defense Intelligence Agency et al. 2012) states that 'during the next 10 years, water problems will contribute to instability in states … [W]ater problems – when combined with poverty, social tensions, environmental degradation, ineffectual leadership, and weak political institutions – contribute to social disruptions that can result in state failure' (US Defense Intelligence Agency et al. 2012: iii). The recent drought in Syria is one such example of how problems of water availability can contribute to conflict and state failure. Conflicts over water resources issues are not new. The Pacific Institute has documented thousands of such conflicts over human history (Gleick 1998).
- **Environmental benefits**: Freshwater extraction rates already exceed environmentally sustainable yields in most nations including Australia and the USA. All ecosystems rely on water for their health, but humanity has diverted over 55% of the entire world's freshwater away from ecosystems. There is now three to six times more water in artificial dams and reservoirs than in natural rivers (UN DESA 2011). Continuing to divert the current high percentage of water from natural ecosystems for human consumption is unsustainable and will result in the further collapse of many ecosystems.

For these reasons, two of the targets of proposed UN Sustainable Development Goal 6 are:

- by 2030, substantially increase water use efficiency (and water recycling) across all sectors and ensure sustainable withdrawals and supply of freshwater to address water scarcity; and
- by 2030, implement integrated water resources management at all levels, including through transboundary cooperation as appropriate (UN n.d.).

13.5 Stakeholder Responsibilities

13.5.1 Urban Water Utilities

In the last 20 years, urban water utilities and their representative water services industry groups have been playing a more proactive role to:

- identify and model climate change scenarios to determine potential risks and vulnerabilities to meeting the water supply–demand gap;
- identify, develop and implement climate change adaptation and mitigation strategies with a particular focus on addressing the energy–water nexus;
- partner with the energy sector to improve understanding of and identification of win–win strategies to address the energy–water nexus in cities; and
- develop online tools to help enable the development of climate change adaptation plans or climate change mitigation plans.

These activities represent a significant shift and progress on the water–energy nexus and climate change. But, to date, there has been little or no formal recognition of the potential for adaptation–mitigation nexus synergies by urban water utilities and their representative industry groups, despite the significant body of work on energy–water nexus synergies.

The authors have not been able to find any examples of water utilities formally taking this integrated adaptation/mitigation approach. Nor have we been able to find any formal recognition of these synergies in adaptation/mitigation by urban water sector industry groups. Hence, a logical next step for the urban water services sector and their industry groups is to start to integrate their climate change adaptation and mitigation plans and strategies using the insights from Section 13.3 of this chapter and the insights summarised in Table 13.3. Water utilities can benefit from taking this approach not simply for their own operations but also for their water efficiency programmes with end users. Emphasising the adaptation/mitigation synergy co-benefits for end users of water can help to motivate end users to make greater investments (compared to BAU) through this better understanding of the multiple climate mitigation and sustainable development co-benefits of climate change adaptation investments in the sector and vice versa. See Table 13.3 for examples of such adaptation/mitigation synergies for water utilities, water end users, water treatment plants and urban planners.

Many water utilities, businesses and households are trying to adapt and mitigate climate change, but they are doing so in a wider context of energy, resource consumption and embodied carbon. Businesses and households are often unable to easily quantify and make decisions about their wider supply-chain energy, water, carbon footprints. National, state and city-level governments have a role to improve information, decision-making, planning and policies to reduce unsustainable urban and regional energy, water and carbon flows and transition to a low-carbon, resilient and sustainable future.

13.5.2 City Governments and Planning Authorities

Cities and their governments are capable of significantly helping urban water utilities and water end users improve how they adapt to and mitigate climate change, through their responsibilities for urban planning, regulation, programmes and leadership. Large flows of energy, water and resources move through cities; hence, since they are being planned, built and retrofitted, there is much scope to influence the type and quantity of these flows. Many aspects of urban planning influence urban water usage. To date, there has been little work investigating how urban water policy can best achieve optimal climate change adaptation and mitigation objectives simultaneously. Quantitative performance measures for cities on urban water adaptation and mitigation, and their synergies or conflicts, are not well developed or systematically used or applied. An effort to benchmark urban water climate change mitigation and adaptation performance would help to enable a more systematic approach to be taken. This is increasingly being recognised. Overall, there is a significant shift starting to occur in cities, as outlined in Table 13.3.

There is much, therefore, that city governments can do to help enable this shift through urban planning and other governance powers. But whether cities transition to low-carbon resilient futures also depends on whether or not a similar transition is achieved regionally and nationally. Hence, to truly take an integrated adaptation/mitigation approach for the urban water sector, national and subnational government policy leadership is needed.

Table 13.3. *The opportunity to realise positive urban water climate change adaptation/mitigation linkages*

Where have cities been?	Where are cities going?
Planning and strategies, programmes, regulation: urban water climate change adaptation and mitigation done in isolation.	Integration of low-carbon resilient planning and strategy development, maximising urban climate change adaptation/mitigation synergies and avoiding adaptation/mitigation conflicts.
Standards and codes related to urban water climate change adaptation and mitigation are made in isolation (e.g. water efficiency and energy efficiency codes and standards).	Standards and codes related to urban water climate change adaptation and mitigation, such as building codes, planning codes, minimum energy and water efficiency performance standards, are better integrated. For example, the Californian Energy Commission now has the authority to set codes and standards for the energy–water efficiency nexus in California.
Urban climate/energy and climate/water policies are developed in isolation as separate issues.	Integrated urban climate/energy/water policy development to maximise adaptation/mitigation synergies. For example, The Californian Global Warming Solutions Act 2006[1] included a cross-sector working group tasked with investigating the climate–energy–water nexus, particularly water-related energy consumption.
Urban energy supply and water supply are seen as separate and distinct from each other, with energy and water policy developed largely in isolation from each other.	Recognition that urban water supply systems play an integral role in urban energy systems through: (i) electricity supply (hydroelectric dams, in pipe turbines), and (ii) electricity storage (pumped storage hydro constitutes 99% of global energy storage). Recognition that energy systems consume significant amounts of water, and different decarbonisation pathways either reduce or increase this.
Urban water supply and treatment systems are dependent on external electricity sources from the 'energy sector'.	Urban water supply and treatment systems increasingly becoming more energy efficient and self-sufficient in energy generation, and providing excess electricity back into the grid. For example, there are already water treatment plants powered 100% by on-site renewable energy.
Urban water and energy supply utilities are rewarded for the water and energy they sell, rather than for optimal lowest-cost integrated resource management and planning.	Urban water and energy supply utilities incentivised by government instead to implement lowest-cost demand management and supply-side investments to meet energy and water supply–demand gaps.
Tools currently in use for climate change adaptation or mitigation or urban water	Tools updated to enable urban water utilities to better optimise climate change

[1] *Global Warming Solutions Act*, Assembly Bill (AB) 32; Cal Health & Saf Code §§ 38500–38599 (2006).

Table 13.3. (*cont.*)

Where have cities been?	Where are cities going?
utilities and water end users are developed in isolation from each other.	mitigation/adaptation synergies in their decision-making.
Communication of water efficiency or energy efficiency opportunities to water/energy end users (business, households) are in isolation from each other.	Development of communication and behaviour change tools that map and explain win–win energy–water efficiency nexus opportunities to business and households.
Education and training in urban climate change adaptation and mitigation are in isolation from each other.	Education and training in urban climate change adaptation–mitigation nexus.
Research on urban water climate change adaptation and mitigation is undertaken in isolation.	There has been a significant body of research into the energy–water nexus to better understand opportunities for better integrated approaches to climate change adaptation and mitigation (Kenway et al. 2008; Retamal et al. 2010).
	There has been a body of research in water-sensitive urban design, investigating how to take integrated optimised adaptation/mitigation strategies at the precinct and urban scale through water-sensitive cities. It investigates, for instance, how to both reduce risks of urban flooding and reduce the urban heat island effect through green urban development (Wong et al. 2013).

13.5.3 *Role of National and State/Provincial Governments and Policy Reform*

13.5.3.1 *A National Urban Water Climate Change Adaptation–Mitigation Nexus Policy Reform Agenda*

As explained in the previous chapters on the energy sector, there are significant risks of inappropriate adaptation and mitigation responses from BAU water and energy policies in many countries:

- Current water policies in many countries risk improving water practices in ways that inadvertently increase energy intensity and greenhouse gas emissions. Some examples are: desalination plants, irrigation piping, water tanks, and water treatment and recycling, unless these are powered by low-carbon energy sources.
- Current energy policies also risk undercutting water policies and long-term food security in many countries by significantly increasing the water intensity of energy production and potentially threatening the long-term groundwater quality, which is essential to food security. For example, current energy policy in some countries allows the rapid expansion of the coal seam gas sector, which in the long term will result in significant amounts of groundwater being extracted.

These risks exist because water policy in most countries is developed by siloed government departments not adequately considering the interlinkages between water, energy and

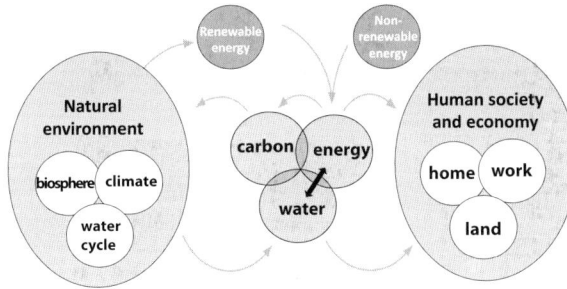

Figure 13.2 Water, energy and carbon are intrinsically interlinked with the natural environment (left) and human society and economy (right).
Source: Adapted from PMSEIC (2010).

climate change mitigation and adaptation. To date, these issues have been largely explored in what is known as the climate–energy–water and food nexus literature. In essence, national and subnational governments need to evolve their national water policy frameworks and water markets to optimise and incentivise investments in climate change adaptation/mitigation.

13.5.3.2 Addressing the Climate–Energy–Water–Soil–Food Nexus

Nations and their cities need to chart a new course that simultaneously:

- improves water productivity without significantly increasing energy costs and greenhouse gas emissions;
- meets national energy needs without increasing the water intensity of energy production or threatening food security;
- enables nations to achieve ambitious greenhouse gas targets in ways that improve energy and water productivity while enhancing food security.

This is because water usage is intrinsically interlinked with energy, food, soil, land, the natural environment and climate systems, as illustrated in Figure 13.2 (PMSEIC 2010).

13.5.3.3 Harmonising Markets for Carbon, Energy and Water to Ensure More Efficient Allocation of Resources

While energy and water markets exist and carbon markets are emerging, the linkages between them require that the markets and their non-market environments operate under consistent guiding principles. Hence, it is recommended that consistent principles for the use and marketing of these finite resources be developed and implemented.

13.5.3.4 Adopting Smarter Networks for Energy and Water Systems to Better Match Supply with Demand

Win–win improvements in energy/water efficiency, and more diverse low-carbon renewable energy production and storage, provide an opportunity to better match energy and water supply and demand in the future. Smart grids, water metering and sensor technologies are opening up new ways to improve access to real-time information to better manage and optimise urban energy/water supply and demand (PMSEIC 2010). This can help delay the need for new energy and water supply infrastructure.

13.5.3.5 Other Government Mechanisms

Other government mechanisms to assist include setting targets, lifting standards, and implementing planning and environmental assessment requirements and incentives. Other examples include:

- setting national and subnational greenhouse gas reduction targets, and water efficiency/ productivity and energy efficiency/productivity targets, supported by national energy and water productivity strategies, which can help to stimulate and motivate action and policy reform;
- establishing and setting targets for urban energy and water demand management strategies, whcih can reduce peak energy and water demand;
- developing integrated urban water and energy resource management plans based on thorough research and analysis for major cities, which can identify clearly optimal climate change mitigation and adaptation strategies for the next few decades;
- developing project-level environmental impact assessments and strategic environmental assessment processes that require urban water-related development proponents to have to consider climate change risks and adaptation and mitigation opportunities, synergies and conflicts;
- ratcheting up building code standards to improve energy and water efficiency standards for new buildings;
- reviewing and streamlining economic incentives to encourage uptake of appliances and equipment which improve energy and water efficiency for business and households, and harmonising local and national rebates, incentives and regulations in these areas of energy and water efficiency (PMSEIC 2010);
- using two- or three-tiered inverted water and electricity tariffs for residential and business energy and water consumer pricing to discourage over-consumption relative to the business service or industrial process involved;
- funding government- or utility-led campaigns, community awareness-raising and grant programmes to upgrade residential building stock and encourage businesses to be more energy and water efficient; and
- investigating a 'climate change levy', via regulating a minute increase to water prices to raise revenues to invest in projects to improve the sustainability resilience and efficiency of urban water systems.

13.5.3.6 Including Urban Water Climate Change Policy Reform as Part of a Broader Strategy to Reinvent Resilient Low-Carbon Cities

To address the energy, water and carbon issues that face cities and towns, it is optimal if these agendas are included as part of a broader national sustainable/resilient cities and towns initiative. A number of simultaneous actions are recommended, in particular:

- reducing demand for limited natural resources, such as water, simultaneously with a reduction of greenhouse gas emissions;
- transitioning to more resilient water-sensitive urban design approaches and relevant policy reforms to underpin this; and
- developing systems that are resilient to shocks and investing in infrastructure and technological innovations in energy and water supply (PMSEIC 2010).

See the Technologies for Decarbonising the Electricity Sector section of this book (Chapters 2 through 8) for policies to help encourage and enable decarbonisation of the water sector.

13.5.3.7 Enhancing Skills and Knowledge

The above recommendations require not only improving skills and knowledge to address the climate–energy–water–soil–food nexus but also urban water climate change adaptation and mitigation integration. To address this, the following is recommended:

- the development of a publicly funded integrated climate change adaptation/mitigation assessment tool for water utilities, building on the existing climate change adaptation and mitigation tools;
- the development of publicly funded free online integrated climate change adaptation/ mitigation guides for the urban water supply sector and water end users; and
- embedding education and training modules on climate change adaptation/mitigation integration for the water sector into the water professional community and formal education systems (PMSEIC 2010).

These would be best done in partnership with relevant peak global and national water professional and industry groups such as the International Water Association, which represents urban water professionals and decision makers internationally.

Such efforts would be assisted by simple reforms to make attention on the climate change mitigation/adaptation nexus in the water sector a mainstream consideration, such as in the Carbon Disclosure Project's questionnaire to water utilities and cities (Carbon Disclosure Project 2015). The questionnaire should include a subsection on what actions are being taken to integrate climate change adaptation and mitigation plans.

13.6 Conclusion

This chapter has addressed the nexus between climate change adaptation and mitigation in the water sector, and sustainable development. The research, practice and examples discussed here have demonstrated that this nexus is strong, and that it is crucial that these aspects be considered together in an integrated way within the policy framework. Doing so will result in significant benefits in terms of improved outcomes for sustainable water and energy resource use, as well as associated greenhouse gas emissions reductions. In addition, there are significant economic and social benefits associated with an integrated approach to these issues. These benefits include vastly reduced capital and operating expenditure for water supply and infrastructure, employment benefits, improved health of citizens and enhanced amenity of our cities and landscapes.

There is a complex array of stakeholders and actors that impact on water as well as complex institutional arrangements which are difficult to change. To bring these stakeholders together in an integrated way is challenging, but it is also essential.

References

ABI (Association of British Insurers) (2009). *Financial Risks of Climate Change*. ABI Research Paper No. 19. London: Association of British Insurers. Available at: www .ipcc.ch/apps/njlite/ar5wg2/njlite_download2.php?id=9144.

Australian Government (n.d.). Rainwater. *YourHome*. Available at: www.yourhome.gov .au/water/rainwater.

Black, M. and King, J. (2009). *The Atlas of Water: Mapping the World's Most Critical Resource*. Oakland, CA: University of California Press.

Carbon Disclosure Project (2015). *Putting a Price on Risk: Carbon Pricing in the Corporate World*. Report. New York: Carbon Disclosure Project. Available at: https://6fefcbb86e61af1b2fc4-c70d8ead6ced550b4d987d7c03fcdd1d.ssl.cf3.rackcdn .com/cms/reports/documents/000/000/918/original/carbon-pricing-in-the-corporate-world.pdf?1472456914.

Chen, M., Xie, P. and Janowiak, J. (2002). Global land precipitation: A 50-yr monthly analysis based on gauge observations. *Journal of Hydrometeorology*, 3, 249–266.

Cohen, R., Ortez, C. and Pinkstaff, C. (2009). *Making Every Drop Work: Increasing Water Efficiency in California's Commercial, Industrial, and Institutional (CII) Sector*. Natural Resources Defense Council. Available at: www.nrdc.org/water/cacii/files/cii .pdf.

Cooley, H. and Gleick, P. (2009). Urban water use efficiencies: Lessons from United States cities. In P. Gleick, ed., *The World's Water: The Biennial Report on Freshwater Resources*, Vol. 8. Oakland, CA: The Pacific Institute, pp. 101–122.

Cooley, H., Christian-Smith, J., Gleick, P. H., Cohen, M. J. and Heberger, M. (2010). *California's Next Million Acre-Feet: Saving Water, Energy, and Money*. Oakland, CA: The Pacific Institute. Available at: http://pacinst.org/app/uploads/2013/02/next_ million_acre_feet3.pdf.

Craig, I., Aravinthan, V., Baillie, C. P. et al. (2007). Evaporation, seepage and water quality management in storage dams: A review of research methods. *Environmental Health*, 7, 84–97.

DNV KEMA Energy (2013). *Large-Scale Electricity Storage*. Available at: www.dnvkema .com/Images/Large-scale-electricity-storage.pdf (site discontinued).

Dobbs, R., Oppenheim, J., Thompson, F., Brinkman, M. and Zornes, M. (2011). *Resource Revolution: Meeting the World's Energy, Materials, Food and Water Needs*. McKinsey Global Institute. Available at: www.mckinsey.com/business-functions/ sustainability/our-insights/resource-revolution.

Dudding, M., Evans, R. and Dillon, P. (2006). *Broad Scale Map of ASR Potential in Melbourne*. Smart Water Fund.

Emelko, M., Silins, U., Bladon, K. and Stone, M. (2011). Implications of land disturbance on drinking water treatability in a changing climate: Demonstrating the need for 'source water supply and protection' strategies. *Water Research*, 42, 461–472.

EPA (US Environmental Protection Agency) (2015). *Adaptation Strategies Guide for Water Utilities*. United States Environmental Protection Agency. Available at:

https://19january2017snapshot.epa.gov/sites/production/files/2015-04/documents/updated_adaptation_strategies_guide_for_water_utilities.pdf.

Fischedick, M., Roy, J., Abdel-Aziz, A. et al. (2014). Industry. In O. Edenhofer, R. Pichs-Madruga, Y. Sokona et al., eds., *Climate Change 2014: Mitigation of Climate Change. Contribution of Working Group III to the Fifth Assessment Report of the Intergovernmental Panel on Climate Change*. Cambridge: Cambridge University Press, pp. 739–810. Available at: www.ipcc.ch/site/assets/uploads/2018/02/ipcc_wg3_ar5_chapter10.pdf.

Fyfe, J., McKibbin, J., Mohr, S., Madden, B., Turner, A. and Ege, C. (2015). *Evaluation of the Environmental Effects of the WELS Scheme*. Report prepared for the Australian Commonwealth Government Department of the Environment. Institute for Sustainable Futures, University of Technology Sydney.

GBC (Green Building Council of Australia) (n.d.). *Case Study: 30 The Bond*. Green Building Council of Australia. Available at: www.gbca.org.au/docs/case%20study%2030%20The%20Bond.pdf.

Gleick, P. H. (1998). Water and conflict. In P. H. Gleick, ed., *The World's Water 1998–1999*. Washington, DC: Island Press, pp. 105–135.

Harisson, P. and Gleick, P. (2014). Water, drought, climate change, and conflict in Syria. *Weather, Climate and Society*, 6, 331–340.

Hawken, P., Lovins, A. and Lovins, L. (1999). *Natural Capitalism: Aqueous Solutions*. London: Earthscan.

Heberger, M., Cooley, H. and Gleick, P. (2014). *Urban Water Conservation and Efficiency Potential in California*. Issue brief IB:14-05-D. Oakland, CA: The Pacific Institute and Natural Resources Defense Council (NRDC). Available at: https://pacinst.org/wp-content/uploads/2014/06/ca-water-urban.pdf.

Hennessy, K., Fitzharris, B., Bates, B. C. et al. (2007). Australia and New Zealand. In M. L. Parry, O. F. Canziani, J. P. Palutikof et al., eds., *Climate Change 2007: Impacts, Adaptation and Vulnerability. Contribution of Working Group II to the Fourth Assessment Report of the Intergovernmental Panel on Climate Change*. Cambridge: Cambridge University Press, pp. 507–540. Available at: www.ipcc.ch/site/assets/uploads/2018/03/ar4_wg2_full_report.pdf.

Hodgkin, T. (2005). *Aquifer Storage Capacities of the Adelaide Region*. Report DWLBC 2004/47. Adelaide: South Australian Government Department of Water, Land and Biodiversity Conservation.

Hussey, K. and Pittock, J. (2012). The energy–water nexus: Managing the links between energy and water for a sustainable future. *Ecology and Society*, 17, 31.

IPCC (Intergovernmental Panel on Climate Change) (2012). *Managing the Risks of Extreme Events and Disasters to Advance Climate Change Adaptation*. Edited by C. B. Field, V. Barros, T. F. Stocker et al. Cambridge: Cambridge University Press. Available at: www.ipcc.ch/report/managing-the-risks-of-extreme-events-and-disasters-to-advance-climate-change-adaptation/.

Jiménez Cisneros, B. E., Oki, T., Arnell, N. W. et al. (2014). Freshwater resources. In C. B. Field, V. R. Barros, D. J. Dokken et al., eds., *Climate Change 2014: Impacts, Adaptation, and Vulnerability. Part A: Global and Sectoral Aspects. Contribution of Working Group II to the Fifth Assessment Report of the Intergovernmental Panel on Climate Change*. Cambridge: Cambridge University Press, pp. 229–269. Available at: www.ipcc.ch/site/assets/uploads/2018/02/WGIIAR5-Chap3_FINAL.pdf.

Kenway, S., Priestly, A., Cook, S. et al. (2008). *Energy Use in the Provision of Urban Water in Australia and New Zealand*. Water for a Healthy Country National Research Flagship Report Series. Australia: Commonwealth Scientific and Industrial Research

Organisation (CSIRO). Available at: www.clw.csiro.au/publications/waterfora healthycountry/2008/wfhc-urban-water-energy.pdf.

Kron, W. and Berz, G. (2007). Flood disasters and climate change: trends and options: A (re-)insurer's view. In J. L. Lozan, H. Grasl, P. Hupfer, L. Menzel and C.-D. Schonwiese, eds., *Global Change: Enough Water for All?* Hamburg: Wissenschaftliche Auswertungen, pp. 268–273.

McDonald, R., Green, P., Balk, D. et al. (2011). Urban growth, climate change, and freshwater availability. *Proceedings of the National Academy of Sciences*, 108, 6312–6317. Available at: www.pnas.org/content/108/15/6312.

McDonald, R. I. and Shemie, D. (2014). *Urban Water Blueprint: Mapping Conservation Solutions to the Global Water Challenge*. Washington, DC: The Nature Conservancy. Available at: http://water.nature.org/waterblueprint/#/section=overview&c=3:6 .31530:-37.17773.

Mitchell, D., Beecher, J., Chesnutt, T. and Pekelney, D. (2008). *Transforming Water: Water Efficiency as Stimulus and Long-Term Investment*. US Alliance to Save Water. Available at: https://digitalscholarship.unlv.edu/cgi/viewcontent.cgi?article=1001& context=water_pubs.

Mohtadib, S., Canec, M. A., Kushnirc, Y. and Colin, P. (2015). Climate change in the Fertile Crescent and implications of the recent Syrian drought. *Proceedings of the National Academy of Sciences*, 112, 3241–3246. Available at: www.pnas.org/con tent/pnas/112/11/3241.full.pdf.

Munich Re Group (2005). *Weather Catastrophes and Climate Change: Is There Still Hope for Us?* Munich: Münchener Rückversicherungs Gesellschaft.

OECD (2008). *Costs of Environmental Policy Inaction: Summary for Policy-makers*. Paris: OECD.

Petheram, C. (2001). Towards a framework for predicting impacts of land-use on recharge. *Australian Journal of Soil Research*, 40, 397–417.

PMSEIC (Australian Prime Minister's Science, Engineering and Innovation Council) (2010). *Challenges at Energy–Water–Carbon Intersections*. Report of PMSEIC working group. Canberra: Prime Minister's Science, Engineering and Innovation Council. Available at: http://web.science.unsw.edu.au/~matthew/FINAL_EnergyWaterCarbon.pdf.

Po-An, S. and Karney, B. (2014). Micro hydroelectric energy recovery in municipal water systems: A case study for Vancouver. *Urban Water Journal*, 12. DOI: 10.1080/ 1573062X.2014.923919.

Postel, S. (1997). *Last Oasis: Facing Water Scarcity*. New York: Worldwatch Institute.

Rajsekhar, D. and Gorelick, S. M. (2017). Increasing drought in Jordan: Climate change and cascading Syrian land-use impacts on reducing transboundary flow. *Science Advances*, 3, e1700581. Available at: http://advances.sciencemag.org/content/3/8/ e1700581.

Retamal, M., Turner, A. and White, S. (2010). The water–energy–climate nexus: Systems thinking and virtuous circles. In C. Howe, J. Smith and J. Henderson, eds., *Climate Change and Water: International Perspectives on Mitigation and Adaptation*. Denver: American Water Works Association and International Water Association Publishing, pp. 99–109.

Scatena, M. and Williamson, D. (1999). *A Potential Role for Artificial Recharge in the Perth Region: A Pre-feasibility Study*. Centre for Groundwater Studies report No. 84. Glen Osmond, South Australia: Centre for Groundwater Studies.

Sharma, K. and Kothari, D. (2016). Floating solar PV potential in large reservoirs in India. *International Journal for Innovative Research in Science & Technology*, 2, 2349–6010.

Smith, H., Sheridan, G., Lane, P. and Haydon, S. (2011). Wildfire effects on water quality in forest catchments: A review with implications for water supply. *Journal of Hydrology*, 396, 170–192.

Sustainability Victoria (2006). *Fact Sheets and Calculation Fact Sheets: Boiler Optimisation*. Melbourne: Victorian State Government.

SWRCB (State Water Resources Control Board) (2016). State's cumulative water savings continue to meet Governor's ongoing water conservation mandate. Media Release from the California State Water Control Board 5 January. Available at: www.swrcb .ca.gov/press_room/press_releases/2016/pr1516_nov_conservation.pdf.

Sydney Water (2007). *Best Practice Guidelines for Water Conservation in Commercial Office Buildings and Shopping Centres*. Sydney: Sydney Water Corporation.

Sydney Water (2009). *Best Practice Guidelines for Water Efficiency in Clubs*. Sydney: Sydney Water Corporation. Available at: www.sydneywater.com.au/web/groups/pub licwebcontent/documents/document/zgrf/mdq1/~edisp/dd_045254.pdf.

Sydney Water (2013). *Climate Change Adaptation Program*. Sydney: Sydney Water Corporation. Available at: www.sydneywater.com.au/web/groups/publicwebcon tent/documents/document/zgrf/mdy5/~edisp/dd_069672.pdf.

Turner, A., Willets, J., Fane, S. et al. (2010). *Guide to Demand Management and Integrated Resource Planning*. Report prepared for the National Water Commission and the Water Services Association of Australia. Sydney: Institute for Sustainable Futures, University of Technology Sydney. Available at: www.researchgate.net/publication/ 271530911_Guide_to_Demand_Management_and_Integrated_Resource_Planning.

Turner, A., White, S., Chong, J., Dickinson, M. A., Cooley, H. and Donnelly, K. (2016). *Managing Drought: Learning from Australia*. Alliance for Water Efficiency, the Institute for Sustainable Futures, University of Technology Sydney and the Pacific Institute for the Metropolitan Water District of Southern California, the San Francisco Public Utilities Commission and the Water Research Foundation. Available at: www.researchgate.net/ publication/297723736_Managing_Drought_Learning_from_Australia.

UN (n.d.). Sustainable Development Goal 6. *Sustainable Development Goals Knowledge Platform*. Available at: https://sustainabledevelopment.un.org/sdg6.

UN DESA (2011). *World Economic and Social Survey, the Great Green Technological Transformation*. United Nations Department of Economic and Social Affairs. Available at: www.un.org/en/development/desa/policy/wess/wess_current/2011wess .pdf.

UNEP (UN Environment Programme) (2016). *Options for Decoupling Economic Growth from Water Use and Water Pollution*. Report of the International Resource Panel Working Group on Sustainable Water Management. UN Environment Programme. Available at: www.resourcepanel.org/reports/options-decoupling-economic-growth- water-use-and-water-pollution.

US Defense Intelligence Agency (2012). *Global Water Security*. Intelligence community assessment ICA 2012-08. US Defense Intelligence Agency. Available at: www.dni .gov/files/documents/Special%20Report_ICA%20Global%20Water%20Security.pdf.

UWSRA (Urban Water Security Research Alliance) (2012). *Reducing Losses in Urban Water Supplies*. Fact sheet. Urban Water Security Research Alliance. Available at: www.urbanwateralliance.org.au/publications/factsheets/UWSRA_Fact_Sheet_6.pdf.

von Weizsacker, E., Hargroves, K., Smith, M., Cheryl, D. and Stasinopoulos, P. (2009). *Factor Five: Transforming the Global Economy through 80% Increase in Resource Productivity*. London: Earthscan.

Wallbridge & Gilbert (2009). *Urban Stormwater Harvesting Options Study*. Technical report C081266. Government of South Australia Stormwater Management Authority.

Available at: www.sma.sa.gov.au/wp-content/uploads/2018/07/UrbanStormwater HarvestingOptionsStudy_WEB.pdf.

Wallis, P. J. (2014). The water impacts of climate change mitigation measures. *Climatic Change*, 125, 209–220.

Williams, A. et al. (2015). Contribution of anthropogenic warming to California drought during 2012–2014. *Geophysical Research Letters*, 42, 6819–6828.

Wong, T. H. F., Allen, R., Brown, R. R. et al. (2013). *blueprint2013: Stormwater Management in a Water Sensitive City*. Melbourne: Cooperative Research Centre for Water Sensitive Cities. Available at: https://watersensitivecities.org.au/wp-con tent/uploads/2016/06/blueprint2013.pdf.

World Bank Group, ESMAP (Energy Sector Management Assistance Program) and SERIS (Solar Energy Research Institute of Singapore) (2018). *Where Sun Meets Water: Floating Solar Market Report – Executive Summary*. Washington, DC: The World Bank. Available at http://documents.worldbank.org/curated/en/579941540407455831/ Where-Sun-Meets-Water-Floating-Solar-Market-Report-Executive-Summary.

WSAA (Water Services Association of Australia) (2012). *Cost of Abatement in the Australian Water Industry*. Occasional paper 28. Sydney: Water Services Association of Australia. Available at: www.wsaa.asn.au/sites/default/files/publication/download/Occasional% 20Paper%2028%20Cost%20carbon%20abatement%20in%20the%20urban%20water %20industry%20May%202012.pdf.

14

National Climate Change Adaptation Case Study: Early Adaptation to Climate Change through Climate-Compatible Development and Adaptation Pathways

TIM CAPON, MARK STAFFORD SMITH AND RUSSELL WISE

Executive Summary

This chapter describes four case studies – three from Australia and one from rural Indonesia – that build the argument that to enhance the potential for multiple benefits, climate adaptation needs to be integrated into development and planning processes.

The case studies demonstrate (1) the early benefits from adaptation to coastal inundation, (2) the importance of considering the distribution of costs and benefits across communities, (3) the low-regrets nature of some early adaptation actions and (4) the synergies between adaptation measures and sustainable development.

Early climate change adaptation helps create resilience by maintaining the valuable options provided by various forms of diversity, flexibility and adaptability – factors that enable people to take advantage of future opportunities.

The multiple benefits are reflected in the global goals for adaptation. The multiple benefits include disaster mitigation, greenhouse gas mitigation, improving management approaches by incorporating flexibility to enable adaptive learning, for example, in areas such as coastal planning and development, and redressing food insecurity, inequalities and gender imbalances in the context of sustainable development.

Adaptation research has generated capabilities and capacity that can help inform adaptation processes. It has advanced through collaborations between researchers, practitioners and decision makers, who have all learned from the practical application of adaptation approaches.

Regulatory and institutional barriers, however, continue to present a challenge to effective climate adaptation. Adaptation pathways approaches can help stakeholders overcome these challenges and integrate adaptation into broader planning approaches.

This chapter identifies precedents from research and application that can provide practical guidance to decision makers seeking to apply adaptation pathways approaches.

14.1 Introduction

The need to understand the strategic imperative of climate adaptation for Australia and how best to realise it was one of the major questions behind the research agenda of the CSIRO (Commonwealth Scientific and Industrial Research Organisation) National Climate

Adaptation Research Flagship. To address this need requires understanding of whether there are net benefits from early action – including whether actions taken to adapt to a changing climate have multiple benefits for society, the environment and the economy – and the mechanisms by which decision makers can be incentivised or enabled to consider and act on climate adaptation. Research into climate adaptation has produced examples across geographic scales, domains (i.e. health, water management and coastal development) and decision-making levels, involving diverse stakeholders in government, business and community, that demonstrate potential net benefits including multiple benefits of early action (see Table 14.1). To realise these multiple benefits, adaptation needs to account for deep uncertainty (Lempert et al. 2003, 2019; Hallegatte et al. 2012), intangible values and the possibility of large-scale change. There is also a need to understand the dependencies and interdependencies between actions, their associated risks and the multiple benefits of taking a more integrated and coordinated approach (e.g. Australian Department of Home Affairs 2018; O'Connell et al. 2019).

Early climate change adaptation can help minimise or avoid some impacts and costs from potential future disruptions to individuals, communities, ecosystems and broader economic and social activities, and can help build the capacity needed to deal with emerging risks. For example, more resilient infrastructure allows economies to reduce the impacts from disasters and to recover faster (Lloyd's of London 2018). Estimates of the economic benefits of investments in resilience suggest the benefits exceed costs by at least 4:1 (UNISDR 2007; UK Government Office for Science 2012; Lloyd's of London 2018). A recent review by the Global Commission on Adaptation (2019) found that the overall rate of return from investments in resilience has benefit–cost ratios ranging from 2:1 to 10:1 or even higher. Early climate change adaptation helps create resilience by maintaining the valuable options provided by various forms of diversity, flexibility and adaptability – these factors enable people to take advantage of future opportunities created by climate change as well as policy, technological and research responses to climate change.

In this chapter, we describe four examples – three from Australia and one from rural Indonesia – that build the argument that to realise the additional benefits, climate adaptation needs to be integrated into development and planning processes. The case studies demonstrate (1) the early benefits from adaptation to coastal inundation, (2) the importance of considering the distribution of costs and benefits across communities, (3) the low-regrets nature of some early adaptation actions and (4) the synergies between adaptation pathways approaches and sustainable development. These case studies demonstrate the economic rationale of early action and the importance of integrating climate adaptation into planning for risk management and sustainable development. This isn't always a matter of simply considering climate change and adaptation costs and benefits within conventional applications of cost–benefit analysis or risk assessment. The case studies reveal the diversity of adaptation contexts, ranging from highly populated urban areas where clearly identifiable decision-making processes exist to ensure the reliable provision of critical infrastructure and services, through to peri-urban and rural environments where affected stakeholders are dispersed over large areas and no single decision-making process exists with the mandate to inform and implement strategic responses.

Table 14.1. *Australia: examples of direct and indirect impacts from key climate changes, and of low-regret adaptation strategies with potential additional benefits*

Climate change phenomena	Forecast change[a]	Potential direct and indirect risks (and opportunities)[b]	Some adaptation actions with mitigation and additional benefits[b,c]
1. Higher average temperature: increases in hot days, and changes to frost regimes	Annual average temperatures increase above the climate of 1986–2005 by: • 2030: 0.5–1.4 °C • 2090: 2.7–5.0 °C (high emissions) • 2090: 1.2–2.6 °C (moderate emissions). Hot days in Melbourne (>35 °C max.) increase from 11 days per year now to 13 days by 2030, and 16–24 days by 2090 (for other cities, see Webb and Hennessy 2015). With fewer than 10 days per year above 35 °C currently, Canberra is expected to have over 10–20 more days per year above 35 °C by 2070 (NSW Office of Environment & Heritage 2014).	**Urban/built environment:** Strengthens urban heat island. Changed heat/cold stress for humans, reduced outdoor activity and active transport, raised obesity levels, raised morbidity and mortality, altered sports regimes, productivity losses (construction, etc.), more school closures. (Note: net effects tend to be negative in warm climate cities, but may have initial net positive effects in cool cities.) Heat stress to materials and equipment, component failures, increased material deterioration rates, increased repairs and maintenance, transport system disruptions. Damage to urban greenspace, further feedback to urban heat island. Increased energy use and water use, higher utility peak loads, reduced equipment lifetimes. **Rural:** Effects on delivery of ecosystem goods and services – agriculture,	*Promote green urban development: cool pavements, permeable landscapes, water-sensitive urban design, shade trees and increased urban canopy.* *Invest in solar photovoltaic (PV)-shaded car parks and shade sails over playground areas.* *Invest in shadeways (i.e. tree canopy along the main urban bike paths) to support active travel in summer.* *Implement urban water efficiency and conservation; store storm water for urban greenspace.* *Improve building and appliance efficiency to reduce peak electricity and water demand, and utilise more distributed PV to contribute to meeting higher peak loads.* *Upgrade maximum and night-time temperature standards on engineering components in old and new infrastructure.* Harden road and rail transport systems. *Enhanced social networks to help people at risk.*

Table 14.1. (*cont.*)

Climate change phenomena	Forecast change[a]	Potential direct and indirect risks (and opportunities)[b]	Some adaptation actions with mitigation and additional benefits[b,c]
		water, biodiversity, etc. (including some opportunities).	Provide shade and shelter for livestock in heatwaves (and frost extremes, where relevant).
		Crop and livestock heat/frost stress, failure of chilling requirements, lost productivity, *changes to value chains, stranded regional infrastructure assets, changed rural community composition.*	Reassess regional infrastructure to be fit for changing crops.
			Explore new insurance instruments.
		Changed inflow regimes for water supplies, insufficient snow to support ski-fields, shortened winter comfortable tourism season inland.	Diversify tourism experiences in snowfields with mountain biking, hiking, camping and sustainable recreational mountain fishing.
		Species change, loss of native species, changed invasive species profile.	*Sustain diverse ecosystems, to support fauna survival in extremes.*
2. More weather extremes (e.g. cyclones, hailstorms, flooding, high winds)	Fewer but more intense cyclones in coastal Northern Australia.	**Urban/built environment:**	*Ensure wind-related building and retrofit standards are met; strengthen these in South East Queensland.*
	20% increase by 2050 in hailstorms in Sydney region, from once every 5–8 years at present (Leslie et al. 2008).	Wind/hail/flood damage to buildings, home evacuations, *increased insurance premiums (or uninsurability), disrupted business operations, volunteer downtime.*	Enforce flood zoning in planning and construction.
	General increase in intensity of extreme rainfall events, even where mean rainfall decreases, leading to increased risks of flooding (magnitude very locality dependent).	Access cut, road damage, transport and resupply disruption.	*Upgrade public infrastructure (including roads), especially for flooding.*
		Risk of deaths and injury, *increased food and services costs.*	*Enhance urban greenspace water management, rain- and storm-water harvesting and recycling.*
		Rural:	*Invest in greater urban forest canopy.*
		Crop and stock losses, commodity supply interruption (e.g. bananas).	*Encourage distributed energy and water systems that are more resilient to disruption.*
		Emergent:	
		Competition for public disaster relief funds and emergency services	

	Impacts	Responses
	capacity, productivity losses, supply and value chain disruption, reduced tax receipts.	*Solar-powered street lighting.* *Choose appropriate crops for flood- or wind-prone areas.* *Invest in agroforestry and tree windbreaks to improve resilience.* *Obtain crop weather insurance.*
3. Increasing drought Time spent in drought likely to increase in central and (especially) southern parts of Australia (see also Hennessy et al. 2008).	**Urban/built environment**: Reduced potable and non-potable water supplies; *water restrictions; loss of plants in gardens and greenspace.* *Increased food prices (at least for specific commodities, especially irrigated), possible impacts on electricity prices.* **Rural**: Crop and stock losses, *diverted feed stocks, soil erosion and loss of long-term productivity.* *Community decline, rural suicides, lower school attendance.* Inadequate environmental water leading to loss of habitats.	*Promote drought-tolerant native gardens and greenspaces combined with urban storm- and rain-water harvesting into urban lakes.* *Long-term incentives and regulations aimed at demand management for water.* *Promote household and commercial water efficiency features that also reduce energy use.* *Promote water use efficiency measures in horticultural and irrigated crops (e.g. polymer covers).* *Develop and plant drought-tolerant crop cultivars, especially in forecast dry years.* *Support drought preparedness and early livestock sales.* *Develop flexible prioritisation procedures to facilitate environmental water flows at critical times.*
4. Rising bushfire risk Harsher fire-weather climate in the future. Increase in 'very high fire danger' days per year of 10–30% by 2030 and of 20–100% by	**Urban/built environment**: Damage to property and assets, impeded access, risk to coal mines, *increased insurance costs.*	*Enforce building construction and maintenance standards.* *Manage gardens (especially natives) to reduce fire risks without increasing water use.*

369

Table 14.1. (*cont.*)

Climate change phenomena	Forecast change[a]	Potential direct *and indirect risks (and opportunities)*[b]	Some adaptation actions with mitigation and additional benefits[b,c]
	2050 under high emissions scenarios (see Lucas et al. 2007).	Deaths and injuries, community *trauma.* *Interrupted businesses, volunteer downtime, increasing risk of fire seasons that overlap with other places that might provide volunteer support or machinery, and reduced seasons for controlled burns.* *Damage to electricity and communications networks, long-term impacts on water catchments.* **Rural**: Loss of crops and stock, loss of rural infrastructure and machinery, *increased insurance costs.* Loss of critical habitat, *effects on endangered species, promotion of fire-dependent species, loss of soil carbon.*	Ensure escape routes are available. Control burns appropriately, to avoid losses of large areas in one go. Implement strategic fire regimes in natural areas that allow burning but manage emissions. Implement fire breaks in rural areas.
5. Rising sea level and storm surge	Sea-level rise of ~0.3 m by 2030 and 0.6–0.98 m by 2100 (under RCP8.5) with a small possibility of larger rises. Potential increase in storm intensities add to storm surge.	**Urban/built environment**: Coastal flooding and damage to or loss of property/infrastructure/ roads, access cut, home evacuations. *Business and tourism interruption, funding diverted to property protection, reduced land values, foreshore development restrictions.*	Planned retreat or protective measures for existing low-lying buildings and precincts. Proactive planning controls on new buildings and precincts within future storm-surge areas. *Construction controls and building standards for properties and infrastructure at potential future risk.*

6. Warming and acidifying ocean Near-coastal sea surface temperatures increase by 0.4–1.0 °C by 2030 and 2–5 °C by 2090. Ocean acidifies by pH 0.08 by 2030 and pH 0.3 by 2090.	Loss of beach amenity, community conflict over values. **Rural**: Loss of mangroves and marshes, loss of foredunes, reduced services for environmental protection from storm surge. **Urban/built environment**: Increased concrete and wood deterioration rates in ports and coastal infrastructure. **Rural**: Changing locations of fisheries, changed travel times for fishing fleets, possible new international tensions on fisheries exploitation. Damage and loss of Great Barrier Reef and other reefs, decline of tourism industry dependent on reefs. Marine pest species invade new areas.	*Consider shorter lifetime buildings in areas at possible future risk, with easily recoverable building materials.* Facilitate landward migration of marshes and mangroves to sustain their protective services (and biodiversity value). Increased construction standards for concrete/wood in coastal applications. *Proactive repairs and maintenance for coastal assets for water/sewer/ energy utilities.* *Reassess location and configuration of new or refurbished ports.* Diversify tourist experiences around Great Barrier Reef via land-based rainforest-related tourism and water-sport-based tourism.

---

[a] Change forecasts sourced from Climate Change in Australia website (Climate Change in Australia n.d.) and Intergovernmental Panel on Climate Change (IPCC) Working Group II Fifth Assessment report (Reisinger et al. 2014) unless otherwise cited. Note that explicit data pertain to particular locations – the website shows how these may vary across Australia, where this understanding is available.

[b] Documentation about the risks, opportunities, adaptation strategies and their costs and benefits may mostly be found somewhere on the CSIRO Climate Adaptation (CSIRO n.d.b) or AdaptNRM (CSIRO n.d.a) websites unless otherwise cited. A further good source is the National Climate Change Adaptation Research Facility publications (NCCARF n.d.).

[c] *Italicised* items in this column have the potential to be implemented in ways that provide emissions reductions benefits, though this is not automatic; all items have potential for some social or environmental benefits.

Notwithstanding the many adaptation examples with economic, social and environmental rationales (Table 14.1), there is growing recognition among decision makers and societies more broadly that the rapidly increasing rate and magnitude of change, and the fundamental changes to biophysical and ecological systems, are creating deep uncertainties and ambiguities which prevailing decision-making tools and processes (particularly economic valuation and assessments) will increasingly struggle to accommodate.[1] This will be further complicated by the need for increasingly novel responses to unprecedented changes, whose effectiveness and viability will often only be able to be determined through purposeful adaptive learning studies such as small-scale pilots. In such contexts, the term 'adaptation pathways' has emerged as a powerful concept, metaphor, analytical framework and planning tool for agencies to more confidently make decisions today that meet near-term imperatives while ensuring options are maintained or created in the future to enable continual adaptive learning in the face of deep uncertainty and intertemporal complexity (Wise et al. 2014). These approaches are reasonably new and are one of the emerging frontiers of adaptation research and development and are briefly explored at the end of this chapter.

14.2 Developing an Evidence Base for Low-Regrets, Multiple-Benefit Climate Adaptation

The objectives of the Paris Agreement[2] include enhancing adaptive capacity, strengthening resilience and reducing vulnerability to climate change, while contributing to sustainable development and ensuring an adequate adaptation response in the context of limiting warming to below 2 °C. These strategic objectives raise key research questions and practical policy questions about the relationship between climate adaptation and sustainable development. To explore this relationship, we need to investigate whether there are net benefits from early action versus delayed action, and whether climate adaptation actions have additional benefits for society, the environment and the economy. Additional benefits include benefits from disaster mitigation, greenhouse gas mitigation and the development of more adaptive approaches in the contexts of coastal planning and development, food insecurity, and other inequalities and gender imbalances (Ellis et al. 2013; Suckall et al. 2015). Research into climate adaptation has produced examples across geographic scales, domains (i.e. health, water management and coastal development) and decision-making levels, involving diverse stakeholders in government, business and community, that demonstrate potential net benefits (including multiple benefits) of early action (Table 14.1).

In this section, we describe four case studies of climate adaptation.

1. 'Rising tides: Adaptation policy alternatives for coastal residential buildings in Australia' (Wang et al. 2015);

---

[1] Organisations and institutional processes also need to be adapted to suit the context of climate change. For example, see Lonsdale (2012) for a discussion of how to build learning organisations for climate adaptation.
[2] *Paris Agreement Under the United Nations Framework Convention on Climate Change*, opened for signature 16 February 2016. Available at: https://unfccc.int/process-and-meetings/the-paris-agreement/the-paris-agreement.

2. 'Costs and coasts: An empirical assessment of physical and institutional climate adaptation pathways' (Fletcher et al. 2013);
3. 'Risk assessment of climate adaptation strategies for extreme wind events in Queensland' (Stewart and Wang 2011);
4. 'Framing the application of adaptation pathways for rural livelihoods and global change in eastern Indonesian islands' (Butler et al. 2014).

These case studies demonstrate the early benefits from adaptation to coastal inundation, the importance of considering the distribution of costs and benefits across communities, the low-regrets nature of some early adaptation actions and the synergies between adaptation actions and sustainable development. These case studies are representative of a growing number of analyses that demonstrate the benefits of early action.

14.2.1 *'Rising Tides: Adaptation Policy Alternatives for Coastal Residential Buildings in Australia'*

Wang et al. (2015) found that adaptation action for coastal inundation can be a no-regrets strategy and demonstrated higher net benefits from early action, but noted that difficult prioritisation decisions will be needed given the limited resources available for adaptation. Wang et al. (2015) compared several alternative national policy stances for adapting coastal residential buildings to storm surges and rising sea levels, including protection via the construction of seawalls, accommodation of climate change impacts by raising floor heights and avoiding impacts by limiting new developments in hazardous areas. These adaptation strategies were compared for the increase in sea levels associated with urban development scenarios (business-as-usual, urban consolidation and inland regional development) and alternative policy stances (baseline, static, reactive and anticipatory).

A protection strategy was modelled using the construction of a seawall as a proxy. The construction costs of building seawalls were estimated based on a minimum fixed cost for any seawall of AUD 1000 per metre, plus a flexible cost component that varies with the square of the height of the wall. This flexible cost was estimated based on a cost of AUD 3000 per metre for building a 6-m-high seawall (Gold Coast City Council 2014). An accommodation strategy was modelled by the implementation of new building standards at a cost of adding 0.2% to the per square-metre cost of a building for every 10-cm increase in floor height, for any new buildings constructed in areas exposed to 100-year events. An avoidance strategy was modelled as avoiding building in areas currently and projected to be exposed to natural hazards under climate change, and the capital costs of this strategy were assumed to be zero.

The business-as-usual projection for population and urban development classified statistical local areas (SLAs) into deciles, distributed projected population increases to SLAs pro rata until they reached the density of the next highest decile, and then redistributed remaining population increases proportionately to unfilled deciles. The projection for urban consolidation distributed population increases to SLAs pro rata until they reached a density of 1000 people per hectare. The inland regional development projection modelled a linear increase in the populations of Toowoomba, Launceston, Bendigo, Ballarat, Albury-Wodonga, Shepparton, Wagga Wagga, Tamworth, Dubbo, Bathurst, Orange, Kalgoorlie and Lismore to accommodate 10 million people in 2100, with other projected population increases distributed according to the business-as-usual projection.

Four different policy stances were modelled:

1. A static policy stance required existing assets to be rebuilt to the same standard before any damage occurred and new buildings constructed based on the building standards and climate information that were available in 2006.
2. A reference stance required new buildings to comply with current building regulations, with hazard levels based on the historical climate information available in 2006 informing the building standards for new buildings.
3. A reactive policy stance specifies that no preventative action is taken until an extreme event occurs but that adaptation actions are based on updated climate information to that time, for example, the height of a seawall built in 2030 would be determined to avoid a 1-in-100-year storm tide using the climate data available in 2030 to define a 1-in-100-year event.
4. An anticipatory policy stance includes adaptation investment based on the design standard expected in another 30 years (up to 2100), based on the A1B climate scenario used by the IPCC (Nakićenović et al. 2000): for example, the height of a seawall built in 2030 is determined to avoid a 1-in-100-year storm tide using the A1B scenario to define a 1-in-100-year event in 2060.

The combination of an anticipatory policy stance and adaptation strategies to protect against storm tide surges as outlined above reduced expected damages to residential housing to around AUD 200 million, with a net benefit of around AUD 4 billion up to 2100 (net present value, AUD 2006, 2.6% discount rate) in comparison with current building standards based on historical climate information. Wang et al. (2015) found that the distribution of damages from storm tide surges is unevenly distributed across Australia, with over 80% of damages expected in Queensland.

While seawalls were chosen to represent a protective strategy, Wang et al. (2015) noted that other protection measures might be preferred in some areas to avoid the loss of amenity and ecological damage associated with seawalls. Seawalls were used to illustrate coastal protection in order to limit the complexity of the modelling, whereas in practice alternatives such as beach nourishment, groynes, artificial reefs, wetlands, living breakwaters or replanting mangroves could be even cheaper and have higher net benefits.

14.2.2 'Costs and Coasts: An Empirical Assessment of Physical and Institutional Climate Adaptation Pathways'

Fletcher et al. (2013) analysed the distribution of costs and benefits for alternative adaptation strategies for responding to the risk of coastal inundation. Using a financial cost–benefit analysis they considered the affordability of adaptation and the consequences for property values but did not include sources of non-market values. This means the results

Table 14.2. *Settlement case study statistics*

Case study	1	2	3	4	5	6
Category	Suburb	Hamlet	Hamlet	Central business district	Suburb	Canal estate
Residential properties modelled	560	312	122	575*	489	2620
Median property value (AUD k)	181	241	169	215	267	290
Median infrastructure value (AUD k)	89	125	88	114	136	141
Subregion population	3900	1800	500	9900	3000	4700
Median household income (AUD k/year)	44	35	38	52	67	52

* 1346 commercial properties were also modelled for case study 4. *Source*: Fletcher et al. (2013).

are indicative but should not be used to directly inform decision-making without also adding other kinds of costs and benefits to the analysis. What this analysis does show, however, is that adaptation actions that can have benefits that exceed costs for a whole community may not be affordable based on the budget available for adaptation. This means that the distribution of costs and benefits across a community is important for assessing the equitability of alternative adaptation strategies. For instance, in some cases a very small number of households might receive a large benefit from a community-level adaptation while most households receive little or no benefit.

Case studies were selected from the residential sector in six Australian coastal communities within Moreton Bay, the Sunshine Coast and Cairns (Table 14.2).

The strategies considered included protection measures such as building seawalls, redesigning infrastructure to accommodate inundation, for example, by raising floor heights, and retreating out of the areas likely to be inundated via the purchase of the properties at risk. The costs of the protection and accommodation strategies were estimated from the literature, with the cost of building a seawall capable of withstanding a 1-in-100-year event in 2050 estimated at AUD 2500 per metre and the cost of raising houses estimated at approximately AUD 40 000 per house. The purchase price of individual properties for the retreat strategy were estimated using a hedonic pricing model. To estimate the costs of inundation, Fletcher et al. (2013) simulated storm surges from 2010 to 2100 by considering damage to buildings estimated using damage curves (Middelmann-Fernandes 2010) and the devaluation of land in the residential sector estimated using hedonic pricing analysis (Rambaldi et al. 2012). Alternative adaptation strategies were simulated and compared the distribution of the costs of implementation with the distribution of the benefits in the form of avoided damages.

Even if benefits exceed costs in the long term, communities may not have the financial capacity to invest in the adaptation measures mentioned above. Accordingly, estimates of costs and benefits were combined with estimates of financial capacity to assess the affordability of adaptation. Table 14.3 shows the adaptation budget, costs and benefits of adaptation for each case study. It shows whether mean household benefits exceed mean

Table 14.3. *The net benefits and affordability of adaptation*

Case study	1	3	4	5	6
Budget (AUD k/property)	34	30	40	52	40
Seawalls					
Cost (AUD k/property)	37	34	6	9	26
Benefit (AUD k/property)	1155	382	155	39	27
Benefit/cost	31.15	11.22	25.73	4.09	1.03
Budget/cost	0.92	0.87	6.65	5.64	1.52
Economic? (Benefit/cost>1)	Yes	Yes	Yes	Yes	Yes
Affordable? (Budget/cost>1)	No	No	Yes	Yes	Yes
Floor heights					
Cost (AUD k/property)	14	7	2	1	6
Benefit (AUD k/property)	736	259	3	1	4
Benefit/cost	53.15	35.61	1.85	0.65	0.58
Budget/cost	2.47	4.08	25.98	40.12	6.61
Economic? (Benefit/cost>1)	Yes	Yes	Yes	No	No
Affordable? (Budget/cost>1)	Yes	Yes	Yes	Yes	Yes
Retreat					
Cost (AUD k/property)	170	59	83	56	209
Benefit (AUD k/property)	1104	378	148	39	28
Benefit/cost	6.47	6.43	1.78	0.71	0.13
Budget/cost	0.20	0.50	0.48	0.92	0.19
Economic? (Benefit/cost>1)	Yes	Yes	Yes	No	No
Affordable? (Budget/cost>1)	No	No	No	No	No

Source: Fletcher et al. (2013).

household costs for each adaptation strategy and assesses the affordability of each strategy based on whether mean household adaptation budgets exceed mean household costs. The adaptation budget per property was estimated based on a contribution of 1.92% of median annual household income per year for 40 years, indexed by a discount rate to 2050, and the adaptation cost per property was estimated based on the length of seawall, the number of properties and median household income.

The affordability of the adaptation strategies differs greatly across the case studies, largely because adaptation costs are very variable, and are affected by their exposure to the ocean, density and household income. For instance, Case Study 2 turned out to receive little benefit (in the form of avoided damage) from any of the three adaptation options evaluated because it is naturally protected from storm surges (it is therefore not included in Table 14.3). Some larger communities like Case Studies 4 and 5 have a relatively low exposure to the ocean and therefore lower costs per property for adaptation, whereas others, like Case Study 6, with a complex and extended exposure to storm surges, can have high adaptation costs per property and therefore relatively low affordability despite having many high-income households. Communities like Case Study 3 have low affordability because of high costs per property, few contributing households and low median household income.

A significant proportion of households in Case Studies 1, 3 and 4 are at risk from a 1-in-100-year event by 2100, and because benefits are estimated in the form of avoided damage, benefits exceed costs for all three adaptation strategies for these case studies. Raising floor heights is the cheapest strategy for all case studies, but as with Case Studies 5 and 6, if most costs are due to land devaluation then the costs would exceed the expected benefits. Building a seawall is expected to have net benefits by 2100 for all case studies except Case Study 2, even when only a small proportion of the community is at risk. Accordingly, we might conclude that building seawalls is justified across all case studies; however, this analysis is based on a financial cost–benefit analysis that does not consider non-market values such as the negative impacts of seawalls on the benefits people enjoy from beaches and the negative effects that seawalls have by displacing the force of waves to neighbouring properties and beaches. To directly inform decision-making, a social cost–benefit analysis needs to consider all costs and benefits. Still, Fletcher et al.'s (2013) financial cost–benefit analysis can be used to demonstrate how the equitability of community-level adaptations such as seawalls differs across the communities within the case studies.

Fletcher et al. (2013) found that for many case studies most people would not receive a net benefit by contributing to a community-level adaptation such as a seawall. The median benefit–cost ratio of adaptation is always lower than the corresponding mean benefit–cost ratio, indicating that in all cases a few households receive a disproportionate benefit from a seawall. For example, in Case Studies 3, 5 and 6, more than half of households receive no benefit from a seawall. In Case Studies 5 and 6, most households receive little or no benefit from a seawall, while a very small number of households receive a very large benefit. Fletcher et al. (2013) classified coastal settlements based on physical risk factors including terrain and building location, the distribution of risk throughout a community and the socio-economic factors that determine the capacity for adaptation. Table 14.4 presents the results of this classification as a framework that can help communities think further about the kinds of adaptation they should investigate, based on the financial costs and benefits, affordability and equity of alternative adaptation strategies.

Table 14.4. *A classification of coastal communities and adaptation strategies*

Economic	Equitable	Affordable	Case study	Action
No	–	–	2	Do nothing
Yes	No	No	3	Retreat/household adaptation, e.g. raised floor heights
		Yes	5, 6	Household adaptation, e.g. raised floor heights
Yes	Yes	No	1	Funding from larger-scale government for community engineering, e.g. seawall
		Yes	4	Local council to fund community engineering, e.g. seawall

Source: Fletcher et al. (2013).

14.2.3 'Risk Assessment of Climate Adaptation Strategies for Extreme Wind Events in Queensland'

Stewart and Wang (2011) investigated adaptation strategies for mitigating the impact of extreme wind events such as tropical cyclones and severe storms on four coastal and urban communities in Queensland; specifically, Cairns, Townsville, Rockhampton and Brisbane. With current design standards based on the assumption of a static climate, the higher wind speeds projected for the region (and increasingly already experienced) under climate change will lead to increased damage to residential buildings in these cities. This vulnerability could be reduced by updating building standards to cope with higher wind speeds (i.e. Australian Standards AS4055–2006 and AS1170.2-2011) (Adaptation Strategy 1); retrofitting pre-1980 buildings to current standards (Adaptation Strategy 2); or repairing pre-1980 wind-damaged houses to current standards (Adaptation Strategy 3). Adaptation Strategy 1 involves changing the Australian Standard, 'Wind Loads for Houses', so that new constructions and alterations are designed to resist 50% higher wind pressures.

These three adaptation strategies were assessed in terms of the avoided damage for climate scenarios ranging from 'no change to moderate change' (25% reduction in cyclone frequency, 10% increase in wind speeds) to 'significant change' (no change in cyclone frequency, 20% increase in wind speeds), and a shift in the distribution of cyclones towards Brisbane by 2100. Spatial and temporal stochastic[3] simulation methods were used to take into account wind field characteristics, uncertainty in the vulnerability of houses to wind, costs of adaptation, timing of adaptation, discount rates, future growth in new housing and increases in wind speeds over time. The results are summarised in Tables 14.5 and 14.6.

Without any changes to building standards, there is a high likelihood of costly increases to wind damage from climate change, with a 90% chance that the increase in total damages from the moderate and significant wind change scenarios would exceed AUD 1.6 billion and AUD 5.9 billion by 2100, respectively. Whereas Adaptation Strategies 2 and 3 often give net losses, Adaptation Strategy 1, with new constructions and alterations designed to resist 50% higher wind pressures, can significantly reduce wind vulnerability. Wind losses up to 2100 could be reduced by up to AUD 10.5 billion for a cost of AUD 2.2 billion. The results indicate a net present value for Brisbane about 10 times higher than for the other three cities combined, and, given their populations, if urban areas in the Sunshine Coast and Gold Coast were included in the analysis, the benefits for South East Queensland alone could be as high as AUD 15 billion by 2100.

Importantly, even without any climate change, increasing building standards can still be cost-effective. There is a 97.6% likelihood that Adaptation Strategy 1 would result in a mean net present value of AUD 202.7 million for Brisbane foreshore locations. Stewart and Wang (2011) noted that this strategy makes sense as a no-regrets policy. Further, there is evidence of benefits from early adaptation action, since net benefits in the form of avoided damage accumulate over time. At that time, it was found that delaying action until 2020 or 2030 would have a lower net benefit compared to immediate implementation.

---

[3] Stochastic means having a random probability distribution or pattern that may be analysed statistically but may not be predicted precisely.

Table 14.5. *Summary of net benefit from implementing adaptation options for foreshore houses in Brisbane, Cairns and Townsville, for moderate climate change (there are no foreshore houses as defined in Rockhampton). Adaptations 2 and 3 are considered only for pre-1980 houses in cyclonic areas*

Adaptation action applied to foreshore housing		Adaptation 1		Adaptation 2		Adaptation 3	
		Average net benefit (millions)	Pr(net benefit > 0)*	Average net benefit (millions)	Pr(net benefit > 0)*	Average net benefit (millions)	Pr(net benefit > 0)*
Brisbane	2030	38	58.9%	–	–	–	–
	2050	142	90.0%	–	–	–	–
	2070	240	98.0%	–	–	–	–
	2100	340	100%	–	–	–	–
Cairns	2030	–9	1.0%	–28	10.2%	–0.4	0.4%
	2050	–13	1.7%	–19	16.0%	–0.2	1.3%
	2070	–14	2.1%	–15	19.0%	0.1	2.8%
	2100	–15	2.5%	–13	20.3%	0.4	5.7%
Townsville	2030	–4	12.9%	–4	16.0%	–0.2	0.4%
	2050	–5	24.5%	0	28.3%	–0.1	1.3%
	2070	–6	34.3%	2	33.3%	0.1	2.8%
	2100	–6	44.8%	2	25.8%	0.2	5.7%

* The probability the net benefit is greater than zero. This reflects the uncertainties in the assessment.
Source: Stewart and Wang (2011).

14.2.4 Framing the Application of Adaptation Pathways for Rural Livelihoods and Global Change in Eastern Indonesian Islands

Butler et al. (2014) consider whether low-regret, multiple-benefit strategies can be integrated with human development by applying an adaptation pathways approach.

Evidence from interview and focus groups conducted in the Nusa Tenggara Barat Province in eastern Indonesia, in 2012, highlighted that the vulnerability of poor rural communities is influenced by 20 interacting drivers of change, including climate variability and change, local unemployment, inefficient development investment, fuel and energy prices, and land, water and food availability. For example, in the Nusa Tenggara Barat Province, by 2100, rising sea temperatures may result in more extreme weather events and increase sea level by 1 m, and by 2080, rainfall will be concentrated into fewer events, while population could grow from 4.5 million in 2010 by 41% to 6.37 million by 2050. Additionally, the surveys revealed local contexts characterised by high levels of poverty and inequality that are resilient to change due to corruption, traditional norms and practices, inefficient development investment, power imbalances and fatalism. These findings indicated that climate adaptation cannot be separated from the cultural, political, economic, environmental and development contexts and that, to reduce poverty, climate variability and change must be considered in conjunction with the other interacting drivers.

Butler et al. (2014) then demonstrated through participatory action research how the collective identification of low-regret, multiple-benefit adaptation strategies is crucial for

Table 14.6. *Summary of net benefit from implementing adaptation options for all houses in Brisbane, Cairns, Townsville and Rockhampton, for moderate climate change. Adaptations 2 and 3 are considered only for pre-1980 houses in cyclonic areas*

Adaptation action applied to all housing		Adaptation 1		Adaptation 2		Adaptation 3	
		Average net benefit (millions)	Pr(net benefit > 0)*	Average net benefit (millions)	Pr(net benefit > 0)*	Average net benefit (millions)	Pr(net benefit > 0)*
Brisbane	2030	−299	21.0%	−	−	−	−
	2050	413	43.3%	−	−	−	−
	2070	1228	61.2%	−	−	−	−
	2100	2149	79.1%	−	−	−	−
Cairns	2030	−40	11.8%	−198	1.6%	−1.7	0.4%
	2050	−22	22.2%	−236	2.2%	−0.4	1.4%
	2070	8	31.4%	−226	2.6%	1.7	3.1%
	2100	42	41.0%	−221	2.8%	4.5	6.7%
Townsville	2030	−42	12.9%	−173	1.0%	−1.8	0.4%
	2050	−13	24.5%	−238	1.2%	−0.4	1.4%
	2070	25	34.3%	−229	1.6%	2.1	3.1%
	2100	69	44.8%	−224	1.7%	5.6	6.7%
Rockhampton	2030	−14	17.1%	−167	0.5%	−0.8	0.2%
	2050	11	32.8%	−172	1.2%	0.6	1.3%
	2070	38	45.8%	−167	1.4%	2.6	2.5%
	2100	69	60.0%	−164	1.5%	5.4	5.8%

* The probability the net benefit is greater than zero. This reflects the uncertainties in the assessment.
Source: Stewart and Wang (2011).

these communities to be able to deal adaptively with the inevitable yet uncertain declines in availability of land, water and food. These are caused by the interacting changes in climate, population and ecosystem integrity along with the contested and ineffective social, economic and political processes and corruption that contributed to, and were exacerbated by, these scarcities. Butler et al. (2014) do this by showing how: (1) low-regret measures are relatively low cost and provide relatively large benefits under the range of possible projected futures (i.e. by considering the types of interactions that characterise such complex systems and ensuring available options are diverse and flexible); and (2) options that provide or demonstrate multiple additional benefits can enhance the acceptance of such measures where there are many competing stakeholder interests or contested perspectives. Further to these beneficial outcomes of adopting low-regrets, multiple-benefits approaches, Butler et al. (2014) also demonstrate how the adoption of a pathways approach – involving the adoption of livelihoods analysis, participatory scenario planning and experimental adaptive co-management – can also help build the capabilities and capacity of stakeholders to understand and address proximate and systemic causes of vulnerability and contested decision-making (see also Butler et al. 2015, 2016 for further evidence).

While the Australian case studies presented in this chapter provide evidence of examples where climate adaptation options can be identified that are low-regrets or have multiple benefits, the example of Nusa Tenggara Barat Province in eastern Indonesia demonstrates the necessity of tackling climate adaptation in conjunction with other sources of vulnerability and by being conscious of broader development goals. This is an argument in favour of mainstreaming climate adaptation into development planning and risk management. Synergies exist between approaches to adaptation and sustainable development, and although there may be challenging trade-offs that need to be managed, if synergies are to be identified and realised, then we need to consider early and integrated approaches to climate adaptation.

14.3 Adaptation Pathways

Markets and decision-making processes involving strategy development, planning and policy choices around economic development and conservation all rely upon rules. However, these rules have not evolved to accommodate the complex and systemic risks associated with climate change. This means that the benefits of acting early and the multiple benefits of climate-compatible development will remain unrealised if adaptation challenges are ignored and development and planning focuses on short-term considerations. For example, this pattern of risk governance – ignoring the systemic and deeply uncertain and ambiguous nature of climate and disaster risks – is likely to result in coastal communities that lock themselves into increasingly costly futures by continually building defences, repairing damaged infrastructure and losing beaches, dunes and estuaries as a result of a closed-system, reductionist and probability-based approach to risk assessment and management. In contrast, planning for climate impacts such as sea-level rise, increasingly severe storms, flooding and inundation requires the coordinated and systematic responses of many stakeholders across jurisdictions and levels of government to ensure that essential services are provided in the short term while maintaining options for the future. Existing decision-making processes and tools and the training, skills and capabilities of those responsible for assessing and managing risks are all limited for dealing with these challenges.

The adaptation pathways concept provides an umbrella conceptual and analytical framework for several approaches and related developments (such as participatory co-design, visioning, scenario planning, values analysis, systemic thinking, real options analysis and adaptive learning) that have sought to integrate climate adaptation into broader planning processes. Case studies such as those presented above stimulated research efforts in Australia, and around the world, to increase efforts into better understanding the more systemic drivers of vulnerability and how adaptation can more effectively (proactively and strategically) address these. In particular, efforts have been shifting into developing the ideas, concepts, models, processes and frameworks that promote different ways of thinking (i.e. diagnosis and framing of adaptation) and decision-making (i.e. planning and implementation). This is important because institutional and cultural changes are required to adapt effectively to changes in system dynamics that are novel, uncertain, span temporal and spatial scales and cross jurisdictional boundaries.

These shifts in research and development have coincided with, or responded to, the growing calls from stakeholders around the world for new approaches that can support and enable them to more holistically and adaptively accommodate the deep uncertainties in knowledge, the highly ambiguous nature of the goals and the high levels of distribution in authority and power (i.e. across jurisdictional boundaries and economic sectors) associated with their climate adaptation planning and decision-making. These experiences stimulated CSIRO's Climate Adaptation Flagship to increase efforts into better understanding these institutional and values dimensions to the challenges of adaptation, particularly in contexts of deep uncertainty and intertemporal complexity (e.g. Wise et al. 2014). Or, possibly more accurately, additional efforts were shifted into developing the different ways of thinking, sense-making and decision-making that more holistically and adaptively consider and account for the interdependencies between knowledge, societal values and systems of formal and informal rules (Gorddard et al. 2016). Efforts have principally drawn upon thinking and approaches developed in cognate disciplines and domains of application spanning disaster risk reduction, sustainable development, integrated coastal zone management, catchment and water resource management, food security, governance of epidemics and the energy–climate–food–water nexus.

The disciplinary and theoretical perspectives, methodologies and methods for doing this are many and diverse, drawing primarily on theoretical developments in connected areas around decision-making under uncertainty (Lempert et al. 2003; Ben-Haim 2006; Lempert et al. 2006; Stern 2006, 2013; Weitzman 2011; Jeuland and Whittington 2013), dynamic sustainabilities (Leach et al. 2010), pathways for resilience, adaptation and transformation, for innovation, adaptive management and adaptive governance (Pahl-Wostl 2009; Jacobsson and Bergek 2011; Ison et al. 2013), social psychology and behavioural economics, and theories of societal change from evolutionary, development and institutional economics, adaptive governance, articulation theory, path creation and activity fields (Ostrom 1990; North 1992, 1994; Ostrom et al. 1994; Smith et al. 2005; Newig et al. 2007; Voß et al. 2007; Scott 2008; Dopfer and Potts 2009; Kingston and Caballero 2009; Potts et al. 2010; Ostrom 2011; Downing 2012; Nelson 2012; Juma 2014). These diverse perspectives have contributed to the development of theoretical and practical frameworks that can be used to support adaptation efforts.

Table 14.7 lists some of the key conceptual frameworks used to support climate adaptation decision-making to support decision-making in contexts of high uncertainty in knowledge, highly distributed power (many decision makers across jurisdictional boundaries) and where goals are contested or ambiguous or to promote adaptation as change in the decision context.

For practical guidance on applying adaptation pathways approaches we recommend reading further about some of the lessons learned from recent experiences applying such approaches, including Park et al. (2012), Bloemen et al. (2017), Bosomworth et al. (2017), Ramm et al. (2018), Stephens et al. (2018) and Haasnoot et al. (2019). Prominent examples of the adaptation pathways approach being applied in Australasia include the National Coastal Adaptation Decision Pathways programme (Lin et al. 2017) and the City of Melbourne 2017 Climate Change Adaptation Strategy (City of Melbourne 2017).

Table 14.7. *Key conceptual frameworks used to support climate adaptation decision-making*

Conceptual framework	Further reading
Pathways as a metaphor for promoting flexibility in planning	Yohe and Leichenko (2010); Stafford Smith et al. (2011); Barnett et al. (2013)
'Resilience, adaptation, transformation pathways'	Pelling (2011)
'Structured learning approaches' to integrating top-down and bottom-up planning	Butler et al. (2015)
'Decision-focused adaptation pathways'	Reeder and Ranger (2011); Moss and Martin (2012); Siebentritt et al. (2014); Bosomworth et al. (2015); Young and Hall (2015)
'Dynamic adaptive policy pathways'	Haasnoot et al. (2013); Kwakkel et al. (2015)
'Adaptation pathways for addressing systemic drivers of vulnerability'	Downing (2012); Butler et al. (2014); Maru et al. (2014); Rosenzweig and Solecki (2014); Wise et al. (2014); Fazey et al. (2015)
'Adaptation services' as recognising values not previously included	Lavorel et al. (2015)
'Values, rules, knowledge' model – a decision context-focused approach to adaptation	Gorddard et al. (2016)
Diagnostic model of operating environments	Leith et al. (2014)

14.4 Conclusion

Adaptation research has provided several case studies that demonstrate the benefits of considering the challenges of climate adaptation early, in order to ensure the best chance of realising the benefits of early action and enabling adaptation that keeps options open. This research has also generated the capabilities and capacity, such as data, tools, models and processes, that enable the kinds of research that is embodied in these case studies to be replicated, and has also revealed general principles and lessons that can help inform adaptation processes. Regulatory and institutional barriers continue to present a challenge to effective climate adaptation. This is especially true given the uneven distribution of costs and benefits and the limited availability of risk management processes that can deal with the uncertainty and intertemporal complexity of adaptation challenges. Adaptation pathways approaches, as illustrated in this chapter and in the following sectoral chapters, provide one way forward, by helping stakeholders integrate adaptation into broader planning approaches.

References

Australian Department of Home Affairs (2018). *Profiling Australia's Vulnerability: The Interconnected Causes and Cascading Effects of Systemic Disaster Risk*. Australia: Commonwealth Government of Australia. Available at: www.aidr.org.au/media/6682/national-resilience-taskforce-profiling-australias-vulnerability.pdf.

Barnett, J., Waters, E., Pendergast, S. and Puleston, A. (2013). *Barriers to Adaptation to Sea-Level Rise*. Final report. Gold Coast, Australia: National Climate Change Adaptation Research Facility. Available at: https://apo.org.au/sites/default/files/resource-files/2013-05/apo-nid33956.pdf.

Ben-Haim, Y. (2006). *Info-Gap Decision Theory: Decisions under Severe Uncertainty*, 2nd ed. London: Academic Press.

Bloemen, P., Reeder, T., Zevenbergen, C., Rijke, J. and Kingsborough. A. (2017). Lessons learned from applying adaptation pathways in flood risk management and challenges for the further development of this approach. *Mitigation and Adaptation Strategies for Global Change*, 23, 1083–1108; online 2017.

Bosomworth, K., Harwood, A., Leith, P. and Wallis, P. (2015). *Adaptation Pathways: A Playbook for Developing Robust Options for Climate Change Adaptation in Natural Resource Management*. Southern Slopes Climate Change Adaptation Research Partnership (SCARP). Hobart: RMIT University, University of Tasmania and Monash University.

Bosomworth, K., Leith, P., Harwood, A. and Wallis, P. J. (2017). What's the problem in adaptation pathways planning? The potential of a diagnostic problem-structuring approach. *Environmental Science & Policy*, 76, 23–28.

Butler, J. R. A., Suadnya, W., Puspadi, K. et al. (2014). Framing the application of adaptation pathways for rural livelihoods and global change in eastern Indonesian islands. *Global Environmental Change*, 28, 368–382.

Butler, J. R. A., Wise, R. M., Skewes, T. D. et al. (2015). Integrating top-down and bottom-up adaptation planning to build adaptive capacity: A structured learning approach. *Coastal Management*, 43, 346–364.

Butler, J. R. A., Suadnya, W., Yanuartati, Y. et al. (2016). Priming adaptation pathways through adaptive co-management: Design and evaluation for developing countries. *Climate Risk Management*, 12, 1–16.

City of Melbourne (2017). *Climate Adaptation Strategy Refresh 2017*. City of Melbourne. Available at: www.melbourne.vic.gov.au/sitecollectiondocuments/climate-change-adaptation-strategy-refresh-2017.pdf.

Climate Change in Australia (n.d.). *Climate Change in Australia*. Available at: www.climatechangeinaustralia.gov.au.

CSIRO (Commonwealth Scientific and Industrial Research Organisation) (n.d.a). *AdaptNRM*. Available at: https://adaptnrm.csiro.au/.

CSIRO (n.d.b). *Climate Adaptation*. Available at: https://research.csiro.au/climate.

Dopfer, K. and Potts, J. (2009). On the theory of economic evolution. *Evolutionary and Institutional Economics Review*, 6, 23–44.

Downing, T. E. (2012). Views of the frontiers in climate change adaptation economics. *WIREs Climate Change*, 3, 161–170.

Ellis, K., Cambray, A. and Lemma, A. (2013). *Drivers and Challenges for Climate Compatible Development*. Climate and Development Knowledge Network. Available at: https://cdkn.org/resource/drivers-and-challenges-for-climate-compatible-development/?loclang=en_gb.

Fazey, I., Wise, R. M., Lyon, C., Câmpeanu, C., Moug, P. and Davies, T. E. (2015). Past and future adaptation pathways. *Climate and Development*, 8, 1–19.

Fletcher, C. S., Taylor, B. M., Rambaldi, A. N. et al. (2013). *Costs and Coasts: An Empirical Assessment of Physical and Institutional Climate Adaptation Pathways*. Final report. Gold Coast, Australia: National Climate Change Adaptation Research Facility. Available at: https://nccarf.edu.au/costs-and-coasts-empirical-assessment-physical-and-institutional-climate-adaptation/.

Global Commission on Adaptation (2019). *Adapt Now: A Global Call for Leadership on Climate Resilience. Global Commission on Adaptation*. Global Centre on Adaptation. Available at: https://gca.org/wp-content/uploads/2019/09/GlobalCommission_ Report_FINAL.pdf.

Gold Coast City Council (2014). *The A-line Seawall*. Available at: www.goldcoast.qld.gov .au/thegoldcoast/gold-coast-seawalls-40670.html.

Gorddard, R., Colloff, M., Wise, R. M., Ware, D. and Dunlop, M. (2016). Values rules and knowledge: Adaptation as change in the decision context. *Environmental Science & Policy*, 57, 60–69. Available at: www.sciencedirect.com/science/article/pii/ S1462901115301210.

Haasnoot, M., Kwakkel, J. H., Walker, W. E. and ter Maat, J. (2013). Dynamic adaptive policy pathways: A method for crafting robust decisions for a deeply uncertain world. *Global Environmental Change*, 23, 485–498.

Haasnoot, M., Brown, S., Scussolini, P., Jimenez, J. A., Vafeidis, A. T. and Nicholls, R. J. (2019). Generic adaptation pathways for coastal archetypes under uncertain sea-level rise. *Environmental Research Communications*, 1, 071006.

Hallegatte, S., Shar, A., Lempert, R., Brown, C. and Gill, S. (2012). *Investment Decision Making under Deep Uncertainty: Application to Climate Change*. Policy Research Working Paper 6193. The World Bank. doi.org/10.1596/1813-9450-6193.

Hennessy, K., Fawcett, R., Kirono, D. et al. (2008). *An Assessment of the Impact of Climate Change on the Nature and Frequency of Exceptional Climatic Events: Report to Australian Government*. Canberra: Bureau of Meteorology and Commonwealth Scientific and Industrial Research Organisation (CSIRO).

Ison, R., Blackmore, C. and Iaquinto, B. L. (2013). Towards systemic and adaptive governance: Exploring the revealing and concealing aspects of contemporary social-learning metaphors. *Ecological Economics*, 87, 34–42.

Jacobsson, S. and Bergek, A. (2011). Innovation system analyses and sustainability transitions: Contributions and suggestions for research. *Environmental Innovation and Societal Transitions*, 1, 41–57.

Jeuland, M. and Whittington, D. (2013). *Water Resources Planning under Climate Change: A 'Real Options' Application to Investment Planning in the Blue Nile*. Discussion Paper Series EfD DP 13-05. Environment for Development Resources for the Future. Available at: www.jstor.org/stable/resrep14977?seq=1#metadata_info_tab_contents.

Juma, C. (2014). Complexity, innovation, and development: Schumpeter revisited. *Policy and Complex Systems*, 1, 4–21.

Kingston, C. and Caballero, G. (2009). Comparing theories of institutional change. *Journal of Institutional Economics*, 5, 151–180.

Kwakkel, J. H., Haasnoot, M. and Walker, W. E. (2015). Developing dynamic adaptive policy pathways: A computer-assisted approach for developing adaptive strategies for a deeply uncertain world. *Climatic Change*, 132, 373–386.

Lavorel, S., McIntyre, S., Colloff, M. et al. (2015). Ecological mechanisms underpinning climate adaptation services. *Global Change Biology*, 21, 12–31.

Leach, M., Scoones, I. and Stirling, A. (2010). Governing epidemics in an age of complexity: Narratives, politics and pathways to sustainability. *Global Environmental Change*, 20, 369–377.

Leith, P., O'Toole, K., Haward, M., Coffey, B., Rees, C. and Ogier, E. (2014). Analysis of operating environments: A diagnostic model for linking science, society and policy for sustainability. *Environmental Science & Policy*, 39, 162–171.

Lempert, R. J. (2019). Robust decision making (RDM). In V. A. W. J. Marchau, W. E. Walker, P. J. T. M. Bloemen and S. W. Popper, eds., *Decision Making under Deep*

Uncertainty: From Theory to Practice. Cham: Springer International Publishing, pp. 23–51.

Lempert, R. J., Popper, S. W. and Bankes, S. C. (2003). *Shaping the Next One Hundred Years: New Methods for Quantitative, Long-Term Policy Analysis*. Report prepared for the RAND Pardee Centre, Santa Monica. Available at: www.rand.org/pubs/monograph_reports/2007/MR1626.pdf.

Lempert, R. J., Groves, D. G., Popper, S. W. and Bankes, S. C. (2006). A general, analytic method for generating robust strategies and narrative scenarios. *Management Science*, 52, 514–528.

Leslie, L. M., Leplastrier, M. and Buckley, B. W. (2008). Estimating future trends in severe hailstorms over the Sydney Basin: A climate modelling study. *Atmospheric Research*, 87, 37–51.

Lin, B. B., Capon, T., Langston, A. et al. (2017). Adaptation pathways in coastal case studies: Lessons learned and future directions. *Coastal Management*, 45, 384–405.

Lloyd's of London (2018). *Innovative Finance for Resilient Infrastructure: Innovation Report 2018 – Understanding Risk*. Lloyd's of London and the Centre for Global Disaster Protection. Available at: www.lloyds.com/news-and-risk-insight/risk-reports/library/understanding-risk/innovative-finance-for-resilient-infrastructure.

Lonsdale, K. (2012). *Beyond Tools: Building Learning Organisations to Adapt to a Changing Climate*. Final report, Victorian Centre for Climate Change Adaptation Research Visiting Fellowship. Melbourne: Victorian Centre for Climate Change Adaptation Research. Available at: www.vcccar.org.au/sites/default/files/publications/VCCCAR_Final_Report_Kate_Lonsdale_forweb_150713.pdf.

Lucas, C., Hennessy, K., Mills, G. and Bathols, J. (2007). *Bushfire Weather in Southeast Australia: Recent Trends and Projected Climate Change Impacts*. Consultancy report prepared for the Climate Institute of Australia. Melbourne: Bushfire Cooperative Research Centre. Available at: https://publications.csiro.au/rpr/download?pid=proci te:5910842c-f62e-4006-b88f-1055d8e981fa&dsid=DS1.

Maru, Y. T., Stafford Smith, M., Sparrow, A., Pinho, P. F. and Dube, O. P. (2014). A linked vulnerability and resilience framework for adaptation pathways in remote disadvantaged communities. *Global Environmental Change*, 28, 337–350.

Middelmann-Fernandes, M. (2010). Flood damage estimation beyond stage–damage functions: An Australian example. *Journal of Flood Risk Management*, 3, 88–96.

Moss, A. and Martin, S. (2012). *Flexible Adaptation Pathways*. Policy brief. Scotland: ClimateXChange. Available at: www.climatexchange.org.uk/media/1595/flexible_adaptation_pathways.pdf.

Nakićenović, N., Davidson, O., Davis, G. et al. (2000). Summary for policymakers. In N. Nakićenović, O. Davidson, G. Davis et al., *Emissions Scenarios: A Special Report of Working Group III of the Intergovernmental Panel on Climate Change*. Cambridge: Cambridge University Press, pp. 1–20. Available at: www.ipcc.ch/site/assets/uploads/2018/03/emissions_scenarios-1.pdf.

NCCARF (National Climate Change Adaptation Research Facility) (n.d.). NCCARF Publications. *NCCARF*. Available at: https://nccarf.edu.au/nccarf-publications/.

Nelson, R. (2012). Why Schumpeter has had so little influence on today's main line economics, and why this may be changing. *Journal of Evolutionary Economics*, 22, 901–916.

Newig, J., Voß, J.-P. and Monstadt, J. (2007). Governance for sustainable development in the face of ambivalence, uncertainty and distributed power: An introduction [editorial]. *Journal of Environmental Policy and Planning*, 9, 185–192.

North, D. C. (1992). Institutions and economic theory. *American Economist*, 36(1), 3–6.

North, D. C. (1994). Institutional change: A framework of analysis. *Economic History*, 9412001. Available at: http://ideas.repec.org/p/wpa/wuwpeh/9412001.html.

NSW Office of Environment & Heritage (2014). *Australian Capital Territory Climate Change Snapshot*. Sydney, New South Wales. Available at: www.environment.act .gov.au/__data/assets/pdf_file/0009/671274/ACTsnapshot_WEB.pdf.

O'Connell, D., Maru, Y., Grigg, N. et al. (2019). *Resilience, Adaptation Pathways and Transformation Approach. A Guide for Designing, Implementing and Assessing Interventions for Sustainable Futures*, version 2. Canberra: Commonwealth Scientific and Industrial Research Organisation (CSIRO). Available at: https://research.csiro.au/eap/rapta/.

Ostrom, E. (1990). *Governing the Commons: The Evolution of Institutions for Collective Action*. New York: Cambridge University Press.

Ostrom, E. (2011). Background on the Institutional Analysis and Development Framework. *Policy Studies Journal*, 39, 7–27.

Ostrom, E., Gardner, R. and Walker, J. (1994). *Rules, Games, and Common-Pool Resources*. Ann Arbor, MI: The University of Michigan Press.

Pahl-Wostl, C. (2009). A conceptual framework for analysing adaptive capacity and multi-level learning processes in resource governance regimes. *Global Environmental Change*, 19, 354–365.

Park, S. E., Marshall, N. A., Jakku, E. et al. (2012). Informing adaptation responses to climate change through theories of transformation. *Global Environmental Change*, 22, 115–126.

Pelling, M. (2011). *Adaptation to Climate Change: From Resilience to Transformation*. London: Routledge.

Potts, J., Foster, J. and Straton, A. (2010). An entrepreneurial model of economic and environmental co-evolution. *Ecological Economics*, 70, 375–383.

Rambaldi. A. N., Fletcher, C. S., Collins, K. and McAllister, R. R. J. (2012). Housing shadow prices in an inundation-prone suburb. *Urban Studies*, 50, 1889–1905.

Ramm, T. D., Watson, C. S. and White, C. J. (2018). Strategic adaptation pathway planning to manage sea-level rise and changing coastal flood risk. *Environmental Science & Policy*, 87, 92–101.

Reeder, T. and Ranger, N. (2011). *How Do You Adapt in an Uncertain World? Lessons from the Thames Estuary 2100 Project*. World Resources Report Uncertainty Series. Washington, DC: World Resources Institute. Available at: https://wriorg.s3 .amazonaws.com/s3fs-public/uploads/wrr_reeder_and_ranger_uncertainty.pdf.

Reisinger, A., Kitching, R. L., Chiew, F. et al. (2014). Australasia. In V. R. Barros, C. B. Field, D. J. Dokken et al., eds., *Climate Change 2014: Impacts, Adaptation, and Vulnerability. Part B: Regional Aspects. Contribution of Working Group II to the Fifth Assessment Report of the Intergovernmental Panel on Climate Change*. Cambridge: Cambridge University Press, pp. 1371–1438. Available at: www.ipcc .ch/site/assets/uploads/2018/02/WGIIAR5-Chap25_FINAL.pdf.

Rosenzweig, C. and Solecki, W. (2014). Hurricane Sandy and adaptation pathways in New York: Lessons from a first-responder city. *Global Environmental Change*, 28, 395–408.

Scott, W. (2008). Approaching adulthood: The maturing of institutional theory. *Theory and Society*, 37(5), 427–442.

Siebentritt, M., Halsey, N. and Stafford Smith, M. (2014*). Regional Climate Change Adaptation Plan for the Eyre Peninsula*. South Australia: Eyre Peninsula Integrated Climate Change Agreement Committee. Available at: www.naturalresources.sa.gov

.au/files/sharedassets/eyre_peninsula/corporate/climate-change-adaptation-2014-plan
.pdf.

Smith, A., Stirling, A. and Berkhout, F. (2005). The governance of sustainable socio-technical transitions. *Research Policy*, 34, 1491–1510.

Stafford Smith, M. D., Horrocks, L., Harvey, A. B. and Hamilton, C. (2011). Rethinking adaptation for a four degree world. *Philosophical Transactions of the Royal Society A*, 369, 196–216.

Stephens, S. A., Bell, R. G. and Lawrence, J. (2018). Developing signals to trigger adaptation to sea-level rise. *Environmental Research Letters*, 13, 104004.

Stern, N. (2006). *The Stern Review: The Economics of Climate Change.* Cambridge: Cambridge University Press.

Stern, N. (2013). The structure of economic modeling of the potential impacts of climate change: Grafting gross underestimation of risk onto already narrow science models. *Journal of Economic Literature*, 51, 838–59.

Stewart, M. G. and Wang, X. (2011). *Risk Assessment of Climate Adaptation Strategies for Extreme Wind Events in Queensland.* Canberra: Commonwealth Scientific and Industrial Research Organisation (CSIRO). Available at: https://publications.csiro.au/rpr/download?pid=csiro:EP112958&dsid=DS1.

Suckall, N., Stringer, L. and Tompkins, E. (2015). Presenting triple-wins? Assessing projects that deliver adaptation, mitigation and development co-benefits in rural Sub-Saharan Africa. *Ambio*, 44, 34–41.

UK Government Office for Science (2012). *Reducing Risks of Future Disasters: Priorities for Decision Makers.* UK Government Office for Science. Available at: www.gov.uk/government/publications/reducing-risk-of-future-disasters-priorities-for-decision-makers.

UNISDR (UN International Strategy for Disaster Reduction) (2007). *Costs and Benefits of Disaster Risk Reduction.* High level dialogue information note No. 3. Available at: www.unisdr.org/files/1084_InfoNote3HLdialogueCostsandBenefits.pdf.

Voß, J.-P., Newig, J., Kastens, B., Monstadt, J. and Nölting, B. (2007). Steering for sustainable development: A typology of problems and strategies with respect to ambivalence, uncertainty and distributed power. *Journal of Environmental Policy and Planning*, 9, 193–212.

Wang, C.-H., Baynes, T., McFallan, S. et al. (2015). Rising tides: Adaptation policy alternatives for coastal residential buildings in Australia. *Structure and Infrastructure Engineering: Maintenance, Management, Life-Cycle Design and Performance*, 12, 463–476.

Webb, L. B. and Hennessy, K. (2015). *Climate Change in Australia: Projections for Selected Australian Cities.* Commonwealth Scientific and Industrial Research Organisation (CSIRO) and Bureau of Meteorology. Available at: www.climatechangeinaustralia.gov.au/en/publications-library/brochures/.

Weitzman, M. L. (2011). Fat-tailed uncertainty in the economics of catastrophic climate change. *Review of Environmental Economics and Policy, Symposium on Fat Tails*, 5, 275–292. Symposium originally held at Harvard University, Cambridge, MA.

Wise, R. M., Fazey, I., Stafford Smith, M. et al. (2014). Reconceptualising adaptation to climate change as part of pathways of change and response. *Global Environmental Change*, 28, 325–336.

Yohe, G. and Leichenko, R. (2010). Adopting a risk-based approach. *Annals of the New York Academy of Sciences*, 1196, 29–40.

Young, K. and Hall, J. (2015). Introducing system interdependency into infrastructure appraisal: From projects to portfolios to pathways. *Infrastructure Complexity*, 2, 1–18.

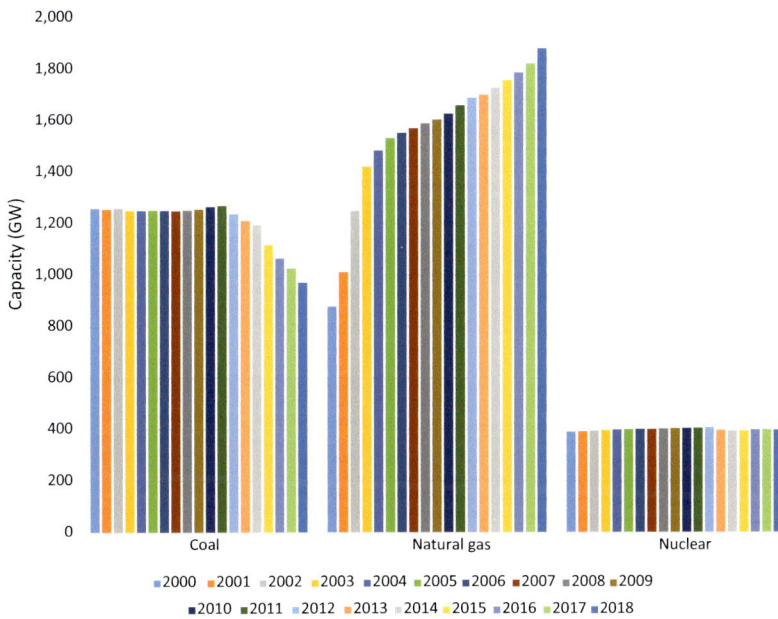

Figure 1.2 US generator capacity by fossil fuel source.

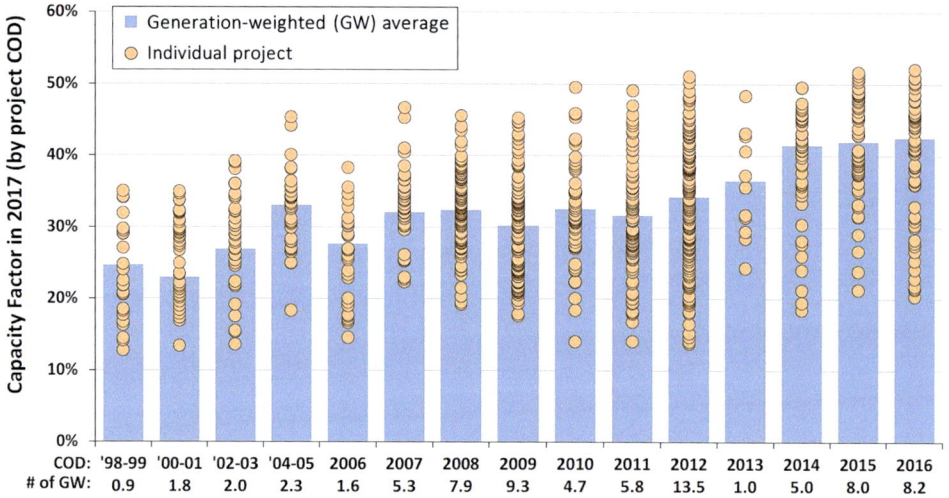

Figure 2.4 Calendar year 2017 capacity factors by commercial operation date (COD) for US wind farms. Also shown along the x axis is the installed capacity of wind that reached COD in each year.

Figure 2.5 Levelised cost of energy of wind in USA.

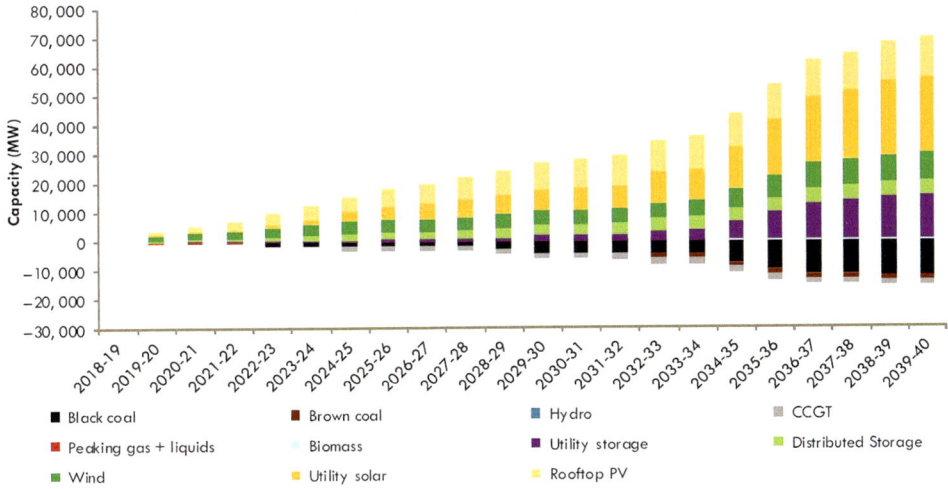

Figure 2.6 Relative change in installed capacity in the neutral scenario..

Legend:
- Black coal
- Peaking gas + liquids
- Wind
- Brown coal
- Biomass
- Utility solar
- Hydro
- Utility storage
- Rooftop PV
- CCGT
- Distributed Storage

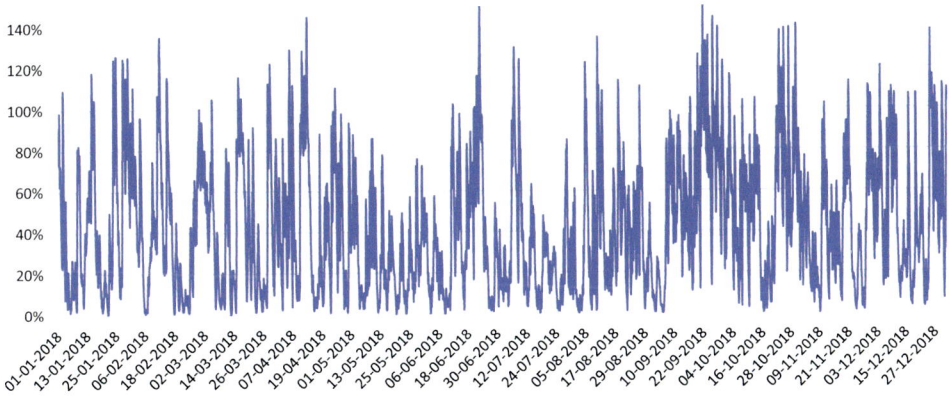

Figure 2.7 Fraction of Danish demand from wind generation, hourly data, December 2014.

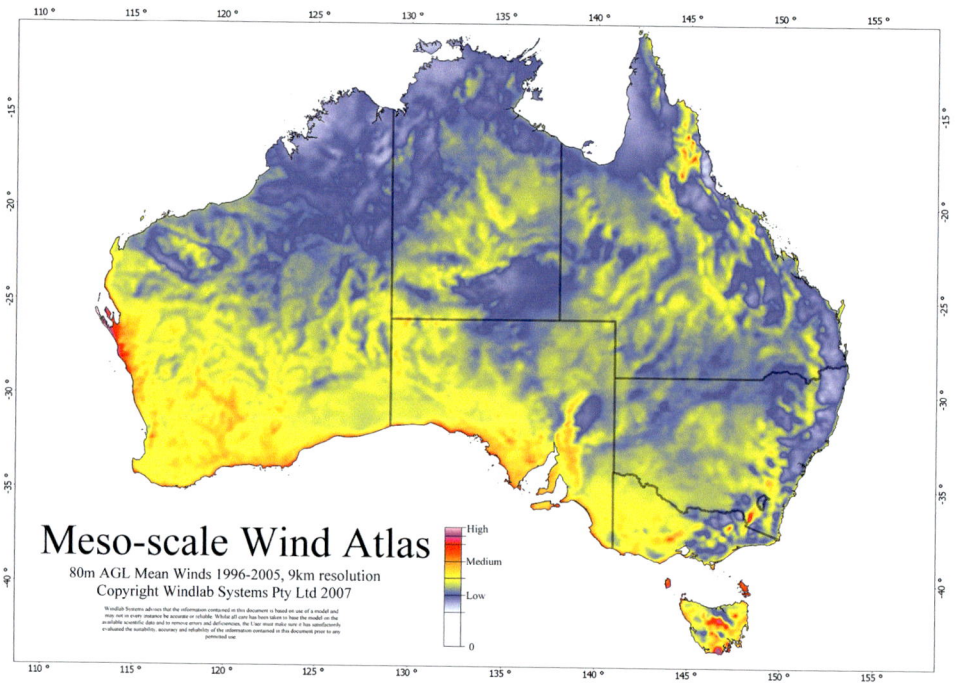

Meso-scale Wind Atlas

80m AGL Mean Winds 1996-2005, 9km resolution
Copyright Windlab Systems Pty Ltd 2007

High

Medium

Low

0

Figure 2.8 Australian Wind Atlas.

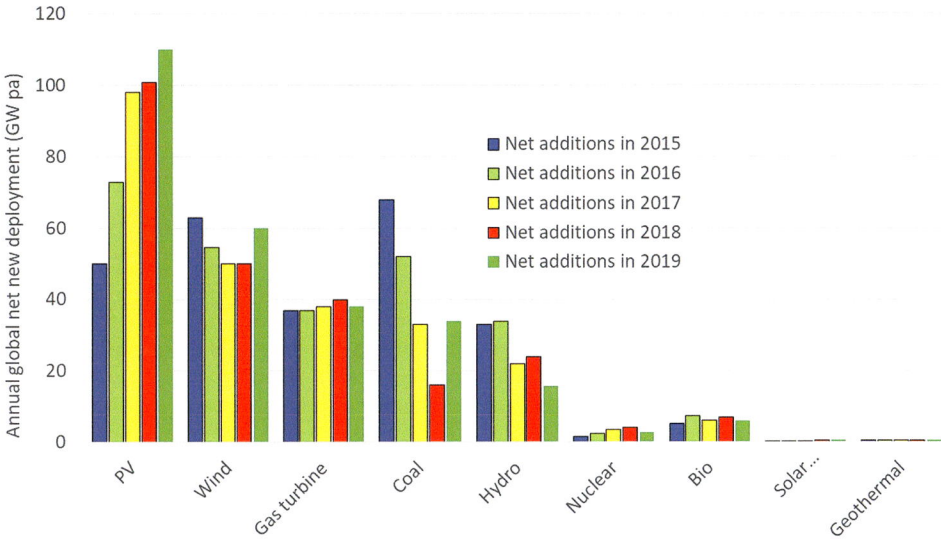

Figure 3.2 New generation capacity added in the year 2015 to 2019 by technology type. Most net new generation capacity is from wind and PV power.

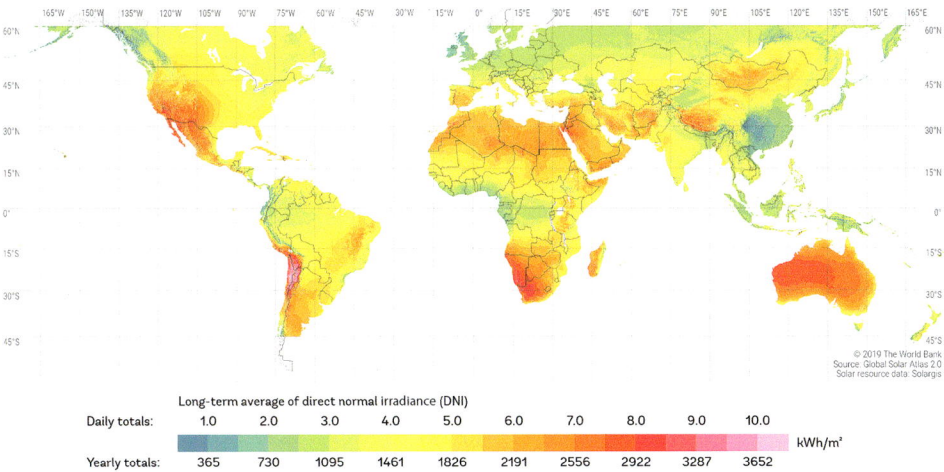

Figure 4.1 Direct normal irradiance (DNI, kWh/m² per year) at sites around the world. Optimal locations for CSP are Chile/Bolivia/Argentina, Australia, South Africa/Namibia and USA/Mexico. Other excellent locations include Spain, India, China, Morocco and Israel.

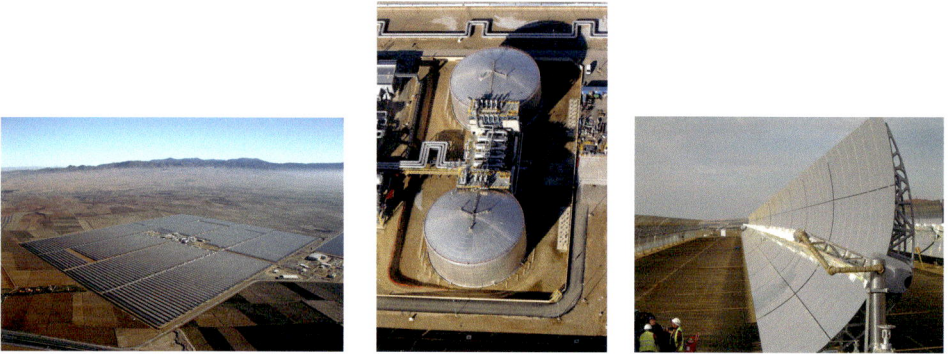

Figure 4.5 Andasol 1, a Spanish 50-MWe trough system with molten salt thermal energy storage. Left: aerial view; centre: molten salt storage tanks; right: one of the parabolic trough collectors.

Figure 4.6 Gemasolar, at Fuentes de Andalucía in Spain. This 19.9-MWe system has 15 hours of thermal energy storage, allowing continuous 24-hour operation in summer months.

Figure 4.7 Khi Solar One. Left: 140-m² heliostats; centre: view of solar field, taken from the receiver; right: the tower, incorporating three cavity receivers.

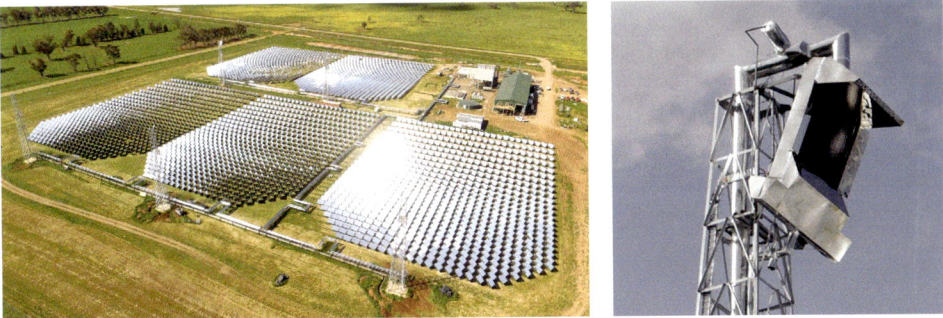

Figure 4.8 Jemalong Solar Thermal Station (left), developed by Vast Solar in central New South Wales, Australia. This modular system has five 1.25-MWth towers, 27-m high (right), with sodium as the HTM, connected to a single 1.1-MWe steam turbine.

Figure 4.9 Left: the Noor III plant near Ouarzazate, Morocco, has a 250-m-high tower and 7400 heliostats, and is co-located with two large parabolic trough plants. Right: Sundrop Farms, Port Augusta, a CSP plant installed to provide combined power and desalinated water for a commercial tomato greenhouse in South Australia.

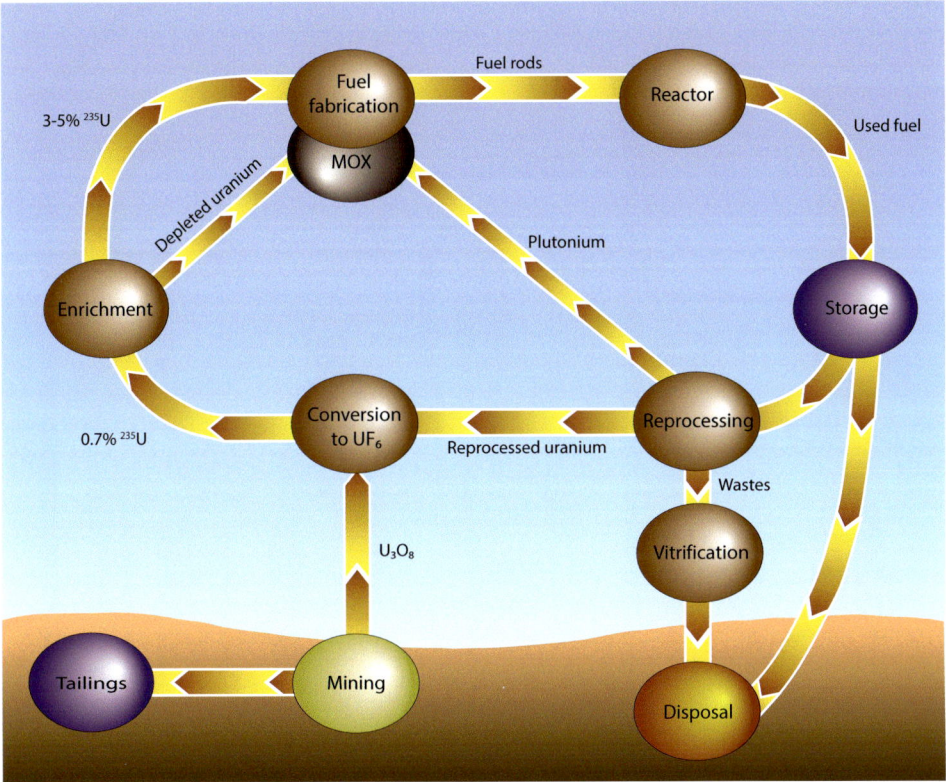

Figure 5.1 Overview of the nuclear fuel cycle.

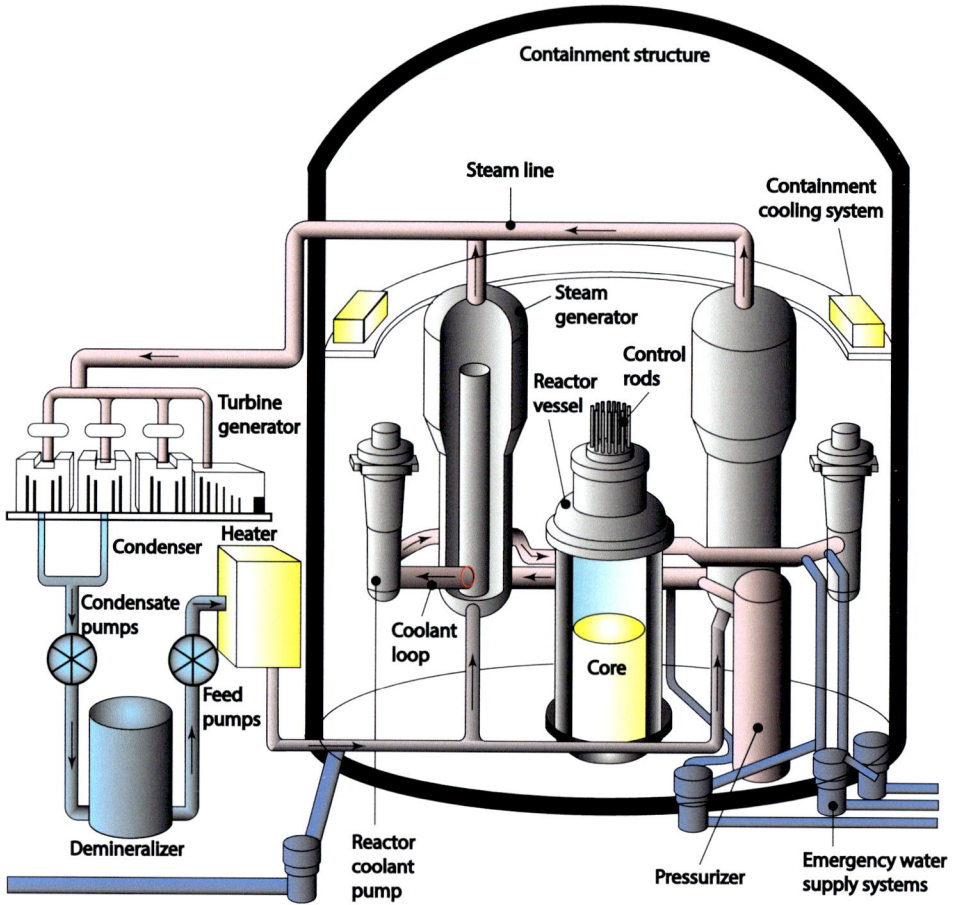

Figure 5.2 Pressurised water reactor (PWR).

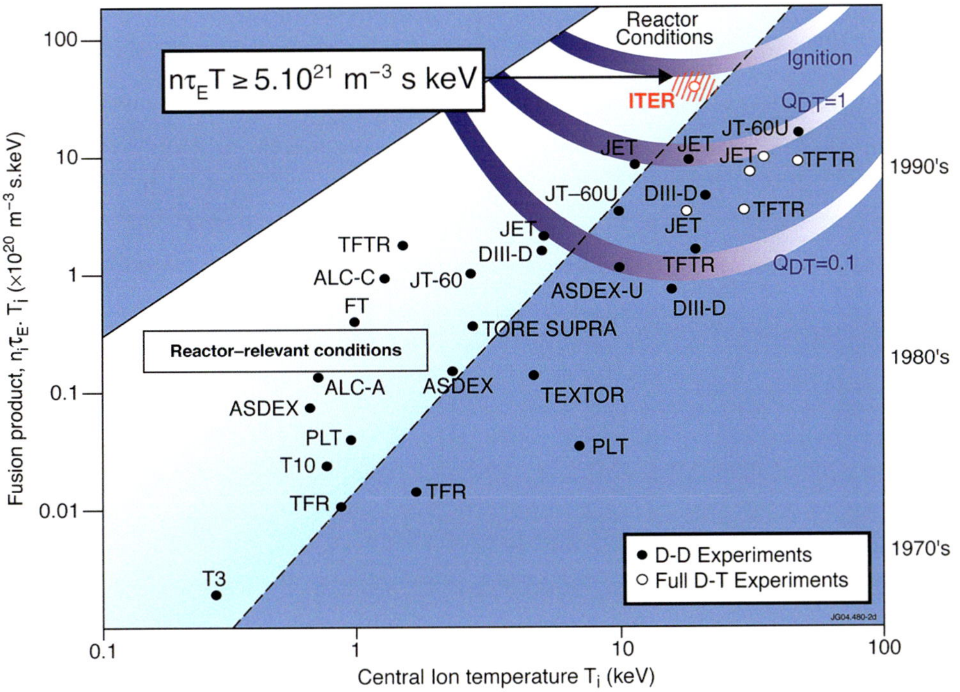

Figure 5.3 The triple product for fusion breakeven (left vertical axis) as a function of plasma temperature (horizontal axis in kiloelectron volts: 10 keV ~ 100 000 000 °C). The progress of fusion experiments around the world over time (right-hand axis) is shown leading to projected breakeven (Q_{DT}) by the ITER programme.

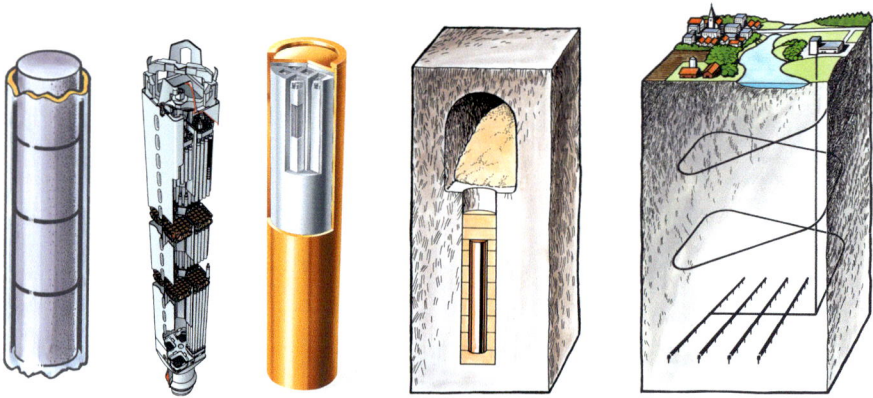

Figure 5.4 Swedish KBS-3 system for deep geological disposal of used fuel.

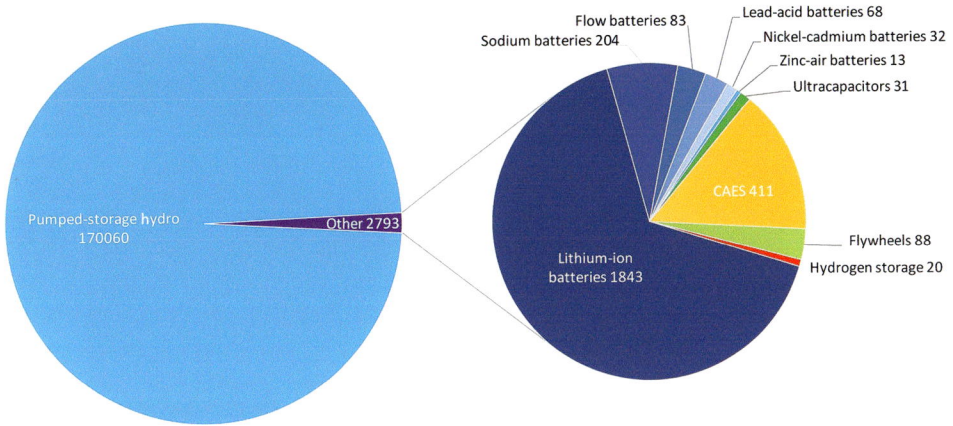

Figure 7.1 Overwhelmingly, energy storage installed in global electricity grids is provided by PSH. Total global installed grid-scale energy storage capacity (power capacity, megawatts (MW)) by technology type.

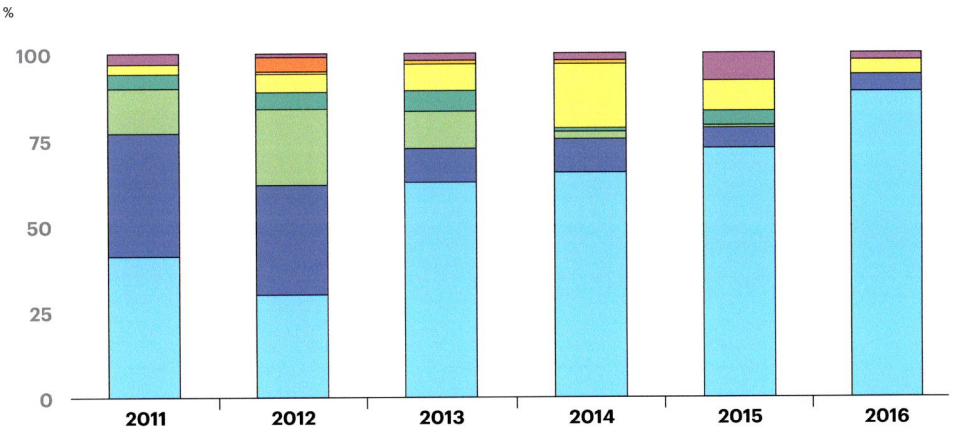

Figure 7.2 For non-PSH storage, battery storage will dominate the growth of energy storage capacity over the years ahead.

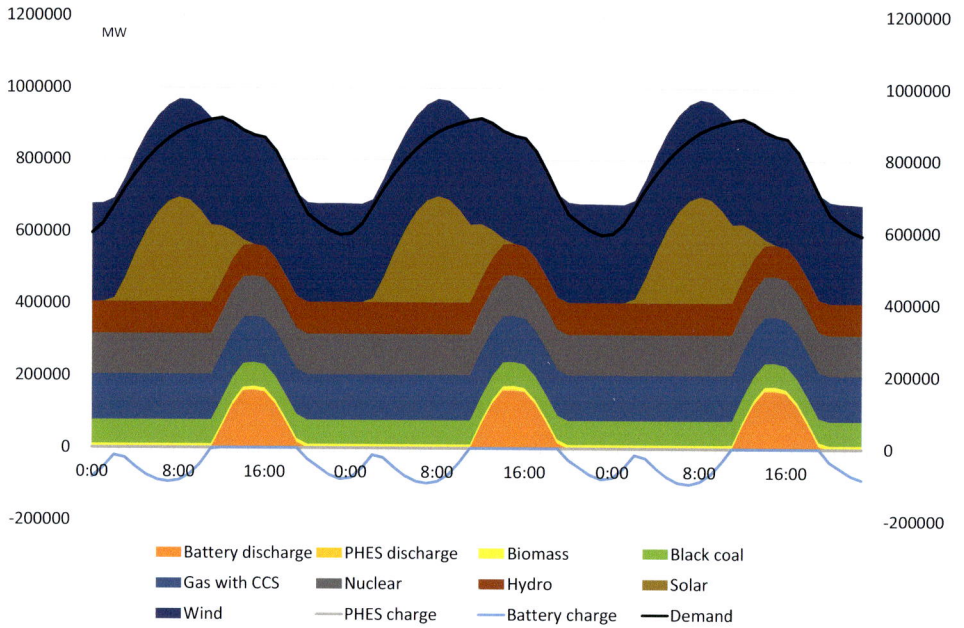

1200000
MW

1000000

800000

600000

400000

200000

0

-200000

1200000

1000000

800000

600000

400000

200000

0

-200000

0:00   8:00   16:00   0:00   8:00   16:00   0:00   8:00   16:00

- Battery discharge
- PHES discharge
- Biomass
- Black coal
- Gas with CCS
- Nuclear
- Hydro
- Solar
- Wind
- PHES charge
- Battery charge
- Demand

Figure 7.3 An example energy demand profile and generation mix from the North American region during spring, showing the diurnal energy storage cycle that provides medium-term energy balancing.

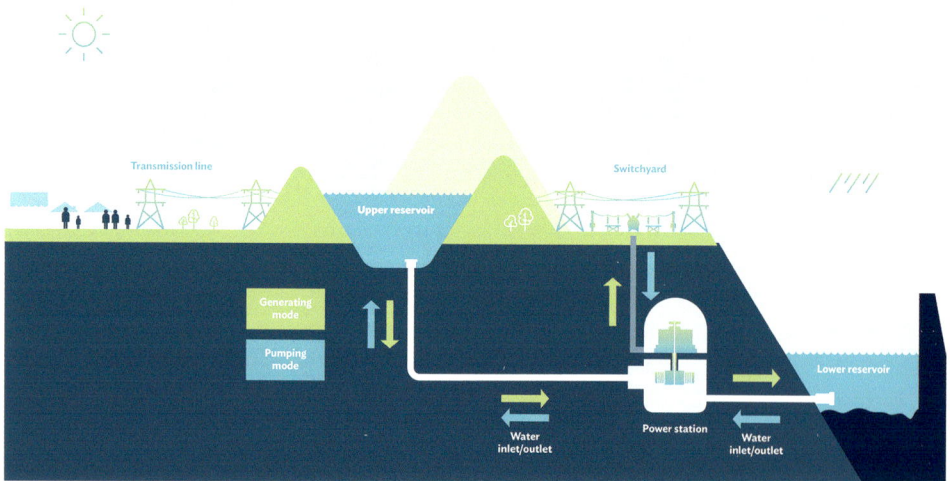

Figure 7.4 Pumped storage hydro schematic.

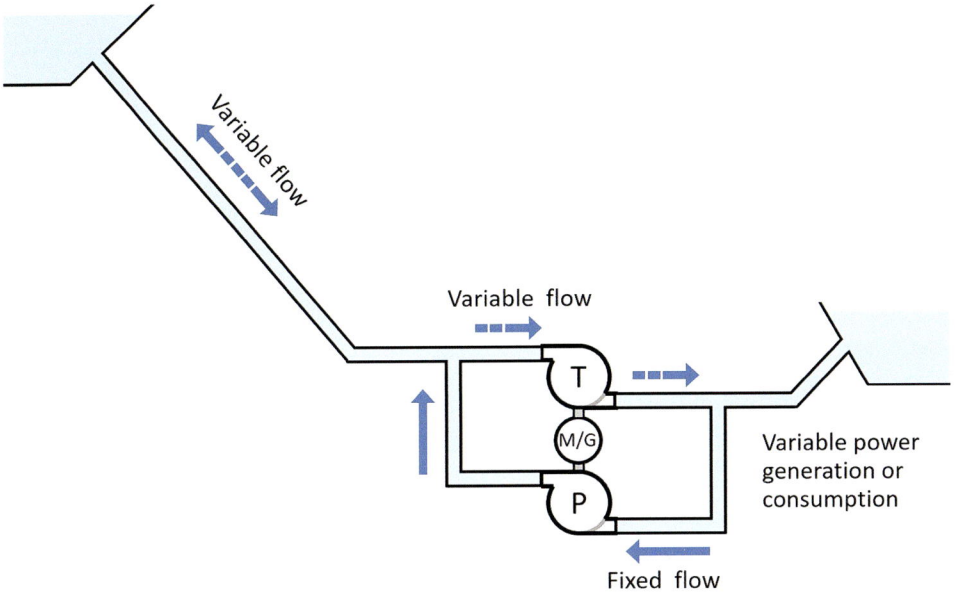

Figure 7.8 Ternary sets, utilising a single synchronous machine with integrated pump and turbine and a hydraulic short circuit, afford maximum flexibility and seamless changeover from pumping to turbining mode.

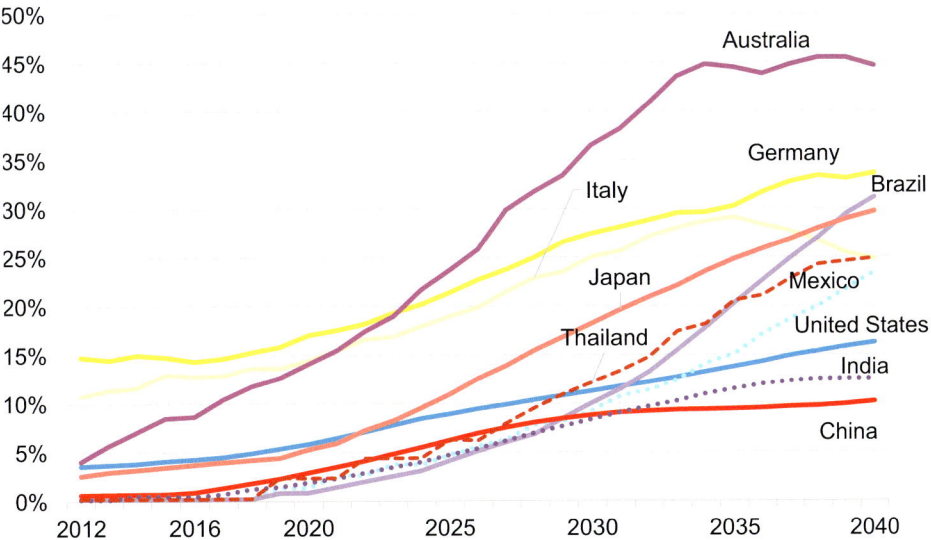

Figure 7.9 Many grids globally will demonstrate increasingly high levels of decentralisation over the decades ahead.

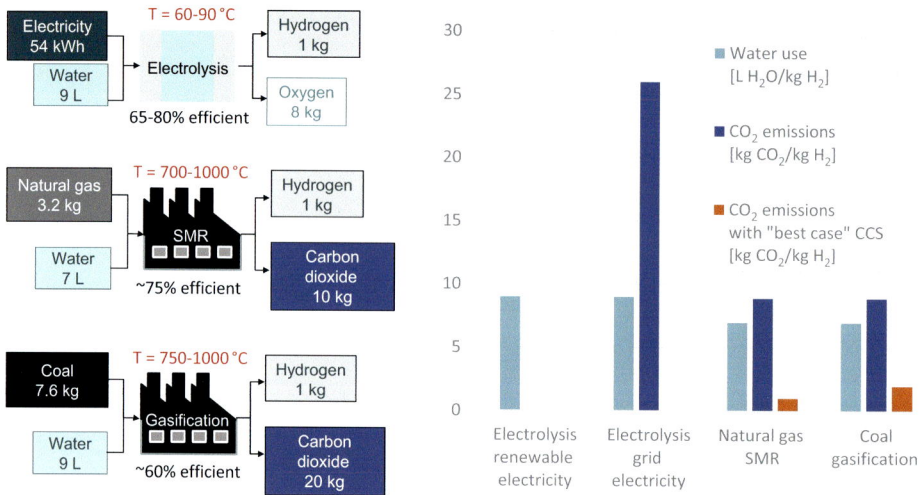

Figure 8.1 Comparison of SMR, coal gasification and electrolysis hydrogen production technologies, showing process inputs, temperatures and outputs. Bar chart compares water usage and CO$_2$ emission intensities for the different processes with and without 'best-case' CCS, equivalent to 90% CO$_2$ capture and retention rate.

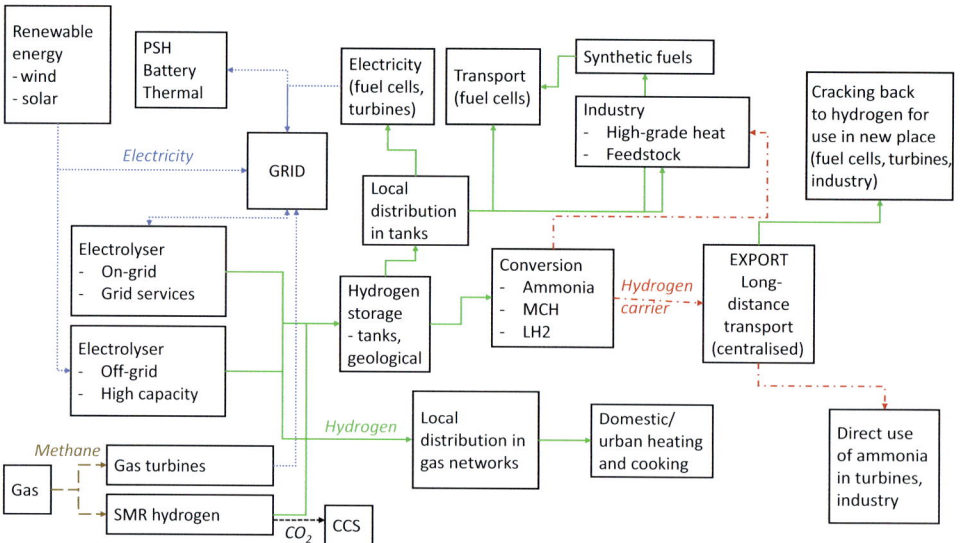

Figure 8.2 Schematic showing the possible ways that hydrogen technologies could be integrated into the larger energy system.

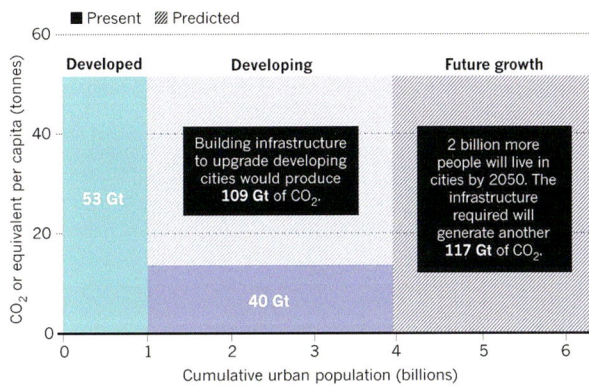

Figure 11.1 Estimated carbon emissions from future urban development.

Mitigation strategies	Emergency risk reduction	Insurance	Urban planning and zoning regulations	Design guidelines	Neighbourhood watch programmes and safety nets	Education and capacity building	Health and livelihoods	Resilient energy installations	Water and wastewater adaptive management	Inland and coastal flood protection	Climate proof infrastructure (for example, transportation)	Wetland restoration	Green roofs and walls	Green space and bioswales
Urban design and form	+		+	−[a]	++			+	+[b]	++	+ −[c]	−−	+	−−[d]
Modal shift, mobility services, traffic optimisation	+		++	−[e]			+[f]			−	++			−−
High-efficiency, low-emissions, smaller vehicles	+	−		−				−−[g]			+			
Low-energy demanding, heat-resistant architecture	+	+		++ −[h]		+	++[i]	+	+[j]				+ −[k]	++
High-efficiency appliances and equipment									+					
Energy efficient and low-carbon urban industries							+	+	+	+				++
High performance operation of buildings				+				+	+[l]	+[m]			+[n]	
Reducing Urban Heat Islands (such as white and green surfaces, green infrastructure)		+	++[o]	++		+	+	−−			+	++	++	++[p]
Infrastructure-integrated renewable energy systems generation	++			+				++[q]			−−[r]			
Fuel switch to low(er) carbon generation	+		+					++	−[s]		−		+	
Affordable low-carbon, durable construction materials; timber infrastructure	−[t]		+	+				++[u]			+ −[v]			
Carbon capture and utilisation in construction materials														
Lifestyle, behaviour, choices, sustainable consumption and production, sharing economy, circular economy	+		+[w]			+	++	+	+					+[x]

[a]Urban design for optimised adaptation and mitigation may coincide or compromise each other. [b]Building orientation, height, and spacing can help reduce need for cooling units. [c]Flood protection may compromise urban design best serving mitigation purposes. [d]Maximising compact urban design can reduce green space areas. [e]Urban designs best serving disaster risk reduction or adaptation needs may compromise the energy efficiency of the transport system. [f]Traffic optimisation results in improved air quality; modal shift typically results in more activity, that is, health gains. [g]Increased vehicular air conditioning will increase transport emissions. [h]In heat-prone regions design guidelines may prioritise the availability of mechanical cooling to reduce health risks, exacerbating emissions. [i]Very high-efficiency buildings with heat recovery ventilation have major health and welfare benefits. [j]High-efficiency buildings often also manage water resources efficiently. [k]In heat-prone regions design guidelines may prioritise the availability of mechanical cooling to reduce health risks, exacerbating emissions, but otherwise the synergies are dominant. [l]High-performance operation of buildings will increase the efficiency of mechanical cooling. [m]Highperformance operation typically also extends to better water management. [n]Green roofs will improve energy efficiency and operation of building. [o]Enhances climate security resilience against extreme events. [p]Green space will reduce urban heat islands and reduce risk of flooding. [q]Renewable energy reduces risk of power loss during extreme events. [r]Energy dependency on pumping water from flooding. [s]Some small-scale energy generation technologies require water resources. [t]Timber infrastructure may be less resilient to disasters than conventional ones. [u]Utilising lightweight construction and phase-change materials (PCMs), solar heat can be absorbed by PCMs, in turn improving thermal regulation of buildings while also reducing energy, heating and cooling. [v]Climate-proof infrastructure could utilise timber; in other cases it needs to rely on concrete. [w]Incorporating institutions and stakeholders into planning can improve lifestyle choices of city as a whole. Integrated approaches encourage more stakeholders to engage in the project, as multiple sectors and institutions are impacted by the adaptation and mitigation efforts . [x]Experiencing biodiversity has been proven to improve life quality and environmental consciousness.

Figure 11.3 Key interactions between urban mitigation and adaptation strategies.

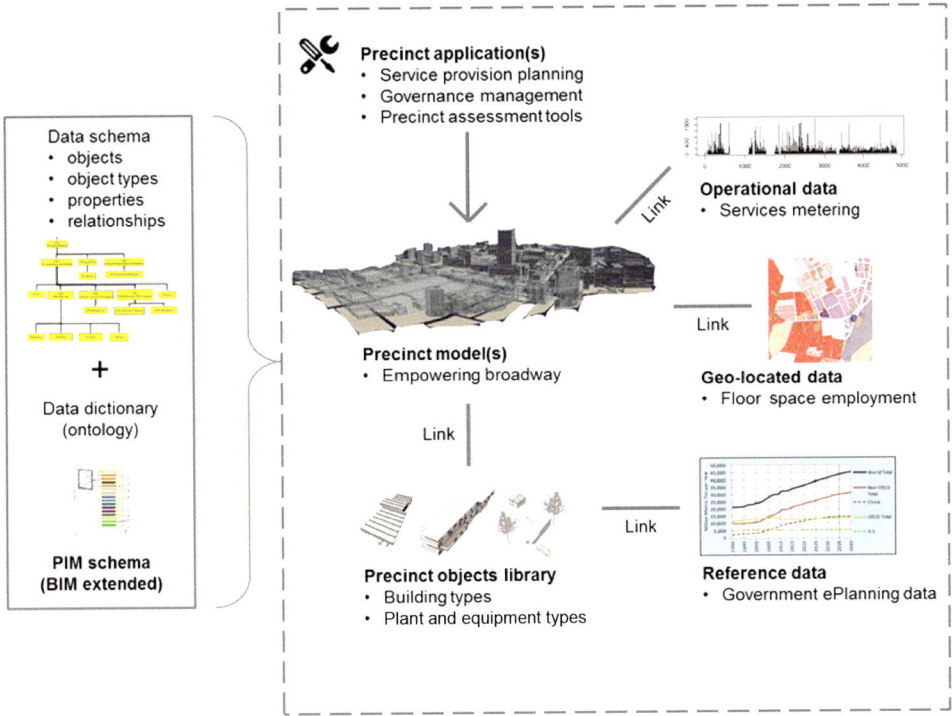

Figure 12.3 Operational structure of the PIM platform.

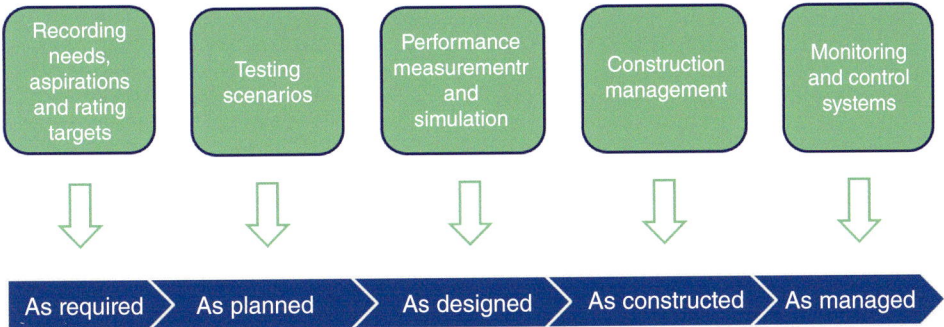

PIM provides a definitive repository of information at all stages in PRECINCT design and management based on open standards

| Recording needs, aspirations and rating targets | Testing scenarios | Performance measurementr and simulation | Construction management | Monitoring and control systems |

As required → As planned → As designed → As constructed → As managed

Figure 12.4 Life-cycle management of precincts using PIM.

Figure 12.5 Mutopia representation of Docklands and Fishermans Bend precincts, Melbourne (with the CBD in the background).

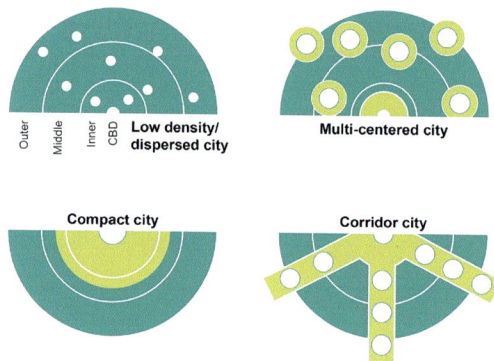

Figure 12.6 Archetypal urban forms.

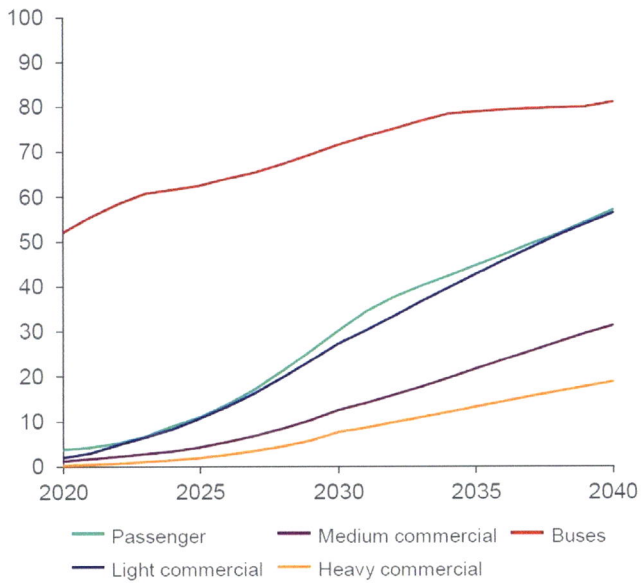

Figure 15.1 Global electric vehicle share of new vehicle sales by segment (% of new sales).

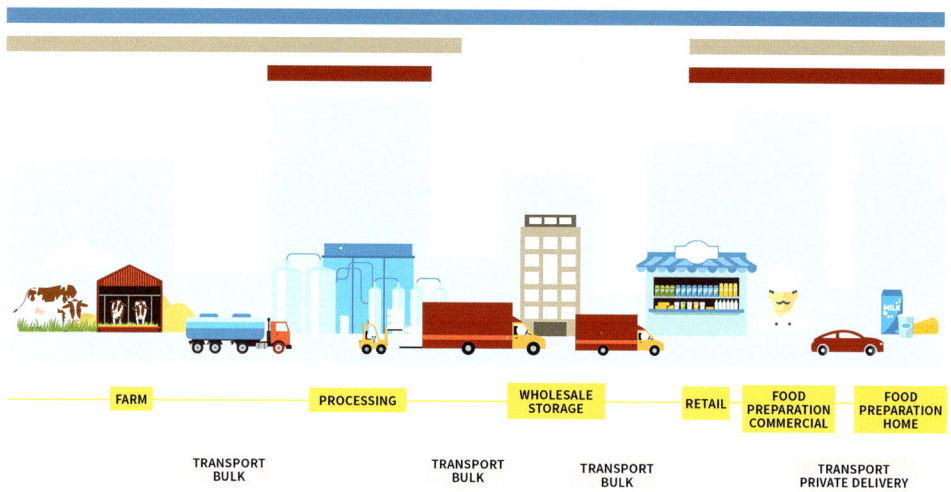

Legend:
- Primary energy use
- Extension of shelf life by preservation/refrigeration
- Dewatering and drying
- Process heat and cooking

Figure 16.1 Example of the 'value chain' approach applied to the 'farm to plate' system by the Australian Alliance for Energy Productivity. This highlights the interdependence of the participants on each other, and opportunities to optimise overall system productivity.

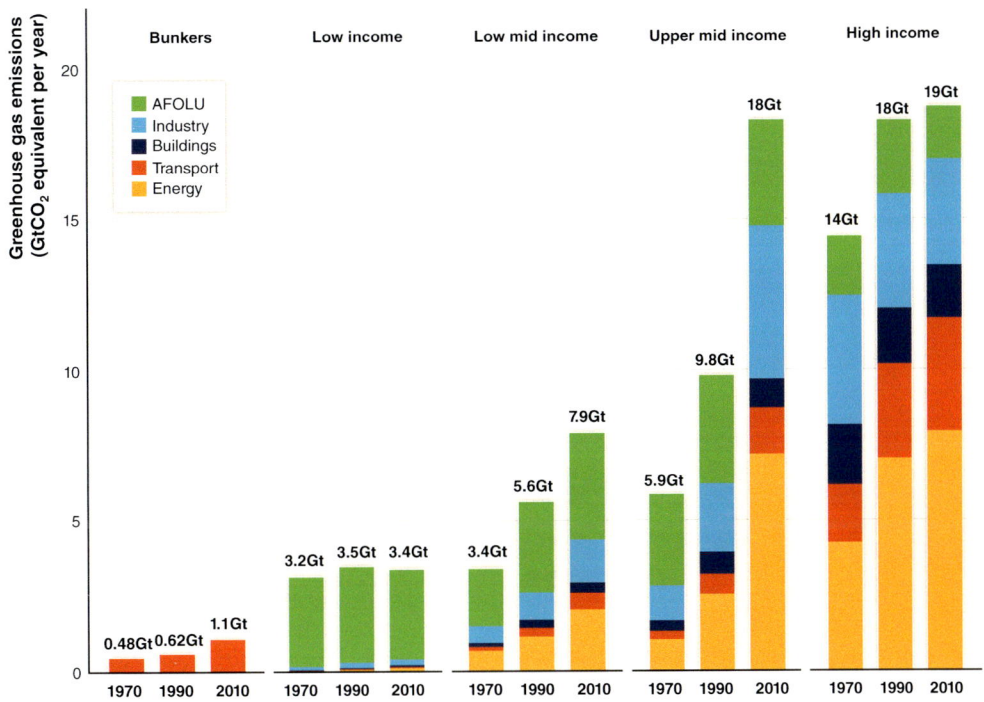

Figure 17.1 Percentage greenhouse gas emissions by sector, for low- to high-income countries.

(a) Pre-agricultural era carbon stocks

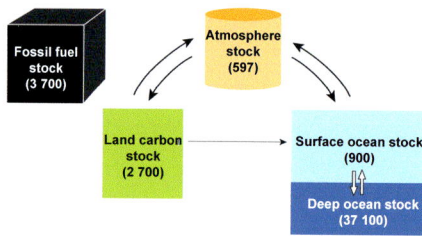

Fossil fuel stock (3 700)

Atmosphere stock (597)

Land carbon stock (2 700)

Surface ocean stock (900)

Deep ocean stock (37 100)

(b) Pre-industrial era changes in carbon stocks and flows

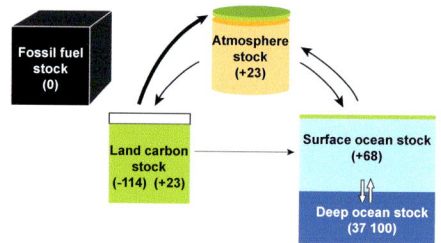

Fossil fuel stock (0)

Atmosphere stock (+23)

Land carbon stock (-114) (+23)

Surface ocean stock (+68)

Deep ocean stock (37 100)

(c) Contemporary era changes in carbon stocks and flows

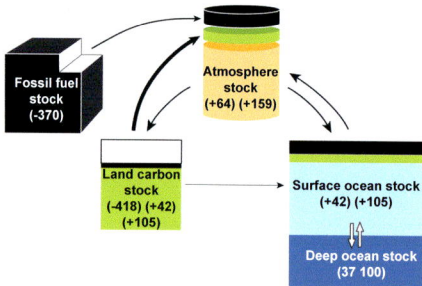

Fossil fuel stock (-370)

Atmosphere stock (+64) (+159)

Land carbon stock (-418) (+42) (+105)

Surface ocean stock (+42) (+105)

Deep ocean stock (37 100)

(d) Hypothetical re-filling of the land stock

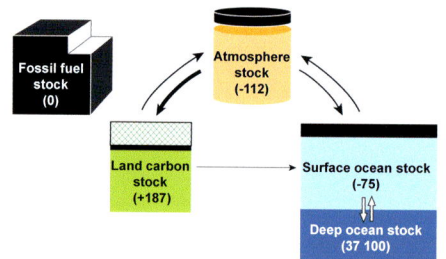

Fossil fuel stock (0)

Atmosphere stock (-112)

Land carbon stock (+187)

Surface ocean stock (-75)

Deep ocean stock (37 100)

Figure 17.2 Changes in the primary stocks of the global carbon cycle.

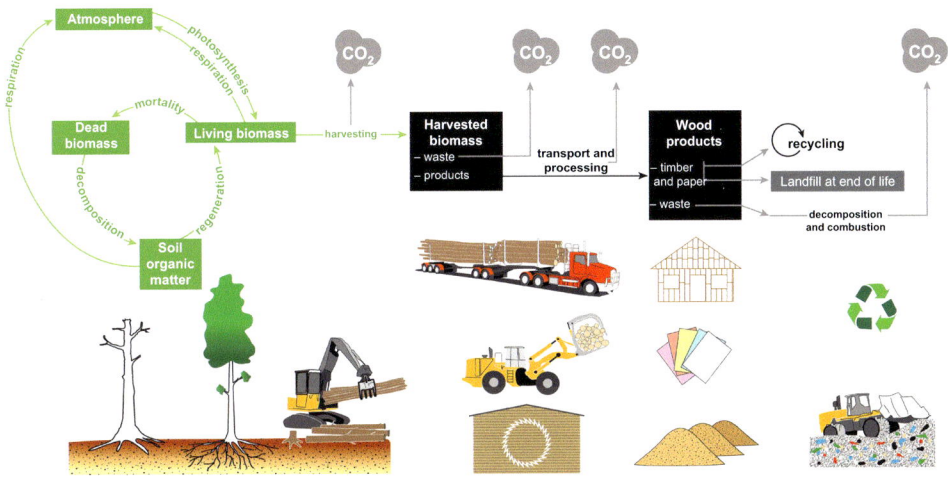

Figure 18.1 The carbon cycle in the forest and harvested wood system.

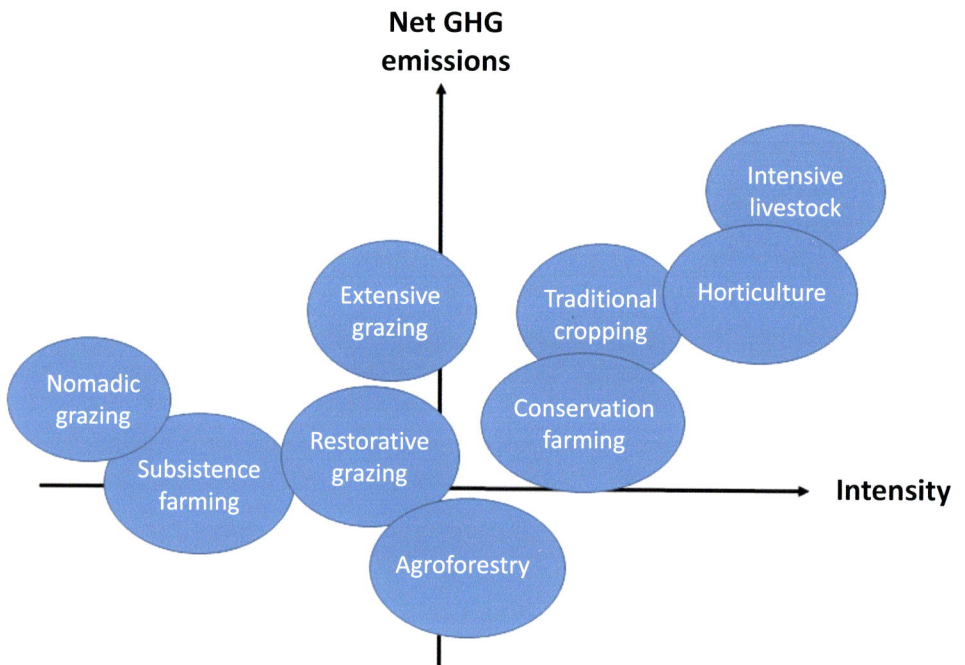

Figure 19.1 General relationship between agricultural intensity and net GHG emissions for a range of agricultural systems.

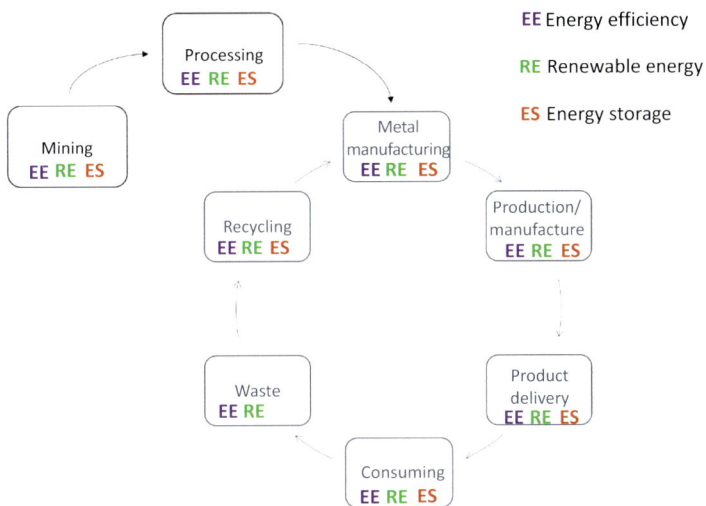

Figure 20.1 Decarbonising the mining and metals supply chains through an integrated approach using innovations in energy efficiency (EE), renewable energy (RE) and energy storage (ES).

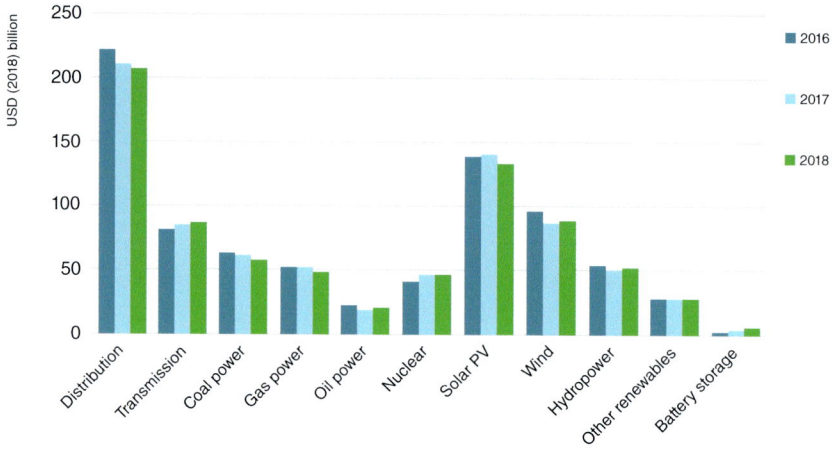

Figure 23.1 Global investment in the power sector by technology.

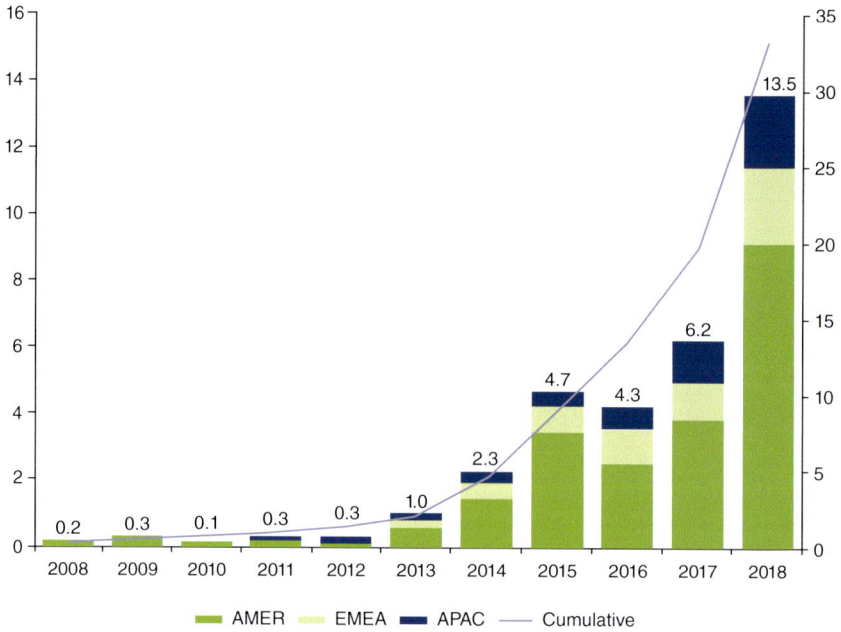

Figure 23.2 Global volume of corporate PPAs signed, by region, 2008–2018 GW.

15

Transport

MICHAEL H. SMITH, PETER STASINOPOULOS, ALAN PEARS AND ESHAN AHUJA

Executive Summary

The chapter discusses the role of the transport sector in climate change noting that it accounts for some 18% of greenhouse gases (GHGs), with emissions rising rapidly.

Transport infrastructure is also highly vulnerable to the effects of climate change including high temperature impact on roads – melting surfaces – expansion and warping of rail tracks and damage to coastal infrastructure from sea-level rise. More resilient infrastructure and better emergency planning are necessary.

It is possible to decarbonise the transport sector rapidly with many no-regrets and profitable measures, largely through electrification utilising renewable energy.

The adoption of battery-electric and hydrogen fuel cell cars and light vehicles can greatly reduce GHGs as the energy system itself is decarbonised and rooftop solar energy grows, making significant cost savings. Many countries have set timetables for phasing out internal combustion engine cars.

The adoption of electrically powered buses is already well advanced with more than 50% of new buses purchased now battery electric or hydrogen fuel cell powered.

There is also scope for significant reduction of energy demand and emissions reduction in this sector by 2030 through mode shift, city design that encourages walking and cycling, e-biking and better renewably powered public transport, as well as car-sharing schemes.

Heavy road transport and shipping are moving to low-carbon propulsion through low-carbon hydrogen or ammonia with battery-electric and fuel cell technology also being trialled.

Air transport is more challenging but battery-electric shorthaul aircraft are already under trial and hydrogen or biofuels may substitute for fossil fuels for larger and longer-range aircraft.

The change to electric vehicles will necessitate increased electricity generation but this can be partially offset by smart 'off-peak' charging and the adoption of vehicle-to-grid technology, whereby car and other vehicle batteries can serve to help reduce peak electricity demand in grids as well as powering households and commercial buildings.

Integrated planning encompassing several arms of policy is required to achieve optimum results in transforming the transport sector. These include urban planning, transport planning, establishing appropriate legal and regulatory conditions, facilitating investment in electric-vehicle charging infrastructure and ensuring availability of finance and providing incentives to facilitate transition to electrification,

The chapter concludes that, with integrated planning and action, decarbonisation of the transport sector by 2050 is feasible.

15.1 Introduction

The transport sector is a large contributor to greenhouse gases (GHGs) and its related infrastructure is also vulnerable to the effects of climate change. It is critically connected to the wider issue of decarbonising the energy system, as the main means of decarbonising transport is through electrification. This applies to all land-based modes including rail, road transport, urban delivery systems, cars, motorcycles, bicycles and scooters. Electric and hybrid systems are already being trialled for sea transport and air transport, while hydrogen and its derivative, ammonia, show great promise as replacement fuels for some transport modes. Decarbonising the sector calls for coordinated government and industry actions impacting all sectors and human lifestyles.

The transport sector is highly vulnerable to the effects of climate change, particularly temperature rises and extreme weather events that can flood infrastructure and damage ports and other coastal installations, so adaptation strategies are important in this sector. This chapter deals with mitigation, through such means as electrification utilising renewable energy, but also discusses impacts and possible adaptations, and policy responses. Mitigation and adaptation measures need to be considered together for best results.

The transport sector is responsible for over 11% of total global energy use – equivalent to about 18% of global carbon dioxide emissions – and is one of the sectors with the fastest growing rate of emissions. If no additional action is taken, emissions from transport are forecast to double by 2050 (Sims et al. 2014).

Mitigation consistent with keeping warming levels under 1.5 °C is possible (Luderer et al. 2018) by halving final energy use by the transport sector by 2030, phasing out of fossil fuel passenger vehicle sales by 2035–2050 (Kuramochi et al. 2017), behaviour change, modal shifts, rapid decarbonisation of the electricity supply sector and development of zero-emission hydrogen fuel sources (Grubler and Wilson 2014; Breyer et al. 2017; Jacobson 2017; Jacobson et al. 2017).

The economic, business competitiveness, social equity, air quality and health benefits of improved fuel efficiency, electrification of transport systems and behaviour change are significant, making the business case for transitioning to zero-emissions transport sector strong.

As governments invest in better public transport and cycling infrastructure, and as vehicle- and bike-sharing schemes increase, more households will be able to meet their transport mobility needs with less recourse to individual motor vehicles.

Corporates and businesses, as well as households, can reduce operational costs of their car, vehicle, van or truck fleets by as much as 70% by shifting to efficient electric vehicles. The Climate Group in its Progress and Insights Report 2021 notes that members of its

Electric Vehicle 100 (EV100) initiative had doubled electric vehicle use over the previous year to 169 000 with commitments to achieve 4.8 million by 2030 (The Climate Group 2021).

15.2 Why Is Action on Climate Change a Priority for the Transport Sector?

Transport systems need to be designed and planned to be resilient to extreme weather events, but, as the Intergovernmental Panel on Climate Change (IPCC) points out in its Fifth Assessment: 'The literature on urban transport and climate change focuses more on mitigation, with less attention to vulnerability, impacts, and adaptation. Existing studies on impacts are often limited to the short-term demand side, particularly in passenger transport' (Revi et al. 2014: 559).

The transport sector is particularly vulnerable to climate change because of its extensive fixed infrastructure and the associated investments. Some of the vulnerabilities of the sector include the risks of:

- rail accidents due to train tracks warping under extreme heat and disruptions of rail lines due to flooding;
- airport tarmacs and roads melting due to higher temperatures;
- hailstorms damaging transport vehicles;
- damage and flooding to ports and sea transport vessels from extreme weather events;
- loss of ice roads in Alaska, Canada, Northern Europe and Russia (EPA 2013); and
- flooding of subway systems.

In the longer term, sea-level rises pose significant risks to coastal transport systems and related infrastructure in many countries around the world.

While decision makers are increasingly aware of these climate-change-related risks, still many businesses, cities and nations do not have adequate emergency plans in place to deal with extreme weather events should they occur. For instance, in 2011, a survey of over 550 companies around the world revealed that 51% of business supply-chain-related transport was affected by adverse weather over the previous year. Over that year, 49% of businesses lost productivity from such disruption, while their costs increased 38% and their revenue decreased 32% (Zurich Financial Services Group 2011).

In the USA in 2012, Hurricane Sandy flooded the New York subway system, causing the worst damage ever to the subway and stopping train services for several days. The hurricane also caused cancellations of thousands of airline flights. When Hurricane Katrina bore down on New Orleans in 2005, many, especially poor, people were stuck in the city because of lack of adequate public transport for evacuation (Schwartz and Litman 2008). In 2017, Hurricane Harvey shut down Houston, Texas, a major transport hub, and damaged over a million cars.

Transportation systems and related infrastructure thus also have a vital role to play in ensuring adaptive capacity and resilience in extreme weather events by enabling:

- the quick evacuation of millions of people if a hurricane is forecast to hit populated coastal areas;

- the rapid movement of people to higher ground in times of extreme flooding; and
- the prompt evacuation of people from places at risk of being surrounded by mega-fires.

Clearly, there is already a need to improve the resilience of transportation systems and related infrastructure to deal with already occurring extreme weather events. This need will become more urgent as the climate warms, increasing the risks of more intense extreme weather events. Planners, policy-makers, standards boards and companies need to design transport systems that are able to withstand still more intense extreme weather events in the future.

There is a range of adaptation options for the transport sector. However, if climate change is not mitigated, some areas of longer-term adaptation for transport systems and related infrastructure will be prohibitively expensive, including: rebuilding and adapting coastal transport systems vulnerable to sea-level rises in coming centuries; retrofitting underground metro systems to make them less vulnerable to flooding risks (e.g. Japan, London); and building seawalls and other barriers for low-lying cities.

Most of the literature on the costs of climate change adaptation focuses on the potential costs for adaptation to 2050 or 2100 but, if climate change is not mitigated, beyond 2100, infrastructure-related costs in coastal cities, including for transport, will increasingly dominate adaptation costs, especially as sea levels rise beyond 1 metre. Hence the importance of focusing on rapid mitigation for this and all other sectors, to avoid having to adapt to worst-case scenarios, which are expected to be prohibitively expensive. It is in the interests of all key actors in the transport and related infrastructure sector to play their parts to achieve rapid decarbonisation as well as adaptation.

There are strong economic, social and environmental benefits for most stakeholders to take action in the transport sector. It is in the transport sector that stakeholders have perhaps the most capacity to reduce GHG emissions because it is one of the largest emitting sectors. According to the 2014 IPCC Assessment, 'The transport sector produced $7.0\,GtCO_2e$ (gigatonnes of carbon dioxide equivalent) of direct GHG emissions (including non-CO_2 gases) in 2010 and hence was responsible for approximately 23% of total energy-related CO_2 emissions ($6.7\,GtCO_2$)' (Sims et al. 2014: 603). The transport sector is also the sector with the most rapidly increasing emissions.

15.3 What Potential Is There for Climate Change Mitigation?

A range of global and national studies show that there is greater potential to decarbonise the transport sector to contribute to national and global efforts to reduce emissions than previously understood even 5 years ago. The technologies now exist to decarbonise the electricity grid and simultaneously decarbonise the transport sector by 2045–2050, but significant actions need to be started. Extensive studies (see Box 15.1) (Sims et al. 2014) show that it is technically and economically possible to decarbonise the transport sector with many no-regrets and profitable measures by 2050.

These studies evidence how to achieve decarbonisation in the transport sector with the following common elements.

Box 15.1 **Global and National Transport Sector Decarbonisation Case Studies**

In 2019, Gota et al. showed that achieving rapid decarbonisation in the transport sector is more technically and politically possible because of a range of technical and policy innovations (Gota et al. 2019). In particular, they emphasised the potential of 'Avoid' and 'Mode Shift' measures by improving user information and, in turn, increasing the potential for sharing and connecting low-emission modes (Fulton et al. 2017). This can be complemented by vehicle restrictions (e.g. quotas) as used in China; another strategy is to curtail motorisation (Kenworthy 2017). Regulations in Singapore have resulted in zero growth of personal four-wheelers (LTA 2017). Such restrictions can be complemented by internal combustion engine (ICE) phase-out policies, which are non-coercive in the short term but have potential to guide near-term manufacturer decisions in the context of a long-term commitment. Several governments and major manufacturers have already announced ICE phase-out dates.

It is technically and economically feasible for transport sectors to transition to a low-carbon state by 2050, as the following studies show:

- **USA**: The US Department of Energy (2013) found that the US transport sector can eliminate oil dependence and reduce GHG emissions by more than 80% by 2050 if significant barriers can be overcome. This outcome requires a combination of using electric and hydrogen technologies to increase the fuel efficiency of vehicles, and transport demand management.
- **California**: In 2005, California set a mandatory target to reduce GHG emissions to 80% below 1990 levels by 2050 (Yang et al. 2015). They calculated that the target could be met at a cost of 0.03–0.5% of discounted cumulative gross state product. The required changes include major investments in renewable electricity generation, biofuel and hydrogen production; vehicle improvements through electric and fuel cell drivetrain technologies; and carbon capture.
- **Europe**: Ambel and Earl (2018) showed that it is technically possible to fully decarbonise the transport sector by reducing demand for carbon-intensive forms of transport and by shifting to cleaner transport modes to deliver zero-emission mobility.
- **Japan**: Japan set a target to reduce GHG emissions by 80% below 1990 levels by 2050, when annual transport GHG emissions are expected to have halved to 122 Mt (megatonnes) due to a declining population and the adoption of low-carbon technologies. Oshiro and Masui (2015) found that a high carbon tax of USD 530–750/tCO$_2$e can further reduce these annual emissions to 42–60 Mt, by driving growth in the market's shares of renewable electricity generation, battery electric vehicles and fuel cell electric vehicles.

15.3.1 Halving Energy Demand by 2030

The International Energy Agency (IEA) has found that it is technically possible to reduce energy demand in the transport sector by 50% by 2030 (IEA 2012). This is achieved largely through mode shift and through making these lower-carbon modes more efficient and powered by renewables as, by 2030, electric vehicles are likely to make up only 20–25% of total vehicles in usage. So, halving energy demand by 2030 is achieved by a combination of the following:

- **Designing cities to not be automobile dependent and to be relatively compact to avoid urban sprawl**: Urban planning to reduce passenger travel demand has been identified as a key measure that cities can adopt to reduce emissions. It helps make walking, cycling and public transport convenient. In Chapter 11, Figure 11.2 shows the difference in land area of car-dependent Atlanta, USA, and low-carbon Barcelona, Spain.

- **Encouraging active transport**: Active transport is the combination of cycling and walking with public transport. Copenhagen (Denmark), Amsterdam, Utrecht, Nijmegen, Groningen and Eindhoven (Netherlands), Malmo (Sweden), Antwerp (Belgium), Friedberg (Germany), Portland (Oregon, USA), Bordeaux (France), Seville and Barcelona (Spain), Tokyo (Japan) and Berlin (Germany) are demonstrating what is possible through making the effort to reform urban and transport planning and invest in active transport infrastructure to significantly lift the percentage of commutes by walking and cycling combined with public transport. For instance, the Netherlands has shown that it is possible for as much as 25% of all trips to be taken by bicycle.

- **Supporting the shift to lower-carbon emissions by encouraging bike-share and car-share schemes**: More than 1000 cities have constructed bicycle-share schemes and over 600 have implemented car-share schemes in the last 20 years (Larsen 2013).

- **Investing in low-emission public transport systems to encourage mode shifts to public transport (light rail, buses and rail)**: Compared with conventional passenger cars, the energy and GHG intensities per passenger are 4–6 times lower for buses and 3–7 times lower for light rail during peak times (Chester et al. 2009). A number of cities – including Shanghai (China), Mumbai (India), Madrid (Spain), Copenhagen (Denmark), Amsterdam (Netherlands), Friedberg (Germany), Tokyo (Japan) and Curitiba (Brazil) – are demonstrating what is possible through investing in public transport. Curitiba (Brazil) and Bogota (Colombia) are model examples of the 164 cities worldwide which have built bus rapid transit systems, carrying close to 33 million passengers a day. There is a significant difference in per capita emissions from transport across cities around the world due to the relatively high use of low-carbon transport options such as public transport instead of cars. For instance, China, through investing in metro rail systems and urban bus systems, implementing quotas on the number of new car licences, and investing in facilities for electric scooters and bicycles, has managed to stop the growth in uptake of cars in some of its largest cities, such as Beijing and Shanghai (Yang et al. 2014).

- **Shifting modes to more efficient forms of freight transport**: Shifting part of the freight journey from trucks to rail or shipping results in large energy savings of 75% and 85%, respectively (von Weizsacker et al. 2009).

- **Achieving still more emissions reductions from mode shifts by improving the efficiency of buses, rail and shipping**: It is possible to significantly improve the efficiency of buses via transitioning to electric drivetrains. For instance, the US National Renewable Energy Laboratory has found that the fuel economy of electric buses is around four times higher than that of compressed natural gas and diesel buses operated on equivalent routes (Eudy et al. 2016).

- **Improving efficiency and shifting modes in air travel**: There is potential to improve fuel efficiency by as much as 50% by 2050 in the air transport sector through further improvements in flight management and logistics, weight reduction, aerodynamics, engine efficiency and the upgrade of fleets (IATA 2009). Much research is currently under way into use of renewable energy, including algal fuels, to provide a non-land-based source of renewable energy for air transport. Rapid advances are being made in the design and operational ability of short-haul all-electric aircraft and hybrid longer-haul aircraft to enable aircraft to be powered by renewables.

- **Reducing energy demand via transitioning to electric vehicles**: Electric vehicles are 70–80% more efficient than ICE vehicles (RAP 2017). With improved energy efficiency, it is then possible to more easily power most transport systems with zero-carbon electricity sources.

15.3.2 Decarbonisation via Electrification Combined with Renewables, Hydro and Nuclear Electrical Power Sources

As well as reducing energy demand, it is now technically and economically possible to decarbonise the energy used in transportation systems principally because of the potential to electrify most forms of transport or utilise other zero-emission fuels, such as hydrogen. Electrification enables a variety of low-carbon energy sources, such as renewables and nuclear energy, to power transport vehicles and systems. For instance, as in Box 15.1, numerous studies in many countries show that it is technically and economically possible to decarbonise the transport sector by 2050. Many countries are committing to sourcing 100% renewable energy to power their transport sectors by dates between 2020 and 2050. This is because technologies exist now to enable much of the transport sector to, over time, increasingly use electricity from renewable and low-carbon technologies, and use hydrogen fuels generated from renewables.

15.3.3 Combining Household Rooftop Solar Photovoltaics with Electric Scooters/ E-Bikes/E-Motorcycles/E-Cars to Achieve Zero Operational Cost Transport

As solar photovoltaic (PV) prices have fallen, it is now cost-effective in many countries for households to purchase solar PV systems to charge their electric scooters, e-bikes, e-motorbikes or electric cars. There are now over 300 million electric motorbikes/scooters/ bikes globally in use, which could, in principle, be powered by renewable energy. China had 250 million e-bikes in 2017 (Newman et al. 2017). E-motorcycle and e-scooter markets are likely to grow at 4–6 million per annum in coming years (Navigant Research 2013). There is exponential growth in sales of low-emission vehicles, such as all-electric-powered vehicles, which can all be powered by zero-emission electricity sources. The cost per kilowatt-hour for a battery used in a standard electric vehicle has fallen from USD 1000 in 2010 to 156/kWh in 2019 (S&P Global 2020). Electric car ranges have improved to over 250 km on a single charge for most models.

The total cost of owning an electric vehicle – including charging and maintenance – has fallen below conventional car ownership in Europe in 2018. Four countries – namely Norway, China, Japan and the USA – already have over a million electric cars in use as of 2020. Japan has invested heavily in electric car charging stations. These now outnumber petrol stations in Japan. A growing list of countries, including Norway, Italy, Austria, Denmark, Japan, Korea, the Netherlands, Spain and Portugal have established targets for electric vehicle sales.

Over 70 car manufacturers globally are already manufacturing all-electric cars, including a number of vehicle manufacturers that manufacture only electric vehicles, such as Tesla (USA-based), BYD Auto (China-based) and Luxgen Motor Co., Ltd (Taiwan-based). The majority of global car manufacturers have publicly committed to move to all-electric vehicle production over varying time frames.

15.3.4 Utilising Renewables to Power Business Delivery Vans, Utes, Trucks and Semi-Trailers, Bus and Rail

In 2017, major corporations, including Unilever and Ikea, initiated the EV100 initiative (also discussed in Chapter 16) to accelerate the shift to electric vehicles and away from petrol- and diesel-powered transportation. Ikea has committed to transitioning to 100% electric vehicles for its delivery of all products in five major global cities, including New York, London and Amsterdam. All of Ikea's deliveries in Shanghai are now by 100% electric delivery vehicles. Electric vehicles have significantly cheaper operational costs and therefore the business case for corporations investing in electric delivery vehicles is strong.

All-electric utility trucks (utes) and truck models are being sold by companies such as BYD, Volvo, Renault Trucks, Mitsubishi Fuso, Mercedes-Benz. Rivian, and Tesla are also planning on providing electric utes and trucks to market post 2020. The major global trucking company, Cummins, in addition to Volvo, DAF, MAN, Freightliner and Tesla, is aiming to produce all-electric large-scale semi-trailer trucks that can be powered by zero-carbon electricity sources.

As buses shift to electric power, it is possible to charge them overnight at bus depots with solar PV systems and batteries or decarbonised grids with pumped hydro and storage. Bloomberg New Energy Finance (BNEF) predicts that, by 2025, the majority of the global bus fleets will have begun their conversion to all-electric models due to the strong business case. E-buses have much lower operating costs and are already cheaper, on the basis of total cost of ownership, than conventional buses (BNEF 2018). As of 2018, over 50% of all new buses being procured globally were electric (BNEF 2018), as shown in Figure 15.1.

Rail systems already are powered, significantly, by electricity. So, both rail and buses can be powered by low-carbon sources such as renewable, hydroelectric and nuclear power. Many regions rich in hydroelectricity, such as Norway, Sweden and Brazil, already power their railways with a significant percentage of hydroelectricity. For instance, the Austrian and Swiss railways already run on 93% and 75% renewable power, respectively, mainly hydroelectricity (Energy Matters 2014). The Netherlands achieved its target for its rail network to be 100% powered by a mixture of wind and hydroelectric renewable energy by 2018 a year early, in 2017.

In Germany, Deutsche Bahn, the second largest transport company in the world, has publicly committed to raising the percentage of wind, hydro and solar energy to power its

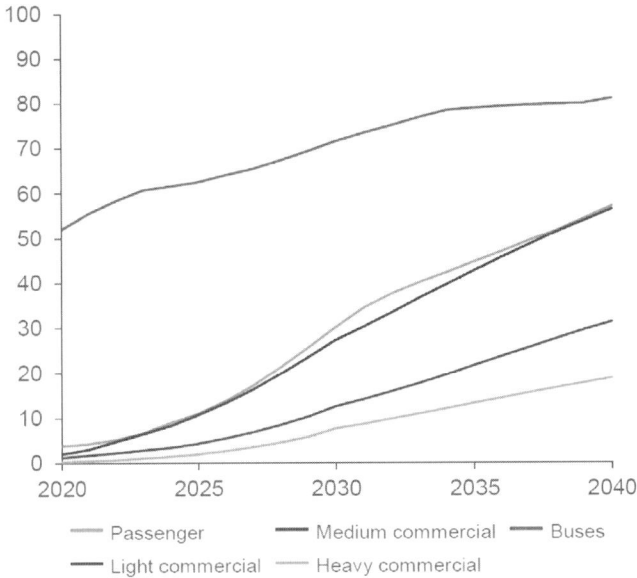

Figure 15.1 Global electric vehicle share of new vehicle sales by segment (% of new sales).
Source: BNEF (2018). For a colour version of this figure, please see the colour plate section.

trains from 28% in 2014 to 100% by 2050, because its customers want it to. In the UK, EDF, the major French company that owns and runs nuclear power, are working with the UK rail network to supply zero-emissions electricity to contribute to powering UK rail (Topham 2013). Scotland is aiming for its rail sector to be largely powered by renewables by 2030. France has over 50% of its rail network electrified. In the next few years, electrification on secondary lines is expected to continue until all rail infrastructure in France has been electrified, thus making it technically possible for it all to be powered by low-carbon sources (Vinot and Coussy 2009). Indian Railways is aiming to secure 25% of its energy requirements from renewable sources by 2025 (Baker 2017).

15.3.5 Utilising Low-Carbon Fuels, Renewables and Electrical Storage to Power Air Travel

The aircraft sector is investing significantly in low-carbon fuels and becoming increasingly powered by electricity, as demonstrated by the examples below:

- Norway's OSM Aviation has ordered, in 2019, 60 all-electric aeroplanes from Colorado-based electric aircraft manufacturer Bye Aerospace. Norway plans to make all short-haul flights of 1.5 hours or less using electric aircraft by 2040 (Agence France-Presse 2018).
- Avinor, the public operator of Norwegian airports, is planning to launch a tender to test a commercial short-haul route by an electric-powered aeroplane in 2025 (Agence France-Presse 2018).

- Zunum Aero, a start-up partly financed by Boeing, plans to have an electric aircraft available by 2022 (Deshayes 2018).
- Airbus, Rolls Royce and Siemens have tested a demonstrator hybrid model from 2017 to 2020 (Airbus n.d., 2017). These developments have the potential to reduce greenhouse gas emissions and also reduce noise levels.

Research is being undertaken into how best to power aircraft entirely from renewables and batteries, as illustrated by the famous Solar Impulse aircraft, which completed a circumnavigation of the globe using only solar power and electric batteries. In the future, this technology could help protect the airline industry from fluctuating fossil fuel prices, which historically have affected profit margins significantly.

15.3.6 Decarbonising Sea Travel

The majority of governments have agreed to a target of 50% GHG emissions reductions in the maritime industry by 2050. Still greater reductions will be needed to achieve the Paris Climate Change Agreement targets for this sector. Manufacturers of passenger ships, cargo ships and ferries are developing ammonia fuel cell (Gallucci 2021), hydrogen fuel cell, hybrid electric and full alternatives to reduce reliance on diesel and heavy fuels. YARA International, a Norwegian chemical company, is working with technology group Kongsberg to produce a zero-emissions electric ship. Norway launched the first hydrogen ferry in 2018 and is aiming to have 60 hydrogen electric fuel cell ferries by 2022. Maersk, the world's largest shipping company, committed in 2018 to reaching carbon neutrality by 2050. Using a 2007 baseline, Maersk has cut carbon dioxide emissions by 46% and is committed to sourcing only carbon-neutral vessels from 2030 to meet its 2050 target.

Pursuing these forms of mitigation in the transport sector does not just improve fuel efficiency, cut costs and cut GHG emissions, they can also lead to numerous adaptation co-benefits. In addition, some adaptation strategies also have mitigation and sustainable development co-benefits. These are considered below.

15.4 Mitigation Strategies with Adaptation and Sustainable Development Synergies

15.4.1 Improving Energy, Food and Water Security

15.4.1.1 Improving Electricity Grid Security Against Heatwaves by Zero-Emission Vehicles with Vehicle-to-Grid Capability

Global warming is forecast to lead to greater intensity of heatwaves and higher extreme summer temperatures, causing a 30-fold increase in air-conditioning electricity demand by 2100. This trend risks higher peak electricity demand periods in summers. Zero-emission vehicles with vehicle-to-grid (V2G) capability, which allow the vehicles' batteries to store and sell electricity back to the grid at peak demand periods, have the potential to contribute significantly to meeting summer peak electricity demand. According to the US Energy Department's Lawrence Berkeley National Laboratory data, if 25% of the US vehicle fleet consisted of plug-in electric vehicles with V2G capability, the total energy storage capacity would be almost 1000 gigawatt-hours, equal to the total electricity-generating capacity of the USA (V2G-Sim n.d.; Casey 2019).

The automotive industry is moving quickly to offer electric vehicles with V2G capability; the Nissan LEAF 2.0 is already on the market at an affordable price. Renault is trialling V2G capability with their new generation Zoe. Honda, Toyota and Hyundai are exploring the potential for V2G in their electric vehicles. The Hyundai fuel cell vehicles have capability for their fuel cells to be used to enable V2G capability. Cars and other vehicles with V2G capability can have batteries up to three times the storage of a household battery, and so offer significant benefits to grid stability. Some large vehicles such as school buses, fire trucks and rubbish trucks are idle for significant periods, so offer significant potential for grid support while meeting their primary functions.

15.4.1.2 Improving Food Security through Reducing Pressures on Food Prices by Reducing Demand for Oil and Biofuels

International food prices have become increasingly correlated with oil prices for several reasons, including that large-scale food production and processing is increasingly energy-intensive; as a result, when fossil fuel prices go up, biofuels become more competitive, so more investment goes into biofuels, which reduces land for agriculture and raises food prices. In the decade 2005–2014, this mechanism caused significant food price spikes that resulted in food riots and deaths in many countries (Tenenbaum 2008). As outlined above, there is potential to reduce demand for oil, and thus long-term oil prices, through fuel efficiency, behaviour change and mode shifting; and the low-carbon electrification of transport systems.

15.4.1.3 Improving Water Security through Reducing Dependency on Water-Intensive Transport Fuels

Compared to business as usual this century, the water intensity of transport fuels can be significantly reduced by reducing demand for biofuels and petrol through sustainable transport behaviour change, mode shifting and the low-carbon electrification of transport systems using water-efficient renewables, such as solar and wind power (i.e. not biofuels).

15.4.2 Addressing Risks from Higher Temperatures

15.4.2.1 Reducing Urban Heat Island Impacts and the Impact of Air-Conditioning Electricity Demand on the Grid: Solar PV Car Parks with Electric Vehicle Charging Stations

Black asphalt car parks contribute significantly to urban heat island effects, raising temperatures to dangerous levels in the peak of summer for those using the car parks. These hot car park surfaces radiate heat, which also increases the air-conditioning load on the adjoining commercial buildings, increasing peak summer electricity demand. On very hot days, this effect increases the risk of air-conditioning electricity demand of buildings exceeding the capacity of the grid, resulting in electricity blackouts. Investing in solar PV structures for commercial and government-run car parks offers multiple benefits. Solar car park structures with electric vehicle charging stations: provide shade over car parks, reducing radiant heat and localised urban heat island impacts; reduce risks of heat stress to those utilising car parks; and reduce air-conditioning loads of adjoining commercial buildings, while helping those adjoining commercial buildings to meet 100% of their electricity needs from on-site solar generation. They can also harvest rainwater for watering of ovals and greenery and provide shading, which has been shown to improve electric vehicle driving range because it reduces the need for drivers to turn on the air-conditioning system, which drains electric vehicle batteries.

For these reasons, investment in solar PV car parks is now growing in many markets globally. Financing via energy performance contracts or power purchase agreements are further enabling greater investment in solar PV car parks.

15.4.2.2 *Reducing Risks of Higher Air Pollution through Reducing Transport-Related Air Pollution and Reducing Risks of Higher Temperatures/Drought*

Mitigation measures, such as reducing demand for fossil-fuel-based transport and transitioning from fossil fuel to low-carbon-based transport systems, reduces urban air pollution/ozone production and black carbon emissions (Ramanathan and Carmichael 2008). This outcome alone justifies the investment costs for cities already battling air pollution. Mitigation also reduces risks of high temperatures, further helping to reduce air pollution levels, as higher temperatures can result in higher urban air pollution through a few mechanisms – for instance, higher temperatures increase the rate of ozone smog formation in major cities. In 2013–2014, California's drought caused several temperature inversions over Los Angeles, with a layer of warmer air trapping cooler air below, concentrating pollution near the ground. High-pressure systems have resulted in fewer storms, causing less circulation and more pollution stagnating over the city.

15.4.2.3 *Reducing Risk of Heatwaves That Cause Blackouts and Stop Low-Carbon Train Services*

The USA, the UK, Australia, Holland, India and China have all seen train services lose their electricity due to blackouts caused by heatwaves. A wide range of climate change mitigation measures can reduce the risks of power failures or blackouts, thus helping cities and countries reduce the risks of failure of electrified transport systems, thus improving resilience. These measures include:

- investing in energy efficiency to reduce peak electricity demand;
- decentralising energy supply, with a wider variety of low-carbon energy sources as well as energy storage, so that cities and transport systems are not so dependent on electricity from only a few highly centralised sources;
- a transition to electric vehicles with V2G technologies that, when idle, can be used to rapidly supply the grid to help reduce the negative impacts of a power failure to the grid;

- solar-powered LED (light-emitting diode) street and traffic lighting so that, if there is a power failure, streets are still well lit to prevent accidents at night.

15.4.2.4 Maintaining Train Services by Stopping Higher Temperatures Warping Train Tracks

On warm days, rails in direct sunshine can be as much as 20 °C above air temperature. This condition results in rails expanding and warping. To prevent this warping, several steps need to be implemented, including:

- as already done in many countries, painting train tracks with white reflective paint to reduce the heat absorbed, lower the temperature by 5–10 °C and reduce the risks of warping, thereby additionally reducing the need for as many 'water trains' to cool train tracks during heatwaves;
- checking the stability of the tracks and, where necessary, replenishing the ballast that surrounds the sleepers and re-tensing continuously welded rail;
- measuring and monitoring rail temperatures by installing probes that provide alerts when track temperatures rise.

These measures help to prevent rail buckling and thus reduce train accidents, save lives and prevent the need for replacing steel rail track, which saves energy and reduces emissions.

15.4.2.5 Maintaining Low-Carbon Road-Based Transport Services by Stopping Higher Temperatures Melting Road Surfaces

Road surfaces made of asphalt will start melting at 50 °C. There are many cases of asphalt road surfaces starting to melt once air temperatures reach over 45 °C if the road is in direct sunlight and there is minimal cooling wind. To address this degradation, local governments put gravel over the road or use asphalt surfaces made using polymer-modified binders, which raise the softening point of the asphalt to around 80 °C. These measures help to avoid the need to use water to cool the surface of roads to prevent melting during heatwaves.

15.5 Integrated Sustainable Urban and Transport Planning and Policy

As Sims et al. (2014) explain, despite there now being many cost-effective, low-carbon, resilient solutions to improve transport systems, 'Financial, institutional, cultural, and legal barriers constrain low-carbon resilient technology uptake and behavioural change. All of these barriers include the high investment costs needed to build low-emissions transport systems, the slow turnover of stock and infrastructure, and the limited impact of a carbon price on petroleum fuels already heavily taxed' (Sims et al. 2014: 604). These barriers can be overcome by governments implementing a mix of integrated urban and transport planning, targeted government behavioural change programmes, investment in active travel and public transport infrastructure, encouragement of bike- and car-sharing schemes, support for the electrification of transport, strong and clear electric vehicle targets and national policies supported by budget cost neutral targeted grants, and access to low-

Sustainable urban transportation systems

Policies and regulations — Advanced technologies

Command and control | Market based | Land use planning | ITS | Engines and fuels

Command and control
- Demand: Vehicle use restrictions; Vehicle quotas; Peak and local restrictions
- Supply: Roads and streets; Public transit; Pedestrian and cycle; Restricted vehicle zones; Restricted parking; HOV

Market based
- Demand: Car sharing; Various taxes; Rebates and incentives; Transit subsidies; Road pricing; Congestion metering; Parking pricing; ALS
- Supply: Pivate transit

Land use planning
- TOD
- HDD
- UGB

ITS
- Intelligent infrastructure: ATMS; ATIS; ETC and ERP; APTS
- Intelligent vehicle: AVCS; Drive by wire; Navigation assistance; Mayday systems

Engines and fuels
- CNG LNG
- Biodiesel
- LPG
- Ethanol and bioethanol
- Hydrogen fuel cells

⟵ **Institutional capacity, environmental mapcat, socio-economic impact, financial feasibility** ⟶

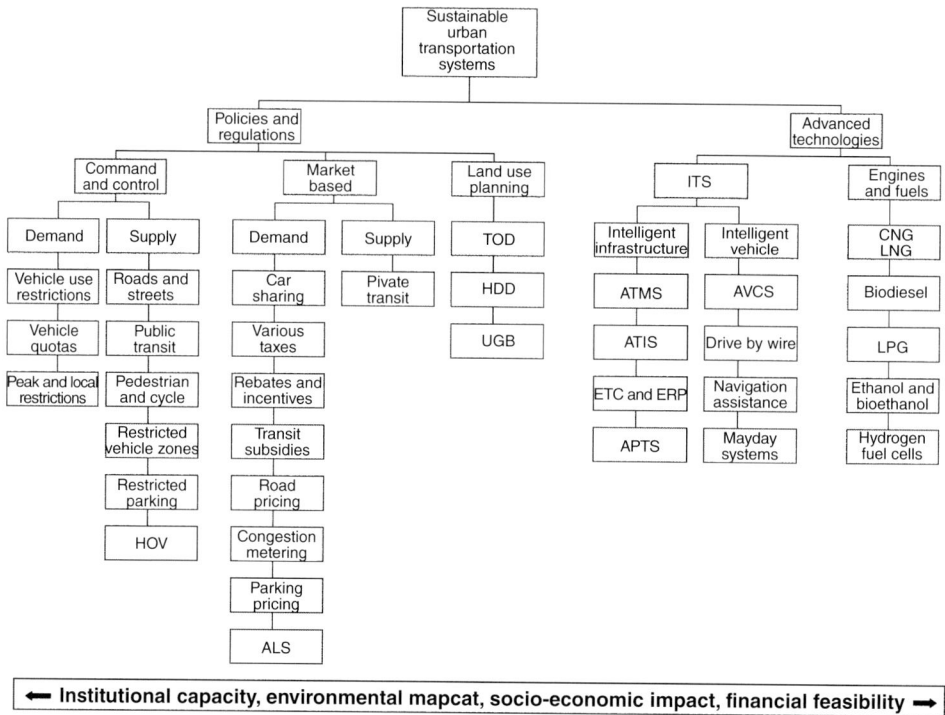

Figure 15.2 Framework of transportation policies that can lead to behavioural measures, technological advances and infrastructural changes.

Note: ALS: area licensing scheme; APTS: advanced public transportation system; ATIS: advanced traveller information system; ATMS: advanced traffic management system; AVCS: advanced vehicle control system; CNG: compressed natural gas; ERP: electronic road pricing; ETC: electronic toll collection; HDD: high-density development; HOV: high-occupancy vehicle; ITS: intelligent transportation system; LNG: liquefied natural gas; LPG: liquefied petroleum gas; TOD: transit-oriented development; UGB: urban growth boundaries.

Source: Moavenzadeh and Markow (2007).

interest climate financing options (i.e. private–public partnerships, climate bonds and green bonds); see Figure 15.2. Nations currently have transport-sector-related policies, incentives, regulation and research and development programmes.

Nations need to take an integrated approach to sustainable transport policy reform that reviews all current policy and incentive settings, and transitions to create a better sustainable transport policy framework to lay a foundation for the zero-carbon transition for the transport sector. Generally, climate change mitigation and adaptation outcomes in transport are more effectively achieved through integrated sustainable development policies (Dhar and Shukla 2015; Garg et al. 2015), which value triple bottom-line outcomes.

15.6 Policy Options

As discussed in Chapter 12, increased use of public transport and active travel (walking, bicycling) helps to reduce emissions while having health and lifestyle co-benefits.

Governments can help with integrated planning to facilitate these developments through appropriate regulatory structures and incentives. A key component of planning should be electrification of transport at all levels, ranging from heavy transport, rail and buses, to cars, motorcycles, bicycles and scooters, and the energy should come from renewable sources. Sharing schemes for bicycles and cars can contribute and are popular. Even in the USA, one of the most car-dependent nations in the world, most trips are of 6 miles or less (Office of Energy Efficiency and Renewable Energy 2018) so bicycles or e-bikes can substitute for second cars in families. Very fast trains can substitute for aircraft on some routes while electric aeroplanes can progressively replace fossil fuel models, at least for short routes.

Some governments are setting phase-out dates for the sale of ICE cars, and vehicle manufacturers are transitioning to electric drivetrains with firm target dates. Governments can assist with planning infrastructure for charging to overcome concerns about longer-distance journeys by electric vehicles.

The cost of batteries has kept the purchase prices of electric vehicles generally above those of ICE cars. As battery prices decline, this gap will close. According to industry sources, this is expected to be around 2025 (BNEF 2017). In the meantime, savings on operational costs – low fuel and maintenance costs – plus government incentives (where they exist) can make electric vehicles attractive, and a new business model – 'buy the car, lease the battery' – can reduce 'sticker shock', as batteries can be around 33% of the price of the car. Battery warranties of 6–8 years and falling replacement costs can reduce anxiety about battery life. There is a good case for governments to encourage take-up of electric vehicles, as they reduce air pollution, are cheaper to operate and, where powered by renewable energy, contribute to reducing GHGs. Where V2G technology is in place they can also provide grid backup power. Governments can offer access to low-interest finance, grants, lower registration fees and traffic privileges such as use of bus routes to encourage the transition. Financial packages can bridge the price gap between electric vehicles and ICE cars. These can be phased out as prices reach parity and operational costs strongly favour electric vehicles.

As of 2019, 50% of new bus sales were electric-powered (BNEF 2019; Green Car Congress 2019). The cost of ownership of electric buses is now lower than for diesel. Vendors are offering attractive packages including loans of vehicles for trials and 'turn-key' business and financing models where the vendor provides infrastructure and mainten-ance services, paid back over time by operational savings and by being able to repurpose and sell bus batteries for a second life in the stationary energy sector. There is a strong case for accelerating the electrification of commercial transport generally as it is responsible for 50% of GHG emissions from the transport sector.

To maximise the GHG reduction benefits of electrification of transport, nations need to simultaneously transition to 100% zero-carbon forms of electricity supply. Governments can also encourage households and businesses to invest in on-site solar PV systems to enable them to charge their electric vehicles with 100% renewable energy while waiting for the rest of the grid to be decarbonised.

15.7 Conclusion

The transport sector has a major role in GHG emissions, so offers considerable scope for their reduction. The sector can rapidly reduce emissions through an integrated approach to encourage greater use of public and active transport, combined with rapid decarbonisation of the electricity grid and take-up of electric bikes, scooters and cars. Electrification of buses is moving rapidly and prospects for the transition of commercial vehicles to electric power are highly favourable, with major vehicle builders set to launch e-trucks and heavy transports. At the same time as mitigation strategies are being pursued, significant attention needs to be paid to adaptation measures, especially where there are synergies between mitigation and adaptation.

References

Agence France-Presse (2018). Norway aims for all short-haul flights to be 100% electric by 2040. *The Guardian*. 18 January. Available at: www.theguardian.com/world/2018/jan/18/norway-aims-for-all-short-haul-flights-to-be-100-electric-by-2040.

Airbus (n.d.). E-Fan X: A giant leap towards zero-emission flight. *Airbus*. Available at: www.airbus.com/innovation/zero-emission/electric-flight/e-fan-x.html.

Airbus (2017). Airbus, Rolls-Royce, and Siemens team up for electric future: Partnership launches E-Fan X hybrid-electric flight demonstrator. *Airbus.com*. 28 November. Available at: www.airbus.com/newsroom/press-releases/en/2017/11/airbus–rolls–royce–and–siemens-team-up-for-electric-future-par.html.

Ambel, C. and Earl, T. (2018*). How to Decarbonise European Transport by 2050*. In-house analysis by Transport and Environment. European Union Commission. Available at: www.transportenvironment.org/sites/te/files/publications/2018_11_2050_synthesis_report_transport_decarbonisation.pdf.

Baker, J. (2017). Indian Railways: Blazing a trail towards renewable energy. *Railway Technology*. 10 November. Available at: www.railway-technology.com/features/indian-railways-blazing-trail-towards-renewable-energy.

BNEF (Bloomberg New Energy Finance) (2017). Electric cars to reach price parity by 2025. *BloombergNEF*. 23 June. Available at: https://about.bnef.com/blog/electric-cars-reach-price-parity-2025/.

BNEF (2018). Electric buses in cities: Driving towards cleaner air and lower CO2. *BloombergNEF*. 10 April. Available at: https://about.bnef.com/blog/electric-buses-cities-driving-towards-cleaner-air-lower-co2/.

BNEF (2019). Electric transport revolution set to spread rapidly into light and medium commercial vehicle market. *BloombergNEF*. 15 May. Available at: https://about.bnef.com/blog/electric-transport-revolution-set-spread-rapidly-light-medium-commercial-vehicle-market/.

Breyer, C., Bogdanov, D., Gulagi, A. et al. (2017). On the role of solar photovoltaics in global energy transition scenarios. *Progress in Photovoltaics: Research and Applications*, 25, 727–745.

Casey, T. (2019). Electric vehicle-to-grid technology gears up for the mass market (#CleanTechnica interview). *CleanTechnica*. 26 January. Available at: https://cleantechnica.com/2019/01/26/electric-vehicle-to-grid-technology-gears-up-for-the-mass-market-cleantechnica-interview.

Chester, M., Horvath, A. and Garnaut, R. (2009). Environmental assessment of passenger transportation should include infrastructure and supply chains. *Environmental Research Letters*, 4, 024008.

Deshayes, P.-H. (2018). Faced with global warming, aviation aims to turn green. *Phys.org*. 8 April. Available at: https://phys.org/news/2018-04-global-aviation-aims-green .html.

Dhar, S. and Shukla, P. (2015). Low carbon scenarios for transport in India: Co-benefits analysis. *Energy Policy*, 81, 186–198.

Energy Matters (2014). Netherlands trains to run on 100% green energy by 2018. *Energy Matters*. 21 May. Available at: www.energymatters.com.au/renewable-news/em4312.

EPA (US Environmental Protection Agency) (2013). Climate impacts on transportation. *US Environmental Protection Agency*. Available at: https://19january2017snapshot .epa.gov/climate-impacts/climate-impacts-transportation_.html.

Eudy, L., Prohaska, R., Kelly, K. and Post, M. (2016). *Foothill Transit Battery Electric Bus Demonstration Results*. Golden, CO: National Renewable Energy Laboratory. Available at: www.nrel.gov/docs/fy16osti/65274.pdf.

Fulton, L. (2017). Can we reach 100 million plug-in electric vehicles by 2050? A new GFEI report says it is possible, but will be challenging. *Global Fuel Economy Initiative*. 31 May. Available at: www.globalfueleconomy.org/blog/2017/may/can-we-reach-100-million-plug-in-electric-vehicles-by-2050-a-new-gfei-report-says-it-is-possible-but-will-be-challenging.

Gallucci, M. (2021). Why the shipping industry is betting big on ammonia. *IEEE Spectrum*. 23 February. Available at: https://spectrum.ieee.org/transportation/marine/why-the-shipping-industry-is-betting-big-on-ammonia.

Garg, A., Naswa, P. and Shukla, P. (2015). Energy infrastructure in India: Profile and risks under climate change. *Energy Policy*, 81, 226–238.

Gota, S., Huizenga, C., Peet, K., Medimorec, N. and Bakker, S. (2019). Decarbonising transport to achieve Paris Agreement targets. *Energy Efficiency*, 12, 363–386.

Green Car Congress (2019). BloombergNEF: Electrics to take 57% of global passenger car sales, 81% of municipal bus sales by 2040. *Green Car Congress*. 16 May. Available at: www.greencarcongress.com/2019/05/20190516-bnef.html.

Grubler, A. and Wilson, C., eds. (2014). *Energy Technology Innovation: Learning from Historical Successes and Failures*. Cambridge: Cambridge University Press.

IATA (International Air Transport Association) (2009). *The IATA Technology Roadmap Report*. International Air Transport Association. Available at: www.iata.org/en/pro grams/environment/technology-roadmap/.

IEA (International Energy Agency) (2012). *Technology Roadmap: Fuel Economy of Road Vehicles*. Paris: International Energy Agency. Available at: https://webstore.iea.org/ technology-roadmap-fuel-economy-of-road-vehicles.

Jacobson, M. Z. (2017). Roadmaps to transition countries to 100% clean, renewable energy for all purposes to curtail global warming, air pollution, and energy risk. *Earth's Future*, 5, 948–952.

Jacobson, M. Z., Delucchi, M. A., Bauer, Z. A. F. et al. (2017). 100% clean and renewable wind, water, and sunlight all-sector energy roadmaps for 139 countries of the world. *Joule*, 1, 108–121.

Kenworthy, J. R. (2017). Is automobile dependence in emerging cities an irresistible force? Perspectives from São Paulo, Taipei, Prague, Mumbai, Shanghai, Beijing, and Guangzhou. *Sustainability*, 9, 1953.

Kuramochi, T., Höhne, N., Schaeffer, M. et al. (2017). Ten key short-term sectoral benchmarks to limit warming to 1.5 °C. *Climate Policy*, 18, 1–19.

Larsen, J. (2013). Bike-sharing programs hit the streets in over 500 cities worldwide. *Earth Policy Institute*. 25 April. Available at: www.earth-policy.org/plan_b_updates/2013/update112.

LTA (Land Transport Authority) (2017). Certificate of entitlement quota for November 2017 to January 2018. *Land Transport Authority*. Available at: www.lta.gov.sg/content/ltagov/en/newsroom/2017/10/2/certificate-of-entitlement-quota-for-novem ber-2017-to-january-2018-and-vehicle-growth-rate-from-february-2018.html.

Luderer, G., Vrontisi, Z., Bertram, C. et al. (2018). Residual fossil CO_2 emissions in 1.5–2 °C pathways. *Nature Climate Change*, 8, 626–633.

Moavenzadeh, F. and Markow, M. (2007). *Moving Millions: Transport Strategies for Sustainable Development in Megacities*. Dordrecht: Springer.

Navigant Research (2013). Sales of electric motorcycles and scooters will reach 6 million annually by 2023. *Navigant Research*. 13 May. Available at: www.navigantresearch.com/newsroom/sales-of-electric-motorcycles-and-scooters-will-reach-6-million-annually-by-2023 (site discontinued).

Newman, P. (2017). Decoupling economic growth from fossil fuels. *Modern Economy*, 8, 791–805.

Office of Energy Efficiency and Renewable Energy (US Department of Energy) (2018). FOTW #1042, August 13, 2018: In 2017 nearly 60% of all vehicle trips were less than six miles. *Energy.gov*. 13 August. Available at: www.energy.gov/eere/vehicles/articles/fotw-1042-august-13-2018-2017-nearly-60-all-vehicle-trips-were-less-six-miles.

Oshiro, K. and Masui, T. (2015). Diffusion of low emission vehicles and their impact on CO_2 emission reduction in Japan. *Energy Policy*, 81I, 215–225.

Ramanathan, V. and Carmichael, G. (2008). Global and regional climate changes due to black carbon. *Nature Geoscience*, 4, 221–227.

RAP (The Regulatory Assistance Project) (2017). *Getting from Here to There: Regulatory Considerations for Transportation Electrification*. The Regulatory Assistance Project. Available at: www.raponline.org/wp-content/uploads/2017/06/RAP-regula tory-considerations-transportation-electrification-2017-may.pdf.

Revi, A., Satterthwaite, D. E., Aragón, F. et al. (2014). Urban areas. In C. B. Field, V. R. Barros, D. J. Dokken et al., eds., *Climate Change 2014: Impacts, Adaptation, and Vulnerability. Part A: Global and Sectoral Aspects. Contribution of the Working Group II to the Fifth Assessment Report of the Intergovernmental Panel on Climate Change*. Cambridge: Cambridge University Press, pp. 535–612. Available at: www.ipcc.ch/site/assets/uploads/2018/02/WGIIAR5-Chap8_FINAL.pdf.

Schwartz, M. and Litman, T. (2008). Evacuation station: The use of public transportation in emergency management planning. *ITE Journal*. Available at: www.vtpi.org/evacuation.pdf.

Sims, R., Schaeffer, R., Creutzig, F. et al. (2014). Transport. In O. Edenhofer, R. Pichs-Madruga, Y. Sokona et al., eds., *Climate Change 2014: Mitigation of Climate Change. Contribution of Working Group III to the Fifth Assessment Report of the Intergovernmental Panel on Climate Change*. Cambridge: Cambridge University Press, pp. 599–670. Available at: www.ipcc.ch/site/assets/uploads/2018/02/ipcc_wg3_ar5_chapter8.pdf.

S&P Global (2020). As battery costs plummet, lithium-ion innovation hits limits, experts say. *S&P Global*. 14 May. Available at www.spglobal.com/marketintelligence/en/news-insights/latest-news-headlines/as-battery-costs-plummet-lithium-ion-innovation-hits-limits-experts-say-58613238.

Tenenbaum, D. J. (2008). Food vs. fuel: Diversion of crops could cause more hunger. *Environmental Health Perspectives*, 116, A254–A257. Available at: www.ncbi.nlm.nih.gov/pmc/articles/PMC2430252/.

The Climate Group (2021). EV100 progress and insights report. *The Climate Group*. Available at: www.theclimategroup.org/ev100-annual-report-2021.

Topham, G. (2013). EDF £3bn deal with Network Rail makes trains 'mainly nuclear powered'. *The Guardian*. 12 January. Available at: www.theguardian.com/uk/2013/jan/11/edf-3bn-network-rail-nuclear.

US Department of Energy (2013). *Transportation Energy Futures Study*. US Department of Energy. Available at: www.energy.gov/eere/analysis/transportation-energy-futures-study.

V2G-Sim (n.d.). Vehicle–grid integration. *V2G-Sim*. Available at: http://v2gsim.lbl.gov/background/vehicle-grid-integration.

Vinot, S. and Coussy, P. (2009). Greenhouse gas emissions and the transport sector. *IFP Panorama*. Available (in French) at: https://inis.iaea.org/collection/NCLCollectionStore/_Public/42/013/42013960.pdf?r=1.

von Weizsacker, E., Hargroves, K., Smith, M., Desha, C. and Stasinopoulos, P. (2009). *Factor Five: Transforming the Global Economy through 80% Improvements in Resource Productivity*. London: Earthscan.

Yang, J., Liu, Y., Qin, P. and Liu, A. A. (2014). *A Review of Beijing's Vehicle Lottery: Short-Term Effects on Vehicle Growth, Congestion, and Fuel Consumption*. Environment for Development Discussion Paper Series EfD DP 14-01. Shanghai: Environment for Development.

Yang, C., Yeh, S., Zakerinia, S. and McCollum, D. (2015). Achieving California's 80% greenhouse gas reduction target in 2050: Technology, policy and scenario analysis using CA-TIMES energy economic systems model. *Energy Policy*, 77, 118–130.

Zurich Financial Services Group (2011). *Supply Chain Resilience 2011*. Zurich: Business Continuity Institute. Available at: www.cips.org/Documents/Resources/Knowledge%20Summary/BCI%20Supply%20Chain%20Resilience%202011%20Public%20Version.pdf.

16

Industry and Manufacturing

MICHAEL H. SMITH, ALAN PEARS, PETER STASINOPOULOS, ALI HASANBEIGI
AND ESHAN AHUJA

Executive Summary

Industry is a major contributor to climate change, both through emissions from its own activities and its influence on emissions from activities upstream and downstream. Many industrial sites, and their supply chains and customers, are vulnerable to the physical consequences of climate change as well as to policy and consumer responses to the climate challenge.

Profits from industrial production are dependent on consumer demand for products and services, and the ways those products and services are provided. Industry is undergoing disruptive changes. Powerful forces such as digitalisation, dematerialisation, decentralisation, optimisation, electrification, efficiency improvement and the emergence of circular economies may influence levels of production, as well as emission intensity of activity. Industrial activity is spreading across other sectors as modular and distributed production technologies evolve and new business models emerge. Industrial firms are facing increasing pressure from regulators, investors and customers to develop and implement strategies to demonstrate corporate responsibility and dramatically cut climate impacts. Emissions from industrial energy use, particularly for high-temperature heat, and process emissions are generally considered difficult to reduce, and substantial capital is invested in existing facilities, so industry faces significant transition challenges. Competition from non-traditional competitors and innovation increases the risk from failure to cut emissions. However, there is enormous potential to capture multiple benefits through aggressive, innovative decarbonisation, utilising energy efficiency and demand management, renewable energy, advanced electrification, carbon capture utilisation and storage and carbon offsets. New business strategies can target growth markets and involve cooperation along supply chains to reduce life-cycle emissions and capture greater business value.

The economic productivity and business competitiveness case for action on climate is strong: it cuts business operational costs and reduces extreme weather risk exposure, while strategically positioning manufacturing companies for fast-growing markets in low-carbon resilient products and services.

This is being increasingly recognised by industry where numerous manufacturers have committed to greenhouse gas reduction targets.

This industry leadership needs to be complemented by effective government policy. The chapter finishes by overviewing a suite of policies that national and subnational government policy-makers can consider, supporting the transition to a zero-carbon resilient industrial sector.

16.1 Introduction

This chapter begins with a discussion of the rapidly changing roles of industrial activities in the global economy, its contribution to climate-changing emissions, and the significant climate-related risks it faces. It questions some widely held assumptions about future trends in demand for industrial output, and the nature of industrial activity: people want services they value, not materials or products. Disruptive solutions combined with increasingly powerful drivers of change create risks for many existing industries: incremental change will not be an adequate response. Comprehensive business strategies that involve supply-chain and life-cycle solutions are outlined. They focus on the capture of opportunity and value, not just the reduction of risk. Specific measures that can cut industrial emissions and capture value are described, along with the broader benefits they may allow businesses to capture. These types of climate action can improve businesses' bottom lines. For instance, DuPont has saved USD 5 billion between 1990 and 2014 from energy efficiency (Vassallo and Smith 2010), stating that 'Upwards of 40% of industry's energy efficiency improvement opportunities can be realized through low or no-cost projects' (Vassallo and Smith 2010). DuPont is not alone. The Dow Chemical Company has saved USD 27 billion in energy efficiency initiatives from 1990 to 2014 (Almaguer 2015). Studies show that companies taking smart action on climate deliver double the financial returns of their rivals (Carrington 2011).

16.2 Context

Industrial progress plays a key role in shaping our future, but it is also shaped by many powerful forces. These forces are reshaping industrial activity within a broader shift towards a global services-based economy, driven by digitalisation, dematerialisation and virtualisation (replacement of physical products, activities and movement by digital and communication-based services). Services now comprise over two-thirds of the global economy, with industry's share declining to around a quarter in 2015. Around 90% of jobs in high-income OECD countries are in the services sector (Deloitte 2018). The energy and emission intensity of the services sector is typically much lower than that of manufacturing, so economic growth based on shifting from industrial activity to services can also be achieved while reducing emissions.

Global carbon dioxide (CO_2) and non-CO_2 emissions from industry, including purchased electricity, process and waste treatment emissions, comprise approximately 30% of total climate impact (IPCC 2013; IEA 2019a). Emissions from combustion of fossil fuels and electricity consumed in industry dominate, but process emissions, for example from cement production, and waste-related emissions are substantial. Emission intensity varies widely across industrial sectors. Metal and mineral processing can have very high energy intensity, while recycling and high-value manufacturing can have quite low emissions per unit of value added. Food processing is a substantial contributor to industry emissions.

But industry's impact is far more pervasive, as its decisions, products and activities influence emissions from mining or harvesting, through operations to end-of-life of equipment and infrastructure. Industrial activity is also blurring across industry sectors and the broader economy. Modular, decentralised production technologies are being integrated with upstream and downstream activities. Agriculture may be combined with processing and tourism activities. A shop may incorporate a micro-brewery or in-store bakery. A design business may have a 3D printer for prototyping, custom and low-volume production. Within industry, roles are evolving rapidly as work is reshaped by smart mechanisation, digitalisation, advanced data analytics, communications and structural change, driven by changing production technologies such as robotics and artificial intelligence (Deloitte 2019).

Industry is a very diverse sector of the economy. Emission and energy intensities vary widely, as does economic and employment contribution. Some industries focus on exports, while others operate largely within local economies. For example, in Australia in 2017–2018, the metal products industry's final energy intensity per unit of gross value added was 2.7 times the industry average, while that of the machinery and equipment industry was a fiftieth of that average (ABS 2019; DoEE 2019). So, the emission impacts of growth of different industries varies widely. It is possible to achieve economic growth while reducing emissions if activity shifts from a high-emission-intensity industry sector to a low-intensity sector.

The impacts of industry on climate, and the environment more broadly, are coming under increasing scrutiny. At the same time, industry is becoming more vulnerable to extreme climate events, flow-on effects from policy changes and shifts in consumer expectations. Innovative business models and technologies are reframing consumer preferences and disrupting markets.

The demand for virgin materials, a major element of industrial energy consumption and emissions, is changing. On one hand, population growth and traditional economic development models grow demand, while dematerialisation, recycling, material substitution, metallurgy, circular economy models, advanced design, optimisation and other factors mentioned earlier reduce demand. Reduction in food waste and changing diets are influencing food processing and agricultural emissions.

At a fundamental level, consumers do not want specific products or energy: they want services that they perceive to be valuable or essential (Pears 2007). Perceptions of value, and how much energy or equipment is needed or desired, are influenced by past experience and technological and social context. In our modern world, these perceptions can change quickly. For example, the need for a car to access a workplace may be replaced by internet communication or electric bike, leading to a dramatic change in the need for materials and transport fuel.

16.2.1 Key Climate and Energy Issues for Industry

Climate change and technological change (within production of materials, goods and consumer preferences, as well as energy supply and use) are fundamental drivers of industry activity and structure.

Humanity's slow response to the messages from climate science now means that industry must transform to very low, or even negative, emission business models as quickly as possible. At the same time, it must manage the direct, upstream and downstream physical impacts of climate change. This may result in devaluation of existing plant and equipment. Investments in incremental change or business as usual will lead to stranded assets, adverse financial impacts and reputational damage. At the same time, there is extreme uncertainty about which options are likely to be 'winners' in a time of disruptive change and volatile politics. So all paths face significant business risks, and many involve transformation of business basics.

This situation makes development of effective policy to influence industrial emissions quite difficult, and individual industrial firms can face challenging barriers to change, such as loss of asset value of fossil-fuel-powered industrial processes, business models producing internal-combustion-engine-powered cars or low profit margins.

That said, increasing numbers of manufacturers have committed to greenhouse gas reduction targets as part of the We Mean Business 'Renewable Energy 100' (RE100), 'Energy Productivity 100' (EP100) and 'Electric Vehicle 100' (EV100) initiatives. By doing so, leading manufacturers are increasingly working to transition to 100% renewable electricity, transition their fleets to electric vehicles and double energy productivity by 2030 from a 2005 baseline.

One useful framework for businesses to apply was produced by the Task Force on Climate-related Financial Disclosures (TCFD 2017), a global group of financial, prudential and banking regulators. This framework, shown in Figure 23.3, summarises the main business risks and opportunities, and considers their varied nature, from legal to reputation, from transitional to long term, across all aspects of a business. Central banks and regulators are encouraging, and beginning to require, businesses to formally and thoroughly address the impacts of climate issues on shareholder value and business viability. This has major implications for industry, as there will be increasing accountability and higher expectations of management to develop and implement practical, profitable business models for transition to, and operation in, a low- or zero-carbon economy.

In practical terms, this requires a business to:

- understand its climate-related risks and opportunities, including disruption of supply chains, sudden government policy changes, changes in consumer behaviour driven by extreme events or ongoing change and potential competitors that may emerge to take advantage of new market opportunities and technologies;
- identify short- and long-term strategies to mitigate or offset relevant emissions and business impacts;
- implement effective measures and strategies to drive necessary changes within the organisation and its supply chains, and maintain revenue and profit;

- understand the climate impacts and vulnerabilities of its supply chains and customers (and potentially its customers' customers), as well as issues over the whole life cycle of materials and products, and how these may impact on its business model; and
- consider the extent to which it should take responsibility for indirect (upstream and downstream) emissions, commonly described as Scope 3 emissions. For example, the Australian Government's voluntary *Climate Active* scheme requires participants to identify and take action on 'significant' Scope 3 emissions (Climateactive 2020). A major resources company BHP (2018) has committed to address its Scope 3 emissions.

For many firms, the scale of emissions upstream and/or downstream from their own activities may far exceed those they directly control. While some may argue that these emissions are the problems of those businesses or customers, the reality is that every element in a supply chain relies on all others in order to satisfy end consumers and governments, to make profits and to operate within the fundamental limits of the environment. As climate pressures build, coordinated, cooperative action will be essential as community pressures and public policy shift. A key issue is that these factors can change very quickly, after an extreme wildfire or storm, after an election or driven by an innovation. Early movers and those with well-designed contingency strategies will be better positioned.

16.3 Understanding and Framing the Context

Industrial firms face serious challenges in adapting to a zero-carbon world. The proportion of overall consumer expenditure on goods and services that flows to industrial businesses is often relatively small, and expenditure on energy and environmental management in industry is even smaller. When energy comprises less than 5% of business input costs in many cases, and is seen as an 'essential input' to production, it is no surprise that energy and its flow-on impacts are low priorities when managing change within many industrial firms. But, given that fossil fuel use and industrial process emissions must be dramatically reduced, they are taking on a business significance much greater than just avoided cost, as reflected in the TCFD's work, discussed earlier.

Many industries produce commodities and items with little differentiation, and these comprise a relatively small proportion of the total wholesale or retail price of goods or services. Many materials and manufactured components are inputs to the production of goods that are assembled in low energy intensity 'light manufacturing' facilities and may not be very visible within final products sold to consumers. Hence, materials, components and sub-assemblies such as motors or compressors tend to compete among suppliers on price, more than high-value attributes. Profit margins can be relatively low. Many consumer decisions that impact on industrial energy and environmental outcomes involve little or no consideration of the energy, resources and environmental impacts associated with the production of items they buy. As economies move towards a low-carbon future, differentiated products, such as 'green' steel, may have higher value especially if their environmental contribution, as well as intrinsic worth, are well marketed.

A time of disruptive change brings both opportunities and risks – and it is often possible to transform risk into opportunity, or to reduce adverse impacts. For example, an open-cut mine may be converted to a pumped hydroelectricity storage, and may have access to good renewable energy resources (e.g. the Kidston Pumped Storage Project in North Queensland, Australia (Australian Renewable Energy Agency 2020)). A retired coal power station or industrial site may have access to major power lines and a skilled workforce. Repurposing a site may defer site rehabilitation costs. An old industrial site may become valuable land as urban development occurs. A substance traditionally seen as a waste may have significant value, and redirecting it may also reduce waste management costs. For example. one dairy processing plant found that its waste effluent contained valuable pharmaceuticals (Meat and Livestock Australia 2011).

A change in response to one issue may bring other benefits, or may facilitate new business opportunities. For example, as discussed in Chapter 12, improving a building's energy efficiency may improve the occupant health and productivity, enhance the asset value, extend the heating and cooling equipment lifetime, reduce the amount of water used in cooling towers and make the building more resilient to extreme temperatures and less vulnerable to power interruptions. Skills gained and intellectual property created in making the change may open new business opportunities.

Within this challenging context, the following sections map out options for industry to transition towards a zero-carbon economy.

16.3.1 Understanding the Systems and Services

It is important to understand what the fundamental services your customers (and, if appropriate, their customers) value, and how you contribute to delivery of those services. For example, an aluminium can competes with other kinds of containers (and post-mix services and reusable containers) that provide drinks that, in turn, slake thirst and/or provide enjoyment or status or health outcomes. Aluminium may also compete with other materials in car manufacture to reduce weight (and fuel consumption). While steel is commonly seen as an essential for building, infrastructure and goods (see Table 16.1), it is increasingly competing with other alternatives, such as avoiding the need for construction (e.g. online shopping, repurposing existing buildings) or substituting with other materials such as engineered timber or plastics.

16.3.2 Understanding the Systems That Deliver the Services

The systems that deliver these services include physical supply chains, business networks, financial systems, cultural networks, etc. Underpinning most systems are many unstated assumptions and 'groupthink', often linked to past and existing practices, sunk capital (human and intellectual as well as physical and financial), boundaries of networks, and the like.

If we step back from a specific site or process, and consider the end-user service provided by a value chain, there may be potential to reduce demand for virgin materials

Table 16.1. *Global steel demand by category and examples of underlying services*

Product category	Percentage of 2011 steel demand	Examples of services
Construction	51.2	Shelter, workplaces, transport infrastructure
Mechanical machinery	14.5	Production of goods
Metal products	12.5	Storage, tools, containers
Automotive	12.0	Mobility, freight movement on roads, access to services
Other transport	4.8	Mobility, freight movement
Electrical equipment	3.0	Energy supply and management
Domestic appliances	2.0	Cleaning, entertainment, comfort, sales
Total	**1518 MtCO$_2$ equivalent**	

Source: World Steel Association (2012).

through virtualisation, dematerialisation, system optimisation, reuse, recycling, material substitution and circular economy models.

We can also redesign production systems. Distributed production and energy supply systems offer important options in some areas. For example, micro-breweries, in-store bakeries linked to tourism, 'dark' kitchens (kitchens shared by several restaurants, cooking food purely for delivery, not eat-in customers) with bicycle delivery services and other means of capturing greater business value are out-competing traditional 'industrial' facilities. New business models can reduce overall value chain costs and/or create higher 'perceived value' among customers. They may create disruptive threats to existing industries.

16.3.3 Understanding the Scale and Nature of System Elements' Influence on Emissions and Climate Impacts, and Their Vulnerability to Climate Change Factors

These include physical, social, political, financial dimensions of climate change itself and the ways people, businesses and governments may respond.

An assessment handbook published under the Australian Government's *Energy Efficiency Opportunities* programme (DRET 2011) provides a comprehensive guide for an organisation or site to develop a strategy based on this approach. This industrial programme delivered large energy savings at an average cost of *minus* AUD 95 per tonne of avoided emissions – that is a *benefit* of AUD 95/tonne of avoided emissions equivalent to about a 2-year simple payback period, compared with the carbon price *paid* by government under its Emission Reduction Fund of around AUD 15/tonne of avoided emissions (ACIL Tasman 2013: 53).

Reports by the Australian Alliance for Energy Productivity (A2EP 2020a) apply 'systems and services' approaches across 'value chains', including 'farm to plate' and

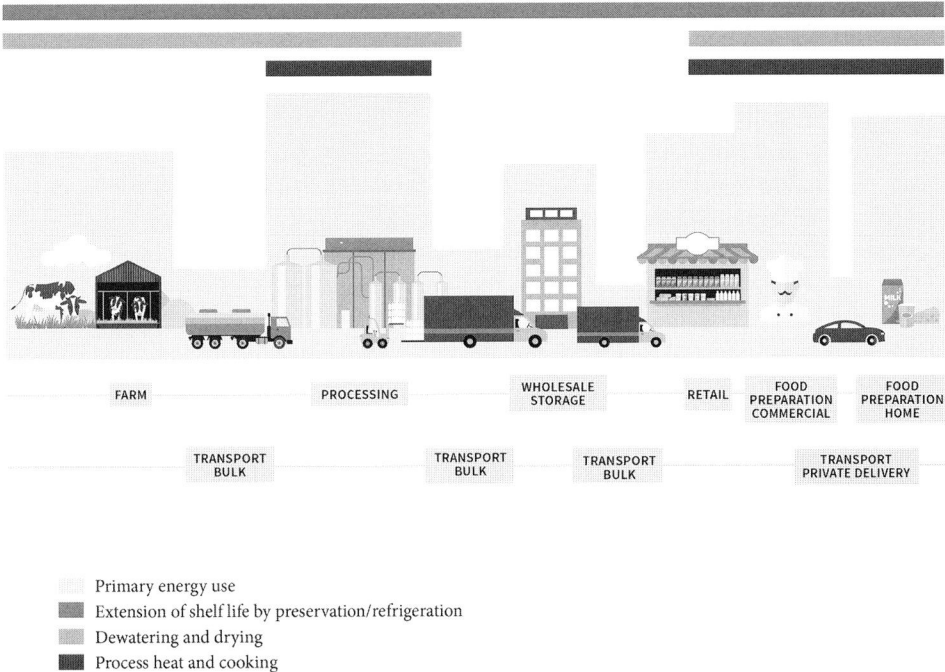

Figure 16.1 Example of the 'value chain' approach applied to the 'farm to plate' system by the Australian Alliance for Energy Productivity. This highlights the interdependence of the participants on each other, and opportunities to optimise overall system productivity.
Source: A2EP (2020a). For a colour version of this figure, please see the colour plate section.

'raw materials to shelter' that help industry to think beyond individual processes and sites, and capture benefits from emerging technologies and practices. Figure 16.1 illustrates this approach. It also highlights the scale of primary energy use (which correlates fairly well with carbon emissions) in the elements of the value chain. While food processing is the biggest energy user, retail, commercial, transport and consumer energy use are comparable.

16.3.4 Identifying Key Drivers, Influencers, Constraints, Boundaries, Beliefs, Assumptions and Priorities

Most decision makers believe they are rational and logical. But this is always within a bounded reality. Adoption of change often requires reframing of perceptions of reality, and development of a practical transition path. It also requires consideration of the broader context within which a business operates.

A range of key actors must be considered, all of which can contribute to or slow achievement of a low-carbon resilient future for this sector. These include:

- industrial and manufacturing companies and their designers, through the process of designing and building manufacturing plants and products;

- financiers: institutional investors; bankers, whose priorities and investment tools often influence the direction of this sector; and insurers who carefully assess long-term risk influence investment decisions;
- industry associations, who reflect and shape industry attitudes and government policies;
- government policy-makers, product standards authorities and regulators, whose actions can address major market and informational failures, and who can implement policies needed to underpin change; and
- consumers, who need to be encouraged and assisted with good information and incentives to purchase the lower-carbon, more resilient manufactured products.

16.3.5 Framing Proposed Change in Ways That Address Business Priorities – or Provide a Path to Change Them

Each group within an organisation, and each organisation within a supply chain, has its own conscious and unconscious priorities. Their understanding of what they do and why they do it is shaped by the factors described above. Change is more likely if it has 'perceived value' to the decision maker(s), who may have priorities different from those of the overall organisation. Growth in output is often valued more than reduction in costs and improvements in efficiency. A manager on a fixed-term contract may heavily discount the value of benefits beyond that time frame.

The International Energy Agency (IEA) has conducted research into the 'multiple benefits' of energy efficiency measures, which are often ignored in evaluation of energy-saving options (IEA 2014). Yet the IEA found that these extra benefits were often worth far more to a business than the value of energy saved.

An example of this was a project conducted by the Australian Alliance for Energy Productivity with a supermarket chain (A2EP 2017). Energy is a very small proportion of overall business cost for a supermarket. But food loss in the supply chain is a major issue, as it involves substantial costs that are important to management, such as loss of potentially saleable goods, unnecessary labour and reputational risk from sale of poor-quality food. Temperature sensors that sent real-time data to 'the cloud' were placed in samples of food at the farm. These provided real-time information on food condition and location. It was now possible to identify when and where temperatures deviated from requirements.

This meant remedial action could be taken by contacting the custodian, and the reasons for deviation identified, such as faulty refrigeration or poor handling practices. It was now possible to identify equipment that required repair or upgrade, target staff training, optimise scheduling of deliveries, manage staff availability and identify and address points of vulnerability to hot weather. This approach saved a lot of money, energy and emissions, as it focused attention on inefficient and faulty equipment, avoided the need to re-cool food that had warmed up and reduced organic food waste. More than that, it opened up a whole new way of operating the business and engaging with its supply chain and customers.

16.4 Climate Response Strategies Addressing Challenges, Risks, Opportunities, Synergies and Nexuses

Many climate actions are simplistically categorised into mitigation or adaptation. Many measures achieve both kinds of outcomes. Further, they also often capture synergies or multiple benefits across a range of areas. As shown in the example above, one climate action can cut emissions, improve business productivity, improve health and reduce vulnerability to changing climate. Indeed, climate and energy actions can contribute to many of the UN's Sustainable Development Goals (SDGs). For example, efficient agricultural equipment, food preservation and cooking help reduce hunger, while an efficient lamp, tablet computer and access to the internet can facilitate education.

In this section, we identify a number of potentially useful climate and energy measures and strategies for industry, and point out how they can contribute to business success and broader social, economic and environmental improvement.

16.4.1 Climate-Driven Risks and Opportunities

Emissions of greenhouse gases are 'energising' climate, and ongoing change will continue, even if atmospheric concentrations are reduced. So, while aggressively cutting emissions, industry must also adapt. Extreme weather events will become more frequent and more intense. Climate zones are shifting and variability is increasing.

The impacts on existing facilities, supply chains and sources of inputs should be evaluated, so that adaptation strategies can be formulated. Planning of new facilities should factor in a range of climate scenarios.

Business impacts could include outcomes such as:

- flooding, inundation and fire damage to sites, infrastructure or communities;
- damage to buildings, plant, infrastructure (on-site and essential services) and sources of process inputs, including agricultural produce;
- interruptions to activities, supply from traditional sources and communications;
- effects on staff activity and operation as a result of extreme conditions of heat, cold or wind, causing thermal stress, disease and shifting beyond designed operating conditions;
- changes to costs, quality, reliability and availability of energy and water;
- increased operating costs, lower production;
- flow-on impacts, such as price changes in one commodity rapidly impacting on other commodity prices (Dobbs et al. 2013).

The following section explores strategies that can incorporate consideration of these potential impacts into constructive actions.

16.5 Climate Change Mitigation and Adaptation Nexus Opportunities

A wise risk management approach for the sector is to identify and implement both profitable and cost-minimising synergistic mitigation and adaptation opportunities to reduce carbon liability and improve the resilience of plants and supply chains. Manufactured product and business model designs should be reviewed to target growing markets for low- and zero-emission products and services that also help customers to adapt to changing climate. Such products and business models are likely to be more profitable in the long term. Industrial and manufacturing companies can simultaneously reduce emissions and exposure to climate-change-related supply chain shocks and input cost rises, while improving energy, water and materials security and reducing vulnerability to extreme weather events through the following strategies.

16.5.1 Strategy 1: Innovating Low-Carbon, Resilient Products

Manufacturers can increase profits by strategically positioning themselves for the rapidly growing global markets in low-carbon, resilient products (Hargroves and Smith 2005). Manufacturers have a unique ability then to contribute to simultaneously improving resilience and reducing greenhouse gas emissions throughout the economy via their products and services, while differentiating them and deepening their customer relationships. This approach responds to life-cycle analysis that shows over 60% of total life-cycle energy and water use occurs over the 5–30 years of customer usage. So, operational energy and water efficiency of manufactured products during the 'use phase' play critical roles in enabling the rest of society to play their parts in mitigating and adapting to climate change. Also, product manufacturers can design products to maximise recyclability to move to a circular economy, with all the mitigation and adaptation co-benefits that entails.

Finally, product designers and manufacturers can play a critical role in developing innovatively designed eco-efficient renewable energy, energy storage, zero-emission vehicles, zero-emission buildings, and agricultural and industrial equipment products that are critical to bringing down the costs of transitioning to a low-carbon future and improving energy, food and water security, as discussed in earlier chapters.

16.5.2 Strategy 2: Transitioning to Low-Carbon Energy Security

Manufacturing companies are increasingly insulating themselves from the risks of rising energy prices and disruptions from extreme weather events to energy supply chains, improving energy security by simultaneously (i) improving energy efficiency, (ii) reducing and reusing waste heat and (iii) meeting their own energy needs with renewable energy options.

Improving Energy Efficiency. The technical potential to improve the energy efficiency of traditional industrial plants is around 25% (Cullen et al. 2010). New plants and processes can save more.

The potential exists to improve motor-driven system energy efficiency (e.g. pumps, fans, conveyors) by 30–50% (IEA 2007). This may be conservative as advances in the 'industrial internet', known as the 'internet of things' or 'Industry 4.0', are already enabling greater real-time control and optimisation of industrial plants to improve the efficiency of motor-driven systems as well as other energy- and water-using systems (A2EP 2020b). In practice, many manufacturing companies achieve greater than 25% efficiency improvement. The Dow Chemical Company has reduced production energy intensity by 40%, saving USD 27 billion between 1990 and 2015 from improved energy efficiency (Almaguer 2015). Through process innovation, additional energy efficiency reductions of 20–50% in energy intensity may potentially be realised, depending on the energy sector (Cullen et al. 2010).

Reducing Waste Heat Losses and Utilising Waste Heat Recovery. The industrial sector currently loses significant amounts of energy as waste heat. Often, heat is produced and distributed at temperatures far higher than required by processes, often as steam. Modular equipment, often using electricity instead of gas, can deliver heat at suitable temperatures more flexibly and with lower distribution losses (Jutsen et al. 2017).

Opportunities to reduce heat losses include better insulation of equipment and piping. There are numerous and widely documented opportunities for waste heat recovery and reuse for medium- to high-temperature manufacturing process applications where the waste heat is high temperature and high quality (US DOE 2008; Smith 2014). And in low- to medium-temperature cases, waste heat recovery and reuse is also possible by using heat pump technologies to increase the temperature and quality of a waste heat stream to a higher, more useful temperature, and to recover latent heat by condensing water vapour from humid air streams. Utilising heat pumps (which can be powered by renewable energy) enables the quality of waste heat to be improved and for it to be reused in manufacturing subsectors with low- to medium-temperature industrial processes such as chemicals and pharmaceuticals, food and beverage and paper and pulp manufacturing (US DOE 2004).

Renewable Energy to Meet Industrial Electricity Demand. The cost of renewable energy, energy efficiency improvement and smart energy management is declining, while performance is improving and the range of applications is expanding. For many, these options now provide the lowest-cost outcome, though barriers include lack of access to best-practice equipment and supply chains, limited access to finance and institutional and political barriers. To illustrate this shift in economics within the Asia-Pacific region, APEC's *Outlook* comments, 'The additional USD 12 trillion of capital investment required in the 2DC [2 degrees Celsius scenario] (compared with the BAU [business as usual]) is more than offset by USD 16 trillion in fuel savings resulting from demand reduction' (APERC 2019: 3).

As explained in Chapters 2–8, renewable energy (such as wind and solar photovoltaic (PV) energy) can provide stable and reliable electricity around the clock if an adequate means for energy storage is also utilised, while solar thermal, hydro generation and geothermal provide continuous energy generation. As a result, around the world different manufacturing plants are demonstrating that it is possible to meet manufacturing plants' electricity needs with renewable energy. In 2019, Nucor Steel began using renewable electricity for one of its steel plants in Sedalia, USA (Walton 2019), and the Tesla Motors factory in Nevada, USA, will run entirely on a combination of solar energy and power from a nearby wind farm and geothermal electricity plant (Klender 2019).

The science of using low-carbon, renewable energy to power electrolysis to produce other metals is rapidly being developed to produce iron (Sadoway 2014; Lord 2020), copper, nickel and titanium (Boston Metal 2020). For instance, Boston Metal's metal oxide electrolysis process is simpler, modular and scalable, so the initial capital cost barrier is reduced compared with traditional blast furnaces.

Hydropower was one of the main sources of energy for the first industrial revolution, and has long been used to manufacture aluminium through electrolysis. It can power many types of metal manufacturing.

In some chemical industry subsectors, and the paper and pulp sector, it is possible for some plants to meet most of their energy needs through cogeneration. Cogeneration systems can be powered by biogas, biomass and other low-carbon renewable fuels.

To summarise, practical and cost-effective options for zero net emission industrial energy include:

- energy efficiency improvements;
- adoption of high efficiency, flexible electric technologies, with demand management and response capability, combined with use of on-site renewable electricity generation complemented by purchase of remotely sourced renewable electricity;
- purchase of firmed (i.e. supply reliability matched to user requirements) renewable electricity or hydropower via long-term power purchase agreements from third parties or energy utilities;
- demand management, demand response and targeted energy efficiency (A2EP 2020a);
- a shift away from use of coal and fossil gas, particularly boilers with lengthy steam distribution systems, unless it can be teamed with carbon capture, use and storage (CCUS);
- use of renewable energy and electricity, and renewably sourced hydrogen, for heat, especially for very high-temperature industrial heat (COAG Energy Council 2019);
- consideration of CCUS, noting that applications of CCUS in fossil fuel production can be controversial, and development of large-scale technologies requires substantial development and investment. However, the IEA sees a potentially important role for it in industry, especially steel, cement, chemicals and (both fossil-fuel-based and renewable) hydrogen production, and considers industrial CCUS, at an estimated USD 15–40/tonne of captured CO_2, likely to be significantly cheaper than CCUS for power generation (IEA 2019b); and
- purchase of carbon offsets for emissions that cannot be avoided. A commitment to achieve net zero emissions using a mix of mitigation actions and offsets also builds in a price on carbon emissions to influence decision-making within an organisation, as each decision either cuts emissions or increases the number of carbon offsets that must be purchased.

Renewable Energy to Meet Industrial Process Heat Demand. Process heating accounts for roughly two-thirds of the total energy used in industrial sectors (EESI 2011). In some sectors, such as glass manufacturing, it can be as much as 80%.

A logical approach to review industrial heat involves consideration of issues such as:

- What service(s) does the process, and the overall value chain, provide?
- Can virtual solutions or alternative materials deliver these services?
- Can product wastage be reduced, such as through improved quality control, production and supply chain management?
- Can the need for heat be avoided in part or all of a process?
- What temperatures are really required to drive processes?
- Can plant layout, flexible operation or use of localised heat sources minimise distribution losses?
- Can waste heat from the process be utilised on-site or nearby?
- Can 'waste heat' from other sources (including neighbouring sites) be utilised, either at its present temperature or upgraded with a heat pump or 'topped up'?
- Can zero-emission sources of heat be utilised, as discussed below? Examples include bioenergy, renewable hydrogen or other combustible gases and electric technologies such as heat pumps, induction, electromagnetic radiation and resistance-using heating elements.

Much of industry's gas is used to deliver relatively low-temperature heat, often by diluting high-temperature steam. This heat could be provided at suitable temperatures by industrial heat pumps, waste heat recovery and/or other options. Gas systems may also use surprisingly large amounts of electricity to run pumps, fans and other equipment: this electricity could be used more productively. Limited monitoring of gas use and analysis of its efficiency means that the potential for cost-effective emissions reduction through efficiency improvement and switching to high-efficiency electric technologies is not well understood at many sites. Rapid changes in performance of technologies, energy prices and business models support innovation.

A decision to buy a new gas or coal boiler or heater instead of a high-efficiency alternative solution locks in higher emissions over the life of that equipment – and makes it relatively more expensive to shift to a zero-emission energy option at the end of that equipment's life. Integration of that equipment into the process, and staff familiarity with its operation, would also create a barrier to change.

While there is much enthusiasm for production of 'zero-emission' hydrogen as a replacement for fossil gas, the reality is not simple. Costs must be substantially reduced before it is competitive (COAG Energy Council 2019). A likely transition pathway to green hydrogen (produced by water splitting using electrolysis) is via production from fossil gas and even coal, which creates even higher emissions than direct use of gas or coal over the next decade or two, unless they are integrated with CCUS. It seems likely that large-scale CCUS from hydrogen production could be more than a decade away, and its cost is uncertain, though cheaper than CCUS for fossil fuel power stations. Several studies suggest it may be necessary if several major industrial sectors such as cement and chemicals are to approach zero-emission production (see IEA 2019b).

Large amounts of high-temperature heat are used in certain industries, such as cement, metals, chemicals and brick production. There have been many studies of ways of reducing emissions from these activities, for example Beyond Zero Emissions (2017), A2EP (2020a) and Fischedick et al. (2014), and examples of some of these activities are presented later in this chapter.

Across national and global industrial process heat demand, close to a half of industrial process heat is required at low to medium temperatures (50–200 °C) which can relatively easily and cost-effectively be provided by a range of renewable energy sources listed below, as outlined by Lovegrove et al. (2015):

- solar thermal technologies can be provided by simple arrays of evacuated tube collectors, and large non-concentrating solar thermal systems have numerous applications in industrial process heat and in agriculture/crop drying (Kempener et al. 2015);
- combustion of biomass or solid waste: combustors are typically combined with steam boilers and could thus substitute for any gas use for process heat in the range 100–350 °C;
- gasification of biomass or solid waste, to produce a renewable gas mixture that can directly replace natural gas in appropriately retuned gas combustion systems;
- solar-thermal-driven gasification of solid materials provides an identical product to conventional gasification, but with the final gas being partly solar derived and partly derived from the original solid in energy content;
- anaerobic digestion in tanks or covered ponds is a proven approach and can be fed with wastes such as sewerage, or effluent from operations such as feedlots or abattoirs, to help provide food processes with biogas to replace natural gas to run their plants;
- direct renewable electricity generation (via for example solar PV or wind systems) can be used to operate electric heaters or heat pumps; and
- geothermal heat sources: there are a range of interesting geothermal prospects under investigation. Most attention has been paid to applications in renewable electricity production. However, underground resources range from temperatures around 100 °C up to 250 °C and a significant number of process heat applications may be addressed within this range.

Renewable technologies also exist that can provide industrial process heat at 250–800 °C (Lovegrove et al. 2015). For instance, Bill Gates has funded a solar thermal start-up, Heliogen, which has produced concentrated solar energy to temperatures greater than 1000 °C. Heliogen can replace the use of fossil fuels in critical industrial processes, including the production of cement, steel and petrochemicals (Heliogen n.d.).

16.5.3 Strategy 3: Improving Materials Security through Transitioning to the Circular Economy

Manufacturing and industry rely on increasingly complex global supply chains for inputs and revenue streams. For instance, in 2011, a study found that '51% of business supply chains were affected by adverse weather over the past year. Over the previous year, 49% of businesses lost productivity from such disruption, while their cost increased by 38% and their revenue decreased by 32%' (Zurich Financial Services Group 2018). The financial benefits of proactively improving 'input materials' security are great, and include insulating against disruption and price shocks, which are significant because resource input costs are as much as 40% of costs in some manufacturing subsectors in Europe. Hence, the financial imperative to improve resource efficiency, recycle and diversify supply chains to source more recycled content is clear. Studies show that if European manufacturers alone did this and thereby contributed to the more circular flow of resources known as the circular economy, it could result in annual cost savings of as much as USD 600 billion per annum (WEF 2014). This is why research has found a correlation between improved resource productivity and increased competitiveness of manufacturing companies (UNEP 2011). It is possible for manufacturers to improve material security and improve profits by:

- reviewing company material input requirements and risks to supply chains due to climate change, extreme weather events, price shocks and political and regulatory risks;
- improving materials efficiency and designing for remanufacture (Stahel 1981);
- closing the loop on materials, increasing recycling and diversifying supply chains to include more local recycled content.

This can achieve significant energy savings, over the life cycle, through remanufacturing and using recycled materials when manufacturing products: examples of potential energy savings are: steel (62–74%), aluminium (95%), copper (70–85%), lead (60–80%), zinc (60–75%), magnesium (97%), paper (65%) and plastics (40–50%) (BMRA n.d.; Ditze and Scharf 2008).

16.5.4 Strategy 4: Improving Food Security and Reducing Food Waste through Supply Chain Innovation

Leading food and beverage processing and manufacturing companies have a critical role to play in improving food security. They can work with and influence the food and water security practices of farmers and suppliers as well as playing a key role to reduce food waste in supply chains. Major droughts around the world have had negative impacts on the food manufacturing sector; from California, USA, to South Africa, to São Paolo, Brazil, to Mediterranean countries, to Australia, with higher primary food commodity input costs (Smith and Dyer 2015). Food and beverage manufacturers, who rely on ingredients from the agricultural sector, therefore have a strong financial interest in working with agricultural suppliers to help improve supply chain food and water security, reduce waste and improve resilience to extreme weather events.

16.6 Examples of Industry Sectors

16.6.1 How Much Production of 'New' Material Is Needed?

As discussed earlier in this chapter, consumers do not want metals, materials or even products. They want valued useful services, which may be provided without physical products, with less material or with non-traditional materials. So, discussion of future options for energy- and/or emission-intensive industries should begin with analysis of the services provided by the industry sector to its downstream customer chain through to the end users of the relevant services, and possible alternative ways of providing those services in low- or zero-carbon ways.

As noted earlier, circular economy models can support recovery and reprocessing of existing materials, which often involves much less energy consumption and emissions. Capture of high-value substances from what has traditionally been seen as 'waste' can also be a useful source of material.

There are substantial stores of materials in landfills, derelict equipment and buildings, 'housefill' and 'garage fill', that could be recovered for reprocessing. Indeed, depending on economics, there can even be a case for accelerated recovery of materials, for example by retiring old, inefficient vehicles, appliances and equipment. This can be done via incentives (e.g. 'cash for clunkers' schemes), strict registration renewal standards (as applied in Japan) or age limits. This can eliminate inefficient or faulty equipment from the stock to save energy and water while making more recycled resources available to substitute for virgin materials.

The industries involved in these approaches to provision of materials and products tend to be more geographically spread, and to involve a wider range of skills; thus, they can support a more diversified workforce and activity in more locations. These models can also involve shorter, more diversified and resilient supply chains.

16.6.2 *'Light' Industries*

Much of the focus on industrial energy and emissions is on energy- or emission-intensive industries. But many of the decisions that influence the amounts and types of materials used across society are made by low-emission-intensity 'light manufacturers', and within supply chains downstream from them, which are much more engaged with end customers. These businesses design products and services, and deliver them. In doing so, they choose materials and influence the operational energy and emissions of their chains of customers.

To the extent that these firms demand low-priced inputs over low environmental impact options, they shape the scale and nature of upstream emission-intensive industries. This is reflected in the 'value chain' approach applied by the Australian Alliance for Energy Productivity, discussed earlier in this chapter (A2EP 2020a). These 'intermediary' businesses, between heavy industry and end consumers, therefore play a critically important role in determining the overall scale of carbon emissions and energy use, both in mining and processing, and in operation of equipment by consumers.

Material inputs comprise a relatively small proportion of their input costs, which are often dominated by labour, capital and product distribution costs. So a shift to low-carbon, more sustainable inputs will often make little difference to the retail price of their output. Further, if a change in materials allows them to convince customers of 'higher value' they may be able to increase profits and offset higher input costs. Given this situation, relatively simply policy changes, new incentives (or cost penalties such as a carbon price), disruptive innovations or a combination of improved information and changed consumer preferences, could have major impacts on business models of heavy industries and demand for high-temperature heat.

As concern about climate change and other environmental issues rises, potential competition presents an increasing risk to the traditional business models of energy- and emission-intensive businesses. Much more study of this risk, and its implications, is needed.

It is risky to assume that forecasts of future demand for many energy- or emission-intensive materials will grow or follow past trends.

16.6.3 Steel

Steel production is a major contributor to global climate impact. Table 16.1 shows a breakdown of the main services now provided by steel. Over half of steel is now used in construction, much of it combined with cement in the form of reinforced concrete. One study has suggested that, if the developing world grows using these materials for buildings and infrastructure, 35–60% of the global carbon budget available to 2050 would be spent (Müller et al. 2013). So the future of substantial steel production is linked to that of cement and building techniques, and to climate policy.

Automotive vehicles use 12% of all steel. A typical car is idle for around 95% of the time, so an enormous amount of financial capital, urban space and material is poorly utilised. To what extent might virtual travel, micro-mobility solutions, public transport, the sharing economy and autonomous vehicles affect this demand, especially in high-density cities? To what extent might these factors offset or outweigh growth in demand for cars in growing economies?

Progress on lightweight, high-strength steels and alloys means less steel is needed for a given service. Sophisticated design optimises steel usage, while alternative design approaches such as tensile structures can use much less steel, and can justify more expensive, stronger cables, including carbon fibre or premium steel. Steel competes with other materials: fibre-reinforced concrete, plastics, carbon fibre, timber or other metals could replace some steel. Alternatives to reinforced concrete in building and infrastructure would also impact on demand for steel.

The Intergovernmental Panel on Climate Change (IPCC) (Fischedick et al. 2014) suggests that significant metal losses in the production and supply chain could be reduced, and that optimised commercial building structural design and extension of building life could cut steel usage in that area by up to three-quarters.

So future demand for steel may differ from present forecasts. Sourcing may shift to more distributed recycling, or smaller, modular steel production options.

The declining cost of renewable hydrogen from concentrated renewable energy sources near iron ore mines, such as in Northern Australia, could lead to substantial production of steel there (Lord 2020). Transporting iron or steel instead of iron ore would reduce transport costs and emissions.

A number of methods are being developed to achieve zero-carbon, resilient steel products, including:

- **Hydrogen steel reduction**: Reduction of iron ore with renewable hydrogen offers the possibility of near zero-emissions steel. Four European steel producers – ArcelorMittal, Salzgitter, Handelsblatt and Voestalpine – are working on replacing coal with zero-carbon hydrogen. ArcelorMittal aims for carbon neutrality by 2050 partly by using hydrogen to reduce steel. A Swedish consortium working on hydrogen-based steel production estimates that at today's prices it would cost 20–30% more than conventional steel, with this difference falling over time (Vogl et al. 2018). Lord (2018) describes several development projects and has mapped out an ambitious scenario for steel production in Northern Australia (Lord 2020), while the Australian Government has developed a comprehensive hydrogen strategy that includes use of hydrogen for steel production (COAG Energy Council 2019).
- **Molten oxide electrolysis**: According to the World Steel Association: 'Molten oxide electrolysis works by passing an electric current through molten slag fed with iron oxide. The iron oxide breaks down into liquid iron and oxygen gas. No carbon dioxide is produced' (World Steel Association 2012: 21). If this is powered by zero-carbon energy sources, this would be a zero-carbon form of making steel. In Europe, ArcelorMittal is developing this approach, leading the EU-funded Siderwin project to develop a pilot plant to demonstrate it. A team at Massachusetts Institute of Technology, USA, has been developing this approach through their spin-off company US Boston Metal (Sadoway 2014; Boston Metal 2020); Boston Metal proposes a modular and scalable approach that involves much less initial capital than a conventional blast furnace.
- **Zero-carbon electric-arc-furnace steel plants**: Steel made from electric arc furnaces powered by electricity can be made zero emissions now if these plants are powered by zero-emission electricity sources. Several firms already do this around the world including Nucor Steel, USA (see Lord 2018).

It is possible to reduce emissions in all existing steel manufacturing plants, as international studies report that significant energy efficiency potential exists, of 22–31% for basic oxygen furnace steelmaking, 22–47% for electric-arc-furnace steelmaking and up to 90% for continuous casting and rolling (US DOE 2010). Investing in energy efficiency opportunities can also yield significant co-benefits and additional cost savings through increased product quality, yield and productivity, and reduced air pollution emissions (Fischedick et al. 2014). These specific energy efficiency and waste heat recovery opportunities are covered in a detailed guide by Worrell et al. (2010), and other guides are available on the International Institute for Industrial Productivity global database for steel (IIP 2015), and in publications such as von Weizsacker et al.'s *Factor Five: Transforming the Global Economy through 80% Improvements in Resource Productivity* (von Weizsacker et al. 2009).

Using renewable sources of reductants can also reduce emissions in steelmaking. The Commonwealth Scientific and Industrial Research Organisation (CSIRO) has found that:

32–58 per cent of coal and coke could be replaced with bio-based charcoal, significantly reducing the net greenhouse gas emissions produced by the industry. This substitution could occur without substantially modifying the steelworks. This is one of the technologies that make up our low-emission integrated steel making process (ISP). The other is the dry granulation of slag, which reduces water usage and recovers waste heat and converts the large volume of by-product slag into a substitute for low emission cement production. Together in this integrated process they could more than halve the steel industry's greenhouse gas emissions without increasing production costs. *(Jahanshahi and Xie 2012)*

16.6.4 Cement Manufacturing

Cement manufacturing, with an estimated 2.2 Gt (gigatonnes) of global annual direct CO_2 emissions from thermal energy consumption and production processes (IEA 2019a), contributes 7% of global greenhouse gas emissions and is the third largest energy-consuming and CO_2-emitting industry sector, after iron/steel and chemicals/plastics. Carbon dioxide emissions from chemical breakdown of inputs and heating limestone to make clinker make up 65% of direct CO_2 emissions from cement production. Consequently, there is a global trend to reduce clinker use and develop substitutes for it, as well as developing new types of cement production that emits no CO_2. There is significant potential to cut emissions in this sector by 2050 through the following innovations:

- Geopolymer cement, which has been commercialised by Zeobond Pty Ltd, can provide 80% emissions reductions (Deventer et al. 2012). This can be used for approximately 40–50% of cement applications, such as cement slabs for buildings, or for use as concrete in roads/footpaths. For other applications, the regulatory and standards hurdles have been (or are in the process of being) addressed in Australia (von Weizsacker et al. 2009; Smith et al. 2010; Deventer et al. 2012).
- There are other efforts around the world to develop lower-carbon cements. For instance, a Swiss team has developed limestone calcined clay cement (LC3) which uses raw materials that are readily available globally, namely calcined clay and ground limestone. They claim to have performance levels as good as, or better than, Portland cement, while reducing CO_2 emission intensity by 40% (Hicks 2014).
- Given that process emissions dominate climate impact, there has been a focus on development of CCUS techniques. The IEA (2019b) estimates CCUS could avoid 0.4 Gt of CO_2 annually in 2060. While most CCUS involves removal of CO_2, then transport and storage, some processes react CO_2 with rock containing magnesium or calcium minerals. This can produce useful building materials. The IEA (2018, 2019b) has also reported that CO_2 can replace water in curing cement.

Hence, HeidelbergCement, the fourth largest global cement manufacturer, has publicly committed to transitioning to net zero emissions by 2050. The IEA (2019b) sees materials efficiency, followed by reduced clinker-to-cement ratios, CCUS and fuel and feedstock switching as major contributors to cement emissions reduction.

Within the building sector, innovation is providing alternatives to concrete in almost every area. Cross-laminated timber construction insulated structural timber panels, insulated timber party walls in apartments, engineered timber and composite beams, plastics and prefabricated construction are among the innovations reducing demand for concrete

(Lippke et al. 2011). It is possible to further reduce the amount of cement used in cities through improved holistic design; for instance, utilising lightweight materials and insulation for residential foundations and flooring as an alternative to concrete slabs (Empower Construction 2018). City governments can encourage the use of nature strips and green urban spaces with shade trees and permeable pathways, instead of concrete, to reduce the urban heat island effect. Instead of using cement-filled trapezoidal storm-water drains, water-sensitive urban design approaches with permeable surfaces can be used to restore the natural hydrology of cities.

It is also possible to significantly reduce greenhouse gas emissions in the cement sector through energy efficiency and waste heat recovery strategies, increasing use of waste feedstocks, utilising solar thermal energy for cement manufacture process heat and utilising CCUS for traditional Portland cement processes, as summarised by the International Institute of Industrial Productivity (IIP 2015).

For a detailed description of energy efficiency and waste heat recovery opportunities, the International Institute for Industrial Productivity provides a database of relevant international publications and tools (IIP 2015). Lawrence Berkeley National Laboratories Industrial Energy Efficiency Group also provides detailed guides and tools for cement industries around the world on how to cost-effectively reduce emissions.

16.6.5 Chemicals and Plastics Manufacturing

According to the IPCC (Fischedick et al. 2014), the chemicals industry emits 15% of direct industrial emissions, but it is also a substantial electricity consumer. A small number of processes dominate this sector's emissions. These include:

ethylene, ammonia, nitric acid, adipic acid and caprolactam used in producing plastics, fertilizer, and synthetic fibres. Emissions arise both from the use of energy in production and from the venting of by-products from the chemical processes. The synthesis of chlorine in chlor-alkali electrolysis is responsible for about 40% of the electricity demand of the chemical industry. *(Fischedick et al. 2014: 759)*

Where chemical products replace more emissions-intensive options, these emissions may be offset by reductions in emissions from other sectors.

For the chemicals and plastic sector, there is significant potential to decarbonise by 2050, through an integrated approach. There is significant potential to reduce the amount of chemicals and plastics used by modern society through:

- sustainable agriculture and dietary behaviour change – eating fresh, organic food, fruit and vegetables rather than heavily processed food has been found to be better for people's health and the planet while using less chemicals, plastics and energy over the life cycle of its production and consumption (Katz and Meller 2014);
- recycling urban organic and sewerage waste streams (as alternatives to artificial fertilisers) (Scialabba and Müller-Lindenlauf 2010);
- reducing chemical usage in industry; and

- replacing plastic office and indoor furniture with timber, hemp or bamboo indoor and office furniture.

Further decarbonisation and higher energy security for the sector can be achieved through a range of strategies including energy efficiency improvements, cogeneration and renewable energy. The chemical and plastics industry can significantly improve energy efficiency through a combination of:

- achieving a 20–40% reduction in greenhouse gas emissions by improved catalysis by 2050 (IEA 2013: 19);
- cutting energy usage in energy-intensive chemical processes such as (i) distillation, (ii) evaporation and (iii) drying by using more energy-efficient forms of chemical and physical separation;
- cutting energy usage in plastics and plastic moulding through strategies such as:
  - improving extruder drive system energy efficiency: by investing in higher-efficiency motors with variable-speed drives and correctly sizing such motors;
  - reducing waste heat losses by insulating the extruder barrel;
  - saving energy in mould closing and clamping systems, by better matching load requirements by using control systems and variable-speed drives;
  - improving the energy efficiency of auxiliary equipment, by optimising the efficiency of existing motors, pumps, air compressors or upgrading them with newer models; and
  - upgrading equipment: for instance, fully electric injection moulding tends to be significantly more energy efficient than hydraulic injection moulding (PACIA 2008);
- utilising low-carbon-powered cogeneration: US studies also show that cogeneration and tri-generation could generate 1.8 times the total amount of existing electricity consumption in US chemical plants, enabling chemical companies to sell electricity back to the grid at a profit (Khrushch et al. 1999); and
- utilising power purchase agreements for power plants with 100% renewable energy: the Dow Chemical Company in the USA is rapidly increasing its purchasing of renewable energy for its plants through power purchase agreements.

Studies show that, in use, the chemicals and plastic sector's products and services, such as insulation products and use of lightweight composite plastic parts in transport vehicle systems, save three times the energy reported in building and transport sectors, through reducing heat losses and enabling less weight to be transported, than the entire amount now used in the production and disposal of all chemical products (ICCA 2009). Producing products which enable greater energy efficiency in other parts of the economy position chemical companies to profit from a transition to a more energy-efficient economy. The following are just some of the enabling technologies the chemical sector has, and is pioneering, to help create a low-carbon future through 'green chemistry':

- ultra-light, strong and recyclable composite plastics for 'light-weighting' transport vehicles;
- novel catalysts enabling the development of second-generation biofuels and high-value materials, as have been pioneered by the Canadian catalyst company Iogen (Iogen 2015);

- longer-lasting rechargeable batteries that can be recycled at end of life;
- carbon sponges that promise a low-energy way to capture carbon;
- replacing wooden crates/storage pallets for shipping with long-lasting strong plastic pallets, reducing demand for timber, thus reducing pressure on deforestation; and
- using recycled plastics to make products that need to last a long time, such as local government park benches and school chairs.

16.7 Government Policy Options to Underpin the Transition

Government and policy-makers have key roles to play to help ensure that policies underpin, rather than undercut, businesses' efforts to improve their climate change mitigation and adaptation performance while also unlocking new sources of productivity growth. There is a clear role for government here because of split incentives (e.g. short-term profits versus long-term sustainability), lack of access to capital and skills for small to medium-sized enterprises, imperfect information for consumers and insufficient research, development and deployment (R&D&D) funding to overcome risk aversion for innovation, as well as the need for structural adjustment and worker retraining packages. Policy options to consider (Smith 2015) include:

- **Economic instruments**, such as putting a price on carbon, carbon emissions trading schemes and cost-reflective energy pricing.
- **Industry policy**, such as the development of and implementation of 'Advanced (low-carbon) Manufacturing' industry policy and associated R&D&D initiatives.
- **Investment in R&D&D policy and programmes** to de-risk take-up and adoption of key enabling technologies for the decarbonisation and climate adaptation of this sector.
- **Structural adjustment** policy, programmes and packages, especially for trade-exposed industry subsectors or where traditional regionally based industrial plants need assistance to transition to the zero-carbon, resilient economy.
- **Access to flexible low-interest financing**, potentially by the creation of government-backed 'Green banks' or 'clean energy financing facilities'.
- **Regulatory instruments**, such as minimum energy and water efficiency performance standards (MEPS) for industrial equipment. Most countries' MEPS schemes apply only to residential appliances, commercial lighting and perhaps motors, but there are many other types of energy- and water-using industrial equipment. Expanding MEPS schemes to cover industrial and manufacturing equipment comprehensively would save many billions of dollars per annum globally in reduced energy and water costs to industry. Other regulatory levers include product standards and product stewardship requirements.
- **Providing information** via creation of sustainability-related opportunities, benchmarking and information-sharing programmes involving the largest 100–1000 manufacturing companies. Ideally, national governments would initiate a high-profile sustainable business programme with one central web portal, with information, resources, education and training resources for all business sectors on how to achieve (and finance) a transition to a low-carbon sustainable and resilient future. Such a programme would require companies above a certain size to complete regular reviews and assessments of their climate change

mitigation and/or adaptation strategies or sustainability performance annually. The State Government of Massachusetts has required annual assessments of sustainable energy, water, materials and chemicals performance. Similarly, national top 500–1000 company energy efficiency opportunity assessment programmes, which require those companies to improve their energy efficiency, have been running in most OECD countries for at least 15 years. Hence, there is an opportunity to build on these programmes and expand them to become low-carbon sustainable business opportunity assessment programmes.

- **Circular economy policies**: It is important for governments to recognise and encourage the key role of manufacturing in developing the circular economy via improved incentives to design for remanufacturing and recyclability and to improve resource productivity. As this chapter showed earlier, incentives and regulations to encourage a transition to the circular economy can both help reduce greenhouse gas emissions and build resilience to disruptions to supply chains from extreme weather events and related price shocks. Several national waste/resource policy reforms are heading in this direction. The European Union's thematic strategies on sustainable use of natural resources (Europa 2005a) and on waste prevention and recycling (Europa 2005b), Japan's Sound Material-Cycle Society (Environment Agency, Japan 2000) and China's circular economy policy reforms provide examples. Common elements of these integrated policies are:
  - ○ targeting primarily the environmental impacts rather than material use per se;
  - ○ taking an integrated life-cycle approach;
  - ○ building partnerships with stakeholders, rather than using command-and-control approaches (OECD 2008);
  - ○ building markets for recycled products and materials;
  - ○ increasing use of economic instruments: for instance, increased landfill levies or restriction of landfilling of recyclables and incentives for recovery of resources from existing landfills and other sources.

These integrated circular economy policies normally place strong emphasis on material efficiency, redesign and reuse of products, recycling of end-of-life materials and products (i.e. considering end-of-life materials and products as resources rather than waste) and environmentally sound management of residues (management standards). This also provides revenue to government, which can fund incentives to improve overall resource productivity in the economy.

- **Sustainable consumption programmes and policies**: An educated population that makes informed consumer choices is essential to support a rapid market shift to low-carbon, resilient products. Government has a key role to play through improved product labelling, product design and efficiency standards. Also, public education on the health, social and economic co-benefits of reducing demand for high-carbon-intensity products and increasing demand for low-carbon products and services is needed.

16.8 Conclusion

Incremental change in industry will be insufficient to meet the urgent global imperative to cut emissions to net zero while coping with the impacts of ongoing climate change.

Traditional industry also faces new challenges from disruptive technological innovation and emerging business models within industry itself, and across all sectors. Risks for industrial firms include physical impacts of climate change together with policy, regulator, consumer and investor responses. The threat of loss of asset value is real and growing.

However, threat is balanced by opportunity. Comprehensive and creative business strategies offer the potential to cut costs, manage transitional stresses, optimise business practices, capture additional value and build new revenue streams. This will involve the application of life-cycle analysis, application of circular economy models and adoption of disruptive technologies and business models. All options involve risks, but business as usual and incremental change involve most risk.

References

A2EP (Australian Alliance for Energy Productivity) (2017). *Food Cold Chain Optimisation: Improving Energy Productivity Using Real Time Food Condition Monitoring through the Chain*. Australian Alliance for Energy Productivity (A2EP). Available at: https://iifiir.org/en/fridoc/food-cold-chain-optimisation-improv ing-energy-productivity-using-real-4770.

A2EP (2020a). *Innovation: The Next Wave*. Australian Alliance for Energy Productivity (A2EP). Available at: https://a2ep.org.au/our-work/innovation-the-next-wave.

A2EP (2020b). *Transforming Energy Productivity in Manufacturing*. Australian Alliance for Energy Productivity (A2EP). Available at: www.2xep.org.au/transforming-energy.html.

ABS (Australian Bureau of Statistics) (2019). 5206.0 – Australian National Accounts: National income, expenditure and product. Table 37: Industry Gross Value Added, Chain volume measures, Annual. *Australian Bureau of Statistics*. Available at: www.abs.gov.au/AUSSTATS/abs@.nsf/DetailsPage/5206.0Mar%202018? OpenDocument.

ACIL Tasman (2013). *Energy Efficiency Opportunities Program End of First Full Five-Year Cycle Evaluation: Final Report*. Canberra: Department of Resources Energy and Tourism.

Almaguer, J. A. (2015). The Dow Chemical Company: Energy management case study. In A. Rossiter and B. Jones, eds., *Energy Management and Efficiency for the Process Industries*. Wiley, pp. 25–36. Available at: https://onlinelibrary.wiley.com/doi/book/10.1002/9781119033226.

APERC (Asia Pacific Energy Research Centre) (2019). *APEC Energy Supply and Demand Outlook*, 7th ed., Vol. 1. Tokyo, Japan: Asia Pacific Energy Research Centre (APEC) and The Institute of Energy Economics, Japan (IEEJ). Available at: www.apec.org/Publications/2019/05/APEC-Energy-Demand-and-Supply-Outlook-7th-Edition—Volume-I.

Australian Renewable Energy Agency (2020). Kidston Pumped Storage Project. *ARENA*. Available at: https://arena.gov.au/projects/kidston-pumped-storage-project/.

Beyond Zero Emissions (2017). *Zero Carbon Industry Plan: Rethinking Cement*. Zero Carbon Australia. Available at: https://bze.org.au/research/manufacturing-industrial-processes/rethinking-cement/.

BHP (2018). Addressing greenhouse gas emissions beyond our operations. *BHP*. 27 August. Available at: www.bhp.com/media-and-insights/prospects/2018/08/addressing-greenhouse-gas-emissions-beyond-our-operations/.

BMRA (British Metal Recycling Association) (n.d.). Cool facts about metals recycling. *BMRA*. Available at: www.recyclemetals.org/about-metal-recycling/cool-facts.html.

Boston Metal (2020). *Boston Metal*. Available at: www.bostonmetal.com/home/.

Carrington, D. (2011). Why low carbon means high profit – eventually. *The Guardian*. 14 September. Available at: www.theguardian.com/environment/damian-carrington-blog/2011/sep/14/carbon-green-economy-emissions.

Climateactive (2020). *ClimateActive.org.au*. Available at: www.climateactive.org.au/.

COAG (Commonwealth of Australian Governments) Energy Council (2019). *Australia's National Hydrogen Strategy*. Canberra: Council of Australian Governments. Available at: www.industry.gov.au/sites/default/files/2019-11/australias-national-hydrogen-strategy.pdf.

Cullen, J., Milford, R. and Allwood, J. (2010). Options for achieving a 50% cut in industrial carbon emissions by 2050. *Environmental Science & Technology*, 44, 1888–1894.

Deloitte (2018). The services powerhouse: Increasingly vital to world economic growth. *Deloitte Insights*. 12 July. Available at: www2.deloitte.com/us/en/insights/multi media/infographics/trade-in-services-economy-growth-infographic.html.

Deloitte (2019). The future of work in manufacturing: What will jobs look like in the digital era? *Deloitte Insights*. 14 April. Available at: www2.deloitte.com/us/en/insights/industry/manufacturing/future-of-work-manufacturing-jobs-in-digital-era.html.

Deventer, J. S. J., Provis, J. L. and Duxson, P. (2012). Technical and commercial progress in the adoption of geopolymer cement. *Minerals Engineering*, 29, 89–104.

Ditze, A. and Scharf, C. (2008). *Recycling of Magnesium*. Clausthal-Zellerfeld: Papierflieger Verl.

Dobbs, R., Oppenheim, J. and Thompson, F. (2013). *Resource Revolution: Tracking Global Commodity Markets*. McKinsey Global Institute. Available at: www .mckinsey.com/business-functions/sustainability/our-insights/resource-revolution-tracking-global-commodity-markets.

DoEE (Australian Department of the Environment and Energy) (2019). Table F: Australian energy statistics. In *Australian Energy Update 2019*. Australian Government Department of the Environment and Energy. Available at: www.energy.gov.au/publi cations/australian-energy-update-2019.

DRET (Australian Department of Resources, Energy and Tourism) (2011). *Energy Efficiency Opportunities: Assessment Handbook*. Australian Government Department of Resources, Energy and Tourism. Available at: www.energy.gov.au/publications/energy-efficiency-opportunities-assessment-handbook.

EESI (Environmental and Energy Study Institute) (2011). *Solar Thermal Energy for Industrial Uses*. Issue brief. Environmental and Energy Study Institute. Available at: www.eesi.org/files/solar_thermal_120111.pdf.

Empower Construction (2018). Fast, efficient and feature packed alternative to concrete slabs. *Empower Construction*. 2 February. Available at: www.empowerconstruction .com.au/news/fast-efficient-and-feature-packed-alternative-to-concrete-slabs.

Environment Agency, Japan (2000). *The Challenge to Establish the Recycling-Based Society: The Basic Law for Establishing the Recycling-Based Society Enacted*. Tokyo: Government of Japan.

Europa (2005a). *Thematic Strategy on the Sustainable Use of Natural Resources*. Munich: European Commission. Available at: www.eea.europa.eu/policy-documents/the matic-strategy-on-the-sustainable.

Europa (2005b). *Thematic Strategy on the Prevention and Recycling of Waste*. Munich: European Commission. Available at: www.eea.europa.eu/policy-documents/the matic-strategy-on-the-prevention.

Fischedick, M., Roy, J., Abdel-Aziz, A. et al. (2014). Industry. In O. Edenhofer, R. Pichs-Madruga, Y. Sokona et al., eds., *Climate Change 2014: Mitigation of Climate Change. Contribution of Working Group III to the Fifth Assessment Report of the Intergovernmental Panel on Climate Change*. Cambridge: Cambridge University Press, pp. 739–810. Available at: www.ipcc.ch/site/assets/uploads/2018/02/ipcc_wg3_ar5_chapter10.pdf.

Hargroves, K. and Smith, M. (2005). *The Natural Advantage of Nations: Business Opportunities, Innovation and Governance in the 21st Century*. London: The Natural Edge Project, Earthscan.

Heliogen (n.d.). *Heliogen*. Available at: https://heliogen.com/.

Hicks, J. (2014). Green cement to help reduce carbon emissions. *Forbes*. 23 June. Available at: www.forbes.com/sites/jenniferhicks/2014/06/23/green-cement-to-help-reduce-carbon-emissions/.

ICCA (International Council of Chemical Associations) (2009). *Innovations for Greenhouse Gas Reductions: A Lifecycle Quantification of Carbon Abatement Solutions Enabled by the Chemical Industry*. International Council of Chemical Associations. Available at: www.americanchemistry.com/Policy/Energy/Climate-Study/Innovations-for-Greenhouse-Gas-Reductions.pdf.

IEA (International Energy Agency) (2007). *Energy Technologies at the Cutting Edge*. Paris: International Energy Agency. Available at: www.iea.org/reports/energy-technologies-at-the-cutting-edge.

IEA (2013). *Technology Roadmap: Energy and GHG Reductions in the Chemical Industry via Catalytic Processes*. Paris: International Energy Agency. Available at: www.iea.org/reports/technology-roadmap-energy-and-ghg-reductions-in-the-chemical-industry-via-catalytic-processes.

IEA (2014). *Capturing the Multiple Benefits of Energy Efficiency*. Paris: International Energy Agency. Available at: webstore.iea.org/capturing-the-multiple-benefits-of-energy-efficiency.

IEA (2018). *Technology Roadmap: Low Carbon Transition in the Cement Industry*. Paris: International Energy Agency.

IEA (2019a). Global CO_2 emissions by sector, 2017. *IEA.org*. 26 November. Available at: www.iea.org/data-and-statistics/charts/global-co2-emissions-by-sector-2017.

IEA (2019b). *Transforming Industry through CCUS*. Paris: International Energy Agency.

IIP (Institute for Industrial Productivity) (2015). Iron and steel. *Industrial Efficiency Technology Database* [database]. Available at: www.iipinetwork.org/wp-content/Ietd/content/iron-and-steel.html.

Iogen (2015). Cellulosic ethanol process. *Iogen Corporation*. Available at: www.iogen.ca/cellulosic_ethanol/index.html.

IPCC (2013). Summary for policymakers. In T. F. Stocker, D. Quin, G.-K. Pattner et al., eds., *Climate Change 2013: The Physical Science Basis. Contribution of Working Group I to the Fifth Assessment Report of the Intergovernmental Panel on Climate Change*. Cambridge: Cambridge University Press. Available at: www.ipcc.ch/site/assets/uploads/2018/02/WG1AR5_SPM_FINAL.pdf.

Jahanshahi, S. and Xie, D. (2012). Current status and future direction of CSIRO's dry slag granulation process with waste heat recovery. Paper presented at 5th International Congress on the Science and Technology of Steelmaking (ICS 2012), Dresden, 1–3 October. Available at: https://publications.csiro.au/rpr/pub?list=BRO&pid=csiro:EP125951&sb=RECENT&n=2&rpp=25&page=1&tr=9&dr=all&dc4.creator=xie,%20dongsheng.

Jutsen, J., Pears, A. and Hutton, L. (2017). *High Temperature Heat Pumps for the Australian Food Industry: Opportunities Assessment*. Sydney: Australian Alliance for Energy Productivity (A2EP). Available at: www.airah.org.au/Content_Files/ Industryresearch/19-09-17_A2EP_HT_Heat_pump_report.pdf.

Katz, D. L. and Meller, S. (2014). Can we say what diet is best for health? *Annual Review of Public Health*, 35, 83–103. Available at: www.researchgate.net/publication/ 260914255_Can_We_Say_What_Diet_Is_Best_for_Health.

Kempener, R., Burch, J., Brunner, C., Navntoft, C. and Mugnier, D. (2015). *Solar Heat for Industrial Processes*. Technology Brief E21. IEA (International Energy Agency)-ETSAP (Energy Technology Systems Analysis Programme) and IRENA (International Renewable Energy Agency). Available at: www.irena.org/publica tions/2015/Jan/Solar-Heat-for-Industrial-Processes.

Khrushch, M., Worrell, E., Price, L., Martin, N. and Einstein, D. (1999). *Carbon Emissions Reduction Potential in the US Chemicals and Pulp and Paper Industries by Applying CHP Technologies*. Report No. LBNL-43739. Berkeley, CA: Ernest Orlando Lawrence Berkeley National Laboratory, Environmental Energy Technologies Division. Available at: www.researchgate.net/publication/315754511_Carbon_ Emissions_Reduction_Potentials_in_Pulp_and_Paper_Mills_by_Applying_ Cogeneration_Technologies.

Klender, J. (2019). Tesla resumes Gigafactory 1 solar panel installations. *Teslarati*. 25 October. Available at www.teslarati.com/tesla-resumes-gigafactory-1-solar-panel-installations/.

Lippke, B., Oneil, E., Harrison, R., Skog, K., Gustavsson, L. and Sathre, R. (2011). Life cycle impacts of forest management and wood utilization on carbon mitigation: Knowns and unknowns. *Future Science: Carbon Management*, 2, 303–333.

Lord, M. (2018). *Electrifying Industry: Zero Carbon Industry Plan*. Melbourne: Beyond Zero Emissions. Available at: https://bze.org.au/wp-content/uploads/2020/12/electri fying-industry-bze-report-2018.pdf.

Lord, M. (2020). *From Mining to Making: Australia's Future in Zero-Emissions Metal*. Melbourne: Energy Transition Hub, University of Melbourne. Available at: www .energy-transition-hub.org/resource/mining-making-australias-future-zero-emissions-metal.

Lovegrove, K., Edwards, S., Jacobson, N. et al. (2015). *Renewable Energy Options for Australian Industrial Gas Users: Background Technical Report*. Background technical report ITP/A0142 rev. 2.0. Canberra: IT Power (Australia) Pty Ltd and ARENA (Australian Renewable Energy Agency). Available at: https://itpau.com.au/wp-con tent/uploads/2018/08/ITP_REOptionsForIndustrialGas_TechReport.compressed.pdf.

Meat and Livestock Australia (2011). *MLA Bioactives Workshop October 2011: Final Report*. Meat and Livestock Australia. Available at: www.mla.com.au/download/ finalreports?itemId=2084.

Müller, D. B., Liu, G., Løvik, A. N. et al. (2013). Carbon emissions of infrastructure development. *Environmental Science & Technology*, 47, 11739–11746.

OECD (2008). *Costs of Environmental Policy Inaction: Summary for Policy-makers*. Paris: OECD.

PACIA (Plastics and Chemicals Industry Association) (2008). *Sustainability Leadership Framework for Plastics and Chemicals Industries*. Melbourne: Plastics and Chemicals Industry Association.

Pears, A. (2007). Imagining Australia's energy services futures. *Futures*, 39, 253–271.

Sadoway, D. R. (2014). *A Technical Feasibility Study of Steelmaking by Molten Oxide Electrolysis*. Fact sheet 9956. Cambridge, MA: Massachusetts Institute of Technology (MIT). Available at: https://doi.org/10.2172/974198.

Scialabba, N. and Müller-Lindenlauf, M. (2010). Organic agriculture and climate change. *Renewable Agriculture and Food Systems*, 25, 158–169.

Smith, M. (2014). *Assessing Climate Change Risks and Opportunities: Industrials, Materials and Manufacturing Sector*. Canberra: The Investor Group on Climate Change (IGCC) and The Australian National University (ANU). Available at: https://igcc.org.au/wp-content/uploads/2020/06/Assessing-Climate-Change-Risks-and-Opportunities-for-Investors.pdf.

Smith, M. (2015). Industry and manufacturing. In D. Lindenmayer, S. Dovers and S. Morton, eds., *Ten Commitments Revisited: Securing Australia's Future Environment*. Collingwood: Commonwealth Scientific and Industrial Research Organisation Publishing.

Smith, M. and Dyer, G. (2015). Strategies to mainstream climate change, energy, water and food security nexus knowledge and skills. In J. Pittock, K. Hussey and S. Dovers, eds., *Climate, Energy and Water*. New York: Cambridge University Press, pp. 231–252.

Smith, M., Hargroves, K., Desha, C. and Stasinopoulos, P. (2010). Factor 5 in eco-cement: Zeobond Pty Ltd. *Ecos Magazine*, 21, 149. Available at: www.ecosmagazine.com/?act=view_file&file_id=EC149p21.pdf.

Stahel, W. (1981). *Jobs for Tomorrow: The Potential for Substituting Manpower for Energy*. New York: Vantage Press.

TCFD (Task Force on Climate-related Financial Disclosures) (2017). *Recommendations of the Task Force on Climate-related Financial Disclosures*. Final report. Task Force on Climate-related Financial Disclosures. Available at: www.fsb-tcfd.org/publications/recommendations-report/.

UNEP (UN Environment Programme) (2011). *Towards a Green Economy: Pathways to Sustainable Development and Poverty Eradication*. UN Environment Programme. Available at: https://sustainabledevelopment.un.org/index.php?page=view&type=400&nr=126&menu=35.

US DOE (Department of Energy) (2004). *Waste Heat Reduction and Recovery for Improving Furnace Efficiency Productivity and Emissions Performance: A Best Practices Process Heating Technical Brief*. United States Department of Energy. Available at: www.energy.gov/sites/prod/files/2014/05/f15/35876.pdf.

US DOE (2008). *Waste Heat Recovery: Technology and Opportunities in U.S. Industry*. United States Department of Energy. Available at: www.osti.gov/biblio/1218716-waste-heat-recovery-technology-opportunities-industry.

US DOE (2010). *Steel Industry Technology Roadmap*. United States Department of Energy.

Vassallo, D. and Smith, C. (2010). *Optimizing the Success of Industrial Energy Efficiency Improvement Programs by Driving Culture Change*. DuPont Energy Efficiency Culture White Paper.

Vogl, V., Åhman, M. and Nilsson, L. J. (2018). Assessment of hydrogen direct reduction for fossil-free steelmaking. *Journal of Cleaner Production*, 203, 736–745.

von Weizsacker, E., Hargroves, K., Smith, M., Desha, C. and Stasinopoulos, P. (2009). *Factor Five: Transforming the Global Economy through 80% Improvements in Resource Productivity*. London: Earthscan.

Walton, E. (2019). Nucor to use clean energy to power mill. *Sedalia Democrat*. 14 December. Available at: www.sedaliademocrat.com/news/nucor-to-use-clean-energy-to-power-mill/article_7c222da4-1e44-11ea-93b3-037d5308eeae.html.

WEF (World Economic Forum) (2014). *Towards the Circular Economy: Accelerating the Scale-up across Global Supply Chains*. Geneva: World Economic Forum. Available at: www3.weforum.org/docs/WEF_ENV_TowardsCircularEconomy_Report_2014.pdf.

World Steel Association (2012). *Sustainable Steel: At the Core of a Green Economy.* World Steel Association. Available at: www.worldsteel.org/en/dam/jcr:5b246502-df29-4d8b-92bb-afb2dc27ed4f/Sustainable-steel-at-the-core-of-a-green-economy.pdf.

Worrell, E., Blinde, P., Neelis, M., Blomen, E. and Masanet, E. (2010). *Energy Efficiency Improvement and Cost Saving Opportunities for the U.S. Iron and Steel Industry: An ENERGY STAR Guide for Energy and Plant Managers.* Berkeley, CA: Enerst Orlando Lawrence Berkeley National Laboratory, Environmental Energy Technologies Division. Available at: www.energystar.gov/sites/default/files/buildings/tools/Iron_Steel_Guide.pdf.

Zurich Financial Services Group (2018). *Supply Chain Resilience 2011.* Zurich: Business Continuity Institute. Available at: www.thebci.org/uploads/assets/uploaded/6bd728bd-bf0e-4eb7-b15fa67164eb9484.pdf.

Land Use, Forests and Agriculture

17

Land Use

Executive Summary

Globally, the land-use sector produces the second highest level of greenhouse gas emissions, after the energy sector, and the highest in some lesser-developed regions. The land can also be a major sink for anthropogenic emissions.

The land sector is at risk because plant growth and distribution are dependent on climate conditions and their variability. Agricultural and forest productivity are affected by the combined pressures of climate change, demand for land resources for food production, deforestation and degradation of natural systems, and population pressure. Of the world's forests, around 30% are in a degraded condition having undergone human modification, and only 36% remain in primary condition.

Shifts in plant and animal distributions and ecosystems occur in response to changes in climate conditions and the incidence and severity of extreme events: higher temperatures, reduction in water availability in some areas and flooding in others, droughts and wildfires.

The global carbon cycle is closed, with a fixed total quantity, but this carbon is continually exchanged between the land, oceans and the atmosphere. The combustion of fossil fuels adds carbon dioxide (CO_2) to the atmosphere, and about half these emissions due to human activities remain as an elevated atmospheric CO_2 concentration, lasting beyond human time spans and with some impacts occurring on timescales of millennia.

The land sector has many opportunities for mitigation, including options that are cost-effective and can be implemented quickly. Reduction in emissions from both land-use activities and fossil fuel use are needed.

Priorities for mitigation of, and adaptation to, climate change are: first, avoiding emissions by protecting carbon stocks in natural ecosystems through ceasing deforestation and conserving forests; second, sequestering carbon through restoration of degraded ecosystems; and third, reducing emissions through transferring wood production to plantations on existing cleared land, improving the efficiency of processing of harvested wood including reducing waste, producing higher-value wood products, substituting emissions-intensive building materials with wood and recycling.

The chapter describes a comprehensive carbon accounting system that accounts separately for stocks and flows of carbon that have different longevity and reversibility: biocarbon (plant biomass, animals and soil organic matter), which is constantly exchanged between the land, atmosphere and oceans; and geocarbon (oil, gas, coal, limestone, carbonate rocks), which remains static until disturbed by human activity and results in a one-way flow to the atmosphere over human timescales.

In carbon accounting policies, land-based abatement activities have been suggested as 'offsetting' emissions from combustion of fossil fuels. However, this is a false assertion as land-based activities replace the carbon previously lost from the land sector. This is an important mitigation activity in its own right, but it does not replace the carbon lost through fossil fuel combustion. Additionally, the stock of carbon that can be sequestered by the land sector is limited by its productive capacity, which is determined by the availability of water and nutrients and suitable climate conditions.

Mechanisms for action in the land-use sector are described, which require changes to planning frameworks to take into account mitigation and adaptation; and a multi-stakeholder approach is presented encompassing a wide range of mechanisms. Examples of adaptation options are provided in the specific chapters on forests and agriculture.

17.1 Introduction

The goal of keeping global warming to well below 2 °C (over pre-industrial levels) was set with the aim of preventing dangerous anthropogenic interference with the climate system that would expose humanity to the risk of major changes in climate and extreme events resulting in reduced well-being. The basis for setting this target was to allow natural ecosystems and human modified systems to adapt to climate change. Changes in the global carbon cycle due to human activities demonstrate the risks of changes in the land sector because the climate and carbon cycle have a positive feedback. Changes in climate affect photosynthesis, respiration and organic matter decomposition, all of which influence the global land–atmosphere carbon flux. Changes in this flux that increase emissions contribute to the climate change problem. Systems of accounting for the stocks (how much is in living and dead biomass, soil carbon, etc.) and flows of carbon (how much moves from one stock to another in a given time period) must ensure that mitigation and adaptation benefits are maximised. This is achieved by prioritising activities from among the range of potential opportunities for mitigation. Mechanisms for action on mitigation and adaptation rely on developing new planning frameworks that incorporate a changing climate, adaptive man-agement and reliance on renewable resources across all levels of society.

Globally, the land-use sector is the second-largest emitting sector after energy, and higher than the transport sector. In low- to middle-income countries, relative emissions from the land sector are far higher, often being the highest emitting sector (see Figure 17.1). Emissions in the land sector are due to degradation and deforestation, with around 30% of the world's forests being in a degraded condition due to human modification, 34% of forests having been cleared, and only 36% remaining in a primary condition (Mackey et al. 2020). Human disturbance of the land sector has resulted in the loss of about 150 GtC (gigatonnes of carbon), equivalent to 35% of the accumulated anthropogenic emissions (Houghton 2007, 2008; FAO 2010; IPCC 2014). Net emissions of carbon from the land

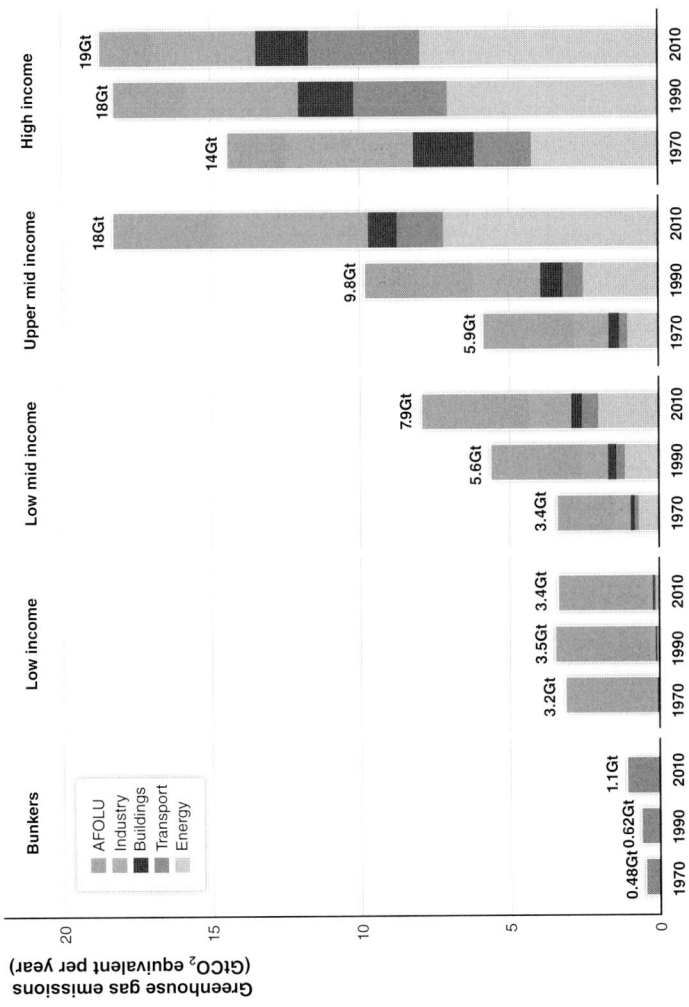

Figure 17.1 Percentage greenhouse gas emissions by sector, for low- to high-income countries.

Note: AFOLU = agriculture, forestry and other land use. For a colour version of this figure, please see the colour plate section.

Source: Figure Ts.3 (fragment) from Edenhofer et al. (2014).

sector due to land use, land-use change and forestry represented 12% of the annual increase in atmospheric carbon dioxide (CO_2) concentration by 2000–2009 (Friedlingstein et al. 2010; IPCC 2014). The land sector thus has many opportunities for mitigation, including options that are cost-effective and can be implemented rapidly (Richards and Stokes 2004; Benndorf et al. 2007; Nabuurs et al. 2007; IPCC 2014; Mackey et al. 2020).

17.2 Climate Change Risks

The land sector is at high risk because of the combined effects of climate change, increasing demand for agricultural and forest commodities, increasing population pressure for land and resources and over-exploitation and degradation of natural resources. These effects reduce the provision of ecosystem services, such as water supply, carbon storage, biodiversity and soil fertility, and hence lead to the loss of productive capacity of the land.

The land sector is vulnerable to climate change because plant establishment, growth and species distributions and survival are highly dependent on climate variables. Vulnerability of ecosystems to climate change depends on the sensitivity of specific ecosystems to climate conditions and their capacity to adapt to changes. Climate change involves both shifts in average climatic conditions and increased climatic variability. Changes in average climatic conditions generally involve higher temperatures, but are region-specific in terms of changes in diurnal and seasonal patterns of temperature and precipitation. Increased climatic variability is manifest particularly in a higher frequency and magnitude of extreme events, such as heatwaves, droughts, wildfires, flooding, heavy snowfalls, hurricanes and cyclones. Changes in these variables affect crop yields, livestock and forest productivity, and the distribution and survival of species including pest and disease organisms.

The impact of climate change is highly variable regionally, for different ecosystems and human production systems. Quantitative spatial scenarios of climate change risks to ecosystems globally were analysed using outputs from general circulation models as inputs to a dynamic global vegetation model (Scholze et al. 2006). This modelling predicted that there was a 44% risk of the land sector becoming a carbon source this century under a scenario of more than 3 °C warming, implying a positive climate feedback. A high risk of forest loss is predicted for Eurasia, eastern China, Canada, Central America and Amazonia, with forest extensions into the Arctic and semi-arid savannas; more frequent wildfires in Amazonia, boreal forests and many semi-arid regions including Australia; more run-off north of 50° N and in tropical Africa and north-western South America; and less run-off in West Africa, Central America, southern Europe and the eastern USA. Such changes in climate patterns would likely result in biome shifts. Some biome shifts are predicted even for global warming of less than 2 °C (Scholze et al. 2006). Risks to agricultural and forest production occur because of the effect of higher temperatures, reduced water availability or flooding, and an increased frequency of extreme events on the growth and survival of plants, animals, pest and disease organisms.

Solutions to these risks of climate change in the land sector can be achieved through management activities that have both mitigation and adaptation benefits. With scarce resources, it is important to identify forms of investment and management actions that act synergistically to produce multiple benefits. To date, most discussions have focused on strategies either to mitigate or adapt to climate change in the land sector. However, it is important to consider potential synergies where both activities are possible simultaneously as integrated response options based on land management, as this can significantly increase efficiencies and positive outcomes (Smith et al. 2019).

17.3 Opportunities for Mitigation

The land sector offers opportunities for climate change mitigation; but there are limits. Removals of carbon from the atmosphere by uptake in the land sector represent replacing the stock that was previously depleted. Emissions reductions are required from both the fossil fuel and land sectors if the Paris Agreement[1] goals are to be met. The role of the land sector as both a sink and source in the global carbon cycle and the historical changes among the global reservoirs are described in Box 17.1 and illustrated in Figure 17.2. Cycles of the other main greenhouse gases of methane and nitrous oxide are described in Boxes 17.2 and 17.3.

Potential mitigation activities should be assessed in terms of the magnitude, time frame and cost of the changes in carbon stocks. From the perspective of the carbon cycle, it is the total amount of carbon and the length of time it is stored in the land sector that influence the carbon stock in the atmosphere. Assessment of these criteria allows ranking of carbon stocks in reservoirs and prioritising activities according to their relative benefit for climate change mitigation. Prioritisation should be: first, avoiding emissions; second, sequestration; and third, reducing emissions (Mackey et al. 2013). Such an assessment of criteria and prioritisation will ensure that incentives for changing land management practices have the maximum benefit for climate change mitigation.

The land-use sector has a key role in mitigation of climate change, through management of both biophysical processes and human activities. As Figure 17.1 shows, emissions from the land-use sector contribute the majority of greenhouse gas emissions for many low- and lower-middle-income countries. Opportunities to create change should be viewed in the context of the significant contribution of the land sector to national economies and sustainable development through impact on economic growth, employment, development goals and food security.

17.3.1 Avoiding Emissions

The most effective form of climate change mitigation is avoiding carbon emissions from all sources; both fossil fuel and land use. Avoiding emissions is more effective than removal of CO_2 from an elevated concentration in the atmosphere. This conclusion is based on the understanding of the lifetime of the airborne fraction of a pulse of CO_2 (Mackey et al. 2013).

---

[1] *Paris Agreement Under the United Nations Framework Convention on Climate Change*, opened for signature 16 February 2016. Available at: https://unfccc.int/process-and-meetings/the-paris-agreement/the-paris-agreement.

Box 17.1 **Changes in the Global Cycles of Carbon Due to Human Activities**

The major reservoirs of the global carbon cycle are the atmosphere, land, oceans and fossil fuels (Figure 17.2). The amount of carbon in the Earth system is fixed, and it is stored in and transferred between these reservoirs in different forms. The carbon cycle is thus a closed cycle.

The land and ocean reservoirs exchange carbon with the atmosphere in the form of two-way flows. By contrast, the fossil fuel reservoir did not have an active exchange of carbon within the cycle until human activities caused combustion and emissions of CO_2, which is a one-way flow to the atmosphere (on human timescales).

Time periods of change in the carbon cycle correspond to major phase shifts in the magnitude of the human environmental footprint in terms of land clearing and use of fossil fuels:

(a) Pre-agricultural era (>8000 BCE), with no distinguishable net change in the carbon stocks of reservoirs.

(b) Pre-industrial era (8000 BCE to 1850 CE), with a depletion in land carbon due to human land-use activities.

(c) Contemporary era (1850 CE to present), with depletion of land carbon and fossil fuel carbon and increase in atmospheric and ocean carbon.

(d) Hypothetical and unachievable case of refilling the land stock. This case would occur if all previously cleared and degraded land was returned to its pre-agricultural carbon stock, and there were zero emissions from the fossil fuel stock. Sequestration in the land stock is limited

(a) Pre-agricultural era carbon stocks

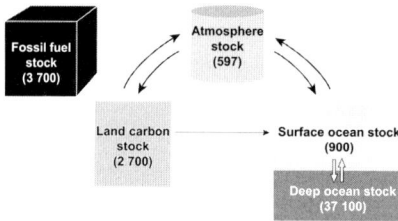

Fossil fuel stock (3 700)
Atmosphere stock (597)
Land carbon stock (2 700) → Surface ocean stock (900)
Deep ocean stock (37 100)

(b) Pre-industrial era changes in carbon stocks and flows

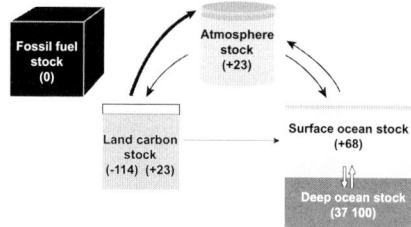

Fossil fuel stock (0)
Atmosphere stock (+23)
Land carbon stock (-114) (+23)
Surface ocean stock (+68)
Deep ocean stock (37 100)

(c) Contemporary era changes in carbon stocks and flows

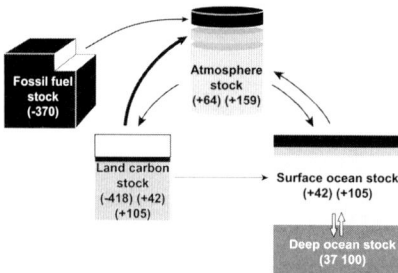

Fossil fuel stock (-370)
Atmosphere stock (+64) (+159)
Land carbon stock (-418) (+42) (+105) → Surface ocean stock (+42) (+105)
Deep ocean stock (37 100)

(d) Hypothetical re-filling of the land stock

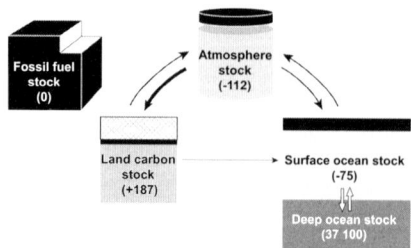

Fossil fuel stock (0)
Atmosphere stock (-112)
Land carbon stock (+187) → Surface ocean stock (-75)
Deep ocean stock (37 100)

Figure 17.2 Changes in the primary stocks of the global carbon cycle.
Source: Mackey et al. (2013). For a colour version of this figure, please see the colour plate section.

to the amount of carbon that was depleted from previous land use. A small amount of fossil fuel emissions potentially could be stored as a result of the CO_2 fertilisation effect.

The numbers in parentheses (GtC) are indicative estimates of the carbon stocks in (a), and changes in the stock in (b) to (d). Arrows represent the direction of carbon flows between reservoirs over the era. Arrows in (a) represent the natural carbon cycle and arrows in (b) to (d) represent the impact of human activities. The impact of human activities of land-use change and combustion of fossil fuels in transferring carbon stocks between reservoirs is shown as colour-coded slices (not drawn to scale) (Mackey et al. 2013).

Currently, the land and oceans act as sinks, with greater removals of CO_2 from the atmosphere than emissions. Globally, the land and oceans had removed an estimated 56% of all CO_2 emitted from human activities from 1958 to 2010 (GCP 2011). The land is currently acting as a sink, even though emissions from deforestation and degradation are continuing, because of a number of factors that are resulting in increased net plant productivity. Plant productivity is increased by an elevated atmospheric CO_2 concentration that increases rates of photosynthesis (the CO_2 fertilisation effect), although this effect is constrained by nutrient availability and, often, water availability; the lengthening of the growing season due to increases in temperature in northern temperate regions; and nitrogen deposition in northern temperature regions. However, there are competing processes that potentially increase or decrease regional land carbon stocks, particularly related to changes in water availability in different regions. Projections of the land and ocean carbon sinks have indicated that future climate change would lead to a reduced efficiency of these sinks to absorb anthropogenic carbon emissions, leading to a larger fraction of the atmospheric CO_2 remaining airborne, and thus amplifying global warming (Friedlingstein et al. 2006).

The residence time of a unit of CO_2 in the atmosphere is complex and depends on different processes that take up CO_2, each of which has different time constants. About 60% of the CO_2 emitted into the atmosphere is removed on a timescale of 100 years by the land and ocean, in approximately equal proportions. The longevity of the last 20–35% of emissions in the atmosphere is estimated to be 2–20 millennia. These proportions vary depending on other climate variables. Final removal of CO_2 from the atmosphere occurs only when it has completely dissolved in the deep ocean, which requires the concurrent dissolution of carbonates from ocean sediments (about 5000 to 10 000 years) and enhanced weathering of silicate rocks (about 100 000 years) (Archer et al. 2009). Even if scenario (d) was achieved, after 100 years there would still be an extra 134 GtC in the atmosphere compared with the pre-agricultural era due to fossil fuel emissions (Plattner et al. 2008; Le Quéré et al. 2009, 2013).

The essence of the climate change problem is that permanent removals from the atmosphere as carbon deposits in ocean sediments are not sufficient to neutralise fossil fuel stock losses which are not recoverable on a human timescale (Feely et al. 2004; Archer et al. 2009). Much of the fossil fuel emissions that are transferred into the biosphere and oceans are stored there only temporarily; only an estimated 0.1 GtC per year is removed permanently as ocean sediments (Feely et al. 2004), representing only 1.9% of the current annual increase in atmospheric concentrations.

Box 17.2 **The Methane Cycle**

Methane (CH_4) is also a key greenhouse gas associated with land-based systems, especially agriculture. The natural methane cycle consists largely of methane from biological sources, but also some releases from geological reservoirs. Natural biological sources include methane from decomposition in wetlands, tundra and other low-oxygen environments, from wild ruminant animals, from termites and from fires. Human activities have increased methane emissions particularly via increased ruminant livestock, increased waste from non-ruminant livestock such as pigs, increased paddy rice area and production, increased fires from land clearing, landfills and waste water facilities and increased fugitive emissions associated with fossil fuel extraction. The dominant sink of methane in the atmosphere is via oxidisation by hydroxyl radicals, decomposing to water vapour and other chemicals. Methane can also be consumed by soil organisms, particularly in dry soil. Methane has a comparatively short residence time in the atmosphere (approximately 10 years) and so reductions in emissions could relatively rapidly reduce this component of atmospheric warming. In agricultural systems, methane essentially is an unwanted energy loss from productivity, and so a range of approaches have been tried to reduce this loss, for example capturing methane from intensive livestock systems and combusting the gas to produce energy.

Box 17.3 **The Nitrous Oxide Cycle**

Nitrous oxide (N_2O) is a further major greenhouse gas produced in land-based systems such as agriculture and forestry. Nitrous oxide is part of the natural nitrogen cycle, occurring in trace amounts in the atmosphere. Human activities such as increased application of nitrogen fertiliser to crops, forages and forests to increase growth rates and protein content have accelerated and increased the size of the nitrogen cycle. Particularly high nitrous dioxide emissions arise from wet, warm and carbon-rich soils. Because the loss of nitrogen from human-managed systems represents a loss of an expensive input, a range of management and technological methods have been developed to reduce loss rates such as more closely matching nitrogen demand with supply and use of nitrification inhibitors. Nitrous oxide has a long residence time in the atmosphere (about 120 years), and it also reduces stratospheric ozone levels, increasing ultraviolet damage to plants and animals.

Avoided emissions from reducing deforestation and degradation represent some of the most rapidly implemented and cost-effective forms of mitigation activities. The mitigation value of protecting natural ecosystems from deforestation and degradation is due to the longevity of the accumulated carbon stocks in the ecosystem; not their present net rate of CO_2 uptake.

As avoided emissions are the most effective form of mitigation, it is important that regulations for recognising and rewarding mitigation activities include ecosystem protection and conservation management. Protection of carbon-dense ecosystems is critical, including forests, mangroves and peatlands. Natural ecosystems are resilient: that is, they have the capacity to persist when subjected to disturbance and environmental change without human inputs. This resilience governs the stability of the ecosystem's carbon stock. Accumulation of carbon over long time periods is thus greater in natural rather than modified ecosystems (unless conditions with artificial inputs are employed) (Thompson et al. 2009). Additionally, protection of natural ecosystems has co-benefits for sustainable land management, biodiversity conservation and sustainable livelihoods.

17.3.2 Sequestration

Sequestration in the land sector represents refilling the reservoir or repaying the debt from past deforestation and degradation. Deforestation has resulted in approximately half the world's forests being cleared (Archer 2005; MEA 2005). Only about 36% of the world's forests are in a primary condition (with minimal human disturbance) (FAO 2015). The remaining forest has been degraded to some degree by land-use activities, such as logging and soil disturbance, which result in depletion of organic carbon stocks and emissions of CO_2. Emissions from forest degradation are highly variable and poorly quantified, but estimates indicate that they are approximately half that of deforestation (Asner et al. 2010).

Mechanisms for sequestration in the land sector include reforestation, afforestation and restoration, which increase current carbon stocks. Benefits of these sequestration mechanisms include the capacity for relatively rapid change at low cost, and with a limited need for major technological development (Olsson et al. 2019). Sequestration has been used in scenarios of a zero-carbon economy to offset the remaining low level of emissions, particularly from agriculture and mining (ClimateWorks Australia et al. 2014; IPCC 2018). From a practical implementation and economic perspective, sequestration in the land sector requires financing and this is most likely achievable through a 'polluter pays' system, whether as direct carbon credits or a redistribution of finances by government.

The potential carbon stock that can be sequestered is limited by the area of land available and the capacity of the land to take up carbon. Not all the land previously cleared can be reforested when there are other competing land uses, particularly food production. Thus, trade-offs occur in the use of land for restoration and carbon storage compared with production of food and fibre, and in some cases water resources (IPCC 2018; Olsson et al. 2019). Additionally, existing land degradation, particularly of soils, means that in some areas the land carbon stock cannot be refilled. The capacity of the land to store carbon is determined mainly by climatic factors, but locally modified by substrate and topographic factors that affect plant growth and soil organic matter content. Ultimately, the capacity is limited to the amount of carbon previously depleted by land use, without additional human inputs of resources.

The total potential for sequestration in the land sector through reforestation, if it replaced all the carbon that had been released by land use up to the early 2000s, was estimated to reduce the atmospheric CO_2 concentration at the end of the century by 40–70 ppm (parts per million). Conversely, complete deforestation would increase atmospheric CO_2 concentration by an estimated 130–290 ppm (House et al. 2002). In comparison, the projected range of elevated atmospheric CO_2 concentration by the end of the century, under a range of fossil fuel emissions scenarios, is 170–600 ppm above 2000 levels (Prentice et al. 2001).

These estimates highlight the modest capacity for sequestration in the land sector to reduce atmospheric CO_2 concentrations compared with potential emissions from land-use activities and combustion of fossil fuels.

17.3.3 Reducing Emissions

Current land-use activities that create emissions can be modified to reduce these emissions. The following are examples of activities that can be changed to reduce emissions from the land-use sectors.

Forest sector:

- reducing rates of deforestation by reducing energy poverty, illegal logging and demand for forest products and land conversion by lowering pressures to expand agriculture, biofuel production, mines and urbanisation;
- maximising carbon storage in native forests by conserving native forests and restoring degraded native forest;
- improving carbon storage in harvested forest systems by increasing rotation lengths, reducing waste in harvesting and improving silvicultural and processing methods to produce high-quality timber;
- growing trees in plantations, which increases efficiency of production and reduces wastage, although care is required in species selection and maintaining biodiversity and sustainable production;
- improving efficiency of processing wood products to reduce waste and increase productivity;
- producing high-value-added wood products which have a mitigation benefit by increasing the longevity of the carbon storage in the products;
- maximising substitution benefits of wood products by using high-value timber instead of greenhouse-gas-intensive materials such as steel, aluminium and cement; and
- increasing recycling of wood and paper products.

Similar options for reducing emissions exist for the agricultural sector: increasing efficiencies of production, reducing emissions from waste, increasing above- and below-ground carbon stocks, improving value chains and enhancing strategic decision-making capabilities (Chapter 19).

17.4 Synergies between Mitigation and Adaptation Activities and Sustainable Development

The global challenges of population growth, resource scarcity and degradation, combined with the impacts of climate change that are already locked in due to the current atmospheric CO_2 concentration, as well as due to future increases, mean that land management policies and actions must support the combined objectives of sustainable development, mitigation and adaptation. Significant potential to reduce greenhouse gas emissions cost-effectively is derived from activities that provide synergistic mitigation and adaptation benefits.

Additionally, these activities often have co-benefits that enhance the resilience of ecosystems and the capacity for sustainable development. Defining a sustainable development pathway requires accounting for natural assets, identifying the drivers of change in these assets and allowing trade-offs to be evaluated.

Strategies for investment and policy interventions can be directed at both the supply side and demand side of production. Supply-side strategies aim to increase productivity, including new technologies and practices to increase crop and livestock yields, while restoring and protecting ecosystem services. Demand-side strategies aim to reduce pressures on natural resources, including reducing food loss and waste, changing diets and product specifications and sourcing renewable, non-carbon energy systems rather than biofuels (Global Commission on the Economy and Climate 2014).

The dual pressures of expanding agricultural production and minimising degradation of natural resources, particularly forests, need not be conflicting. Increasing agricultural production can be achieved through natural resource restoration to increase productivity, investment in research and development, technological advances, expanded agricultural credit and rural infrastructure, all of which do not necessitate deforestation. This process is exemplified by Brazil, where deforestation was reduced from the mid-2000s yet agricultural productivity continued to increase, due to the combined factors of the ability to quantify deforestation by remote sensing, improved law enforcement to reduce forest clearing, and agricultural finance conditioned on compliance with anti-deforestation policies. However, recent changes in government policies have reversed this trend and allowed increasing deforestation (Silva Junior et al. 2021).

The potential for strategies in the agriculture and forest sectors to help achieve cost-effective mitigation and adaptation outcomes is acknowledged in the literature and by some country climate action plans, such as the Ethiopian Climate Resilient Green Economy Strategy, the Rwandan Green Growth Strategy and the Kenyan Climate Smart Agriculture and Agroforestry Strategy. These national strategies identify and specify significant 'big wins' that simultaneously achieve sustainable development, mitigation and adaptation. Improving crop and livestock production, and the use of perennials, agroforestry practices and conservation tillage to increase food yields, also reduces erosion and improves food security, farmer incomes, soil health, water retention and resilience to climate variability. The capacity to gain from these synergies is due to the strong climate–energy–water–food production nexus in the land-use sector.

In addition to the sustainable development benefits from creating more resilient and productive agricultural and forest systems, other low-carbon strategies also offer ways to revitalise rural and regional economies. Strategies include payments to landholders for provision and maintenance of ecosystem services, carbon sequestration and sources of renewable energy.

17.5 Carbon Accounting Systems to Maximise Mitigation/Adaptation Benefits

Systems used for measuring, monitoring and reporting carbon stocks and stock changes nationally and internationally are important mechanisms for maximising the benefit of land-use activities for mitigation and adaptation. The rules and methods for these accounting systems influence whether reported emissions reductions truly reflect the net increase in atmospheric CO_2 concentration. Understanding the science of the global carbon cycle of stocks and flows enables the design of accounting systems that allow the identification of the opportunities and limitations of different mitigation activities.

Carbon accounting systems should be comprehensive of reservoirs, lands and activities, and separate the measurement, monitoring and reporting of carbon stock changes in each reservoir. This allows transparency of the origins and transfers of stocks and understanding of their impact within the global carbon cycle. In terms of national accounting to identify sources of emissions, it is the changes in stocks and flows due to anthropogenic activities that are causing climate change and so are relevant for identifying impacts and policies for abatement.

Carbon stocks in the land sector are referred to as biocarbon, and carbon stocks in fossil fuels are geocarbon (Ajani et al. 2013). These reservoirs differ fundamentally in the characteristics of carbon density, stability and restoration time of their carbon stocks and the reversibility of stock losses; that is, whether the stock can be restored and over what time period. Because of these different characteristics, the disturbance of different primary reservoirs by human activities will influence the climate to differing degrees (Prentice et al. 2001). Stocks of carbon in different reservoirs are not equivalent even though the carbon atoms are exchangeable irrespective of the source.

Geocarbon reservoirs consist of coal, oil, gas, limestone, other carbonate rocks and methane clathrates. The reservoirs are generally stable and inert without human intervention, except for fugitive emissions from gas deposits and volcanic activity. Stock losses from geocarbon reservoirs are effectively irreversible over timescales relevant to climate change and human society. Biocarbon reservoirs consist of biomass in plants and animals, and soil organic matter, and are spatially located in terrestrial, aquatic and marine ecosystems. Differentiating ecosystem types between natural, semi-natural and managed is important to account for the differing qualities of their carbon stocks, in terms of the density, stability and restoration capacity (Ajani et al. 2013). Natural ecosystems are generally more carbon-dense and resilient than managed systems due to their biodiversity and capacity for adaptation and regeneration (Thompson et al. 2009).

Carbon atoms in the biocarbon reservoir are constantly being exchanged with carbon in the atmosphere and ocean. The magnitude of the exchange of carbon is large and highly variable, with diurnal, seasonal, annual and multi-annual patterns, mainly related to climate variability. For example, the biocarbon reservoir has oscillated by more than 1 GtC during the past half-century due to the effects of climate variability on the relative rates of photosynthesis and respiration (WBGU 1998). However, some carbon stocks in the land sector can be considered a stable quantity in a dynamic equilibrium, where carbon atoms are exchanged but the stock remains at the same magnitude with fluctuations over time. By contrast, a trend in the exchange rate represents a change in stock over a time period. The impact of land management activities on atmospheric CO_2 concentration is best assessed as changes in carbon stocks between reservoirs over given time periods; not rates of change as annual flows.

Currently, national greenhouse gas inventories report information about anthropogenic flows, that is, greenhouse gas emissions to and removals from the atmosphere, in accordance with policies and compliance targets determined by the United Nations Framework

Convention on Climate Change (UNFCCC) negotiations. Reporting flows are appropriate for the fossil fuel sector, where flows are one-way, stock changes are almost entirely anthropogenic and stocks are stable in the absence of human perturbation. For the land sector, however, the land–atmosphere interaction is different because flows are two-way, with emissions to and removals from the atmosphere. Flows occur due to natural processes as well as anthropogenic disturbances. Often, the difficulty for accounting is distinguishing the processes impacting these flows. Flow-based inventories obscure fundamental differences between natural ecosystems, managed systems and fossil fuels in terms of the density, stability and restoration capacity of their carbon stocks. Additionally, the objective of climate change mitigation and the goal of the UNFCCC – reduction in the atmospheric concentration of CO_2 – is a stock change, and this is reported on a global basis. A disjuncture exists between the global information on stock change and the national information about annual flows in inventory systems that are used to assess targets. Hence, at the national level, carbon flow information needs complementing with information about carbon stocks in fossil fuels and ecosystems to provide policy-makers and the public with a complete set of information. Quantification and reporting of information about changes in stocks of geocarbon and biocarbon will help communication about prioritising activities for climate change mitigation and the inevitable trade-off decisions.

The current flow-based inventory has caused some unintended consequences because of incomplete accounts and lack of differentiation of stock longevities and stabilities. Such problems are demonstrated by the current forest inventory used in the Australian *State of the Environment* Report to assess environmental condition against criteria and indicators of sustainability. The native forest sector is reported as a net carbon sink, but the accounting included carbon sequestration in regenerating forest post-harvesting, but excluded emissions from harvesting prior to 1971, and carbon flows due to impacts on soil and coarse woody debris and non-timber vegetation (Dean et al. 2012). Adding a stock account to the existing flow inventories will create an opportunity to include all activities and to disaggregate the land sector carbon information by an ecosystem ranking based on differences in carbon stock density, stability and restoration capacity.

The comprehensive carbon accounting framework of carbon stocks and flows described by Ajani and Comisari (2014) and Keith et al. (2021) has been adopted in the revised System of Environmental-Economic Accounting – Ecosystem Accounting, and endorsed by the United Nations Statistics Division (UNSD 2021). This system of accounting is more general than the national greenhouse gas inventories and provides information at multiple scales which meet the growing demand of multiple objectives. Comprehensive carbon stock accounting was demonstrated to be feasible and populated with indicative estimates at a national scale (Ajani and Comisari 2014), and quantified at a regional scale (Keith et al. 2019). These stock accounts provide a complementary extension to the existing national accounts that use a flow-based system. These carbon accounts are based on the same principles as the national accounts of comprehensiveness and linkages across multiple stocks and flows, and so allow the carbon accounts to be fully integrated with the national economic information system.

Comprehensive accounts that identify sources due to anthropogenic activities are important to support national mitigation policies by providing comparable information on carbon stocks in fossil fuels and all ecosystems and land uses across the country. Mitigation policies aim to address climate change and improve the environment by rewarding abatement at the lowest cost and, as such, attract a diverse range of competing abatement activities in the energy, land and marine sectors. Scenario analysis can be used to examine policy options expressed as stock changes over periods of time, and hence understand the implications of abatement options for agriculture, forest, natural ecosystem and water assets. By linking the carbon accounts with economic information, the economic and employment impacts of these options can be estimated.

Current international mitigation policy is being guided by a desire to limit warming to a 2 °C increase in order to prevent dangerous anthropogenic interference with the climate system, with ambition for a 1.5°C limit. To help quantify the implications of this 2 °C limit, the 'budget approach' has been recommended, but not universally agreed upon politically. This approach calculates the global budget of cumulative CO_2 emissions that can be released to the atmosphere if total atmospheric concentrations are to be kept at a level that would limit warming to 1.5 °C or 2 °C. The estimated budget for a 66% probability of meeting the 1.5 °C global warming target is 114 GtC, which represents approximately 11 years of annual emissions. These figures were reported in 2018 and have been updated over the last two decades (WBGU 1998; Graßl et al. 2003; Allen et al. 2009; Schellnhuber et al. 2009; Steffen and Hughes 2013; IPCC 2018). Comprehensive carbon accounting is required to provide information about attribution of the changes in this total stock of carbon in the atmosphere.

17.6 Mechanisms for Action

Traditional planning frameworks are based on the assumption of a constant climate and cheap energy. However, there is an urgent need to update these frameworks to include both mitigation and adaptation considerations. Strategic planning combined with adaptive management is a powerful tool to guide future action, identify implementing roles and monitor outcomes. Different 'user groups' can play important roles.

At the scale of regional land management, carbon accounts provide important information for monitoring environmental health and condition. Long-term storage of carbon in vegetation and soil can be predicted from information about current carbon stocks in ecosystems combined with their carbon carrying capacity under conditions of environmental variability and natural disturbance regimes but without human activities. Maintaining vegetation cover and soil organic matter are key components of sustainable land management. As per the other sectors, there are a range of key actors and levers that can be used to achieve a transition to sustainable development, climate change mitigation and adaptation simultaneously for the land sector, including the following listed in Table 17.1.

Both the information in carbon accounts and the policy goals are important for assessing the most effective actions for mitigation and adaptation. Actions under the policy context of 'offsets' are becoming increasingly popular and are thought to provide solutions for many land management decisions. However, many issues are involved that require careful consideration of the carbon accounting; these are discussed in Box 17.4.

Table 17.1. *Potential stakeholders and levers that could be employed to achieve synergistic benefits for sustainable development, climate change mitigation and adaptation*

Key stakeholders	Potential options
Government	Economic incentives (carbon pricing, emissions trading schemes, markets for ecosystem services), licensing, regulation, standards and certification, research and development, best-practice information, funding stakeholder engagement processes, education and training. Land use and planning and funding regional natural resource management (NRM) multi-stakeholder groups. Restoring and managing ecosystem services sustainably. Community leadership and environmental stewardship. Ensuring rule of law and fair trade to eliminate illegal logging imports. Integrity systems to reduce risks of corruption.
Regional NRM multi-stakeholder groups	Sharing and improving science-based approaches to NRM in forestry and agriculture. Sharing best-practice knowledge in a local context.
Industry groups	Providing policy advice to government. Producing sustainability frameworks and action plans with their members to communicate and share the latest best practice in a framework with sustainability indicators to help members improve their practices. For example, the 2008 Australian Plantation, Timber Products and Paper Council (A3P) Sustainability Action Plan and Framework. Providing innovation portals to share the latest results of research, development and deployment; for example, the Australian New South Wales Farmers Federation Agricultural Innovators initiative and web portal and the New South Wales Farmers Federation Energy and Agriculture information web portal.
Landholders: forestry and agriculture	Leadership to provide evidence that sustainable forestry and agriculture is achievable and profitable.
Non-governmental organisations (NGOs)	Third-party certification schemes for sustainable and fair trade of agricultural and forestry products to build markets for these products. Advocacy and campaigns. Communication and knowledge-sharing of best practice.
Science/research	Earth systems (climate, soils, water, biodiversity) science in global, regional and local contexts to improve decision-making by landholders and government policy-makers. Research into science-based sustainable agriculture and forestry, such as agro-ecology. Optimising whole-of-farm, whole-of-forestry approaches for environmental sustainability, reduced greenhouse gas emissions, productivity and resilience.
Consumers	Consumers play a critical role by choosing certified sustainable agricultural and forestry products, reducing food waste and choosing to eat low-carbon diets.
Urban and peri-urban farming	There is significant potential for urban and peri-urban agriculture.

Box 17.4 **The Conceptual Framework of Offsets**

At a broad conceptual level, there are two competing perspectives on land sector offsets: one that opposes the use of land sector abatement to offset fossil fuel emissions and another that encourages the use of offsets. The historical context of these differing perspectives originates from the negotiations of the land use, land-use change and forestry (LULUCF) sector that came after the Kyoto Protocol targets for emissions reduction had been set. Hence, accounting for the land sector was used as a means of offsetting fossil fuel emissions without changing the targets (Schulze et al. 2002). This was seen by some as lessening the incentive to reduce fossil fuel emissions (Höhne et al. 2007). Offsets were used as financial instruments to reduce net emissions.

Opponents of land sector offsets generally base their position on four arguments. First, removals of carbon by uptake in the land sector represent the repayment of past carbon losses (a debt) from deforestation and degradation, which are equivalent to 35% of the accumulated CO_2 in the atmosphere. Hence, using land sector abatement to offset fossil fuel emissions is the equivalent of double claiming on a debt repayment: that is, trying to use a single repayment to satisfy two debts (the historical land sector debt and the current debt from fossil fuel combustion).

Second, there are qualitative differences in the nature of biological and geological carbon stocks that make it inappropriate to use land sector abatement to offset fossil fuel emissions. This relates to the longevity and reversibility of carbon stocks. Carbon losses from biological stocks are reversible on human-relevant timescales, while losses from geocarbon reservoirs are not (Matthews and Caldeira 2008). Offsetting geocarbon for biocarbon embodies an incorrect assumption that reservoirs and their stocks of carbon are uniform and interchangeable from a climate perspective. Both the magnitude and longevity of stocks must be considered. Long-lived carbon stocks have a different influence on atmospheric CO_2 concentration compared with short-lived stocks (Ajani et al. 2013).

Third, the land sector does not have the capacity to offset fossil fuel emissions continuing for a long time. Emissions from fossil fuel and cement production were 9.7 ± 0.5 Gt in 2019 (GCP 2020). At this rate, a theoretical global return of all land to natural carbon stocks (an addition of about 150 Gt) would offset 15.5 years of fossil fuel emissions. However, the capacity of the land sector to store more carbon is finite and ecosystems cannot return to their full carbon carrying capacity because of competing demands for land, in terms of production of food and fibre, and soil resources in some areas have been permanently degraded. Allowing offsets in the land sector can delay action on reducing emissions from fossil fuels if it is used solely to meet emissions reduction targets.

Fourth, using land sector abatement to offset fossil fuel emissions can have perverse outcomes. For example, encouraging land sector abatement can result in the planting of fast-growing plant species that have negative landscape and biodiversity impacts, or can lead to activities that fail to reduce atmospheric CO_2 concentrations in the long term.

Supporters of land sector offsets argue that the opposition to offsets is based on a misconception about the relevant baseline against which to judge the climate benefits of the land sector abatement activity. The climate benefit of any activity should be judged against a counterfactual in which no direct policy measures are introduced to address climate change. If, in the absence of direct policy measures to address climate change, the relevant land sector stock remained depleted, action to replenish the stock would provide a real and additional climate benefit. The opposing view is based on a sunk cost fallacy: it involves judging the merits of a current decision on the basis of a past sunk cost (the original depletion of the stock).

According to supporters of land sector offsets, this same logic answers the opponents' claims about the reversibility of land versus geological carbon stocks. The relevant issue is not whether a stock can physically be replenished but whether, in the absence of additional policy intervention, it will be.

The response given by supporters of land sector offsets to the third and fourth arguments against them is that neither are reasons to prohibit their use. That land sector offsets cannot be used to fully or even substantially offset fossil fuel use is a statement of fact. However, it is not relevant to the decision of whether to allow land sector offsets. The primary argument for using land sector offsets is that land sector abatement is often cheaper than abatement in other sectors. If abatement from two sources provides equivalent climate benefits, and there is a choice between the two, the rational solution is to select the cheapest. The fact that there may be limits on the amount of abatement that a source or sink can provide is not relevant to the abatement decision.

Perverse impact risks are a material issue with many types of land sector abatement and most carbon offset schemes have mechanisms for dealing with them. In some cases, this can result in particular types of land sector abatement being prohibited. For example, policy-makers may choose to prohibit the use of invasive species for reforestation. However, stopping all land sector offsets on the basis of risks posed by a few, when those risks can be managed, is irrational.

These differing arguments illustrate the problems that can occur when opinions become polarised. If the common goals of climate change mitigation and sustainable land management are sought, then it should be possible to develop policies to penalise activities that cause emissions and incentivise activities that reduce emissions or sequester carbon. The issue becomes one of policy, to develop funding sources or markets, which do not have to be in the form of direct market offsets, and without creating perverse outcomes. There are real differences between biocarbon and geocarbon stocks in terms of their longevity and irreversibility. Separating the accounting and targets for geocarbon and biocarbon would help to solve the current disagreements about offsets because the transfers of stocks between reservoirs would be transparent. Avoiding and reducing land carbon emissions is an integral part of a comprehensive approach to solving the climate change problem. Many land sector mitigation activities are highly desirable because they are cheap, can be implemented quickly and have co-benefits for land management. But they should be considered as additional to, and not an alternative to, reducing fossil fuel emissions.

17.7 Conclusion

Mitigation activities in the land sector are critical for achieving the goal of keeping global warming to less than 2 °C over pre-industrial levels. Additionally, the positive feedback between carbon storage in the land sector, and the risks of loss due to changed climate conditions affecting productivity, exacerbate the global warming problem. However, land-use activities should not be considered as 'offsets' for reducing fossil fuel emissions. To account for the stocks and flows of carbon adequately to inform policy decisions, the different qualities of carbon stocks, in terms of the density, stability and restoration capacity, should be accounted separately as biocarbon and geocarbon. The land sector has many opportunities for mitigation including options that are cost-effective, can be implemented quickly and which operate on either the supply side or the demand side of productivity. A multi-stakeholder approach is needed to achieve a transition to sustainable development, climate change mitigation and adaptation simultaneously, and many co-benefits emerge from this nexus.

References

Ajani, J. and Comisari, P. (2014). *Towards a Comprehensive and Fully Integrated Stock and Flow Framework for Carbon Accounting in Australia*. Discussion Paper. Australia: The Australian National University (ANU). Available at: https://coombs-forum.crawford.anu.edu.au/sites/default/files/publication/coombs_forum_crawford_anu_edu_au/2014-09/carbon_accounting_discussion_paper_revised_sept_2014.pdf.

Ajani, J., Keith, H., Blakers, M., Mackey, B. G. and King, H. P. (2013). Comprehensive carbon stock and flow accounting: A national framework to support climate change mitigation policy. *Ecological Economics*, 89, 61–72.

Allen, M. R., Frame, D. J., Huntingford, C. et al. (2009). Warming caused by cumulative carbon emissions towards the trillionth tonne. *Nature*, 458, 1163–1166.

Archer, D. (2005). Fate of fossil fuel CO_2 in geologic time. *Journal of Geophysical Research*, 110, 1–6.

Archer, D., Eby, M., Brovkin, V. et al. (2009). Atmospheric lifetime of fossil fuel carbon dioxide. *Annual Review of Earth and Planetary Sciences*, 37, 117–134.

Asner, G. P., Powell, G. V. N., Mascaro, J. et al. (2010). High resolution forest carbon stocks and emissions in the Amazon. *Proceedings of the National Academy of Sciences*, 107, 16739–16742.

Benndorf, R., Federici, S., Forner, C. et al. (2007). Including land use, land-use change, and forestry in future climate change, agreements: Thinking outside the box. *Environmental Science & Policy*, 10, 283–294.

ClimateWorks Australia, ANU (Australian National University), CSIRO (Commonwealth Scientific and Industrial Research Organisation) and CoPS (Centre for Policy Studies) (2014). *Pathways to Deep Decarbonisation in 2050: How Australia Can Prosper in a Low Carbon World*. Technical report. Melbourne: ClimateWorks Australia. Available at: www.climateworksaustralia.org/wp-content/uploads/2014/09/climate works_pdd2050_technicalreport_20140923-1.pdf.

Dean, C., Wardell-Johnson, G. and Kirkpatrick, J. B. (2012). Are there any circumstances in which logging primary wet-eucalypt forest will not add to the global carbon burden? *Agricultural and Forest Meteorology*, 161, 156–169.

Edenhofer, O., Pichs-Madruga, R., Sokona, Y. et al. (2014). Technical summary. In O. Edenhofer, R. Pichs-Madruga, Y. Sokona et al., eds., *Climate Change 2014: Mitigation of Climate Change. Contribution of Working Group III to the Fifth Assessment Report of the Intergovernmental Panel on Climate Change*. Cambridge: Cambridge University Press, pp. 33–107. Available at: www.ipcc.ch/site/assets/uploads/2018/02/ipcc_wg3_ar5_technical-summary.pdf.

FAO (Food and Agriculture Organization) (2010). *Global Forest Resources Assessment 2010: Main Report*. FAO Forestry Paper 163. Rome: Food and Agriculture Organization of the UN. Available at: www.fao.org/3/i1757e/i1757e00.htm.

FAO (2015) *Global Forest Resources Assessment 2015: How Are the World's Forests Changing?*, 2nd ed. Food and Agricultural Organization of the United Nations.

Feely, R. A., Sabine, C. L., Lee, K. et al. (2004). Impact of anthropogenic CO_2 in the $CaCO_3$ system in the oceans. *Science*, 305, 362–366.

Friedlingstein, P., Cox, P., Betts, R. et al. (2006). Climate–carbon cycle feedback analysis: Results from the C^4MIP model intercomparison. *Journal of Climate*, 19, 3337–3353.

Friedlingstein, P., Houghton, R., Marland, G. et al. (2010). Update on CO_2 emissions. *Nature Geoscience*, 3, 811–812.

GCP (Global Carbon Project) (2020). *Carbon Budget 2020*. Available at: www.globalcarbonproject.org/carbonbudget/archive/2011/CarbonBudget_2011.pdf.

Global Commission on the Economy and Climate (2014). *Better Growth, Better Climate: The New Climate Economy Report*. Synthesis Report. Washington, DC: The Global Commission on the Economy and Climate. Available at: https://newclimateeconomy.report/2016/wp-content/uploads/sites/2/2014/08/BetterGrowth-BetterClimate_NCE_Synthesis-Report_web.pdf.

Graβl, H., Kokott, J., Kulessa, M. et al. (2003). *Climate Protection Strategies for the 21st Century: Kyoto and Beyond*. Special Report. Berlin: German Advisory Council on Global Change (WBGU). Available at: www.gci.org.uk/Documents/wbgu_sn2003_engl.pdf.

Höhne, N., Wartmann, S., Herold, A. and Freibauer, A. (2007). The rules for land use, land use change and forestry under the Kyoto Protocol: Lessons learned for the future climate negotiations. *Environmental Science & Policy*, 10, 353–369.

Houghton, R. A. (2007). Balancing the global carbon budget. *Annual Review Earth and Planetary Science*, 35, 313–347.

Houghton, R. A. (2008). Carbon flux to the atmosphere from land-use changes: 1850–2005. In *TRENDS: A Compendium of Data on Global Change*. Oak Ridge, TN: Carbon Dioxide Information Analysis Center, Oak Ridge National Laboratory, U.S. Department of Energy. Available at: https://cdiac.ess-dive.lbl.gov/trends/land use/houghton/houghton.html.

House, J. I., Prentice, I. C. and Le Quéré, C. (2002). Maximum impacts of future reforestation or deforestation on atmospheric CO_2. *Global Change Biology*, 8, 1047–1052.

IPCC (2014). *Climate Change 2014: Synthesis Report. Contribution of Working Groups I, II and III to the Fifth Assessment Report of the Intergovernmental Panel on Climate Change*. Edited by R. K. Pachauri and L. A. Meyer. Geneva: IPCC. Available at: www.ipcc.ch/report/ar5/syr/.

IPCC (2018). *Global Warming of 1.5 °C: An IPCC Special Report on the Impacts of Global Warming of 1.5 °C Above Pre-Industrial Levels and Related Global Greenhouse Gas Emission Pathways, in the Context of Strengthening the Global Response to the Threat of Climate Change, Sustainable Development, and Efforts to Eradicate Poverty*. Edited by V. Masson-Delmotte, P. Zhai, H.-O. Pörtner et al. Cambridge: Cambridge University Press. Available at: www.ipcc.ch/sr15/.

Keith, H., Vardon, M., Stein, J. A. and Lindenmayer, D. (2019). Contribution of native forests to climate change mitigation: A common approach to carbon accounting that aligns results from environmental-economic accounting with rules for emissions reduction. *Environmental Science & Policy*, 93, 189–199.

Keith, H., Vardon, M., Obst, C. et al. (2021). Evaluating nature-based solutions for climate mitigation and conservation requires comprehensive carbon accounting. *Science of the Total Environment*, 769, 14434.

Le Quéré, C. (2009). Trends in the sources and sinks of carbon dioxide. *Nature Geoscience*, 2, 831–836.

Le Quéré, C., Peters, G. P., Andres, R. J. et al. (2013). Global carbon budget 2013. *Earth System Science Data*, 6, 689–760.

Mackey, B., Prentice, I. C., Steffen, W. et al. (2013). Untangling the confusion around land carbon science and climate mitigation policy. *Nature Climate Change*, 3, 552–557.

Mackey, B., Kormos, C., Keith, H. et al. (2020). Understanding the importance of primary tropical forest protection as a mitigation strategy. *Mitigation and Adaptation Strategies for Global Change*, 25, 763–787. Available at: https://doi.org/10.1007/s11027-019-09891-4.

Matthews, H. D. and Caldeira, K. (2008). Stabilizing climate requires near-zero emissions. *Geophysics Research Letters*, 35, L04705.

MEA (Millennium Ecosystem Assessment) (2005). *Ecosystems and Human Well-being: Biodiversity Synthesis*. Washington, DC: World Resources Institute. Available at: www.millenniumassessment.org/documents/document.354.aspx.pdf.

Nabuurs, G. J., Masera, O., Andrasko, K. et al. (2007). Forestry. In: B. Metz, O. R. Davidson, P. R. Bosch, R. Dave and L. A. Meyer, eds., *Climate Change 2007: Mitigation. Contribution of Working Group III to the Fourth Assessment Report of the Intergovernmental Panel on Climate Change*. Cambridge: Cambridge University Press, pp. 541–581. Available at: www.ipcc.ch/site/assets/uploads/2018/02/ar4-wg3-chapter9-1.pdf.

Olsson, L., Barbosa, H., Bhadwal, S. et al. (2019). Land degradation. In P. R. Shukla, J. Skea, E. C. Buendia et al., eds., *Climate Change and Land: An IPCC Special Report on Climate Change, Desertification, Land Degradation, Sustainable Land Management, Food Security, and Greenhouse Gas Fluxes in Terrestrial Ecosystems*. Available at: www.ipcc.ch/srccl/chapter/chapter-4/.

Plattner, G. K., Knutti, R., Joos, F. et al. (2008). Long-term climate commitments projected with climate–carbon cycle models. *Journal of Climate*, 21, 2721–2751.

Prentice, I. C., Farquhar, G. D., Fasham, M. J. R. et al. (2001). The carbon cycle and atmospheric carbon dioxide. In J. T. Houghton, Y. Ding, D. J. Griggs et al., eds., *Climate Change 2001: The Scientific Basis. Contribution of Working Group I to the Third Assessment Report of the Intergovernmental Panel on Climate Change*. Cambridge: Cambridge University Press, pp. 185–237. Available at: www.ipcc.ch/site/assets/uploads/2018/02/TAR-03.pdf.

Richards, K. R. and Stokes, C. (2004). A review of forest carbon sequestration cost studies: A dozen years of research. *Climatic Change*, 63, 1–48.

Schellnhuber, H. J., Messnre, D., Leggewie, C. et al. (2009). *Solving the Climate Dilemma: The Budget Approach*. Special Report. Berlin: German Advisory Council on Global Change. Available at: www.wbgu.de/en/publications/publication/special-report-2009.

Scholze M., Knorr, W., Arnell, N. W. and Prentice, I. C. (2006). A climate-change risk analysis for world ecosystems. *Proceedings of the National Academy of Sciences*, 35, 13116–13120.

Schulze, E.-D., Valentini, R. and Sanz, M.-J. (2002). The long way from Kyoto to Marrakesh: Implications of the Kyoto Protocol negotiations for global ecology. *Global Change Biology*, 8, 505–518.

Silva Junior, C. H. L., Pessôa, A. C. M., Carvalho, N. S. et al. The Brazilian Amazon deforestation rate in 2020 is the greatest of the decade. *Nature Ecology and Evolution*, 5, 144–145. Available at: https://doi.org/10.1038/s41559-020-01368-x.

Smith, P., Nkem, J., Calvin, K. et al. (2019). Interlinkages between desertification, land degradation, food security and GHG fluxes: Synergies, trade-offs and integrated response options. In P. R. Shukla, J. Skea, E. C. Buendia et al., eds., *Climate*

Change and Land: An IPCC Special Report on Climate Change, Desertification, Land Degradation, Sustainable Land Management, Food Security, and Greenhouse Gas Fluxes in Terrestrial Ecosystems. In press. Available at: www.ipcc.ch/srccl/chapter/chapter-6/.

Steffen, W. and Hughes, L. (2013). *The Critical Decade 2013: Climate Change Science, Risks and Responses*. Canberra: ACT Climate Commission Secretariat. Available at: https://researchers.mq.edu.au/en/publications/the-critical-decade-2013-climate-change-science-risks-and-respons.

Thompson, I., Mackey, B., McNulty, S. and Mosseler, A. (2009). *Forest Resilience, Biodiversity and Climate Change. A Synthesis of the Biodiversity/Resilience/Stability Relationship in Forest Ecosystems*. CBD Technical Series No. 43. Montreal: Secretariat of the Convention in Biological Diversity. Available at: www.cbd.int/doc/publications/cbd-ts-43-en.pdf.

UNSD (United Nations Statistics Division) (2021). System of Environmental–Economic Accounting – Ecosystem Accounting: Final draft. Available at: https://unstats.un.org/unsd/statcom/52nd-session/documents/BG-3f-SEEA-EA_Final_draft-E.pdf.

WBGU (German Advisory Council on Global Change) (1998). *The Accounting of Biological Sinks and Sources under the Kyoto Protocol: A Step Forwards or Backwards for Global Environmental Protection?* Special report. Berlin: German Advisory Council on Global Change. Available at: www.wbgu.de/fileadmin/user_upload/wbgu/publikationen/sondergutachten/sg1998/pdf/wbgu_sn1998_engl.pdf.

18

Forests

HEATHER KEITH, ANDREW MACINTOSH, BRENDAN MACKEY
AND MICHAEL H. SMITH

Executive Summary

Action on climate change is a priority for this sector because rates of forest degradation and deforestation contribute about 12% to global greenhouse gas emissions. Reducing carbon loss caused by harvesting forests, and protecting forests to allow them to continue growing, as well as restoration and reforestation on previously deforested land, provide major global climate change mitigation strategies.

The Intergovernmental Panel on Climate Change (IPCC) Fifth Assessment Report estimated that about 65% of the total mitigation potential in the forest sector is located in the tropics, and of this total about 50% could be achieved by reducing deforestation.

If strong action is not taken on climate, the forest sector faces serious risks from climate change such as increased risks of more intense and frequent mega-fires, drought and loss of ecosystem resilience due to species loss. Long-term loss of forest resources due to climate change may directly affect 90% of the 1.2 billion forest-dependent people who live in extreme poverty.

There are many strategies to simultaneously help reduce rates of deforestation and degradation of native forests and encourage restoration and reforestation to maximise the forest sector's contribution to climate mitigation, while supporting forest resilience and adaptation to climate change. There are many climate change adaptation and ecosystem resilience co-benefits from applying forest sector climate mitigation strategies, such as reducing flood risks, improving water quality and reducing soil erosion and desertification. Others include conserving biodiversity and heritage values, and avoiding the effects of changed vegetation on local climate conditions, such as reduced precipitation and increased temperature.

There is growing recognition of synergies between adaptation and mitigation strategies for the forest sector.

Effective governance and policy are critical for supporting and incentivising efforts to conserve native forests, stop illegal logging and invest in restoration of native forests and plantation forests, but pose challenges especially for developing countries with less mature governance systems.

Alternative policy development frameworks are discussed, noting that success or otherwise of policies is significantly influenced by political and economic contexts. Reducing

Emissions from Deforestation and forest Degradation (REDD) and REDD+ schemes and policy frameworks including the LEAF–BEDS (Legality, Effectiveness, Acceptability, Feasibility–Beneficiaries, Elites, Detractors, Society) model are discussed.

18.1 Introduction

The forest sector includes all landscapes that are dominated by forest ecosystems. Various definitions of forest exist based on scientific, political and economic criteria. From a scientific perspective, forests can be defined by the structure of the vegetation, in terms of the dominant plant form of trees with canopy height greater than 10 m and the foliage cover of greater than 30% density (Carnahan 1976). The height and density of a forest canopy is highly significant, as these serve to modify the internal forest microenvironment buffering it from external perturbations, including climatic variability. Ecologically, natural forests can be distinguished from plantation forests, which are crops of trees harvested on a regular basis as input to the industrial production of wood and pulp-based products. Like other crops, plantations require ongoing human management inputs at all stages of their development. Natural forests are also used as a source of woody biomass that is harvested as input to industrial processes, but differ from plantations in that they rely primarily on natural processes for their regeneration. The different types of forests, therefore, can be managed for a range of goals, from conservation to production, and occur in a range of conditions, from natural, to semi-natural, to plantation.

The role of forests in mitigation and adaptation is investigated, including the ways in which climate change impacts the forest sector. About 34% of the world's natural forest cover (61.5×10^6 km^2) has been lost. Of the world's remaining area of forests (40.1×10^6 km^2), approximately 57% is subject to industrial logging or designated for multiple uses including wood production, 7% is plantation and around 36% (14.5×10^6 km^2) is primary forest, as reported in the first decade this century (Archer 2005; MEA 2005; FAO 2010; Mackey et al. 2015). The rate of deforestation as loss of natural forest was 9.5 million hectares (ha) of forest per year on average from 1990 to 2015, with the rate decreasing from 13 million ha per year in the first decade to 6.5 million ha per year in the last decade (FAO 2015). Much of the forest that remains has been degraded to some degree by land-use activities, such as logging, construction of roads and infrastructure, and associated soil disturbance, all of which result in depletion of biomass and soil organic carbon stocks and emissions of carbon dioxide (CO_2) (Mackey et al. 2020).

Net emissions from deforestation and degradation are about 12% of global anthropogenic CO_2 emissions for 2000–2007 and this is a decline since the 1990s. Gross emissions are at least twice this amount and equivalent to approximately one-third of fossil fuel emissions (IPCC 2014). Data for gross emissions and removals are derived from Pan et al. (2011), Baccini et al. (2012), Houghton (2012). Emissions from forest degradation are highly variable and poorly quantified, but estimates indicate that they range from half to greater than the total emissions from deforestation (Asner et al. 2010; Houghton 2013). Areas of forest degraded are larger than that deforested in many regions, such as the

Amazon (Berenguer et al. 2014). These current patterns of forest management are unsustainable, resulting in ongoing deforestation and degradation due to industrial logging, swidden farming (shifting or 'slash-and-burn' agriculture) and conversion for industrial ranching and agriculture.

The forest sector now has opportunities to contribute significantly to the mitigation solutions to climate change because carbon stocks in forests have been reduced to such an extent by past land use. In the tropics alone, it is estimated that forest protection could generate mitigation benefits of at least 3 GtC (gigatonnes of carbon) per year for decades based on avoided emissions from ceasing deforestation and degradation, along with the additional sequestration from regrowth of previously logged and cleared forest (Houghton et al. 2015). However, this kind of mitigation action at the scale required is not a straightforward matter. National governments can have development strategies that are contingent on clearing or exploiting forests. In many cases, governments have signed long-term logging contracts with multinational corporations. Particularly in the developing world, most tropical forests are home to Indigenous peoples who seek development opportunities. Increasingly, forest protection is in conflict with the land-use pressures arising from increasing demand for the production of food, fibre and energy. Furthermore, there is considerable polarisation of opinion about forest management strategies and the benefits of protection versus production, which hinder policies and actions for change towards sustainable development.

Despite the difficulties, there is growing agreement among the international community and in most countries with significant forested lands that forests provide important ecosystem services for people, that the current situation is unsustainable and that action is needed. In addition to their climate change mitigation benefits, forests provide a range of critical ecosystem services, including the quality and supply of freshwater, the conservation of biodiversity, provision of clean air, regulation of regional precipitation regimes and reducing the risk of flooding and soil erosion during extreme weather events (Mackey et al. 2015). Forests provide the raw materials for wood, paper and energy products essential to current human way of life. Forest resources are integral to the sustainable livelihoods of much of the world's population in developing countries, with biomass the main energy source for about 2 billion people and accounting for one-third of all energy consumption in these countries (UNDP 2000). Degradation of these forest resources threatens both livelihoods of individual people and health of the planet. A primary reason for both the high value placed on forest ecosystem services and the pressure from other land uses is that all require the same environmental conditions for plant growth in terms of rainfall, temperature, fertile soil and sunlight.

The risks from climate change hinge on our collective success or failure to mitigate greenhouse gas (GHG) emissions over the coming decades. The less successful our mitigation efforts, the greater the climate risks, and the more need there will be for adaptation. Forests are significant reservoirs of carbon, so protecting and restoring them can be an important mitigation action. Additionally, healthy forests provide valuable ecosystem services that can assist human communities adapt to climate change. However, forest ecosystems are also being impacted by a rapidly changing climate, interacting with prevailing land-use activities that are undermining their resilience and adaptive capacity. Therefore, significant changes to forest management are needed using an ecosystem-based approach, which optimises the synergistic effects of mitigation by reducing emissions and increasing sequestration, improves sustainable land management and livelihoods, and assists forests and forest-dependent communities to cope with the future risks of climate change (Locatelli et al. 2011). Here, we identify opportunities for forest management pathways that deliver mitigation and adaptation in the context of sustainable development goals.

18.2 Climate Change Risks Related to Forests

Risks of climate change to forests include both the impacts of gradual changes in average climate conditions and the potential increase in intensity and frequency of extreme climate events. Although trees are long-lived organisms and many changes in forests occur over long time frames, impacts of current climate change are apparent now and the risks of future more serious climate change are profound. The impacts of climate change on forest ecosystems encompass a spectrum of plant physiological processes, ecosystem responses, ecosystem services, wood product industries and local community livelihoods.

The risks of climate change to forest ecosystems and resources are significant, whether in terms of the goods and services provided by forests, human well-being, survival of other species or the commodities required by society. At the beginning of this century, the FAO estimated that the long-term loss of forest resources due to climate change may directly affect 90% of the 1.2 billion forest-dependent people who live in extreme poverty (FAO 2004). Given the range of risks from climate change, it is critical to reduce and manage these risks both to forests ('adaptation for forests') and to the services provided by forests used by humanity ('adaptation of forests for people').

Examples of the risks of climate change on forests include the following:

- **Risks of change in natural ecosystems: composition, distribution, physiology and phenology**. Distributions of species are influenced by temperature in relation to latitude and altitude, and by rainfall in relation to continentality and topography. Hence, composition of ecosystems will change as competitive relationships among species and their physiological and phenological characteristics change under different conditions (Hughes 2000). Rising temperatures over time will result eventually in a change in the taxonomic composition of the forest ecosystem as conditions become less than optimum for the dominant tree species, and other species migrate in and out-compete. Regions with a decreasing rainfall regime may change to the point where there is insufficient water available to support tree cover of sufficient height and density to be classified as forest, in which case the landscape will develop into a different ecosystem type. In regions with an increasing or stable rainfall regime, forests are likely to persist although they may change in composition.
- **Risks of loss of resilience, extinction and greater vulnerability of biodiversity**, due to inability to move or adapt in response to climate change, especially when fragmentation

and absence of sufficient corridors or restricted habitat conditions limit movement (Steffen et al. 2009; Thompson et al. 2009; Hughes 2012).

- **Risks of forests becoming a net carbon source**. Many climate models project that, under high-emission scenarios, rising temperatures, drought and fires will lead to a reversal of forests from being a carbon sink to becoming a net carbon source by 2100 (Sitch et al. 2008; Bowman et al. 2009; Allen et al. 2010). This reversal from a sink to a source of carbon has already occurred during periods of drought in Australia (Keith et al. 2012), Europe (Cias et al. 2005), North America (Breshears et al. 2005) and the Amazon (Phillips et al. 2009; Phillips et al. 2010).

- **Risks of reduced productivity and increased tree mortality** due to higher temperatures and heatwaves combined with more intense drought and shifts in rainfall patterns (Williams et al. 2010, 2013; Anderegg et al. 2013). Higher rates of tree mortality and dieback have been documented under conditions of increased temperature and intensity of drought in temperate regions of western North America (van Mantgem et al. 2009), Canada (Michaelian et al. 2011), southern Europe (Carnicer et al. 2011), Australia (Fensham et al. 2009; Keith et al. 2012) and in tropical forests (Kraft et al. 2010; Phillips et al. 2010).

- **Risks of increased incidence of pests, diseases and weeds as indirect effects of climate change** (Hellman et al. 2008; Brasier and Webber 2010). Changes in climate conditions affect occurrence, productivity and competition between exotic and natural organisms, and often negatively impact productivity and survival of forests. For instance, the mountain pine beetle has negatively impacted more than 11 million ha of forest in Canada and the western USA since the late 1990s (FAO 2010). Outbreaks of the pest are exacerbated by higher minimum temperatures in winter and maximum temperatures in summer, as well as reduced summer precipitation (Kurz et al. 2008).

- **Risks of increased frequency and severity of wildfires as indirect effects of climate change** (Williams et al. 2001; Cary et al. 2012). The magnitude and timing of potential climate-driven changes in fire regimes are likely to negatively impact productivity and survival of forests. Changes to fire regimes will be highly variable regionally, depending on regional changes in temperature and rainfall, their seasonality and extreme events. Westerling et al. (2011) predicted a shift in fire–climate–vegetation relationships in the Greater Yellowstone ecosystem of conifer forest in Wyoming by mid-century because the increased fire frequency and extent would be inconsistent with persistence of the current suite of conifer species.

- **Risks of crossing irreversible ecosystem tipping points**. Of the nine tipping elements short-listed by Lenton et al. (2008), two involve forest ecosystems: the boreal forest and Amazon forest. Predicted impacts of climate change in the boreal forest biome show non-linear shifts in tree cover and transitions to tundra and steppe (Scheffer et al. 2012). In the Amazon, the risk of severe drought episodes, together with land-use change and forest fire, are predicted to cause significant changes to forest extent under a 3–4 °C level of warming (Betts et al. 2004; Malhi et al. 2009; Nobre and Borma 2009; Smith et al. 2014).

- **Risks of greater pressures for deforestation due to climate change reducing agricultural productivity**. This could result in greater demand for land for agriculture to improve food security resulting in greater pressures for deforestation. Already clearing of forests for agriculture is a major driver of deforestation globally. Given the range of risks from climate change for agricultural systems (see Chapter 19), there is a risk that this could increase agriculture-related deforestation.

• **Risks of loss of economic value of forests due to climate change** altering the tree species that can be grown commercially in certain regions. For instance, this shift could reduce the economic value of European forested lands by 14–50% by 2100, owing to the decline of economically valuable species that can be grown in future European climates (Hanewinkel et al. 2012).

Many aspects of managing carbon stocks in forest ecosystems include benefits for both mitigation and adaptation, as well as co-benefits that enhance the resilience of ecosystems and the capacity for sustainable development. To understand how best to mitigate and adapt to climate change in the forest sector, it is important to understand the following key issues:

(a) the processes driving the carbon cycle in forest systems and the reasons for changes in carbon stocks (Box 18.1 and Chapter 17);
(b) the current drivers responsible for GHG emissions in the sector (Section 18.3);
(c) the main strategies available for mitigation, adaptation and sustainable development, and the potential for synergies or conflict (Section 18.4);
(d) the co-benefits for productivity, health, society and the environment derived from pursuing such strategies, which will help to motivate and build political will, industry and public support for necessary policy reform (Section 18.4.5); and
(e) the policy and institutional reform options needed to enable climate change mitigation and adaptation in this sector, while improving environmental and social sustainability (Section 18.5).

Evaluation of the net mitigation benefit of different strategies for management of native forests – at the simplest level, conservation versus harvesting for forest products – is based on the carbon storage and net anthropogenic emissions. In a conserved native forest, the carbon stock is at a maximum given the environmental conditions and disturbance regimes that characterise the landscape. Stability of these natural carbon stocks is conferred by the capacity for resilience and self-regeneration of natural ecosystem processes. In a harvested forest system, carbon is also stored in wood and paper products and landfill. Wood products can have a lower embodied energy than other construction products, and thus their substitution can avoid emissions. Additionally, biomass can be used as feedstock for bioenergy, which can be substituted for fossil fuel energy. Quantifying carbon accounts in these systems requires estimating the magnitude of each of these stocks, their longevities and their emissions due to combustion and decomposition.

In the past, there have been attempts to provide guidance on the relative benefits of these mitigation strategies (conservation versus harvested forest) in the absence of context. More recent research has highlighted that the benefits of these strategies is sensitive to contextual factors, including the nature of the forests, market factors, alternative land uses and policy institutions (Macintosh et al. 2015). An important issue is the time frame for assessment:

Box 18.1 **Understanding Processes of the Carbon Cycle in Forest and Harvested Wood Systems**

Carbon dioxide is removed from the atmosphere through the process of photosynthesis in plants, whereby CO_2 and water vapour are absorbed through the stomata in leaves and produce carbohydrates and oxygen (Law et al. 2002). A proportion of the carbon is emitted through autotrophic respiration, and the remainder is stored in plant biomass and accumulates as plants grow. Carbon is transferred between biomass components within the ecosystem; from living to dead plant biomass, falling branches and trees, insect and animal consumption, deposition as dung, mortality of animals, transfers to fungal hyphae, exudation and mortality of roots. Dead biomass represents input to soil organic matter through decomposition. Organic carbon in soils is transferred between different physical and chemical forms during decomposition and incorporation into the soil matrix. Soil organisms, including invertebrates and microorganisms, consume organic matter and emit CO_2 through heterotrophic respiration.

   The net exchange of carbon between the atmosphere and the forest is the difference between the fluxes of photosynthesis and combined autotrophic and heterotrophic respiration; and this represents the change in carbon stock. In all forest ecosystems, either conserved or managed for production, natural disturbance regimes will affect the net carbon exchange in a stochastic manner.

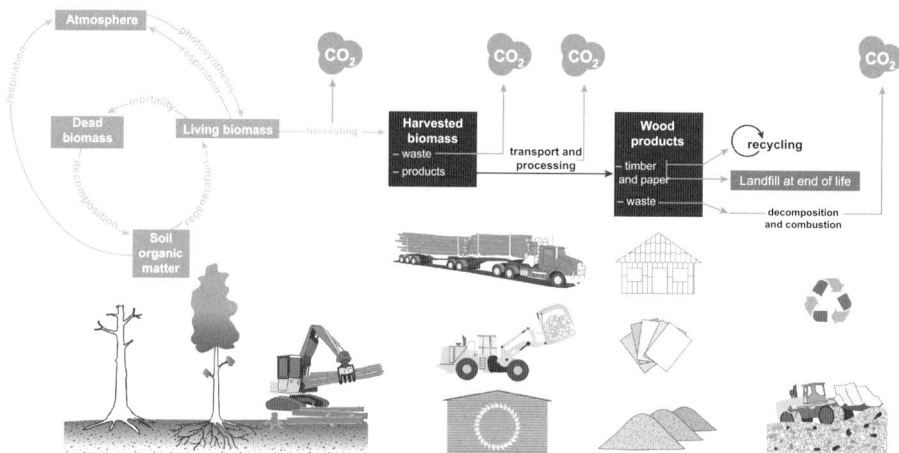

Figure 18.1 The carbon cycle in the forest and harvested wood system.
Source: Courtesy of the authors. For a colour version of this figure, please see the colour plate section.

the relevant timescale for effective climate mitigation activities is within the next few decades. A temporal imbalance between carbon emissions and carbon sequestration rates in the order of many decades may not provide a net mitigation benefit. Forest mitigation activities also have the potential to provide either co-benefits or perverse impacts. Hence, it is important that any forest mitigation strategy be developed to suit the relevant ecological, social, economic and political conditions.

In Australia, several studies have found that the conservation of native forests can generate significant reductions in net emissions (Macintosh 2011, 2012a; Keith et al. 2015; Macintosh et al. 2015). For example, Keith et al. (2015) looked at the net emission outcomes from forest conservation relative to a reference case of commercial harvesting in two case study systems of mixed eucalypt forest on the South Coast of New South Wales and mountain ash forest in the Central Highlands of Victoria. Testing simulations with a wide range of parameter values reported in the literature demonstrated that none of the scenarios could increase the carbon stock in the harvested forest system sufficiently to exceed the increase in carbon stock that could be achieved by changing the management of native forest to conservation over a 100-year time frame.

18.3 Drivers of Change in Forest Carbon Stocks

The exchange of carbon between the forest and atmosphere is large in magnitude and highly variable, with temporal patterns ranging from hours to decades, but the carbon stock remains in a stable state as a dynamic equilibrium. Change in the forest carbon stock only occurs if there is a trend in the net exchange with the atmosphere. There are many processes related to disturbance events and human activities that create an imbalance in the net exchange, and thus drive change in forest carbon stocks.

The carbon stock in a natural ecosystem under regimes of natural disturbance events is referred to as the carbon carrying capacity. This is distinguished from the current carbon stock in a system subject to human activities. The carbon sequestration potential is calculated as the difference between the carbon carrying capacity and the current carbon stock of the system under current conditions of disturbance and/or human activities.

Human activities are drivers of change in forest carbon stocks due to various forms of deforestation and degradation. The reasons for these activities include the provision of goods and services, in the form of energy; wood and paper products; and land for other activities, such as cropping, grazing, settlements or mining. Land conversion to farmland, biofuel or vegetable oil crop production is the single biggest cause of deforestation globally. In many countries, government subsidies and incentives for these activities that cause deforestation are over 100 times greater than the current levels of financing to reward countries for not harvesting their forests (Table 18.1). Global consumption of forest products is indicative of the main drivers of market demand: 50% is used for energy production, 30% is used for wood products, 10% is used for paper products and the remainder is unaccounted for (FAO Statistics 2007–2013: FAO n.d.).

The following are some of the main drivers of change in forest carbon stocks due to human activities:

- **Deforestation for change in land use** to provide cleared land for agriculture, grazing and settlements has been the primary cause of emissions from the land sector. An estimated 13 million ha of forests are cleared each year (FAO 2013a), releasing about

Table 18.1. *Reducing Emissions from Deforestation and Forest Degradation (REDD+) finance received compared with domestic expenditure on biofuel and agricultural subsidies (average annual USD million)*

	REDD+ finance (2006–2014)	Agricultural subsidies (2010–2012 annual average)	Biofuel subsidies (2009)
Brazil	158	11 082	2700
Chile	0	709	n/a
China	9	160 023	500
Indonesia	165	27 072	79
Mexico	12	7880	n/a
Total	**346**	**206 766**	**3279**

Source: REDD+ finance (Barnard and Nakhooda 2014); Agricultural subsidies (OECD 2014); Biofuel subsidies (Gerasimchuk et al. 2012).

1.5 GtC (equivalent to 5.5 $GtCO_2$) into the atmosphere (Gustavsson et al. 2006). The primary, immediate effect of deforestation for land conversion is the loss of carbon from the woody biomass of trees that are cleared and then typically burnt (van der Werf et al. 2009). An additional effect is the loss of carbon from the soil, and particularly peat, due to mechanical disturbance, higher temperatures, enhanced rates of respiration and reduced organic inputs. Degradation of the soil, including loss of organic matter, compaction and erosion means that the restoration potential of the vegetation and carbon stocks may be hindered.

• **Deforestation due to energy poverty**. Approximately half of the timber harvested annually is used for energy production by people who do not have electricity, mainly in Africa and Asia (FAO 2013b). In this case, deforestation is a form of unintentional land-use change resulting from the need for wood resources rather than conversion to another land use.

• **Deforestation for conversion to biofuels** is a particular case of plantations where rotation lengths are even shorter than for wood production. The biomass harvested from the plantation is used as feedstock for combustion as biofuel to substitute for fossil fuels. Biofuel as an energy source is renewable, but it is not carbon neutral. The loss of carbon stock when the original primary forest was cleared from the land must be taken into account. Additionally, there are emissions associated with plantation establishment, harvesting, maintenance, transport, processing and possibly fertiliser use. To demonstrate mitigation benefit, biofuel production must generate and use additional carbon, that is, carbon that plants would not otherwise absorb (Searchinger 2010; Hudiburg et al. 2011; Schulze et al. 2012).

• **Deforestation for palm oil production** is increasingly responsible for deforestation in Indonesia and Malaysia (Carlson et al. 2012; see Box 18.2). Deforestation for palm oil plantation expansion in Kalimantan, Indonesia, alone is projected to contribute 18–22% (0.12–0.15 GtC per year) of Indonesia's 2020 CO_2 equivalent (CO_2e) emissions (Carlson

---

Box 18.2 **Reducing Global Demand for Deforestation Related to Palm Oil Production**

Rapidly rising demand for palm oil has, and is forecast to, contribute significantly to deforestation in Indonesia and Malaysia (Carlson et al. 2012). Demand for palm oil is high for production of packaged foods, personal care and cleaning products. Its versatility is due to an exceptionally high melting point and very high saturation levels. The combination of these properties means it can be used without creating trans fats. Some vegetable oils have one or the other property but not both. Hence, products from most other vegetable oils result in the formation of trans fats. Export revenues from palm oil are so high that REDD+ payments of USD 10 per tCO_2e avoided are not sufficient to stop most palm-oil-related deforestation (Carlson et al. 2012).

Mechanisms to reduce forest conversion for palm oil include both increasing REDD+ payments and steps to curb palm oil demand. Increasing the carbon price for REDD+ will reduce the incentive for forest conversion for palm oil. Reducing demand for palm oil products that are derived from deforestation can be achieved by certification within the food supply chain to inform consumers of product sources. The Roundtable on Sustainable Palm Oil (RSPO) certifies sustainable palm oil production where it is not contributing to deforestation. The global market for certified palm oil is now over 13% of total production and growing rapidly. Major supermarket chains in Europe, the UK, USA and Australia have made commitments to source palm oil only from sustainably certified sources.

To reduce pressure on the world's tropical forests further, research is ongoing to find a cheap alternative to palm oil, which has the same properties of reduced trans fats and longer food storage time. Researchers at the University of Bath have found a possible alternative to palm oil, an oily yeast, *Metschnikowia pulcherrima* – previously used in South Africa's wine industry. The yeast can grow on most organic feedstocks and the land requirement for commercial yeast production would be significantly less than palm oil (from 10 to possibly 100 times less). To be viable, the cost of yeast production would need to be USD 800–900 per tonne of oil, but this has not yet been achieved (Carlson et al. 2012). Hence, more research is needed to reduce the costs and find other alternatives to palm oil.

---

et al. 2012). A significant percentage of exported palm oil globally comes from plantations that illegally replaced forests.

- **Remnant forest conversion to plantations or harvested regrowth forests**. In the Kyoto Protocol's first commitment period, it was not compulsory for developed countries to report emissions and removals associated with the conversion of remnant forests to plantations or harvested regrowth forests because the land use was not changed. The forest, and its carbon stock, may have been fundamentally altered but the land use was still considered to be a forest. In the Protocol's second commitment period, all developed countries had to report emissions and removals from forest management, including those associated with the conversion of remnant forests to plantation and harvested regrowth systems. How forest management will be accounted for in the post-2020 regime is still uncertain. Ideally, the accounting framework should be comprehensive, capturing all anthropogenic impacts on forest systems (Harmon et al. 1990; Schlamadinger and Marland 1999; Keith et al. 2015).

- **Degradation due to forest harvesting** results in depletion of ecosystem carbon stocks and significant CO_2 emissions, even though the land-use type is not changed. Global assessment in 2010 estimated that 57% (2.337 billion ha) of remaining forest is subject to industrial logging and degradation (FAO 2010). The carbon stock in regrowth forest averaged over the harvesting rotation is commonly about half, or in the range 30–70%, of the stock in the primary forest. Even when the stocks of carbon in the wood products and waste components of the system are included, the total stock is less than in the primary forest (Harmon et al. 1990; Krankina and Harmon 2006; Dean et al. 2012; Keith et al. 2015; Macintosh et al. 2015). Emissions from forest degradation, including from commercial logging, are estimated to have a similar magnitude as emissions from deforestation in many regions (Asner et al. 2010; Macintosh 2011). Forest biomass is also harvested to provide bioenergy. Bioenergy does not necessarily substitute for fossil fuels in the marketplace. The greater supply of energy sources may create a larger market, or bioenergy may displace other forms of renewable energy rather than fossil fuel forms. Degradation is difficult to detect and monitor on large scales and from remote sensing because land cover change is not involved (Harris et al. 2012). Globally, the carbon loss due to forest degradation has yet to be calculated but estimates for tropical forest degradation range between 0.6 GtC and 1.47 GtC per year (ISU 2015). In general, allowing logged forests to regrow could remove 1–3 GtC from the atmosphere (Houghton 2013).

- **Environmental burden shifting** has become a factor driving deforestation in a globalised world. While some countries may appear to be making progress in increasing their domestic stock of forests, they may be simultaneously shifting the environmental and social burden of deforestation to other countries. For instance, the total area forested in the EU has grown in the last decade, while at the same time Europe is the main market for around 50% of forest and agricultural products derived from illegal deforestation.

- **Corruption and illegal logging** are major drivers of deforestation. Transparency International (TI 2006) studies show that, in countries where excessive corruption prevails, the destruction of natural resources, such as local forests, for private gain is high. This is highlighted in Chapter 22, which addresses how nations, companies and communities can reduce the levels of corruption.

- **Natural disturbance events** such as wildfires, pests, disease, droughts, cyclones and storms result in mortality of vegetation and sometimes combustion or decomposition, with consequent emissions of CO_2. However, under a natural regime of these disturbance events, the emissions at the time of the event are replaced by the uptake of CO_2 in the regrowing vegetation. Regaining the balance of carbon stocks occurs over variable time frames depending on the disturbance type and ecosystem, and may take years to decades. As an example, the losses and gains of forest carbon due to a wildfire were monitored in mountain ash forest in Victoria. An average of 10% of the biomass carbon stock was emitted through combustion during the fire, and this amount would be replaced within about 10 years of regrowth (Keith et al. 2014a). Long-term changes in the forest carbon stock would occur only when the disturbance regime was changed; that is, the frequency, intensity or magnitude of the events was changed. This new regime would result in a difference between rates of emissions and removals of CO_2 over time; that is, the net exchange with the atmosphere. For example, predicted increases in severity of fire regimes under conditions of climate change, with higher temperatures and more extreme droughts, potentially could reduce forest carbon stocks (Westerling et al. 2011). Additionally, conditions under a changed climate are likely to intensify positive feedback processes between stress factors, such as drought and pest outbreaks, resulting in reduced plant growth and/or increased mortality, and hence loss of carbon stock.

Clearly, many drivers of deforestation and degradation result in carbon stock losses from the forest sector. Hence, a holistic approach to forest management is needed, which addresses and reverses the fundamental drivers for deforestation and degradation, and is supported by policy and institutional reforms to create the right incentives for change.

18.4 Strategies to Simultaneously Achieve Climate Change Mitigation and Adaptation Goals

Management actions can reduce, but not eliminate, the risks of impacts on ecosystems due to climate change, as well as increase the inherent capacity of ecosystems to adapt (Settele et al. 2014). Goals for mitigation and adaptation in the forest sector can be achieved through a more holistic, landscape-wide approach to land management activities, which delivers mitigation outcomes by increasing long-term stable carbon storage, promotes adaptation by increasing the resilience of ecosystems, protects and restores land productivity, generates co-benefits for other ecosystem services such as water quality and supply, and helps to improve sustainable livelihoods.

Management of forests for the objective of GHG mitigation requires strategies to maximise long-term stable biocarbon stocks by conserving natural forests and allowing them to continue growing without being perturbed by harvesting of biomass (Harmon et al. 1990; Keith et al. 2014b). However, there is significant demand for wood and paper products and bioenergy, which necessitates the harvesting of some biomass. One strategy to optimise trade-offs between competing forest uses is to conserve existing native forests, while harvesting wood and fibre from industrial plantations, which can provide high volumes of wood from a small fraction of the available land. For example, an average plantation produces 5–40 m^3 of usable wood per hectare per year, versus less than 3 m^3 from a natural forest. In 2010, about 60% of global wood supply came from plantations that represent 6–10% of forest cover globally (Paquette and Messier 2010). Promoting a shift to plantation forests on existing cleared land as the primary source of solid wood and wood fibre could reduce the pressure on natural forests and allow them to be prioritised for other ecosystem services including mitigation, biodiversity conservation and water quality.

The various adaptation responses required in forests depend on the management context. For conservation forests, the best response is to remove other threatening processes so that the natural resilience and adaptive capacity derived from their biodiversity can operate optimally. For plantation forests, strategic planning is needed; for example, the genetic material planted should be appropriate for changing climatic conditions over the coming rotation period.

A holistic approach to forest management includes strategies for mitigation and adaptation, as well as sustainable development; we outline four main strategies and 12 additional sub-strategies for the forest sector.

18.4.1 Strategy 1: Reducing Rates of Deforestation

Primary forests in an intact state store the highest carbon densities because most living biomass carbon is found in large, old trees (Stephenson et al. 2014) and in undisturbed soil stocks. The primary mitigation value of forest ecosystems is the accumulated, standing stock of carbon. The short-term flux of carbon between forests and the atmosphere through photosynthesis and respiration is of secondary importance, as this net exchange varies on a seasonal and annual basis in response to short-term fluctuations associated with prevailing weather conditions. A common misunderstanding is that natural forests should be replaced with fast-growing plantations as younger trees have a faster rate of carbon uptake and therefore function as a larger sink. However, this ignores the fact that even old growth forests have been shown to continue to function as sinks (Pan et al. 2011; Bellassen and Luyssaert 2014), and furthermore the loss of carbon stock from clearing the natural ecosystem must be accounted for; this takes hundreds of years to replenish (Harmon et al. 1990). Additionally, the standing stock of carbon in a plantation is much less than in a natural forest (Dean et al. 2012; Keith et al. 2015).

Protecting primary forests by avoiding deforestation will significantly reduce GHG emissions from the land sector. The Intergovernmental Panel on Climate Change (IPCC) Fifth Assessment Report estimated that about 65% of the total mitigation potential in the forest sector is located in the tropics, and of this total about 50% could be achieved by reducing deforestation. The reduction in emissions was estimated to be up to 13.8 $GtCO_2e$ per year at prices on the carbon market up to USD 100 per tCO_2 (Smith et al. 2014). According to the Stern Review, the estimated cost of avoiding deforestation entirely in eight countries collectively responsible for 70% of land-use emissions is between USD 1 and 2 per tCO_2 (Stern 2006, based on Grieg-Gran 2006). Stopping deforestation of natural forests was one of the key recommendations in the global action plan of the World Resources Institute's *New Carbon Economy* report, to be achieved by strengthening investment in forest protection that is linked to performance (WRI 2014).

A number of countries have implemented total or partial bans on deforestation, such as China, New Zealand, the Philippines, Costa Rica, Sri Lanka, Thailand and Vietnam. In most OECD countries, the area of forest has remained stable or has slightly increased. Some countries like China, South Korea and Costa Rica have invested significantly in forest restoration and expanded their forests (Chong 2005). A substantial number of developing countries, which host most of the world's tropical forests, have reduced rates of deforestation and increased afforestation in the last decade. Examples of successful projects to reduce deforestation and promote reforestation occur in Guyana, Madagascar, Kenya, India, Mexico, Vietnam, Costa Rica, Tanzania, Mozambique and El Salvador (Boucher et al. 2014). Progress in these countries is due to a combination of implementation of comprehensive sets of policies and programmes (Section 18.5), including REDD+; payment for ecosystem services provided by forests; stronger law enforcement; governance reforms; changes in social, economic and market forces; and combining environmental actions with social and economic developments. As a result of these positive efforts (Boucher et al. 2014), combined with economic and market forces, the annual pace of deforestation fell by 19% – from 18 million ha to 13 million ha – in the first decade of this century (FAO 2010).

It should be noted though that nearly all these successes are partial ones. In some of these countries, such as Brazil, deforestation rates have started to increase as commodity prices have risen and the government has granted amnesty to illegal logging pre-2008. This has led to many assuming there will be another future amnesty for illegal logging that has been done or will be in the future (Fearnside 2015). In addition, there is the issue of environmental burden shifting. If reduced deforestation in one country is simply replaced by 'leakages' such as the transfer of deforestation and related emissions elsewhere as a result of globalised commodity production, then the gains in terms of climate change mitigation are negated (Boucher et al. 2014). Rates of deforestation are closely related to economic and market forces, as well as government policies, and so it can be difficult to attribute multiple causes to changes in trends of deforestation.

Strategies to reduce emissions that are focused on the fundamental drivers of deforestation are suggested as being more effective than detailed measurement and accounting of carbon in forests under changed management conditions. Providing incentives for policy and legal reforms in developing countries, such as forest governance and land tenure, are essential changes required to reduce forest loss (Daviet et al. 2009; Dooley 2014).

18.4.1.1 Strategy 1.1: Reducing Pressures for Land Conversion

Land conversion to non-forest uses is the biggest driver for deforestation globally.

- **Reducing pressures to convert forests into farmland** by improving the productivity of agricultural systems (as explained in Chapter 19). Improving efficiency of the food production chain, thus increasing supply, will lower the relative price of agricultural commodities and reduce the demand for expansion of agricultural land. There are many mechanisms that have potential to improve efficiency of agricultural production, including: selective breeding, cultivation, genetic modification and related technologies to improve yields from livestock and crops; increasing nutrient inputs; improving access to technology in developing countries; improving transportation and marketing; and reducing food wastage. Some degraded land can be restored to return it to agricultural production, as currently about a third of all previously arable land has been degraded and is not used. Improved sustainability of farming practices of small-scale farmers can reduce the need to regularly slash and burn new areas of forest to create new farmland.

Maintenance of forests or agroforestry on farms could be encouraged if there is a price on carbon to provide an income for maintaining existing (and restoring) forest systems.

- **Reducing pressures to convert coastal forests and wetlands into aquaculture**. Coastal wetlands, particularly mangrove forests, have large carbon stocks in their biomass and anaerobic sediments. Pressure for deforestation and draining of wetlands for development in coastal areas, particularly for aquaculture, has resulted in large emissions due to this land-use change. In Indonesia, 40% of mangrove forests have been cleared, where the ecosystem has an average carbon stock of 1080 tC/ha, and a total stock of 3.14 GtC in the 2.9 million ha of mangroves. The average annual emissions from clearing mangroves are 0.19 GtCO$_2$e per year, which is 10–31% of the national annual emissions from the land-use sector. The export revenue from aquaculture, mainly shrimp products, is USD 1.5 billion per year. Ceasing deforestation of mangrove forests would be an economically viable abatement activity if the price of carbon was USD 10/tCO$_2$e, providing a revenue of USD 1.9 billion (Murdiyarso et al. 2015). Conservation of mangrove forests, and other coastal wetland ecosystems, are good examples of the mitigation–adaptation–sustainable development nexus. Abatement is economically viable with a modest price for carbon, compared with the existing industry. As an adaptation measure, natural coastal ecosystems are the best form of protection against rising sea levels. Coastal ecosystems provide many additional services that improve local community livelihoods, including protecting near-shore fisheries.

- **Reducing pressures to convert forests into crops for biofuel production**. Most of the world's increased demand for biofuels this century has come from national mandates to use higher percentages of biofuels to reduce fossil fuel emissions and improve fuel security. The bulk of mandates, as of 2015, come from the 27 countries of the EU, but 13 countries in the Americas, 12 in the Asia-Pacific and 8 in Africa also have mandates or targets in place (Smith et al. 2014). As Chapter 15 of this book has shown, it is technically possible for the global transport sector (with the exception of air transport) to achieve a transition to low or zero net carbon without biofuels, through a combination of improved fuel efficiency, modal shifts, electrification and decarbonisation of that electricity supply. As the energy section of this book has overviewed, there are promising alternatives to wood/crop/land-based biofuel production, such as algal fuels and other second-generation biofuel production that could provide alternative fuels for air travel without requiring significant amounts of land.

- **Reducing pressures on forests from urbanisation**. Urbanisation requires land that previously was used for forests or farming. Displacement of farms creates pressures to move and convert new forests into farms. This negative impact of urbanisation can be mitigated by reducing the rate of population increase and increasing urban density. Increased urban density can be achieved without significantly increasing the embodied energy of cities through improved urban planning and mainstreaming the use of alternative building and construction materials to replace high GHG-emissions-intensity methods of making steel, cement and aluminium (Chapter 16).

18.4.1.2 Strategy 1.2: Reducing Energy Poverty

A significant percentage of harvested forest products is used for energy production (~1900 million m^3) by over one billion people who are without electricity for cooking, heating and light (FAO 2014). Almost 76% of the population in Africa (over 500 million people) and 51% of the population in developing nations in Asia – India, Bangladesh, Pakistan – use wood as the only household cooking fuel (Seifert 2013). Moreover, indoor air pollution from burning wood is responsible for more than 1.5 million premature deaths each year, half of them children under the age of five, the rest women. Enabling technologies such as energy-efficient cooking stoves and lighting using solar power, or other renewable energy sources, can be implemented to reduce pressure on wood from forests (see Chapter 4).

18.4.1.3 Strategy 1.3: Reducing Global Demand for Timber and Paper Products from Natural Forests

Approximately 30% of harvested forests are used for timber products and 10% for paper products (FAO Statistics, 2007–2013: FAO n.d.). Hence, reducing global demand for wood products is an important mechanism to reduce deforestation. This is already occurring due to reduced demand in developed countries for paper with the rise of digital technologies and information storage systems, and the increased supply of recycled wood and paper. The following mechanisms can be used to eliminate, replace or reduce demand for wood products:

- **Reducing demand for paper through digitalisation and virtualisation**: for example, e-commerce and digital publishing. The paperless office – digital storage of information instead of printed storage, email instead of letters – in principle can reduce demand for paper. In developed countries, digital technologies have contributed to a stagnation and, more recently, decline in demand for paper and paperboard products since the mid-2000s (Mery et al. 2010; Macintosh 2013). While paper and paperboard consumption has continued to grow in developing countries, the spread of digital technology could assist in reversing this trend.
- **Eliminating junk mail and unnecessary packaging**. These strategies have enabled many developed nations to reduce demand for paper consumption over the last decade. Total paper consumption has decreased by about 20% over the last decade in North America and Europe (Environmental Paper Network 2018).
- **Making paper out of alternative feedstocks**. Until 1843, all paper was obtained from hemp, cotton and linen materials, and none from wood fibre. In the twentieth century, wood pulp was the primary feedstock for paper and paperboard. Since the early 2000s, however, the use of recovered fibre pulp (recycled paper and paperboard) has increased substantially, rising from approximately 10 million tonnes to almost 80 million tonnes in a decade. This growth in the use of recovered fibre pulp has allowed pulp consumption and paper and paperboard production to grow with minimal growth in wood fibre inputs (Macintosh 2013). Looking forward, there is the prospect of further increasing the use of recovered fibre pulp and using alternative biomass feedstocks, including crops like kenaf. Kenaf produces more fibre per hectare per year, absorbs more carbon and is easier to pulp

than wood fibres, thus also saving on energy inputs to processing. Additionally, agricultural wastes can provide viable biomass feedstocks for pulp fibre. While some agricultural wastes are best returned to the soil, wastes that would be burnt or disposed of in landfill could be used as feedstocks. Use of this biomass waste provides farmers with extra income and reduces air pollution as the waste is no longer burnt.

- **Replacing timber freight pallets with plastic that lasts up to 200 years**. Timber freight pallets use 11% of all timber products globally. Plastic pallets are lighter, longer lasting and more durable, provide less of a medium for pathogens and are also designed to use less storage space than timber pallets (Lee 2004).

18.4.1.4 Strategy 1.4: Reducing Deforestation by Reducing Illegal Logging

Illegal logging constitutes over 50% of all forestry activities in tropical forests (including the Amazon Basin, Central Africa and South East Asia), and 15–30% of all wood traded globally. The trade in illegally harvested timber is very lucrative, estimated to be worth between USD 30 and USD 100 billion annually. It also depresses prices of legally harvested timber by at least 7%, reducing margins for legally produced products and reducing government revenue. Other factors contributing to illegal logging include palm oil investment and poverty, where the latter leads to slash-and-burn agriculture and deforestation for biomass energy sources. An integrated approach is needed to address illegal logging that:

- reduces poverty and energy poverty;
- improves agricultural productivity in developing tropical countries so as to break the cycle of slash-and-burn agriculture;
- increases the demand for legally sourced wood products and reduces the demand for illegally sourced wood products (for example, through certification and bans on trade in illegal timber products);
- reduces the demand for deforestation for palm oil production;
- increases the capacity to identify illegal logging and enforce forest laws in developing countries.

18.4.2 Strategy 2: Maximising Carbon Storage in Native Forests

18.4.2.1 Strategy 2.1: Conserving Native Forests

Conservation of existing carbon stocks in forest vegetation and soil is an important mechanism for mitigation. Implementing conservation mechanisms requires natural resource planning to prioritise forest types for the provision of a range of ecosystem services, including carbon storage, water resources, wood production, bioenergy production and biodiversity conservation. Some combinations of these services are compatible and enhancing, while other services are conflicting and detrimental. Condition or state of the forest can be classified as natural, semi-natural or plantation, and this influences their current services and capacity to change. Identifying and quantifying forest condition, the services provided and their compatibility is the first stage in planning. Application of the

Environmental Economic Accounting – Ecosystem Accounts, is allowing progress in natural resource management (United Nations Statistical Commission 2021).

18.4.2.2 Strategy 2.2: Restoring Degraded Native Forests

Two billion hectares of land offer opportunities for restoration across the world, in both tropical and temperate countries. The International Union for Conservation of Nature (IUCN) 'Bonn Challenge', established at the invitation of the Government of Germany and IUCN in 2011, had the global goal of restoring 150 million ha of deforested and degraded lands by 2020, and 350 million ha by 2030. An assessment of progress in 2018 across 19 countries showed that approximately half the area of their country commitments had been brought under restoration (Dave et al. 2019). This approach is aimed at facilitating the implementation of various international commitments including the United Nations Framework Convention on Climate Change (UNFCCC) REDD+ goal. This could also help to sequester an additional 1 GtCO$_2$e per year, reducing the current emissions gap by 11–17%. This challenge is supported by the Global Partnership on Forest Landscape Restoration, with IUCN as its secretariat. IUCN published in 2015 the report *Synergies between Climate Mitigation and Adaptation in Forest Landscape Restoration*, outlining how forest landscape restoration has been achieved recently in seven countries to realise both mitigation, adaptation and sustainable development outcomes (Rizvi et al. 2015).

18.4.3 Strategy 3: Improving Carbon Storage in Harvested Forests Systems

Management of forests for production of harvested wood products should prioritise the use of plantations, which provide the most efficient system of production (Paquette and Messier 2010). Where land-use decisions prioritise native secondary forest for production of harvested wood products, then management strategies must seek to optimise mitigation and adaptation goals by maximising carbon storage, but within the constraints of production. Such changes in forest management practices should not be in the form of a subsidised system that is substituting for more efficient and economically viable plantations.

Harvested forest systems store 30–70% less carbon than primary forests because the trees are younger, biomass waste after harvesting decomposes and soil disturbance enhances respiration rates and erosion (Krankina and Harmon 2006; Cyr et al. 2009; Bryan et al. 2010; Carlson et al. 2010; Keith et al. 2015; Macintosh et al. 2015).

18.4.3.1 Strategy 3.1: Increasing Rotation Lengths

Maximising carbon stocks in the standing forest is a major component of reducing emissions. Biomass carbon stocks increase with forest age, and so in a production forest the average carbon stock is higher when the rotation length is greater. Waiting longer before harvesting may delay economic return but could be compensated by the increased value of the carbon stock and potentially higher quality timber in older, larger trees.

18.4.3.2 Strategy 3.2: Reducing Waste in Harvesting

Reducing waste in harvesting would improve efficiency in the utilisation of wood resources and reduce emissions from the combustion or decomposition of the waste material. However, removal of all biomass off-site is not recommended because the export of nutrients would result in loss of soil fertility, and loss of habitat in the litter and woody debris.

18.4.3.3 Strategy 3.3: Improving Soil Conservation Practices

Improving soil conservation practices would increase biomass productivity and soil carbon content. Practices include erosion control, minimising compaction, maintaining vegetation and/or litter cover and minimising mechanical disturbance.

18.4.3.4 Strategy 3.4: Improving Silvicultural and Processing Methods to Produce Long-Lived Wood Products

This includes selecting species that can be grown efficiently in plantations and can adapt to changing climate conditions, integrating plantation–production systems, processing methods and facilities for different wood types and products and developing new engineered wood products. Long-lived wood products have a mitigation benefit compared with many current wood products by increasing the longevity of the carbon storage in the products. Global statistics of wood products show that the lifetime of products is very short compared with the lifetime of a tree. Harvested wood products used for paper consumption amounted to 510 million m^3 in 2007, with a lifetime of 1–10 years, and the amount of solid wood products was 688 million m^3, with a lifetime of 30– 90 years. Only about 4% of the carbon in trees ends up in solid wood products (Le Quéré et al. 2013; Keith et al. 2015; Macintosh et al. 2015). Greater investment is required in developing new wood products, processing facilities and markets for products with greater longevity.

18.4.3.5 Strategy 3.5: Maximising Substitution Benefits of Wood Products

Harvested wood products are not carbon neutral because of the loss of carbon stock in the original forest and/or the forgone sequestration associated with preventing the forest from reaching its carbon carrying capacity. There are also fossil fuel emissions in wood production processes (Ingerson 2011). However, the use of wood products derived from sustainably managed plantations as a substitute for more emissions-intensive products such as concrete, steel, aluminium and wood products from native forests can reduce GHG emissions (although not if these other products are produced using non-carbon renewable energy sources).

18.4.4 Strategy 4: Reducing Emissions from Waste

Waste biomass material is produced during harvesting, during the processing of timber and paper products and at the products' end of life. CO_2 is emitted from the waste material during combustion or decomposition. Increasing the proportion of harvested wood that is transferred to long-lived products will not only increase profits but also reduce emissions and the amount of harvested wood required to meet demand. Efficiency in processing can also include generation of energy from waste material that is used as an energy source in production and/or sold back to the grid.

18.4.4.1 Strategy 4.1: Increasing Recycling of Wood and Paper Products

Recycling of products and materials reduces demand for harvesting additional wood and reduces the amount of processing waste, thus reducing GHG emissions (Acuff and Kaffine 2013). Currently, more than half the paper used in Europe, North America and some Asian countries is recycled. Recycling at a pulp and paper mill reduces the energy intensity of paper production by around 50%, thus reducing GHG emissions. The rate of recycling is increasing, particularly with the advent of new technologies. Printers are now available that will de-ink and reprint on paper up to five times, thus moving paper recycling from the pulp mill into offices, and reducing GHG emissions of paper production by over 70%.

18.4.5 Co-benefits, Perverse Impacts and the Nexus with Adaptation

All forest-related mitigation activities involve trade-offs and, mostly, at least some costs: resources must be expended or opportunities forgone in order to realise the mitigation opportunities. In some cases, the mitigation activity will result in greater social benefits than costs. However, even in these cases, there will be costs incurred somewhere. How the costs and benefits of forest-related mitigation activities are distributed and managed will usually be critical to the success of policy efforts. The nexus between adaptation and mitigation is highlighted in Box 18.3, giving details of national and international strategies in the forest sector.

Inevitable costs are not unique to the forest sector. However, one of the factors that sets forest-related mitigation activities apart from activities in other sectors is the prevalence of spill-over effects (positive and negative externalities), co-benefits and perverse impacts. Mitigation actions to prevent deforestation and conserve forests have many co-benefits for land management: reducing flood risks, improving water quality, reducing soil erosion and desertification, reducing the spread of dryland salinity, conserving biodiversity and heritage values and avoiding the effects of changed vegetation on changing local climate conditions (forest loss can reduce precipitation and increase temperature variability). Some of these co-benefits accrue to forest owners. However, in most cases, a substantial proportion of the benefits accrue to the community; they are both co-benefits and positive externalities. Consequently, if financial incentives are provided to forest owners to prevent deforestation for carbon purposes only, there is a risk that the levels of conservation will be suboptimal. Forest owners may fail to protect the forests, even when it is socially beneficial, because they are unable to internalise the broader societal benefits.

In contrast, policies to prevent deforestation can have perverse impacts, including leakage, theft, gaming and corruption. Leakage refers to the risk that efforts to prevent deforestation in one area increase deforestation elsewhere, either by directly displacing the activity or via market forces. By providing incentives to conserve forests, there is also a risk

---

Box 18.3 Growing Recognition of the Adaptation–Mitigation Nexus in the Forest Sector: International and National Climate Change Strategies

There is growing recognition internationally that forest-related mitigation projects, such as REDD+ projects, also have the potential to support the adaptation of forests to climate change by reducing the pressures of human activity on them, by connecting forested areas and by conserving biodiversity hotspots. A REDD+ project's carbon storage has a higher chance of lasting if it incorporates adaptation measures for communities and forest ecosystems. If adaptation is not considered, the negative impacts of climate change could jeopardise project outcomes. Integrating adaptation measures can also increase local people's acceptance of, and interest in, the project, because adaptation focuses on immediate local needs, whereas mitigation addresses longer-term global benefits. At the same time, if an adaptation project also contributes to climate change mitigation, the local community may be able to benefit from the carbon funding and capacity-building inherent in international instruments such as REDD+. Furthermore, donors may prefer supporting adaptation projects that also have global mitigation benefits and vice versa.

The need for combined mitigation and adaptation actions is recognised formally by an increasing number of nations including Cameroon (Chia et al. 2015), Rwanda (Republic of Rwanda 2011), Ethiopia (FDRE 2011) and Kenya (FAO 2011).

---

that this may encourage some officials to divert scarce resources to non-productive activities in order to extract rents. This might involve the misappropriation of protected forests from traditional owners, threatening to clear forests in order to extract payments or the bribery of officials so that particular forest owners receive payments. These activities not only undermine the objective of reducing deforestation but can also stifle economic growth, deprive vulnerable communities of vital economic and cultural resources and undermine confidence in government.

Similar issues arise in relation to reforestation and afforestation, maximising carbon storage in forest reserves and improving carbon storage in harvested forests. The expansion of forest cover can have all of the co-benefits associated with the avoidance of deforestation but, depending on how it is done, it can also have adverse effects. These include negative impacts on local economies due to reductions in employment stemming from the change in land use, reductions in catchment run-off from the expansion of forest cover, increases or changes in water pollution due to forest-related fertiliser use, losses of biodiversity and heritage values, increases in bushfire risk and upward pressure on food prices. Widespread transformation of land for mitigation purposes, in the form of fast-growing plantations in areas where the species did not occur previously, or the conversion of previously unculti-vated land to bioenergy plantations, will lead to negative impacts on ecosystems and biodiversity (Settele et al. 2014). Ideally, both the positive and negative effects of reforest-ation and afforestation programmes will be managed in a way that maximises social well-being. In practice, achieving this is difficult, as it requires the balancing of competing interests in imbalanced political decision-making processes.

The potential for forest-related mitigation activities to have spill-over effects, co-benefits and perverse impacts means that it is important for them to be integrated with adaptation strategies and gain the benefits of synergies (Ravindranath 2007; Guariguata et al. 2008; Locatelli et al. 2011). For example, financial incentives to encourage mitigation activities can diversify the income streams of forest owners and forest-dependent communities and, in so doing, make them more resilient to the effects of climate change. In other cases, mitigation and adaptation imperatives can be in conflict. For example, the desire to maximise carbon sequestration can lead to the planting of fast-growing invasive species, which threaten biodiversity and further weaken the capacity of threatened species to adapt to climate change (Mackey et al. 2020).

18.4.6 Outcomes from These Strategies

Management of carbon stocks and flows in the forest sector is critical for mitigating the climate change problem, allowing natural ecosystems and human production systems to adapt to climate change and ensuring available resources for sustainable development. The estimated mitigation benefit from ceasing deforestation and forest degradation is a reduction in emissions by 1.4 GtC per year. Allowing logged forests to regrow to their carbon carrying capacity could remove an additional 1–3 GtC per year from the atmosphere (Houghton 2013). Maximising carbon stocks in forests by avoiding deforestation and degradation and allowing forest regeneration, as well as through restoration and reforestation, could reduce emissions by 3.45–3.86 GtC per year, representing 24–33% of all annual mitigation activities globally (ISU 2015).

18.5 Governance and Policy Solutions for Forest Mitigation and Adaptation

There are two main governance and policy issues associated with forest-related mitigation and adaptation issues: first, the distribution of power and responsibility between international/transnational bodies and national/subnational governments; and second, the nature of the domestic policy measures that are necessary to achieve the desired forest mitigation and adaptation objectives.

18.5.1 Multilevel Governance for Forests

For over 40 years, repeated attempts have been made to negotiate effective international agreements to promote the sustainable management and conservation of forests. While progress has been made in some areas, the defining element of the history of international forest agreements has been underachievement (Humphreys 2008). The limiting issue from the beginning has been the nexus between poverty, global inequality and forest conservation. Developing countries have been unwilling to conserve forests and alter forest management practices in order to achieve global environmental objectives unless developed countries provide financial assistance and compensate them for forgone development opportunities. Developed countries, mostly, have been unwilling to oblige. This dynamic lies behind the relative ineffectiveness of the International Tropical Timber Agreements of 1983, 1994 and 2006 and collaborative processes like the United Nations Forum on Forests, and the repeated reliance on non-legally binding agreements like the Forest Principles outcome of the United Nations Conference on Environment and Development in 1992 (*Non-Legally Binding Authoritative Statement of Principles for a Global Consensus on the Management, Conservation and Sustainable Development of All Types of Forests* of 1992) (Humphreys 2008).

This historical impasse explains the enthusiastic reception that proposals for REDD and REDD+ have received in many quarters (Macintosh 2012b). At the heart of all REDD proposals is the notion that developing countries will be compensated by developed countries for reducing GHG emissions from deforestation and forest degradation. The REDD+ pro-posal merely adds the provision of compensation for the maintenance and enhancement of forest carbon stocks; an addition that was included so as to ensure that developing countries that had already taken steps to conserve forests would not miss out on financial assistance.

Much of the hype surrounding REDD and REDD+ has centred on market-based proposals, under which developing countries (or regions within them) that reduce deforest-ation and forest degradation emissions below a preset baseline (or reference level) would receive credits that could be sold to, and used by, developed countries to meet their mitigation obligations. One of the attractions of market-based proposals is that govern-ments in developed countries do not necessarily have to finance the purchase of credits from developing countries. By linking domestic carbon markets to REDD and REDD+ markets, developed country governments can rely on private entities with carbon liabilities to purchase REDD credits when it is in their interests to do so. Simultaneously, market-based REDD and REDD+ schemes could: (a) provide the necessary financial resources to prompt developing countries to change their forest management practices; (b) lower the social cost in developed countries of meeting mitigation obligations; and (c) lower the abatement costs of private entities with carbon liabilities in developed countries.

While offering many potential benefits, an international market-based REDD scheme has proven difficult to negotiate. Many groups have opposed a market-based REDD scheme because it will not reduce global GHG emissions; could increase global emissions if additionality, leakage and permanence risks are poorly managed; and could lead to a range of perverse impacts, including the abuse of Indigenous peoples, corruption and adverse biodiversity outcomes. Apart from the opposing voices, there have proven to be substantial practical obstacles to creating an effective market-based scheme, none more significant than the lack of institutional capacity in developing countries and disagreements over the methods for determining baselines.

In an effort to address these concerns, it was agreed at the 16th Conference of the Parties (COP 16) to the UNFCCC (the Cancun Agreements) that REDD+ would be introduced in phases, although rules and methodologies were not finalised at COP 25 in Madrid in 2019:

... beginning with the development of national strategies or action plans, policies and measures, and capacity-building, followed by the implementation of national policies and measures and national strategies or action plans that could involve further capacity-building, technology development and transfer and results-based demonstration activities, and evolving into results-based actions that should be fully measured, reported and verified. *(UNFCCC 2010: FCCCC/CP/2010/7/Add.1/ paragraph 73)*

The parties also agreed to a number of principles and safeguards on REDD+ activities to ensure environmental integrity, promote co-benefits and guard against perverse impacts, including that parties should promote and support:

- actions to address the risks of reversals (permanence);
- actions to reduce displacement of emissions (leakage);
- respect for the knowledge and rights of Indigenous peoples and members of local communities;
- the full and effective participation of relevant stakeholders, in particular Indigenous peoples and local communities, in REDD+ activities; and
- actions consistent with the conservation of natural forests and biodiversity conservation and ensuring that REDD+ actions are 'not used for the conversion of natural forests, but are instead used to incentivise the protection and conservation of natural forests and their ecosystem services, and to enhance other social and environmental benefits' *(UNFCCC 2010: FCCCC/CP/2010/7/Add.1/Appendix 1/paragraph 2)*.

In conjunction with the UNFCCC negotiations, a number of countries and international organisations have been investing in REDD+ capacity-building and results-based demonstration projects, where financial payments are made conditional on the delivery of outcomes. As flagged in the Cancun Agreements, results-based financing is viewed as the stepping stone to any market-based scheme. At COP 19 in Warsaw in late 2014, the Warsaw Framework for REDD+ was agreed that again emphasised the importance of results-based financing and stressed that such finance was likely to come from a variety of sources: public and private, bilateral and multilateral, including the Green Climate Fund.

The progress made in REDD+ in recent years has illustrated the role that international and transnational bodies can, and indeed have to, play in forest-related mitigation and adaptation activities. Solutions to forest management issues in developing countries are unlikely to be found in the absence of international involvement. Many developing countries do not have the domestic institutional and financial capacity to resolve forest management challenges alone. Having said this, the history of underachievement in international forest agreements, the protracted nature of the UNFCCC REDD+ negotiations and the high-profile failure of a number of early REDD+ projects, demonstrate the magnitude of the obstacles to success. Australia, a developed country with mature and stable political institutions and a high level of technological capacity, has struggled for decades to control deforestation and improve forest management (Macintosh 2012b). Achieving the same in developing countries with less mature and stable institutions is orders of magnitude more difficult.

Similar lessons concerning the importance of international governance structures in forest management and the obstacles to success have been drawn from the attempts to control illegal logging. Consistent with the objects of the International Tropical Timber Agreements and other processes, a number of developed countries have taken steps to curb the importation of illegally sourced wood products. Amongst the most well known of these measures are the Lacey Act provisions of the United States Code,[1] which, following amendments in 2008, have made it unlawful for anybody to:

import, export, transport, sell, receive, acquire, or purchase in interstate or foreign commerce . . . any plant;

(a) taken, possessed, transported, or sold in violation of any law or regulation of any State, or any foreign law, that protects plants or that regulates:

  (i) the theft of plants;

  (ii) the taking of plants from a park, forest reserve, or other officially protected area;

  (iii) the taking of plants from an officially designated area;

  (iv) the taking of plants without, or contrary to, required authorisation;

(b) taken, possessed, transported, or sold without the payment of appropriate royalties, taxes, or stumpage fees required for the plant by any law or regulation of any State or any foreign law; or

(c) taken, possessed, transported, or sold in violation of any limitation under any law or regulation of any State, or under any foreign law, governing the export or trans-shipment of plants.

(16 USC § 3372)

The Lacey Act also makes it unlawful for any person to import any plant unless the person files an import declaration containing, among other things, the scientific name of the plant, value of the importation, quantity of the plant and name of the country from which the plant was harvested. A 'plant' is defined for these purposes as 'any wild member of the plant kingdom, including roots, seeds, parts, or products thereof, and including trees from either natural or planted forest stands'. Since their introduction, enforcement actions have been taken on a number of occasions in relation to these provisions, demonstrating the seriousness with which compliance is taken by US authorities. The provisions have also prompted changes in the purchasing practices of wood importers, as the provisions require them to exercise 'due care' when engaging in these activities.

Australia followed the Lacey Act model in crafting the *Illegal Logging Prohibition Act 2012* (Cth).[2] The Act creates offences for importing illegally sourced wood products and processing illegally logged timber and establishes due diligence requirements for those engaged in importing wood products or processing raw logs. Compliance with the due diligence requirements prescribed in the *Illegal Logging Prohibition Regulation 2012* (Cth)[3] provides importers and wood processors with a degree of protection against prosecution for the importation of illegal logs and products.

The EU has a more comprehensive approach to curtailing the importation of illegal wood products, known as the Forest Law Enforcement Governance and Trade (FLEGT) Action Plan, which was commenced in 2003. The main objectives of the FLEGT Action

---

[1] *Lacey Act of 1900*, 16 USC §§ 3371–3378 (2008). Available at: www.law.cornell.edu/uscode/text/16/3371.

[2] *Illegal Logging Prohibition Act 2012* (Cth). Available at: www.legislation.gov.au/Details/C2012A00166.

[3] *Illegal Logging Prohibition Regulation 2012* (Cth). Available at: www.legislation.gov.au/Details/F2018C00885.

Plan are to tackle illegal logging and improve forest governance in timber-exporting countries. Like the plant provisions of the Lacey Act, the Action Plan includes regulations (introduced in 2010) that prohibit the placing on the EU market for the first time of illegally harvested timber and products derived from such timber, and a requirement for EU traders to exercise due diligence when placing wood and wood products on the market. There are also record-keeping requirements to facilitate traceability and programmes to change government and corporate purchasing practices. However, the centrepiece of the plan is a licensing system based on voluntary partnership agreements (VPAs) between exporting countries and the EU. Under the VPAs, the exporting countries agree, with EU assistance, to implement reforms to their forest governance systems, including by establishing product stewardship and export licensing processes. The exporting countries are given a period to implement the reforms, after which the EU only accepts licenced wood products from the country. Six countries have concluded VPAs: Ghana, Republic of Congo, Cameroon, Indonesia, Central African Republic and Liberia. Voluntary partnership agreement negotiations are ongoing with several other countries, including Côte d'Ivoire, Democratic Republic of the Congo, Gabon, Guyana, Honduras, Laos, Malaysia, Thailand and Vietnam.

These types of consumer-country-led efforts have been complemented by non-government accreditation schemes, the most well known of which are the Programme for the Endorsement of Forest Certification (PEFC) and Forest Stewardship Council (FSC). The total area of certified forest worldwide reported in 2021 is 224 million ha under FSC (FSC n.d.) and 320 million ha under PEFC (PEFC n.d.), which is about 2.4% of total forest area managed for industrial logging. By comparison, in 1991 there were virtually no forests certified. So, in combination, these programmes appear to have made inroads into improving practices in developed countries, especially Canada, and reducing the prevalence of illegal logging and trade in illegal wood products in particular areas. However, their global impact seems to have been limited (Hoare 2015). The reasons for this relate to changes in global wood product markets. Growth in demand from developing countries like China has diluted the influence of the developed country bans and certification schemes. There has also been a marked increase in the proportion of illegal timber being derived from land-use change (conversion of forest for agriculture) and an increase in the proportion of timber being supplied by small-scale producers who lie beyond the reach of developing country regulators. A persistent constraint on progress has been the lack of institutional capacity and corruption in developing countries, and the resistance of their governance systems to change.

18.5.2 Domestic Policy Measures for Sustainable Forest Management

Discussion of policy solutions for forest-related mitigation and adaptation issues often focuses on the design and implementation of policy instruments: for example, which types of policy instruments should be used, how they should be calibrated and what the implementation challenges are that need to be managed? These issues are of considerable importance. However, in order for these issues to be of relevance, the countries or regions in question must have the institutional maturity and stability that is necessary to support the creation and implementation of effective public policy instruments.

Legal institutions perform three basic functions in policy processes: they set the boundaries on government policy-making (which policies can be made and how they can be made); they determine the initial allocation of property rights between policy actors (who owns or controls the relevant resources); and they provide the tools by which policy-makers can modify the behaviour of relevant actors, including by rearranging property rights. In countries with immature institutions, all three of these functions can be inhibited. In forest management, one of the most prevalent problems concerns the absence of clearly defined property rights over forest resources and the means of enforcing them. Owing to these issues, many forests in developing countries are a de facto open-access resource that illegal operators are able to exploit almost without restraint. The absence of formal property rights in the land or forests not only limits the capacity to control access to forest resources but also strips away the incentive to invest in post-harvest regeneration and management.

The ability to resolve these issues is impeded by a collection of interrelated constraints (Macintosh and Wilkinson 2015). The creation of stable legal institutions requires consensus among political and economic elites. Yet any attempt to establish them will result in a reallocation of resources that benefits some and disadvantages others. Those threatened by change typically use their resources to obstruct progress. Further, any attempt to rearrange institutional and property right arrangements requires resources, something that developing countries lack. This situation is exacerbated by the fact that the state typically requires stable institutions in order to raise revenue. The nature of these constraints explains the difficulties that have impeded international efforts to address forest management issues. Yet it also points to the necessity for continued international engagement and assistance to help developing countries overcome these obstacles to progress.

Where there is sufficient institutional stability to support the creation of functional policies, the critical issues for policy-makers concern the choice of instrument types and mixes, their calibration and implementation. Policy instruments may be substantive (altering the allocation of economic resources or income distribution so as to directly change well-being), or procedural (informing policy processes; managing intra- and inter-government and society interactions; and evaluating implementation) (Howlett 2004).

Policy instruments are generally categorised as the choice approach (where instrument categories reflect the choices available to decision makers), and the resource approach (where instrument categories are based on the government resources they employ) (Howlett 1991).

A four-part typology for a choice approach is described by Hamilton and Macintosh (2008) that may be used to address forest management and climate change issues: information instruments, voluntary instruments, market-based instruments and regulation (Table 18.2).

The relative merits of individual policy instruments are debated, typified by the heated dispute between advocates of market-based instruments and regulatory approaches, and the seemingly eternal emissions trading scheme versus carbon tax debate. The focus of research has moved beyond the study of individual instruments and is now centred more

Table 18.2. *Four-part choice-based typology of forest-relevant policy instruments*

Instrument type	Description	Examples
Information	Mechanisms that aim to improve outcomes by improving people's awareness and understanding of relevant environment issues and building their capacity to respond to environmental threats.	Government dissemination of information on the carbon intensity of wood and non-wood projects Provision of advice on management and wood processing practices
Voluntary	Mechanisms that aim to protect the environment where relevant economic agents are able to decide whether or not to participate; that is, involvement in the programme is voluntary and no direct penalties are imposed on non-participants, although incentives may be used to encourage participation.	Unilateral initiatives by forest owners to improve practices Provision of subsidies or other incentives to protect forests or improve management practices
Market-based	Mechanisms that force economic agents to internalise all or part of the social costs associated with environmentally harmful activities and that rely on market forces to promote efficiency. In doing so, they seek to impose additional costs on producers that harm the environment and reward those that improve environment outcomes, while utilising market forces to improve the allocation of resources.	Carbon taxes Emissions trading schemes Improvement of property rights over forest lands and resources
Regulation	Mechanisms that impose legally enforceable restrictions on economic agents to realise environmental protection objectives. They are sometimes referred to as 'command-and-control' mechanisms because they prohibit or mandate certain actions (the command) and use the threat of punishment to motivate compliance (the control).	Prohibitions on deforestation Requirements to obtain approval to engage activities that result in deforestation or forest degradation Requirements to carry out particular management activities (weeding and pest control) or use specified techniques/practices

Source: Hamilton and Macintosh (2008).

on how and why instruments are combined into instrument mixes, particularly in the context of networked policy settings and multilevel governance structures (Howlett 2004; Howlett and Rayner 2007). One of the chief contentions from this literature is that debates about the technical merits of different instruments, made without reference to the context in which policy is being made, are fruitless because they ignore the variables that influence

Table 18.3. *The LEAF–BEDS model of policy-making*

Variable	Explanation
Variables that influence policy-making	
Legality	The formal legal institutions that determine what is legally permissible
Effectiveness	The specific factual, institutional and network context that determines the nature of the problem and how it can be resolved
Acceptability	The institutional, ideological and network factors that determine what is acceptable to relevant policy actors
Feasibility	Governmental capacity, or the resource factors that determine what is known and feasible
Preferences of groups that determine acceptability	
Beneficiaries	Those who obtain special benefits above those enjoyed by the general public
Elites	Privileged policy actors inside and outside government that exert disproportionate influence on policy-making
Detractors	Those who incur special costs above those borne by the general public
Society	The general public, or those that neither gain any special benefits nor incur any special costs

Source: Macintosh and Wilkinson (2015).

and constrain the choices of policy-makers. It is not the theoretical merits of policy instruments that matter, but whether they can be created and implemented, and how they are likely to perform in the practical context in question.

The importance of context can be conceptualised using the LEAF–BEDS model of policy-making (Macintosh and Wilkinson 2015). This complexity-theory-based model divides the variables that influence policy-making into: legality (L), effectiveness (E), acceptability (A) and feasibility (F). Acceptability is presented as being a function of the preferences of four groups: beneficiaries (B), elites (E), detractors (D) and society (S) (Table 18.3). Policy outputs and outcomes are shaped and constrained by the LEAF–BEDS variables. In this context, constraints are defined as any variable that materially restricts the options available to policy-makers or their ability to pursue their personal policy preferences. To constitute a constraint, the variable must be able to prevent, or provide a substantial disincentive to, the pursuit of an option or desired course of action. The model's constraints stem from factors related to legality, acceptability and feasibility.

As the LEAF–BEDS model suggests, policy-making is a bounded game in which the merits of the instruments being considered are only one of the factors that influences choices. The choices of policy design or implementation are bounded by technical attributes, legal institutions, acceptability and feasibility constraints, and shaped by all four of the variables: legality, effectiveness, acceptability and feasibility. Given this, the 'optimal' policy instruments and instrument mixes are those that are able to balance the competing interests, and work within the constraints, to achieve the desired procedural and substantive objectives.

The experiences with deforestation in the Australian context highlight the extent to which instrument choice, calibration and implementation is shaped by these variables. Traditionally, environmental and land-use matters have been the domain of the states rather than the federal government and, for much of the twentieth century, they had contradictory policies on deforestation, with measures to conserve native vegetation existing alongside regulatory and policy programmes that promoted deforestation. Reform of these structures began in the mid-1980s, when significant changes were made to the land clearing regulations in South Australia and Western Australia. The Australian Government also launched a number of information and voluntary 'beneficiary pays' measures – where the government pays landholders to alter their practices – including the National Tree Program in 1982, the One Billion Trees and Save the Bush programmes in 1989 (Macintosh 2011).

While an important first step, the reforms of the 1980s did little to curb the rates of deforestation. Land clearing laws remained lax in most jurisdictions, particularly in the states facing the greatest deforestation pressures: Queensland and New South Wales. Similarly, the non-regulatory programmes may have generated localised benefits but they were not on a scale that was capable of driving significant changes in nationwide land-use practices (Macintosh 2011).

The turning point for deforestation came in 1995, where regulatory reform began in Queensland and New South Wales. In Queensland, the state government started by issuing guidelines to control broad-scale tree clearing on leasehold land, the dominant form of land tenure in Queensland's rangelands. After that, it went through three further rounds of regulatory tightening, followed by a marked loosening. The tightening started with the passage of the *Vegetation Management Act 1999* (Qld[4]), which extended the vegetation clearing restrictions to freehold land. Flaws in the regime ensured that high rates of clearing continued after the *Vegetation Management Act* came into operation. Because of this, the Queensland Government placed a moratorium on clearing applications in May 2003 and, a year later, a new regime commenced that aimed to end broad-scale clearing of remnant vegetation by 31 December 2006.

While the changes of 2003–2004 were significant, they left room for continued vegetation clearing. Of particular concern was the fact that regrowth vegetation remained vulnerable to widespread removal. This was partially addressed in April 2009, when a moratorium was placed on the clearing of high-value regrowth vegetation and native vegetation adjacent to watercourses in the 'priority' Great Barrier Reef catchments, with a permanent change in October 2009.

The impact of these regulatory reforms varied. The 1995 changes failed. Queensland's rate of woody vegetation clearing did not decline in response to the reforms; the changes failed to reduce even the rate of clearing on leasehold land, which the legislation was specifically designed to address. This was a product of a combination of factors, including a lack of enforcement, the inclusion of various exemptions for deforestation on leasehold land and the fact that clearing on freehold agricultural land was largely unregulated.

---

[4] *Vegetation Management Act 1999*. Available at: www.legislation.qld.gov.au/view/html/inforce/current/act-1999-090.

The regulatory regime that was introduced in 1999 encountered similar problems. If anything, it probably resulted in an increase in emissions in the late 1990s and 2000, which was offset by falls in subsequent years because it brought forward planned clearing. This was a product of gaps in the regime and the fact that there was a sizeable delay between the time the relevant legislation was passed (8 December 1999) and when it commenced (15 September 2000). The delay provided a window in which there was panic pre-emptive clearing by landholders.

The reforms of 2003–2004 and 2009 were far more successful, a fact reflected in a sharp decline in both remnant clearing and regrowth clearing that occurred after 2006. However, a counter-reaction emerged. The four phases of deforestation reform in Queensland were all introduced by centre-left Labor governments. The success of the 2003–2004 and 2009 reforms pushed landholders to pressure the conservative opposition to pledge to loosen the laws if they came to office. They did so in 2012 and honoured their pre-election pledge, introducing legislative changes in 2013 that sought to reduce regulatory obstacles to deforestation and enable broad-scale deforestation approvals to be provided for 'high-value agricultural clearing' and 'irrigated high-value agriculture clearing'. The reforms also allowed vegetation clearing to occur, in particular circumstances, without approval, under self-assessable codes and area management plans. There is evidence to suggest that, as a result of the regulatory loosening, Queensland's rates of deforestation have increased since 2013.

Over the periods of regulatory tightening and the post-2012 loosening, the federal government has continued to offer financial inducements to promote the conservation and management of native vegetation, including in Queensland. It is purported to regulate deforestation if it poses a threat to various matters of national environmental significance. In 2015, the federal government introduced a new scheme that offers landholders carbon credits for stopping the re-clearing of forest. The scheme was not extended to remnant forest because of the technical difficulty of setting baselines; that is, determining whether, in the counterfactual, the forest would be cleared.

This history of deforestation policy-making in Queensland illustrates how the LEAF–BEDS variables and constraints both drive and hinder reform. The periods of regulatory tightening reflected the ideological preferences of the Labor governments and the pressure applied by environment groups (benefactors) and the broader community to curb deforestation. It was also a product of the declining economic power of the landholder groups (detractors) engaged in deforestation. For much of the past 40 years, farmers' terms of trade have been declining and deforestation has progressively moved into more and more marginal areas. As the economic returns from deforestation waned, so did the political influence of those wanting to do it. Why then did the conservatives decide to loosen the regulations? The answer lies in the influence of the landholders (detractors) over conservative political elites and the ideological preferences of these elites; they placed agricultural expansion and a belief in its virtues above conservation.

The choice of regulation as the primary instrument of Labor state governments can also be explained by the LEAF–BEDS model. Policy-makers opted for regulation over the alternatives because of feasibility constraints (the state government did not have the financial means to buy out landholders) and the knowledge it was likely to be effective (it would achieve the desired environmental objective). The use of 'beneficiary pays' measures throughout this period is explained by the low political costs associated with their use (detractors tend not to oppose voluntary subsidies) and the desire to be seen to be reacting to a perceived problem.

18.6 Conclusion

Action on forest management to protect carbon stocks, reduce emissions and increase sequestration is a critical component of climate change mitigation. Management activities for both mitigation and adaptation are synergistic and so have multiple benefits. The key issue for policy and implementation is prioritisation of strategies based on scientifically robust methods for carbon accounting and evaluation. The first priority is protection of carbon stocks in existing native forests by reducing rates of deforestation and degradation. The second priority is maximising carbon storage in native forests through conservation and restoration. The third priority is maximising carbon storage in harvested forest systems, and the fourth is reducing emissions from waste. The nexus of protecting carbon stocks in forests for climate mitigation and protecting biodiversity and ecosystem integrity with the goals of sustainable development demonstrate the importance of integrated policies.

Effective governance and policy to support forest management must take account of the context. There are no universally optimal policy solutions for dealing with deforestation and forest degradation. Policy approaches that are worthwhile in one context might be inappropriate in another. What is likely to matter most is the extent to which the constraints on policy-making limit choices and whether and how different instruments and approaches can be used to manipulate the constraints and take advantage of opportunities.

References

Acuff, K. and Kaffine, D. T. (2013). Greenhouse gas emissions, waste and recycling policy. *Journal of Environmental Economics and Management*, 65, 74–86.

Allen, C. D., Macalady, A. K., Chenchouni, H. et al. (2010). A global overview of drought and heat-induced tree mortality reveals emerging climate change risks for forests. *Forest Ecology and Management*, 259, 660–684.

Anderegg, W. R. L., Kane, J. M. and Anderegg, L. D. L. (2013). Consequences of widespread tree mortality triggered by drought and temperature stress. *Nature Climate Change*, 3, 30–36.

Archer, D. (2005). Fate of fossil fuel CO_2 in geologic time. *Journal of Geophysical Research*, 110, 1–6.

Asner, G. P., Powell, G. V. N., Mascaro, J. et al. (2010). High resolution forest carbon stocks and emissions in the Amazon. *Proceedings of the National Academy of Sciences*, 107, 16739–16742.

Baccini, A., Goetz, S. J., Walker, W. S. et al. (2012). Estimated carbon dioxide emissions from tropical deforestation improved by carbon-density maps. *Nature Climate Change*, 2, 182–185.

Barnard, S. and Nakhooda, S. (2014). *The Effectiveness of Climate Finance: A Review of the Scaling-up Renewable Energy Program.* Working Paper 396. Overseas Development Institute (ODI). Available at: www.odi.org/sites/odi.org.uk/files/odi-assets/publications-opinion-files/9075.pdf.

Bellassen, V. and Luyssaert, S. (2014). Managing forests in uncertain times. *Nature*, 506, 153–155.

Berenguer, E., Ferreira, J., Gardner, T. A. et al. (2014). A large-scale field assessment of carbon stocks in human-modified tropical forests. *Global Change Biology*, 20, 3713–3726.

Betts, R., Cox, P., Collins, M., Harris, P. and Huntingford, C. (2004). The role of ecosystem–atmosphere interactions in simulated Amazonian precipitation decrease and forest dieback under global climate warming. *Theoretical and Applied Climatology*, 78, 157–175.

Boucher, D., Elias, P., Faires, J. and Smith, S. (2014). *Deforestation Success Stories: Tropical Nations Where Forest Protection and Reforestation Policies Have Worked.* Tropical Forest and Climate Initiative of the Union of Concerned Scientists. Available at: www.ucsusa.org/global_warming/solutions/stop-deforestation/deforestation-success-stories.html#.V20e8vl96M8.

Bowman, D., Balch, J. K., Artaxo, P. et al. (2009). Fire in the Earth system. *Science*, 324, 481–484.

Brasier, C. and Webber, J. (2010). Plant pathology: Sudden larch death. *Nature*, 466, 824–825.

Breshears, D. D., Cobb, N. S., Rich, P. M. et al. (2005). Regional vegetation die-off in response to global-change-type drought. *Proceedings of the National Academy of Sciences*, 102, 15144–15148.

Bryan, J., Shearman, P., Ash, J. and Kirkpatrick, J. B. (2010). Estimating rainforest biomass stocks and carbon loss from deforestation and degradation in Papua New Guinea 1972–2002: Best estimates, uncertainties and research needs. *Journal of Environmental Management*, 91, 995–1001.

Carlson, M., Chen, J., Elgie, S. et al. (2010). Maintaining the roles of Canada's forests and peatlands in climate regulation. *The Forestry Chronicle*, 86, 1–10.

Carlson, K. M., Curran, L. M., Ratnasari, D. et al. (2012). Committed carbon emissions, deforestation, and community land conversion from oil palm plantation expansion in West Kalimantan, Indonesia. *Proceedings of the National Academy of Sciences*, 109, 7559–7564.

Carnahan, J. A. (1976). Natural vegetation: Map with accompanying booklet commentary. In *Atlas of Australian Resources*, Second series. Canberra: Geographic Section, Department of National Development.

Carnicer, J., Coll, M., Ninyerola, M., Pons, X., Sanchez, G. and Penuelas, J. (2011). Widespread crown condition decline, food web disruption, and amplified tree mortality with increased climate change-type drought. *Proceedings of the National Academy of Sciences*, 108, 1474–1478.

Cary, G. J., Bradstock, R. A., Gill and A. M. and Williams, R. J. (2012). Global change and fire regimes in Australia. In R. A. Bradstock, G. A. Malcolm and R. J, Williams, eds., *Flammable Australia: Fire Regimes, Biodiversity and Ecosystems in a Changing World*. Melbourne: Commonwealth Scientific and Industrial Research Organisation (CSIRO) Publishing, pp. 149–170.

Chia, E. L., Somorin, O. A., Sonwa, D. J., Bele, Y. M. and Tiani, M. A. (2015). Forest–climate nexus: Linking adaptation and mitigation in Cameroon's climate policy process. *Climate and Development*, 7, 85–96.

Chong, S. K. (2005). Anmyeon-do Recreation Forest: A millennium of management. In P. Durst, C. Brown, H. D. Tacio and M. Ishikawa, eds., *In Search of Excellence: Exemplary Forest Management in Asia and the Pacific*. Bangkok: Asia-Pacific Forestry Commission, Food and Agriculture Organization (FAO) Regional Office for Asia and the Pacific, pp. 251–259.

Cias, P., Reichstein, M., Viovy, N. et al. (2005). Europe-wide reduction in primary productivity caused by the heat and drought in 2003. *Nature*, 437, 529–533.

Cyr, D., Gauthier, S., Bergeron, Y. and Carcaillet, C. (2009). Forest management is driving the eastern North American boreal forest outside its natural range of variability. *Frontiers in Ecology and the Environment*, 7, 519–524.

Dave, R., Saint-Laurent, C., Murray, L. et al. (2019). *Second Bonn Challenge Progress Report: Application of the Barometer in 2018*. Gland: International Union for Conservation of Nature (IUCN). Available at: https://portals.iucn.org/library/sites/library/files/documents/2019-018-En.pdf.

Daviet, F., Goers, L. and Austin, K. (2009). *Forests in the Balance Sheet: Lessons from Developed Country Land Use Change and Forestry Greenhouse Gas Accounting and Reporting Practices*. Working Paper. Washington, DC: World Resources Institute. Available at: www.wri.org/publication/forests-balance-sheet.

Dean, C., Wardell-Johnson, G. W. and Kirkpatrick, J. B. (2012). Are there circumstances in which logging primary wet-eucalypt forest will not add to the global carbon burden? *Agricultural and Forest Meteorology*, 161, 156–169.

Dooley, K. (2014). *Misleading Numbers: The Case for Separating Land and Fossil Fuel Based Carbon Emissions*. Report. Brussels: Fern. Available at: www.fern.org/misleadingnumbers.

Environmental Paper Network (2018). *The State of the Global Paper Industry*. Available at: https://environmentalpaper.org/wp-content/uploads/2018/04/StateOfTheGlobalPaperIndustry2018_FullReport-Final-1.pdf.

FAO (Food and Agriculture Organization) (n.d.). FAOSTAT [data resource]. *FAO.org*. Available at: www.fao.org/faostat/en/#data/FO.

FAO (2004). *The State of Food and Agriculture 2003–04. Agricultural Biotechnology: Meeting the Needs of the Poor?* Rome: Food and Agriculture Organization of the UN. Available at: www.fao.org/3/a-y5160e.pdf.

FAO (2010). *Global Forest Resources Assessment 2010*. FAO Forestry Paper 163. Rome: Food and Agriculture Organization of the UN. Available at: www.fao.org/3/a-i1757e.pdf.

FAO (2011). *Transitioning Towards Climate-Smart Agriculture in Kenya*. Rome: Food and Agriculture Organization of the UN. Available at: www.fao.org/3/a-i4259e.pdf.

FAO (2013a). *Forestry*. Fact sheet. Rome: Food and Agriculture Organization of the UN. Available at: www.fao.org/docrep/014/am859e/am859e08.pdf.

FAO (2013b). *Forest Product Statistics: 2013 Global Forest Products Facts and Figures*. Rome: Food and Agriculture Organization of the UN. Available at: www.ipcinfo.org/fileadmin/user_upload/newsroom/docs/FactsFigures2013_En.pdf.

FAO (2014). *Forest Product Statistics: 2014 Global Forest Products Facts and Figures*. Rome: Food and Agriculture Organization of the UN. Available at: www.fao.org/forestry/44134-01f63334f207ac6e086bfe48fe7c7e986.pdf.

FAO (2015). *Global Forest Resources Assessment 2015*. Rome: Food and Agriculture Organization of the UN. Available at: www.fao.org/forest-resources-assessment/past-assessments/fra-2015/en/.

FDRE (Federal Democratic Republic of Ethiopia) (2011). *Ethiopia's Climate-Resilient Green Economy – Green Economy Strategy*. Federal Democratic Republic of Ethiopia. Available at: www.undp.org/content/dam/ethiopia/docs/Ethiopia%20CRGE.pdf.

Fearnside, P. (2015). What lies behind the recent surge of Amazon deforestation. *Yale Environment 360*. 9 March. Available at: http://e360.yale.edu/feature/what_lies_behind_the_recent_surge_of_amazon_deforestation/2854/.

Fensham, R. J., Fairfax, R. J. and Ward, D. P. (2009). Drought induced tree death in savanna. *Global Change Biology*, 15, 380–387.

FSC (Forest Stewardship Council) (n.d). Facts & figures. *FSC*. Available at: https://fsc.org/en/facts-figures.

Gerasimchuk, I, Bridle, R., Beaton, C. and Charles, C. (2012). *State of Play on Biofuel Subsidies: Are Policies Ready to Shift?* Report. Canada: International Institute for Sustainable Development (IISD). Available at: www.iisd.org/gsi/sites/default/files/bf_stateplay_2012.pdf.

Grieg-Gran, M. (2006). *The Cost of Avoiding Deforestation: Report Prepared for the Stern Review of the Economics of Climate Change*. London: International Institute for Environment and Development (IIED).

Guariguata, M., Cornelius, J., Locatelli, B., Forner, C. and Sanchez-Azofeifa, G. (2008). Mitigation needs adaptation: Tropical forestry and climate change. *Mitigation and Adaptation Strategies for Global Change*, 13, 793–808.

Gustavsson, L., Pingoud, K. and Sathre, R. (2006). Carbon dioxide balance of wood substitution: Comparing concrete- and wood-framed buildings. *Mitigation and Adaptation Strategies for Global Change*, 11, 667–691.

Hamilton, C. and Macintosh, A. (2008). Human ecology: Environmental protection and ecology. In S. Jorgensen, ed., *Encyclopedia of Ecology*. Elsevier, pp. 1342–1350.

Hanewinkel, M., Cullmann, D. A., Schelhaas, M. J., Nabuurs, G. J. and Zimmermann, N. E. (2012). Climate change may cause severe loss in economic value of European forest land. *Nature Climate Change*, 3, 203–207.

Harmon, M. E., Ferrell, W. K. and Franklin, J. F. (1990). Effects on carbon storage of conversion of old-growth forests to young forests. *Science*, 247, 699–703.

Harris, N., Brown, S., Hagen, S. et al. (2012). Baseline map of carbon emissions from deforestation in tropical regions. *Science*, 336, 1573–1576.

Hellman, J. J., Byers, J. E., Bierwagen, B. G. and Dukes, J. S. (2008). Five potential consequences of climate change for invasive species. *Conservation Biology*, 22, 534–543.

Hoare, A. (2015). *Tackling Illegal Logging and the Related Trade: What Progress and Where Next?* Chatham House. Available at: www.chathamhouse.org/publication/tackling-illegal-logging-and-related-trade-what-progress-and-where-next.

Houghton, R. A. (2012). Carbon emissions and the drivers of deforestation and forest degradation in the tropics. *Current Opinion in Environmental Sustainability*, 4, 597–603.

Houghton, R. A. (2013). The emissions of carbon from deforestation and degradation in the tropics: Past trends and future potential. *Carbon Management*, 4, 539–546.

Houghton, R. A., Byers, B. and Nassikas, A. A. (2015). A role for tropical forests in stabilizing atmospheric CO_2. *Nature Climate Change*, 5, 1022–1023.

Howlett, M. (1991). Policy instruments, policy styles, and policy implementation: National approaches to theories of instrument choice. *Policy Studies Journal*, 19, 1–21.

Howlett, M. (2004). Beyond good and evil in policy implementation: Instrument mixes, implementation styles, and second generation theories of policy instrument choice. *Policy and Society*, 23, 1–17.

Howlett, M. and Rayner, J. (2007). Design principles for policy mixes: Cohesion and coherence in 'new governance arrangements'. *Policy and Society*, 26, 1–18.

Hudiburg, T., Law, B. E., Wirth, C. and Luyssaert, S. (2011). Regional carbon dioxide implications of forest bioenergy production. *Nature Climate Change*, 1, 419–423.

Hughes, L. (2000). Biological consequences of global warming: Is the signal already apparent? *Tree*, 15, 56–61.

Hughes, L. (2012). Can Australian biodiversity adapt to climate change? In D. Lunney and P. Hutchings, eds., *Wildlife and Climate Change: Towards Robust Strategies for Australian Fauna*. Sydney: Royal Zoological Society of Australia, pp. 8–10.

Humphreys, D. (2008). The politics of 'avoided deforestation': Historical context and contemporary issues. *International Forestry Review*, 10, 433–442.

Ingerson, A. (2011). Carbon storage potential of harvested wood: Summary and policy implications. *Mitigation and Adaptation Strategies for Global Change*, 16, 307–323.

IPCC (2014). *Climate Change 2014: Synthesis Report. Contribution of Working Groups I, II and III to the Fifth Assessment Report of the Intergovernmental Panel on Climate Change*. Edited by R. K. Pachauri and L. A. Meyer. Geneva: IPCC. Available at: www.ipcc.ch/report/ar5/syr/.

ISU (International Sustainability Unit) (2015). *Tropical Forests: A Review*. London: The Prince's Charities International Sustainability Unit. Available at: www.pcfisu.org/resources/.

Keith, H., van Gorsel, E., Jacobsen, K. L. and Cleugh, H. A. (2012). Dynamics of carbon exchange in a *Eucalyptus* forest in response to interacting disturbance factors. *Agricultural and Forest Meteorology*, 153, 67–81.

Keith, H., Lindenmayer, D. B., Mackey, B. G. et al. (2014a). Accounting for biomass carbon stock change due to wildfire in temperate forest landscapes in Australia. *PLoS One*, e107126.

Keith, H., Lindenmayer, D. B., Mackey, B. G. et al. (2014b). Managing temperate forests for carbon storage: Impacts of logging versus forest protection on carbon stocks. *Ecosphere*, 5, 75.

Keith, H., Lindenmayer, D. B., Macintosh, A. and Mackey, B. G. (2015). Under what circumstances do wood products from native forests benefit climate change mitigation? *PLoS One*, 10, e0139640.

Kraft, N. J. B., Metz, M. R., Condit, R. S. and Chave, J. (2010). The relationship between wood density and mortality in a global tropical forest data set. *New Phytologist*, 188, 1124–1136.

Krankina, O. N. and Harmon, M. E. (2006). Forest management strategies for carbon storage. In H. Salwasser and M. Cloughsey, eds., *Forests, Carbon and Climate Change: A Synthesis of Science Findings*. Portland, OR: Oregon Forest Research Institute, pp. 79–92.

Kurz, W. A., Dymond, C. C., Stinson, G. et al. (2008). Mountain pine beetle and forest carbon feedback to climate change. *Nature*, 452, 987–990.

Law, B. E., Falge, E., Gu, L., Baldocchi, D. D. and Bakwin, P. (2002). Environmental controls over carbon dioxide and water vapour exchange of terrestrial vegetation. *Agricultural and Forest Meteorology*, 113, 97–120.

Le Quéré, C., Peters, G. P., Andres, R. J. et al. (2013). Global carbon budget 2013. *Earth System Science Data*, 6, 689–760.

Lee, S. G. (2004). A simplified life cycle assessment of re-usable and single-use bulk transit packaging. *Packaging Technology and Science*, 17, 67–83.

Lenton, T. M., Held, H., Kriegler, E. et al. (2008). Tipping elements in the Earth's climate system. *Proceedings of the National Academy of Sciences*, 105, 1786–1793.

Locatelli, B., Evans, V., Wardell, A., Andrade, A. and Vignola, R. (2011). Forests and climate change in Latin America: Linking adaptation and mitigation. *Forests*, 2, 431–450.

Macintosh, A. (2011). *Potential Carbon Credits from Reducing Native Forest Harvesting in Australia*. CLP Working Paper 2011/1. Canberra: Australian National University (ANU) Centre for Climate Law and Policy.

Macintosh, A. (2012a). *Tasmanian Forests Intergovernmental Agreement: An Assessment of Its Carbon Value*. Canberra: Australian National University (ANU) Centre for Climate Law and Policy.

Macintosh, A. (2012b). The Australia clause and REDD: A cautionary tale. *Climatic Change*, 112, 169–188.

Macintosh, A. (2013). *The Australian Native Forest Sector: The Causes of the Decline and Prospects for the Future*. Technical brief No. 21. Canberra: The Australia Institute. Available at: www.tai.org.au/sites/default/files/TB%2021%20State%20of%20the%20native%20forest%20industry_1_3.pdf.

Macintosh, A. and Wilkinson, D. (2015). Complexity theory and the constraints on environmental policy-making. *Journal of Environmental Law*, 28, 65–93.

Macintosh, A., Keith, H. and Lindenmayer, D. (2015). Rethinking forest carbon assessments to account for policy institutions. *Nature Climate Change*, 5, 946–952.

Mackey, B. G., DellaSala, D. A., Kormos, C. et al. (2015). Policy options for the world's primary forests in multilateral environmental agreements. *Conservation Letters*, 8, 139–147.

Mackey, B., Kormos, C., Keith, H. et al. (2020). Understanding the importance of primary tropical forest protection as a mitigation strategy. *Mitigation and Adaptation Strategies for Global Change*, 25, 763–787. Available at:)https://doi.org/10.1007/s11027-019-09891-4.

Malhi, Y., Aragao, L. E. O. C., Galbraith, D. et al. (2009). Exploring the likelihood and mechanism of a climate-change-induced dieback of the Amazon rainforest. *Proceedings of the National Academy of Sciences*, 106, 20610–20615.

MEA (Millennium Ecosystem Assessment) (2005). *Ecosystems and Human Well-being: Biodiversity Synthesis*. Washington, DC: World Resources Institute. Available at: www.millenniumassessment.org/documents/document.354.aspx.pdf.

Mery, G., Katila, P., Galloway, G. et al., eds. (2010). *Forests and Society: Responding to Global Drivers of Change*. International Union of Forest Research Organizations. Available at: www.iufro.org/science/special/wfse/forests-society-global-drivers/.

Michaelian, M., Hogg, E. H., Hall, R. J. and Arsenault, E. (2011). Massive mortality of aspen following severe drought along the southern edge of the Canadian boreal forest. *Global Change Biology*, 17, 2084–2094.

Murdiyarso, D., Purbopuspito J., Kauffman J. B. et al. (2015). The potential of Indonesian mangrove forests for global climate change mitigation. *Nature Climate Change*, 5, 1089–1092.

Nobre, C. A. and Borma, L. D. S. (2009). 'Tipping points' for the Amazon forest. *Current Opinion in Environmental Sustainability*, 1, 28–36.

OECD (2014). *Agricultural Policy Monitoring and Evaluation 2014*. Report. Paris: OECD. Available at: www.oecd-ilibrary.org/agriculture-and-food/agricultural-policy-monitoring-and-evaluation-2014_agr_pol-2014-en.

Pan, Y, Birdsey, R. A., Fang, J. et al. (2011). A large and persistent carbon sink in the world's forests. *Science*, 333, 988–993.

Paquette, A. and Messier, C. (2010). The role of plantations in managing the world's forests in the Anthropocene. *Frontiers in Ecology and the Environment*, 8, 27–34.

PEFC (Programme for the Endorsement of Forest Certification) (n.d.). Programme for the Endorsement of Forest Certification. Available at: www.pefc.org/.

Phillips, O. L., Aragão, L. E., Lewis, S. L. et al. (2009). Drought sensitivity of the Amazon rainforest. *Science*, 323, 1344–1347.

Phillips, O. L., van der Heijden, G., Lewis, S. L. et al. (2010). Drought–mortality relationships for tropical forests. *New Phytologist*, 187, 631–646.

Ravindranath, N. H. (2007). Adaptation and mitigation synergy in the forest sector. *Mitigation and Adaptation Strategies for Global Change*, 12, 843–853.

Republic of Rwanda (2011). *Green Growth and Climate Resilience – National Strategy for Climate Change and Low Carbon Development*. Government of Rwanda. Available at: https://greengrowthknowledge.org/national-documents/rwanda-green-growth-and-climate-resilience-national-strategy-climate-change-and.

Rizvi, A. R., Baig, S., Barrow, E. and Kumar, C. (2015). *Synergies between Climate Mitigation and Adaptation in Forest Landscape Restoration*. Gland: International Union for Conservation of Nature (IUCN). Available at: https://portals.iucn.org/library/sites/library/files/documents/2015-013.pdf.

Scheffer, M., Hirota, M., Holmgren, M., van Nes, E. H. and Chapin, F. S. (2012). Thresholds for boreal biome transitions. *Proceedings of the National Academy of Sciences*, 109, 21384–21389.

Schlamadinger, B. and Marland, G. (1999). Net effect of forest harvest on CO_2 emissions to the atmosphere: A sensitivity analysis on the influence of time. *Tellus B: Chemical and Physical Meteorology*, 51, 314–325.

Schulze, E.-D., Körner, C., Law, B. E., Haberl, H. and Luyssaert, S. (2012). Large-scale bioenergy from additional harvest of forest biomass is neither sustainable nor greenhouse gas neutral. *Global Change Biology Bioenergy*, 4, 611–616.

Searchinger, T. (2010). Biofuels and the need for additional carbon. *Environmental Research Letters*, 5, 1–10.

Seifert, T. (2013). *Bioenergy from Wood: Sustainable Production in the Tropics*. Dordrecht: Springer Netherlands.

Settele, J., Scholes, R., Betts, R. et al. (2014). Terrestrial and inland water systems. In C. B. Field, V. R. Barros, D. J. Dokken et al., eds., *Climate Change 2014: Impacts, Adaptation, and Vulnerability. Part A: Global and Sectoral Aspects. Contribution of Working Group II to the Fifth Assessment Report of the Intergovernmental Panel on Climate Change*. Cambridge: Cambridge University Press, pp. 271–359. Available at: www.ipcc.ch/site/assets/uploads/2018/02/WGIIAR5-Chap4_FINAL.pdf.

Sitch, S., Huntingford, C., Gedney, N. et al. (2008). Evaluation of the terrestrial carbon cycle, future plant geography and climate carbon cycle feedbacks using five Dynamic Global Vegetation Models (DGVMs). *Global Change Biology*, 14, 2015–2039.

Smith, P., Bustamante, M., Ahammad, H. et al. (2014). Agriculture, forestry and other land use (AFOLU). In O. Edenhofer, R. Pichs-Madruga, Y. Sokona et al., eds., *Climate Change 2014: Mitigation of Climate Change. Contribution of Working Group III to the Fifth Assessment Report of the Intergovernmental Panel on Climate Change*. Cambridge: Cambridge University Press, pp. 816–922. Available at: www.ipcc.ch/site/assets/uploads/2018/02/ipcc_wg3_ar5_chapter11.pdf.

Steffen, W., Burbidge, A., Hughes, L. et al. (2009). *Australia's Biodiversity and Climate Change*. Melbourne: Commonwealth Scientific and Industrial Research Organisation (CSIRO) Publishing.

Stephenson, N. L., Das, A. J., Condit, R. et al. (2014). Rate of tree carbon accumulation increases continuously with tree size. *Nature*, 507, 90–93.

Stern, N. (2006). *The Stern Review: The Economics of Climate Change.* Cambridge: Cambridge University Press. Available at: https://webarchive.nationalarchives.gov .uk/20100407172811/http://www.hm-treasury.gov.uk/stern_review_report.htm.

Thompson, I., Mackey, B., McNulty, S. and Mosseler, A. (2009). *Forest Resilience, Biodiversity and Climate Change. A Synthesis of the Biodiversity/Resilience/ Stability Relationship in Forest Ecosystems.* CBD Technical Series No. 43. Montreal: Secretariat of the Convention in Biological Diversity. Available at: www .cbd.int/doc/publications/cbd-ts-43-en.pdf.

TI (Transparency International) (2006). *Corruption and the Environment.* Transparency International. Available at: www.earth.columbia.edu/sitefiles/file/education/docu ments/progs_of_study/TransparencyInternationalfinalreport1may06.doc.

United Nations Statistical Commission (2021). *System of Environmental Economic Accounting: Ecosystem Accounting.* New York: United Nations Statistical Commission. Available at: https://unstats.un.org/unsd/statcom/52nd-session/docu ments/BG-3f-SEEA-EA_Final_draft-E.pdf.

UNDP (UN Development Programme) (2000). *World Energy Assessment: Energy and the Challenge of Sustainability.* New York: UN Development Programme, UN Department of Economics and Social Affairs and World Energy Council. Available at: www.undp.org/content/dam/aplaws/publication/en/publications/environ ment-energy/www-ee-library/sustainable-energy/world-energy-assessment-energy- and-the-challenge-of-sustainability/World%20Energy%20Assessment-2000.pdf.

UNFCCC (UN Framework Convention on Climate Change) (2010). *Report of the Conference of the Parties on its Sixteenth Session, held in Cancun. Decisions Adopted by the Conference of the Parties* 1/CP.16 The Cancun Agreements: Outcome of the work of the Ad Hoc Working Group on Long-term Cooperative Action under the Convention. Available at: https://unfccc.int/sites/default/files/ resource/docs/2010/cop16/eng/07a01.pdf.

van der Werf, G. R., Morton, D. C., DeFries, R. S. et al. (2009). CO_2 emissions from forest loss. *Nature Geoscience*, 2, 737–738.

van Mantgem, P. J., Stephenson, N. L., Byrne, J. C. et al. (2009). Widespread increase of tree mortality rates in the western United States. *Science*, 323, 521–524.

Westerling, A. L., Turner, M. G., Smithwick, E. A. H., Romme, W. H. and Ryan, M. G. (2011). Continued warming could transform Greater Yellowstone fire regimes by mid-21st century. *Proceedings of the National Academy of Sciences*, 108, 13165–13170.

Williams, A. A. J., Karoly, D. J. and Tapper, N. (2001). The sensitivity of Australian fires danger to climate change. *Climatic Change*, 49, 171–191.

Williams, A. P., Allen, C. D., Millar, C. I. et al. (2010). Forest responses to increasing aridity and warmth in the southwestern United States. *Proceedings of the National Academy of Sciences*, 107, 21289–21294.

Williams, A. P., Allen, C. D., Macalady, A. K. et al. (2013). Temperature as a potent driver of regional forest drought stress and tree mortality. *Nature Climate Change*, 3, 292–297.

WRI (World Resources Institute) (2014). *Better Growth, Better Climate. The New Climate Economy Report.* The Global Commission on the Economy and Climate. Washington, DC: The Global Commission on the Economy and Climate, World Resources Institute. Available at: www.newclimateeconomy.report/2014.

19

Agriculture

MARK HOWDEN

Executive Summary

Food and agriculture are critical to sustainable and prosperous societies and both need to be a core part of addressing climate change. This reflects both the large contribution that food and agriculture make to global greenhouse gas (GHG) emissions (approximately 29% of anthropogenic emissions) and the large-scale vulnerability of food and agriculture to ongoing and projected climate changes, especially in developing countries and those with vulnerable populations. Globally, food and agriculture have been particularly exposed to climate variability, and while this exposure has provided some adaptive capacity, other evidence suggests that adaptation to current climate variability and to emerging climate change is far from complete and that there is a growing gap between the current state of adaptation and where it rationally should be. The broadly negative climate changes projected over the next decades are likely to add to the challenges of this sector and will interact strongly with other expectations, such as: doubling of production, provision of emission reduction options and maintenance of biodiversity and ecosystem services, among others. This chapter identifies the key GHG emissions from agriculture and food systems and how they can be reduced in tandem with ongoing adaptation to climate change: the fundamentals of successful low-carbon, resilient and prosperous agriculture systems. A key theme is understanding the interactions between the different potential activities and the opportunities from synergistic changes, including how policy changes could influence the uptake of improved agricultural practices and result in more proactive value chain management. However, to date, food and agriculture has received far less policy support and research and development focus than other, often smaller and less vulnerable, sectors. This chapter paints a clear picture of the opportunities, over the next decades, for effective outcomes in reducing emissions and for meeting growing food security needs, as well as recognising the opportunities that arise through consideration of the interaction between these. Addressing the interactions between climate change and food and agricultural systems is a challenge we can't afford to fail and an opportunity we can't afford not to grasp.

19.1 Introduction

Terrestrial agriculture is not only the source of the majority of human food, but is also a major economic and livelihood activity, especially in the developed world – including acting as a method of savings and insurance – and it is often tightly embedded within local cultures (Rivera-Ferre et al. 2016). Agriculture is highly diverse, ranging from low-input subsistence farming through to highly industrialised production systems; with varying ranges of labour, technology and knowledge, natural and economic capital and energy inputs (Rivera-Ferre et al. 2016). Agriculture occupies approximately 38.4% of the Earth's land area, with approximately 68% of this being permanent pastures used for livestock and the remainder used for annual and permanent crops (Arneth et al. 2019). Agriculture uses approximately 25–33% of the global terrestrial potential net primary productivity, doubling since 1910 (Arneth et al. 2019), and consequently it has increasingly significant impacts on global energy, water and nutrient cycles. Overuse such as through excessive cultivation, nutrient depletion, land clearing or overgrazing degrades the natural resource base and its capacity to provide ecosystem services including agricultural production (Olsson et al. 2019). Often, these degradation processes reduce the above-ground and below-ground carbon stocks, resulting in net emissions of carbon dioxide to the atmosphere – thus contributing to climate change. These degradation processes also change fluxes of water and energy between the land surface and the atmosphere, generally resulting in increases in local temperatures and in some instances changes in local patterns of rainfall. Agriculture also generates other greenhouse gas (GHG) emissions, such as methane emissions from ruminant livestock and nitrous oxide from fertiliser use, collectively resulting in global emissions of between 15% and 23% (Vermeulen et al. 2012; IPCC 2019) and further emissions arising from pre-farm, on-farm and post-farm activities. The total emissions from the food system (or value chain) are consequently high – about 21–37% of the global total (IPCC 2019). However, agricultural lands also have significant potential to act as a net sink for GHGs under some circumstances (IPCC 2019). The size of these various sources and sinks has significant seasonal and geographical uncertainty and variability. Consequently, emission reduction (mitigation) in agriculture and food systems should be a high priority, both as a topic of research and implementation and as part of global attempts to achieve the Paris Agreement[1] goals.

Agricultural systems not only affect the climate but are also profoundly affected by climate in many ways. Climate demonstrably impacts on choice of production system, production potential and variability, product quality, areas planted and preferred soil types, management systems and technologies, input levels and their costs, product prices, natural resource management (including water), human health, businesses along value chains, regional economies and social fabric. Thus, it follows that if the climate changes, all of these elements may be affected; either a little or a lot. This will sometimes lead to positive outcomes and sometimes negative ones. These climate change impacts will arise from increases *inter alia* in average temperature, temperature and rainfall extremes, storm and hail risk, potential evaporation and sea-level rise (Porter et al. 2014; Mbow et al. 2019). There are particularly likely to be spatially variable changes in rainfall (i.e. some areas receiving more rainfall while others become drier: Naumann et al. 2018) and cold

---

[1] *Paris Agreement Under the United Nations Framework Convention on Climate Change*, opened for signature 16 February 2016. Available at: https://unfccc.int/process-and-meetings/the-paris-agreement/the-paris-agreement.

temperature extremes (e.g. Crimp et al. 2019) depending on location. These will in turn affect the growing environment of crops and pastures, drought frequency, flood risk, frost risk and bushfire risk, among other things. Hotter and drier conditions will generally accelerate desertification processes and tend to reduce current arable land area in the subtropical to temperate zones, but warming is also expected to open up land to agriculture in the higher latitudes that is currently too cold, although this may be limited by suitable soil types and land-use conflicts (Mbow et al. 2019).

These changing climate conditions will shift spatially and temporally as climate change progresses. They will also interact with increases in atmospheric carbon dioxide (CO_2) concentrations, which generally make crops and pastures more water, radiation and nitrogen efficient, but at some expense in terms of quality metrics, reducing the nutritional value of food (Porter et al. 2014). The aggregate impacts of these changes already appear to be affecting agricultural production, often reducing production growth and profitability below what it would otherwise be, but in a smaller number of cases increasing it (Porter et al. 2014; Hughes et al. 2017, 2019). These changes are anticipated to accelerate and become more net negative as climate change progresses.

The significance and broad scale of these anticipated impacts make climate adaptation a high priority for agriculture production systems, regions or value chains (e.g. Howden et al. 2007). Importantly, many climate adaptations are a priori expected to have some impact on net GHG emissions and it will become increasingly important to avoid perverse adaptations (i.e. those which increase net emissions significantly). Equally, emission reduction strategies (mitigation) can affect adaptation options, sometimes enhancing them and sometimes having negative impacts now or in the future by restricting options. Consequently, climate change mitigation strategies themselves need to take climate impacts and adaptation into consideration and require both supply-side and demand-side strategies (e.g. Vermeulen et al. 2012; Smith et al. 2019).

In addition to interactions between climate adaptation and mitigation, there will be interactions between these and the multiple other drivers impacting on agriculture. These drivers include interests in increasing biodiversity, enhancing ecosystem services, reducing unwanted externalities such as nitrogen pollution, changing product specifications (e.g. food miles or contaminants), the long-term reduction in the terms of trade for farmers, reduced arable land area from degradation, desertification and urban expansion and less water for irrigation due to competition from urban areas and industry (e.g. Mbow et al. 2019). Importantly, climate issues will interact with the increasing demands being placed on agriculture: approximate doubling of food production for an additional 2–3 billion, wealthier people buying greater amounts of more resource-intensive foods, more cereal production for animal feed and more biofuel needed to reduce GHG emissions (Ziska et al. 2012). Meeting these multiple challenges will require new and effective options for decision makers across agricultural systems globally.

In this chapter we assess climate mitigation and then adaptation options for agriculture and the interactions between these, with the intent of better identifying win–win options that align with sustainable enhancement in productivity as well as a range of other co-benefits. We also discuss governance and policy options that can address barriers to effective action.

19.2 Agricultural GHG Emissions and Their Net Reduction

The agriculture sector globally is a significant emitter of GHGs via both direct and indirect emissions. Globally, direct emissions are 6.2 ± 1.4 Gt (gigatonnes) CO_2 equivalents (CO_2e) per year (9–14% of total anthropogenic emissions: IPCC 2019) with the contribution of methane (primarily from the digestive tracts of livestock and paddy rice-farming) being twice as much as nitrous oxide. The energy in methane emitted by ruminants is about 5–6% of the gross energy of feed eaten (Charmley et al. 2016) and consequently there has been a long-term interest in reducing these emissions so as to enhance existing animal production levels globally (e.g. Blaxter and Clapperton 1965). Similarly, farm productivity can be enhanced in some situations via capturing of methane produced by livestock manure, which can be used to produce energy (electricity and heat) and fertiliser.

Nitrous oxide (N_2O) is formed particularly in wet soils where there is carbon available for microbial consumption. Nitrogen emitted from soils in the form of nitrous oxide is lost to plant and animal production (nitrogen being a critical, limiting and expensive nutrient in many situations). Consequently, reducing nitrous oxide losses is also aligned with enhanced sustainable agriculture practices.

Direct GHG emissions also arise from on-farm use of fuel, for example with the use of tractors, irrigation pumping and machinery for milking, generating CO_2 emissions and trace amounts of other GHGs. Fuel can be a large cost for many industrialised farming systems and achieving efficiencies in use can increase profits.

Land-use change is also a significant net source of emissions from agricultural activity, contributing 4.9 ± 2.5 GtCO$_2$e/year (5–14%; IPCC 2019). Significant emissions arise from ongoing land-clearing activities where biomass carbon is transferred to the atmosphere via burning or decomposition. Soil carbon can also be a major contributor to either GHG emissions or mitigation in agricultural systems. For example, the conversion of land from a forest or grassland to a cultivated cropping system can result in the loss of up to 50–60% of soil carbon. This loss is important not only in GHG emission terms, as it has been known for many decades that increasing levels of soil carbon usually results in more productive soils, arising from improved water infiltration, higher water-holding capacity, better structure, improved nutrient status and higher levels of biological activity, leading to enhanced, sustainable production (e.g. Meyer et al. 2015).

Lastly, emissions can also arise from pre-farm and post-farm processes, with these being between 2.6 GtCO$_2$e/year and 5.2 GtCO$_2$e/year (5–10%: IPCC 2019). These sources include fossil fuel use, embodied in many agricultural inputs (fertiliser, herbicides, diesel, water, etc.) and also arise from storage, transport, processing and packaging. The emissions for these different components (as well as those associated with on-farm fuel-based emissions) are often counted in other inventory sectors such as energy, industrial activity and transport and so are often not immediately attributed to agriculture and food systems. Importantly, all of these components are costs (and hence there is a tendency to want to reduce these where practicable) and reducing these costs also tends to reduce emissions. Consequently, there is a broad alignment between efficient agricultural production systems and their associated value chains and with reductions in net GHG emissions.

Consequently, agriculture and food systems, when viewed broadly, are responsible for 10.8–19.1 GtCO$_2$e/year (21–37% of anthropogenic emissions – with best estimates of 29%: IPCC 2019). This is similar to the emissions associated with electricity and heat generation (31%) and about double those associated with the transport sector, which are about 15% of global emissions.

In contrast with other economic sectors, which have feasible mitigation pathways (e.g. renewable electricity generation) that could, if implemented, markedly reduce net emissions, agriculture currently has relatively few options. For example, decades of research have gone into the reduction of methane emissions from ruminant livestock and reducing loss of nitrogen (including losses from nitrous oxide) but with only marginal success to date, and that mainly in more intensive systems, often with inadequate whole-life-cycle emissions accounting (e.g. Eckard et al. 2010). Some of these emission reduction options and their effectiveness for different agricultural sectors are outlined in Table 19.1. Hence, in the absence of effective and implementable breakthroughs, due to projected shrinking emissions from the electricity and other sectors, agriculture is likely to become a larger proportion of the global GHG emissions over time. This means that making progress on ways to reduce GHG emissions from the agricultural sector is likely to increase in importance over time. Importantly, such activities can be aligned with agendas to increase agricultural production, production efficiency, profitability and sustainable use of natural resources and as a consequence reduce business risk (e.g. Smith et al. 2019). Unfortunately, this alignment has often been obscured by ideological agendas (e.g. Arbuckle et al. 2015), limiting progress.

The pressures to increase aggregate agricultural production are likely to rise markedly over the next decades due to the projected approximate doubling of food demand by 2050 (FAO 2018). This will occur at the same time as increased expectations of reliability in both the quantity and quality of produce. One response which has historically been used is to increase the area of production – also called extensification. However, this can have significant consequences in terms of GHG emissions, particularly from reductions in above- and below-ground carbon stores arising from the process of land clearing or conversion from grazing to cropping (e.g. Johnson et al. 2014; Figure 19.1). Extensification can also have significant consequences in terms of biodiversity loss and social and economic changes, among others. Variations in environmental, climatic, social, economic and institutional factors result in some extensification pathways being more likely to have fewer unwanted impacts than others (e.g. Johnson et al. 2014). Clearly, there are also limits to extensification, due the limited availability of suitable land and sometimes of infrastructure.

Hence, there is also growing interest in intensification of the use of existing agricultural lands as a way to operate within some of these constraints. Intensification can take many forms but generally involves aiming to increase the production per unit area, or value of production per unit area, by increasing the level of material and/or management inputs into existing systems or through changes in the agricultural system itself (Gregory et al. 2002). Examples of the latter include changing from grazing to cropping or from dryland cropping to irrigated cropping or horticulture. There can be quite different general relationships between the intensity of agriculture activities and the GHG emissions arising from them (Figure 19.1). Intensification can provide higher returns, higher value products and reduced GHG emissions

Table 19.1. *Key agricultural systems and their main sources of GHG emissions and broad emission reduction options*

Agricultural system	Key emission sources and sinks	Broad mitigation options
Extensive grazing	The primary emissions source is enteric methane, with additional methane and nitrous oxide emissions from manure. For this and the other systems below, CO_2 and other GHGs can be emitted from fossil fuels consumed on-farm, during transportation and in storage and processing, and from emissions embodied in various other inputs into cropping systems such as in fertilisers, machinery and energy used on-farm and in herbicides, pesticides and fungicides as well as through some management actions such application of lime and from burning practices. Soil carbon can be lost or gained as a result of various management practices, as can biomass carbon including via inclusion of agroforestry components.	Direct emission reductions can arise from use of methane inhibitors such as 3-nitrooxypropanol (3-NOP) (Jayanegara et al. 2018) and from livestock breeding (e.g. Barwick et al. 2019) A broad range of management activities that enhance pasture and/or animal productivity or efficiency can reduce emissions per unit product but not necessarily total emissions. Examples include: stocking rate adjustment, improved animal husbandry, improved reproductive performance, improved pasture management including increasing legume components, agroforestry or woody weed management and supplementary feeding (Ghahramani et al. 2019). Improvements can also occur across the value chain (e.g. renewable energy for processing).
Dryland cropping	The key direct emissions are nitrous oxide from fertilised soils with a small amount of various emissions from burning crop residues. Carbon dioxide emissions can arise from soil carbon rundown following conversion to cropping but some practices can also slowly rebuild soil carbon (Smith et al. 2019). As noted above, there can be additional pre-farm, on-farm and off-farm emissions.	Emission reduction strategies for nitrous oxide include application of nitrification inhibitors such as dicyandiamide (DCD) or 3-, 4-dimethylpyrazole phosphate (DMPP), matching nitrogen fertiliser applications to crop demand, avoiding fertiliser application onto warm, waterlogged soils with copious organic matter. Care is needed that inhibitor use does not increase emissions post-crop (Scheer et al. 2017). Currently, most inhibitors are not cost-effective in most dryland cropping systems.
Horticulture and irrigated cropping	Key direct emissions are nitrous oxide from soils and methane emissions from paddy rice, with additional methane and nitrous oxide emissions from crop residue burning. As noted above there can be additional pre-farm, on-farm and off-farm emissions including from	Key emission reduction options include those related to cropping (as above) but also in particular to more efficient irrigation systems and infrastructure as well as semi-saturated culture, alternate wetting

Table 19.1. (*cont.*)

Agricultural system	Key emission sources and sinks	Broad mitigation options
	infrastructure, control of growing environment, growing media production, water supply and energy use.	and drying system and low-methane varieties in rice-farming systems.
Intensive animal production	Key direct emissions are from enteric methane, with additional methane, nitrous oxide and CO_2 emissions from manure management, electricity used for milking, temperature-control, lighting and ventilation systems. As noted above there can be additional pre-farm, on-farm and off-farm emissions including chemicals used in husbandry and pasture management, water provision and in feedstuffs such as grain concentrates.	Key emission reduction options include those in the extensive grazing systems above but also include a range of additional diet additions including dietary oils, probiotics and enzymes and treatments such as 3-NOP. Waste products such as manure can be used to partly replace fertiliser use in nearby cropping systems (replacing their embodied emissions) and can increase soil carbon levels but also can be a source of nitrous oxide, which can be reduced with application of chemicals such as DCD.

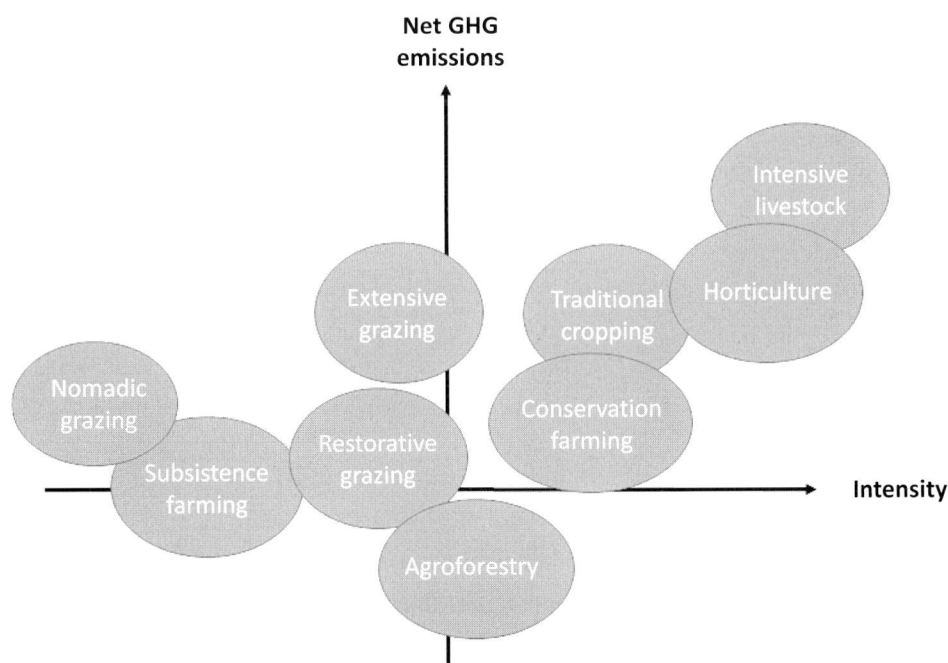

Figure 19.1 General relationship between agricultural intensity and net GHG emissions for a range of agricultural systems.

Source: Created by the author. For a colour version of this figure, please see the colour plate section.

per unit product but it can also increase price-based risk (Lin et al. 2008), increase total GHG emissions, increase off-site externalities such as nutrient pollution and biodiversity loss and it can homogenise production systems, exposing them to increased risk of various disruptions (e.g. Alauddin and Quiggin 2008). Risks also arise from higher dependence on allied infrastructure (e.g. electricity for production systems, heating/cooling/ventilation/cleaning, fossil fuel use and equipment manufacturing). Nevertheless, intensification has been attributed with reducing large amounts of land clearing (also termed 'land sparing') and associated environmental and social costs (Bennetzen et al. 2016). Emission reductions could also be achieved via reductions in food loss and waste (currently estimated to be 25–30% of food produced: Arneth et al. 2019), which would reduce the emissions across the value chain.

The trajectories of intensification and extensification are both highly stable and different when viewed across regions (Figure 19.2). Europe has a strong intensification trajectory for crop systems with increased yields from a smaller area cropped over time. The USA and Asia have a weaker but consistent intensification trajectory with increased yields to a similar degree but only slightly reduced (USA) or increased (Asia) cropped area. In contrast, both Africa and Australia, where cereal cropping tends to be subject to higher climatic risk, have clear extensification trajectories with a relatively smaller increase in yields but increased production historically occurring primarily via increased cropped area. Clearly, altering these trajectories is not a simple process: they are highly stable over periods of decades.

Some farmers have identified that intensification does not consistently increase return on investment given climate and price volatility and reducing terms of trade. Hence, they are designing low-input systems that reduce financial outlays and reduce risk. One example of

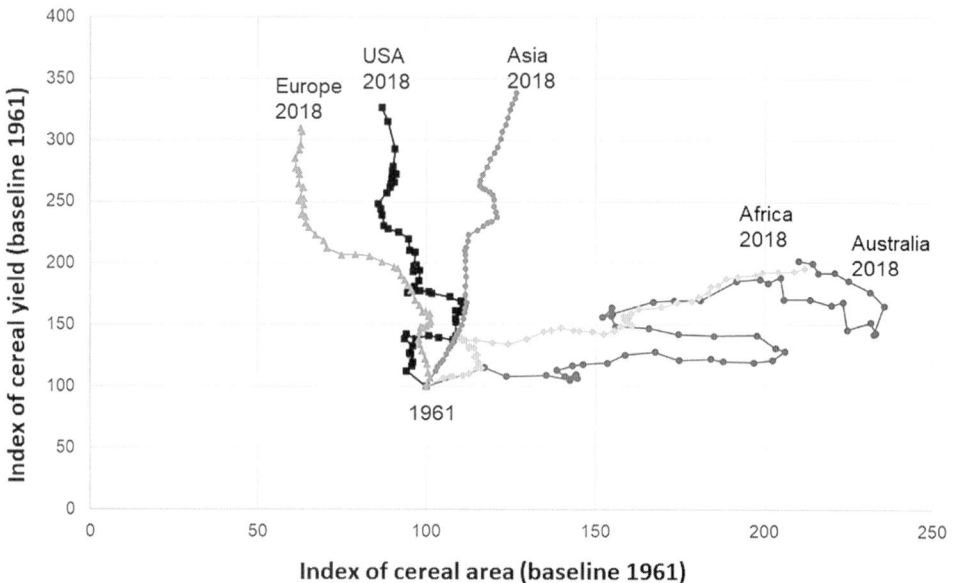

Figure 19.2 Trajectories of intensification and extensification of cereal cropping in Europe, the USA, Asia, Africa and Australia. To reduce noise, the data are presented as a 5-year running mean. Changes in cereal crop yield and area are standardised against 1961 values. *Source*: Data from FAOStat (FAO n.d.).

this is pasture cropping, where cool-season crops such as wheat, barley and oats are sown into summer-growing, relatively lightly grazed native grass pastures using direct drilling techniques with ultra-low levels of applied fertiliser (Millar and Badgery 2009). In some climate regions, this approach can more effectively use the available climate and soil resources, increasing overall production and profit but not increasing risk (Thomas et al. 2014). Pasture cropping appears to also increase soil cover (hence reduce soil erosion risk), increase surface soil carbon, increase biodiversity and reduce unwanted subsoil drainage, which can increase dryland salinisation risk. This example shows the importance of farmer-based innovation.

19.3 Climate Change Impacts and Adaptation in Agricultural Systems

The fundamental task of agriculture is to match plant or livestock genetics and management actions with the available resources and environment (of which climate is a key part) so as to achieve as best as possible the desired balance between risk and productivity (and profitability in enterprises with a focus on financial returns). The consequences of not doing this matching well are either underperformance or incurring unnecessary risk. Neither of these are desirable. Given the importance of climate in this equation and the pervasiveness of the impacts of climate change across agriculture and food value chains (Mbow et al. 2019), and conversely the substantial impacts of agriculture and food on climate, this section addresses the critical issue of how to adapt agriculture while simultaneously reducing net GHG emissions (Smith et al. 2019).

The huge diversity of agricultural activities successfully implemented in diverse local environments across the globe suggests that an equivalent diversity of climate adaptation options exists. For example, in wheat-based cropping systems alone, there are many potential adaptations which include, among other options, changes in varieties or species (e.g. interleaving winter crops with summer crops), planting times, crop density and row spacing, nutrient management and rotations, crop residue and soil water management, pests, disease and weed management and altering the mix of livestock (e.g. Porter et al. 2014; Mbow et al. 2019). These adaptations can be implemented singly or in combination. Comprehensive reviews of these adaptation options dealing with the grains, cotton, sugar, rice, horticulture and vegetables, viticulture, broadacre livestock, intensive livestock and agroforestry sectors are available globally (e.g. Mbow et al. 2019) and for different nations (e.g. Stokes and Howden 2010, for Australia). This section briefly addresses climate change impacts, adaptation options and their implications for emission reduction in relation to extensive grazing, dryland cropping, horticulture and irrigated cropping and intensive livestock systems. This intersection of sustainable food production, climate adaptation and emission reduction is often termed 'climate-smart agriculture' (Lipper et al. 2014).

19.3.1 Extensive Grazing

Extensive grazing systems tend to occur where some climate, soil or other factor limits the adoption of cropping activities. In particular, extensive grazing systems in many parts of the world have evolved as a way of addressing problematic climate variability. Nevertheless, even given this background they are often evaluated as being susceptible to climate changes (Porter et al. 2014; Rivera-Ferre et al. 2016). For example, in Australia, broadacre grazing systems have already suffered a 22% loss in productivity due to climate changes already experienced (Hughes et al. 2019). Table 19.2 outlines the key climate change adaptation options for extensive grazing systems and their mitigation implications, including co-benefits.

19.3.2 Dryland Cropping

As noted above, there are many potential climate change adaptation options for dryland cropping systems and their use could provide substantial productivity benefits under climate change (Challinor et al. 2014). Table 19.3 outlines key climate change adaptation options for dryland cropping and their mitigation implications, including co-benefits.

19.3.3 Horticulture and Irrigated Cropping

The fundamental premise of irrigation is that it reduces risks associated with unreliable rainfall and enhances productivity. This appears to be a sensible strategy for modest climate changes but under more significant changes can result in almost complete industry failure when there is not water to be allocated, as has already happened in various countries (South Africa, Australia) in association with climate change. Table 19.4 outlines the key climate change adaptation options for horticulture, viticulture and irrigated cropping and their mitigation implications, including co-benefits.

19.3.4 Intensive Livestock Systems

Akin to intensive cropping and horticulture systems, intensive livestock systems generally aim to manage the inputs and production environment of livestock systems to increase productivity and efficiency and to reduce various risks. Table 19.5 outlines the key climate change adaptation options for intensive livestock and their mitigation implications, including co-benefits.

19.4 Integration of Emission Reduction, Climate Adaptation and Food Security Goals

The above sections have assessed different agricultural systems in terms of their key emissions and emission reduction options, key options to adapt to climate change and the implications of these for emission reduction. This section seeks to assess a range of emission reduction options in different sectors and integrate the above discussions along with the implications of projected climate change (e.g. Smith et al. 2019). Critically, most agricultural emission reduction activities involve trade-offs, and usually at least some costs. However, they can also result in many benefits including enhanced climate change adaptation. How the costs, benefits and co-benefits of emission reduction activities are distributed and managed will usually be critical to the success of policy efforts (e.g.

Table 19.2. *Key climate change adaptation options for extensive grazing systems and their mitigation implications, including co-benefits*

Adaptation option	Mitigation implications including co-benefits
Progressively recalculate and adjust safe stocking rates and pasture utilisation levels, including strategic spelling, taking into account observed and projected climate change and seasonal climate forecasts	Likely to reduce methane emissions per unit product due to improved per animal productivity, but effects on net emissions depend on the specific adjustments made. Large and diverse co-benefits including avoidance of severe degradation events, which may lose substantial above- and below-ground soil carbon to the atmosphere.
Select animal lines that are resistant to higher temperatures but maintain production	Likely to reduce methane emissions per unit product due to improved per animal productivity and capacity to smooth out production variation.
Modify timing of mating, weaning and supplementation based on seasonal conditions in order to reduce exposure of offspring to climatic extremes	Minor impacts on emissions apart from those associated with improved per animal productivity but potentially substantial impacts on sustainability of the natural resource base.
Improve nutrient management using sown legumes and phosphate fertilisation where appropriate	Lower on-farm nitrous oxide emissions and embodied emissions and likely increased soil carbon levels in many situations. A range of co-benefits arising from increased pasture productivity such as potentially increased soil cover. Issues include risks from soil acidification and sometimes increased financial risk.
Observe climate-induced changes in vegetation and modify management accordingly	Minor emission impacts associated with better sustainable management of biomass carbon stores. Co-benefits from reduced degradation risk.
Provide extra shade using trees and constructed shelters	Minor impacts on emissions apart from those associated with improved per animal productivity but potentially substantial impacts on animal welfare.
Effective control of woody weeds taking into account regional biodiversity needs	Impacts on carbon stocks could be either positive or negative depending on the situation. Potentially large, low opportunity cost carbon sequestration opportunities but at the cost of long-term production and with some risk of losing large proportions of the carbon stock to events such as wildfire.
Improve pest management	Minor impacts on emissions apart from those associated with improved per animal productivity but improvements in animal health and welfare.
Improve on-property water management, particularly for pasture irrigation	Potentially reduced emissions from more efficient systems but in some cases, moving

Table 19.2. (*cont.*)

Adaptation option	Mitigation implications including co-benefits
	from artesian gravity drains to pumped, piped systems could increase net GHG emissions. Co-benefits include leaving water available for other uses.
Develop software to assist proactive decision-making at the on-farm scale	Minimal impact on emissions by itself but potential overall improvement to productivity and hence lowered emissions per unit product.
Expand routine record-keeping of weather, pest and diseases, weed invasions, inputs and outputs	Minimal impact on emissions by itself but potential overall improvement to productivity.
Diversify on-farm production and consider alternate land uses	The emission consequences are highly dependent on the changes made. Aligning land use with the climate, soil and other assets available is the fundamental function of agriculture. So, doing this well is crucial.

Source: Modified from Stokes et al. (2010).

Guariguata et al. 2008). For example, land-use policies to prevent emissions and increase sequestration can have perverse impacts, including leakage, gaming and corruption. Leakage refers to the risk that abatement efforts in one area increases emissions elsewhere, either by directly displacing the activity or via market forces. Similarly, financial incentives to encourage emission reduction activities can diversify the income streams of landowners and dependent communities and, in doing so, make them more able to cope with the effects of climate change. In other cases, mitigation and climate adaptation imperatives can be in conflict. For example, maximising carbon sequestration can be achieved via planting fast-growing species. However, these often use large amounts of water, reducing catchment yields and they perform best on good soils and thus may compete with agricultural food production. Additionally, they usually have little biodiversity value and, moreover, often being invasive, can compete with native species, weakening the capacity of threatened species and vulnerable communities to adapt to climate change (Hurlbert et al. 2019). Table 19.6 outlines extensive and intensive livestock emission reduction management and policy options and their implications for climate adaptation and exposure to climate change, while Table 19.7 summarises the major emission reduction options for extensive cropping systems and horticulture, how they interact with other issues and, similarly, how this affects climate adaptation and exposure to climate change.

19.5 Governance and Policy Solutions for Agricultural Adaptation and Mitigation

There is a range of policy adaptations that form the basis for the decision environment within which farm and enterprise adaptations take place. These include industry and

Table 19.3. *Key climate change adaptation options for dryland cropping and their mitigation implications, including co-benefits*

Adaptation option	Mitigation implications including co-benefits
Develop further climate risk amelioration approaches (e.g. zero-tillage and other minimum disturbance techniques, retaining residue, extending fallows, row spacing, planting density, staggering planting times, erosion control infrastructure, controlled traffic)	These adaptations tend to reduce fuel-based emissions and store slightly more carbon in the surface soil than would otherwise have occurred due to higher biomass production, reduced biomass carbon losses and protection of the soil surface. More efficient use of applied nitrogen may reduce nitrous oxide emissions. Co-benefits include reduced degradation, improved production and profitability and reduced risk. Issues can include increased deep drainage of water (hence leaching and salinisation risks) and capital costs.
Alter planting rules to be more opportunistic depending on environmental conditions (e.g. soil moisture), climate (e.g. frost risk) and markets	Could be reduced risk of soil carbon loss from degradation events if done effectively. Generally positive in terms of managing financial and natural resource degradation risks. Issues include that this type of operation can increase psychological stress on farmers or may increase emissions if not implemented well due to failed plantings.
Select varieties from the global gene pool with high levels of CO_2 responsiveness, appropriate thermal time and vernalisation requirements, heat shock resistance, drought tolerance and recovery, high protein levels, resistance to new pest and diseases and perhaps that set flowers in hot/windy conditions	Likely higher and more reliable biomass generation provides the feedstock for increased soil carbon. More efficient crops reduce inputs and embodied emissions in those.
Maximise utility of seasonal climate forecasts by research, development and extension that combines them with on-ground measurements (i.e. soil moisture, nitrogen), market information and systems modelling	Likely minor impacts on emissions apart from those noted above arising from effective agronomic decision-making. Important role in avoiding large-scale degradation events (e.g. McKeon et al. 2009).
Research and revise soil fertility management (fertiliser application, type and timing, increase legume phase in rotations) on an ongoing basis.	Reduced nitrous oxide and embodied emissions from more efficient fertiliser use regimes. Reduced inputs generally lowers risk.
Further develop area-wide management operations, integrated pest management and other innovative pest, disease and weed adaptations	Lowers levels of inputs with high levels of embodied emissions (e.g. pesticides). A range of human, animal and natural system co-benefits.
Forward selling and crop insurance	Few impacts on emissions.
Explore transformation options in the cropping zones that can provide positive production, environmental and social outcomes from these major changes	Depends markedly on the transformation. Those that move cropping land to extensive grazing could result in net reductions in emissions. Transformation of high rainfall grazing lands to cropping is likely to result in marked increases in CO_2 emissions via soil carbon loss. Transformations are highly context-sensitive in terms of co-benefits.

Source: Modified from Howden et al. (2010).

Table 19.4. *Key climate change adaptation options for horticulture, viticulture and irrigated cropping and their mitigation implications, including co-benefits*

Adaptation option	Mitigation implications including co-benefits
Agronomic management (seeding, harvest, layout, rotations, pruning suited to the new conditions)	Overall, by increasing productivity and reducing risk, these practices reduce GHG emissions per unit product but not necessarily total emissions.
Evaluate genetic environment by management interactions so as to guide conventional breeding (heat and drought tolerance, increased water use efficiency, improved quality) and biotechnology options	Again, by increasing productivity and reducing risk, these adaptation practices reduce GHG emissions per unit product but not necessarily total emissions. Co-benefits arise from more reliable systems with fewer inputs.
Further improve water distribution systems (to reduce leakage and evaporation), irrigation practices such as water application methods, irrigation scheduling and moisture monitoring to increase efficiency of use and enhance yields	Reductions in emissions via reduced pumping and lower overall water use per unit product. Co-benefits include reduced off-site impacts such as dryland salinisation and more water potentially available for other uses.
Identify crop-specific CO_2 response and seek to maximise the benefits of this	For some crops this will mean increased nitrogen fertilisation with attendant increases in on-farm nitrous oxide emissions and embodied emissions. Co-benefits include increased water use efficiency from elevated CO_2.
Tailored climate-sensitive pest, disease and weed management	Likely higher and more reliable biomass generation provides the feedstock for increased soil carbon. More efficient crops reduce inputs and embodied emissions in those such as in pesticides.
Use climate forecasts (multi-day to seasonal) to manage risk	Likely minor impacts on emissions apart from those noted above arising from more effective agronomic decision-making. Important co-benefits in aligning farm inputs with demand, increasing overall efficiency and reducing various farm-level and value chain risks.
Reassess location to manage climate risk	Depends markedly on the transformational change. Re-establishment is likely to be associated with significant emissions associated with infrastructure development, land clearing and change in soil carbon levels (if changing from productive woody or grassland vegetation). Transformations are highly context-sensitive in terms of co-benefits.
Develop and modify markets and value chains for new crops and new cropping schedules. Using geographic dispersion of production units to reduce risk	There appear to be a large number of options to reduce GHG emissions along supply chains (Lim-Camacho 2016) and for consumers to exercise informed choice in relation to products with different environmental footprints. There could be a range of co-benefits to consumers arising from climate adaptations, such as produce available for longer seasons.
Climate derivatives and insurance	Minor impacts on emissions, but via management of risk could encourage flexible and season-appropriate management, which would tend to reduce degradation risk.

Source: Modified from Bange et al. (2010), Gaydon et al. (2010), Webb and Whetton (2010) and Webb et al. (2010).

Table 19.5. *Key climate change adaptation options for intensive livestock and their mitigation implications, including co-benefits*

Adaptation option	Mitigation implications including co-benefits
Selection of drought-tolerant pasture species	Potential for increased soil carbon inputs arising from enhanced productivity and also by reducing methane emissions per unit product by providing livestock with improved feed. Co-benefits include improved soil cover, which could reduce water erosion.
Fodder conservation and use strategies	Reduced emission per unit product but not necessarily total emissions. Strategic use of fodder and supplementary feeding can smooth out production and financial variation and improve product quality.
Forward contracting and insurance including for supplementary feed	Minor impacts on emissions but via management of risk could encourage flexible and season-appropriate management, which would tend to reduce degradation risk.
Selection for heat-tolerant but productive livestock	This is likely to increase per animal productivity in warm climates leading to possible reductions in emissions per unit product. Co-benefits include improvements in animal welfare. Issues could arise from a tendency to ignore other strategies such as provision of shade.
Changed feeding regimes	This is likely to increase per animal productivity in warm climates leading to possible reductions in emissions per unit product.
Agistment	Some emissions in the process of transporting livestock. Co-benefits arise from avoiding overgrazing and degradation.
Improved irrigation efficiency	Reductions in emissions via reduced pumping and lower overall water use per unit product. Co-benefits include reduced off-site impacts such as dryland salinisation, reduced eutrophication and more water potentially available for other uses.
Improved soil water infiltration and water-holding capacity	Improved pasture productivity could increase soil carbon levels slightly.
Use climate forecasts (multi-day to seasonal) to manage risks and opportunities	Likely minor impacts on emissions apart from those noted above arising from more effective agronomic decision-making. Important co-benefits in aligning farm inputs with demand, increasing overall efficiency and reducing various farm-level and value chain risks.
Climate-controlled animal housing	Emissions associated with the housing and any active cooling (or heating) systems which are fossil-fuel-based. Co-benefits include improved animal welfare, higher productivity and improved product quality.
Renewable and bio-based power generation to avoid off-site power outages	Reduction of emissions via substitution for fossil-fuel-based energy. Co-benefits can include cost savings.

Source: Modified from Miller et al. (2010).

Table 19.6. *Extensive and intensive livestock emission reduction management and policy options and their implications for climate adaptation and exposure to climate change*

Emission reduction option	Implications and interaction with other options	Implications of climate change (if any)
Reduction in stocking rate (animals per hectare)	In the absence of productivity-enhancing mitigation strategies (e.g. dietary oils), this will tend to reduce overall farm productivity and profitability although increases in per animal productivity often occur. If currently overstocked, this strategy can reduce degradation risk and improve the natural resource base. Lower stocking rates can increase carbon stocks in many grazing systems but also reduce them in some South American systems (de Oliveira Silva et al. 2016). It can also reduce the palatability of pastures.	Reduced pasture productivity in many subtropical and temperate regions may force reductions in stocking rates anyway (Ghahramani and Moore 2015), in which case mitigation and adaptation actions are broadly aligned. In some equatorial and high-latitude regions where forage production may increase, this mitigation strategy will be in tension with potential stocking rate increases.
Improved animal husbandry (especially disease, pest and heat management)	This increases animal productivity through improved feed conversion efficiency, welfare and often farm-level profitability (Waghorn and Hegarty 2011). This option would usually be integrated with other strategies such as pasture improvement or management to maximise return on investment. Improved housing for intensive livestock can be integrated into intensification strategies that improve reliability and overall productivity. These strategies can significantly reduce GHG emissions per unit product.	This will be a key adaptation to projected increases in heat stress and possible changes in pest and disease distribution and severity such as bluetongue. Hence, adaptation and mitigation elements are aligned. The increased environmental control with improved housing can reduce climate change impacts but potentially at the risk of increased GHG emissions from cooling/heating systems if they are powered from fossil fuel sources.
Improved reproductive performance	This increases total system output in many grassland grazing systems, and is a key goal in many livestock improvement programmes. It will often be dependent on improved husbandry and pasture management.	More challenging climate conditions can impact on reproductive performance and so this is likely to also be considered as an adaptation option. Higher reproductive performance can, however, result in increased exposure to climate risk, often impacting on the resource base, potentially reducing carbon stocks.

Table 19.6. (*cont.*)

Emission reduction option	Implications and interaction with other options	Implications of climate change (if any)
Improved forage management and agroforestry	This can increase the standing stock of carbon and also increase (or at least maintain) soil carbon, as well as potentially delivering a range of other ecosystem services. In addition, it can reduce methane yields through improving diet quality (e.g. Verchot et al. 2007).	Improved forage management is a core adaptation strategy in many systems and hence mitigation and adaptation goals are likely to be strongly aligned. Agroforestry is an option that may be particularly important for smallholder farms.
Increased legume component	This can increase intake but reduce methane emissions, but sometimes at the expense of increasing nitrous oxide emissions. In some situations, increasing legume content in pastures can increase soil carbon. This option is often undertaken in tandem with other improvements.	Increased atmospheric CO_2 concentrations are likely to reduce forage protein content, while potentially more extreme weather can lower forage digestibility, in both cases placing a premium on having adequate legumes in pastures.
Woody weed management	This usually attempts to limit the density of woody vegetation and hence above-ground carbon. Implications for soil carbon can be mixed. Management on a landscape matrix basis may enable win–win opportunities (Moore et al. 2001).	Projected climate changes and CO_2 increases are likely to require enhanced woody weed management especially in tropical and subtropical zones (Howden et al. 2001). In some circumstances, adaptation and mitigation goals may require trade-offs.
Supplementary feeding	Grain and other feed supplements (e.g. molasses) can reduce methane yields and enhance production, and if used strategically can protect the above- and below-ground carbon stores (Thornton and Herrero 2010). Urea and phosphorus supplements can enable overgrazing in droughts and hence further damage the natural resource base.	Given projected increases in climate variability, supplementary feeding could become a more standard part of livestock farming in many regions.
Enhanced robustness and efficiency of livestock value chain	Improved input and output management and including externalities as part of food-footprint-type approaches could require systemic change	Projected increases in climate variability may require buffering strategies across value chains and spatially as well as closer attention to

Table 19.6. (*cont.*)

Emission reduction option	Implications and interaction with other options	Implications of climate change (if any)
	in farming systems, bringing into play several of the above strategies (Garnett 2011). Improving livestock value chains is seen as a key poverty and nutritional insecurity alleviation strategy in many developing countries.	meeting market specifications. Hence, mitigation and adaptation are likely to be broadly aligned.

regional development policies, stewardship programmes, infrastructure development, industry capacity development programmes and other policies such as those relating to drought support, rural adjustment, quarantine, trade and finance, among many others. There is often substantial variation in the broad policy approaches used by different jurisdictions within a country and these can be difficult to change significantly. For example, in many countries there is a culture of active avoidance of more interventionist policy in the agricultural sector, and the majority of policy support is delivered via infrastructure provision and supporting the knowledge and innovation system, with any payments to producers being modest, targeted and direct. In contrast, countries such as the USA, Japan and those in the EU have a long history of more interventionist policies but also much higher levels of producer support (about 20% in the EU but 43% in Japan and 62% in Norway). In the EU, the majority of funding goes directly to producers, mostly in response to broad achievement of 'stewardship' goals, which include climate adaptation and promoting resource efficiency and the transition to a low-carbon economy. A small proportion goes to investments in knowledge and infrastructure. Developing countries often have much more limited policy portfolios and even some which appear to be perverse, often involving subsidising production, maintaining regional populations or export limitations in times of shortage. A rational response from an agricultural producer could well vary depending on the type of policy environment they find themselves in. Nevertheless, there is a growing view that maintaining a flexible research and development base to inform policy adaptations as well as farm-level changes is essential to deliver potential adaptation and mitigation benefits (e.g. Porter et al. 2014).

One typology of policy approaches is that of Hamilton and Macintosh (2008), who provide a four-part typology: information instruments, voluntary instruments, market-based instruments and regulation. Table 19.8 provides examples of the types of instruments that might be used to address agricultural production and sustainability issues.

The underlying rationale for integrating or 'mainstreaming' climate change considerations into existing government policies and initiatives (e.g. economics, regional development, agriculture and food, energy, water resources, health, etc.) is now well established,

Table 19.7. *Key emission reduction options for extensive cropping systems and horticulture, their interactions with other issues and the implications for climate adaptation and exposure to climate change*

Emission reduction option	Implications and interaction with other options	Implications of climate change
Reduce nitrous oxide emissions	The key options are (1) via optimisation of fertiliser management to increase in-crop uptake of soil nitrogen and to reduce excess nitrogen at the end of the crop; and (2) nitrification inhibitors. Both of these can reduce the cost of the nitrogen applied but the latter particularly is not yet cost-effective without subsidy. Reduced nitrogen leaching has benefits in terms of soil acidification and off-site impacts such as waterway eutrophication.	Nutrient demands of crops will tend to change, approximately in proportion, with growing conditions. Nitrous oxide emissions tend to be higher in warm, wet conditions with high carbon availability thus increasing the need for best management practice. Rising CO_2 will add a further demand for nutrients to meet the enhanced growth potential of most crops.
Reduce stubble burning and stubble retention after harvest	This can protect the soil from wind and water erosion, reduce soil evaporation via increased soil cover and increase the carbon supply for soil organisms, and hence biological activity, which tends to increase water infiltration and nutrient availability. All of these tend to increase productivity. In some cases, biomass retention can increase disease risk (e.g. take-all disease).	Most climate changes will tend to increase the importance of these management options. They will provide benefits in hotter and drier scenarios via improved water use efficiency, less wind erosion risk and in wetter and more extreme rainfall scenarios will reduce water-based erosion risks and nutrient loss.
Reduce on-farm fossil fuel use	Use of zero-tillage, stubble retention, precision agriculture, horticulture and viticulture, reduced soil compaction practices, integrated weed, pest and disease management can all reduce on-farm diesel use.	All of these practices can have some element of climate adaptation under different scenarios. For example, disease management can progressively optimise treatments as climate changes progress.
Reduce levels of embedded emissions in inputs	Inputs such as fertiliser, herbicide, pesticide, fungicide and fuel are all expensive, so optimising these tends to be aligned with both mitigation and efficient farm financial management. However, reducing inputs too far can reduce biomass carbon inputs into the system, reducing soil carbon stores over time.	Problematic climate changes are likely to drive adoption of lower risk farm practices and these often involve lower levels of farm inputs, and hence a tendency towards lower GHG emissions.
Reduce post-farm emissions	There can be opportunities to improve value chains via	Emissions from storage, transport and processing of crop products

Table 19.7. (*cont.*)

Emission reduction option	Implications and interaction with other options	Implications of climate change
	streamlining of storage, transport and processing. For example, reducing post-harvest losses effectively reduce the GHG emissions per unit final product.	are, everything else being equal, proportional to crop production, so could go either up or down depending on climate change.
Increase soil carbon levels	Increased soil carbon can enhance water infiltration and soil water-holding capacity, enhance soil physical and chemical fertility and reduce wind and water erosion, and hence is aligned with many natural resource management and productivity goals. However, enhanced soil carbon is intimately linked with higher nitrogen levels, which can be expensive if gained via man-made fertiliser inputs rather than biological ones.	Warmer temperatures are likely to result in soil carbon loss, everything else being equal, as they reduce biomass inputs and increase decomposition rates. This will be particularly noticeable in high-carbon soils in cool, wet climates (Crowther et al. 2016). Hence, even effective soil carbon management strategies may not result in a net increase in carbon.

with substantial analysis of the interactive nature of climate adaptation, GHG emission reduction, natural resource degradation and achievement of sustainable development (e.g. Hurlbert et al. 2019; Smith et al. 2019). This will need to be done in a very dynamic environment where rapid change is the norm: rapid climate changes, rapid change in energy systems, rapid change in consumer preferences, such as for plant-based meat substitutes, rapid change in agricultural technologies and management and rapid changes resulting from unexpected impacts such as the coronavirus COVID-19. The challenge now is to implement this more integrated policy and management approach, framed as a series of experiments that interact with the dynamic environment and, in parallel with this, to establish an effective monitoring programme to learn what works, what does not and why.

19.6 Conclusion

Food and agriculture are critical to sustainable and prosperous societies and need to be a core part of addressing climate change. This is reflected in the pre-eminent place that food and agriculture hold in the Paris Agreement and other global agreements such as the UN Convention to Combat Desertification. It also reflects both the large contribution they make to global GHG emissions (21–37% of anthropogenic emissions) and the large-scale vulnerability of food and agriculture to ongoing and projected climate changes, particularly

Table 19.8. *Four-part choice-based typology of agriculture-relevant climate change policy instruments*

Instrument type	Description	Examples
Information	Mechanisms that aim to improve outcomes by improving awareness and understanding of relevant sustainable production and climate change issues and building capacity to respond to environmental threats.	Research, development and extension (also known as RD&E) on seasonal climate forecasts and how to use them effectively, climate change projection information, identifying climate-smart agriculture options that reduce net GHG emissions and risk management, consumer education about GHG and climate adaptation implications of dietary choices
Voluntary	Mechanisms that aim to promote sustainable agriculture through voluntary involvement in programmes where no direct penalties are imposed on non-participants, although incentives may be used to encourage participation.	Agroforestry and carbon forestry, soil carbon building, improved climate risk management, involvement in Landcare programmes, GHG and climate adaptation labelling on products.
Market-based	Mechanisms that force farmers and agribusiness to internalise all or part of the social costs associated with environmentally harmful activities and that rely on market forces to promote efficiency. In doing so, they seek to impose additional costs on producers and actions that harm the environment and reward those that improve environment outcomes, while utilising market forces to improve the allocation of resources.	GHG or carbon taxes or prices, emissions trading schemes, industry structural adjustment, improvement of property rights over agricultural lands and resources including carbon stocks, GHG footprint information at point of sale, alternative products.
Regulation	Mechanisms that impose legally enforceable restrictions on farmers and agribusiness to realise environmental protection objectives. They are sometimes referred to as 'command-and-control' mechanisms because they prohibit or mandate certain actions (the command) and use the threat of punishment to motivate compliance (the control).	Land clearing controls, vegetation management regulation, fire regulation, pollution and pesticide or herbicide residue regulation, permit requirements to carry out particular management activities (weeding and pest control) or to use specified techniques/practices.

Source: Hamilton and Macintosh (2008).

in developing countries and in those with vulnerable populations. However, to date, food and agriculture have received far less policy and research support and/or development focus than other, smaller and less vulnerable, sectors. This chapter has painted a clear picture of the opportunities for effective outcomes in reducing emissions and for meeting growing food security needs, and in particular the opportunities that arise through recognition of the interaction between these. Addressing the interactions between climate change and food and agricultural systems is a challenge we can't afford to fail and an opportunity we can't afford not to grasp.

Acknowledgements

I would like to acknowledge Dr Craig Strong and Helen King for helpful discussions when framing the paper, Dr Steven Crimp for reviewing an earlier draft and Juliet Meyer for preparing the reference list.

References

Alauddin, M. and Quiggin, J. (2008). Agricultural intensification, irrigation and the environment in South Asia: Issues and policy options. *Ecological Economics*, 65, 111–124.

Arbuckle, J. G., Morton, L. W. and Hobbs, J. (2015). Understanding farmer perspectives on climate change adaptation and mitigation: The roles of trust in sources of climate information, climate change beliefs, and perceived risk. *Environment and Behavior*, 47, 205–234.

Arneth, A., Denton, F., Agus, F. et al. (2019). Framing and context. In P. R. Shukla, J. Skea, E. C. Buendia et al., eds., *Climate Change and Land: An IPCC Special Report on Climate Change, Desertification, Land Degradation, Sustainable Land Management, Food Security, and Greenhouse Gas Fluxes in Terrestrial Ecosystems*. In press. Available at: www.ipcc.ch/site/assets/uploads/sites/4/2019/12/04_Chapter-1 .pdf.

Bange, M. P., Constable, G. A., McRae, D. and Roth, G. (2010). Cotton. In C. Stokes and S. Howden, eds., *Adapting Agriculture to Climate Change: Preparing Australian Agriculture, Forestry and Fisheries for the Future*. Clayton, Victoria: Commonwealth Scientific and Industrial Research Organisation (CSIRO) Publishing, pp. 49–66.

Barwick, S. A., Henzell, A. L., Herd, R. M., Walmsley, B. J. and Arthur, P. F. (2019). Methods and consequences of including reduction in greenhouse gas emission in beef cattle multiple-trait selection. *Genetics Selection Evolution*, 51, 18.

Bennetzen, E. H., Smith, P. and Porter, J. R. (2016). Decoupling of greenhouse gas emissions from global agricultural production: 1970–2050. *Global Change Biology*, 22, 763–781.

Blaxter, K. L. and Clapperton, J. L. (1965). Prediction of the amount of methane produced by ruminants. *British Journal of Nutrition*, 19, 511–522.

Challinor, A. J., Watson, J., Lobell, D. B., Howden, S. M., Smith, D. R. and Chhetri, N. (2014). A meta-analysis of crop yield under climate change and adaptation. *Nature Climate Change*, 4, 287–291.

Charmley, E., Williams, S. R. O., Moate, P. J. et al. (2016). A universal equation to predict methane production of forage-fed cattle in Australia. *Animal Production Science*, 56, 169.

Crimp, S., Jin, H., Kokic, P., Bakar, S. and Nicholls, N. (2019). Possible future changes in South East Australian frost frequency: An inter-comparison of statistical downscaling approaches. *Climate Dynamics*, 52, 1247–1262.

Crowther, T. W., Todd-Brown, K. E. O., Rowe, C. W. et al. (2016). Quantifying global soil carbon losses in response to warming. *Nature*, 540, 104–108.

de Oliveira Silva, R., Barioni, L. G., Hall, J. A. J. et al. (2016). Increasing beef production could lower greenhouse gas emissions in Brazil if decoupled from deforestation. *Nature Climate Change*, 6, 493–497.

Eckard, R. J., Grainger, C. and de Klein, C. A. M. (2010). Options for the abatement of methane and nitrous oxide from ruminant production: A review. *Livestock Science*, 130, 47–56.

FAO (Food and Agriculture Organization) (n.d.). FAOSTAT [data resource]. FAO.org. Available at: www.fao.org/faostat/en/#data/FO.

FAO (2018). *The Future of Food and Agriculture: Alternative Pathways to 2050*. Rome: Food and Agriculture Organization of the United Nations. Available at: www.fao .org/global-perspectives-studies/resources/detail/en/c/1157074/.

Garnett, T. (2011). Where are the best opportunities for reducing greenhouse gas emissions in the food system (including the food chain)? *Food Policy*, 36, S23–S32.

Gaydon, D., Beecher, H. G., Reinke, R., Crimp, S. and Howden, S. M. (2010). Rice. In C. Stokes and S. Howden, eds., *Adapting Agriculture to Climate Change: Preparing Australian Agriculture, Forestry and Fisheries for the Future*. Clayton, Victoria: Commonwealth Scientific and Industrial Research Organisation (CSIRO) Publishing, pp. 67–84.

Ghahramani, A. and Moore, A. D. (2015). Systemic adaptations to climate change in southern Australian grasslands and livestock: Production, profitability, methane emission and ecosystem function. *Agricultural Systems*, 133, 158–166.

Ghahramani, A., Howden, S. M., del Prado, A. et al. (2019). Climate change impact, adaptation, and mitigation in temperate grazing systems: A review. *Sustainability*, 11, 7224.

Gregory, P. J., Ingram, J. S. I., Andersson, R. et al. (2002). Environmental consequences of alternative practices for intensifying crop production. *Agriculture, Ecosystems & Environment*, 88, 279–290.

Guariguata, M. R., Cornelius, J. P., Locatelli, B., Forner, C. and Sánchez-Azofeifa, G. A. (2008). Mitigation needs adaptation: Tropical forestry and climate change. *Mitigation and Adaptation Strategies for Global Change*, 13, 793–808.

Hamilton, C. and Macintosh, A. (2008). Human ecology: Environmental protection and ecology. In S. Jorgensen, ed., *Encyclopedia of Ecology*. Elsevier, pp. 1342–1350.

Howden, S. M., Moore, J. L., McKeon, G. M. and Carter, J. O. (2001). Global change and the mulga woodlands of southwest Queensland: Greenhouse gas emissions, impacts, and adaptation. *Environment International*, 27, 161–166.

Howden, S. M., Soussana, J. F., Tubiello, F. N., Chhetri, N., Dunlop, M. and Meinke, H. (2007). Adapting agriculture to climate change. *Proceedings of the National Academy of Sciences*, 104, 19691–19696.

Howden, S. M., Gifford, R. M. and Meinke, H. (2010). Grains. In C. Stokes and S. Howden, eds., *Adapting Agriculture to Climate Change: Preparing Australian Agriculture, Forestry and Fisheries for the Future*. Clayton, Victoria:

Commonwealth Scientific and Industrial Research Organisation (CSIRO) Publishing, pp. 21–48.

Hughes, N., Galeano, D. and Hatfield-Dobbs, S. (2019). The effects of drought and climate variability on Australian farms. *ABARES Insights*, 6, 11.

Hughes, N., Lawson, K. and Valle, H. (2017). *Farm Performance and Climate: Climate Adjusted Productivity on Broadacre Cropping Farms.* Research report 17.4. Canberra, Australia: Australian Bureau of Agricultural and Resource Economics and Sciences. Available at: http://data.daff.gov.au/data/warehouse/9aas/2017/FarmPerformanceClimate/FarmPerformanceClimate_v1.0.0.pdf.

Hurlbert, M., Krishnaswamy, J., Davin, E. et al. (2019). Risk management and decision making in relation to sustainable development. In P. R. Shukla, J. Skea, E. C. Buendia et al., eds., *Climate Change and Land: An IPCC Special Report on Climate Change, Desertification, Land Degradation, Sustainable Land Management, Food Security, and Greenhouse Gas Fluxes in Terrestrial Ecosystems.* In press. Available at: www.ipcc.ch/site/assets/uploads/sites/4/2019/11/10_Chapter-7.pdf.

IPCC (Intergovernmental Panel on Climate Change) (2019). *Climate Change and Land: An IPCC Special Report on Climate Change, Desertification, Land Degradation, Sustainable Land Management, Food Security, and Greenhouse Gas Fluxes in Terrestrial Ecosystems.* In press. Edited by P. R. Shukla, J. Skea, E. C. Buendia et al. Available at: www.ipcc.ch/srccl/.

Jayanegara, A., Sarwono, K. A., Kondo, M. et al. (2018). Use of 3-nitrooxypropanol as feed additive for mitigating enteric methane emissions from ruminants: A meta-analysis. *Italian Journal of Animal Science*, 17, 650–656.

Johnson, J. A., Runge, C. F., Senauer, B., Foley, J. and Polasky, S. (2014). Global agriculture and carbon trade-offs. *Proceedings of the National Academy of Sciences*, 111, 12342–12347.

Lim-Camacho, L., Crimp, S., Ridoutt, B. et al. (2016). *Adaptive Value Chain Approaches: Understanding Adaptation in Food Value Chains.* Australia: Commonwealth Scientific and Industrial Research Organisation (CSIRO). Available at: https://research.csiro.au/climatesmartagriculture/wp-content/uploads/sites/248/2019/07/CSIRO_AVC_FInal-Report_v1.4_single.pdf.

Lin, B. B., Perfecto, I. and Vandermeer, J. (2008). Synergies between agricultural intensification and climate change could create surprising vulnerabilities for crops. *BioScience*, 58, 847–854.

Lipper, L., Thornton, P., Campbell, B. M. et al. (2014). Climate-smart agriculture for food security. *Nature Climate Change*, 4, 1068–1072.

Mbow, C., Rosenzweig, C., Barioni, L. G. et al. (2019). Food security. In P. R. Shukla, J. Skea, E. C. Buendia et al., eds., *Climate Change and Land: An IPCC Special Report on Climate Change, Desertification, Land Degradation, Sustainable Land Management, Food Security, and Greenhouse Gas Fluxes in Terrestrial Ecosystems.* In press. Available at: www.ipcc.ch/site/assets/uploads/sites/4/2020/02/SRCCL-Chapter-5.pdf.

McKeon, G. M., Stone, G. S., Syktus, J. I. et al. (2009). Climate change impacts on northern Australian rangeland livestock carrying capacity: A review of issues. *The Rangeland Journal*, 31, 1–29.

Meyer, R., Cullen, B. R., Johnson, I. R. and Eckard, R. J. (2015). Process modelling to assess the sequestration and productivity benefits of soil carbon for pasture. *Agriculture, Ecosystems & Environment*, 213, 272–280.

Millar, G. D. and Badgery, W. B. (2009). Pasture cropping: A new approach to integrate crop and livestock farming systems. *Animal Production Science*, 49, 777.

Miller, C. J., Howden, S. M. and Jones, R. N. (2010). Intensive livestock industries. In C. Stokes and S. Howden, eds., *Adapting Agriculture to Climate Change: Preparing Australian Agriculture, Forestry and Fisheries for the Future*. Clayton, Victoria: Commonwealth Scientific and Industrial Research Organisation (CSIRO) Publishing, pp. 171–185.

Moore, J. L., Howden, S. M., McKeon, G. M., Carter, J. O. and Scanlan, J. C. (2001). The dynamics of grazed woodlands in southwest Queensland, Australia, and their effect on greenhouse gas emissions. *Environment International*, 27, 147–153.

Naumann, G., Alfieri, L., Wyser, K. et al. (2018). Global changes in drought conditions under different levels of warming. *Geophysical Research Letters*, 45, 3285–3296.

Olsson, L., Barbosa, H., Bhadwal, S. et al. (2019). Land degradation. In P. R. Shukla, J. Skea, E. C. Buendia et al., eds., *Climate Change and Land: An IPCC Special Report on Climate Change, Desertification, Land Degradation, Sustainable Land Management, Food Security, and Greenhouse Gas Fluxes in Terrestrial Ecosystems*. In press. Available at: www.ipcc.ch/srccl/chapter/chapter-4/.

Porter, J. R., Xie, L., Challinor, A. J. et al. (2014). Food security and food production systems. In C. B. Field, V. R. Barros, D. J. Dokken et al., eds., *Climate Change 2014: Impacts, Adaptation, and Vulnerability. Part A: Global and Sectoral Aspects. Contribution of Working Group II to the Fifth Assessment Report of the Intergovernmental Panel on Climate Change*. Cambridge: Cambridge University Press, pp. 485–533. Available at: www.ipcc.ch/site/assets/uploads/2018/02/WGIIAR5-Chap7_FINAL.pdf.

Rivera-Ferre, M. G., López-i-Gelats, F., Howden, M., Smith, P., Morton, J. F. and Herrero, M. (2016). Re-framing the climate change debate in the livestock sector: Mitigation and adaptation options. *Wiley Interdisciplinary Reviews: Climate Change*, 7, 869–892.

Scheer, C., Rowlings, D., Firrell, M. et al. (2017). Nitrification inhibitors can increase post-harvest nitrous oxide emissions in an intensive vegetable production system. *Scientific Reports*, 7, 43677.

Smith, P., Nkem, J., Calvin, K. et al. (2019). Interlinkages between desertification, land degradation, food security and GHG fluxes: Synergies, trade-offs and integrated response options. In P. R. Shukla, J. Skea, E. C. Buendia et al., eds., *Climate Change and Land: An IPCC Special Report on Climate Change, Desertification, Land Degradation, Sustainable Land Management, Food Security, and Greenhouse Gas Fluxes in Terrestrial Ecosystems*. In press. Available at: www.ipcc.ch/srccl/chapter/chapter-6/.

Stokes, C. and Howden, M., eds. (2010). *Adapting Agriculture to Climate Change: Preparing Australian Agriculture, Forestry and Fisheries for the Future*. Clayton, Victoria: Commonwealth Scientific and Industrial Research Organisation (CSIRO) Publishing.

Stokes, C., Howden, S. and Ash, A. (2010). Adapting livestock production systems to climate change. *Recent Advances in Animal Nutrition*, 2009, 115–133.

Thomas, D. T., Lawes, R. A., Descheemaeker, K. and Moore, A. D. (2014). Selection of crop cultivars suited to the location combined with astute management can reduce crop yield penalties in pasture cropping systems. *Crop and Pasture Science*, 65, 1022.

Thornton, P. K. and Herrero, M. (2010). Potential for reduced methane and carbon dioxide emissions from livestock and pasture management in the tropics. *Proceedings of the National Academy of Sciences*, 107, 19667–19672.

Verchot, L. V., Van Noordwijk, M., Kandji, S. et al. (2007). Climate change: Linking adaptation and mitigation through agroforestry. *Mitigation and Adaptation Strategies for Global Change*, 12, 901–918.

Vermeulen, S. J., Campbell, B. M. and Ingram, J. S. I. (2012). Climate change and food systems. *Annual Review of Environment and Resources*, 37, 195–222.

Waghorn, G. C. and Hegarty, R. S. (2011). Lowering ruminant methane emissions through improved feed conversion efficiency. *Animal Feed Science and Technology*, 166, 291–301.

Webb, L. and Whetton, P. (2010). Horticulture. In C. Stokes and S. Howden, eds., *Adapting Agriculture to Climate Change: Preparing Australian Agriculture, Forestry and Fisheries for the Future*. Clayton, Victoria: Commonwealth Scientific and Industrial Research Organisation (CSIRO) Publishing, pp. 119–136.

Webb, L., Dunn, G. M. and Barlow, E. (2010). Winegrapes. In C. Stokes and S. Howden, eds., *Adapting Agriculture to Climate Change: Preparing Australian Agriculture, Forestry and Fisheries for the Future*. Clayton, Victoria: Commonwealth Scientific and Industrial Research Organisation (CSIRO) Publishing, pp. 101–118.

Ziska, L. H., Bunce, J. A., Shimono, H. et al. (2012). Food security and climate change: On the potential to adapt global crop production by active selection to rising atmospheric carbon dioxide. *Proceedings of the Royal Society B: Biological Sciences*, 279, 4097–4105.

Mining, Metals, Oil and Gas

20

Mining, Metals, Oil and Gas

MICHAEL H. SMITH, JANE HODGKINSON, ALAN PEARS AND
PETER STASINOPOULOS

Executive Summary

Mining, minerals and metals production and recycling contribute >12% of global green-house gas emissions, and the oil, coal and gas sector contributes over 50%.

A transition to a low-carbon economy will increase mineral commodity demands by up to 10-fold by 2050. Improving the quality of lives in developing countries requires affordable, sustainable low-carbon buildings for 9 billion people by 2050, further increasing mineral and metal resource demands. Mining mineral ores is critical to manufacturing low-carbon technologies including LEDs, hybrid/battery-powered transport vehicles, solar photovoltaic (PV) panels, wind turbines, batteries, fuel cells, cogeneration systems, carbon capture and storage and nuclear power.

Successful transition by the mining, minerals processing and metals sectors will ensure they can sustainably provide much-needed resources. Many technologies needed for the low-carbon economy are more metal-intensive compared to traditional fossil fuel approaches of delivering equivalent energy services. The projected increase in demand provides a major business opportunity for the sector, in turn providing a driver for the required investment.

Investment is needed primarily to move towards low-carbon mining and improved efficiency and carbon footprint reduction of metals recycling.

Low-carbon resource extraction can take advantage of climate mitigation measures such as on-site solar PV panels, wind-generated power and batteries, which are becoming more economic, and the use of power purchase agreements and energy performance contracts for renewable energy; in turn, the cost of energy to the sector will decrease.

In the oil and gas sector, a reduction in flaring, venting and fugitive emissions, combined with transitioning oil and gas companies into 'low-carbon energy services' via investing in low-carbon energy supply and storage technologies, will provide environmental and economic benefits.

Adaptation of the sector to cope with extreme weather events is critical to ensuring the materials can be delivered to low-carbon technology producers. Extreme weather can disrupt supply chains and resources markets, so measures to maintain supply are essential. Open-cut mines are vulnerable to extreme rainfall rendering them unproductive. The sector requires significant volumes of water, which is problematic during extreme drought.

Heat stress, fire and smoke hazards are a significant issue at many mines, exacerbated by global warming. Reducing exposure to such risks through an integrated adaptation–mitigation approach reduces operational, maintenance and insurance costs.

Adaptation–mitigation strategies can have co-benefits, improving the business case for investment. This chapter reviews tools to help the sector simultaneously achieve both climate mitigation and adaptation cost-effectively.

20.1 Introduction

As the world moves away from fossil fuels and takes up low-carbon technologies, resources to build the technology will increase demand for many mineral and metals commodities by up to 10-fold by 2050 (Vidal et al. 2013; Speirs et al. 2014; World Bank 2017; Hodgkinson and Smith 2018). Coupled with this, developing countries are aiming to sustainably improve living conditions for 9 billion people by 2050 (UNEP 2011), while the population of the world continues to grow, further increasing metals demands.

While the concept of switching to battery- and hydrogen-powered vehicles or wind-generated power, for example, is considered a 'green' option, the production of the infrastructure and equipment requires mining and processing. The mining, minerals and metals processing and recycling industries produce greenhouse gas emissions. Therefore, the way in which the sector meets future resource demands must be considered seriously, to ensure the industry can provide materials designed to reduce a problem, and, in doing so, not inflate it.

This chapter discusses how the next mineral and metals mining boom is likely to be driven by strong action on climate, as the low-carbon economy transition takes place. It also considers how that boom provides a number of challenges and opportunities for the sectors and how oil and gas companies have the opportunity to broaden their portfolios into renewable energies and energy storage technologies to provide zero-emission 'energy services' (Pinkl 2019). The chapter also presents how the mining, oil, gas and petrochemicals industries may take on adaptive strategies to make them more resilient to climate extremes (Smith 2013a, 2013b), to ensure reliable production and delivery to low-carbon technology manufacturers. Barriers to change may stem from a lack of integration of adaptation–mitigation measures within companies, and while minimised capex (capital expenditure) may also prevent suitable investment, the sector's investment cycles can be heavily influenced by fluctuating commodity prices driven by changes in demand. All of this is coupled with a broad lack of information or knowledge about adaptation and mitigation through many parts of the industry. Nevertheless, many new low carbon footprint methods and processes are available or are being researched but, as the industry is globally dispersed, for each locality with its own drivers and challenges, transition may not appear straightforward.

This chapter considers some alternative energy options and re-engineering opportunities available to these industries, such as in the processes of comminution and froth flotation, which are presently energy intensive. Examples are provided of activities already taking place where win–win opportunities have been found to reduce risk and improve productivity. The role that policy-makers play will be very important for ensuring that innovative practice is encouraged and the development of the sector is sustainable within, and as a critical supplier to, the low-carbon economy.

20.2 A Driver for the Next Minerals Mining Boom

Modelling studies show that efforts to achieve decarbonisation by 2050 in line with the Paris Agreement,[1] and also to achieve the UN's Sustainable Development Goals to reduce global poverty, result in demand growing 5–10-fold for many metal commodities from extractive mining and metal recycling (Vidal et al. 2013; Speirs et al. 2014; World Bank 2017; Hodgkinson and Smith 2018). The UN Environment Programme (UNEP) International Resource Panel (IRP) has found that significant new mining discoveries and investments will be needed, combined with a substantial increase in speciality metal recycling, to meet the rising demand for the transition to the low-carbon economy (UNEP 2011). This is because metals are critical inputs to enable manufacturing of the required low-carbon technologies including:

- **Energy efficiency technologies**. Light-emitting diodes (LEDs), energy-efficient appliances, heat pumps and induction stove cooktops all are largely manufactured from metals (in addition to glass ceramics and a range of other chemicals that require mined resources). For instance, LED lighting manufacture requires multiple materials including arsenic, gallium, indium and the rare-earth elements cerium, europium, gadolinium, lanthanum, terbium and yttrium (Wilburn 2012). Most of the world's supplies of these materials are by-products from aluminium, copper, lead and zinc production (Wilburn 2012). Rising demand for LED lighting has already required the opening of new mines (Wilburn 2012).
- **Solar photovoltaic (PV) systems**. A wide range of mined resources are needed to manufacture solar PV panels including: for the solar cells, cadmium, copper, gallium, indium, molybdenum, selenium, silica and tellurium; for the semiconductor chips, arsenic and boron minerals; and for the frames, bauxite (aluminium), coal and iron ore (steel). According to the World Bank, limiting global warming to 2 °C will increase metal demand for solar PV manufacture by 300% (World Bank 2017).
- **Wind power systems**. Turbines and constituents require, for example, iron ore, steel and other metals, alloys and composites, bauxite (aluminium), metallurgical coal, molybdenum and zinc (for galvanising). According to the World Bank, limiting global warming to 2 °C will increase metal demand for wind power systems manufacture by 250% (World Bank 2017).
- **Electric and hybrid electric transport vehicles**. A wide range of metals and other resources are needed to manufacture hybrid cars, including: for the body of the car,

---

[1] *Paris Agreement Under the United Nations Framework Convention on Climate Change*, opened for signature 16 February 2016. Available at: https://unfccc.int/process-and-meetings/the-paris-agreement/the-paris-agreement.

chromium, manganese, molybdenum, bauxite (aluminium), iron ore and thermal coal; for wiring and circuitry, copper, gold, platinum and tungsten; and for batteries, cadmium, cobalt, lead, lithium, nickel and rare-earth oxides. Electric vehicles also require batteries that need lithium, lead, zinc, graphite and numerous rare-earth metals. The World Bank finds that limiting global warming to 2 °C will increase metal demand for batteries by 1200% by 2050 above business as usual (World Bank 2017).

As a result, industry leaders and academic experts predict that strong action between now and 2030 to achieve the Paris Climate Change Agreement and UN's Sustainable Development Goals will drive the next minerals mining boom (Hodgkinson and Smith 2018). For instance:

- Tom Butler, head of the International Council on Mining and Metals, stated in 2018 that 'demand for renewable energy and batteries will drive a new resources boom for copper and lithium' (Latimer 2018). Therefore, mining and resources companies can strategically position themselves for these positive and broad growth markets (ICMM 2012a).
- In 2019, Swiss-based mining giant Glencore, one of the major global coal-mining companies, announced that it would cap coal production at current levels, and instead focus on new mining investments in commodities including copper, cobalt, nickel, vanadium and zinc, in recognition of the fact that it is these metals that will be increasingly needed to manufacture clean technologies to meet the Paris Climate Change Agreement targets (Glencore Ulan Coal n.d.).

Speirs et al. (2014) reported that, in the period 1971–2011, low-carbon technologies contributed to significant increases in demand for many metals, including gallium (increased by 1600%), tellurium (1000%), lithium (800%) and cobalt (400%). In 2011, Deng et al. calculated that achieving 100% renewable energy by 2050 would result in requirements for approximately 3200 million tonnes of steel, 310 million tonnes of aluminium and 40 million tonnes of copper to build the latest generations of wind and solar facilities (Deng et al. 2011). This corresponded to a 5–18% annual cumulative increase in the global production of these metals for the next 40 years (Deng et al. 2011) equivalent to up to a 10-fold increase in demand for many metal commodities. Required solar and wind facilities were reported to need 15 times more concrete, 90 times more aluminium, and 50 times more iron, copper and glass than fossil fuel or nuclear energy plants. Other ferrous, base and minor metals would also be needed. The likely effect of a transition to high penetration of electric vehicles would be a 2–7-fold increase in demand for lithium by 2030 (Speirs et al. 2014) and a 10-fold increase in lithium demand per annum by 2050 compared to 2012 levels (Mohr et al. 2012). In addition, studies suggested a 50–600% per annum increase in demand by 2030 for different types of specialist metals such as gallium (400–600%), neodymium (150–380%) and indium (300%) (Speirs et al. 2014).

Many studies suggest that demand can only be met through increased metals recycling. Recycling, compared with producing metal from a new resource, reduces intense energy, water and carbon dioxide (CO_2) production. Recycling diversifies metal supply chains' increasing resilience to supply disruption from extreme weather events. Globally, an increase in metal recycling will provide mitigation and adaptation opportunities (Ditze and Scharf 2008). While base metal recycling is becoming more common, rates of recycling speciality metals in 2011 were just 1% (UNEP 2011). The UNEP IRP has published reports that provide evidence of significant technical potential for further improvement in recycling levels (UNEP 2011, 2013). There is an opportunity for resources companies to diversify and utilise their metallurgical expertise to expand into the metal recycling sector.

While the 'carbon bubble' literature shows that, to meet the 2 °C Paris Agreement target, 'a third of oil reserves, half of gas reserves and over 80 per cent of current coal reserves . . .' should remain unused (McGlade and Ekins 2015: 187), these studies ignore the fact that the resources are expected to continue being used as feedstocks to meet growing demands. Fossil fuels are currently used widely for non-energy requirements including a significant percentage of the world's demand for chemicals and plastics. Global demand is increasing at over 10% in many Asian countries. If global demand for chemicals was to be met by biological sources this would use up a significant percentage of the world's arable land and freshwater. Research, development and deployment (R&D&D) is needed to enable the production of zero-carbon footprint chemicals and plastics from fossil sources. The use of oil, gas and coal for chemical and plastic production is well understood (Nalbandian 2014). Other critically important materials for the low-carbon economy include graphite and graphene. Graphene can be produced from coal and methods to also produce graphite from coal is the subject of research (Wu et al. 2013; Ye 2013). Critical elements for low-carbon technologies, including rare-earth elements, are also being sought from coal and coal waste in the USA, Russia, China and Australia (Hower et al. 2016; Hodgkinson and Grigorescu 2019; Zhao et al. 2019). The remainder of this chapter focuses on low-carbon resilient pathways for the metals supply chain. Both offshore and onshore oil and gas wells, infrastructure and pipelines are exposed significantly to damage from extreme weather events (Climate Commission 2013).

20.3 Decarbonising the Metals Supply Chain

Decarbonising the mining and metals supply chain is technically possible (Kirk and Lund 2018) and financially of interest. It will help to lower energy input costs and improve resilience, becoming less dependent on volatile diesel fuel supply chains that can be disrupted by extreme weather events. Energy input costs for mining companies have increased over the last decade in major mining countries such as Australia, Chile and South Africa, due to a combination of rising diesel prices and declining ore grade concentrations. The amount of electricity used in mining sectors at remote sites can be significant and can account for up to 30% of total mining costs. In 2015, the Australian resources sector used 28.2 TWh (terawatt-hours), or 11% of all electricity in Australia, and approximately 45% (12.4 TWh) of that consumed was off-grid (Department of Industry, Innovation and Science 2015). Improving energy security through improved efficiency and the introduction of renewable energy and improved energy storage, especially for

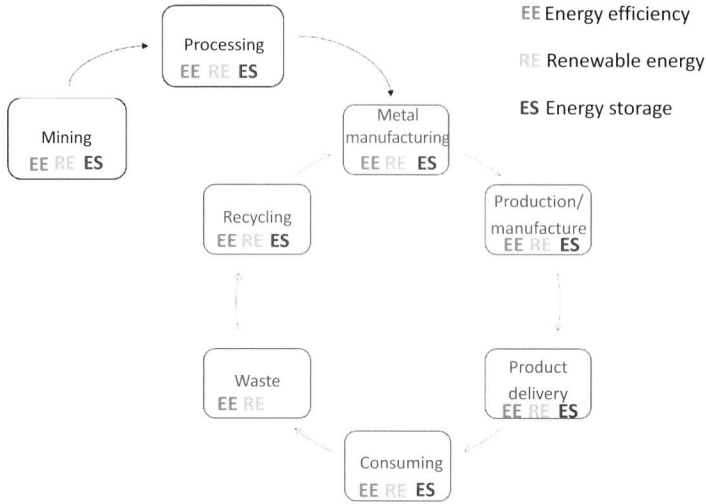

Figure 20.1 Decarbonising the mining and metals supply chains through an integrated approach using innovations in energy efficiency (EE), renewable energy (RE) and energy storage (ES).

Source: Courtesy of the authors. For a colour version of this figure, please see the colour plate section.

off-grid mining and mineral processing, is now being considered for most new mining and mineral processing projects, as companies try to cut operational energy costs (Figure 20.1). The use of renewable energy to process metals is growing and electricity bills at refineries are expected to be cut by as much as 30% (Lord 2020).

20.3.1 Energy Efficiency Improvement Opportunities

The financial benefits for mining, mineral processing and metal manufacturing companies investing in energy efficiency improvements at existing and new mines and manufacturing plants are well documented. Companies can reduce energy costs by 10–20% below business-as-usual trends in most existing mines (ClimateWorks Australia and DRET 2012) and manufacturing plants through energy efficiency. Companies may be able to achieve a 30–50% reduction in energy intensity at new mines (Pokrajcic and Morrison 2008; Smith 2013b) and metal manufacturing plants and supply chains (Cullen et al. 2010). The most effective ways in which mining and mineral processing can reduce energy intensity are improving the energy efficiency of the materials movement, ventilation, comminution, froth flotation and minerals separation, general equipment and mine site accommodation and facilities, as well as transportation (see Box 20.1). The energy intensity of most industrial processes is found to be at least 50% greater than that of the theoretical minimum determined by the laws of thermodynamics (Worrell et al. 2009). Ways to reduce energy intensity in metal manufacturing are covered in Chapter 16.

Box 20.1 **Strategies to Improve Energy Efficiency of Mining and Mineral Processing**

The major ways mining and mineral processing are able to reduce energy intensity and improve energy efficiency are in materials movement, ventilation, comminution, froth flotation and minerals separation, general equipment and mine site accommodation and facilities. Improving energy efficiency in these areas will cut energy costs and improve energy productivity (Stadler et al. 2015). Powering mine and mineral processing sites with renewable energies makes production more cost-effective, and yields adaptation co-benefits such as improved water efficiency.

Materials Movement Energy Efficiency Opportunities

For the mining sector, one of the largest areas of energy efficiency opportunity lies in materials movement. Opportunities include:

- improving the fuel efficiency of haul trucks (Department of Resources, Energy and Tourism n.d.);
- improving the efficiency of electric draglines currently used widely to remove overburden (Rathmann and Heuer 2007). Rio Tinto's Blair Athol coal mine refurbished its dragline and achieved a 25% energy efficiency saving (Rathmann and Heuer 2007); and
- using overburden slushers that use two winches to drag large buckets across overburden, instead of electric draglines, and are 60% more energy efficient (Bosmin n.d.).

When able to adapt or upgrade open-cut or underground mine materials movement equipment, significant energy cost savings can be achieved through:

- investing in lightweight, hybrid diesel electric trucks that improve fuel efficiency – one of the benefits of hybrid trucks is that they can recover energy (up to 10% is claimed) by using their regenerative braking, which diesel haul trucks are traditionally unable to do;
- investing in the new generation of fully electric underground drilling, scooping and materials movement vehicles that are more productive and energy efficient, and emit no diesel fumes, enabling underground ventilation energy costs to be reduced by up to 90% (Vella 2013); and
- using hybrid or fully electric trolley trucks/underground vehicles that use tram power lines to access or feed in electricity, saving fuel. Their regenerative braking enables some energy to be recovered on descent into the mine.

Other options include use of the following:

- conveyor-belt-based systems of materials movement: the Clausthal University of Technology's study on 'Energy-Efficient Conveyor Technology and Climate Protection' concluded that 'conveyor belts achieve a much better energy scorecard, requiring only about 20% of the energy needed by heavy-duty trucks' (Tudeshki 2009: 31); conveyor belts are being used more widely as they can be incorporated into mobile, in-pit crusher/conveyor systems; and
- mobile in-pit crusher/conveyor systems, which are very energy efficient: as Norgate and Haque from Australia's CSIRO (Commonwealth Scientific and Industrial Research Organisation) found, 'Potential greenhouse gas savings of 100 000–133 000 tonnes of CO_2 per annum were

reported for these IPCC [in-pit crusher/conveyor] systems compared to a conventional shovel/truck operation' (Norgate and Haque 2010).

Comminution: Crushing and Grinding Energy Efficiency Opportunities

Comminution grinds particles to a small enough size to allow separation of the valuable mineral from ore bodies by using, for example, froth flotation. Comminution can be as little as 1% energy efficient, and thus contributes significantly to the energy intensity of mineral processing. Comminution energy consumption is forecast to increase significantly over time, especially for gold and copper mining as ore grades continue to decline, if the following steps are not implemented:

- **Smart blasting, ore sorting and pre-concentration of ores before crushing and grinding**. This means less material is fed into the comminution process, so less material has to be handled per unit of product.
- **Optimising the selection of target product size(s)**. This is critical, as size affects comminution energy intensity.
- **Optimising particle size reduction in grinding**. Screens and filtering devices can be applied to achieve greater consistency of 'feed' for the grinding process to reduce energy losses in the grinding process.
- **Use of flexible comminution circuits**. An ore body with varying concentrations, when passed through one standard comminution circuit, must lead to inefficiencies. If a mine has a variable ore body, investing in more than one crusher and grinding mill will more efficiently process relatively high- and low-grade mineral ores separately.
- **Use of more efficient comminution equipment**. Studies show energy intensity can be reduced by up to 30% simply by using the latest, more energy-efficient, comminution equipment.
- **Improving the energy efficiency of motor systems used to drive crushing and grinding mills with variable-speed drives**. For example, Barrick Gold Corporation achieved 4.4% energy efficiency improvements in their motor systems (Buckingham et al. 2011).

Froth Flotation Energy Efficiency Opportunities (Metal Ore Mining and Processing)

Opportunities exist to maximise froth flotation energy efficiency in existing plants (Table 20.1). The design of new froth flotation systems can be also improved in the design phase. Design, layout and equipment selection should all be performed with the view to make the most energy-efficient circuit possible (Murphy 2013).

Ventilation and Air-Conditioning Energy Efficiency Opportunities: Relevant for All Types of Mining Subsectors

Ways in which energy efficiency can be improved and diesel-related fumes can be reduced include:

- investment in electric-powered underground equipment to replace diesel-powered underground mining machines (Government of Canada, Natural Resources Canada 2013);
- increasing motor-driven fan-system energy efficiency to improve ventilation energy efficiency;

Table 20.1. *Energy-saving opportunities in an existing flotation plant*

Process	Maintenance	Re-engineering
Feed size optimisation	Correct drivetrain alignment	Alternative flotation mechanisms
Pulp density optimisation	Correct drive-belt tensioning	Pipe rerouting
Airflow rate optimisation	Maintaining lubrication schedule	Sump redesign
Impeller speed optimisation		Resizing of pumps
Revision of control philosophy	Wear component replacement	Assess high-efficiency motors
		Assess variable-speed drives

Source: Murphy (2013).

- reducing the need for air conditioning in underground mines by providing staff with lightweight personal cooling units (as used by athletes competing in endurance sports);
- reducing the numbers of personnel in underground mines through implementation of remotely controlled operations to reduce demand for air conditioning;
- recovery of some of the potential energy of cold water, which flows down into underground mines, for air-conditioning use (Whillier 1977).

More detailed coverage of energy efficiency opportunities is available: DRET (2010, 2011), Hargroves et al. (2014) and EEC (2016).

20.3.2 Renewable Energy and Energy Storage

Mining sites utilise electricity for grinding, some on-site transportation (conveyors, electric haul trucks), air conditioning and ventilation and typical staff housing electricity-consuming appliances. Renewable energy and on-site storage can now meet these on-site energy needs, reduce greenhouse gas emissions and energy costs and improve energy security, which is valuable for mines in remote locations that may be affected by extreme weather events. The costs for batteries and renewable energy have continued to fall steeply since 2008 and are forecast to fall further. The International Renewable Energy Agency (IRENA) expects there to be a further reduction of 60% by 2025 driven by growing supply chains and economies of scale. Some R&D&D funding bodies, such as the Australian Renewable Energy Agency (ARENA), have funded projects that aim to reduce costs further. For example, ARENA funded the development of modular drop-in and removable solar power systems, which are now commercially available, specifically to meet the needs of remotely located mining companies. Innovations such as these suit the need for fast adaptation that can provide energy efficiently as the mine grows and develops over its lifetime (Engineers Australia 2016).

Renewable energy firms are finding solutions to overcome the barrier that upfront costs present to mining companies wishing to adopt renewable energy. One such option is the power purchase agreement (PPA) under which the mining firm pays fixed annual fees once the project begins to produce power. This provides a critical enabler that allows renewable energy to compete with diesel power systems. For instance, the USD 40 million off-grid solar power station and battery storage facility at Sandfire Resources' DeGrussa Copper Mine in Western Australia has a 6-year PPA with upfront costs of less than AUD 1 million. This combination of modular solar PV and battery systems with PPAs is predicted to become the mainstream vehicle for mining companies in Australia to utilise renewable energy. Some financiers are offering investments in solar PV systems with flexible-term PPAs of as little as 5 years, with the possibility of extension. If the contract is not renewed, the PV system is shifted to another place (Boyle 2016). This is helping to enable significant uptake of renewables in mining in several parts of the world. For instance, in northern Chile and southern Peru, 11 of 13 large solar PV facilities provide electricity for the mining industry (IEA 2017). Evans and Peck (2011) reported that wind power has been cost-competitive in many mining regions. In 2011, Australian electricity produced from solar PV systems was shown to be more cost-effective than electricity from diesel (Bearsley 2016). While the economics fluctuate, infrastructure options have developed further, providing benefits to adopting solar PV systems into micro and mini grids and hybrid power systems, especially valuable to remote communities (ARENA and PowerWater 2019). More recently, production of small energy loads from new installations of PV systems compared with solar towers, diesel and wind turbines is shown to be economically effective (Abu-Hamdeh and Alnefaie 2019). Globally, many mining and mineral processing companies are utilising renewable energy, and several mining and renewable energy websites such as 'Energy and Mines'[2] and 'The Energy Platform'[3] indicate that more companies are ready to do so.

As costs of renewable energy and batteries continue to fall and as alternative financing agreement options become available, the potential grows for improving uptake of renewable energy to meet the energy needs of metal manufacturers. To date, this has already led to 50-MW (megawatts) to >100-MW renewable solar PV projects being built to meet metal manufacturers' energy needs in Townsville and Broken Hill in Australia. The Tesla Motors battery factory in Nevada, USA, is expected to run entirely on a combination of power from a nearby wind farm, geothermal electricity and solar energy (Klender 2019). Sanjeev Gupta, CEO of the GFG Alliance group of companies, is investing heavily in solar PV technology and batteries to power the Arrium group's steel mills in Australia (Evans 2018). In Sweden, the AUD 80 million HYBRIT project seeks to create zero-carbon steel with hydrogen made by renewable energy. Sweden holds 90% of the EU's iron ore reserves and has committed to invest in R&D&D zero-carbon steel (BZE 2018).

Hydroelectric power (HEP or hydropower) has long been used to power some metals manufacturing plants. It was one of the main sources of energy for the first 'industrial revolution' over 500 years ago and again later in the early 1800s. In more recent times,

---

[2] energyandmines.com.   [3] www.clydeco.com/blog/energy.

hydropower has been used widely to manufacture aluminium through electrolysis (Reuters 2019). Methods to use low-carbon energy to power electrolysis to produce other metals is rapidly being developed, to produce iron (Ahman et al. 2018), copper, nickel (Ivanhoe Mines Ltd 2018) and titanium (Chen et al. 2000). Hydropower and renewable-energy-generated electricity can also power arc furnaces. Arc furnaces are used in the production of steel, silicon, calcium carbide, ferroalloys and other non-ferrous alloys (BZE 2018).

Many metal manufacturing processes do not need just electricity. They also require medium- to high-temperature process heat input. Such heat may be produced without generating CO_2 emissions using renewable energy technologies (Eglinton et al. 2013; Lovegrove et al. 2015); such technologies are not commercially competitive at medium to high temperatures, but greater R&D&D and other policies can bring down costs.

Finally, falling costs for renewable energy and storage, via batteries and pumped storage hydro, now enable legacy mining sites to be given a second life as clean energy generators. For instance, BHP Billiton hired Rocky Mountain Institute in 2017 to assess this potential for their North American sites. The Rocky Mountain Institute identified 0.5 GW (gigawatts) of potential. For most of the sites, because of their location, solar PV technology emerged as the largest opportunity, with a few being well suited for wind development (RMI 2017).

20.4 Mitigation Options in the Oil and Gas Sector

20.4.1 Reducing Natural Gas Flaring

Investment in the reduction of natural gas flaring offers a significant mitigation, development, public and environmental health 'win' opportunity. Flaring, venting and fugitive emissions cause huge economic loss to a company and prevent the resource from being used for energy and transport needs (Global Methane Initiative 2011). Further, reduction of flaring provides significant public and environmental health benefits from improved air quality (IRIN 2007). Studies have shown significant economic, health and climate change mitigation benefits from investing in reducing flaring for both Russia and Nigeria (Carbon Limits 2013). Flash gas compressors process gas that would have been otherwise flared, helping to reduce emissions and increase value from the resource (Budde et al. 2014). Chevron has identified additional activities that may eliminate 80% of flares (Chevron Australia 2009). Many countries and companies joined the World Bank's global 'zero gas flaring by 2030' initiative to collectively share knowledge of best practice to reduce flaring. The low-carbon scenario of not flaring not only appreciably reduces emissions but also generates higher economic returns from oil and gas resources. The low-carbon emission pathway is financially self-sustaining, as the economic returns greatly outstrip capital investment costs for most years to 2035 (Cervigni et al. 2013) (Figure 20.2).

20.4.2 Reducing Fugitive Emissions

Natural gas is often described as a 'transition' fuel between coal and renewables. However, new natural gas combined-cycle power plants reduce climate impacts only when methane

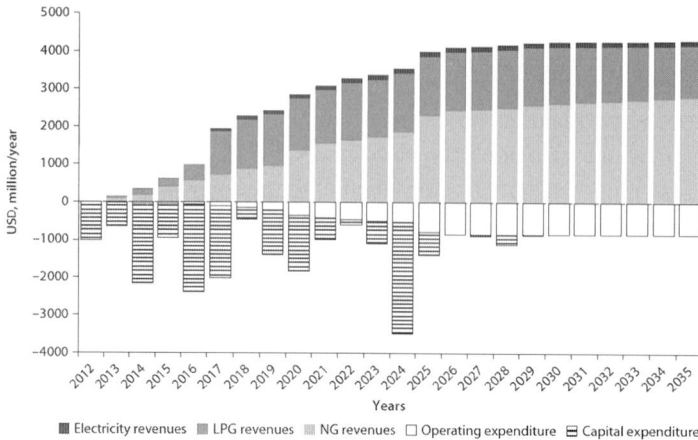

Figure 20.2 Revenues and costs for low-carbon transition for Nigerian oil and gas sector.
Note: NG = natural gas; LPG = liquid petroleum gas.
Source: Cervigni et al. (2013).

leakage remains under 3%. Leakage must remain under 1% to have an immediate climate benefit (Caulton et al. 2014). Methane is a potent greenhouse gas, converting to CO_2 after several years. Methane leakage in unconventional natural gas fields and processing plants can be cut by up to 80% (Dittrick 2012). These opportunities lie mainly in the design and construction phase and at the end of life. For instance, if concrete capstones of unconventional gas wells are built of insufficient integrity, this can allow methane leakage over future centuries. Regulation with effective compliance enforcement is needed to ensure decommissioning will prevent all future leakage.

20.4.3 *Investing in Carbon Capture and Storage (CCS) to Enhance Oil Extraction*

Some technologies and expertise to perform carbon capture and storage (CCS) have been derived from innovations made by the oil and gas sector. Carbon dioxide is captured and pumped back into deposits to enhance oil recovery, some of which becomes trapped and affectively sequestered underground. This is economically feasible under suitable conditions in some oil and gas fields, and saline formations. Investing in CCS provides an opportunity to help cover the cost of CO_2 capture for oil recovery.

20.4.4 *Transitioning from Natural Gas to Biogas*

Biogas (biomethane or biopropane) is produced and captured from waste water, landfill, agricultural or forestry waste sources, often situated near end users. Biogas can be injected into the existing natural gas distribution network. Biogas potential is significant (Energy Networks Australia et al. 2017). In Canada, for example, an estimated 1300 billion cubic feet of biogas could be produced annually, representing approximately 50% of domestic gas consumption (Canadian Gas Association 2014), and in the UK it could meet 30–50% of natural gas demand (National Grid 2016).

20.4.5 Transitioning from Natural Gas to Creating Hydrogen Gas

Hydrogen is seen as a major energy carrier in the future. If produced by zero-carbon energy sources, hydrogen has the potential to be a zero-emissions carrier to contribute to mitigating climate change. This is covered in Chapter 8.

20.4.6 Key Role: Low-Carbon Chemical and Plastic Feedstocks

The most valuable end products for the oil and gas sector are value-added petrochemical products and speciality chemicals. Per unit of oil and gas, it is possible for the sector to earn up to 10 times more from processed chemicals than from simply selling oil and natural gas.

A low-carbon economy will continue to need chemicals and plastics. To produce them solely from land-based biomass sources will use significant arable land and water, making food production almost impossible. While new methods to produce plastics are being explored (Section 20.4.8), it is essential that chemicals and plastics can also be produced by a net zero-carbon petrochemical sector, within which carbon omissions are offset through sequestration and planting options, or within which closed-loop factories are introduced. The International Energy Agency has shown that, by 2050, the petrochemical sector can cut energy demand and greenhouse gas emissions by 40% through improved catalysts alone (IEA 2013: 5). Energy savings and greenhouse gas reduction could also be achieved.

20.4.7 Combining Feedstocks to Create Polymers That Store CO_2

A modern approach to polymer processing is to combine traditional (oil, gas, coal) feed-stocks with CO_2 to synthesise polymers and high-value chemicals at low temperatures and pressures (Global CCS Institute and Parsons Brinckerhoff 2011). The resulting polymer is versatile as it may be hard, soft, transparent or opaque.

20.4.8 Converting Methane Emissions into Thermoplastic Products

Newlight Technologies has developed proprietary chemical-processing methods that capture waste methane and combine it with air to form a resin ('AirCarbon'), essentially sequestering the carbon in a thermoplastic material that performs like a petroleum-based plastic (Zhou 2015).

20.4.9 Diversifying into the Renewable Energy and Energy Storage Sector

Most oil and gas major companies acknowledge that the zero-emission transportation sector transition will mean that oil consumption will peak and then fall this century. By definition, if natural gas is a transition fuel, then at some point in the future, to achieve a net zero emission economy, society will need to transition from its current large dependency on natural gas to reduce demand for it. So, the best strategy to ensure long-term sustainability for oil and gas sector companies is for them to utilise their profits to diversify into the rapidly growing markets for renewable energy, energy storage and zero-emission vehicles, in order to maintain and increase share value. This is recognised by Royal Dutch Shell, Total, Eni and Equinor, who have embarked on their transition from oil companies to low-carbon energy services supply companies (Pinkl 2019). For instance:

- Italian petroleum major Eni plans to diversify to progressively decarbonise its energy mix including solar projects in Italy, Algeria, Pakistan and Egypt;
- Total, the French oil major, has a formal commitment to 20% low-carbon assets, a fifth of its portfolio in low-carbon businesses, by 2035 (FT 2016). In 2016, the company invested USD 1.1 billion to buy the battery maker Saft Groupe SA, complementing its 2011 purchase of a majority stake in the solar-panel maker SunPower Corp;
- Shell is investing heavily in the Indian renewable energy market, and in 2019 acquired a 20% stake in India's Orb Energy solar provider (Hodges 2019);
- Equinor has committed to 'grow renewable energy capacity tenfold by 2026, developing as a global offshore wind major, and strengthen its industry leading position on carbon efficient production, aiming to reach carbon neutral global operations by 2030' (Equinor 2020).

To conclude: there are many ways that the sector can both cut emissions and strategically position oil and gas companies as energy services and low-carbon chemical and plastics companies, to improve their sustainable competitive advantage in the long term.

20.5 Supply Chain Resilience: Adaptation–Mitigation Co-benefits

As metals production will need to increase substantially to manufacture low-carbon technologies at a speed and scale sufficient to achieve the Paris Climate Change Agreement and UN's Sustainable Development Goals, the metals supply chain will need to improve its resilience to extreme weather events to guarantee supply. Developing and implementing adaptation and mitigation strategies involves upfront costs, so it is important to identify smart strategies that simultaneously improve resilience and reduce greenhouse gas emissions. To better understand and map climate change adaptation–mitigation nexus opportunities thoroughly for the sector, it is first vital to understand the major areas of risk that extreme weather and ongoing climate change pose to the sector. This also provides empirical evidence of why action on climate change needs to be a priority for decision makers in this sector.

The metals supply chain is particularly vulnerable to climate change because it is a capital-intensive sector with remotely situated mines, long supply chains and fixed assets that require significant energy and water resources. For instance, analysed data

(Acclimatise 2010) of mining and mineral processing companies from the Carbon Disclosure Project showed that:

- 51% of mining and mineral processing companies experienced negative impacts from extreme weather events;
- 81% of mining and mineral processing companies' physical assets were exposed to risks by extreme weather events; and
- 53% of mining and mineral processing companies were vulnerable to water stress and restrictions.

The metals supply chain is vulnerable to disruptions due to more intense extreme weather events causing water shortages, reduced rainfall and resulting drought and dust, as well as storms, flood, increased precipitation and rising temperatures (Eschner and Hodgkinson 2014). Specific risks include:

- **disruption of supply chains and transport routes** as mines are often in remote regions, thus requiring long supply chains that can make metal supply chains more vulnerable to extreme weather events;
- **extreme flooding events leading to loss of production in mines and mineral process-ing plants** as flooding of open-cut and underground mines causes significant costs to mining and can make mining impossible during the period of inundation, thus reducing overall productivity and increasing costs for damage repair. For instance, total costs to the mining sector in Queensland, Australia, of flooding events in early 2011 have been estimated at AUD 2.5 billion (World Bank and Queensland Reconstruction Authority 2011), which was greater than the negative impact of the global financial crisis on the Australian mining sector;
- **extreme droughts reducing water availability, posing a risk to production** as mining and mineral processing methods are water intensive, so the challenges of greater water variability and scarcity that climate extremes and climate change pose bring significant investment and operational risks. The mining sector often operates in relatively dry climates and thus is vulnerable to even small changes in water availability, which will worsen as climate changes; and
- **heat stress leading to loss of productivity** from an increase in the number of days per annum over 35 °C. In Australia, Africa and South America, where a significant percent-age of global mining and mineral processing occurs, the number of days over 35 °C is expected to increase significantly by 2030 and more so by 2070 without global action to reduce greenhouse gas emissions (Smith 2013a).

Risk management activities typically use knowledge of risk based on past, current and known factors. Climate adaptation to reduce risk, however, must rely on some factors that are not well constrained, some unknowns, potential projections and a suite of past data embedded with influences of the changes already occurring in the climate system. Future scenario modelling (Greiner et al. 2014) is able to provide a good indication of what may be expected, so vulnerability assessments are possible, to provide risk and hazard reduction planning. Maladaptation can be avoided by selecting options that have considered the needs and activities of others in the mining area and along the value chain in addition to activities that have considered climate mitigation. Adaptation may be physical changes to the infrastructure to reduce physical risk, may require alternative storage sites and changes to methods and practices at the site or it may instead focus on recovery and reduction in downtime through faster recovery time after an event.

In a study that investigated the drivers for adaptation by Australian mining and mineral processing companies, as reported by companies to the Carbon Disclosure Project (2003 to 2014), it was found that their reasons varied over time (Eschner and Hodgkinson 2014). For example, prior to 2010, protection of assets and sustaining production were some of the most popular reasons to adapt, but after 2010, market demand, reputation and consumer expectations or media response ('praise' or 'shame' of specific activities) became the greatest drivers for adaptation. Over the long term, however, from 2003 to 2014, a clear, increasing trend was evident: that both asset protection and sustaining production were drivers for adaptation activities taking place (Eschner and Hodgkinson 2014). Conversely, the main reasons for adaptation not taking place were found to be uncertainties in regulation, the market and in climate change generally, with a reported lack of information on expected changes or impacts (Eschner and Hodgkinson 2014). A more recent assessment of adaptation in mining from a global perspective identified some similar drivers to those in Australia, but also assessed the type of changes being made (Loginova and Batterbury 2019). Incremental adaptation was found to prevail, whereas transformational and transitional adaptation were less popular although active in some places. Other drivers of adaptation activity cited included societal pressure, interactions with Indigenous people and responses to recent events.

20.5.1 Typical Approaches to Adaptation

Although a large proportion of mining and oil and gas companies have yet to perform dedicated climate adaptation, of those that have, the most frequently cited approaches have been internal management activities (Smith 2013a, 2013b). Although a wide range of approaches for climate change management (adaptation or mitigation) are available for both mining and oil and gas companies (Smith 2013a, 2013b), the most frequently cited within the Carbon Disclosure Project (2003–2014) were:

- internal management typically responsible for determining strategies for climate change adaptation and mitigation;
- internal vulnerability assessments and reviews;
- planning – such as planning operations that avoid or reduce impacts, planning response to emergencies and planning to reduce hazards for people and operations;
- adaptation to take advantage of opportunities and maximise mitigation co-benefits;
- enhancement of infrastructure and increased resilience;
- engagement with stakeholders, value chain and public to identify opportunities to adapt;

- engagement with organisations that can provide climate change information to better understand vulnerabilities and expectations and perform vulnerability assessments;
- implementation of natural resource management strategies.

Other strategies include use of past experiences as a guide for decision-making (although this may not consider continuing changes to climate patterns); diversification of business activities including routes, providers or customers; workshops to engage stakeholders and seek feedback; hedging and enhanced insurance cover; and hiring consultants to manage and develop the strategy (Eschner and Hodgkinson 2014).

Leading oil and gas companies are starting to manage their climate-change-related risks through adaptation strategies that exceed present compliance requirements to reduce risks of negative shocks to operational performance, revenues and business continuity (Smith 2013b). Limits to adaptation include water scarcity due to climate change; increasing hurricane intensity and frequency, and sea-level rise and storm surges that negatively impact petroleum refineries and petrochemical plants built next to ports.

20.5.2 Mine and Pit-to-Port Vulnerability Assessments

Many companies perform internal management activities to implement vulnerability assessments and determine strategies for climate change adaptation and mitigation. This is the first step towards effective adaptation. A planned approach requires an assessment of the mine site or value chain's susceptibility, vulnerability and resilience, in addition to cooperative discussions with supply chain members during the process to assess from pit to port where adaptation will be most valuable. Some companies might only perform vulnerability assessments of the mine site itself. If so, it should be borne in mind that bottlenecks or interruptions caused by extreme events may have influences both up and down the value chain, having a flow-on impact to the mine site. Some mining companies have successfully employed an external company or consultant to perform a vulnerability assessment as the first stage in identifying adaptation strategies. Mavrommatis et al. (2019) presented Climate Change Economic Risk in the Mining Industry, which is a multi-risk assessment methodology to select the most cost-effective adaptation from available strategies to reduce hazards caused by climate-change-related extreme weather events.

20.5.3 Regional and Site-Level Modelling

Nelson and Schuchard (n.d.) describe how regional and local modelling has been undertaken by companies to identify both risks and opportunities. For example:

- 'Exxaro used downscaled general circulation models to assess climate change impacts for both operations and communities where employees are located. The study examined both natural climate hazards and inherent vulnerability of existing infrastructure, population, and socioeconomic activities' (Nelson and Schuchard n.d.: 6).
- 'Vale commissioned the National Institute for Space Research of Brazil to assess vulnerability under different climate change scenarios in northern and southern Brazil

and their effects on factors such as water availability and biodiversity' (Nelson and Schuchard n.d.: 6).

20.5.4 Planning Adaptation in Ways That Maximise Mitigation Co-benefits

A vulnerability assessment may identify critical risks and priority areas requiring resilience enhancement, and may also identify opportunities that adaptation can enhance further. Planning adaptation activities needs to consider available capitals such as financial, human, social, natural or physical (Loechel et al. 2010, 2011). By considering expected future changes and available capital, a targeted approach may provide the most suitable level of expenditure and risk reduction. When risks have been identified through a vulnerability assessment, planning to reduce hazards or take advantage of opportunities such as mitigation and energy reduction and, where required, enhancement of infrastructure, can be carried out.

20.5.5 Specific Adaptation Strategies with Mitigation Co-benefits

20.5.5.1 Reducing Fuel Supply Disruption Due to Extreme Weather

Extreme weather events can damage or block roads or rail, making it impossible for diesel supply to get to the mine and mineral processing site. It is possible to improve energy efficiency using alternative power sources such as localised geothermal, solar, wind and renewably powered cogeneration, each of which also provide mitigation options. According to Pike Research (REM 2012),

worldwide investment by the mining industry on renewable energy and energy conservation will reach approximately $8.4 billion by 2016 and nearly $20 billion by 2020. Under a more aggressive scenario, in which the global economy expands more rapidly and policy mandates pertaining to climate change are more robust, spending could reach $15.6 billion by 2016 and $30 billion annually by 2020, the cleantech market intelligence firm forecasts. (*Pike Research quoted in REM 2012*)

The payback times for the renewable energy developments discussed can range from 3.5 to 8 years. However, PPAs are increasingly available to help spread the cost of installing renewable energy. Importantly, modular renewable energy equipment can be relocated if a mine no longer needs it. Additionally, if mines are located near oil and gas projects, gas-powered cogeneration systems are more efficient and have lower emissions than diesel generators.

20.5.5.2 Reducing Drought and Water Shortage Risks by Improving Water Security

Drought can lead to insufficient water supplies for resource companies to produce and process ore and coal; it can also lead to mine site degradation caused by vegetation loss and erosion, and excess dust and extreme heat can increase the risk of fire. Dust suppression requires well-managed water usage. Drought will increase the need for earlier (wet season) water storage local to the mine to reduce pumping needs. Drought-resilient rehabilitation would be valuable at a mine site as it will reduce erosion and dust while the foliage can provide a carbon sink.

Mines can treat and process water for recycling at the mine site. Recycled water may be used to reduce dust, for drilling fluid and for processing of materials, and with each use may need to meet a different water quality standard. A technology called Virtual Curtain has been developed to remove metal contaminants from mine waste water, leaving a semi-solid by-product that is cheaper, less time-consuming and less energy-intensive to treat and dispose of than contaminated water (Douglas et al. 2010). The resulting water is of rainwater quality and can be safely discharged or used at the mine site. The Virtual Curtain (VCL n.d.) uses minerals to trap a variety of contaminants typically found in mine water such as cadmium, arsenic and iron. The process does not need expensive infrastructure or complicated chemistry.

Risk-based approaches that include uncertainty in hydrology and commodity market projections (hydro-economics) may help protect mining projects against drought (Ossa-Moreno et al. 2018). Preparing for drought through storage of water in aquifers provides long-term local water supply security, reducing the need to pump or cart water great distances. A further important point to note in relation to water is the need to adopt an 'adaptation mindset'. This means being ever ready for a flood, even after taking steps to focus on water efficiency during drought. This strategy can greatly assist mining companies operating in remote, dry areas where, when rain falls, it is heavy. For instance, Fortescue Metals Group Pty Ltd has invested significantly in managed aquifer capture, storage and recharge for their mines in north-west Australia (FMG 2019).

Using water more efficiently is also important, and many energy-saving measures (climate change mitigation) discussed in Section 20.4 also have water-saving (climate change adaptation) co-benefits. Water- and energy-saving examples include:

- using conveyor materials movement technologies instead of haul trucks, which is up to 80% more energy efficient and reduces water requirements for dust suppression on haul truck roads;
- smart blasting, pre-concentration and gangue rejection reduces both the amount of energy and water needed to crush and grind ores;
- using energy-efficient comminution and froth flotation technologies, which generally also use less water;
- cleaning and recycling of mine and processing water to reduce the need for purchasing or extracting additional water;
- implementing water-sharing schemes that utilise or provide 'spare' water to other mines or users such as agriculture; and
- reducing water losses through evaporation from mining solution ponds leaching pads by up to half by using impermeable white or reflective plastic covers (ICMM 2012b).

20.5.5.3 Reducing Flood Potential

Site and regional water management can reduce the potential for widespread flooding and long-term downtime. Mines can perform local and regional weather monitoring to ensure on-site storage has been emptied and is ready for the new influx of water, while excess water is safely released or diverted. This can provide extra storage for a dry period, avoiding the need to use excess energy to pump or cart water back to the mine. Keeping pits free from excess water will also reduce the need to use energy for pumping when releasing the water. Water in pits can also become contaminated and require treatment prior to releasing it. Anglo American Metallurgical Coal introduced their 'Rain Immunization Project', for example, to reduce the number of days of production lost to rain and to increase their water-pumping capacity (Hodgkinson et al. 2014: 1670).

Other options for reducing the impact of potential flooding include the following:

- Increasing the height of levées or stream diversions that reduce overland flow into pits allows water to continue flowing through or past the site, preventing the need for pumping water from a pit and saving the energy on pumping and decontaminating pit water.
- Ensuring a mine site is resilient to additional water flow through the process of hotspot mapping may also help a mine adapt and reduce impacts of future events. Using, for example, the CRATER (climate-related adaptation from terrain evaluation results) method (Hodgkinson et al. 2013) provides mines with a visual assessment, using multiple criteria evaluation to aid decision-making, providing prioritisation on what area to spend time and money for reducing risk/improving resilience. This approach can avoid downtime, damage, HSE (health, safety and environment) issues and infrastructure washout. It also allows the mine to produce 'what if' scenarios and identify adaptation most suited to the specific mine, avoiding activities that will unnecessarily use energy, haulage and resources.
- Ensuring stockpiles are accessible during a flood to fill clients' orders will protect reputation, save legal costs and safeguard revenue. From a mitigation perspective, safe storage of stockpiles ahead of an extreme event provides product from a suitably located source as opposed to what may be a more distant location where floods have not occurred, avoiding the need for excessive transport to client. BHP Billiton Mitsubishi Alliance, for example, successfully instigated 'Project Noah', which assured delivery to clients by relocation of stockpiles prior to extreme events (Hodgkinson et al. 2014).
- Reducing susceptibility to flooding may also include strategies such as relocating equipment to areas of the mine less prone to ponding or flooding, and water-sharing schemes that harvest water during floods and store water for times of drought. Improved rail resilience at bridges and through flood-prone areas, preventing the need for long diversions, delays and excess fuel usage, are other measures. Construction of alternative, raised or more resilient roads that may reduce the energy and cost required for repairing roads, in addition to reducing downtime while they cannot be used, are also good strategies.

20.5.5.4 Reducing Cyclone and Extreme Storm Risk

Relocation, adaptation or replacement of infrastructure and equipment found to be in vulnerable areas that may be subject to damage from high winds and lightning strike may be possible; alternatively, real-time weather warnings can be distributed to all mine staff to implement safety procedures, including evacuation procedures. In some mines, remote control of equipment has been employed to reduce hazard exposure. Automated or remote shutdown of some equipment during storms can help prevent damage. Automated systems can also reduce the number of personnel at the site, saving fly-in-fly-out costs, stresses and potential delay issues. Closedown of equipment when mine staff are evacuated will also prevent the use of energy when not needed. Where extreme rainfall has led to flooding, undercutting the toes of open faces and spoil heaps can be monitored, to prepare for rockfall or landslip. Subsided areas that have developed over underground mines will alter surface flow, which can lead to unexpected flooding in areas of the mine or local community. Surface mapping using CRATER (or other mapping programs) will identify potential hazards as mining and subsidence progresses, and this can be resolved through installation of drainage or by moving threatened infrastructure.

20.5.5.5 Reducing Heat Risk

The numbers of days and nights of maximum temperatures are increasing, which in some places is expected to increase the risk of bushfire which, in turn, can lead to mine fires, a risk especially for coal mines. Stockpiles can be monitored using thermal imagers to assess potential for spoil heap fires, and with sprinkler and hose systems installed to reduce heat if needed. Bushfires around the mine site can be monitored and managed with fire-fighting equipment and trained personnel at the mine and, importantly, evacuation procedures can be put in place. Preventing mine fires does not just help improve resilience but also mitigation, as bushland and coal mine fires emit greenhouse gas emissions.

20.5.5.6 Reducing Heat Stress and Rising Energy Costs from Higher Temperatures

Extreme heat has health impacts for mine workers. There are numerous ways mines can adapt to reduce this risk with significant climate change mitigation co-benefits including:

- installing cooling systems in trucks and other mining vehicles and in on-site buildings – the use of renewable energies in such cooling systems provides a mitigation option;
- the use of fire-retardant vegetation and reduction of fuel loads, with appropriate fire-breaks around the mine, will reduce fire risk and reduce chances of fires causing coal mines to start burning and releasing greenhouse gas emissions.

20.6 Barriers to Change

The major market, information and institutional failures and barriers that have prevented investment in the 'win–win' adaptation and mitigation measures outlined above are typically:

- the perception by decision makers that action on climate change and sustainable development is an overhead rather than a productivity-enhancing investment;

- a broad lack of climate adaptation and sustainability integration in mining company strategy;
- the focus of the industry and investors on minimising capex on capital-intensive mining projects, which leads to neglecting opportunities to reduce operating energy and water expenditure (opex) by up to 50%;
- fluctuating mineral commodity prices creating company and investor uncertainty, making it more difficult for decision makers to confidently commit internal funds and for external investor capital to be accessed; and
- lack of skills and knowledge about climate adaptation and mitigation.

Energy Transition Hub (Lord 2020) identified some barriers to change, including that: existing supply chain relationships could create barriers to newcomers into mining chains, creating a risk for expanding to alternative suppliers in the case of an extreme event; investment in and demonstration of new options may be difficult beyond a pilot scale if it risks current production; and that there may be competition with countries that already have cheap renewable energy such as Canada and Norway.

20.7 Addressing Barriers to Change

20.7.1 Changing Perceptions of Decision Makers: Productivity Co-benefits of Action on Climate Change

'. . . every one percent improvement in productivity translates to a $170 million saving'. Andrew MacKenzie, CEO of BHP Billiton.

(Mining 2014)

Historically, there has been a negative attitude from some mining sector decision makers towards climate change policy reform: but this is now changing. As of April 2016, the Mining Association of Canada announced its support for a price on carbon (MAC 2016). Having said that, there is still a widely held fear that a price on carbon could harm mining sector productivity, profit margins and international competitiveness. Hence, it is very important to address this concern. There is empirical evidence that implementation of a price on carbon emissions, complemented by well-designed policies, can enhance rather than reduce international competitiveness. This has been researched by Dr Michael Scott from the University of Queensland. In a case study, he has shown how using smart blasting, pre-concentration of mineral ores and energy-efficient comminution strategies at an Australian gold mine would cut energy costs and greenhouse gas emissions by 12% while providing other benefits in improved energy productivity, reducing energy costs per tonne of gold produced (Scott 2014). These financial savings were found to exceed the carbon emission costs of AUD 20 per tonne should they be imposed on the gold mine. Research in South Africa also showed that it is possible for mining companies to maintain or increase profits even with rising energy and carbon costs, through implementation of smart investment in energy efficiency strategies (Chavalala and Nhamo 2014). The research also showed that the South African Government's decision to apply a carbon price in 2015 was one of three main factors influencing a gold mining company's decision to invest in cost-effective energy-efficient technology improvements. The International Council on Mining and Metals supports a price on carbon as a market incentive that will encourage innovation. A range of 28 mining companies such as Vale, Exxaro Resources, Harmony Gold Mining, Anglo American Platinum and Sibanye Gold already use an internal price on carbon as a tool to help make investment decisions (Carbon Disclosure Project 2015).

20.7.1.1 Use of Carbon Price to Retrofit Mines and Design Low-Carbon New Mines

Climate-change-related revenues should be used to help manage a transition to a low-carbon future. As the International Council on Mining and Metals states:

Emission reduction policies and measures provide new sources of revenues for governments. Such revenues should be directed towards two specific areas: (1) supporting the development of climate friendly technologies; and (2) helping 'exposed' economic sectors and populations adjust to the costs associated with a carbon limited future. *(ICMM 2013: 8)*

The Mining Association of Canada, in April 2016, formally supported, for the first time, a price on carbon if some of the revenue from it was recycled to help the mining sector invest to improve efficiency and cut its greenhouse gas emissions.

20.7.1.2 Resource Rent and Profit-Based Taxes

Governments can manage competitiveness issues regarding resource rent taxes by designing them to be super profits taxes. Governments can also manage potential negative impacts on mining competitiveness by (i) implementing effective revenue recycling (see above), (ii) allowing free permits under an ETS or exceptions to a price on carbon and (iii) implementing border tax adjustments (ICMM 2011).

20.7.1.3 Mitigation and Adaptation Boost to Productivity

Mining productivity literature provides evidence that there is value in mining companies transitioning to a low-carbon economy through energy efficiency improvements, as well as investments in on-site renewable energy, because many of these investments can unlock new sources of labour, capital, energy and resource productivity (Stadler et al. 2015).

Improving energy efficiency of comminution will be essential this century to maintain and improve productivity to deal with the declining mineral concentrations of ore being mined. High-grade ore typically requires less energy to process than to extract the same value from lower-grade ore. As Figure 20.3 shows, a major factor in the decline of mining productivity in Australia since the 1970s has been the fall in mineral ore concentrations or grade and their corresponding increase in required energy and water costs (Bye 2011). Hence, it is imperative for mining companies to improve energy and water efficiency, in order to cut rising energy and water costs to process the declining ore grades sustainably. As the discussion of climate change mitigation measures earlier in this chapter showed, there is potential to reduce energy requirements and costs by 10–30% below the business-as-usual trends of existing mines. Recent Australian studies suggest that energy savings of up to 50% per tonne of product, below business-as-usual energy usage trends, are feasible in new 'greenfield' mining-to-mill

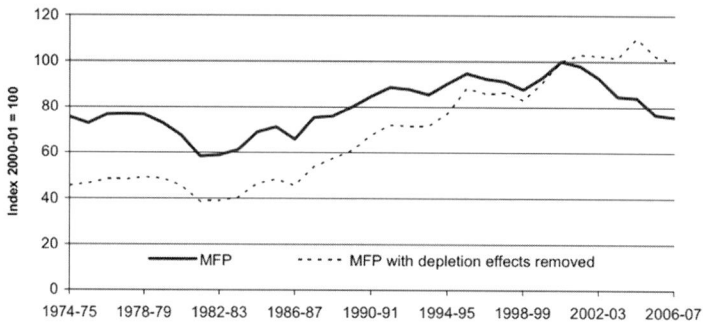

Figure 20.3 Mining multifactor productivity (MFP) and MFP with mineral ore grade depletion effects removed.
Source: Topp et al. (2008).

operations. Since energy cost rises are inevitable, due to the historic trend of falling ore body concentrations, investment in climate change mitigation measures like energy efficiency opportunities is essential for the mining sector to improve productivity. Revenues from a price on carbon can be recycled to help the mining sector retrofit its mines to improve energy efficiency and diversify into renewable energy supply.

Improving energy efficiency of new resource developments can reduce a mine's operational expenditure (opex), enabling companies to pay off their capex and more quickly boost profits and the capital productivity of such investments. Recent analysis of large resource developments in Australia have found that, if profit margins on a long-term asset are 5% and energy is 20% of operational costs, a 25% improvement in energy efficiency will double the profit of the life of that project. Some more energy-productive mining technologies also offer significant labour productivity savings (Stadler et al. 2015), such as:

- **in underground mining**, by investing in automated electric-powered underground drilling, scooping and materials movement machines. Electric-powered underground equipment can replace use of diesel, eliminating fumes and the subsequent need for high ventilation energy by up to 90%; and
- **in open-cut mining**, most mines use haul trucks to transport ore for processing. In-pit crushers and conveyors can yield significant energy, water and labour productivity gains. BHP Billiton informed UK investors of their decision to move towards truckless mines through investing in in-pit crusher/conveyor systems in 2012. BHP Billiton executive Marcus Randolph stated: 'When you run a truck, it takes 10 to 11 employees for every truck ... It takes 4½ to 5 to run it, all the crews that do the maintenance on it, all the camp people that do the camp cleaning and cooking and everything else ... If you go truckless [and use input crushers and conveyors] you get rid of all of them. You do this at a time when you see increasing diesel prices, carbon taxes, a number of reasons why getting rid of trucks or using fewer trucks is desirable.' Mr Randolph said the technology was already viable in mines with soft ground that did not require blasting but he said it could be adapted to also work in mines that did require blasting (Ker 2012).

- Mining company Vale SA (Vale) reported that they are replacing trucks at some mines with in-pit crusher/conveyor technology. In their largest Brazilian iron ore mines, they are replacing trucks with 37 kilometres of conveyor belts and a new railway line to reduce materials-movement-related energy costs by 77% (Vale 2013). Vale states that: 'Compared with conventional methods, the truckless system and ore processing using natural moisture will together cut S11D's annual greenhouse gas emissions by 50%, or 130 000 metric tonnes of CO_2 equivalent. In addition, the main equipment used at the project will be powered by electricity. Only crawler dozers, motor graders and other auxiliary equipment will run on diesel' (Vale 2013: 30).

Numerous studies show that mining companies that are proactive on environmental sustainability and climate change adaptation activities tend to outperform the rest of the market (see Chapter 23).

The energy intensity in the petroleum sector from 1980 to 2015 rose in OECD countries by approximately one-third, due to maturation of oil and gas fields and the trend to exploit more remote resources. Multiple energy efficiency opportunities in petroleum refineries (Worrell and Galitsky 2005) have been developed including the use of heat and power cogeneration, improved heat integration, combustion optimisation, control of air and steam leaks and efficient use of electrical devices (Bernstein et al. 2007).

20.7.1.4 Institutional Investors Calling for Productivity-Enhancing Action on Climate Change

Due to the above evidence, and evidence cited in Chapter 23, action on climate change to assess and manage climate risks is increasingly being recognised as a prudential and fiduciary duty by institutional investors, CEOs and boards and their regulators. As of 2017,

- The investment community is increasingly reviewing how well mining companies manage their risks across the challenges and opportunities of climate change, carbon pricing and energy costs. The Investor Group on Climate Change published a report that identifies and discusses the climate-related risks and opportunities for investors for the mining and mineral processing industry sector (Smith 2013a) and the oil and gas sector (Smith 2013b).
- Non-governmental organisations (NGOs), representing institutional investment funds and investors (such as the Carbon Disclosure Project), are raising public and institutional awareness of climate mitigation and adaptation needs and options (Thomsen et al. 2016). They are putting pressure on mining and mineral processing industry sector companies to publicly report their climate-change-related risks, and their progress on implementing climate change adaptation and mitigation strategies.

20.8 The Role of Government, Policy-Makers and Key Actors

Mining and mineral processing involves making long-term capital-intensive investments. So institutional investors and, increasingly, mining company CEOs are looking for

bipartisan certainty and consistency on climate change and energy policy from governments. They are looking for governments to take a robust empirical approach to address
market failures and design policies that encourage private-sector innovation. And they are
looking for governments to recognise and help address barriers to enabling mining sectors
to genuinely become low-carbon emitters, contributing to national transitions to sustainable
development by constructively focusing on the measures outlined in Section 20.7.

Despite some oil and gas companies already taking action, there is an uneven response
by the sector to the Paris Climate Change Agreement. Institutional investors are concerned
about risks of holding oil and gas company stocks due to their potential carbon and
environmental liabilities. Hence, institutional investors have withdrawn USD 5 trillion
from the fossil fuel sector in the last 5 years. Institutional investors have published two
requests for information and action, in 2014 and 2016, respectively, and are also formally
lodging requests to shareholders on their exposure to carbon liabilities and stranded assets.
They have also published guidance documents on minimum governance standards for
unconventional natural gas companies and their projects (Regnan 2015). Two examples
of bodies with interests in transition to a low-carbon economy are IPIECA (originally
designated the International Petroleum Industry Environmental Conservation Association)
and Energy Networks Australia. The IPIECA is a global not-for-profit oil and gas industry
association for environmental and social issues. It has issued:

- the *Low Carbon Pathways* report and *Paris Puzzle* series of factsheets, addressing
  implications and opportunities of the Paris Climate Change Agreement for its members
  (IPIECA 2015);
- an Energy Efficiency and Greenhouse Gas Compendium, to assist oil and gas companies
  identify and implement energy efficiency opportunities in oil and gas extraction
  (IPIECA-IOGP 2014); and
- guides for 'addressing adaptation to climate change' (IPIECA 2013) and water management for the sector.

In Australia, Energy Networks Australia (ENA) is the national body representing
Australia's electricity transmission and distribution and gas distribution networks.
Members of ENA provide more than 16 million electricity and gas connections to almost
every home and business across Australia. It has partnered with Australia's leading
scientific body, the CSIRO, to develop an Electricity Network Transformation Roadmap.

20.8.1 *Mitigation/Adaptation Innovation and Best Practice*

Encouraging innovation and ensuring best practice can be achieved by:

- implementing a price on carbon, and recycling that revenue to fund R&D&D and mine
  site upgrades, to help the mining sector transition to a low-carbon economy, as
  outlined above;
- government-backed climate-change-financing institutions helping mining companies to
  finance demonstration projects to assist the transition to a low-carbon future. There is

often some cultural resistance to change in any business, and therefore also a need for government-backed financing agencies. The Clean Energy Finance Corporation in Australia, for example, assists with reducing the risk of mining companies investing for the first time in technical innovations. Through funding demonstration projects, such bodies help to establish the business case for emerging low-carbon technologies and thereby expedite the pathway to commercialisation;

- supporting research in mining sector risks and vulnerabilities to climate change. Except for Australia, Canada and the USA, few nations have detailed granular studies on how climate change poses risks to their respective mining sectors;
- supporting greater research to better understand the risks and potential costs/loss of productivity from climate change to the mining sector. The negative financial impact of major extreme weather events such as flooding on mining can be greater than the worst economic shocks. For instance, the negative impacts of extreme flooding of mines on mining companies in Queensland, Australia, in 2011 were far greater than the negative financial impact of the global financial crisis (World Bank and Queensland Reconstruction Authority 2011);
- supporting and funding research into the integration of climate change mitigation, climate change adaptation and sustainable development for mining. There is a need for greater research to better analyse and cost the synergies and productivity co-benefits of climate change mitigation–adaptation nexus strategies for this sector;
- investing in R&D&D. Supporting collaborative R&D&D and innovation clusters such as the CRC ORE (Cooperative Research Centres Optimising Resource Extraction: CRC ORE Ltd 2011) in Australia, which focuses on helping mining companies achieve a step change in mining comminution eco-efficiency to cut energy costs and related emissions from mining, crushing and grinding ores. Supporting R&D&D of low-carbon energy supply options for mining operations is required. For example, this is being done by Australian Renewable Energy Agency (ARENA) who are collaborating with many mining companies and other research and development institutions to create hybrid renewable energy systems to power mining operations around the clock in remote areas of Australia (ARENA 2018). There are also many innovations arising from 'Big Data' that can assist the mining sector improve productivity while also mitigating emissions and better adapting to climate change (Stadler et al. 2015);
- addressing information, knowledge and skills gaps. Improving knowledge and skills of mining climate change adaptation, mitigation and sustainable development best practice is imperative. For instance, UNEP and the United Nations Development Programme (UNDP), along with researchers in Australia, have developed the sustainable development series for mining: leading practice guides to inform industry and stakeholders across many aspects of environmental and social sustainable development subjects (UNDP 2018; UNDP and UN Environment 2018; UNEP 2020). Ideally, one additional guide on climate change mitigation and adaptation would be done within this series to improve rapid capacity-building and knowledge. To assist this, co-authors of this chapter have researched and published on many aspects of these topics. Some of the tools that are available to assist mining companies are presented in Box 20.2.

Box 20.2 **What Tools Are Available?**

Here we present a range of existing tools to help mining companies get started. These include:

- Climate scenario mapping to better understand likely changes due to climate change for specific mining regions. The CSIRO '*Climate Scenarios*' for mining regions of Australia are an example of this.
- CRATER, a tool for decision-making to identify hotspots at a mine and to help guide adaptation planning.
- Mine-to-mill comminution optimisation packages that help optimise eco-efficient design.
- Online web portals integrating the latest in applications of renewable energy technologies in mining including 'Energy and Mines'[4] and 'The Energy Platform'.[5]
- Best-practice guides focused on climate change mitigation (COAG et al. 2014) or adaptation strategies for mining companies and the entire mining supply chain (Nelson and Schuchard n.d.; Hodgkinson et al. 2010; Mason et al. 2013).
- Mining-based sustainable development guides. The Australian Government and UNDP, for example, developed a range of high-profile extensive sustainable development and mining publications as guides for mining globally (e.g. UNDP 2018; UNDP and UN Environment 2018; UNEP 2020).
- Guides for investors and mining, oil and gas companies, which present climate risk, adaptation and mitigation opportunities (Smith 2013a, 2013b, 2013c) to overcome. One of the key barriers to mining companies investing in climate change adaptation and mitigation is that investors expect rapid returns while adaptation/mitigation strategies are a longer-term investment. Valuable information on climate risk is presented in these guides for institutional investors to help overcome this barrier (Smith 2013a, 2013b, 2013c).

20.8.2 Sustainable Development

Governments can mandate requirements for companies to submit environmental and social sustainable development programmes as part of the mine development approval process, including applications for significant operational and process changes during the life of the mine. Examples are given in the following subsections.

20.8.2.1 Managing and Optimising Energy Use and Reducing CO_2 Emissions

Governments have a responsibility and long-term financial interest in encouraging and ensuring that mining companies make suitable efforts to better manage energy and reduce greenhouse gas emissions on-site. Members of the International Council on Mining and Metals have all committed to being proactive in reducing overall greenhouse gas emissions and reporting on progress. It might also be beneficial if governments were to require mining companies to review energy and water efficiency and renewable energy opportunities in the design phase of new mining projects and show why they have selected the options being adopted.

---

[4] energyandmines.com.   [5] www.clydeco.com/blog/energy.

20.8.2.2 *Managing Water Resources*

Governments have a responsibility to ensure that water that is extracted during any activity including mining is done so at environmentally sustainable rates and is not impacting on other water users of environmental assets. If governments are to effectively manage this responsibility, they will require surface and groundwater best-practice environmental management standards to be in place. Adherence to these standards would need to be strictly monitored and have appropriate penalties and consequences should they be compromised, in addition to suitable mitigation plans to reduce the negative impacts of compromise. Such standards should require that mining entities will ensure that the quality and quantity of mine water, leach pad drainage and storm-water discharge to the environment meet established effluent discharge guideline values. If it has not already been done, governments should review and, if needed, fund research that improves assessment and reduction of mining-related risks to aquifer systems and water quality to help ensure better decision-making processes in water management.

20.8.2.3 *Managing Pollution and Waste Risks*

Governments have responsibilities to ensure that mining entities plan for extreme weather implications on risks from water-leaching or percolating waste dumps, tailings storage areas and leach pads. Governments also need to regularly check and evaluate a company's efforts, in addition to evaluating mitigation plans, should an environmental accident occur.

20.8.2.4 *Avoiding and Minimising Adverse Effects*

This may be achieved by identifying, reducing, monitoring and addressing potential risks and actual impacts on biodiversity throughout the mining cycle. Additionally, it can be achieved by requiring mining entities to conduct continuous monitoring and reporting in accordance with national standards and conditions of the operating permit. Such requirements should include publication by governments of the company's performance assessments, while the company is held accountable for any accidents or negative impacts. Also, a requirement that mining entities commission independent expert reviews and report to governments prior to development approval when changes in design are proposed, and at regular intervals during the operating phase, would minimise potential adverse effects on natural assets.

20.8.2.5 *Sustainable Development and Intergenerational Equity*

The act of mining constitutes a fundamentally unsustainable one-time opportunity for companies, communities, regions and nations to generate financial wealth from the extraction of a non-renewable resource. Economists argue that it is thus appropriate to ensure that this one-off sale is done in a way that provides for future generations who will not be able to sell off the same resource. Options include super profits taxes and resource rent taxes on miners and royalty payments that are flexible to ensure wealth is shared in profitable times, but do not punish mining companies when commodity prices are low. Given the risks involved in the capital-intensive sector, if it has to be a choice between super profits or mandatory royalties, mining companies tend to prefer a profits-based tax system rather than set royalties that are independent of how well they perform financially. Over the last decade, many governments of resource-rich countries have implemented a range of resource rent taxes or super profit taxes combined with creating sovereign wealth investment funds, to ensure the one-off wealth that is obtained from mining a national resource is used to help a nation achieve development, social and environmental goals. For instance, during the commodities boom of 2010–2013, many resource-rich countries reformed policy to ensure more of the wealth was returned to national governments and communities through a combination of resource rent taxes, requirements to value-add before exporting products or reforms to a government's ability to own a percentage of new mining projects (PricewaterhouseCoopers 2012; Verrender 2012).

20.8.2.6 Managing the Boom–Bust Mining and Resources Cycle

One of the benefits of governments implementing a resource rent super profits tax is that it helps insulate national governments from large negative impacts on government budgets when there is a downturn in mining commodity prices. During mining and resources booms, national governments can invest resource rent tax revenue into a sovereign wealth fund. The Government of Norway, for example (The Treasury 2012) has done this, to transform the one-off sale of their resources into an ongoing investment fund to ensure the proceeds can help their transition to a low-carbon, resilient sustainable future.

20.8.2.7 Social Licence to Operate: Providing Open and Transparent Data on Tax and Royalty Flows, Showing How the Money Is Being Distributed and Used

Governments and the mining sector can work together to help ensure transparency on these issues to help build public trust. To ensure there is trust in mining initiatives, it is important to provide open and transparent data on tax and royalty flows and show how (and for what) the benefits have been distributed at the local, regional and national level. It is in mining companies' interests to participate in such initiatives to maintain community trust and their social licence to operate.

20.9 Conclusion

This chapter presents some adaptation–mitigation opportunities that provide sustainable development strategies for the resources and metals recycling sector's transition to a low-carbon economy. There is a key role for governments and policy-makers to implement effective incentives, policies and R&D&D projects to encourage investors and resource sector companies to rise to the challenge. Technological advances can already provide economic solutions; coupled with the increased demand for the sector's products, the industries are well placed to become a sector that can be both climate friendly and climate resilient, with improved productivity to meet the rising demands for low-carbon technology products.

References

Abu-Hamdeh, N. and Alnefaie, K. (2019). Techno-economic comparison of solar power tower system/photovoltaic system/wind turbine/diesel generator in supplying electrical energy in small loads. *Journal of Taibah University for Science*, 13, 216–224.

Acclimatise (2010). *Building Business Resilience to Inevitable Climate Change*. Carbon Disclosure Project Report. Oxford: Global Mining. Available at: https://climate-adapt .eea.europa.eu/metadata/publications/building-business-resilience-to-inevitable-cli mate-change-the-adaptation-challenge.

Ahman, M., Vogl, V., Nyqvist, B. et al. (2018). *Hydrogen Steelmaking for a Low-Carbon Economy: A Joint LU-SEI Working Paper for the HYBRIT Project*. EESS report 109, SEI Working Paper WP 2018-07. Stockholm: Stockholm Environment Institute and Lund University.

ARENA (Australian Renewable Energy Agency) (2018). *Hybrid Power Generation for Australian Off-Grid Mines*. Handbook. Available at: https://arena.gov.au/assets/2018/ 06/hybrid-power-generation-australian-off-grid-mines.pdf.

ARENA and PowerWater (2019). *Solar/Diesel Mini-Grid Handbook*, 2nd ed. Available at: https://arena.gov.au/assets/2019/10/solar-diesel-mini-grid-handbook-powerwater.pdf.

Bearsley, C. (2016). Can renewable energy lower your cost of production? *AusIMM Bulletin*. December. Available at: https://search.informit.org/doi/epdf/10.3316/ielapa .591632547615963.

Bernstein, L., Roy, J., Delhotal, K. C., Harmisch, J., Matsuhashi, R., Price, L., Tanaka, K., Worrell, E., Yamba, F., Fenqi Z. et al. (2007). Industry. In: B. Metz, O. R. Davidson, P. R. Bosch, R. Dave and L. A. Meyer, eds., *Climate Change 2007: Mitigation. Contribution of Working Group III to the Fourth Assessment Report of the Intergovernmental Panel on Climate Change*. Cambridge: Cambridge University Press, pp. 447–496. Available at: www.cambridge.org/core/books/climate-change-2007-mitigation-of-climate-change/industry/ADD4528468FA9A9C8A4E93A14C85 4C7F.

Bosmin (n.d.). *Overburden Slushers: The Future in Open-Cut Mining*. Technical Bulletin 2. Camira, Queensland: Bosmin. Available at: www.bosmin.com/OS/osbrochure2 .pdf.

Boyle, N. (2016). Lightsource: Adapting renewables to fit with mining's business model. Presented at the Energy and Mines World Congress, Toronto, 21–22 November.

Buckingham, L., Dupont, J., Steiger, J., Blain, B. and Brits, C. (2011). Improving energy efficiency in barrack grinding circuits. Paper presented at the Fifth International Conference on Autogenous and Semiautogenous Grinding Technology (SAG 2011), Vancouver, 25–28 September. Available at: www.ceecthefuture.org/wp-con tent/uploads/2013/01/Improving-Energy-Efficiency-in-Barrick-Grinding-Circuits3.pdf? dl=1.

Budde, F., Günther, C. and Shah, M. (2014). *When Gas Gets Tight: Next Steps for the Middle East Petrochemical Industry*. McKinsey & Company. Available at: www .mckinsey.com/industries/oil-and-gas/our-insights/when-gas-gets-tight-next-steps-for-the-middle-east-petrochemical-industry.

Bye, A. (2011). Case studies demonstrating value from geometallurgy initiatives. In *GeoMet 2011: Proceedings of The First AusIMM International Geometallurgy Conference 2011*. Carlton, Victoria: AusIMM The Minerals Institute, pp. 9–30.

BZE (Beyond Zero Emissions) (2018). *Electrifying Industry: Zero Carbon Industry Plan*. Melbourne Beyond Zero Emissions. Available at: https://bze.org.au/wp-content/ uploads/2020/12/electrifying-industry-bze-report-2018.pdf.

Canadian Gas Association (2014). *Renewable Natural Gas Technology Roadmap for Canada*. Government of Canada. Available at: https://biogasassociation.ca/images/uploads/documents/2017/rng/The_Renewable_Natural_Gas_Technology_Roadmap.pdf.

Carbon Disclosure Project (CDP) (2015). *Putting a Price on Risk: Carbon Pricing in the Corporate World*. Report. New York: Carbon Disclosure Project (CDP). Available at: https://6fefcbb86e61af1b2fc4-c70d8ead6ced550b4d987d7c03fcdd1d.ssl.cf3.rackcdn.com/cms/reports/documents/000/000/918/original/carbon-pricing-in-the-corporate-world.pdf?1472456914.

Carbon Limits (2013). *Associated Petroleum Gas Flaring Study for Russia, Kazakhstan, Turkmenistan and Azerbaijan: Final Report*. Carbon Limits. Available at: www.ebrd.com/downloads/sector/sei/ap-gas-flaring-study-final-report.pdf.

Caulton, D. R., Shepson, P. B., Santoro, R. L. et al. (2014). Toward a better understanding and quantification of methane emissions from shale gas development. *Proceedings of the National Academy of Sciences*, 111, 6237–6242. Available at: www.pnas.org/content/111/17/6237.

Cervigni, R., Rogers, J. and Henrion, M. (2013). *Low-Carbon Development Opportunities for Nigeria*. Washington, DC: World Bank. Available at: http://documents.worldbank.org/curated/en/290751468145147306/Low-carbon-development-opportunities-for-Nigeria.

Chavalala, B. and Nhamo, G. (2014). Clean and energy efficient technology as green economy transition mechanism in South African gold mining: Case of Kusasalethu. *Environmental Economics*, 5, 52–61. Available at: http://businessperspectives.org/journals_free/ee/2014/ee_2014_01_Nhamo1.pdf.

Chen, G. Z., Fray D. J. and Farthing T. W. (2000). Direct electrochemical reduction of titanium dioxide to titanium in molten calcium chloride. *Nature*, 407, 361–364.

Chevron Australia (2009). *The Value of Partnership*. Corporate Responsibility Report. Chevron Australia.

Climate Commission (2013). *The Angry Summer*. Canberra: Australian Government. Available at: www.coolaustralia.org/the-climate-commissions-angry-summer-report/.

ClimateWorks Australia and DRET (Australian Department of Resources, Energy and Tourism) (2012). *Inputs to the Energy Savings Initiative Modelling from the Industrial Energy Efficiency Data Analysis Project*. Commonwealth Government of Australia Department of Resources, Energy and Tourism. Available at: www.climateworksaustralia.org/wp-content/uploads/2012/07/climateworks_esi_ieedap_report_jul2012_small.pdf.

COAG (Council of Australian Governments) Energy Efficiency Exchange, Smith, M. and Pears, A. (2014). *Mining and Minerals Processing Sector: Analysis of Climate Change Risk, Adaptation and Mitigation Strategies*. An Educational Module: Skills for the Carbon Challenge. Canberra: Australian Department of Industry, Innovation, Climate Change, Science, Research and Tertiary Education (DIICCSRTE) and The Australian National University (ANU).

CRC ORE Ltd (2011). *CRC ORE Annual Report 2010–2011: Transforming Resource Extraction*. St Lucia: The University of Queensland. Available at: https://issuu.com/melraassina/docs/crc_ore_annual_report_2010-11.

Cullen, J., Milford, R. and Allwood, J. (2010). Options for achieving a 50% cut in industrial carbon emissions by 2050. *Environmental Science & Technology*, 44, 1888–1894.

Deng, Y., Cornelissen, S. and Klaus, S. (2011). *The Energy Report: 100% Renewable Energy by 2050*. World Wildlife Fund (WWF), with Ecofys and the Office for Metropolitan Architecture (OMA). Available at: https://assets.panda.org/downloads/101223_energy_report_final_print_2.pdf.

Department of Industry, Innovation and Science (2015). *Energy in Australia*. Canberra: Office of the Chief Economist, Commonwealth Government of Australia.

Department of Resources, Energy and Tourism (n.d.). *Analyses of Diesel Use for Mine Haul and Transport Operations*. Case study. Energy Efficiency Exchange. Available at: www.energy.gov.au/sites/default/files/analyses_of_diesel_use_for_mine_haul_and_transport_operations.pdf.

Dittrick, P. (2012). Investor groups ask industry to cut methane emissions. *Oil & Gas Journal*. Available at: www.ogj.com/articles/print/vol-110/issue-6c/general-interest/investor-groups-ask-industry.html.

Ditze, A. and Scharf, C. (2008). *Recycling of Magnesium*. Clausthal-Zellerfeld, Germany: Papierflieger Verl.

Douglas, G., Wendling, L., Pleysier, R. and Trefry, M. (2010). Hydrotalcite formation for contaminant removal from ranger mine process water. *Mine Water and the Environment*, 29, 108–115.

DRET (Australian Department of Resources, Energy and Tourism) (2010). *Energy Efficiency Opportunities – Energy–Mass Balance: Mining*. Australian Government Department of Resources, Energy and Tourism. Available at: www.energy.gov.au/sites/default/files/energy-mass_balance_mining.pdf.

DRET (2011). *Energy Efficiency Opportunities: Assessment Handbook*. Australian Government Department of Resources, Energy and Tourism. Available at: www.energy.gov.au/publications/energy-efficiency-opportunities-assessment-handbook.

EEC (Energy Efficiency Council) (2016). *Australian Energy Efficiency Policy Handbook: Save Energy, Grow the Economy*. Energy Efficiency Council. Available at: www.eec.org.au/uploads/Documents/Platofrm%20Documents/Australian%20Energy%20Efficiency%20Policy%20Handbook%20–%20July%202016.pdf.

Eglinton, T., Hinkley, J., Beath, A. and Dell'Amico, M. (2013). Potential applications of concentrated solar thermal technologies in the Australian minerals processing and extractive metallurgical industry. *Journal of the Minerals, Metals and Materials Society*, 65, 1710–1720.

Energy Networks Australia, APPEA (Australian Petroleum Production and Exploration Association), APGA (Australian Pipelines and Gas Association), Gas Energy Australia and GAMAA (Gas Appliance Manufacturers Association of Australia) (2017). *Gas Vision 2050: Reliable, Secure Energy and Cost-Effective Carbon Reduction*. Energy Networks Australia. Available at: www.energynetworks.com.au/assets/uploads/gasvision2050_march2017.pdf.

Engineers Australia (2016). Mines explore modular and movable solar systems. *Engineers Australia*. 16 December. Available at: https://portal.engineersaustralia.org.au/news/mines-explore-modular-and-movable-solar-systems.

Equinor (2020). Equinor sets ambition to reduce net carbon intensity by at least 50% by 2050. *Equinor*. Available at: www.equinor.com/en/news/2020-02-06-climate-roadmap.html.

Eschner, S. and Hodgkinson, J. (2014). *The Costs and Benefits of Adapting to Climate and Weather Extremes: Australian Mining Chain Business Examples*. Report EP146903. Commonwealth Scientific and Industrial Research Organisation (CSIRO).

Evans, S. (2018). Sanjeev Gupta steps up $1.37b renewable energy build near Whyalla steelworks. *Australian Financial Review*. 15 August. Available at: www.afr.com/companies/energy/sanjeev-gupta-steps-up-137b-renewable-energy-build-near-why alla-steelworks-20180815-h13zcc.

Evans and Peck (2011). *Assessment of the Potential for Renewable Energy Projects in the Pilbara*. West Perth: Evans and Peck.

FMG (Fortescue Metals Group) (2019). Responsible environmental management. *FMGL. com.au*. Available at: www.fmgl.com.au/workingresponsibly/environment.

FT (Financial Times) (2016). Total aims to be 20% low-carbon by 2035. *Financial Times*. Available at: www.ft.com/content/04985ba4-21c8-11e6-aa98-db1e01fabc0c.

Glencore Ulan Coal (n.d.). Bobadeen irrigation scheme. *Glencore Ulan Coal*. Available at: www.ulancoal.com.au/en/environment/biodiversity/Pages/bobadeen-irrigation-scheme.aspx.

Global CCS Institute and Parsons Brinckerhoff (2011). *Accelerating the Uptake of CCS: Industrial Use of Captured Carbon Dioxide*. Global CCS Institute and Parsons Brinckerhoff. Available at: www.globalccsinstitute.com/archive/hub/publications/14026/accelerating-uptake-ccs-industrial-use-captured-carbon-diox ide.pdf.

Global Methane Initiative (2011). *Oil and Gas Systems Methane: Reducing Emissions, Advancing Recovery and Use*. Global Methane Initiative. Available at: www .globalmethane.org/documents/oil-gas_fs_eng.pdf.

Government of Canada, Natural Resources Canada (2013). Energy efficiency in mining. *Natural Resources Canada*. Available at: www.nrcan.gc.ca/mining-materials/mining/green-mining-innovation/energy-efficiency-mining/18312.

Greiner, R., Puig, J., Huchery, C., Collier, N. and Garnett, S. T. (2014). Scenario modelling to support industry planning and decision making. *Environmental Modelling & Software*, 55, 120–131.

Hargroves, K., Gockowiak, K., M'Keague, F. and Desha, C. (2014). *An Overview of Energy Efficiency Opportunities in Mining and Metallurgy Engineering*. The University of Adelaide and Queensland University of Technology (The Natural Edge Project). Available at: https://cms.qut.edu.au/__data/assets/pdf_file/0004/533065/flatpack8-an-overview-of-energy-efficiency-opportunities-in-mining-and-metallurgy-engineering.pdf.

Hodges, J. (2019). Shell takes 20% stake in Indian solar firm Orb Energy. *Bloomberg*. 3 October. Available at: www.bloomberg.com/news/articles/2019-10-02/shell-takes-20-stake-in-indian-solar-firm-orb-energy.

Hodgkinson, J. and Smith, M. (2018). Climate change and sustainability as drivers for the next mining and metals boom: The need for climate-smart mining and recycling. *Resources Policy*, DOI: 10.1016/j.resourpol.2018.05.016.

Hodgkinson, J. H. and Grigorescu, M. (2019). Strategic elements in the Fort Cooper Coal Measures: Potential rare earth elements and other multi-product targets. *Australian Journal of Earth Science*, 67, 305–319.

Hodgkinson, J., Littleboy, A., Howden, M., Moffat, K. and Loechel, B. (2010). *Climate Adaptation in the Australian Mining and Exploration Industries*. Climate Adaptation Flagship Working Paper 5. Commonwealth Scientific and Industrial Research Organisation (CSIRO). Available at: https://research.csiro.au/climate/wp-content/uploads/sites/54/2016/03/5_WorkingPaper5_CAF_pdf-Standard.pdf.

Hodgkinson, J., Grigorescu, M. and Alehossein, H. (2013). *Preparing a Mine for Both Drought and Flood – Stage 1: A Vulnerability and Adaptive Capacity Study*. ACARP C21041. Commonwealth Scientific and Industrial Research Organisation (CSIRO)

Study Report EP132938. Commonwealth Scientific and Industrial Research Organisation (CSIRO). Available at: https://publications.csiro.au/rpr/download?pid=csiro:EP132938&dsid=DS2.

Hodgkinson, J., Hobday, A. and Pinkard, E. (2014). Climate adaptation in Australia's resource-extraction industries: Ready or not? *Regional Environmental Change*, 14, 1663–1678.

Hower, J. C., Granite, E. J., Mayfield, D. B., Lewis, A. S. and Finkelman, R. B. (2016). Notes on contributions to the science of rare earth element enrichment in coal and coal combustion byproducts. *Minerals*, 6, 32.

ICMM (International Council on Mining and Metals) (2011). *Competitiveness Implications for Mining and Metals*. InBrief Report. London: International Council on Mining and Metals (ICMM). Available at: www.iisd.org/system/files/publications/icmm_competitiveness_implications_mining.pdf.

ICMM (2012a). *The Role of Minerals and Metals in a Low Carbon Economy*. Mining's Sustainable Development InBrief Report. London: International Council on Mining and Metals (ICMM). Available at: www.extractiveshub.org/servefile/getFile/id/2872.

ICMM (2012b). *Water Management in Mining: A Selection of Case Studies*. Climate Change Report. London: International Council on Mining and Metals (ICMM). Available at: http://icmm.uat.byng.uk.net/website/publications/pdfs/water/water-management-in-mining_case-studies.

ICMM (2013). *Options in Recycling Revenues Generated through Carbon Pricing: How 16 Governments Invest Their Carbon Revenues*. Climate Change Report. London: International Council on Mining and Metals (ICMM). Available at: http://icmm.uat.byng.uk.net/en-gb/publications/climate-change/options-in-recycling-revenues-generated-through-carbon-pricing.

IEA (International Energy Agency) (2013). *Technology Roadmap: Energy and GHG Reductions in the Chemical Industry via Catalytic Processes*. Paris: International Energy Agency. Available at: www.iea.org/reports/technology-roadmap-energy-and-ghg-reductions-in-the-chemical-industry-via-catalytic-processes.

IEA (2017). *Renewable Energy for Industry*. Paris: International Energy Agency (IEA). Available at: webstore.iea.org/insights-series-2017-renewable-energy-for-industry.

IPIECA (International Petroleum Industry Environmental Conservation Association) (2013). *Addressing Adaptation in the Oil and Gas Industry*. London: IPIECA. Available at: www.ipieca.org/news/addressing-adaptation-in-the-oil-and-gas-industry/.

IPIECA (2015). Natural gas: Into the future (the Paris puzzle). *Ipieca.org*. 15 April. Available at: www.ipieca.org/resources/fact-sheet/natural-gas-into-the-future-the-paris-puzzle/.

IPIECA-IOGP (2014). IPIECA-IOGP pre-recorded webinars on GHG and energy efficiency. *Ipieca.org*. Available at: www.ipieca.org/energyefficiency.

IRIN (Integrated Regional Information Networks) (2007). Gas flaring wrecking Delta communities. *The New Humanitarian*. 12 December. Available at: www.irinnews.org/report/75824/nigeria-gas-flaring-wrecking-delta-communities.

Ivanhoe Mines Ltd (2018). Ivanhoe Mines and Zinjun virtually triple hydroelectric output to support the Kamoa-Kakula Copper Project. *Globe Newswire*. 8 January. Available at: www.globenewswire.com/news-release/2018/01/08/1284934/0/en/Ivanhoe-Mines-and-Zijin-virtually-triple-hydroelectric-output-to-support-the-Kamoa-Kakula-Copper-Project.html.

Ker, P. (2012). BHP aims to dump trucks. *The Sydney Morning Herald*. 1 November. Available at: www.smh.com.au/business/bhp-aims-to-dump-trucks-20121031-28k9h .html#ixzz2k8EjoxUf.

Kirk, T. and Lund, J. (2018). *Decarbonisation Pathways for Mines: A Headlamp in the Darkness*. Snowmass, CO: Rocky Mountain Institute. Available at: www.rmi. org/wp-content/uploads/2018/08/RMI_Decarbonization_Pathways_for_Mines_2018 .pdf.

Klender, J. (2019). Tesla resumes Gigafactory 1 solar panel installations. *Teslarati*. 25 October. Available at: www.teslarati.com/tesla-resumes-gigafactory-1-solar-panel-installations/.

Latimer, C. (2018). Climate change and renewables driving new mining boom, mining chief says. *The Sydney Morning Herald*. 30 October. Available at: www.smh.com.au/ business/the-economy/climate-change-and-renewables-driving-new-mining-boom-mining-chief-says-20181029-p50cm5.html.

Loechel, B., Hodgkinson, J., Moffat, K., Crimp, S., Littleboy, A. and Howden, M. (2010). *Goldfields-Esperance Regional Mining Climate Vulnerability Workshop*. Report on workshop outcomes. CSIRO Report EP106666. Pullenvale, Queensland, Australia: Commonwealth Scientific and Industrial Research Organisation (CSIRO). Available at: https://publications.csiro.au/rpr/pub?list=BRO&pid=csiro:EP106666.

Loechel, B., Hodgkinson, J. and Moffat, K. (2011). *Pilbara Regional Mining Climate Change Adaptation Workshop*. Report on workshop outcomes. CSIRO Report EP118134. Canberra: Commonwealth Scientific and Industrial Research Organisation (CSIRO). Available at: https://publications.csiro.au/rpr/download?pid= csiro:EP118134&dsid=DS1.

Loginova, J. and Batterbury, S. P. J. (2019). Incremental, transitional and transformational adaptation to climate change in resource extraction regions. *Global Sustainability*, 2, 1–12.

Lord, M. (2020). *From Mining to Making: Australia's Future in Zero-Emissions Metal*. Melbourne: Energy Transition Hub, University of Melbourne. Available at: www .energy-transition-hub.org/resource/mining-making-australias-future-zero-emissions-metal.

Lovegrove, K., Edwards, S., Jacobson, N. et al. (2015). *Renewable Energy Options for Australian Industrial Gas Users*. Background technical report ITP/A0142 rev. 2.0. Canberra: IT Power (Australia) Pty Ltd and ARENA (Australian Renewable Energy Agency). Available at: https://itpau.com.au/wp-content/uploads/2018/08/ITP_ REOptionsForIndustrialGas_TechReport.compressed.pdf.

MAC (Mining Association of Canada) (2016). Mining industry supports carbon price to address climate change. *Mining.ca*. 13 April. Available at: https://mining.ca/press-releases/mining-industry-supports-carbon-price-address-climate-change/.

Mason, L., Unger, C., Lederwasch, A., Razian, H., Wynne, L. and Giurco, D. (2013). *Adapting to Climate Risks and Extreme Weather: A Guide for Mining and Minerals Industry Professionals*. Synthesis and Integrative Research Final report. Gold Coast, Australia: National Climate Change Adaptation Research Facility (NCCARF) and Griffith University.

Mavrommatis, E., Damigos, D. and Mirasgedis, S. (2019). Towards a comprehensive framework for climate change multi-risk assessment in the mining industry. *Infrastructures*, 4, 38.

McGlade, C. and Ekins, P. (2015). The geographical distribution of fossil fuels unused when limiting global warming to 2 °C. *Nature*, 517, 187–190.

Mining (2014). BHP Billiton CEO: Automation could save the mining sector billions of dollars. *Mining Global.* 29 April. Available at: www.miningglobal.com/machinery/bhp-billiton-ceo-automation-could-save-mining-sector-billions-dollars.

Mohr, S., Mudd, G. and Giurco, D. (2012). Lithium resources and production: Critical assessment and global projections. *Minerals*, 2, 65–84.

Murphy, B. (2013). Flotation energy consumption and opportunities for improvement. In *MetPlant 2013 Conference: Technical Program.* Perth, Australia: AusIMM The Minerals Institute.

Nalbandian, H. (2014). *Non-Fuel Uses of Coal.* CCC/236. International Energy Agency Clean Coal Centre. Available at: www.usea.org/sites/default/files/052014_Non-fuel%20uses%20of%20coal_ccc236.pdf.

National Grid (2016). The future of gas: Supply of renewable gas. *National Grid.* Available at: https://cadentgas.com/nggdwsdev/media/Downloads/Future%20of%20gas/The-future-of-gas-Feb-16.pdf.

Nelson, J. and Schuchard, R. (n.d.). *Adapting to Climate Change: A Guide for the Mining Industry.* BSR industry brief. Business for Social Responsibility (BSR). Available at: www.bsr.org/reports/BSR_Climate_Adaptation_Issue_Brief_Mining.pdf.

Norgate, T. and Haque, N. (2010). Energy and greenhouse gas impacts of mining and mineral processing operations. *Journal of Cleaner Production*, 18, 266–274.

Ossa-Moreno, J., McIntyre, N., Ali, S. et al. (2018). The hydro-economics of mining. *Ecological Economics*, 145, 368–379.

Pinkl, M. (2019). The renewable energy strategies of oil majors: From oil to energy? *Energy Strategy Reviews*, 26, 100370. Available at: www.sciencedirect.com/science/article/pii/S2211467X19300574.

Pokrajcic, Z. and Morrison, R. (2008). A simulation methodology for the design of eco efficient comminution circuits. In D. Wang, C. Sun, F. Wang, L. Zhang and L. Han, eds., *Proceedings of XXIV International Mineral Processing Congress*, Vol. 1, 1st ed. Beijing: XXIV International Mining Processing Congress, pp. 481–495. Available at: http://espace.library.uq.edu.au/view/UQ:167383.

PricewaterhouseCoopers (2012). *Corporate Income Taxes, Mining Royalties and Other Mining Taxes: A Summary of Rates and Rules in Selected Countries.* Global mining industry update. London: PricewaterhouseCoopers LLP. Available at: www.pwc.com/gx/en/energy-utilities-mining/publications/pdf/pwc-gx-miining-taxes-and-royalties.pdf.

Rathmann, B. and Heuer, U. (2007). Refit of an electric shovel or dragline: A cost saving alternative between frequent repairs and the purchase of a new machine. Paper presented at SME Annual Conference and Expo, Denver, 25–28 February. Available at: https://library.e.abb.com/public/e4d71de93711bdcac12576730045d8eb/SME%202007%20REFIT%20OF%20AN%20ELECTRIC%20SHOVEL.pdf.

Regnan (2015). *Unconventional Oil and Gas Best Practice ESG Risk Management Principles and Recommendations 1.0.* Sydney: Regnan. Available at: https://static1.squarespace.com/static/5a29d7422278e7032800bc21/t/5a5c37ef9140b796c4fb85ab/1515993080089/UnconventionalGas_PrinciplesAndRecommendations_v1_2015Nov18.pdf.

REM (Renewable Energy Magazine) (2012). Mining industry ramps up investments in renewables. *Renewable Energy Magazine.* 18 May. Available at: www.renewableenergymagazine.com/article/mining-industry-ramps-up-investments-in-renewables.

Reuters (2019). Sustainability the new battleground for aluminium producers: Andy Home. *Reuters*. 22 October. Available at: www.reuters.com/article/us-metals-aluminium-ahome-column/sustainability-the-new-battleground-for-aluminum-producers-andy-home-idUSKBN1X11E5.

RMI (Rocky Mountain Institute) (2017). *Sunshine for Mines: A Second Life for Legacy Mining Sites*. Insight brief. Snowmass, CO: Rocky Mountain Institute. Available at: https://d231jw5ce53gcq.cloudfront.net/wp-content/uploads/2017/11/RMI_SecondLifeLegacyMiningSites.pdf.

Scott, M. (2014). Transforming resource extraction and its evaluation. Paper presented at the Forum for Doubling Energy Productivity, CRC for Optimisation of Resource Extraction, 3–4 April. Available at: https://na.eventscloud.com/file_uploads/38fca418268d339710dd371a25d7a1ea_Scott_Michael.pdf.

Smith, M. (2013a). *Assessing Climate Change Risks and Opportunities: Mining and Minerals Processing Sector*. Canberra: The Investor Group on Climate Change (IGCC) and The Australian National University (ANU).

Smith, M. (2013b). *Assessing Climate Change Risks and Opportunities for Investors: Oil and Gas Sector*. Canberra: The Investor Group on Climate Change (IGCC) and The Australian National University (ANU).

Smith, M. (2013c). *Mining and Mineral Processing Sector: Climate Change Risk and Opportunity Assessment. An Educational Module to Help Identify and Implement Climate Change Adaptation and Mitigation Opportunities*. Skills for the Carbon Challenge. Canberra: Department of Industry, Innovation, Climate Change, Science, Research and Tertiary Education (DIICCSRTE) and The Australian National University (ANU).

Speirs, J., Gross, R., Contestabile, M., Candelise, C., Houari, Y. and Gross, B. (2014). *Materials Availability for Low Carbon Technologies: An Assessment of the Evidence*. Report by the UK Energy Research Centre Technology & Policy Assessment Function. Available at: https://d2e1qxpsswcpgz.cloudfront.net/uploads/2020/03/materials-availability-for-low-carbon-technologies.pdf.

Stadler, A, Jutsen, J., Musa, F. and Smith, M. (2015). *Doubling Australia's Energy Productivity by 2030: Re-energising the Mining Sector to Improve Its Competitiveness*. Consultation draft version 1.3. Sydney: Australian Alliance to Save Energy. Available at: www.researchgate.net/publication/304909452_Doubling_Energy_Productivity_by_2030_-_Re-Energising_the_Mining_Sector_to_Improve_Its_Competitiveness_-_Full_Report.

The Treasury (Australia) (2012). *The Role of Sovereign Wealth Funds in Managing Resource Booms: A Comparison of Australia and Norway*. Macroeconomic Group, Australian Government. Available at: https://treasury.gov.au/speech/the-role-of-sovereign-wealth-funds-in-managing-resource-booms-a-comparison-of-australia-and-norway.

Thomsen, D. C., Keys, N., Treichel, P., Connor, S., Bygrave, S. and Smith, T. (2016). *Climate Adaptation: Lessons from NGOs: Turning Challenges into Successes and Sharing Lessons*. National Adaptation Network Social, Economic and Institutional Dimensions Briefing document. Gold Coast, Australia: National Climate Change Adaptation Research Facility.

Topp, V., Soames, L., Parham, D. and Bloch, H. (2008). *Productivity in the Mining Industry: Measurement and Interpretation*. Productivity Commission Staff Working Paper. Canberra: Australian government Productivity Commission. Available at: www.pc.gov.au/research/supporting/mining-productivity/mining-productivity.pdf.

Tudeshki, H. H. (2009). Round table at Hannover Messe 2009: Climate-friendly and energy-efficient raw material extraction. *AMS Online*, 03/2009, 31–34.

UNDP (United Nations Development Programme) (2018). *Extracting Good Practices: A Guide for Governments and Partners to Integrate Environment and Human Rights into the Governance of the Mining Sector*. New York: United Nations Development Programme. Available at: www.undp.org/content/dam/undp/library/Sustainable%20Development/Environmental-Governance-Project/Extracting_Good_Practices_Report.pdf.

UNDP and UN Environment (2018). *Managing Mining for Sustainable Development: A Sourcebook*. Bangkok: United Nations Development Programme.

UNEP (UN Environment Programme) (2011). *Recycling Rates of Metals: A Status Report*. A report of the Working Group on Global Metal Flows to the International Resource Panel. Paris: UN Environment Programme (UNEP). Available at: http://wedocs.unep.org/handle/20.500.11822/8702.

UNEP (2013). *Metal Recycling: Opportunities, Limits, Infrastructure*. A report of the Working Group on the Global Metal Flows to the International Resource Panel. Paris: UN Environment Programme (UNEP). Available at: http://apps.unep.org/piwik/download.php?file=/publications/pmtdocuments/-Metal%20Recycling%20Opportunities,%20Limits,%20Infrastructure-2013Metal_recycling.pdf.

UNEP (2020). *Mineral Resource Governance in the 21st Century: Gearing Extractive Industries Towards Sustainable Development*. A Report by the International Resource Panel. Nairobi: United Nations Environment Programme. Available at: www.resourcepanel.org/reports/mineral-resource-governance-21st-century.

Vale (2013). *Carajás S11D Iron Project: A New Impetus to Brazil's Sustainable Development*. Rio de Janeiro: Vale. Available at: www.vale.com/EN/initiatives/innovation/s11d/Documents/book-s11d-2013-en.pdf.

VCL (Virtual Curtain Limited) (n.d.). *Virtual Curtain Limited*. Available at: www.virtualcurtain.com.au/.

Vella, H. (2013). Maximising underground efficiency with energy conscious mining machines. *Mining Technology*. 18 August. Available at: www.mining-technology.com/features/feature-underground-efficiency-energy-conscious-mining-machines/.

Verrender, I. (2012). Resources tax: What you may not know . . .. *The Sydney Morning Herald*. 20 March.

Vidal, O., Goffé, B. and Arndt, N. (2013). Metals for a low-carbon society. *Nature Geoscience*, 6, 894–896.

Whillier, A. (1977). Recovery of energy from the water going down mine shafts. *Journal of the South African Institute of Mining and Metallurgy*, April, 183–200. Available at: www.saimm.co.za/Journal/v077n09p183.pdf.

Wilburn, D. R. (2012). *Byproduct Metals and Rare-Earth Elements Used in the Production of Light-Emitting Diodes: Overview of Principal Sources of Supply and Material Requirements for Selected Markets*. US Geological Survey Scientific Investigations Report 2012–5215. US Geological Survey. Available at: http://pubs.usgs.gov/sir/2012/5215/.

World Bank (2017). *The Growing Role of Minerals and Metals for a Low Carbon Future*. International Bank for Reconstruction and Development and The World Bank. Washington, DC: World Bank. Available at: http://documents.worldbank.org/curated/en/207371500386458722/The-Growing-Role-of-Minerals-and-Metals-for-a-Low-Carbon-Future.

World Bank and Queensland Reconstruction Authority (2011). *Queensland Recovery and Reconstruction in the Aftermath of the 2010/2011 Flood Events and Cyclone Yasi*.

Washington, DC: World Bank. Available at: http://documents.worldbank.org/curated/en/842511468220781111/Queensland-recovery-and-reconstruction-in-the-aftermath-of-the-2010-2011-flood-events-and-cyclone-Yasi.

Worrell, E. and Galitsky, C. (2005). *Energy Efficiency Improvement and Cost Saving Opportunities for Petroleum Refineries*. Berkeley, CA: Ernest Orlando Lawrence Berkeley National Laboratory, University of California.

Worrell, E., Bernstein, L., Roy, J., Price, L. and Hamisch, J. (2009). Industrial energy efficiency and climate change mitigation. *Energy Efficiency*, 2, 109–123.

Wu, Y., Ma, Y., Wang, Y. et al. (2013). Efficient and large scale synthesis of graphene from coal and its film electrical properties studies. *Journal of Nanoscience and Nanotechnology*, 13, 929–932. Available at: www.ncbi.nlm.nih.gov/pubmed/23646544.

Ye, R., Xiang, C., Lin, J. et al. (2013). Coal as an abundant source of graphene dots. *Nature Communications*, 4, 2943.

Zhou, L. (2015). Creating plastic from greenhouse gas. *Smithsonian Magazine*. 1 May. Available at: www.smithsonianmag.com/innovation/creating-plastic-from-green house-gases-180954540/.

Zhao, L., Dai, S., Nechaev, V. P. et al. (2019). Enrichment of critical elements (Nb–Ta–Zr–Ft–REE) within coal and host rocks from the Datanhao mine, Daqingshan Coalfield, Northern China. *Ore Geology Reviews*, 111, 102951.

Addressing Barriers io Change

21

Trade and Climate Change

KAREN HUSSEY AND THOMAS FAUNCE

Executive Summary

The importance of international trade in the transition to decarbonised economies cannot be overstated. Goods and services will need to be traded across borders so that energy and production systems, industry, agriculture, transport and other sectors of human development can all move towards low-carbon solutions. Yet the international system of trade can also be a constraint on decarbonisation, with concerns about international competitiveness, market power and market access lying at the heart of resistance to ambitious climate policy, both domestically and internationally.

Two trade-related constraints, in particular, impede policy initiatives aimed at achieving decarbonised economic systems. The first constraint is the tension between a country's need to remain competitive in the international system of trade and the significant costs that decarbonisation strategies present in the short to immediate term. This tension imposes very real constraints on a state's ability to 'go it alone' in climate change mitigation policy. A related second obstacle lies in the conflicting principles of international trade and investment law arising from instruments of the World Trade Organization and from bilateral and regional arrangements between nations. Such agreements simultaneously promote somewhat contradictory goals: the removal of 'barriers to trade' – including but not limited to 'behind the border' environmental regulation – on the one hand, and the need to promote 'sustainable development' in the international trade system on the other.

In this chapter we explore the tensions that exist between climate policies and the legal frameworks that support international trade and investment. We explore some of the global mechanisms and processes that can be used to overcome or at least lessen some of those tensions, and in the last section we turn to the Paris Agreement, which may offer new global markets and rules that could support the development of new markets and opportunities in the international system of trade, which could in turn support the transition to decarbonised economies. The chapter concludes by identifying five actions that could successfully support the low-carbon agenda: (1) raising awareness of the strong economic benefits from decarbonisation; (2) developing international frameworks for 'green' finance; (3) pursuing global standards and the standardisation of low-carbon technologies; (4) reforming trade law to encourage low-carbon goods and services; and (5) ensuring coherence across international frameworks.

571

21.1 Introduction

Seizing the many opportunities for low-carbon, resilient and prosperous economies described in the previous chapters of this book will require countries to rapidly undergo massive transformation in energy and production systems, industry, agriculture, transport, land use and other sectors of human development. These transformations offer significant trade and investment opportunities for countries, companies and communities that move to exploit them. Yet, clearly, the extent to which a country can seize the opportunities of the shift to 'green' economies depends on their 'low-carbon competitiveness' at the sector level, defined by Srivastav et al. (2018) as the combination of an ability to shift to low-carbon products and processes, an ability to gain and maintain market share and a favourable starting point in terms of current output and scale. Inevitably, the transition to decarbonised economies will need to be pursued amid fierce resistance from vested interests in carbon-based industries and often in the face of significant domestic fiscal constraints and competing policy priorities. Indeed, despite the now overwhelming scientific case supporting their importance, the absence of effective decarbonisation policy measures, such as a global price on carbon emissions, is evidence of very strong, countervailing corporate forces and, relatedly, community resistance.

The centrality of international trade in these countervailing forces cannot be overstated. Concerns about international competitiveness, market power and market access lie at the heart of resistance to ambitious climate policy both domestically and internationally. Despite the hope offered by the Paris Agreement,[1] the inertia and recalcitrance witnessed at subsequent Conference of the Parties meetings (notably COP 25 in Madrid in December 2019) is almost entirely a result of government and corporate concern about the political and economic costs of climate mitigation – costs which will be realised in part through the international system of trade.

Two trade-related constraints, in particular, impede policy initiatives aimed at achieving decarbonised economic systems. The first constraint is the tension between a country's need to remain internationally competitive in the international system of trade and the significant costs that decarbonisation strategies present in the short to immediate term. This tension imposes very real constraints on a state's ability to 'go it alone' in climate change mitigation policy. Prime facie, this constraint is evident in the absence of strong commitments to reduce greenhouse gas (GHG) emissions by all bar a few major economies, but it is also evident in the often complex and pervasive use of environmentally harmful subsidies in domestic settings to shore up domestic industry (Oosterhuis and Ten Brink 2014). It is indirectly evident in the strong resistance to climate policy held by communities that rely on energy-intensive trade-exposed industries for their livelihood.

A related second obstacle lies in the conflicting principles of international trade and investment law arising from instruments of the World Trade Organization (WTO) and from bilateral and regional arrangements between nations. Such agreements simultaneously promote somewhat contradictory goals: the removal of 'barriers to trade' – including but

---

[1] *Paris Agreement Under the United Nations Framework Convention on Climate Change*, opened for signature 16 February 2016. Available at: https://unfccc.int/process-and-meetings/the-paris-agreement/the-paris-agreement.

not limited to 'behind the border' environmental regulation – on the one hand, and the need to promote 'sustainable development' in the international trade system on the other. While the legal doctrine of the WTO permits the use of trade barriers to protect social or environmental outcomes, the realpolitik of international trade disputes means that more ambitious governments may be reluctant to pursue zero-carbon (but potentially trade-distorting) policies for fear of retribution from trade partners. Similarly, the power of the fossil fuel lobby to use global processes – such as investor-state dispute settlement (ISDS) mechanisms – to undermine global and national attempts to pursue decarbonisation provokes additional concerns. Inherent in both these obstacles is the imbalance that exists between developed and developing economies in the international system of trade: not only are developing countries more vulnerable to climate impacts; many are also more vulnerable to the trade impacts of climate change policies (Brack et al. 2000), as well as to the asymmetry of power that characterises international trade negotiations.

Both constraints have a strong global dimension and consequently there is scope for them to be dealt with through global mechanisms. Certainly, a meaningful global agreement on climate mitigation policy would address both, to varying degrees, because a global constraint on carbon would, as trade negotiators are wont to say, 'level the free market playing field'. But the Paris Agreement does not provide for *global climate policy* – it provides for a global framework in which individual countries develop *individual targets and policies* through Nationally Determined Contributions (NDCs), which is understandable in the context of national interest, but does nevertheless reinforce some of the trade–climate challenges described above. It is these challenges that are the focus of this chapter.

In the following section we explore how concerns about trade competitiveness manifest in domestic economies such that they impede attempts to pursue decarbonisation. In Section 21.3, we explore the tensions that exist between climate policies and the legal frameworks that support international trade and investment. In Section 21.4, we explore some of the global mechanisms and processes that can be used to overcome or at least lessen some of those tensions. Finally, in Section 21.4.5, we turn to the Paris Agreement. Perhaps surprisingly, the Paris Agreement does not make direct reference to trade or investment policy – the links are indirect – but it does offer new global markets and rules which could support the development of new markets in the international system of trade, and we provide some preliminary thoughts on those opportunities.

21.2 State Sovereignty, International Trade and the 'Competitiveness Straitjacket'

Clearly, staying within a 2 °C global rise in atmospheric temperature so as to avoid deleterious societal and environmental consequences will require profound changes in prevailing carbon-based socio-economic development frameworks (SDSN and IDDRI 2014) and a reorientation of the underlying principles of the global world order towards environmental sustainability (Faunce et al. 2014). While this challenge is immense, history tells us that such fundamental transformations have been sought and rapidly achieved in the past. The shift away from the mercantilism that dominated the sixteenth through eighteenth centuries towards the paradigm of comparative advantage and international trade that has characterised the nineteenth and twentieth centuries is the most salient example. However, the anti-mercantilism thought that emerged from Western Europe in the late eighteenth and early nineteenth centuries has subsequently entrenched a range of principles and practices that are now problematic in the quest for low-carbon economies. Paehlke (2003) argues this global economic integration forces states to pursue lowest-common-denominator socio-economic policies, in order to maintain competitiveness vis-à-vis other jurisdictions. The ready acceptance of the principles of comparative advantage and the trade liberalisation agenda has, ironically, produced a situation whereby states are severely constrained by the need to be globally competitive in an international system dominated by economic and political corporate power. This constraint has in turn produced a complex array of domestic policies almost all of which contradict the goal of low-carbon economies.

Why is it that, despite the many promises of collective action and indeed the hope garnered from the Paris Agreement, there has not been more progress made by the international community to mitigate climate change? To many, the answer is self-evident: because states are concerned about the short- to medium-term economic and social costs of fundamentally changing the way the corporations that dominate their economies produce, consume and trade goods and services. However, this answer hides two underlying tensions that need to be identified and understood because they clearly lend themselves to correction at the global level and *not* within the confines of a nation state. Both tensions relate to state sovereignty and we refer particularly to the excellent work of Philip Cerny in this section (Cerney 1993, 1994, 1995, 1997).

The first tension lies in the erosion of state sovereignty (including its capacity for governance and regulation) by increasingly integrated world economic markets dominated by multinational corporations with increasingly linear supply chains. These changes have, in turn, altered the power balance between society and capital in favour of the latter. Such multinational corporations have become the primary owners of mobile factors of production and possessors of largely immobile factors of production like land and labour, at the expense of geographically bound actors like the sovereign state (Hulsemeyer 2004; Vitali et al. 2011).

In parallel, and as a result of corporate election funding, lobbying, corruption and media influence, the principal political authority, in the form of the sovereign state, has declined in influence. In other words, the relative power and leverage of sovereign states has altered in favour of business enterprises. As Hulsemeyer (2004: 2) states, 'in an age of global finance and multinational corporations, states can no longer capture the market that they wish to regulate; the latter has outgrown the former in size, and we can think of this as a pendulum swing toward "the market"'. Similarly, Cerny (1995: 610) writes:

the ability of firms, market actors and competing parts of the national state apparatus itself to defend and expand their economic and political turf through activities such as transnational policy network-ing and regulatory arbitrage – the capacity of industrial and financial sectors to whipsaw the state apparatus by pushing state agencies into a process of competitive deregulation or what economists call competition in laxity – has both undermined the control span of the state from without and fragmented it from within.

The result of this pendulum swing has been the emergence of the so-called 'competition state' whose primary task is not, as was traditionally assumed, to serve the interests of its citizens or their environment, but to make domestic economic activities more competitive in global terms for corporations. This in turn has transformed the state from one that provides public goods and protects the domestic economy from external forces (arguably including human-induced climate change), to one that serves to facilitate greater privatisation, commodification, marketisation and deregulation (Cerny 1995, 1997; Hulsemeyer 2004). In many ways, these developments have, in essence, paralysed the state, preventing it from establishing its own, comparatively more stringent environmental – or climate – policy at the state level. Paehlke (2003: 2) describes the trade-off between economic integration and state sovereignty as 'democracy's dilemma':

Democracy's dilemma is this: global economic integration virtually requires some form of corresponding political integration, but the very notion of global government in any form is worrisome, especially perhaps to those with strong liberal democratic instincts. The response to this dilemma amongst those who advance freer global trade and investment opportunities (and the expanded integration of media, communications, travel and immigrations that are bound up in that process) is often disingenuous denial. They imagine, assert, or proceed as if the world can be integrated ever more tightly economically, while each nation at the same time is 'free' to establish its own rules regarding social equity, environmental protection, and all manner of other 'domestic policy concerns'. This is simply not the case.

Another, contrasting perspective is that the global integration required is being achieved in the corporate realm, through takeovers, collusion and capture of the democratic political system. The constraints on the state can be partly dealt with by 'collective action' at the international level, particularly for a challenge like climate change: on the one hand, the more states that sign on to a global climate agreement, the less likely it is that any one state will suffer economic disadvantage, particularly in terms of foreign corporate investment vis-à-vis its competitors; on the other hand, the global extent and impact of climate change necessitates a global response so that significant results can be achieved. In this sense, the erosion of state sovereignty could be viewed as a positive development for the pursuit of international climate cooperation as it encourages collective action. Unfortunately, the very limited real action on climate mitigation after the Paris Agreement suggests Paehlke's analysis is right.

Somewhat paradoxically, the second challenge posed by corporate-eroded state sovereignty lies in the power still embedded in the principle itself (i.e. state sovereignty), which still allows the weakened states to engage – or not – in international efforts on climate change. That is, the sovereign right of states to pull out of negotiations or agreements on the basis of national interest (or anything else for that matter) can be enough to block attempts at progressive international climate policy – either the state chooses to withdraw from the negotiations, or a 'watering down' of the agreement is negotiated to avoid such a withdrawal. We have seen both instances repeated regularly in the context of the UN Framework Convention on Climate Change (UNFCCC) negotiations. So, states can, and do, engage in international climate politics, but on their own terms and driven by their need to be internationally competitive.

But how does the imperative of 'global competitiveness' in the international system of trade affect a state's ability to pursue the decarbonisation strategies articulated in this volume? In a world of relatively open trade, financial deregulation, increased leverage of multinational corporations and increasing impact of information technology, the constraint of competitiveness is manifest in the state's restricted ability to effectively provide all three categories of public goods (following Cerny 1995: 608–609):

- **Regulatory** public goods are those that provide a workable framework for the state system as a whole. These include the establishment (and protection) of private and public property rights, a stable currency, a (limited) system of trade protection and other systems that act as safety nets for potentially system-threatening market failures (i.e. banking and financial services regulation; anti-corporate fraud regulation). Essentially, regulatory goods provide the framework for economic activity to take place in a way that conforms to basic social values such as predictability and certainty according to the rule of law, justice, equity and environmental sustainability (Faunce et al. 2014). In relation to decarbonisation, regulatory public goods would be those, for example, that ensure tradable carbon credits are protected through law, or the establishment of a legal and market framework that enables the penetration of distributed renewable energy gener- ation into existing fossil-fuel-based electricity markets. However, perhaps the most significant opportunity for governments to account for decarbonisation through regula- tory goods lies in their procurement policies – as the largest 'buyer' of goods and services in most countries, shifting the criteria for procurement to include carbon emissions considerations would fundamentally alter the framework for economic activity in that country.
- **Productive or distributive** goods are those state-sponsored or state-sanctioned activities of production and distribution such as direct or indirect provision of public infrastructure and utilities, state involvement in finance capital, and various public subsidies. The 'productive' aspect refers to the provision of infrastructure (technical and non-technical) while the distributive aspect refers to the full or partial ownership of certain industries by government, most notably the utilities industries. In relation to decarbonisation, such goods would include, for example, the provision of subsidies or hard infrastructure to support the uptake of electric vehicles, or the development of energy efficiency standards in the built environment.
- **Redistributive** goods are the goods most often associated with welfare states. These goods include, among others, health care services, unemployment policies, financial assistance for low-income families as well as the vulnerable and disabled, and environ- mental protection. It is through redistributive goods that political authorities try to achieve an allocation of income that broadly reflects the social values of a given society, which would not occur through the market economy alone. Thus, the choice of

redistributive public good, for instance environmental protection versus support for low-income earners, is a normative judgement made by government to reflect the normative values of society. It has been argued that one of the major failures of contemporary economics is its failure to value ecosystem goods (such as fresh air, water, soil and serenity in nature, etc.), and clearly the failure to account for GHG emissions in economic transactions is a case in point. In relation to decarbonisation, redistributive public goods would include the provision of solar panels for public housing and government action to support 'just' transitions for affected communities.

Regrettably, with regard to regulatory public goods, the increased leverage of multi-national corporations has seen governments opt in favour of competitive deregulation in order to attract foreign direct investment (FDI) over competing nation states. The fact that regulatory public goods together create a regulatory framework only adds to the problem as the state's regulatory core is reshaped around economic – that is, competitive – interests (Cerny 1995). A similar fate exists for productive or distributive goods. As the competitiveness of a nation state becomes the primary objective, states focus on the provision of technical infrastructure and high-level education in an effort to attract more FDI, at the expense of goods like basic public education or health. In addition to this, previously state-owned enterprises (such as power plants, telecommunications facilities, water supply utilities and so on) have been 'sold off', with the state abandoning its shares in return for 'quick cash' and the easing of their fiscal crises. Finally, as the 'competition state' struggles to remain competitive in the global economy, the extent to which redistributive public goods can be provided is severely diminished for two reasons: states have less money to spend on redistributive public goods as a result of their fiscal crises; and the redistributive goods are perceived to increase labour costs that in turn reduce the incentives for productive investment (Hulsemeyer 2004). Consequently, the impact of globalisation has meant that intervention by the state is increasingly shaped by the perceived imperatives of international competitiveness rather than by domestic welfare goals, such as mitigating the effects of climate change.

In relation to decarbonisation specifically, there are many countries in which energy-intensive trade-exposed industries dominate the economy (e.g. the USA, Australia, Saudi Arabia), and for those countries competitiveness and decarbonisation are inimical. Recent experience in Australia provides a salient example, with the concerns of a few small but important electorates in fossil-fuel-dependent Queensland 'trumping' the concerns of many millions more people in less exposed electorates in other states and territories. In the 2019 federal election, political parties' approval or otherwise of the Adani coal mine in Queensland became a litmus test in the false dichotomy between 'protecting jobs and growth' and 'progressive climate change policy'.

Table 21.1 outlines some of the key public goods that states provide to either inhibit or enable decarbonisation. The governance challenge captured by this table is thus: in an era dominated by the imperative of international competitiveness, states are motivated to provide those goods that inhibit decarbonisation, and they are disinclined to provide those that enable decarbonisation.

Table 21.1. *The provision of public goods: prohibiting versus enabling the transition to decarbonised economies*

Type of public good	Examples that inhibit decarbonisation	Examples that enable decarbonisation	Level of government(s) with the potential to act
Regulatory	Weak – or not enforced – environmental regulation of carbon-based industries Free trade agreements that prevent state governments from pursuing trade-related environmental measures Tax arrangements that encourage carbon-based infrastructure development based on short-term economic gains Inhibiting capacity of non-governmental organisations to use the courts to challenge carbon-based industrial development Legal acceptance of carbon-based industries' donations to and lobbying of political parties Legal frameworks that make it difficult for low-carbon industries to penetrate existing markets	Reforms to statutory and institutional arrangements to mandate low-carbon inputs and outputs (and enforcement thereof) A nationally consistent constraint on carbon, such as a tax or emissions trading scheme Renewable energy target and energy efficiency standards relating to emissions in sectors such as transport, built environment, agriculture, etc. Microeconomic reforms to ensure consistency across substate boundaries (i.e. competition policy) Border carbon adjustments (if necessary and feasible, see Section 21.3)	National State/provincial
Productive/ distributive	Regulatory and pricing arrangements that don't reflect the full cost (including GHG emissions) of providing critical services such as water, energy and land development Direct or indirect support to the fossil fuel industry through tax breaks, subsidies, land grants and regulatory exemptions Reinforcing or misrepresenting the job provision and other contributions to the economy provided by carbon-intensive industries even when they are small relative to other sectors	Provision of renewable energy infrastructure or related infrastructure such as that to support the roll-out of electric vehicles Fossil fuel divestment (by large funds, cities) Standards, subsidies and regulatory arrangements to support low-carbon process and production techniques and technologies Research and development activities that encourage the development and uptake of low-carbon technologies Educational arrangements (qualifications, training providers, standards, etc.) to ensure the labour force is equipped to deal with a low-carbon economy	National State/provincial

Redistributive	Poorly targeted subsidies (direct or indirect) that encourage inefficient use of energy by consumers	Reforms to tax systems to protect vulnerable members of society from possible climate-related increases in the cost of utilities (electricity, water, sewerage, etc.)	National
	Carbon-based industries' media and political attacks on low-carbon competitors (wind farms, solar arrays) and on the scientific evidence supporting the urgent need for decarbonisation of the economy	Structural adjustment funds and processes to support those industries that will be phased out	State/provincial
		Rebates to low-income earners to better insulate homes	Local
		Support for the scientific basis behind the urgent need for decarbonisation of the economy	

579

21.3 Tensions between Climate Policy and International Trade and Investment Law

The previous section explored the impact of international trade on state competitiveness and consequently on the state's capacity to address climate change. In many ways it is an indirect tension between trade and climate policy. In this section we explore very direct tensions that exist between trade and climate policy; tensions which are captured in the legal frameworks supporting the international system of trade and which have already been realised through legal disputes.

The Marrakesh Agreement Establishing the World Trade Organization, signed in 1994 (the WTO Agreement),[2] ostensibly seeks to strike a balance between, on the one hand, the expansion of production and trade in goods and services, and, on the other hand, sustainable development and the protection of the environment. Indeed, WTO members recognise, in the first recital of the preamble of the WTO Agreement, that:

their relations in the field of trade and economic endeavour should be conducted with a view to raising standards of living, ensuring full employment and a large and steadily growing volume of real income and effective demand, and expanding the production of and trade in goods and services, while allowing for the optimal use of the world's resources in accordance with the objective of sustainable development, seeking both to protect and preserve the environment and to enhance the means for doing so in a manner consistent with their respective needs and concerns at different levels of economic development.

This book provides a 'how to' model for governments to 'decarbonise' their economies, and while the type of reforms and interventions vary enormously from one sector to another (i.e. addressing energy efficiency draws on vastly different policy instruments to, say, shifting from coal-fired power stations to decentralised, distributed energy generation), the public goods identified in Table 21.1 go some way to describing the types of strategies that governments can and will pursue. However, the pursuit of a low-carbon economy by a state will intersect with the trade system in one of two ways:

- some of the policy instruments used to encourage the adoption of low-carbon technologies or process and production techniques could be deemed contrary to WTO law; and/or
- states will be encouraged to use some kind of trade-related environmental measure, such as carbon footprinting certification or border carbon adjustments, to ensure a level playing field between domestic and imported producers, which could similarly be deemed contrary to WTO law (Cosbey 2016).

Article XX of the General Agreement on Tariffs and Trade (GATT) 1994[3] on general exceptions occupies an important role within the GATT as an expression of the balance the drafters intended between certain agreed non-trade social objectives and the objectives of trade law. Its crafting shows the care that was taken to ensure the autonomy of the members

---

[2] *Marrakesh Agreement Establishing the World Trade Organization*, opened for signature 15 April 1994, 1867 UNTS 3. Available at: www.wto.org/english/docs_e/legal_e/04-wto_e.htm.
[3] *General Agreement on Tariffs and Trade 1994*, initially signed 30 October 1947, revised in 1994. Available at: www.wto.org/english/docs_e/legal_e/06-gatt.pdf.

to pursue such non-trade objectives while remaining within the framework of an agreed set of rules of conduct. In specific circumstances, the pursuit of those social objectives – which include environmental protection – may breach even such fundamental principles as non-discrimination. Two exceptions are of particular relevance to decarbonisation: paragraphs (b) and (g) of Article XX. Pursuant to these two paragraphs, WTO members may adopt policy measures that are inconsistent with even such fundamental principles as non-discrimination in the pursuit of the protection of human, animal or plant life or health (paragraph (b)), or the conservation of exhaustible natural resources (paragraph (g)). For example, this principle was upheld in the case of Australian 'plain packaging' laws for cigarette packets, which were deemed to be a legitimate public health measure (WTO 2018a).

The chapeau of Article XX of the GATT 1994 reads:

Subject to the requirement that such measures are not applied in a manner which would constitute a means of arbitrary or unjustifiable discrimination between countries where the same conditions prevail, or a disguised restriction on international trade, nothing in this Agreement shall be construed to prevent the adoption or enforcement by any contracting party of measures . . .

It is important to note that Article XX exceptions apply to 'this Agreement'. The narrowest meaning proposed in the literature equates 'this Agreement' with the GATT 1994. In other words, Article XX applies not only to some but to all provisions of the GATT 1994. Accordingly, the general exceptions apply to all provisions of the GATT 1994, including rules on subsidies (such as for renewable energy technologies) and countervailing duties.

It is indeed in relation to the use of subsidies that countries have traditionally 'fallen foul' of WTO law, and the shift to decarbonise economies will be no different. The provision of subsidies to support climate-friendly industries is already evident across economies, most obviously in support for renewable energy generation through the use of feed-in tariffs. However, subsidies could also take the form of grants of land or cash, low-interest loans, preferential tax arrangements, price floors or premiums, or prescriptive or mandatory government procurement rules (Cosbey 2016). Certainly, the need for such subsidies to support decarbonisation is obvious: nascent low-carbon industries need financial support so as to glean some competitive advantage over incumbent, carbon-intensive industries. However, the use of subsidies – and the distortion they can create – is not straightforward, from an economic or environmental perspective. For example, subsidies can sometimes be used simply to protect less efficient firms irrespective of their 'environmental' credentials, and similarly they may be prohibitive against the growth of new 'green' industries in developing countries. As Cosbey (2016: 3) explains, 'The contest for market share in the emerging sectors is definitely tilted towards larger economies, both because of their superior ability to support infant industries and because they provide supported firms with a larger domestic market for their products.'

In addition to subsidies, another measure that is used frequently to support environmental outcomes is eco-labelling and third-party certification. Until now, the overwhelming majority of eco-labelling has been non-state-sanctioned; that is, not made mandatory by governments but encouraged by consumer demand and market forces, such as the ability for a firm to receive a premium price for their product and/or to increase their market share. For example, eco-labels to support the goals of sustainable forestry (such as the Forest Stewardship Council), sustainable development (e.g. Fair Trade certification) and sustainable fishing (e.g. the Marine Stewardship Council) have gained in prominence to the extent that some argue they have become de facto mandatory in practice (Botterill and Daugbjerg 2011). Importantly, because these eco-labels are voluntary and not state-sanctioned, they do not contravene WTO law. However, as efforts to address climate change intensify, it is possible that some jurisdictions might introduce mandatory 'carbon footprint' labelling, which would potentially clash with WTO law. Hitherto, though, efforts to establish carbon footprinting have largely failed, owing to challenges in establishing a sound methodology to measure the carbon embedded in products, and the potentially deleterious impact of such labelling on less-developed countries (Cosbey 2016).

Beyond the introduction of subsidies and carbon footprint labelling, another trade-related environmental measure that could support the shift to decarbonised economies is so-called 'border carbon adjustments' (BCAs), also known as border tax adjustments. Responding precisely to the competitiveness challenges described in Section 21.2, BCAs attempt to 'level the playing field' by allowing carbon-taxing countries to impose fees on imports manufactured in non-carbon-taxing countries. Aside from addressing competitiveness concerns for domestic industries, BCAs could also address the problem of carbon 'leakage', that is, carbon-intensive industries moving production to jurisdictions with no or less stringent climate policy, thus mitigating any intended reductions in GHG emissions. So too, the use of BCAs – especially by significant markets such as the EU or the USA – could encourage countries with limited or no climate policy to change tack, on the grounds that it would hurt their export industries not to.

So far, several significant markets have explored the possibility of introducing BCAs, including the USA, but it has been the EU's 'Green Deal' announced in 2019 that has come closest to indicating firm support for the measure, with Ursula von der Leyen, President of the European Commission, articulating her ambition to introduce a Carbon Border Tax so as 'to avoid carbon leakage' and in such a way as to be 'fully compliant with World Trade Organization rules' (von der Leyen 2019: 5). Nevertheless, the idea of BCAs is not new, and there has been significant scholarly and policy work dedicated to understanding the ways in which they could work. Morris (2018) and Cosbey (2016) describe several critical issues to consider, including the scope of the products that ought to be covered by the BCA; the legality of the programme vis-à-vis WTO law; the accessibility of information to support administrators in their implementation and monitoring of the scheme; the administrative costs associated with administering the BCA programme, including determining which responsibilities would fall to which agencies to administer the programme; determining whether or how to reduce or suspend BCAs under certain conditions, such as a change in the climate policy within the trading partner country; agreeing on how stakeholders might appeal determinations made by the BCA-administering body in a way that is transparent, timely and cost-effective; and ensuring the system can take account of developments in the technologies and policies being used to support climate policies internationally.

There is no consensus on how these issues can be satisfactorily dealt with, and while the opportunity for BCAs to achieve some climate ambitions is gaining traction, there are still very strong reservations held by trade policy experts about the potential impact of BCAs on developing countries, and indeed on the international system of trade itself. Still, if the EU were to proceed with their proposed Carbon Border Tax, it would be a very direct example of the trade system being used to further the decarbonisation agenda, and it would surely herald a new impetus for climate action globally.

Relatedly, and similarly important in creating obstacles to the replacement of carbon-based fuels with renewable energy technologies, are ISDS mechanisms in regional and bilateral trade agreements. In July 2011, for example, Texas-based Mesa Power Group LLC served Canada with an ISDS claim under the North American Free Trade Agreement's (NAFTA) Chapter 11 in connection with Ontario's solar feed-in tariff (FIT) programme. Ontario's FIT programme has also been challenged by Japan and the EU under Article 2.1 of the WTO Agreement on Trade-Related Investment Measures (TRIMs Agreement), which restricts states' freedom to impose domestic content performance requirements despite exceptions relevant to the protection of the environment in paragraphs (b) and (g) of Article XX of GATT 1994 (Faunce et al. 2014). While Mesa Power Group LLC lost their claim and were ordered to pay costs, Japan and the EU's claims were upheld. The challenging by corporations of state action to achieve climate mitigation – such as through FITs or standards – is the most tangible example of the tension between trade and climate policy, and we are likely to see more such examples in the coming years. One of the key challenges in such claims lies is determining what is a legitimate measure to protect human, animal or plant life, rather than a form of protectionism. For example, India recently brought proceedings in the WTO Dispute Settlement Panel (DSP) questioning domestic content requirements and subsidies for renewable energy instituted by the governments of the US states of Washington, California, Montana, Massachusetts, Connecticut, Michigan, Delaware and Minnesota. The DSP found that all of the measures at issue were inconsistent with WTO law because they provide an advantage for the use of domestic products, which amounts to less favourable treatment for like imported products (WTO 2014, 2018b, 2019; ICTSD 2016).

In addition to using ISDS mechanisms to retard climate action, companies could also refer to the Energy Charter Treaty (ECT), which was signed in Lisbon in December 1994 with a Protocol on Energy Efficiency and Related Environmental Aspects (PEEREA) (both in force April 1998). The ECT was designed for the protection of foreign investments (national treatment or most-favoured nation treatment) and to provide non-discriminatory conditions for trade in energy materials, products and energy-related equipment based on WTO rules and to ensure reliable cross-border energy transit flows through pipelines, grids and other means of transport. It, too, could well be at the forefront of efforts by multinational corporations with substantial investments in the electricity grid or in centralised fossil fuel supplies (coal, oil and natural gas) to impede the development of competitive solar fuels technologies. A claim has already been brought against Spain, for example, under the ECT, by a group of 14 investors over that nation's retrospective cuts to solar energy tariffs. Foreign investors also challenged the Italian government, under the

ECT or bilateral investment treaties (BITs), over its efforts to roll back FITs in the country's booming solar energy sector (Faunce et al. 2014) although the tribunal found in favour of Italy.

21.4 Global Mechanisms and Processes to Overcome Trade and Climate Policy Tensions

As we've seen above, the tensions between the international system of trade and climate policy are both direct and indirect. Indirectly, the 'corporate competitiveness straitjacket' makes it difficult for countries to take strong action on climate change. In addition, international trade and investment law – and the norms that underpin the international system of trade – can pose technical, legal and geopolitical challenges that undermine climate action. The fact that these tensions exist at the global level means that they are, arguably, best dealt with at that level and there are a range of global mechanisms and processes that can and ought to be pursued in earnest in the coming years. This section identifies five actions that could successfully support the low-carbon agenda:

- raising awareness of the strong economic benefits from decarbonisation;
- developing international frameworks for 'green' finance;
- pursuing global standards and the standardisation of low-carbon technologies;
- reforming trade law to encourage low-carbon goods and services; and
- ensuring coherence across international frameworks.

21.4.1 Raising Awareness of the Strong Economic Benefits from Decarbonisation

Fears about diminished competitiveness pervade politics and policy-making; fears which are reinforced by the lobbying of energy-intensive trade-exposed industries. There is, however, an emerging body of evidence suggesting that concerns about the negative impacts of stringent environmental regulation on international competitiveness are unfounded. Researchers from the OECD (Albrizio et al. 2014) have created an index that calculates the explicit and implicit cost of environmental policies so as to assess the strictness of environmental regulations across 24 wealthy OECD members between 1990 and 2012 and, consequently, to measure the effect of those regulations on productivity. The study found that the strictness of environmental policies has 'increased significantly' in all the countries over the past two decades but that (Albrizio et al. 2014: 6):

First, stringent environmental policies should not be expected to have detrimental effects on productivity, in particular if policies are well designed. Second, there is no evident trade-off between policy stringency and competition-friendliness. The design and implementation of stringent environmental policies can and should be geared toward paying due attention to barriers to entry and competition, making the greening of the economy consistent with continuing productivity growth.

Clearly, this contradicts the prevailing view that stringent environmental regulation leads almost immediately and inevitably to competitive disadvantage. One possible explanation for the findings is that new regulations have pushed firms to operate more efficiently than would otherwise have been the case – or in other words, it reflects Porter's seminal hypothesis that environmental regulation can encourage innovation and investment and be implemented while promoting strong competition (Porter and Linde 1995). The research also found that market-based environmental policies, such as emissions trading schemes, 'tend to have a more robust positive effect on productivity growth' (Albrizio et al. 2014: 6). However, the impact of stronger environmental regulation is not uniform. The most advanced industries and firms have seen an increase in productivity as a result of stricter rules while 'less productive' firms are likely to see negative effects. In short, those firms that already optimise productivity will gain, while those firms that are less efficient will suffer under strong regulatory arrangements. As the world comes to grips with the enormity of the decarbonisation agenda, these findings are significant and should be shared widely, because they will go a long way to removing the 'competitiveness straitjacket'.

21.4.2 Developing International Frameworks for 'Green' Finance

Scaling up the provision of sustainable, low-carbon investment and finance at the global level is urgently needed to ensure that new investment in infrastructure supports the climate agenda while simultaneously fostering economic development. There is no shortage of capital, but new sources of financing need to be mobilised, especially those that explicitly recognise and reward the 'green' credentials of investments. Early moves to develop Sustainable Finance Roadmaps, which provide capital cheaper to those activities that are 'greener', should be pursued as a matter of priority. Such roadmaps will provide 'pathways and policy signals and set frameworks' to enable the finance sector to contribute to the global decarbonisation agenda by providing the trillions of private investment dollars needed (RIAA 2018). So too, public finance and investment can also catalyse the low-carbon transition and governments need to reconsider their support for investments in GHG-intensive activities and mainstream climate objectives into public procurement and official development assistance (Baron 2016).

21.4.3 Pursuing Global Standards and the Standardisation of Low-Carbon Technologies

The central role played by global standardisation in the economic development of the nineteenth and twentieth century cannot be overstated – standardisation was critical for the electrification of cities, the provision of transboundary transportation, the digitalisation of workplaces and the integration of international financial markets. In short, where global standards can be agreed, economic development and international trade follows. For decarbonisation, the following areas could be particularly amenable to global efforts to standardise:

Energy-Intensive, Trade-Exposed Industries. Some industries are particularly vulnerable to international competitiveness concerns in the context of decarbonisation: steel, basic metals, refined products and other chemicals, aluminium production, paper pulping, concrete and the coal, oil and gas industries more directly. For these industries, the options for pursuing meaningful low-carbon outcomes that don't risk losing international competitiveness are very limited. In this case, the best solution would be a set of international standards overseeing lower-carbon process and production techniques. While politically very difficult to achieve, the chance of industry-led standards may be more feasible than state-led standards given the current impasse in international climate negotiations.

Renewable Energy Technologies. The uptake of low- or zero-carbon technologies is central to decarbonisation, and standardisation at the international, national, regional and organisational levels is critical to that dissemination. The International Renewable Energy Agency's (IRENA) study, *International Standardisation in the Field of Renewable Energy*, identifies over 570 standards in the current renewable energy technologies (RET) landscape, but it identifies a number of gaps, particularly for post-installation aspects of RET, such as operation, maintenance and repair. Perhaps most importantly, IRENA recognises the need for standardisation to be sensitive to the context in which standards are imposed, as inequity between countries can be exacerbated in that eventuality, particularly through the trade system:

A key message from this study is that if standards are to remain of global relevance then the international standardisation route should support all regional, demographic, technical development, societal and environmental aspects of their use. This is particularly relevant in developing countries, where issues of cost, capacity or resource availability limit their involvement in the whole international standards development process. Consequently, international standards may not always consider specific issues relevant to some regions, such as specific climate conditions, infrastructure development or skills available for implementing renewable energy systems. It is therefore important to make use of existing mechanisms, and develop new ones, to ensure the engagement of all stakeholders, particularly in developing countries, in the international standardisation process. This engagement is especially relevant if those stakeholders are to be involved in competitive and inclusive global trade. (IRENA 2013: 5)

The Transport Sector. The next wave of low-carbon investment will almost certainly be in the transport sector, with electric vehicles expected to overtake combustion engine vehicles soon after 2035. However, such progress will demand standardisation of electric vehicles and the infrastructure to support them, from charging stations to professional qualifications to repair electric vehicles.

21.4.4 Reforming Trade Law to Encourage Low-Carbon Goods and Services

The inclusion of environmental provisions in what are fundamentally trade agreements has been important to the achievement of more sustainable outcomes because this 'seepage' of environmental issues into the trade domain has led to the spread of environmental principles from one region to another through economic markets. However, as economic integration has spread it has resulted in a shift in focus away from protection-at-the-border policies onto domestic policies that affect production costs and conditions of sale. In doing so, and as we saw earlier in this chapter, a country's competition, health, social security, minimum wage and environmental policies are open to scrutiny by trading partners due to their likely effect on the cost of production and international competitive position (Pearson 1993). But the scrutiny of states' domestic environmental regulations could in itself trigger the development of more robust environmental governance regimes, through a 'building block' of transnational administrative law, with mutual recognition at its heart. The theoretical and conceptual precedence for this lies in earlier work on the role of normative power – the diffusion of norms by strong actors – in international regimes but the potential introduction of BCAs could see such a building block achieved through more immediate means. Nevertheless, even without the imposition of BCAs, a commitment by the WTO and WTO member countries to support the following would be similarly beneficial: (i) reform of tariffs and quota regimes and removal of non-tariff barriers on environmental goods and services; (ii) adoption of third-party certification schemes as mandatory for key carbon-intensive technologies and products; (iii) introduction of 'sustainability' criteria to assess the validity of any investor–state dispute cases; and (iv) rebadging the International Centre for Trade and Sustainable Development (ICTSD) as the International Centre for Trade and Environmental Sustainability (ICTES) and reorganising the WTO as the World Trade and Environmental Sustainability Organization (WTESO).

21.4.5 Ensuring Coherence across Global Agreements and Policy Frameworks

Finally, global decarbonisation will rely on countries, companies and citizens 'pulling in the same direction', which in turn demands consistency and alignment across all international, national and regional policy frameworks – including, but not limited to, trade agreements. Arguably, the UN's Sustainable Development Goals (SDGs) are supposed to achieve that coherence, but the absence of any legal 'teeth' to the SDGs, and the relative obscurity in which they exist, means that pursuit of the SDG agenda alone is unlikely to be enough. Rather, international policy frameworks covering everything from global investment, crime and corruption, through to more recent commitments such as the Kobe 3R Action Plan and the G7 commitment on Responsible Supply Chains need to be reviewed and reformed to account for decarbonisation ambitions.

In relation to the recent Paris Climate Agreement, and in addition to the establishment of NDCs, carbon credits and offset markets are re-emphasised as a means to achieve mitigation, and rules to support such a market are in the process of being drafted. Such a market will go a long way to encouraging a global response to mitigation in the absence of a global carbon price or trading system. Indeed, even though international trade and investment are not covered in the Paris Agreement, the design of the agreement should provide confidence to investors that decarbonisation is here to stay, which in turn will see the benefits of shifting to green economic activity played out.

21.5 Conclusion

International trade, competitiveness and climate policy are inherently linked, and are centrally concerned with the interplay between state, corporate and community power. Until now, and not withstanding some significant investment in low-carbon energy, that power play has seen genuine efforts to reduce GHG emissions largely fail. Nevertheless, reflecting these challenges and to support the transition to low-carbon economies, there are a number of global priorities that need to be pursued through global processes. In the first instance, the economic, social and environmental opportunities presented by a shift to decarbonisation need to be better communicated to all levels of government and to citizens – it is only when the latter demand a shift that the former will feel empowered to overcome the status they find themselves in. So too, efforts to reinforce global mechanisms and processes in support of decarbonisation need to be intensified – from pursuing global standards for new technologies, through to establishing global frameworks for 'sustainable' finance, and finally to securing some level of coherence between existing, and emerging, international policy frameworks. It would be naive to suggest these efforts are easy, least of all at a time when international institutions are under threat, but such coherence at a global level will be needed if the climate emergency is to be addressed quickly.

Acknowledgements

The authors would like to thank Chris McEwan for his invaluable research assistance in the preparation of this chapter, and Dr Lynette Molyneaux and Peter Dawson for their advice on earlier drafts. We dedicate this chapter to the of memory of Tom Faunce, esteemed colleague and co-author.

References

Albrizio, S. et al. (2014). *Do Environmental Policies Matter for Productivity Growth? Insights from New Cross-Country Measures of Environmental Policies*. OECD Economics Department Working Papers No. 1176. OECD Publishing. Available at: www.oecd-ilibrary.org/economics/do-environmental-policies-matter-for-productivity-growth_5jxrjncjrcxp-en.

Baron, R. (2016). *The Role of Public Procurement in Low-Carbon Innovation*. Paris: OECD.

Botterill, L. and Daugbjerg, C. (2011). Engaging with private sector standards: A case study of GLOBALG.A.P. *Australian Journal of International Affairs*, 65, 488–504.

Brack, D., Grubb, M. and Windram, C. (2000). *International Trade and Climate Change Policies*. London: Earthscan.

Cerny, P. (1993). *Finance and World Politics: Markets, Regimes and States in the Post-Hegemonic Era*. Cheltenham: Edward Elgar.

Cerny, P. (1994). The dynamics of financial globalization: Technology, market structure and policy response. *Policy Sciences*, 27, 319–342.

Cerny, P. (1995). Globalisation and the changing logic of collective action. *International Organisation*, 49, 595–625.

Cerny, P. (1997). Paradoxes of the competition state: The dynamics of political globalization. *Government & Opposition*, 32, 251–274.

Cosbey, A. (2016). The Paris Climate Agreement: What implications for trade? *Commonwealth Trade Hot Topics*, 129. Produced by the Trade Division of the Commonwealth Secretariat. Available at: https://pdfs.semanticscholar.org/7455/cf06abd0200c27c3415a6bda5eb426bc6353.pdf.

Faunce, T. A., Wasson, A. and Crow, K. (2014). Environmental sustainability and global health law: The case study of global artificial photosynthesis. In M. Freeman, S. Hawkes and B. Bennett, eds., *Law and Global Health: Current Legal Issues*, Vol. 16. Oxford: Oxford University Press, pp. 465–477.

Hulsemeyer, A. (2004). *Globalization and Institutional Adjustment: Federalism as Obstacle?* Ashgate Publishing.

ICTSD (International Centre for Trade and Sustainable Development) (2016). NAFTA Tribunal issues ruling in Mesa Power–Canada case. *Bridges*, 20. 4 May. Available at: https://ictsd.iisd.org/bridges-news/bridges/news/nafta-tribunal-issues-ruling-in-mesa-power-canada-case.

IRENA (International Renewable Energy Agency) (2013). *International Standardisation in the Field of Renewable Energy*. Abu Dhabi: International Renewable Energy Agency. Available at: www.irena.org/publications/2013/Mar/International-Standardisation-in-the-Field-of-Renewable-Energy.

Morris, A. C. (2018). *Making Border Carbon Adjustments Work in Law and Practice*. Washington, DC: Tax Policy Centre.

Oosterhuis, F. H. and Ten Brink, P. (2014). *Paying the Polluter: Environmental Subsidies and Their Reform*. Cheltenham: Edward Elgar.

Paehlke, R. C. (2003). *Democracy's Dilemma: Environment, Social Equity and the Global Economy*. Cambridge, MA: MIT Press.

Pearson, C. (1993). The trade–environment nexus: What's new since 1972? In D. Zaelka, P. Orbuch and R. Houseman, eds., *Trade and the Environment: Law, Economics and Policy*. Washington, DC: Centre for International Environmental Law.

Porter, M. and van der Linde, C. (1995). Towards a new conception of the environment–competitiveness relationship. *Journal of Economic Perspectives*, 9, 97–118.

RIAA (Responsible Investment Association Australia) (2018). *Sustainable Finance Roadmaps: Aligning Finance with a Resilient and Sustainable Economy*. Briefing paper. Sydney: Responsible Investment Association Australia. Available at: https://responsibleinvestment.org/wp-content/uploads/2018/07/Sustainable-Finance-Roadmap-BRIEFING-PAPER-FINAL-web.pdf.

SDSN (Sustainable Development Solutions Network) and IDDRI (Institute for Sustainable Development and International Relations) (2014). *Pathways to Deep Decarbonization: 2014 Report*. Sustainable Development Solutions Network and the Institute for Sustainable Development and International Relations. Available at: www.globalccsinstitute.com/archive/hub/publications/184548/pathways-deep-decarbonization-2014-report.pdf.

Srivastav, S., Fankhauser, S. and Kazaglis, A. (2018). Low-carbon competitiveness in Asia. *Economies*, 6, 5.

Vitali, S., Glattfelder, J. B. and Battiston, S. (2011). The network of global corporate control. *PLoS One*, 6, e25995.

von der Leyen, U. (2019). *A Union That Strives for More: My Agenda for Europe*. Political guidelines for the next European Commission 2019–2024. European Union. Available at: www.europarl.europa.eu/resources/library/media/20190716RES57231/20190716RES57231.pdf.

WTO (World Trade Organization) (2014). *DS412: Canada – Certain Measures Affecting the Renewable Energy Generation Sector*. Geneva: World Trade Organization. Available at: www.wto.org/english/tratop_e/dispu_e/cases_e/ds412_e.htm.

WTO (2018a). *Australia – Tobacco Plain Packaging (DS435, 458, 467)*. Geneva: World Trade Organization. Available at: www.wto.org/english/tratop_e/dispu_e/cases_e/1pagesum_e/ds467sum_e.pdf.

WTO (2018b). *DS563: United States – Certain Measures Related to Renewable Energy*. Geneva: World Trade Organization. Available at: www.wto.org/english/tratop_e/dispu_e/cases_e/ds563_e.htm.

WTO (2019). *DS510: United States – Certain Measures Relating to the Renewable Energy Sector*. Geneva: World Trade Organization. Available at: www.wto.org/english/tratop_e/dispu_e/cases_e/ds510_e.htm.

22

Improving the Governance of Governments

KEN COGHILL, BARBARA NORMAN, THOMAS SMITH,
CRISTINA NEESHAM AND ABEL KINYONDO

Executive Summary

This chapter evidences the significant risk of corruption as a constant in any system of government and in particular its significance and potential to undermine local, national and global action on climate change. Here we refer not only to the most egregious corruption such as accepting illegal benefits coincident with making decisions favouring donors, but a spectrum extending to flouting the public trust principle for more subtle reasons. The evidence that corruption can undermine efforts on climate is well established:

- Corrupt states have been among the world's worst environmental performers for decades.
- Corrupt states have a higher risk and probability of suboptimal policy and investment decision-making processes with often a lack of transparency.
- In democracies, if there is corruption, it is easier for certain interests to lobby and buy support from at least one political party, thus making it impossible to achieve bipartisan climate change reform for the long term. For example, to decarbonise national economies, private-sector actors need a level playing field on which to operate. If they invest in low-carbon technologies, they must be able to compete fairly for contracts against established firms and carbon-intensive products. Currently, there are over USD 5.2 trillion per annum globally in fossil fuel industry subsidies, and there are also significant government subsidies to agricultural commodity industries driving deforestation globally. Phasing out these subsidies is essential for being able to achieve the Paris Agreement targets and the UN's Sustainable Development Goals, yet corruption or industry capture of key policy agencies of government can make practical achievement of a level playing field impossible for new low-carbon technologies and related firms.
- Governments themselves are significant procurers of products and services. A government's procurement should align with its obligations under the Paris Climate Change Agreement and be consistent with that. Corrupt states are less likely to ensure their procurement aligns with the requirements of achieving the Paris Climate Change Agreement.

Accordingly, integrity in government is one of the most important preconditions for progress towards a low-carbon future. This chapter outlines strategies that can be implemented to proactively reduce the risk of corruption undermining efforts by nations to meet their Paris Climate Change Agreement commitments.

22.1 Introduction

Corruption is a constant risk in any system of government and remains a huge problem in many countries (OSCE 2012; Ugaz 2015). Corruption can undermine any and all attempts to achieve sustainable development, as by definition resources of all types are alienated from the community's shared public interest in favour of individuals and organisations representing vested interests. The resources alienated may be financial, human or physical. At least one and often all three of these resources are limited in almost all states. Having been diverted from serving the public interest, fewer resources remain to deliver goods and services for sustainable development.

For example, massive funds budgeted to address the effects of climate change on Australia's Great Barrier Reef (GBR) were given to the Great Barrier Reef Foundation (a non-government organisation), which had not so much as sought funding, much less demonstrated a capacity to protect the GBR from degradation (Slezak 2019). Rather, the focus was on scheduling payment of the grant to artificially create the appearance that expenditure occurred in a particular reporting period. The grant was unrelated to emissions reduction and was for vaguely worded adaptation purposes (ANAO 2019; Great Barrier Reef Foundation 2020). These funds are now not available for projects known to reduce the stress on the GBR.

In this chapter, we address the significance of corruption to climate change in particular and identify what is required to best address it. In doing so, we examine factors affecting the effectiveness of international bodies and agreements, with particular reference to the integrity and performance of nation states and subnational levels of government.

22.2 Climate Change: A Global Issue

The 2015 Paris Agreement[1] at the 21st Conference of the Parties (COP 21) is recognition that climate change requires global solutions involving purposeful action from all nations. There remains the risk that the solutions it incorporates could be undermined by a few rogue greenhouse-gas-emitting states as well as corruption at local, state and national government levels. The Transparency International (TI) 2011 report reviewed 'major climate-related corruption risks' and concluded that:

Climate change is arguably the greatest governance challenge the world has ever faced. Addressing it requires a degree of urgency, trust, cooperation and solidarity that tests the limits of conventional mechanisms and institutions to manage public goods. *(TI 2011: xxv)*

To meet that challenge:

---

[1] *Paris Agreement Under the United Nations Framework Convention on Climate Change*, opened for signature 16 February 2016. Available at: https://unfccc.int/process-and-meetings/the-paris-agreement/the-paris-agreement.

A robust system of climate governance – meaning the processes and relationships at the international, national, corporate and local levels to address the causes and effects of climate change – will be essential ... *(TI 2011: xxv)*

In the lead-up to the Paris Conference and since, TI has urged nations to act on three major points:

(1) to tackle climate change, they need to be transparent and give citizens a voice;
(2) 'climate finance' could make or break the Paris Agreement – they shouldn't let a corruption scandal ruin everything;
(3) for the Agreement to count, it needs to be accountable (TI 2015b).

This chapter describes in detail how to address these points.

22.3 Implementing the Paris Climate Change Agreement: The Risk from Corruption

Transparency International demonstrates that the risks of corruption in this area transcend the normal risks; for corruption of government in this area puts at risk not only present but also future government decisions. It extends beyond the usual forms – 'misappropriation of funds, bribery, awarding of contracts, and nepotism' – to new areas such as the corruption of the policy processes by 'distortion of scientific facts, the breach of the principles of fair representation and false claims about the green credentials of consumer products' (TI 2011: xxv–xxvi).

All of these consequences 'can be devastating in a policy area in which uncertainty abounds and trust and cooperation are essential' (TI 2011: xxvi). New unprecedented corruption opportunities will arise from the necessary massive and urgent flows of funds to transition the global economy to a low-carbon resilient future. In his 2015 Encyclical addressing climate change, Pope Francis pointed to the 'need ... to combat corruption more effectively' (Francis 2015: 126–127). The evidence that corruption can undermine efforts on climate is well established.

- Corrupt states have been among the world's worst environmental performers for decades (TI Secretariat 2001). Corrupt states tend to have suboptimal policy and investment decision-making processes with a lack of transparency. If such countries have valuable natural resources (e.g. fossil fuels), the private ownership or state control of these resources are at risk of being concentrated in the hands of a few private companies or public officials whose interests then align with maintaining extraction of these resources irrespective of the Paris Climate Change Agreement.
- Corruption erodes public trust within nations in their public institutions and government as it tends to shift investment away from public goods required for transition to a low-carbon economy, including investment in effective public services and infrastructure, capacity-building and better education and training of citizens to empower them to act (TI Secretariat 2001; TI 2011).

- In democracies, if there is corruption, it is easier for vested interests to make cooperation in reform between political parties virtually impossible to achieve.[2]
- To decarbonise national economies, private-sector actors need a level playing field on which to operate. If they invest in low-carbon technologies, they must be able to compete fairly for contracts against established firms and carbon-intensive products. Corruption or industry capture of key energy-related policy agencies of government can make practical achievement of a level playing field impossible for new low-carbon technologies and related firms.
- To decarbonise and increase resilience, private-sector investment is needed. So, a resilient low-carbon future should be far easier to achieve during prosperous times when there is money to invest than in a recession. Corruption puts this at greater risk. The World Bank and others have identified corruption as the single greatest obstacle to economic development; an obstacle that can cause the GDP growth rate of a country to be 0.5–1.0% lower than that of a similar country with little corruption (Knack and Keefer 1995; Mauro 1995; Kaufmann et al. 1999; World Bank 2011; TI 2015a).
- History shows that, where there is endemic corruption, the interests of the poor and marginalised and the environment are ignored. Inequality itself can fuel corruption (TI 2011). Those most at risk of being adversely affected by climate change are usually the most marginalised voices in the political system. The poor particularly suffer, through lower levels of social services, infrastructure investment biased against projects that aid the poor, higher taxes and/or fewer services and disadvantage in selling their agricultural produce (Rose-Akerman 1999). High and rising corruption increases income inequality and poverty (Gupta et al. 1998), making it harder to achieve both climate goals and the UN's Sustainable Development Goals (SDGs).

Accordingly, integrity in government is one of the most important preconditions for progress towards a low-carbon and sustainable future. To meet the challenges, strong national institutions operating under strong fiduciary and other standards are needed. In many countries, however, such institutions and standards are lacking; furthermore, developing countries tend to lack people with the expertise and experience, and systems and support needed (OECD 2015).

The challenge cannot be underestimated. The interconnection between the need to find the best ways to address climate change and to address corruption is clear. The good news is that the ultimate solutions required are the same – improving the governance of governments by addressing and improving the integrity of our government systems, international, national and local.

---

[2] Vested interests have compelling incentives for: maintaining markets in fossil fuels and subsidies for fossil fuel use, currently in excess of USD 600 billion per annum; maintaining markets for established high-emitting businesses and reducing competition from new low-emitting competitors and innovators; and influencing the decision-making process of governments. As a result, they have a vested interest in 'capturing', or having 'disproportionate influence', over one or more of a country's major national political parties.

22.4 How to Proactively Improve the Governance of Governments

Transparency International argues that 'a dramatic strengthening of governance mechanisms can reduce corruption risk and make climate change policy more effective and more successful' and a 'well-co-ordinated system of accountable decision making is essential' (TI 2011: xxvi).[3]

Transparency International also makes the point that 'Important climate decisions are not deliberated and decided upon only by conferences of state parties in Copenhagen or Cancún', and that perhaps even more critical decisions are made in many national and regional venues, cities and in local government (TI 2011: xxvii).

We focus on strengthening government integrity systems of nations, their states and municipalities. Australia is a useful case study of a nation relying very heavily on fossil fuels, being one of the world's highest per capita users and a major exporter of coking coal, thermal coal and liquified natural gas (LNG). Australian national, state, territory and local democracies struggle to deal with powerful vested interests in those areas and to address climate change issues. Corruption control commissions established at subnational and local government levels aim to identify and address serious corrupt conduct and support integrity systems. However, corruption may remain hidden if integrity systems are inadequate.

In identifying government integrity action that needs to be taken in any nation, it is necessary to first determine the standards and principles that are being applied and should be applied; and the nature, extent and quality of the government integrity systems covering all the levels of government. It is then necessary to compare them with best practice and reform them accordingly.

Are there accepted standards and principles by which the integrity of decisions by national, subnational and municipal governments may be guided and assessed?

We suggest that there are at least the following:

(i) the public trust principle: a universal ethical principle and a legal principle in common law countries;
(ii) environmental protection principles recognised by law;[4]
(iii) the Universal Declaration of Human Rights (UDHR), International Covenant on Civil and Political Rights (ICCPR) and other international agreements and conventions concerning human rights, including the SDGs;
(iv) international law principles concerning damage to other states (International Court of Justice and Weeramantry 1997).

Of the above, the public trust principle is the first that needs to be addressed.

---

[3] As TI note, this is a challenge, bearing in mind that there are 'more than 500 multilateral environmental agreements, many relevant to climate change' (TI 2011: xxvii). They also note the predominance of delegates and observers from polluting nations and lobbyists (TI 2011: xvii–xviii).

[4] For example, Napoleonic Code (France); Law of Mother Earth (Ecuador, Bolivia); Basic Law (Germany: Article 20a commits Germany to protect the environment); The Constitution of the Kingdom of the Netherlands 2008 (Article 21).

22.4.1 The Public Trust Principle

The public trust principle is a simple ethical and legal principle that can be traced back to Plato (Plato's *Republic*).[5,6] It is:

Much the most fundamental of fiduciary relations in our society being that which exists between the community ... and the state and its agencies that serve the community. *(Finn 2016: 31)*

Sir Gerard Brennan, the former Chief Justice of the High Court of Australia, stated the principle:

It has long been an established legal principle that a Member of Parliament holds 'a fiduciary relation towards the public' and 'undertakes and has imposed upon him a public duty and a public trust'. *(Brennan 2013)*

After noting that 'the duties of a public trustee are not identical with the duties of a private trustee', he made the point that they are both subject to 'an analogous limitation', namely that 'all decisions and exercises of power be taken in the interests of the beneficiaries and that duty cannot be subordinated to, or qualified by the interests of the trustee' (Brennan 2013).

Turning to the question of enforcement, he said:

True it is that the fiduciary duties of political officers are often impossible to enforce judicially (the courts will not invalidate a law of the Parliament for failure to secure the public interest [footnote]) – the motivations for political action are often complex – but that does not negate the fiduciary nature of political duty. Power, whether legislative or executive, is reposed in members of the Parliament by the public for exercise in the interests of the public and not primarily for the interests of members or the parties to which they belong. The cry 'whatever it takes' is not consistent with the performance of fiduciary duty. *(Brennan 2013)*

The legal principle has been the basis of a number of legal remedies under UK and Australian common law that address breaches of the principle, and is relevant to sentencing and the interpretation of statutes (which would include the wide statutory powers given to ministers, including planning ministers (Smith 2014a)). On the other hand, in the USA the focus appears to have been on its role in holding state governments responsible for protecting the environment (Conca 2014).[7]

As an ethical and legal principle to guide and protect the integrity of our democratic systems of government, it should be to the forefront of the minds both of holders of public office when they are making decisions and communicating with the people, and us the people when assessing the conduct of those holding public office.

---

[5] For detailed discussions of the principle, see Smith (2014b). See also Thornton (2014) and Lusty (2014).

[6] Plato states '... neither arts nor governments provide for their own interests; but, as we were before saying, they rule and provide for the interests of their subjects who are the weaker and not the stronger – to their good they attend and not to the good of the superior'. Plato (360 BCE) *The Republic*, Book 1. Available at: http://classics.mit.edu/Plato/republic.2.i.html.

[7] For detailed analyses and comparisons, see Sampford et al. (2016). See also Hendriks (2010) and French (2010).

Acknowledgement of the public trust principle (or doctrine) varies between jurisdictions. One of the most explicit recognitions is in Pakistan. The Lahore High Court has stated that the 'Public Trust doctrine ... [is] part of our jurisprudence developed by our Supreme Court' (Lahore High Court 2019: 22–23) and observed that it has been part of the international environmental law since the Stockholm Declaration in 1972 (Lahore High Court 2019: 30).

But the principle seems largely forgotten in politics and civil society in some other jurisdictions, such as Australia. In England, a century ago, the principle was still part of the political and civil society culture, when legal historian Maitland could not read a newspaper without finding someone citing the principle (Smith 2014b). Is it still a public part of the UK political and civil society culture?

In Australian politics and government, while in the late 1970s the principle was referred to and publicly applied (Committee of Inquiry Concerning Public Duty and Private Interest 1979), it has been forgotten and largely supplanted by the 'whatever it takes' approach, prioritising pursuit of power over public interest (cf. Brennan 2013). This facilitates powerful vested interests to gain advantage over the public interest. Further, the public trust/fiduciary duty model, in which the people are the principals and beneficiaries, and those entrusted with power by them are their public trustees, has been replaced with the business model in which the principal is the state, government is its business agent and the people are merely clients (Graycar 2013). Accordingly, public officers entrusted with information, money and decisions may see these as 'business' assets, and information as businesses' intellectual property to be treated as commercial in confidence. There is a severe risk of resistance by public officers to openness and accountability of government even though such reforms are critical to good government. Again, powerful vested interests may have significant advantages over the public interest in climate change policy.

In some countries, including the USA, Canada, India, Philippines and South Africa, the public trust principle has resulted in governments bringing actions for compensation or to prevent use of public property in a manner inconsistent with the principle.

Application of the public trust principle to enforce action on climate change has been successful in the Netherlands. A court has ruled that the Dutch Constitution requires the government to take effective action to mitigate climate change. Article 21 'imposes a duty of care on the State relating to the liveability of the country and the protection and improvement of the living environment' (The Hague District Court 2015: §4.36). This landmark decision required the government to reduce national carbon emissions by at least 25% below 1990 levels by 2010 (The Hague District Court 2015).

The decision of the lower court was confirmed by the Netherlands Supreme Court when it rejected an appeal by the government. In doing so, the Court stated that it

based its judgment on the UN Climate Convention [i.e. United Nations Framework Convention on Climate Change (UNFCCC)] and on the Dutch State's legal duties to protect the life and well-being of citizens in the Netherlands, which obligations are laid down in the European Convention for the Protection of Human Rights and Fundamental Freedoms (the ECHR). *(Netherlands Supreme Court 2019)*

Parties to the UNFCCC each undertake to:

formulate, implement, publish and regularly update national and, where appropriate, regional programmes containing measures to mitigate climate change by addressing anthropogenic emissions by
sources and removals by sinks of all greenhouse gases not controlled by the Montreal Protocol, and
measures to facilitate adequate adaptation to climate change. *(UNFCCC 1992: 5)*

The Netherlands Supreme Court's decision confirmed that

Each country is thus responsible for its own share. That means that a country cannot escape its
own share of the responsibility to take measures by arguing that compared to the rest of the
world, its own emissions are relatively limited in scope and that a reduction of its own emissions
would have very little impact on a global scale. The State is therefore obliged to reduce greenhouse
gas emissions from its territory in proportion to its share of the responsibility. *(Netherlands Supreme
Court 2020)*

While the UNFCCC applies to most nation states, European Convention for the
Protection of Human Rights and Fundamental Freedoms (the ECHR) applies only
throughout Europe: that is, the EU and virtually every other European state. It has no
jurisdiction in the USA, Australia or other states outside Europe. However, Wood (Philip
H. Knight Professor of Law, University of Oregon) advises that the Netherlands Supreme
Court opinion:

is very relevant for two reasons. It does not really matter that the opinion rests on laws unique to
Europe and The Netherlands, because plenty of laws support similar rulings across the world. The
public trust is a framework of fundamental rights, and the principle is manifest in a multitude of
nations. Also, many nations have constitutional provisions expressly guaranteeing the right to a
healthy environment. What is vastly important in the Urgenda decision is: (1) the remedy requiring
absolute reduction of greenhouse gases; and (2) the judges' recognition of the role of the courts in
climate crisis. The court rejects the argument that we must leave this climate crisis to the discretion of
the other two branches (legislative and executive) that brought about this emergency through decades
of policies promoting fossil fuels even in the face of grave warnings that continued pollution would
cause catastrophe. We need judicial dominoes to fall quickly in order to stave off tipping points that
would unravel human civilization and preclude human survival within the lifetimes of young people
today. The law is there – we just need judges to take the reins and use it. *(M. C. Wood, personal
communication, 16 January 2020)*

In the USA, atmospheric trust litigation (ATL) is being used in attempts to force governments to address climate change. At the time of writing (early 2021), in the case *Juliana
v. United States,* the US Ninth Circuit Court of Appeals had dismissed an interlocutory
appeal for lack of legal standing (Our Children's Trust n.d.).

22.4.2 The Public Trust Principle: What Can Be Done?

While the US ATL is unresolved at the time of writing, the Netherlands Supreme Court
decision provides a clear precedent that provides a guide to further action.

However, a major factor influencing effective responses to climate change would be enhancing community education and awareness of the fundamental legal and ethical principle that public office is a public trust, irrespective of the outcome of legal processes. More would become alert to the principle and demand its application. Resuscitation of the principle could be a game changer, as those holding public office would, like the emperor with no clothes, be stripped of the mental, psychological, ethical and legal freedom to subvert the public interest. This reform is as much to the culture of governance as it is to any formal change.

We note that the Global Organization of Parliamentarians against Corruption (GOPAC) recognises that 'to effectively represent citizens, members of parliament carry out their legislative and oversight roles in a way that is demonstrably in the public interest' (GOPAC 2015a). Our public trustees (our elected representatives and our public servants) should take the lead. But ultimately, the responsibility rests with the people. The quality of our democracies is ultimately our responsibility. We would be failing ourselves and our responsibilities if we did not act to revive the public trust principle.

22.4.3 Improving the Governance of Governments: Strengthening National Integrity Systems

The important links in the anti-corruption chain comprise the national integrity system, which is defined as 'the sum of all our institutions, laws, and efforts in stopping corruption' in a particular country (TI Ireland 2009). The national integrity system must be strong and effective to reduce the risk posed by corruption to climate change policies and practices at all levels, from local to global.

While national jurisdictions may differ in their historical and cultural development of the elements that constitute their integrity systems, there are some common patterns that can be observed, such as the existence of a system of (explicit or implicit) norms or rules of conduct, as well as of means to monitor and enforce compliance. Importantly, there is wide agreement that integrity systems include not only structures and institutions of compliance and enforcement, but also a positive culture of fostering ethical behaviour, usually mediated by the shared values of a community, public expectations and professional ideals. In this context, the challenges of climate change raise important issues of interdependent behaviour and collective action, whose solutions rely on shared understandings of the consequences of corrupt behaviours on the future of humankind and its environment.

Beside hard laws and regulations supporting the unequivocal implementation of collective measures mitigating the risks of climate change, national integrity systems may develop 'softer' options, such as codes of ethics and/or codes of conduct within the industries that create prominent impacts on climate change (such as energy, extraction and manufacturing), but also within the government policy-making systems charged with regulatory responses to climate change. The role of an ethical culture in supporting any regulatory frameworks, whether enforceable or not, cannot be overstated. This culture is developed through public debate, social leadership and role modelling – and all sectors of society, from government to industry, from the media to civil society and each individual citizen, have significant roles to play. All involved should be guided at all times by the public office public trust principle.

National integrity systems vary considerably in their ability to produce consistent and effective anti-corruption regulations and means of enforcement. There is no doubt that global success in mitigating the risks of climate change depends on strong national integrity systems worldwide, so that collective action at the global level can be organised and yield desired results. Below we discuss how several essential elements of national integrity systems can be strengthened in order to achieve more cohesion in international climate change response decisions and agreements, and their effects at national policy and practice (implementation) level: the need for an independent judicial branch and the primacy of the rule of law; the features of effective anti-corruption bodies and systems; the role of the Ombudsman, supreme audit organisation (e.g. auditor-general) and parliamentary scrutiny of government; the role of effective parliamentary codes of conduct; the need for best practices in key areas such as receipt of gifts while in public office, political campaign funding and lobbying; as well as effective public interest disclosure (i.e. whistle-blower protection) legislation, policies and practices. The suite of institutional arrangements and powers which support an integrity system is described by TI as pillars of the integrity temple. These pillars include the legislature, political parties, the election management body, the executive, the public sector, law enforcement, the judiciary, the Ombudsman, the audit institution, the corruption control agency, the media, civil society and business (TI 2010). A number of these pillars are discussed below.

21.4.4 *Rule of Law and an Independent Judicial Branch*

The value to society of the rule of law was apparent in ancient Greece (Karayiannis and Hatzis 2007). Since the adoption of the Magna Carta in England in 1215 (John King of England 1215), the rule of law has been increasingly accepted throughout the global community as fundamental to the good governance of society (Bingham of Cornhill KG 2006). The rule of law serves to constrain the executive from corruptly evading its 'fiduciary relation towards the public' in which it 'undertakes and has imposed upon it a public duty and a public trust' (Brennan 2013). The implementation of the rule of law necessarily requires that actions of the executive, and other actors, are judged independently of those accused of breaches: that is, the judicial branch must be independent of influence, whether by the executive or other actors, which corrupts impartial decision-making (Brennan 1996).

22.4.5 *Anti-Corruption Bodies and Systems*

Anti-corruption institutions play a central part in a national integrity system, and this becomes acutely evident in the context of climate change risk mitigation imperatives. The creation and maintenance of anti-corruption bodies at all levels of government within a national jurisdiction is highly desirable. In Australia, for example, the states have led the culture of institutionalising anti-corruption systems. Today, the Crime and Misconduct Commission (Queensland), the Independent Commission Against Corruption (New South Wales) and the Independent Broad-Based Anti-Corruption Commission (Victoria) have well-established policy development and enforcement practices that have significantly altered the culture of professional integrity in the public service.

Corruption is particularly problematic at the interface between policy adoption and implementation, with most cases indicating the subversion of a policy's projected outcomes. The notion of 'walking the talk' is central to maintaining integrity in public policy. Through their investigation and enforcement systems, anti-corruption bodies can therefore ensure that corrupt practices do not take hold. It is important that these institutions are well resourced and equipped to deal with challenging corruption issues and have the statutory powers to carry out their role. This requires support from government (which is difficult to obtain if the government itself resents transparency and scrutiny), the parliament (as legislator and guardian of the democratic system), the judiciary (as protector of the rule of law) and the public (as the locus of social conscience and the beneficiaries of the public trust with whom rests the ultimate responsibility). This is also because the activities and outcomes produced by anti-corruption bodies can also shape integrity cultures more widely, by influencing the development of tacit norms of behaviour in public office, far beyond the boundaries of the law.

22.4.6 Processes and Practices for Transparency and Accountability

Effective action on climate change is at risk unless there are strong integrity system pillars that ensure the availability and examination of information about decisions and actions of the executive. Provisions for freedom of information (or right to information) aim to provide the public with access to information held by the executive; this is important to the objective of accountable government. It enables public access and evaluation of action, or inaction, on climate change. What it fails to do is recognise the benefits to public policy and administration of encouraging and facilitating public engagement in the development and execution of policy (Solomon et al. 2008).

Even more dangerous is the modern day 'burning of the books': the deliberate closure or emasculation of sources of expert information and advice provided by advisory bodies, research organisations and universities, as observed in relation to climate change in a number of countries, including Australia. These policy and budgetary decisions reduce the capability of those states and the global community to identify evidence of climate change and generate appropriate responses.

Protected public interest disclosure and compensation to whistle-blowers are further necessary but not sufficient provisions enhancing the release of information otherwise withheld by the executive to the detriment of the public interest (Devine and Walden 2013).

Parliamentary scrutiny of government policy and practice relies on both effective structural arrangements, particularly committees, and a supportive political culture in which the majority recognises that outcomes are improved by accounting for policy processes, policies and their execution. Likewise, those in the minority should accept an obligation to investigate and challenge executive actions. This is especially important on climate change because 'business-as-usual' policies must be strongly questioned.

Audit institutions must be independent of the executive if they are to fulfil their role of verifying the use of resources and auditing agency performance. They should report to the parliament through public reports (INTOSAI 1998).

The role of the Ombudsman, although focused on administrative actions, includes investigation of the integrity of public service processes and actions. Reports to parliament concerning the execution of climate change policy can be profoundly important.[8]

22.4.7 Codes of Conduct for Members of Parliament

Parliaments form key pillars of integrity systems. Effective functioning requires that parliaments themselves are morally responsible, and accordingly that members of parliament (MPs) conduct themselves to the highest ethical standards, including in addressing climate change.

Unfortunately, there are continuing instances of ethical misconduct by MPs: for example, a series of ethics scandals roiled the US Congress, ending several influential careers in the 1980s and 1990s, and MP expenses scandals rocked the UK parliament in the early twenty-first century (Higham 2012a; Rush and Giddings 2015), as well as Australia's Commonwealth Parliament (e.g. see Anderson and Belot 2017). Ethics scandals have also engulfed various nascent democracies including parliaments in Asia and Africa (Feeney 2005; Reilly 2008; Trease 2010; Faull 2011; Higham 2012b; Kapama 2015).

This leads to increasing use of parliamentary codes of conduct and ethics (codes), to guide and regulate behaviour of members (Coghill and Kinyondo 2015). Codes should result from deliberations by the House's MPs, set standards and guide accepted behaviours for legislative, representative and oversight roles (Pelizzo and Stapenhurst 2004; OSCE 2012). They vary: some are detailed lists of rules embedded in legally binding documents (e.g. in Germany and Latvia), while others simply recite shared values (OSCE 2012; Coghill and Kinyondo 2015). They should be revisited from time to time (Pelizzo and Ang 2008; OSCE 2012; Coghill and Kinyondo 2015).

The failure of MPs to deal with climate change could be the worst ethical scandal ever. Risking the lives of tens of millions of people and the natural resources necessary to sustain life by failing to enact effective policies is an epic moral lapse (Brown 2015). Since the poorest are the most harmed, should it not be considered as an ethical issue? MPs have duties, responsibilities and ethical obligations to others (Brown 2015). Codes potentially have key roles in entrenching fiduciary duty and the public trust principle in MPs' responses to climate change. A code should be complemented by the appointment of an

---

[8] For example, the Australian Ombudsman investigated modelling of climate change by the Australian Bureau of Agriculture and Resource Economics (ABARE) (Commonwealth Ombudsman 1998).

independent parliamentary integrity commissioner who can advise MPs on conduct and may also, as in Queensland, register lobbying (Queensland Integrity Commissioner 2020).

22.4.8 Clear Rules Regarding Receipt of Gifts by Public Officers

Accepting or donating gifts is a particularly difficult area of behaviour to regulate as social acceptability varies so widely between and even within cultural groups. The very act of offering or receiving a gift is known to establish a favourable predisposition to the other person, irrespective of the value of the gift (Malmendier and Schmidt 2012).

However, a total ban on accepting gifts risks setting up even the most ethical public officer for failure. Once a person is tainted as unethical for accepting or offering a gift, no matter how commonplace, reasonable and harmless this social behaviour may be, critics have a tool with which to tar and tarnish the reputation of the individual, and his or her colleagues (Kania 2004). Continuous real-time online disclosure of all gifts greatly reduces risking the appearance of impropriety.

22.4.9 Political Campaign Funding and Payments for Access

Financial or in-kind donations to political parties and candidates are rightly seen as endangering effective policy responses if made by investors, corporations and unions with vested interests in industries which contribute to climate change. This applies whether they contribute directly (e.g. coal-fired generator owners) or indirectly (e.g. financiers) (Market Forces 2019). Currently, there are over USD 5.2 trillion per annum globally in fossil fuel industry subsidies (Irfan 2019), and there are also significant government subsidies to agricultural commodity industries driving deforestation globally (Global Forest Coalition 2018). The extraordinary generosity of fossil fuel businesses to political parties, politicians and candidates is particularly dangerous. It is estimated that fossil fuel interests donated over USD 350 million in 2013 and 2014, which shored up subsidies to the industry of more than 100 times that amount (Oil Change International 2019). Little wonder politicians failed to act to curb emissions!

However, corporations law dictates that company directors must act in the best interests of their specific corporation.[9] That being so, how can any director authorise a donation unless it is intended to either (1) corruptly influence the recipient to act in that particular corporation's interests or (2) improperly divert corporate funds to favour the recipient?

Recent instances of high-wealth individuals in several jurisdictions using their personal assets to campaign at enormous advantage over other candidates and political parties has severely contradicted basic democratic principles. Candidates should campaign on their skills, character and policies rather than personal wealth.

Better regulatory schemes are required, including total bans on donations by corporate and trade union entities, caps on donations by individuals, with online, continuous, real-time disclosure, bans on payment for access to members of the executive and limits on

---

[9] *Corporations Act 2001* (Cth). Available at: www.legislation.gov.au/Details/C2017C00328.

campaign expenditure, including on indirect support through third-party campaigning (Schott et al. 2014).

22.4.10 Clear Rules Preventing Paid Lobbying Post Parliamentary Careers

Lobbying and post-parliamentary careers are linked because of the propensity of former MPs to become lobbyists or otherwise work in the interests of entities, or related entities, for whose interests they formerly had executive responsibility while in government. Examples include senior Opposition Labor Party figures who 'either took up management jobs with mining and energy companies and associations or worked as consultants for them' following Labor's defeat at the 2016 election (Holmes 2016). Former senior Liberal and National Coalition party members are similarly well represented. Most egregiously, fossil fuel consultancies have been formed by cross-party alliances of former Coalition and Labor MPs (Holmes 2016).

Former political staffers are also frequently employed by businesses with which their previous employers dealt (Davies 2015).

The Center for Responsive Politics maintains the extensive Revolving Door database that lists movements of personnel between politics and industry in the USA (Center for Responsive Politics 2019).

Two types of regulation are necessary to reduce the risks of responses to climate change being undermined. First, lobbying should be recorded and disclosed online, continuously and in real time.

Second, former members of the executive should be free to apply the capabilities that they have developed, except that they must not receive any remuneration or other benefit from any entity, or related entity, for the interests of which they formerly had executive responsibility within the previous 5 years. This should extend to former staffers.

22.5 Globalisation of Climate Change Action

Notwithstanding the importance of integrity in individuals and institutions at national and subnational levels, climate change is a global issue without parallel. Avoiding corrupting policies and actions to curb climate change requires that national and subnational actors from the three social sectors – state (public), market and civil society – collaborate and cooperate to protect the home we share.

There is a range of global initiatives that nations and their citizens and organisations can join to help execute this journey to combat corruption and promote open and accountable government. These are summarised in Boxes 22.1 and 22.2.

22.6 Future Global Anti-Corruption and Governance Action

These issues will continue to be negotiated at the national level and executed at the relevant domestic level or levels according to each nation's distribution of legislative and executive powers.

Box 22.1 **Present International Arrangements for Promotion of National Good Governance Rights**

There are specific international conventions, declarations and agreements that declare rights and provide frameworks necessary for addressing the risks of corruption and enhancing good governance – including the direct pursuit of open and accountable governments around the world and the participation of the people in all countries in their political and government systems (Centre for Law and Democracy 2014).

International Conventions

The Universal Declaration of Human Rights (UDHR)[10] (in force 1948): the right to participate in elections (Art. 21), and related rights to equality and dignity and right to freedom of expression (Art. 19), are now part of customary international law. The International Convention on Civil and Political Rights (ICCPR)[11] (which currently has 74 signatories and 168 parties) acknowledges as human rights the right to open and accountable government (Art. 25), and citizens' right to participate in public affairs, the right to vote and equal access to the public service, the right to freedom of expression and the right to access government information (Art. 19) (Parliament of Australia Joint Committee on the Australian Commission for Law Enforcement Integrity n.d.).

Groups of Nations

(a) Charter of the Commonwealth: mutual respect, inclusiveness, transparency, accountability, legitimacy and responsiveness (Commonwealth Secretariat 2015).

(b) Europe and central Asia: Aarhus Convention 1998, on access to information, public participation in decision-making and access to justice in environmental matters.[12]

(c) Council of Europe Convention on Access to Official Documents.[13]

Fortunately, there is an impressive array of concerned and active international and regional organisations with potential to contribute to addressing the issues of national government corruption and of open and accountable national governments globally.

The more successful the efforts to curb corruption, the more difficult it is to defend and sustain subsidies that harm the atmosphere, contribute to climate change and frustrate achieving SDGs. Politicians and public servants are less able to disguise or hide corrupt inducements; business leaders are more wary of offering inducements that risk being disclosed and prosecuted and leading to imprisonment; and there are fewer reasons for subsides that lead to environmental harm and drain public finances.

---

[10] *Universal Declaration of Human Rights*, GA Res 217A (III), UN Doc A/210, opened for signature 10 December 1948. Available at: www.un.org/en/about-us/universal-declaration-of-human-rights.

[11] *International Covenant on Civil and Political Rights*, GA Res 2200A (XXI), opened for signature 19 December 1966. Available at: www.ohchr.org/en/professionalinterest/pages/ccpr.aspx.

[12] *Convention on Access to Information, Public Participation in Decision-making and Access to Justice in Environmental Matters* [*The Aarhus Convention*], UNECE (UN Economic Commission for Europe) adopted 25 June 1998. Available at: www.unece.org/fileadmin/DAM/env/pp/documents/cep43e.pdf.

[13] *Council of Europe Convention on Access to Official Documents*, CETS No. 205, opened for signature 18 June 2009. Available at: http://conventions.coe.int/Treaty/Commun/QueVoulezVous.asp?NT=205&CL=ENG.

Box 22.2 **Present International Arrangements for Direct Action Against Corruption and to Promote National Government Integrity Systems**

Anti-Corruption Focus

International conventions and standards pertaining to integrity and anti-corruption include:[14]

(i) **The United Nations Convention Against Corruption (UNCAC).**[15] As of 2020, the UNCAC has 140 signatories and 187 parties (UN 2020). It requires parties (members) to implement a range of measures to address corruption. Parties are to prepare an action plan in consultation with civil society and accept implementation reviews. The implementation review system consists of two reviews of performance cycles over 5 years.

(ii) **Extractive Industries Transparency Initiative (EITI).** The EITI is an international standard to 'promote the open and accountable management of extractive resources', governed by a board comprising representatives from extractive companies and civil society organisations, governments, institutional investors and international organisations. It has established a standard which sets requirements that must be met by companies and governments to be accepted as EITI-compliant. The standard is to be implemented in some 56 countries. To become a candidate country, a country has to complete four sign-up steps to the satisfaction of the EITI board (EITI 2016).

(iii) **OECD Convention on Combating Bribery of Foreign Public Officials in International Business Transactions.**[16] The Convention establishes standards for criminalising the bribery of public officials in international transactions. Parties are obliged to prosecute those involved in bribery (OECD 2015).

(iv) **G20 Anti-Corruption Action Plan.** The G20 recognises that it has a 'special responsibility to prevent and tackle corruption' (G20 2014:1). Its plan spells out a policy and legal framework based on the UNCAC principles to prevent and address corruption (G20 2014).

(v) **APEC Anti-Corruption and Transparency Working Group.** This Group came into existence in 2004 to advance transparency and fight corruption. It established an Anti-Corruption and Transparency Experts' Task Force (ACT) in 2005, which was upgraded to a Working Group in March 2011. The ACT's roles include promoting cooperation in areas including extradition, legal assistance and enforcement, in particular, recovery and forfeiture. It has 21 member economies, including Australia (APEC Anti-Corruption and Transparency Working Group 2015).

(vi) **Global Organization of Parliamentarians Against Corruption (GOPAC).** This network of parliamentarians was founded in 2002. It has over 700 members. It seeks to address corruption and open and accountable government and so strengthen the integrity of government (GOPAC 2015b).

---

[14] A useful summary of international conventions and standards pertaining to integrity and anti-corruption is available from the Parliament of Australia: Australian Commission for Law Enforcement Integrity. Available at: www.aph.gov.au/Parliamentary_Business/Committees/Joint/Australian_Commission_for_Law_Enforcement_Integrity/Completed_inquiries/2010-13/integrity_inter_op/report/c04.

[15] *United Nations Convention Against Corruption*, GA Res 58/4, UN Doc. A/58/422, adopted 31 October 2003. Available at: https://digitallibrary.un.org/record/505186?ln=en.

[16] *OECD Convention on Combating Bribery of Foreign Officials in International Business Transactions*, opened for signature 17 December 1997. Available at: www.oecd.org/corruption/oecdantibriberyconvention.htm.

Open and Accountable Government Focus

There is one international organisation which is dedicated to this issue: the Open Government Partnership (OGP). It was established in 2011 'to provide an international platform for domestic reformers committed to making their governments more open, accountable, and responsive to citizens' and to 'fighting corruption'. It has over 70 participating countries (OGP 2020).

It has multilateral partnerships with the World Bank, the UN Development Programme, the Inter-American Development Bank and the OECD (OGP 2013).

The Digital 5 (D5) Charter was adopted by five leading OGP member governments in 2014 to 'provide a focused forum to share best practice, identify how to improve the Participants' digital services, collaborate on common projects and to support and champion [their] growing digital economies' (Republic of Estonia et al. 2014).

The OGP has developed a substantial operating platform to enable it, governments and civil society groups in a 'co-creative partnership' to advance the open governance, anti-corruption objective and to monitor their performances. An independent reporting mechanism (IRM) assesses the performance of participating countries in preparing and implementing their biennial action plans, providing progress reports (OGP n.d., 2015a).

The OGP has a steering committee of representatives of governments and civil society organisations in equal numbers. It is chaired by two government co-chairs and two civil society co-chairs (OGP 2015b). There is a small permanent secretariat, the OGP Support Unit, that works closely with the steering committee and serves as a neutral third party between civil society organisations and governments (OGP 2015c). The OGP has also established six working groups focusing on different aspects of open and accountable government. They include 'Anti-Corruption' and 'Openness in Natural Resources', the latter having sustainable development as one of its objectives (OGP 2015d).

There is also an independent active Civil Society Coordination (CSC) team, which supports national civil society actors in their use of the OGP to achieve their objectives (OGP 2014a).

The OGP's *Four-Year Strategy 2015–2018* records its strong growth and its plans. Its flexibility and involvement with civil society would appear to enable it, and the civil society groups of participating nations, to focus on the governance changes they need to ensure that their responses on mitigation of and adaptation to climate change are what is necessary (OGP 2016).

But time is limited. How can more progress be made and how can it be accelerated? Could bringing litigation against polluting nations and states assist to hold them to account?[17] Also, could understanding and acceptance of the proposition that improvements in these areas will promote economic growth be advanced at a higher pace?[18] While such actions may assist in advancing global action, mechanisms for global action are already in place. How can they be best mobilised?

Looking at all the international and regional organisations attempting to help nations in these matters, the OGP may well be the best placed to advance and accelerate the process. It

---

[17] For material that looks at the issues more widely, see for example Transparency International's Climate Change website, available at: www.transparency.org.uk/. Regarding the management of grant funds and finance, see TI (2014a, 2014b).

[18] Cf. speech of Prime Minister Cameron (Cameron 2013) and speech of President Obama (Obama 2014).

appears to already have a flexible and very active consultative, supportive and monitoring structure, and this is what is needed to enable international and regional organisations, civil societies and their national and subnational governments to come together on shared issues, on an equal footing. This enables them not only to strengthen the integrity of their national governance but also to address fundamental long-term challenges we all face, such as climate change.

In the OGP processes, national governments and civil societies collaborate to define their issues and proposals in their National Action Plans, or NAPs. They can pick up an issue that focuses on a particular area of government policy and action and the government integrity action needed. Why not that needed for developing programmes for climate change mitigation and adaptation? The OGP is alive to the relevance and importance of the climate change issue (Marcel 2015; OGP 2015b; Excell 2016) and participating nations have started addressing climate change issues in their action plans – for example, the USA, Mexico, France and the North American Climate, Clean Energy and Environment Partnership (The White House 2016).

A major anti-corruption and governance challenge in addressing climate change will be the successful development and implementation of the Paris Agreement to fund action to mitigate, and to adapt to, climate change. Mitigation funds include the Green Climate Fund (Green Climate Fund n.d.), Reducing Emissions from Deforestation and Forest Degradation (REDD 2015) and green bonds (World Bank n.d.). There are also numerous funds directed to adaptation action. In August 2015, the OECD published a report, *Toolkit to Enhance Access to Adaptation Finance*, in which it emphasised the need to meet necessary standards and safeguards, including 'minimum fiduciary standards from the early stage of development' for new institutions (OECD and GEF 2015: 16). It also notes that some funds have formulated key policies concerning fiduciary risk (e.g. the Adaptation Fund, Global Environment Facility (also an 'open information policy'), the Least Developed Countries Fund and the Green Climate Fund) (Green Bond Principles 2015; OECD 2015: 41–51). The Pilot Program for Climate Resilience relies upon the policies and guidelines of the multilateral development banks around the world that provide funds for it (OECD 2015: 47). Some countries have also started developing regulatory frameworks and fiduciary standards. This includes Bangladesh and Zambia (OECD 2015: 19), their ranks according to the TI Corruption Perception Index being 145 (139 in 2015) and 87 (76 in 2015), respectively (TI 2015a, 2017).

These funds aim to invest very large sums in projects around the world. It will be critical to ensure that all such funds and their support and management are protected from corruption. The risk is high and history suggests that the challenge will be considerable and at times complex. It is unlikely to be sufficient to rely solely on individually negotiated loan arrangements and their requirements and systems, or on specially created national or international bodies to administer the funds. The corrupt will try to find a way around them. It will be necessary to strengthen the openness and accountability of the relevant national governments generally as well as those aspects of government directly affected or involved in the funding processes. Another critical matter is to continue to take steps to ensure that all countries have people with the expertise and experience and the systems needed to address the challenges (as discussed in Chapter 23).

To explore practicalities, consider, for example, Brazil and the Central African Republic, ranked 76 and 145, respectively, on the TI Corruption Perceptions Index 2015 (TI 2015a). In 2018, Brazil's ranking had fallen to 105 (of 180 countries), but Punder argues that this may be a positive, in that '[s]ince the enactment of Brazil's anti-corruption law, the country has finally started to speak publicly about corruption and take the first steps to fight this evil' (Punder 2019; TI 2019a).

Brazil is a participant in the OGP and the UNCAC. Could not Brazil's OGP 2018–20 NAP (NAP4) have been developed in conjunction with the negotiation of any special funding arrangements, and compliance with the NAP's requirements made a condition of instalment payments of the promised funds?

Brazil's NAP4 does at least include a commitment acknowledging climate change.

Open Government and Climate

Commitment 9: Develop, collaboratively, a transparent mechanism for the evaluation of actions and policies related to climate changes.

The commitment is devoted to the improvement of the Climate Policy's management and planning by the evaluation of actions and policies, as well as to the expansion of civic participation.

Agenda 2030 Goal: 13 Take urgent measures to combat climate changes and its impacts; and 16.10 – Ensure public access to information and protect fundamental freedoms, in accordance with national legislation and international agreements. *(OGP 2018)*

Consider also the Central African Republic. It is a member of the UNCAC but not the OGP. Its Corruption Perceptions Index ranking slipped further to 149 in 2018 (TI 2019b). It could be required to revise its NAP to address the risks of corruption involved in such funding programmes, but while that will require civil society input, it will lack the support of the OGP approach. Some countries in that situation do not comply with the criteria for OGP membership, but could an effective international regime require such countries, or the few who are a party to neither OGP nor the UNCAC, to adopt the processes of both as a condition of funding? They would commit to following OGP and UNCAC processes, somewhat like countries wishing to join D5 (see above) agreeing to follow its Charter.

The proposed funding exercises will tend to be big and the issues, measures and tasks involved vary for everyone associated with them, including the OGP. The issues and the measures needed would vary between the participating nations, but there are already examples of action by participants in relevant areas (fighting corruption, strengthening democratic institutions, freedom of information, fiscal transparency, public service delivery, extractive resources transparency) of the OGP to draw on (OGP 2014b). The point should also be made that the tasks will arise in any event because of the ever increasing need to address the mitigation of and adaptation to climate change, and the need for best governance arrangements, including integrity systems, especially in that area. This means that it is an area where the OGP and its present participant countries and civil societies are likely to be dealing with what is required to address the government integrity issues in that area – perhaps starting with assistance from the Natural Resources Working Group (OGP

2015d). At the same time, it is likely to present a significant opportunity for the benefits of the OGP model to extend to other countries and, in the long term, spread the benefits of better government and economic growth around the world. Importantly, doing so will strengthen not only the capacity of the world but also the resilience that is needed to successfully address the present and future challenges of climate change. It should also be noted that the OGP is moving towards participation by countries at the subnational levels of their governments (Heller and Bapna 2015).

22.7 Integrating Anti-Corruption Measures into Subnational and City-Level Governance of Climate Change

As earlier chapters of the book have shown, it will be impossible to achieve global decarbonisation without reducing emissions in cities (see Chapter 11). This section looks at issues that have arisen at the national and subnational level, demonstrating why action is needed and how it can be achieved. What is required is long-term climate-smart urban and regional planning decisions that will enable cities to decarbonise. Cuba demonstrated that major change can be achieved in the face of a sudden large reduction in oil supplies (Morgan 2006). However, corruption of planning authorities can make it impossible to achieve reform quickly enough to avoid dangerous climate change.

22.7.1 Land-Use Decision-Making

Planning for zero-carbon cities will require new pathways for decision-making on land-use activities. This is because we are dealing with a changing environment at a scale and rate not previously planned for. An example is coastal inundation from projected sea-level rise or extreme events (Watson et al. 2015). In the Australian planning system, the consequences of coastal inundation are not necessarily considered, nor are the coastal or ocean impacts insurable. The unpredictability of future climate change means that we will need to move to a 'risk' management approach rather than the traditional zoning of land that never considered eroding land boundaries or shifting land titles (Whetton 2011). This was a significant finding by the Intergovernmental Panel on Climate Change Working Group II (IPCC 2012).

As a consequence of the climate-change-related increased risks to land, the built environment and urban communities, there is now an increasing move to 'adaptive planning'; incorporating a risk management approach into pathways for decision-making (Webb and Beh 2013). This places emphasis on 'learning by doing', recognising that in many respects we are dealing with the unknown in the future. This approach includes continuous monitoring and review. It also includes new tools such as 'scenario planning' as part of the decision-making process, and planning for a coincidence of extreme events (e.g. during the Brisbane 2012 floods, inland flooding coincided with a king sea tide, resulting in record water levels).

Planning for climate change is a 'wicked problem', requiring the contribution of a range of disciplinary perspectives (planning, science, health, economics, design). There is also increasingly an emphasis on co-design and co-production of knowledge and leading practices involving partnership between sustainability scientists, land-use planners and local communities (Norman et al. 2013; NCCARF 2014). It is through this collaborative approach that innovation and learning may find more sustainable solutions on the ground. To facilitate this, new platforms or boundary organisations for innovation and collaboration are emerging; a current example is Canberra Urban & Regional Futures (CURF 2015). Canberra Urban & Regional Futures seeks to better connect leading research with practice and local communities, building pathways towards sustainable futures for cities and regions.

22.7.2 Land-Use Planning

Urbanisation and climate change are two of the most significant drivers of global change. Planning for climate change is concerned with the urbanisation of cities within the context of climate change (Norman 2018). The global urban population has increased from 746 million in 1950 to 3.9 billion in 2014, with an additional 2.5 billion expected by 2050 (UN Department of Economic and Social Affairs, Population Division 2014). The potential impact of urban growth on climate change is significant, with cities consuming 78% of the world's energy and producing more than 60% of all carbon (Norman et al. 2014). The extent and rate of urbanisation varies across nations. The fastest rate of urbanisation is now occurring in the Global South, particularly Asia and Africa, with developed nations like Australia already highly urbanised (IPCC 2014).

As the Secretary-General of the UN[19] has stated, cities will be the pathway to sustainable solutions (Rio+20 2012). Land-use planning will be a key instrument for both mitigation and adaptation to climate change, but it is still at a formative stage with varying degrees of implementation across the nations (IPCC 2012). Planning can actively reduce emissions, through actions such as facilitating renewable energy opportunities including solar panels and wind farms, introducing strategies that encourage a switch from car-based cities to active travel (walkable and cycling cities, light rail) and developing green buildings and precincts. There are also global climate change and cities networks committed to sharing knowledge and leading practice; two examples are the C40 network of global cities – 'cities [that] have emerged as strong and inspiring champions of the kind of ambitious climate action the world needs' (C40 2015) – and the ICLEI global network of local governments (ICLEI Global n.d.).

Such enterprises require maximum integrity.

Recent experience with action on climate change at the local level indicates that there are two important elements that need to be in place. First, a significant influence is the extent to which higher levels of government have supporting frameworks in place (i.e. governance). This has been demonstrated in both developed (e.g. Australia) (NCCARF 2014), and developing nations (e.g. Nepal) (Saito 2013). The second is that 'planning for climate

---

[19] The Secretary-General of the UN was then Ban Ki-moon.

change' will require a much closer connection between sustainability scientists and land-use planners and local communities (Norman et al. 2014).

22.8 Conclusion

This chapter set out to examine opportunities for improving governance to reduce the constant risk of corruption in any system of government and in particular its significance to SDGs and more specifically climate change policy and action. This has implications for societies throughout the world. To return to the three major points raised by TI in advance of COP 21, this chapter has revealed extensive initiatives being taken in recognition of the threats to effective climate action posed by corruption – threats that can range from compromised decision-making affecting land use to grand larceny.

Transparency is essential for its potential to reveal to the people (whether in neighbourhoods or in international communities of interest) both the realities of the nature and effects of climate change and the manner in which power is being exercised in their name to address it (i.e. accountability: see below). This in turn raises the prospect of better-informed decision-making and indeed better forms of decision-making in which political leaders engage their communities.

We have also examined the significant risks of corruption scandals undermining progress. There are encouraging steps towards greatly reducing opportunities for the misuse and diversion of funds needed to tackle climate change. This is especially important given the large sums that need to be committed to aid adaptation and mitigation in poorer and more vulnerable countries. Progress is slow, but it is steady, building on initiatives such as the UNCAC, the OECD, APEC, the G20 initiatives, the EITI and the OGP.

The third point, accountability, is seen to be improving through mechanisms such as UNCAC Action Plans and OGP Action Plans and, most importantly, the mechanisms to monitor and review implementation. Already, OGP has excluded from its membership countries unable to live up to its standards – an encouraging sign.

This chapter has illustrated these points by identifying the potential for improved governance in a key policy area, land use, to have major impacts on integrated approaches to climate change mitigation and adaptation. Underlying these points is the issue of trust, in two related aspects. The first aspect of trust is the entrusted responsibilities which are assigned to members of the political executive. This public trust obliges political leaders, and all other public officers, to always act in the public interest, necessarily the long-term interest, when faced with the existential threat to human civilisation posed by climate change. This has been long recognised as a fundamental ethical principle and is recognised in the law of a number of countries, including the USA, and recently in the Netherlands (The Hague District Court 2015; Netherlands Supreme Court 2020).

The other understanding of trust is the extent to which the people, the beneficiaries of the trust, perceive that those they entrusted with power (MPs and the political executive) have acted in the public interest or have failed to do so. Where corruption is suspected, or occurs, the people's trust in their leaders and public officers will be eroded, and may be severely weakened, impacting on social harmony and increasing the cost of government.

Returning to the first aspect, where corruption does occur, the public trust obligation of our leaders and public offices is undermined and may completely destroy the capacity of government to serve the public interest. Where it affects climate change policy and action, it directly threatens the environmental conditions necessary for human life not only in the nation where the breach of trust occurred but also in every other nation in the world.

More than ever before, with electorates now largely accepting anthropogenic causes of climate change, experiencing its impacts, frustrated by the failure of the political process to address it and turning to the false promises of populists, political leaders must accept their sacred public trust to secure a healthy atmosphere. To do otherwise is to knowingly corrupt the public trust invested in them as public officers.

References

ANAO (Australian National Audit Office) (2019). Award of a $443.3 million grant to the Great Barrier Reef Foundation. *ANAO.gov.au*. 16 January. Available at: www.anao. gov.au/work/performance-audit/award-4433-million-grant-to-the-great-barrier-reef-fo undation#:~:text=2050%20Plan%20activities.-,The%20relevant%20budget%20meas ure%20included%20%24443.3%20million%20for%20a%20partnership,the%20grant %20to%20the%20foundation.

Anderson, S. and Belot, H. (2017). Sussan Ley quits as health minister as Malcolm Turnbull flags political expenses reform. *ABC News*. 14 January. Available at: www .abc.net.au/news/2017-01-13/sussan-ley-tenders-resignation-parliament-expenses-scan dal/8180602.

APEC (Asia-Pacific Economic Cooperation) Anti-Corruption and Transparency Working Group (2015). Anti-corruption and transparency. *Asia-Pacific Economic Cooperation*. Available at: www.apec.org/Groups/SOM-Steering-Committee-on-Economic-and-Technical-Cooperation/Working-Groups/Anti-Corruption-and-Transparency.aspx.

Bingham of Cornhill KG (2006). The rule of law. Sixth Sir David Williams Lecture presented at the Centre for Public Law, 16 November. Available at: www.cpl.law .cam.ac.uk/sites/www.law.cam.ac.uk/files/images/www.cpl.law.cam.ac.uk/legacy/ Media/THE%20RULE%20OF%20LAW%202006.pdf.

Brennan, G. (1996). Judicial independence. Paper presented at the Australian Judicial Conference, The Australian National University, Canberra, 2 November. Available at: www.hcourt.gov.au/assets/publications/speeches/former-justices/brennanj/bren nanj_ajc.htm.

Brennan, G. (2013). Presentation of Accountability Round Table Integrity Awards, Parliament House, Australia, 11 December Available at: www.accountabilityrt.org/ integrity-awards/sir-gerard-brennan-presentation-of-accountability-round-table-integ rity-awards-dec-2013/.

Brown, D. (2015). The worst ethical scandal in Congress: Climate change? *Rock Ethics Institute*. 15 April. Available at: https://archive.thinkprogress.org/the-worst-ethical-scandal-in-congress-climate-change-7b8832b0380e/.

C40 (C40 Cities Climate Leadership Group) (2015). The power of C40 cities. *C40 Cities*. Available at: www.c40.org/cities.

Cameron, D. (2013). Speech presented at Open Government Partnership 2013 [transcript]. Available at: www.gov.uk/government/speeches/pm-speech-at-open-government-partnership-2013.

Center for Responsive Politics (2019). Revolving door. *OpenSecrets.org*. Available at: www.opensecrets.org/revolving.

Centre for Law and Democracy (2014). Briefing Paper on Transparency and Accountability. Briefing paper 47. Centre for Law and Democracy. Available at: www.law-democracy.org/live/briefing-paper-on-transparency-and-accountability/.

Coghill, K. and Kinyondo, A. (2015). Benchmarks for codes of conduct. *The Parliamentarian*, 2015, 172–175.

Committee of Inquiry Concerning Public Duty and Private Interest (1979). *Public Duty and Private Interest: Report of the Committee of Inquiry Established by the Prime Minister on 15 February 1978* [The Bowen Report]. Canberra: Commonwealth Government of Australia. Available at: https://apo.org.au/sites/default/files/resource-files/1979-07/apo-nid33757.pdf.

Commonwealth Ombudsman (1998). Ombudsman releases ABARE investigation report. Media release. *Commonwealth Ombudsman*. 4 February. Available at: www.ombudsman.gov.au/__data/assets/pdf_file/0024/26286/investigation_1998_01.pdf.

Commonwealth Secretariat (2015). Commonwealth Charter. *The Commonwealth*. Available at: https://thecommonwealth.org/about-us/charter.

Conca, J. (2014). Atmospheric trust litigation: Can we sue ourselves over climate change? *Forbes*. 23 November. Available at: www.forbes.com/sites/jamesconca/2014/11/23/atmospheric-trust-litigation-can-we-sue-ourselves-over-climate-change/?sh=7bd200264005.

CURF (Canberra Urban and Regional Futures) (2015). *Canberra Urban and Regional Futures*. Available at: www.curf.com.au.

Davies, A. (2015). CSG industry hires well-connected staffers. *The Sydney Morning Herald*. 25 May. Available at: www.smh.com.au/national/nsw/csg-industry-hires-wellconnected-staffers-20150515-gh2rg3.html.

Devine, T. and Walden, S. (2013). International best practices for whistleblower policies. *Government Accountability Project*. 22 July. Available at: http://whistleblower.org/international-best-practices-for-whistleblower-policies/.

EITI (Extractive Industries Transparency Initiative) (2016). Impacting 49 EITI countries. *EITI Progress Report 2016*. 18 March. Available at: http://progrep.eiti.org/2016/glance/where-implementation-happens.

Excell, C. (2016). Open government and climate action. *OGP Blog*. 27 April. Available at: www.opengovpartnership.org/blog/ogp-webmaster/2016/04/27/open-government-and-climate-action.

Faull, L. (2011). Parliament's ethics committee eviscerates top ANC MP. *Mail and Guardian*. 25 August. Available at: http://mg.co.za/article/2011-08-25-parliaments-ethics-committee-eviscerates-top-anc-mp.

Feeney, S. (2005). The impact of foreign aid on economic growth in Papua New Guinea. *The Journal of Development Studies*, 41, 1092–1117.

Finn, P. (2016). Public trusts and fiduciary relations. In K. Coghill, C. Sampford and T. Smith, eds., *Fiduciary Duty and the Atmospheric Trust*. London: Routledge.

Francis (2015). *Encyclical Letter Laudato Si' of the Holy Father Francis: On Care for Our Common Home*. Rome: Vatican Press. Available at: http://w2.vatican.va/content/dam/francesco/pdf/encyclicals/documents/papa-francesco_20150524_enciclica-laudato-si_en.pdf.

French, R. (2010). Fiduciary duties and indigenous people: In the interface between equitable principles and public law. Available at: www.hcourt.gov.au/assets/publica tions/speeches/current-justices/frenchcj/frenchcj29oct10.pdf.

G20 (2014). *G20 Anti-Corruption Action Plan. G20 Agenda for Action on Combating Corruption, Promoting Market Integrity, and Supporting a Clean Business Environment*. Available at: www.ag.gov.au/Integrity/AntiCorruption/Documents/G20AntiCorruptionActionPlan.pdf.

Global Forest Coalition (2018). Ending subsidies for meat and soy sector is key to halting deforestation, shows new paper. *Global Forest Coalition*. 18 November. Available at: https://globalforestcoalition.org/ending-subsidies-for-meat-and-soy-is-key-to-halting-deforestation/.

GOPAC (Global Organisation of Parliamentarians Against Corruption) (2015a). Parliamentary ethics and conduct. *GOPAC Network*. Available at: http://gopacnetwork.org/programs/parliamentary-ethics-conduct/.

GOPAC (2015b). A global organization of parliamentarians against corruption. *GOPAC Network*. Available at: http://gopacnetwork.org/.

Graycar, A. (2013). Perceptions of corruption in Victoria. Research Paper. Independent Broad-Based Anti-Corruption Commission Victoria. Available at: www.ibac.vic.gov.au/publications-and-resources/article/perceptions-of-corruption-in-victoria.

Great Barrier Reef Foundation (2020). *Great Barrier Reef Foundation*. Available at: www.barrierreef.org/.

Green Bond Principles (2015). Green, social and sustainability bonds. *International Capital Market Association*. Available at: www.icmagroup.org/Regulatory-Policy-and-Market-Practice/green-bonds/.

Green Climate Fund (n.d.). *Green Climate Fund*. Available at: www.greenclimate.fund/.

Gupta, S., Davoodi, H. and Alonso-Terme, R. (1998). *Does Corruption Effect Income Inequality and Poverty?* Washington, DC: International Monetary Fund.

Heller, N. and Bapna, M. (2015). The OGP journey to go subnational. *OGP Blog*. 13 August. Available at: www.opengovpartnership.org/stories/the-ogp-journey-to-go-subnational/.

Hendriks, E. (2010). Common law: Implementing the public trust doctrine in British Columbia. *Water Canada*. Available at: https://poliswaterproject.org/polis-media-highlight/common-law-implementing-public-trust-doctrine-british-columbia-article-water-canada/.

Higham, S. (2012a). Congressional ethics committees protect legislators, critics say. *Washington Post*. 7 October. Available at: www.washingtonpost.com/investigations/congressional-ethics-committees-protect-legislators-critics-say/2012/10/07/a313e59c-e251-11e1-ae7f-d2a13e249eb2_story.html.

Higham, S. (2012b). Parliament ethics committee to continue questioning MP El-Eleimy on Monday. *Ahram Online*. 11 March. Available at: http://english.ahram.org.eg/NewsContent/1/64/36415/Egypt/Politics-/Parliament-ethics-committee-to-continue-questionin.aspx.

Holmes, D. (2016). The fossil-fuelled political economy of Australian elections. *The Conversation*. 22 June. Available at: https://theconversation.com/the-fossil-fuelled-political-economy-of-australian-elections-61394.

ICLEI Global (n.d.). *ICLEI Global*. Available at: www.iclei.org/.

International Court of Justice and Weeramantry, C. (1997). Separate opinion of Vice-President Weeramantry. Gabčíkovo-Nagymaros Project (Hungary/Slovakia). Available at: www.icj-cij.org/en/case/92/judgments.

INTOSAI (International Organization of Supreme Audit Institutions) (1998). *Code of Ethics*. Available at: www.intosai.org/documents/open-access.

IPCC (Intergovernmental Panel on Climate Change) (2012). Summary for Policymakers [Report graphics]. In C. B. Field, V. Barros, T. F. Stocker et al. eds., *Managing the Risks of Extreme Events and Disasters to Advance Climate Change Adaptation. A Special Report of Working Groups I and II of the Intergovernmental Panel on Climate Change*. Cambridge: Cambridge University Press. Available at: www.ipcc.ch/report/managing-the-risks-of-extreme-events-and-disasters-to-advance-climate-change-adaptation/summary-for-policymakers/.

IPCC (2014). *Climate Change 2014: Impacts, Adaptation, and Vulnerability. Part B: Regional Aspects. Contribution of Working Group II to the Fifth Assessment Report of the Intergovernmental Panel on Climate Change*. Edited by A. Reisinger, R. L. Kitching, F. Chiew et al. Cambridge: Cambridge University Press. Available at: www.ipcc.ch/site/assets/uploads/2018/02/WGIIAR5-PartB_FINAL.pdf.

Irfan, U. (2019). Fossil fuels are underpriced by a whopping $5.2 trillion. *Vox*. 17 May. Available at: www.vox.com/2019/5/17/18624740/fossil-fuel-subsidies-climate-imf.

John King of England (1215). *Magna Carta* [British Museum translation]. Runnymede. British Museum item description: Cotton MS Augustus II 106:1215.

Kania, R. R. E. (2004). Ethical acceptability of gratuities: Still saying yes after all these years. *Criminal Justice Ethics*, 23, 54–63.

Kapama, F. (2015). Chenge contest ethics case mode. *Daily News*. 26 February. Available at: www.dailynews.co.tz/news/chenge-contests-ethics-case-mode.aspx.

Karayiannis, A. and Hatzis, A. N. (2007). Morality, social norms and rule of law as transaction cost-saving devices: The case of ancient Athens. *European Journal of Law and Economics*, 33, 621–643. Available at: http://ssrn.com/abstract=1000749.

Kaufmann, D., Kraay, A. and Zoido-Lobatón, P. (1999). *Governance Matters*. World Bank Policy Research Working Paper No. 2196. Washington, DC: World Bank. Available at: http://documents.worldbank.org/curated/en/665731468739470954/Governance-matters.

Knack, S. and Keefer, P. (1995). Institutions and economic performance: Cross-country tests using alternative institutional measures. *Economics and Politics*, 7, 207–227.

Lahore High Court (2019). Writ Petition No. 192069 of 2018. Sheikh Asim Farooq V/S Federation of Pakistan etc. Judgment Sheet. Stereo HCJ DA 38. Available at: http://blogs2.law.columbia.edu/climate-change-litigation/wp-content/uploads/sites/16/non-us-case-documents/2019/20190830_W.P.-No.-1920692018_judgment-1.pdf.

Lusty, D. (2014). Revival of the common law offence of misconduct in public office. *Criminal Law Journal*, 337(38): Available at: www.accountabilityrt.org/wp-content/uploads/2015/02/Lusty-Revival-of-the-Common-Law-Offence-of-Misconduct-in-Public-Office-2014-38-Crim-LJ-337.pdf.

Malmendier, U. and Schmidt, K. (2012). You owe me. Discussion Paper No. 392. Governance and the Efficiency of Economic Systems. Available at: http://epub.ub.uni-muenchen.de/14279/1/392.pdf.

Marcel, M. (2015). Engaging citizens: A game changer for development? *OGP Blog*. 27 February. Available at: www.opengovpartnership.org/stories/engaging-citizens-a-game-changer-for-development/.

Market Forces (2019). Friends in high places: Fossil fuel political donations. *Market Forces*. Available at: www.marketforces.org.au/politicaldonations2019/.

Mauro, P. (1995). Corruption and growth. *Quarterly Journal of Economics*, 110, 681–712.

Morgan, F., dir. (2006). *The Power of Community: How Cuba Survived Peak Oil* [film]. Arthur Morgan Institute for Community Solutions. Available at: www.communitysolution.org/mediaandeducation/films/powerofcommunity/.

NCCARF (National Australian Climate Change Research Facility) (2014). *Supporting Decision-Making for Effective Engagement*. Policy Guidance Brief 3. Gold Coast, Australia: National Climate Change Adaptation Research Facility.

Netherlands Supreme Court (2019). Dutch State to reduce greenhouse gas emissions by 25% by the end of 2020. *de Rechtspraak*. 20 December. Available at: www.rechtspraak.nl/Organisatie-en-contact/Organisatie/Hoge-Raad-der-Nederlanden/Nieuws/Paginas/Dutch-State-to-reduce-greenhouse-gas-emissions-by-25-by-the-end-of-2020.aspx.

Netherlands Supreme Court (2020). Climate case Urgenda. ECLI:NL:HR:2019:2007. *Rechtspraak*. Available at: https://uitspraken.rechtspraak.nl/inziendocument?id= ECLI:NL:HR:2019:2007.

Norman, B. (2018). *Sustainable Pathways for Our Cities and Regions: Planning Within Planetary Boundaries*. London: Routledge.

Norman, B., Steffen, W., Webb, R. et al. (2013). *South East Coastal Adaptation (SECA): Coastal Urban Climate Futures in SE Australia from Wollongong to Lakes Entrance*. Gold Coast, Australia: National Climate Change Adaptation Research Facility.

Norman, B., Steffen, W. and Smith, M. S. (2014). *Cities in Future Earth: A Summary of Key Considerations*. Canberra: Australian Academy of Science.

Obama, B. (2014). Speech presented at Open Government Partnership meeting [transcript]. Available at: www.whitehouse.gov/the-press-office/2014/09/24/remarks-president-obama-open-government-partnership-meeting.

OECD (2015). *Australia: Follow-up to the Phase 3 Report & Recommendations*. OECD. Available at: www.oecd.org/daf/anti-bribery/Australia-Phase-3-Follow-up-Report-ENG.pdf.

OECD and GEF (Global Environment Facility) (2015). *Toolkit to Enhance Access to Adaptation Finance for Developing Countries That Are Vulnerable to Adverse Effects of Climate Change, Including LIDCs, SIDS and African States*. Report to the G20 Climate Finance Study Group. OECD. Available at: www.oecd.org/environment/cc/Toolkit%20to%20Enhance%20Access%20to%20Adaptation%20Finance.pdf.

OGP (Open Government Partnership) (n.d). About the IRM. Available at: www.opengovpartnership.org/process/accountability/about-the-irm/.

OGP (2013). Leading multilateral institutions join forces with the Open Government Partnership. IDB Press release. 30 October. Available at: www.iadb.org/en/news/announcements/2013-10-30/open-government-partnership%2C10626.html.

OGP (2014a). What's in the new OGP action plans? *OGP Blog*. 29 September. Available at: www.opengovpartnership.org/stories/whats-in-the-new-ogp-national-action-plans/.

OGP (2014b). An overview of the OGP European and Asia-Pacific regional summits. Available at: www.opengovpartnership.org/stories/an-overview-of-the-ogp-european-and-asia-pacific-regional-summits/.

OGP (2015a). *Open Government Partnership (OGP) by the Numbers: What the IRM Data Tells Us about OGP Results*. Technical Paper 1. Open Government Partnership. Available at: www.opengovpartnership.org/wp-content/uploads/2019/07/Technical-paper-1_Executive-summary_final.pdf.

OGP (2015b). France to serve as next Open Government co-chair. *OGP Blog*. 29 April. Available at: www.opengovpartnership.org/stories/france-to-serve-as-next-open-government-partnership-co-chair/.

OGP (2015c). *Open Government Support Unit*. Available at: www.opengovpartnership.org/about/ogp-support-unit.

OGP (2015d). Openness in Natural Resources Working Group. Available at: www.opengovpartnership.org/wp-content/uploads/2001/01/ONRWG_Road-Map-Natural-Resource-Openess.pdf.

OGP (2016). *Four-Year Strategy 2015–18*. Open Government Partnership. Available at: www.opengovpartnership.org/wp-content/uploads/2019/06/OGP-4-year-Strategy-FINAL-ONLINE.pdf.

OGP (2018). *Brazil's 4th National Action Plan*. Open Government Partnership. Available at: www.opengovpartnership.org/wp-content/uploads/2019/04/Brazil_Action-Plan_2018-2020_EN.docx.

OGP (2020). Americas Regional Meeting. *OGP*. Available at: www.opengovpartnership.org/americas-regional-meeting/.

Oil Change International (2019). Fossil fuel industry influence in the US. *Oil Change International.* Available at: http://priceofoil.org/fossil-fuel-industry-influence-in-the-u-s/.

OSCE (Organization for Security and Co-operation in Europe) (2012). *Background Study: Professional and Ethical Standards for Parliamentarians.* Organization for Security and Co-operation in Europe. Available at: www.osce.org/odihr/98924.

Our Children's Trust (n.d.). Juliana v. United States youth climate lawsuit. *Our Children's Trust.* Available at: www.ourchildrenstrust.org/juliana-v-us.

Parliament of Australia Joint Committee on the Australian Commission for Law Enforcement Integrity (n.d.). Completed Inquiries 2010–13 CHAPTER 4 – International obligations and regional engagement. Footnote 3 UNODC, UNCAC Signature and Ratification Status as of 24 December 2012. Available at: www.aph .gov.au/Parliamentary_Business/Committees/Joint/Australian_Commission_for_ Law_Enforcement_Integrity/Completed_inquiries/2010-13/integrity_inter_op/report/ footnotes#c04f3.

Pelizzo, R. and Ang, B. (2008). A code of conduct for Indonesia: Problems and perspectives. *Parliamentary Affairs*, 61, 315–333.

Pelizzo, R. and Stapenhurst, R. (2004). *Tools for Legislative Oversight: An Empirical Investigation.* Research Collection School of Social Sciences Paper 44. Available at: https://ink.library.smu.edu.sg/soss_research/44.

Punder, P. (2019). Brazil's falling ranking in the Corruption Perceptions Index. *Corporate Compliance Insights.* 4 February. Available at: www.corporatecomplianceinsights .com/brazils-falling-ranking-in-the-corruption-perceptions-index.

Queensland Integrity Commissioner (2020). About us. *Queensland Integrity Commissioner.* Available at: www.integrity.qld.gov.au/about-us.aspx.

REDD (Reducing Emissions from Deforestation and Forest Degradation) (2015). Our work. *The UN-REDD Programme.* Available at: www.un-redd.org/how-we-work-1.

Reilly, B. (2008). Ethnic conflict in Papua New Guinea. *Asia Pacific Viewpoint*, 49, 12–22.

Republic of Estonia, Israel, Republic of Korea, New Zealand and United Kingdom (2014). D5 Charter. Available at: www.gov.uk/government/uploads/system/uploads/attach ment_data/file/386290/D5Charter_signed.pdf.

Rio+20 (2012). *The Future We Want: Outcome Document of the United Nations Conference on Sustainable Development.* United Nations. Available at: https:// sustainabledevelopment.un.org/content/documents/733FutureWeWant.pdf.

Rose-Akerman, S. (1999). *Corruption and Government: Causes, Consequences, and Reform.* Cambridge: Cambridge University Press.

Rush, M. and Giddings, P. (2015). Written evidence. The Operation of the Parliamentary Standards Act 2009 – Committee on Members' Expenses. Available at: https:// publications.parliament.uk/pa/cm201012/cmselect/cmmemex/writev/1484/m15.htm.

Saito, N. (2013). Mainstreaming climate change adaptation in least developed countries in South and Southeast Asia. *Mitigation and Adaptation Strategies for Global Change*, 18, 825–849.

Sampford, C., Coghill, S. and Smith, T. (2016) *Fiduciary Duty and the Atmospheric Trust.* Abingdon: Routledge.

Schott, K., Tink, A. and Watkins, J. (2014). *Political Donations: Final Report of Panel of Experts.* The Department of Premier and Cabinet. Available at: www.dpc.nsw.gov .au/updates/2014/05/27/panel-of-experts-political-donations.

Slezak, M (2019). Controversial Great Barrier Reef grant did not comply with transparency rules, National Audit Office says. *ABC News.* 17 January. Available at: www.abc.net .au/news/2019-01-16/great-barrier-reef-funding-grant-scrutinised-auditor-general/ 10720928.

Smith, T. (2014a). *Government Secrecy and Urban Planning: The Forgotten Trust and Reform*. Paper presented to the Melbourne University Urban Heritage Conference, Melbourne, 9 October. Available at: www.accountabilityrt.org/wp-content/uploads/2009/11/Smith-T-2014-Lyceum-U3A-Speech-final-_3_.pdf.

Smith, T. (2014b). Integrity in politics? Public office as a public trust? Is there hope? Paper presented to University of the Third Age, Hawthorn, 23 July. Available at: www.accountabilityrt.org/integrity-in-politics-public-office-as-a-public-trust-is-there-hope/.

Solomon, A. M., Webbe, S. and McGann, D. (2008). *The Right to Information: Reviewing Queensland's Freedom of Information Act*. Report by the FOI Independent Review Panel [The Solomon Report]. The State of Queensland Department of Justice and Attorney-General. Available at: https://apo.org.au/sites/default/files/resource-files/2008-06/apo-nid16021.pdf.

The Hague District Court (2015). Zaaknummer C/09/456689 / HA ZA 13-1396 (English translation). Available at: http://elaw.org/system/files/urgenda_0.pdf.

The White House (2016). North American Climate, Clean Energy, and Environment Partnership Action Plan. Obama White House Archives. 29 June. Available at: https://obamawhitehouse.archives.gov/the-press-office/2016/06/29/north-american-climate-clean-energy-and-environment-partnership-action.

Thornton, J. M. S. (2014). Government secrecy and urban planning: The forgotten trust and reform. Address presented to the Melbourne University Urban Heritage Conference. Available at: www.accountabilityrt.org/government-secrecy-and-urban-planning-the-forgotten-trust-and-reform.

TI (Transparency International) (2010). *National Integrity System Assessment Toolkit*. Berlin: Transparency International.

TI (2011). *Global Corruption Report: Climate Change*. London: Earthscan. Available at: www.transparency.org/en/publications/global-corruption-report-climate-change.

TI (2014a). Protecting Climate Finance: An Anti-Corruption Assessment of the Forest Carbon Partnership Facility. Transparency International. Available at: www.transparency.org/en/publications/protecting-climate-finance-an-anti-corruption-assessment-of-the-fcpf.

TI (2014b). Protecting Climate Finance: An Anti-corruption Assessment of the Global Environment Facility's Least Developed Countries Fund and Special Climate Change Fund. Transparency International. Available at: www.transparency.org/whatwedo/publication/protecting_climate_finance_assessment_gef_ldcf_sccf.

TI (2015a). Corruption Perceptions Index 2015. *Transparency International*. Available at: www.transparency.org/cpi2015.

TI (2015b). 3 Conditions for successful COP21 Paris Climate Agreement. *Transparency International*. 3 December. Available at: www.transparency.org/news/feature/3_conditions_for_successful_cop21_paris_climate_agreement.

TI (2017). Corruption Perceptions Index 2016. *Transparency International*. 25 January. Available at: www.transparency.org/en/news/corruption-perceptions-index-2016.

TI (2019a). Brazil. *Transparency International*. Available at: www.transparency.org/country/BRA.

TI (2019b). Central African Republic. *Transparency International*. Available at: www.transparency.org/en/countries/central-african-republic.

TI Ireland (2009). *National Integrity Systems Study*. Transparency International. Available at: http://transparency.ie/resources/NIS.

TI Secretariat (2001). 'Strongest correlation' between corruption and poor environmental performance, study shows. *Transparency International*. 25 January. Available at:

www.transparency.org/news/pressrelease/strongest_correlation_between_corruption_
  and_poor_environmental_performance.
Trease, H. V. (2010). Vanuatu. *The Contemporary Pacific*, 22, 467–476.
Ugaz, J. (2015). More corruption now globally than 20 years ago, says Transparency
  International chair. Available at: https://niajamaica.org/press_release/more-corrup
  tion-now-globally-than-20-years-ago-says-transparency-international-chair/.
UN (2020). United Nations Convention against Corruption: Signature and ratification
  status. UN Office on Drugs and Crime. Available at: www.unodc.org/unodc/en/
  corruption/ratification-status.html.
UN Department of Economic and Social Affairs, Population Division (2014). *World
  Urbanization Prospects: The 2014 Revision. Highlights*. ST/ESA/SER.A/352. New
  York: UN Department of Economic and Social Affairs, Population Division.
  Available at: https://population.un.org/wup/publications/files/wup2014-highlights
  .pdf.
UNFCCC (UN Framework Convention on Climate Change) (1992). Article 4
  Commitments, 1(b). In *United Nations Framework Convention on Climate Change*.
  United Nations. Available at: https://unfccc.int/files/cooperation_and_support/ldc/
  application/pdf/article4.pdf.
Watson, C., Church, J. and King, M. (2015). Sea level is rising fast – and it seems to be
  speeding up. *The Conversation*. 12 May. Available at: https://theconversation.com/
  sea-level-is-rising-fast-and-it-seems-to-be-speeding-up-39253.
Webb, R. and Beh, J. (2013). *Leading Adaptation Practices and Support Strategies for
  Australia: An International and Australian Review of Products and Tools*. Gold
  Coast, Australia: National Climate Change Adaptation Research Facility, Griffith
  University. Available at: https://nccarf.edu.au/leading-adaptation-practices-and-sup
  port-strategies-australia-international-and-australian/.
Whetton, P. (2011). Future Australian climate scenarios. In H. Cleugh, M. S. Smith and
  M. Battaglia, eds., *Climate Change: Science and Solutions for Australia*. Canberra:
  Commonwealth Scientific and Industrial Research Organisation (CSIRO) Publishing,
  pp. 35–44.
World Bank (n.d.). Green bonds. *World Bank*. Available at: https://treasury.worldbank.org/
  en/about/unit/treasury/ibrd/ibrd-green-bonds.
World Bank (2011). World Bank flash: A more open, transparent and accountable World
  Bank. 7 July. *World Bank*. Available at: www.worldbank.org/en/news/press-release/
  2011/07/07/world-bank-flash-more-open-transparent-accountable-world-bank.

23

Financing the Transition

MICHAEL H. SMITH, PABLO BERRUTTI, NATHAN FABIAN AND
NICOLETTE BOELE

Executive Summary

The chapter looks at trends in financing the transition to low-carbon economies, including the growth of an environmental, social and governance (ESG) financial market segment.

The reasons for investing in ESG projects and funds are explored, noting that, as well as helping to avoid catastrophic climate change, such investments often offer superior financial returns. These are helped by the steeply falling costs of renewable energy making for improved business cases.

The recent evolution of the principles of fiduciary duty (of investors) to include the effects of climate change in investment decisions is discussed, pointing to significant changes in Australia, and globally.

The chapter outlines the growing redirection of funds to ESG-oriented investments and the emergence of a diverse range of green financing instruments. It is noted that although not yet dominant in the financial markets, ESG investments and the related funds are no longer niche. They are, however, not yet sufficient to address the needs to hold temperatures within the 2 °C guideline.

The chapter sets out the four steps needed to address this gap in investment, namely:

- government policy change to provide leadership and market confidence – particularly as part of COVID-19 recovery plans;
- improved governance of the financial sector;
- de-risking ESG investments and improving risk/return ratios; and
- improved knowledge and skills in the financial sector.

The chapter concludes by noting that although there are many 'green shoots', progress in financing the transition to low-carbon economies is not yet commensurate with the challenges of climate change.

23.1 Introduction

There is a growing consensus among institutional investors and business leaders that it is in their financial interests to take environmental, social and governance (ESG) principles into account in their decisions, and this is increasingly reflected in legal fiduciary obligations.

These principles include considering the effects of investments on climate change, limiting risk exposures and reducing greenhouse gas emissions.

The 2019 Carbon Disclosure Project's analysis of over 6000 companies concluded that the best performing companies on action on climate change outperformed the market (Carbon Disclosure Project 2019). This reconfirmed the conclusions of previous studies. These facts provide the basis for initiatives like The Climate Action 100+ initiative, whereby a coalition of institutional investors collectively investing over USD 40 trillion are asking the 160 largest global emitters to assess their risk exposure to climate change and develop capital expenditure plans in line with the Paris Climate Agreement targets.

The reasons for this higher performance are well documented (Hargroves and Smith 2005). Companies that are proactive on climate change improve their profitability and competitive advantage through reducing operational costs and strategically positioning their products and services for the fast-growing markets in climate change solutions (UK Government Department for Business Innovation and Skills 2013). Also, the costs of clean technologies have been falling rapidly over the last decade.

The top 100 banks in China, Europe, the USA, Canada and Australia have all issued climate bonds, certified green bonds. As of February 2020, 22 banks have stopped direct financing of new thermal coal-mining projects and 28 banks have stopped direct financing of new coal plants worldwide (BankTrack n.d.). Investment flows to solar, wind and hydropower, in the power sector, made up three-quarters of new capacity added in 2019 (BloombergNEF 2020a). This is because, as the Paris Effect Report found in 2020, 'We are crossing a number of cost tipping points that are creating economic incentive for solar and wind to serve up to 75–90% of power systems (leveraging batteries, hydropower and other flexibility levers)' (SYSTEMIQ 2020).

While these changes are becoming more common, the current speed and scale of the shift by investors and business leaders to invest in a low-carbon resilient future is insufficient to ensure that global warming stays well below 2 °C. There has been strong growth in ESG-linked funds, green loans and other financial instruments, but starting from a small base. Recent studies (BankTrack et al. 2019) indicate that investment in fossil fuels by 33 major banks greatly exceeded their investments in ESG-focused projects, reaching USD 1.9 trillion in 2019 since the 2015 Paris Agreement and USD 3.8 trillion by 2021.

It is estimated that from 2020 to 2030, USD 6.3 trillion will be needed to achieve the UN's Sustainable Development Goals worldwide, increasing to USD 6.9 trillion per annum in investment up to 2030 to make this investment compatible with meeting the Paris Agreement climate goals. This is opening up enormous investment opportunities.

This investment and finance gap means that governments continue to have a key role in accelerating the reallocation of capital by the private sector towards low-carbon resilient investments. The chapter sets out four integrated steps that offer governments a clear path for closing the investment gap and achieving climate and sustainable development objectives.

Governments have the fundamental role of establishing a policy environment that provides long-term predictability, certainty and transparency to investors, business and consumers to encourage them to invest in a low-carbon resilient future.

23.2 Why Is Climate Change and Sustainability a Priority for Investors? The Case for Investing in a Low-Carbon Resilient Economy

There is a growing consensus that it is in the financial interests, and therefore the fiduciary duty, of institutional investors and business leaders to proactively address sustainability and climate change. Good management of environmental, social and governance (ESG) considerations is a proxy for sound business decision-making by quality companies that deliver medium- and long-term value to a range of stakeholders.

The UN-backed Principles of Responsible Investment's (PRI) final report, *Fiduciary Duty in the 21st Century* (PRI 2019), discusses a range of evidence supporting the positive relationship between ESG performance and financial outcomes; for instance:

- A 2014 study of 180 companies showed that those that had voluntarily adopted ESG principles produced superior stock returns (PRI 2019: 18).
- A 2015 study by Fried et al. analysed 2000 empirical studies on the relationship between ESG criteria and investment performance, and found that 90% of the studies identified a non-negative relationship between ESG performance and corporate financial perform- ance, with a clear majority finding a positive relationship (Fried et al. 2015; PRI 2019: 18).
- In 2017, a Bank of America Merrill Lynch study concluded that the stocks in its US portfolio ranked within the top third by ESG scores outperformed stocks in the bottom third by 18% in the 2005–2015 period (PRI 2019: 18).
- The 2019 Carbon Disclosure Project's analysis of over 6000 companies concluded that the best performing companies on action on climate change also outperformed the market. Specifically: 'The STOXX Global Climate Change Leaders Index – which is based on the CDP A List – outperformed the STOXX Global 1800 by 5.4% per annum from December 2011 to July 2018, [demonstrating] that leadership on environmental issues goes hand in hand with being a successful and profitable business' (CDP 2019).

In 2020, major financial management firms, BlackRock and Meketa, separately con- cluded that investment funds have experienced no negative financial impacts from divesting from fossil fuels but, on the contrary evidence of improvement in returns (IEEFA 2021).

Mercer has conducted scenario analysis over three climate change scenarios: 2 °C, 3 °C and 4 °C, respectively, above pre-industrial levels (Mercer 2019). They state: 'In a 2 °C scenario by 2050, there are minor positives . . . for materials, telecoms and consumer staple sectors [as well as renewables]. In 3 °C and 4 °C scenarios, all sectors, apart from renew- ables, have negative return impacts' (Mercer 2019: 35).

The transition of the economy to net zero greenhouse gas emission by 2050 also poses direct and indirect market risks and opportunities. For example, there are risks to investors and fossil fuel companies of being stuck with stranded assets. Some large institutional investors have made specific divestments of companies they have identified as being most at risk of stranded assets; others, like BlackRock (the world's largest asset manager), have set policies to limit investments in thermal coal. In the case of BlackRock, in its

Global investment in the power sector by technology

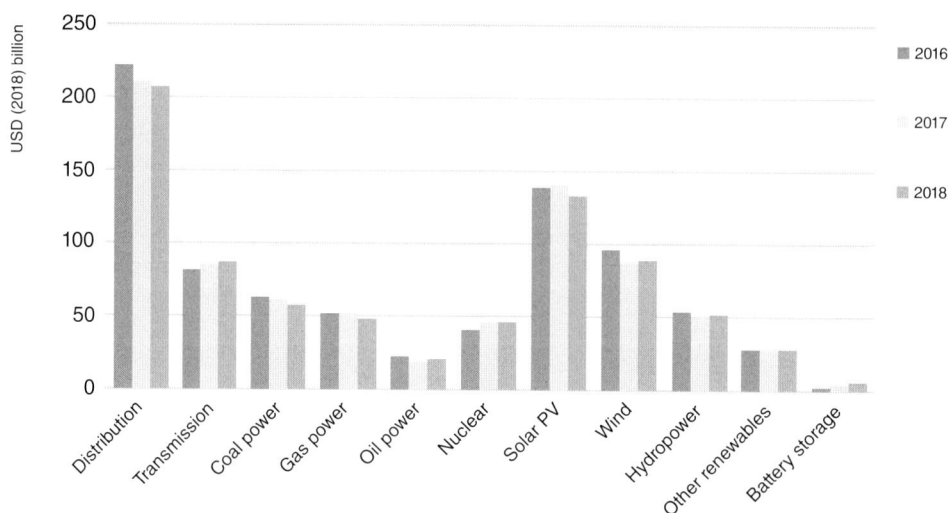

Figure 23.1 Global investment in the power sector by technology.
Note: Gas and oil-fired generation investment includes utility-scale plants as well as small-scale generating sets and engines. Hydropower includes pumped storage hydro. For a colour version of this figure, please see the colour plate section.
Source: IEA analysis with calculations for solar photovoltaic (PV) systems, wind and hydropower (IEA 2019a).

discretionary active investment portfolios (BlackRock n.d.). As reported by IEEFA, BlackRock comments that fossil fuel stocks have underperformed for the last 5 years (to 2021) and forward-looking analysis shows they are exposed to significant regulatory, technological and market risks (IEEFA 2021).

In terms of transition opportunities, the costs of key enabling technologies such as LED (light-emitting diode) lighting, heat pumps, renewable energy and batteries have fallen so much that it is now more profitable for companies to cut greenhouse gas emissions and make the transition to meet their own energy needs with renewable energy (see Figure 23.1).

Over 100 companies (BloombergNEF 2019, 2020b) per year since 2017 are entering into power purchase agreements (PPAs) for renewable energy projects to help meet their energy needs and cut greenhouse gas emissions (see, for example, Figure 23.2). In 2018, Facebook had the largest PPA with solar power agreements equalling 2.1 GW (gigawatts) and wind power 55 MW (megawatts), followed by AT&T (with the equivalent of 820 MW of wind power) and Walmart with 138 MW solar and 533 MW wind power (BloombergNEF 2019: 43).

The RE100 initiative includes more than 260 reporting members who have committed to sourcing 100% renewable energy and includes some of the largest companies in the world. RE100's 2020 progress report found that three-quarters of its members plan to reach 100% renewable energy by 2030, with 53 reporting they have already reached the 100% milestone. PPA's account for 26% of the renewable electricity purchased by RE100 members (RE100 2021: 3).

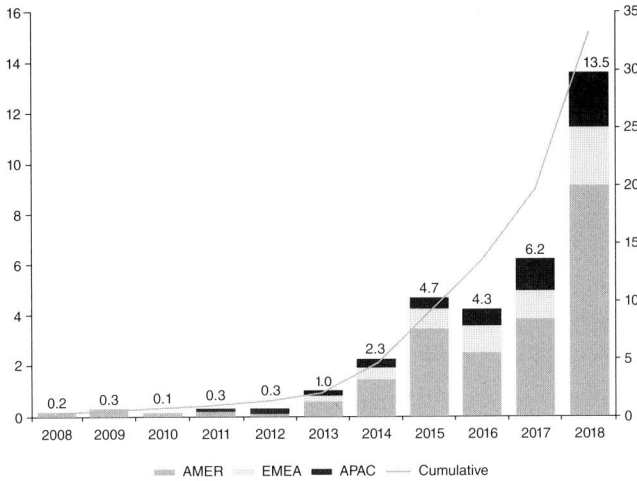

Figure 23.2 Global volume of corporate PPAs signed, by region, 2008–2018 GW.
Note: On-site PPAs with captive installation not included; the Asia Pacific (APAC) number is an estimate; pre-market reform Mexico PPAs are not included; the cumulative total is shown on the right-hand axis; AMER, Americas; EMEA, Europe, Middle East and Africa. For a colour version of this figure, please see the colour plate section.
Source: BloombergNEF (2019: 43).

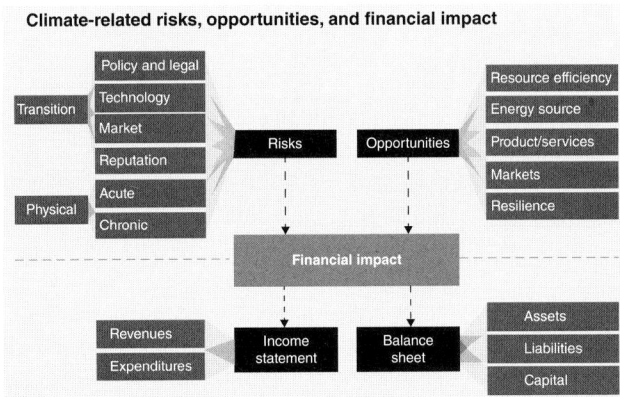

Figure 23.3 Climate change risk management issues relevant for asset owners across asset types.
Source: TCFD (2017: 8).

Taken together, the transition to a low-carbon economy and the influence of the physical impacts of climate change pose manifold risks and opportunities for investors over the short, medium and long term. The Task Force on Climate-related Financial Disclosures (TCFD) identified six types of risk and five opportunities which are likely to influence company performance in the decades ahead (see Figure 23.3). Investors and business leaders are increasingly being expected to make disclosures in line with the TCFD framework due to risks and opportunities described in this section.

In December 2020, the Net Zero Asset Manager initiative was launched with 73 signatories representing USD 32 trillion of assets under management. Its aim is to 'galvanise the asset management industry to commit to a goal of net zero emissions'. The Net Zero Asset Manager initiative mirrors the Net Zero Asset Owner Alliance with 35 investor signatories managing USD 5.6 trillion in assets and the Collective Commitment to Climate Action initiative, which includes 38 bank signatories as of the first half of 2021.

In January 2021, 33 major investors, representing USD 5 trillion assets under management, launched the inaugural 2025 Alliance Target Setting Protocol which requires signatories to report every 5 years against their detailed plans for carbon reduction to achieve net zero emissions by 2025, consistent with a maximum temperature rise of 1.5 °C above pre-industrial levels.

Divestment commitments have also accelerated with the non-government organisation Go Fossil Free tracking; as of early 2021, more than 1300 institutions with more than USD 14.56 trillion in assets have made fossil fuel divestment commitments. However, most of these institutions remain faith-based, educational and philanthropic institutions, with pension and sovereign wealth funds representing only 24% of the institutions who have made commitments.

In 2019 the Global Investor Statement to Governments on Climate Change was signed by 515 investors representing well over USD 35 trillion in assets. The statement notes that the global shift to clean energy is under way but 'much more needs to be done by governments to accelerate the low carbon transition and to improve the resilience of our economy, society and the financial system to climate risks'. It calls on global leaders to take steps to achieve the Paris Agreement goals of keeping global warming well under two degrees Celsius, such as:

- accelerating private-sector investment in the low-carbon transition;
- incorporating Paris-aligned climate scenarios into all policy frameworks;
- putting a meaningful price on carbon;
- phasing out fossil fuel subsidies by set deadlines;
- phasing out thermal coal power worldwide by set deadlines; and
- committing to improve climate-related financial reporting.

23.2.1 The Evolving Nature of Fiduciary Duty

The concept that investors, CEOs and boards have a fiduciary duty to apply a broader lens to risks such as those posed by climate change to business has become increasingly mainstream and required by governments or relevant regulatory agencies. The PRI finds that ESG issues including climate change are material to company financial performance; therefore, a failure to consider them and their financial implications is a failure of fiduciary duty.

Of the top 50 world economies, 48 now have some form of policy designed to help investors consider sustainability risks, opportunities and outcomes (PRI 2019: 13). In the UK, in 2015, Mark Carney, Governor of the Bank of England and Chairman of the Prudential Regulation Committee, identified three main risks stating the following;

There are three broad channels through which climate change can affect financial stability:

First, physical risks: the impacts today on insurance liabilities and the value of financial assets that arise from climate- and weather-related events, such as floods and storms that damage property or disrupt trade;

Second, liability risks: the impacts that could arise tomorrow if parties who have suffered loss or damage from the effects of climate change seek compensation from those they hold responsible. Such claims could come decades in the future, but have the potential to hit carbon extractors and emitters – and, if they have liability cover, their insurers – the hardest;

Finally, transition risks: the financial risks which could result from the process of adjustment towards a lower-carbon economy. Changes in policy, technology and physical risks could prompt a reassessment of the value of a large range of assets as costs and opportunities become apparent. *(Carney 2015)*

In Australia, in 2017, the Australian Prudential Regulatory Authority (APRA), concluded, after a 2-year review, that climate change represents a financial risk to investor portfolios and hence investors have a duty to assess and manage such risks.

In Australia and New Zealand, the legal profession has provided its views of fiduciary duty in the contemporary context of the systemic risk of climate change. Chris Gillies, Olivia Morgan and Scarlet Roberts, of Chapman Tripp, in 2019 provided a legal opinion that was peer reviewed by Alan Galbraith QC. They opined that New Zealand company directors and managed scheme providers are required to take account of climate change considerations in their decision-making (Kalderimis and Swan 2019). This was in line with the 2016 opinion provided by Noel Hutley SC and Sebastian Harford-Davis (of Minter Ellison) in their memorandum of opinion on climate change and directors' duties in Australia, which was updated in 2019 to say:

There are, at the present time, significant and well-publicised risks associated with climate change and global warming that would be regarded by a Court as foreseeable. Such risks require engagement from company directors in affected sectors, particularly in (at least) the banking, insurance, asset ownership/ management, energy, transport, material/buildings, agriculture, food and forest product industries . . .

. . . As time passes, it is increasingly obvious that climate change is and will inevitably affect the economy, and it is increasingly difficult in our view for directors of companies of scale to pretend that climate change will *not* intersect with the interests of their firms. In turn, that means that the exposure of individual directors to 'climate change litigation' is increasing, probably exponentially, with time. *(Hutley and Hartford-Davis 2019: 8–9, italics in original)*

These facts provide the basis for initiatives like The Climate Action 100+, where a coalition of over 500 institutional investors managing over USD 52 trillion worth of assets are asking 160 of the largest and most strategically important global emitters to assess their risk exposure to climate change and develop capex (capital expenditure) plans in line with the Paris Climate Agreement[1] targets. This engagement has resulted in several major mining companies adjusting their business models to diversify investments away from fossil fuels and into the mining of critical metals needed for the low-carbon economy.

---

[1] *Paris Agreement Under the United Nations Framework Convention on Climate Change*, opened for signature 16 February 2016. Available at: https://treaties.un.org/Pages/ViewDetails.aspx?src=TREATY&mtdsg_no=XXVII-7-d&chapter=27&lang=_ en&clang=_en.

Banks are providing an increasingly diverse range of instruments that support green finance, including green bonds, green loans, mortgages and green deposit products. The largest 100 global banks, whether they be in China, Europe, the USA, Canada or Australia, have all issued green bonds. However, these developments must be viewed in the context of the larger picture mentioned earlier, where major global banks continue to invest vast sums in the fossil fuel industry. The World Research Institute, through its Green Targets tool, compared the fossil fuel and renewable energy finance being provided by the world's 50 largest banks and found the vast majority are providing greater funding to fossil fuels than renewables overall when all sectors are included (Pinchot and Christianson 2019; WRI 2019).

As the COVID-19 pandemic has run its course, energy demand has fallen but the Fossil Fuel Finance Report (2021), titled 'Banking on Climate Chaos' states:

Despite a massive global drop in fossil fuel demand and production in 2020, banks' fossil fuel financing still remained above 2016 and 2017 levels. Although overall fossil fuel financing dropped in 2020, bank financing from January to June was the highest of any half year since the adoption of the Paris Agreement, as large energy companies loaded up on cheap debt at the start of the global pandemic, in preparation for expected difficult times.

In the five years since the Paris Agreement was adopted, the world's 60 largest private banks financed fossil fuels with USD 3.8 trillion.

17 of the 60 banks have recently pledged to achieve 'net zero' financed emissions. But our analysis shows that for many of the world's worst funders of fossil fuels, these plans so far are dangerously weak, half-baked,or vague. Even the best overall 'climate impact' commitments are not a substitute for explicit commitments on fossil fuels (and deforestation). 2050 commitments should be met with great skepticism unless they are accompanied by 2021 action on coal, oil, and gas. US and Canadian banks make up only 13 of the 60 but account for about half of fossil fuel financing.

Much of this 3.8 trillion in financing facilitates the expansion of fossil fuel extraction and infrastructure. 39% of total financing went to just 100 key companies with the worst fossil fuel expansion plans. *(Fossil Fuel Finance Report 2021)*

Notwithstanding this later data, the comments of the UN Environmental Programme (UNEP) in its *Sustainable Finance Progress Report* 2019: remain valid. They stated:

There are several indications beyond investment flow volumes that sustainable finance is no longer a niche offering in many areas of the financial markets. There is emerging evidence that a diverse range of larger financial actors around the world are integrating elements of sustainability into financial decision-making in both public and private financial markets. *(UNEP 2019: 10)*

Although encouraging, with a significant acceleration in commitments and actions by the finance sector, the overall picture remains mixed.

23.2.2 What Options Exist to Address the Investment Gap to Stay under Two Degrees?

While these changes are positive, the current speed and scale of the shift by institutional investors and business leaders to invest in a low-carbon resilient future is insufficient to ensure average global warming stays well below 2 °C as required by the Paris Agreement. The IEA *World Energy Outlook 2019* finds that in the Sustainable Development Scenario, around 25% more investment is needed than in the Stated Policies Scenario, with investments in renewable-based power needing to double. In this scenario there is a significant reallocation of investment away from fossil fuels and towards renewable energy, energy efficiency and low-carbon technologies (IEA 2019b).

This 'investment gap' is widely recognised by numerous reports and international bodies including UNCTAD (UN Conference on Trade and Development) (WRI 2014), the IPCC (Intergovernmental Panel on Climate Change), OECD, IMF (International Monetary Fund) and the World Economic Forum (WEF 2013). The UNEP *Sustainable Finance Progress Report 2019* comments that: 'Sustainable finance remains an embryonic topic, but certain areas are experiencing rapid rates of development' (UNEP 2019: 9). It also observes that 'in financial markets, the standout feature is one of sustainable finance becoming increasingly mainstream. Larger actors within the financial system are increasingly aligning themselves with sustainable outcomes' (UNEP 2019: 19).

The investment community has long called for government to lead on clear climate change policy reform, urging for reforms based around near- and longer-term emissions reduction targets and which help create the necessary environment for higher levels of private-sector investment in a low-carbon resilient future. In 2019, the investor statement was signed by 500 investors representing well over USD 35 trillion (IIGCC 2019).

23.3 Encouraging Greater Private-Sector Investment: An Integrated Four-Step Strategy for Government and Key Stakeholders

Despite the trends discussed in this chapter, the scale of the finance gap means that government continues to have a key role in facilitating the reallocation of capital by the private sector towards low-carbon resilient investments. The following four integrated steps offer governments and stakeholders a path for closing the investment gap and achieving climate and sustainable development objectives.

23.3.1 Step 1: Implementing Climate Change Policy Reform to Provide Greater Investment Certainty and Transparency to Support Better-Informed Investment Decisions

The first and most important step is for governments to undertake purposeful climate change policy reform including emissions reduction targets which can achieve net zero emissions no later than 2050. Consistent with the Paris Agreement's review and ratchet mechanism, these targets should become increasingly ambitious as technology and other changes occur, noting that, as earlier chapters illustrate, the necessary technologies are already available. The main challenges lie in policy failure at the national level in many cases, and the global level, as demonstrated in the 2019 Conference of Parties meeting (COP 25) in Madrid.

The challenge and opportunity for governments has been heightened by the COVID-19 pandemic and the prospect of green economic recovery plans. The UN-backed report Are We Building Back Better, found that to February 2021 only 18% of spending by governments in recovery plans was classed as environmentally positive. However, these figures do not fully include green recovery plans for the USA under the new Biden and Harris Administration, which the report notes could exceed USD 2 trillion (O'Callaghan and Murdock 2021).

Investor priorities for climate policy reform were clearly communicated in the previously mentioned *Global Investor Statement to Governments on Climate Change* (IIGCC 2019), which was presented at the UN climate summit in New York in September 2019. Their climate change policy reform priorities were:

- The achievement of the Paris Agreement's goals:
  - updating and strengthening Nationally Determined Contributions to meet the emissions reduction goal of the Paris Agreement, starting the process now and completing it no later than 2021, and focusing swiftly on implementation;
  - formulating and communicating long-term emissions reduction strategies;
  - aligning all climate-related policy frameworks holistically with the goals of the Paris Agreement;
  - supporting a just transition to a low-carbon economy.
- The acceleration of private-sector investment into the low-carbon transition:
  - incorporating Paris-aligned climate scenarios into all relevant policy frameworks and energy transition pathways;
  - putting a meaningful price on carbon;
  - phasing out fossil fuel subsidies by set deadlines;
  - phasing out thermal coal power worldwide by set deadlines.
- A commitment to improve climate-related financial reporting:
  - publicly supporting the Financial Stability Board's (FSB) TCFD recommendations and the extension of its term;
  - committing to implement the TCFD recommendations in their jurisdictions, no later than 2020;
  - requesting that the FSB incorporate the TCFD recommendations into its guidelines;
  - requesting that international standard-setting bodies incorporate the TCFD recommendations into their standards (IIGCC 2019: 2).

As Nick Robbins, former Chair of the UNEP Finance Initiative has explained, policy leadership from governments is critical to:

shift the risk–reward ratio [for investors and business], for example, by setting clear carbon budgets, reforming fossil fuel subsidies, pricing carbon, raising energy efficiency standards, rewarding low-carbon energy supply and building resilience into agriculture and urban infrastructure. *(Robins et al. n.d.)*

Leading governments are increasingly implementing the policies required. According to UNEP's Sustainable Finance Progress Report 2019 produced for the G20 Sustainable

Finance Study Group, 'there is growing evidence that demonstrates the sustainable finance policy over the last year has been characterized by strong growth, increased scope, and greater maturity' (UNEP 2019: 7). According to the PRI, globally, there are now 730 hard and soft law provisions in financial regulations that embed such sustainability consider-ations across some 500 policy instruments, with 97% of these laws having been enacted since 2000 (PRI 2019: 8).

As this book has shown, with few exceptions, positive climate change policy reform is occurring in most countries since the Paris Agreement. China has spent 2 years developing its national emissions trading scheme (announced in 2017) after successful pilot schemes in seven regions (World Bank Group 2019: 35–37). Canada implemented its price on carbon (World Bank Group 2019: 30) and, by 2019, Singapore, in South East Asia, implemented a price on carbon (World Bank Group 2019: 41). Hence, by 2021, over 35% of the world's GDP will be subjected to a price on carbon.

In 2019, there were renewable energy policies and targets in over 130 countries. Such policies improve the risk and return characteristics of low-carbon investments. For example, many governments are using long-term energy contracts as part of the renewable energy reverse auctions to give investors certainty (SYSTEMIQ 2020). This is helping ensure, in the power sector, investment in low-carbon electricity sources now outpaces fossil-fuel-powered electricity generation roughly 3:1 (BloombergNEF 2020a).

Notwithstnding these positive developments, as already indicated, major bank financing of fossil fuels, especially oil and gas developments, continues apace and governments need to consider urgently how this situation can be reversed. In addition to specific climate policies, governments need to incorporate climate goals across agencies. One example where this has not happened is China's Belt and Road Initiative, which is the country's signature foreign policy programme and the largest global infrastructure investment plan ever. Between 2014 and 2017, 91% of energy-sector syndicated loans from six major Chinese banks to Belt and Road Initiative countries were in fossil fuels, with 40% of the lending in 2018 going to coal projects. This is despite China being a global leader in the deployment of renewable energy technology. Encouragingly, other countries and develop-ment institutions have committed to stop financing coal projects, including announcements in 2020 from the second and third largest coal funding countries, Japan and Korea. More recently, China has indicated it will increase environmental standards for Belt and Road projects (Council on Foreign Relations 2021).

23.3.2 Step 2: Improving Governance of the Financial Sector Itself through Clarifying Fiduciary Duty and Establishing More Meaningful Performance Benchmarks

In many jurisdictions, the fiduciary duty of fund trustees has historically been interpreted to preclude the consideration of environmental, social, corporate governance and ethical factors in investment decision-making, as it was assumed that this would undermine the financial returns delivered to beneficiaries. This interpretation originated from the concern that trustees might put their personal ethical values over their fiduciary obligations to their clients.

As already discussed, institutional investors incorporating sustainable development and climate change considerations are acting consistently with their fiduciary duty because climate change risks (transition, physical, liability) and opportunities (new products and markets) will materially affect the financial performance of funds. Moreover, studies indicate (Orlitzky et al. 2003; Fried et al. 2015) that compliance with ESG principles typically has a positive impact on commercial outcomes. Also, as discussed above, greater clarity is being provided by regulators and respected legal voices on fiduciary duties and increased expectations. Box 23.1 presents a more modern understanding of fiduciary duty.

The trend broadly towards the 'stewardship' responsibilities of investors, and more specifically on climate change, is well advanced around the world. Many nations are making reforms in line with this, as exemplified by the following:

- The EU sustainable finance initiative includes sweeping reforms that will see sustainability and climate change goals incorporated into the full finance value chain, from financial advice to institutional investment strategy, and even the capital risk weightings used to regulate bank lending (European Commission n.d.).
- Stewardship codes or equivalent requirements have been introduced in countries including Japan,[2] Malaysia,[3] South Africa[4] and the UK[5] (Generation Foundation et al. 2016: 8). Stewardship codes generally reinforce fiduciary duties but also explicitly call out the importance of ESG issues.
- The French Energy Transition for Green Growth Law,[6] which came into effect on 1 January 2016, strengthens the mandatory carbon disclosure requirements for listed and large unlisted companies, and the ESG disclosure requirements for institutional investors (Generation Foundation et al. 2016: 8).
- Ontario's pension standards legislation (PPA909), effective in 2016, requires pension funds to state in their investment policies whether and how ESG factors are considered in their decision-making processes (Generation Foundation et al. 2016: 8).
- In addition to regulatory-led changes, industry-led initiatives such as the Principles for Responsible Banking, Principles for Responsible Investment and the Principles for Sustainable Insurance are also providing signals to financial services companies on these evolving expectations.

These reforms must address both individual finance institution responsibilities and issues with market structures. One example of a market structural problem is the increased popularity of broad-based passive investment funds which track benchmarks weighted on

---

[2] *Principles for Responsible Investors* [*Stewardship Code*], Japan, 2014. Available at: www.fsa.go.jp/en/refer/councils/ stewardship/20140407.html [English version].
[3] *Malaysian Code for Institutional Investors*, Malaysia, 2014. Available at: www.sc.com.my/api/documentms/download.ashx?id= 9f4e32d3-cb97-4ff5-852a-6cb168a9f936.
[4] *Code for Responsible Investing in South Africa* [*CRISA*], South Africa, 2011. Available at: https://integratedreportingsa.org/the-code-for-responsible-investing-in-south-africa/.
[5] *UK Stewardship Code 2020*, UK, 2020. Available at: www.frc.org.uk/getattachment/5aae591d-d9d3-4cf4-814a-d14e156a1d87/ Stewardship-Code_Dec-19-Final-Corrected.pdf.
[6] *Loi relative à la transition énergétique pour la croissance verte* [*LTECV*] [*The Energy for Green Growth Act*], France, 2015.

Box 23.1 **Modern Fiduciary Duty**

Fiduciary duties (or equivalent obligations) exist to ensure that those who manage other peoples' money act in the interests of beneficiaries, rather than serving their own interests (see PRI 2019). The most important of these duties are:

- **Loyalty**: Fiduciaries should:
  - act honestly and in good faith in the interests of their beneficiaries or their clients;
  - understand and incorporate in their decision-making the sustainability preferences of beneficiaries and/or clients, whether or not these preferences are financially material;
  - impartially balance the conflicting interests of different beneficiaries and clients;
  - avoid conflicts of interest; and
  - not act for the benefit of themselves or third parties.
- **Prudence**: Fiduciaries should act with due care, skill and diligence, investing as an 'ordinary prudent person' would. This includes:
  - incorporating financially material ESG factors into their investment decision-making, consistent with the time frame of the obligation;
  - being an active owner, encouraging high standards of ESG performance in the companies or other entities in which they are invested;
  - supporting the stability and resilience of the financial system.

Fiduciaries should disclose their investment approach to clients and/or beneficiaries including information on how preferences are incorporated into the scheme's investment strategy and the potential risks and benefits of doing so.

---

company size (market capitalisation). In the USA in 2019, 'passive' investment eclipsed 'active' management in size for the first time (Gittelsohn 2019). These funds by definition allocate more capital to incumbent industries, like fossil fuel companies, rather than emerging industries. Part of the European Commission reforms is to create new low-carbon benchmarks, although questions of how to encourage capital to be transitioned from one to the other has not been addressed (European Commission n.d.).

In addition to decisions related to whether or not to finance or provide insurance for a particular asset or activity, financial institutions also have significant influence through existing relationships. Banks, investors and insurance companies have the ability, both formally, through contract terms (banks and insurance companies) and proxy voting (investors) and informally, through relationship management and engagement, to influence the way companies do business. Regulatory reforms and guidance must remove barriers and enhance this important lever for financial services companies to have a positive influence.

Lastly, education is a critical enabler for stronger action on climate finance. While directors and other actors within the finance industry may be aware of changes in these areas, a greater understanding of the science and implications of issues like climate change, along with methods for implementing effective strategies, is required. Regulators can require a demonstration of competence in these areas as they do for other important areas of risk management and strategy. While the importance of investor education is covered under step 4, regulator expectations linked to fiduciary responsibilities are a key driver for rapid adoption.

As these reforms continue to evolve, they must continually be tested against Paris Agreement and the UN's Sustainable Development Goals to ensure they are adequate.

23.3.3 Step 3: De-risk Private- and Public-Sector Investment and Improve the Capital Reward/Risk Ratio – Role of Investment Vehicles, Institutions and Instruments

Traditional investment vehicles, private–public partnerships (PPPs), institutions and instruments have long been used to overcome perceived risks and have been refined by financiers and investors for centuries. These can also be used to help de-risk private-sector investment and improve capital reward/risk ratios for new investments in the green economy. Private–public partnerships and instruments (i.e. bonds, public and private issuance, revolving loan facilities) and institutions (i.e. green banks, clean energy finance corporations and the Global Green Climate Fund) are being adapted in numerous ways. These help to encourage the private sector – including private banks and institutional investors (sovereign wealth funds, pension funds, insurance companies and endowments) – to allocate capital towards low-carbon resilient investments.

23.3.3.1 Investment Instruments: Green Bonds/Climate Bonds/Sustainable Bonds/Green Loans/Sustainability Linked Loans

Green bonds and climate bonds[7] are fixed-income securities issued by the public sector (i.e. governments), or the private sector (i.e. banks or corporations), to raise the necessary capital for a project which contributes to a low-carbon, climate-resilient economy. Green bonds[8] involve the issuing entity guaranteeing to repay the bond over a certain period, plus either a fixed or variable rate of return. They can be asset-backed securities tied to specific green infrastructure projects or 'treasury-style' bonds issued to raise capital that will be allocated across a portfolio of green low-carbon investments.[9] The first green bonds were issued by the European Investment Bank in 2007 and World Bank in 2008. Since then, numerous major banks have followed suit. Green bonds/climate bonds are also issued by governments.

Green bonds (OECD 2015) and climate bonds (Climate Bonds Initiative n.d.) markets are growing rapidly. In 2019, there were USD 694 billion of climate-aligned bonds (Mello 2019). This total was made up of approximately 3590 bonds from 780 issuers

---

[7] The Climate Bonds initiative has created best-practice guides, information kits and the Climate Bonds Standard to assist. The Climate Bonds Standard is a screening tool for investors and governments, which allows them to easily prioritise climate and green bonds with confidence that the funds are being used to deliver climate change solutions. It is available at: www.climatebonds.net/standard.

[8] Voluntary best-practice guidelines called the 'Green Bond Principles' were established by a consortium of investment banks: Bank of America Merrill Lynch, Citi, Crédit Agricole Corporate and Investment Bank, JPMorgan Chase, BNP Paribas, Daiwa, Deutsche Bank, Goldman Sachs, HSBC, Mizuho Securities, Morgan Stanley, Rabobank and SEB.

[9] As theme bonds, climate bonds are similar to the railway bonds of the nineteenth century, the war bonds of the early twentieth century or the highway bonds of the 1960s.

across low-carbon investments in transport, energy, buildings and industry, water, waste and pollution and agriculture and forestry.

The emergence of green and climate bonds is significant and has potential to be transformative. For instance, in 2013, the global bond market was estimated to be USD 78 trillion (Kidney 2013) and is a core part of global financing. The emergence of the 'green' and climate bond markets is therefore important in helping to allocate capital towards activities that support a transition to a low-carbon, climate-resilient world. Governments can issue and incentivise banks to issue green bonds/climate-aligned bonds.

Green bonds can be used by public and private institutions, with many examples, including the following.

- **State governments, cities and institutions**: Massachusetts has issued over USD 3 billion in green bonds to fund land rehabilitation, green space land acquisition, flood prevention, energy efficiency and water efficiency projects (Massachusetts State Treasurer's Office 2015).
- **Cities**: Wuhan Metro has issued two green city bonds to finance seven metro lines (Climate Bonds 2017).
- **Universities**: Monash University, Macquarie University and Australian Catholic University in Australia raised over AUD 200 million through certified green bonds in 2016, 2017 and 2018, respectively.
- **Hospitals**: Tandem Health Partners issued a 32-year green bond to raise CAD 231 million to deliver the public North Island Hospitals Project, in British Columbia (British Columbia Government 2014).
- **Companies**: As part of its commitment to sustainability and to meet its environmental objectives, Vodafone issued a EUR 750 million green bond in May 2019; the company intends to allocate proceeds to a portfolio of eligible green projects including energy efficiency, on-site renewable energy and green buildings (Vodaphone 2019).

Green and sustainability linked loans are also growing rapidly, and offer additional flexibility as they are arranged directly with a bank or financiers rather than having to issue into bond markets. Green loans operate more or less like green bonds with a focus on funding green assets, while sustainability linked loans are used for general corporate purposes but have interest rates which are linked to the sustainability performance of the company.

Australian bank ANZ separates these types of loans into a use of proceeds approach and a sustainability performance linked approach and maps the processes for each (see Figure 23.4).

While smaller than more mature green bonds, both loan types have been growing rapidly, with USD 99 billion issued in 2018. Most of the growth in all sustainability-related debt in 2018 was attributed to sustainability linked loans by Bloomberg (Bloomberg n.d.).

23.3.3.2 Combining Investment Vehicles and Instruments

Cities and regional governments face significant upfront capital barriers to progress large-scale investment in transitioning to low-carbon transport infrastructure, decarbonising the

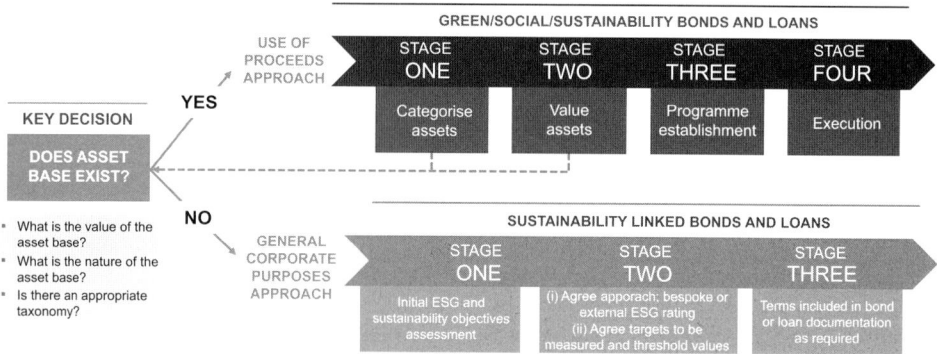

Figure 23.4 Green bonds and sustainability loans.
Source: ANZ (Klyne 2019).

grid, urban/rural water adaptation projects and rural water-efficient infrastructure projects. Historically, such projects were often funded with PPPs to reduce risks and debt to the public sector.

For green PPP projects to be bankable, legal, political, commercial and financial risks need to be appropriately allocated between public and private parties. The International Institute for Sustainable Development's report on the topic, entitled *Green Bonds in Public–Private Partnerships* (Ordonez et al. 2015), provides details on how to do this. The potential to combine, where appropriate, relatively cost-effective financing through green bonds to further leverage private investment and capability through PPPs offers city and regional governments additional options on how to cost-effectively fund the transition to low-carbon, resilient and prosperous economies.

23.3.3.3 Investment Instruments: Revolving Loan Facilities

Financial institutions, corporations, national, state and local governments, universities and community groups can create green revolving loan facilities (RLFs) to support their own internal low-carbon energy efficiency, renewable energy and battery upgrades. Financial institutions and governments can also set up RLFs to support external low-carbon projects in the private sector. Revolving loan facilities start with a fixed pool of internal funds to pay for projects and then some or all the savings are repaid to the RLF. The replenished RLF can then be used to fund additional projects.

Government-sponsored RLFs can use the credit rating and security of the government to typically offer lower interest rates and/or more flexible terms than are available in commercial capital markets. These programmes often focus on financing the upfront cost of:

- energy efficiency upgrades, such as appliances, lighting, insulation, and heating and cooling system upgrades; and
- renewable energy and battery installations.

Depending upon each government's situation and need, RLFs can be capitalised through a variety of sources, including state bond proceeds, treasury investments, ratepayer funds and other special funds. Revolving loan facilities are a widely used proven financing tool to help the transition to low-carbon resilient futures. The Sustainable Endowments Institute, based in the USA, has developed the GRITS software package that enables organisations to effectively manage their rolling loan facilities (Sustainable Endowments Institute n.d.).

23.3.3.4 Investment Institutions

Public or quasi-public purpose-built investment institutions such as green banks are being adopted by governments to complement policies aimed at delivering lower-carbon technologies, enterprises and assets.

The purpose of these investment institutions is to help overcome the 'Valley of Death' phenomenon (Daley et al. 2011) that occurs in the product life cycle. Public grant monies traditionally assist with funding the research, development and even demonstration of many clean energy technologies; however, at pre-commercialisation stage, technologies rely heavily on bulk procurement (often government contracts) and/or high-cost capital to tool up production and bring the product commercially to market. These investment institutions use innovative financing techniques and market development tools in partnership with the private sector to overcome this phenomenon and accelerate deployment of clean energy technologies.

One of the most widely used techniques is investing alongside private counterparts (co-investment), producing multiple benefits to clean energy project proponents, private investors and ultimately the clean technology marketplace, and its consumers. Examples of co-investment include:

- **first loss positions**: government ranks behind other investors upon insolvency or winding up;
- **non-carried interest**: the government's interest is not entitled to profit over and above return of capital or a pre-agreed rate of return;
- **mezzanine or subordinated debt**: government's position on funds lent is subordinate to senior lenders; and
- **capped return**: return on government's capital is capped, allowing other co-investors access to higher up-side on their investment (O'Connor and Chenoweth 2010).

Some investment institutions may also take equity positions in companies, through which they seek to nurture the company's management and share the risks and rewards from deploying its products and services into the marketplace.

In addition to making available *financial* capital – by way of debt or equity – these investment institutions, employing and engaging technology experts, make available their *knowledge and expertise* 'capital' in deals with private-sector lenders and investors. They can bring to the table relevant information about how that technology is applied in a new context or within a new business model, interrogate and challenge the underlying technical and business model assumptions and ultimately improve the risk profile of deals and reduce the cost of capital to the project overall.

Other advantages of co-investment include the ability for governments to recoup and recycle return on funds as well as lowering investment risk. By only co-investing in projects, enterprises or funds that have already attracted private capital, these publicly funded investment institutions can partly mitigate due diligence risks and transaction costs that might be higher where grant funding or sole government investment is the primary source of capital (O'Connor and Chenoweth 2010). Co-investors may include other governments, private banks and institutional investors (sovereign wealth funds, pension funds, insurance companies and endowments).

Investment institutions are not nearly as widespread as they could be but have taken root at a state and country level in the USA, with jurisdictions such as Connecticut, New York, California, Hawaii, Rhode Island and Montgomery County all having created and installed their own investment institutions focused on low-carbon buildings, infrastructure, natural resource management, energy and government utilities.

One of the first investment institutions of its type – founded in Germany in 1948 – was what is now known as KfW. With its origins in a regional municipality in Germany, the bank covers over 90% of its borrowing needs in the capital markets, mainly through bonds that are guaranteed by the federal government. This enables it to provide loans at rates lower than commercial banks in its focus areas of housing and environment, small and medium enterprises, development aid and export and project finance (KfW n.d.).

The UK (Green Investment Bank), Australia (Clean Energy Finance Corporation), Malaysia (Green Technology Financing Corporation) and Japan (Green Finance Corporation) all have investment institutions aimed at stimulating private-sector investment and economic activity, reducing energy costs for ratepayers and expediting the transition to a low-carbon economy.

23.3.4 Step 4: Institutional Investors – Improving Knowledge and Skills

A key finding of the UNEP PRI and UNEP Finance Initiative *Sustainable Finance Fiduciary Duty in the 21st Century* study was that 'Despite significant progress, many investors have yet to fully integrate environmental, social and governance issues into their investment decision-making processes' (PRI 2015: 9). As this book has shown, there are a variety of investment, regulatory and reputational drivers for asset owners to direct their asset consultants and asset managers to perform ESG analysis and design a more climate-balanced portfolio that reduces exposure to high climate risk assets and increases their exposure to low-carbon assets and those well positioned for transition to a low-carbon economy.

A major barrier to improved practices remains the knowledge and competence of finance industry professionals. While this issue is starting to be addressed by universities through integration with existing courses and new dedicated courses, more needs to be done. More concerning, however, is the millions of finance professionals currently working in the industry who were educated at a time when these issues were not part of the curriculum at all. A recent CFA (Chartered Financial Analyst) Institute report found that of over one million LinkedIn profiles assessed, only 1% claimed any sustainability or ESG skills. When surveying CFA Institute charter holders, who carry one of the most prestigious qualifications for investment management professionals, only 11% considered themselves proficient in ESG, with more the 70% wanting training (CFA 2021).

While this is concerning, there are many resources to assist asset owners, asset consultants and asset managers in this regard, as shown in Box 23.2, but, as they are relatively new approaches, many investors and financiers either do not know about them or are not using them. Hence there is a role for government to:

(i) change education standards for investors and companies (including boards); and
(ii) publicly fund research, tool development and the development of effective education and training resources, courses and programmes.

As Box 23.2 shows, there is a wealth of existing resources upon which such courses, tools and training programmes can be based.

23.4 Conclusion

This chapter has reviewed the issues around financing projects and industries that offer environmental, social and governance (ESG) benefits, in particular those that move the world towards a low-carbon sustainable economy and society. There are many encouraging developments in the evolution and development of financial instruments and facilities, in redefining fiduciary duties to more explicitly include ESG principles and in redirecting funds towards building sustainable low-carbon economies. At the same time, the reader's attention is drawn to the slow progress of change in the allocation of financial resources, as illustrated by continuing high levels of investment in the maintenance and expansion of greenhouse-gas-intensive fossil fuel industries, and the accompanying increases in emissions.

There is general agreement that the present (2021) pace of change is not sufficient to enable the achievement of the Paris Agreement goals of keeping temperature rises well below 2 °C compared to pre-industrial levels. Commitments to work to the target of net zero emissions by 2050 have proliferated across governments and many private-sector entities including banks, asset managers and some significant fossil fuel enterprises. In the case of the latter, assertions of good intent to reduce emissions ultimately by transitioning out of high-emitting activities to cleaner ways of delivering the same services have not been matched by the rate of change required to meet the Paris Agreement targets. Rather, some companies and their investors continue to embark on multi-billion-dollar projects in fossil fuel (and related energy) production, especially oil and gas, that would appear to be profitable only if they operate over decades, thus maintaining or increasing greenhouse gas emissions in the longer term.

What is clear is that governments will need to step up to a much stronger leadership role in the transition to a low-carbon economy by setting stronger interim mitigation targets to achieve purposeful emission reductions by 2030 to complement net zero 2050 targets. Such a strengthening might well begin with the phasing out of fossil fuel subsidies and the adoption of policies that provide greater investor certainty for investing in the transition to a zero carbon, resilient and prosperous economy.

Box 23.2 **Sample of Existing Tools/Educational Resources That Guide Governments' Efforts to Improve Education and Training Standards for Investors, Asset Managers, CEOs and Their Boards**

The Global Investor Coalition on Climate Change (GICCC) and UNEP Finance Initiative have published many comprehensive free online 'how to' manuals, explaining step by step how to create more resilient climate-balanced and sustainable investment portfolios, specifically:

- UNEP's Financial Initiative provides guides for investors on how to factor in environmental social and governance issues (UNEP FI and WBCSD 2010).
- The PRI also provides an online academy (PRI n.d.) that anyone can subscribe to and learn how to wisely incorporate ESG considerations into investment decision-making.

There are also specific tools on how to develop a carbon-balanced portfolio.

- The PRI is developing an asset owner climate change strategy guide covering measurement, engagement, investment and avoidance strategies.
- The Asset Owners Disclosure Project (AODP) has published a detailed manual on this subject entitled *Climate Risk Management Best Practice Methodology* (AODP 2013).
- The GICCC has also developed a comprehensive manual (GICCC 2015), providing step-by-step 'how to' guidance for asset owners and asset managers to create a climate-balanced portfolio. This manual recommends the following steps be undertaken:
  - ○ a strategic review to assess portfolios to reduce risks of being caught holding stranded assets, by reducing exposure to (i) fossil fuel and higher carbon-intensive industries and (ii) high-risk assets and investments in property and infrastructure that are in high-risk sea-level rise, flooding or bushfire zones;
  - ○ strategic asset allocation: reviewing and increasing exposure to profitable low-carbon sectors and high-performing 'ESG best of sector' businesses, and conversely reducing exposure to companies that are climate change and sustainable development laggards;
  - ○ reviewing and increasing the percentage of the portfolio involved in projects that achieve climate change mitigation and adaptation;
  - ○ utilising traditional financing models such as green bonds and climate bonds to reduce risk and increase the reward to risk ratio when investing in low-carbon asset classes; and
  - ○ reducing climate vulnerability of existing assets through working with companies to encourage them to make smart investments in climate change adaptation. Most institutional investors invest for the long term in 'blue chip' corporations. Hence it is in their interests to work together to reduce risks of cash flow shocks from asset damage and operational losses due to more intense extreme weather events (GICCC 2015).
- Altiorem is a not-for-profit library and resource centre 'that helps people understand the role finance must play in addressing sustainability challenges and building a better future for all' (Altiorem n.d.). It helps advocates for sustainability make stronger arguments, and 'a new generation of leaders gain a solid understanding of critical sustainability issues and their relevance to business and finance' (Altiorem n.d.).

A consortium of peak institutional investment bodies in partnership with the UNEP Financial Initiative has also published an extensive report in 2014, *Financial Institutions Taking Action on Climate Change* (AIGCC et al. 2014), on examples of climate change leadership from institutional investors, to help institutional investors learn from each other's best practice. In addition, the climate bonds (Climate Bonds Initiative n.d., 2015) and green bonds (OECD 2015; Ordonez et al. 2015; Sustainability Institute and Climate Solutions 2016) initiatives have significant resources online to assist. Finally, as reported above, the 2013 World Economic Forum report, *The Green Investment Report: The Ways and Means to Unlock Private Finance for Green Growth*, lists hundreds of relevant reports and guides to undertaking and implementing private/public financing opportunities (WEF 2013). These tools and resources are critical in assisting the private and public sector to identify new ways to invest in a low-carbon resilient future in ways that are prudent and successful, to build momentum for more such investment in years to come. The UNEP's 2019 *Sustainable Finance Progress Report* provides information on more recent developments (UNEP 2019).

While the interventions discussed in this chapter will aid in scaling up and accelerating the reallocation of capital, a whole-of-government integrated approach to climate change mitigation and adaptation policy is the ultimate way in which governments can send consistent signals to financial markets.

References

AIGCC (Asia Investor Group on Climate Change), IGCC (Investor Group on Climate Change), IIGCC (Institutional Investors Group on Climate Change), PRI (Principles for Responsible Investment) and UNEP FI (UN Environment Programme Finance Initiative (2014). *Financial Institutions Taking Action on Climate Change*. UN Environment Programme Finance Initiative (UNEP FI). Available at: www.unepfi .org/publications/climate-change-publications/technical-advice-for-policy-makers-publications/financial-institutions-taking-action-on-climate-change/.

Altiorem (n.d.). About. *Altiorem.org.* Available at: https://altiorem.org/about/.

AODP (Asset Owners Disclosure Project) (2013). *Climate Risk Management Best Practice Methodology*. London: Asset Owners Disclosure Project. Available at: http:// aodproject.net/wp-content/uploads/2012/03/Asset-Owners-Climate-Change-Best-Practice-Framework-v27.pdf.

BankTrack (n.d.). *BankTrack*. Available at: www.banktrack.org.

BankTrack, RAN (Rainforest Action Network), Indigenous Environment Network, Oil Change International, Sierra Club and Honour the Earth (2019). *Banking on Climate Change: Fossil Fuel Finance Report Card 2019*. Available at: www.banktrack.org/ article/banking_on_climate_change_fossil_fuel_finance_report_card_2019.

BlackRock (n.d.). Sustainability as BlackRock's new standard for investing. *BlackRock.* Available at: www.blackrock.com/au/individual/blackrock-client-letter.

Bloomberg (n.d.). Sustainable debt. *Bloomberg.* Available at: www.bloomberg.com/ impact/products/sustainable-debt/.

BloombergNEF (New Energy Finance) (2019). *Global Trends in Renewable Energy Investment 2019*. Frankfurt School of Finance and Management and Bloomberg New Energy Finance. Available at: https://wedocs.unep.org/bitstream/handle/20 .500.11822/29752/GTR2019.pdf?sequence=1&isAllowed=y.

BloombergNEF (2020a). Solar and wind reach 67% of new power capacity added globally in 2019, while fossil fuels slide to 25%. *BloombergNEF*. Available at: https://about .bnef.com/blog/solar-and-wind-reach-67-of-new-power-capacity-added-globally-in-2019-while-fossil-fuels-slide-to-25/.

BloombergNEF (2020b). Corporate clean energy buying leapt 44% in 2019, sets new record. *BloombergNEF*. 28 January. Available at: https://about.bnef.com/blog/corpor ate-clean-energy-buying-leapt-44-in-2019-sets-new-record/.

British Columbia Government (2014). North Island Hospitals Project green bond issue a first [media release]. Available at https://news.gov.bc.ca/releases/2014FIN0023-000901.

Carney, M. (2015). Breaking the tragedy of the horizon: Climate change and financial stability. Speech given at Lloyd's of London. Available at: www.bankofengland.co .uk/speech/2015/breaking-the-tragedy-of-the-horizon-climate-change-and-financial-stability.

CDP (Carbon Disclosure Project) (2019). World's top green businesses revealed in the CDP A List. *CDP.net*. 22 January. Available at: www.cdp.net/en/articles/companies/ worlds-top-green-businesses-revealed-in-the-cdp-a-list.

CFA (Chartered Financial Analyst) Institute (2021). Future of sustainability in investment management: From ideas to reality. Available at: www.cfainstitute.org/-/media/docu ments/survey/future-of-sustainability.ashx.

Climate Bonds Initiative (n.d.). *Climate Bonds Initiative*. Available at: www.climatebonds .net/.

Climate Bonds Initiative (2015). *Bonds and Climate Change: The State of the Market in 2015*. Climate Bonds Initiative. Available at: www.climatebonds.net/files/files/CBI-HSBC%20report%207July%20JG01.pdf.

Climate Bonds Initiative (2017). *Green Bonds: Opportunities for Cities to Fund Climate Resilient Infrastructure*. Climate Bonds Initiative. Available at: www.climatebonds .net/2017/10/green-city-bonds-opportunities-cities-fund-climate-resilient-infrastruc ture-our-state-market.

Council on Foreign Relations (2021). China's Belt and Road: Implications for the United States. Council on Foreign Relations. Available at: www.cfr.org/report/chinas-belt-and-road-implications-for-the-united-states/.

Daley, J., Edis, T. and Reichl, J. (2011). *Learning the Hard Way: Australian Policies to Reduce Carbon Emissions*. Melbourne: Grattan Institute. Available at: https://grattan .edu.au/report/learning-the-hard-way-australias-policies-to-reduce-emissions/.

European Commission (n.d.). Sustainable finance. *European Commission*. Available at: https://ec.europa.eu/info/business-economy-euro/banking-and-finance/sustainable-finance_en#overview.

Fossil Fuel Finance Report (2021). Banking on Climate Chaos. Available at: https:// reclaimfinance.org/site/wp-content/uploads/2021/03/BOCC__2021_vF.pdf.

Fried, G., Busch, T. and Bassen, A. (2015). ESG and financial performance: Aggregated evidence from more than 2000 empirical studies. *Journal of Sustainable Finance & Investment*, 5, 210–233. Available at: www.tandfonline.com/doi/abs/10.1080/ 20430795.2015.1118917.

Generation Foundation, UNEP FI (Finance Initiative) and PRI (Principles for Responsible Investment) (2016). *Fiduciary Duty in the 21st Century: Statement*. Available at: www.unepfi.org/fileadmin/documents/FiduciaryDutyStatement.pdf.

GICCC (Global Investor Coalition on Climate Change) (2015). *Climate Change Investment Solutions: A Guide for Asset Owners*. Paris: Global Investor Coalition on Climate Change (GICCC). Available at: http://globalinvestorcoalition.org/wp-content/uploads/2015/04/Climate-Change-Investment-Solutions-GuideFINAL.pdf.

Gittelsohn, J. (2019). End of era: Passive equity funds surpass active in epic shift. *Bloomberg*. 12 September. Available at: www.bloomberg.com/news/articles/2019-09-11/passive-u-s-equity-funds-eclipse-active-in-epic-industry-shift.

Hargroves, K. and Smith, M. (2005). *The Natural Advantage of Nations: Business Opportunities, Innovation and Governance in the 21st Century*. London: The Natural Edge Project, Earthscan.

Hutley N. and Hartford-Davis, S. (2019). *Climate Change and Directors' Duties*. Supplementary memorandum of opinion. The Centre for Policy Development. Available at: https://cpd.org.au/wp-content/uploads/2019/03/Noel-Hutley-SC-and-Sebastian-Hartford-Davis-Opinion-2019-and-2016_pdf.pdf.

IEA (International Energy Agency) (2019a). *World Energy Investment 2019*. Paris: International Energy Agency. Available at: www.iea.org/reports/world-energy-invest ment-2019/power-sector#abstract.

IEA (2019b). *World Energy Outlook 2019*. Paris: International Energy Agency.

IEEFA (2021). *Institute for Energy Economics and Financial Analysis*. Available at: www .IEEFA.org.

IIGCC (Institutional Investors Group on Climate Change) (2009). *2009 Investor Statement on the Urgent Need for a Global Agreement on Climate Change*. Institutional Investors Group on Climate Change (IIGCC), Investor Network on Climate Risk (INCR), Investor Group on Climate Change (IGCC) and UNEP FI (UN Environmental Programme Finance Initiative). Available at: www.nbim.no/contentas sets/6e5c230a453d439394c0692b08e2aed4/2009_investor_statement_on_a_global_ agreement.pdf.

IIGCC (2019). *Global Investor Statement to Governments on Climate Change*. Asia Investor Group on Climate Change (AIGCC), The Carbon Disclosure Project (CDP), Institutional Investors Group on Climate Change (IIGCC), Principles for Responsible Investment (PRI) and the UN Environment Finance Initiative (UNEP FI). Available at: https://igcc.org.au/wp-content/uploads/2019/ 06/GLOBAL-INVESTOR-STATEMENT-TO-GOVERNMENTS-ON-CLIMATE-CHANGE.pdf.

Kalderimis, D. and Swan, N. (2019). *Climate Change Risk: Implications for New Zealand Company Directors and Managed Investment Scheme Providers*. Legal opinion. Chapmann Tripp and The Aotearoa Circle. Available at: www.chapmantripp.com/ Publication%20PDFs/Chapman%20Tripp%20Aotearoa%20Circle%20Climate% 20Change%20Risk%20Legal%20Opinion.pdf.

KfW (n.d.). About KfW. *KFW.de*. Available at: www.kfw.de/KfW-Group/About-KfW.

Kidney, S. (2013). 9 Useful facts about the global bond market. *Climate Bonds Initiative*. 27 February. Available at: www.climatebonds.net/2014/05/9-useful-facts-about-global-bond-market.

Klyne, S. (2019). The rise (and rise) of green loans in Australia. *ANZ*. February. Available at: https://institutional.anz.com/insight-and-research/the-rise-and-rise-of-green-loans-in-australia.

Massachusetts State Treasurer's Office (2015). *MassGreenBonds: Investing in a Greener, Greater Commonwealth*. Final Investor Impact Report 2013 Series D. Boston, MA: Massachusetts State Treasurer's Office.

Mello, L. (2019). Shifting pathways: Brazil to review $54 billion infrastructure portfolio in line with international green standards. *Climate Bonds Initiative*. 17 September. Available at: www.climatebonds.net/2019/09/shifting-pathways-brazil-review-54-bil lion-infrastructure-portfolio-line-international-green.

Mercer (2019). *Investing in a Time of Climate Change: The Sequel*. Mercer. Available at: www.mercer.com/our-thinking/wealth/climate-change-the-sequel.html.

O'Callaghan, B. and Murdock, M. (2021). Are we building back better? Available at: https://wedocs.unep.org/bitstream/handle/20.500.11822/35281/AWBBB.pdf.

O'Connor, S. and Chenoweth, J. (2010). *Funding the Transition to Clean Energy Economy*. Melbourne: Australian Conservation Foundation.

OECD (2015). *Green Bonds: Mobilising the Debt Capital Markets for a Low-Carbon Transition*. Paris: OECD. Available at: www.oecd.org/environment/cc/Green%20bonds%20PP%20[f3]%20[lr].pdf.

Ordonez, C. D., Uzsoki, D. and Dorji, S. T. (2015). *Green Bonds in Public–Private Partnerships*. Winnipeg: International Institute for Sustainable Development (IISD). Available at: www.iisd.org/sites/default/files/publications/green-bonds-public-private-partnerships.pdf.

Orlitzky, M., Schmidt, F. L. and Rynes, S. L. (2003). Corporate social and financial performance: A meta analysis. *Organization Studies*, 24, 403–441.

Pinchot, A. and Christianson, G. (2019). How are banks doing on sustainable finance commitments? Not good enough. *World Resources Institute*. 3 October. Available at: www.wri.org/blog/2019/10/how-are-banks-doing-sustainable-finance-commitments-not-good-enough.

PRI (Principles of Responsible Investment) (n.d.). *PRI Academy*. Available at: https://priacademy.org.

PRI (2015). *Fiduciary Duty in the 21st Century*. United Nations Global Compact (UNGC), United Nations Environment Programme Finance Initiative (UNEP FI), Principles for Responsible Investment (PRI) and Inquiry Design of a Sustainable Financial System. Available at: www.unepfi.org/fileadmin/documents/fiduciary_duty_21st_century.pdf.

PRI (2019). *Fiduciary Duty in the 21st Century Final Report*. UN Environmental Programme Finance Initiative (UNEP FI) and Principles for Responsible Investment (PRI). Available at: www.fiduciaryduty21.org/publications.html.

RE100 (2020). Growing renewable power: Companies seizing leadership opportunities. RE100 annual progress and insights report 2020. Available at: www.there100.org/growing-renewable-power-companies-seizing-leadership-opportunities.

Robins., N, HSBC and UNEP FI (UN Environment Programme Finance Initiative) (n.d.). High-level ministerial dialogue on climate finance: Private finance [transcript]. Available at: www.unepfi.org/fileadmin/communications/High-level_Ministerial_Statement.pdf.

Sustainability Institute and Climate Solutions (2016). *Integrated Climate Solutions: Green Bonds*. Sustainability briefing. Durham, NH: Sustainability Institute. Available at: https://scholars.unh.edu/cgi/viewcontent.cgi?article=1015&context=sustainability.

Sustainable Endowments Institute (n.d.). GRITS: The Green Revolving Investment Tracking System. *Sustainable Endowments Institute*. Available at: www.endowmentinstitute.org/our-initiatives/grits/.

SYSTEMIQ (2020). The Paris Effect: How the climate agreement is reshaping the global economy. *SYSTEMIQ*. Available at: www.systemiq.earth/wp-content/uploads/2020/12/The-Paris-Effect_SYSTEMIQ_Full-Report_December-2020.pdf.

TCFD (Task Force on Climate-related Financial Disclosures) (2017). *Recommendations of the Task Force on Climate-related Financial Disclosures: Final Report*. Task Force on Climate-related Financial Disclosures. Available at: www.fsb-tcfd.org/wp-content/uploads/2017/06/FINAL-2017-TCFD-Report-11052018.pdf.

UK Government, Department for Business Innovation and Skills (2013). *Low Carbon Environmental Goods and Services (LCEGS)*. Report for 2009/10. London: UK

Government. Available at: www.gov.uk/government/uploads/system/uploads/attach ment_data/file/224068/bis-13-p143-low-carbon-and-environmental-goods-and-ser vices-report-2011-12.pdf.

UNEP (UN Environmental Programme) (2019). *Sustainable Finance Progress Report*. UN Environment Programme (UNEP). Available at: https://unepinquiry.org/publication/ sustainable-finance-progress-report/.

UNEP FI (UNEP Finance Initiative) and WBCSD (World Business Council for Sustainable Development) (2010). *Translating ESG into Sustainable Business Value: Key Insights for Companies and Investors*. UN Environment Programme Finance Initiative (UNEP FI) and World Business Council for Sustainable Development (WBCSD). Available at: www.unepfi.org/fileadmin/documents/translatingESG.pdf.

Vodaphone (2019). *Green Bond Framework Report*. Available at: www.vodafone.com/ content/dam/vodcom/sustainability/pdfs/green-bond-framework.pdf.

WEF (World Economic Forum) (2013). *The Green Investment Report: The Ways and Means to Unlock Private Finance for Green Growth*. Geneva: World Economic Forum. Available at: http://reports.weforum.org/green-investing-2013.

World Bank Group (2019). *State and Trends of Carbon Pricing 2019*. Washington, DC: World Bank. Available at: https://documents.worldbank.org/en/publication/docu ments-reports/documentdetail/191801559846379845/state-and-trends-of-carbon-pricing-2019.

WRI (World Research Institute) (2014). *Better Growth, Better Climate. The New Climate Economy Report*. The Global Commission on the Economy and Climate. Washington, DC: The Global Commission on the Economy and Climate, World Resources Institute. Available at: www.newclimateeconomy.report/2014.

WRI (2019). Green Targets tool [data resource]. Available at: www.wri.org/finance/banks-sustainable-finance-commitments/.

24

Social Movements for Change

MICHAEL H. SMITH

Executive Summary

The latest climate science suggests that to avoid risks of global warming greater than 1.5–2 °C and irreversible tipping points, rapid greenhouse gas emissions reductions of 5–10% per annum are needed over the next 12 years and beyond. This book has outlined technical, policy/institutional, economic and social solutions to simultaneously mitigate and adapt to climate change to achieve this speed and scale of change.

The technological and policy solutions to achieve this are readily available, and yet nations are not yet committing to targets and policies that will achieve a safe climate. Consequently, a dynamic multi-stakeholder climate change social movement for climate action has formed.

The emerging twenty-first-century climate change social movement is multifaceted and broad-based, led by activists and climate-change-action-oriented non-governmental organisations (NGOs) working in partnership with progressive investor and business groups, coalitions of economists (e.g. the Global Commission on the Economy and Climate), academies of science and the Intergovernmental Panel on Climate Change (IPCC), as well as with educators and existing movements, such as the women's, trade union, social justice, Indigenous and consumer-protection movements.

The broad-based nature of the climate change movement was first demonstrated with the 400 000-strong New York climate change march in September 2014, which brought together UN leaders, unions and social justice and ethnic movements, as well as the environmental movement in New York, along with millions of others marching worldwide.

This bottom-up climate change social movement and its members played a critical role in building new partnerships and coalitions to create the political will for the Paris Climate Change Agreement and UN's Sustainable Development Goals Agreement in 2015. This movement and its key actors are playing a critical role to ensure the Paris Agreement and UN's Sustainable Development Goals are prioritised by decision makers and implemented around the world, as outlined in this chapter.

By providing an overview of the novel climate-change-related NGO partnerships with institutional investors, the insurance sector, business and existing movements that are delivering real change on the ground, the chapter seeks to provide an overview of how you and your business, organisation, government, school or community group can join existing networks to share knowledge and collectively achieve the scale and speed of greenhouse gas emissions reductions required.

24.1 Introduction

This book has outlined technical, policy, institutional, economic and social solutions to mitigate and build resilience to climate change and reduce poverty. The latest climate science suggests that to avoid risks of global warming greater than 1.5–2 °C and irreversible tipping points, rapid greenhouse gas emissions reductions of 5–10% per annum are needed over the next 12 years and beyond. With the technological and policy solutions to achieve this being readily available, and yet nations still not committing to targets and policies that will achieve a safe climate, a multifaceted, dynamic, multi-stakeholder climate change social movement for climate action has formed. Our societies need both individuals who understand and are empowered to embrace changes to their day-to-day lives, and, most importantly, leaders in positions of power to commit to the scale and speed of greenhouse gas reductions required – such as transitioning to 100% renewable energy, regulating fossil fuels and decarbonising the industry, transport and land-use sectors within the next two decades.

As the long-term future well-being of human societies and the ecosystems upon which they depend are at risk from climate change, economy-wide technological and cultural change to a decarbonised future is now required. As the book has shown, to understand how societies can achieve rapid decarbonisation we must acknowledge and better understand how to overcome the barriers to change and to counter the vested interests acting against such change.

One of the barriers to nations adopting appropriate greenhouse gas reduction targets and policy reforms in line with the scale and speed of the problem has been campaigns and lobbying by fossil fuel interests. Earlier chapters have highlighted that rapid decarbonisation poses short-term financial risks to fossil fuel companies whose share market value is significantly based (30–50%) on the asset value of remaining fossil fuel reserves, the more so if carbon capture and storage does not become commercially competitive (Smith 2013c). Several fossil fuel and heavy industry companies have formed 'blocking coalitions' over the past 30 years, which are still effectively preventing efforts to reduce emissions in several major emitting nations (Oreskes and Conway 2010; Maddow 2019). Indeed, strong links between some modern states and fossil fuel extraction, energy and other fossil-fuel-intensive industries have been demonstrated by Newell and Paterson (1998) and Maddow (2019). These companies are very experienced in influencing governments through political party donations, experienced professional lobbyists and media/communication campaigns. These vested interests have, in the past, run media campaigns to delay action on climate change by both (a) disputing the climate science (Oreskes and Conway 2010; Dunlap and McCright 2015) and (b) arguing that if strong action is taken it will reduce business competitiveness, cause job losses and harm economic growth (Smith 2009; Smith et al. 2010). Moreover, they have promoted the mistaken assumption that economic growth is dependent upon growth in the use of fossil fuels. These assumptions, challenged by this book, have driven close links between fossil fuel interests and some modern states (Newell and Paterson 1998: 692–693).

Historically, the funding by some in the fossil fuel industry of think tanks and scientists to deny climate change science (Oreskes and Conway 2010) and promote the idea that fossil fuels are essential to ending poverty and maintaining economic growth reflects the Upton Sinclair adage, 'It is difficult to get a man to understand something, when his salary depends on his not understanding it' (Sinclair 1994). The illegalisation of slavery in Britain and the USA faced strong resistance from a wealthy and powerful industry based on slave labour. Efforts to win women and racial minorities the vote also endured strong resistance, as did efforts to phase out the insecticide DDT (dichlorodiphenyltrichloroethane) and ozone-depleting chemicals, as well as to reduce acid rain in the 1960s, 1970s and 1980s (Smith 2009). In each case, social movements, whether the anti-slavery, suffragette, civil rights or, in more recent times, environmental movement, played a critical role in challenging vested interests, shifting public attitudes and galvanising the moral will of society to achieve change. Such social movements are widely acknowledged as having played critical roles in these advances in human society. This chapter shows that the climate change social movement and its key civil society actors have played critical roles in laying the groundwork to help enable the UN Framework Convention on Climate Change (UNFCCC) Paris Agreement[1] to be achieved and will play a key role in ensuring that its targets are implemented.

24.2 Reframing Macro- and Microeconomic Narratives

Those who do not want action on climate change have played on the fears of workers and business leaders by arguing that action on climate change would harm economic growth and business competitiveness, and lead to large job losses (Smith 2009; Smith et al. 2010). This, plus the lobbying of certain business interests opposed to strong action on climate change, has resulted in many political leaders failing to take on ambitious greenhouse gas reduction targets for fear they would harm economic growth and cause reduced jobs, below business as usual (BAU) (Smith 2009; Smith et al. 2010). These fears are not founded in the latest climate change macro- and microeconomic literature.

24.2.1 Reframing the Macroeconomic Narrative

Since 1997 (Repetto and Austin 1997), a range of studies (Hargroves and Smith 2005; Smith et al. 2010; Smith 2015) have shown that smart action on climate change could boost economic growth by at least USD 25 trillion above BAU by 2030 (Smith 2015). The new non-governmental organisation (NGO) the Global Commission on the Economy and Climate was formed in 2012 to investigate whether it was possible for smart action on

---

[1] *Paris Agreement Under the United Nations Framework Convention on Climate Change*, opened for signature 16 February 2016. Available at: https://unfccc.int/process-and-meetings/the-paris-agreement/the-paris-agreement.

climate change to boost economic growth. Their resulting report in 2014, entitled *Better Growth, Better Climate* (WRI 2014), showed that cutting emissions could generate better growth, with lower air pollution, more liveable and economically efficient cities, more sustainable use of land and greater energy security. In 2018, the Global Commission on the Economy and Climate published *Unlocking the Inclusive Growth Story of the 21st Century: Accelerating Climate Action in Urgent Times*, which confirmed findings (Smith 2015) that strong and smart action on climate can boost the global economy greater than USD 25 trillion above BAU by 2030 (Global Commission on the Economy and Climate 2018). The latest modelling of the OECD also confirms that action on climate change can boost economic and jobs growth above BAU (OECD 2017).

24.2.2 Reframing the Microeconomic Narrative

Just as the evidence outlined above is helping to shift the macroeconomic narrative, the following evidence has helped shift the microeconomic narrative, to show that strong action on climate can help business competitiveness and profitability. First, there is mounting evidence, as shown in the cities and industry chapters (Chapters 11 and 16) and Chapter 23, 'Financing the Transition', that risks of more extreme and intense weather events can be material and affect financial performance for businesses across most sectors (Smith 2013a, 2013b, 2013c, 2014). Second, empirical evidence that companies leading on action on climate change and sustainability, on average, outperform the market, continues to grow. For instance, the 2019 Carbon Disclosure Project's analysis of over 6000 companies concluded that the best-performing companies on action on climate change also outperformed the market (CDP 2019).

These findings are helping to shift the discourse in many countries and, importantly, in many government cabinet, investor and business board meetings, by showing that action on climate is essential to achieving other key priorities for governments and business such as increased productivity growth, innovation and improved risk management. This has created new opportunities for civil society to develop new NGO–investor, NGO–business and NGO–governmental partnerships to help knowledge-share, mainstream stronger commitments and action on climate change, and build new alliances and coalitions to champion purposeful climate change policy reform. A sample of these new partnership models between civil society NGOs and institutional investors, business and other key actors, many of which have emerged over the last decade, are outlined next.

24.3 The Key Role of Civil Society

The climate change social movement is one of the broadest ever created and it is important to provide an overview of some of the contributing NGOs to highlight their breadth and the ways that they complement each other. The following is a sample of the key civil society actors and NGO initiatives which have helped to achieve the UNFCCC Paris Climate Change Agreement, and which will help ensure the intent of the Paris Agreement is implemented.

It is important to recognise that it was climate scientists who first raised the alarm about the risks from ongoing production of more and more greenhouse gas emissions. It is the physical science of climate change and its unmitigated impacts that underpin the need for action. Not surprisingly, those seeking to oppose action on climate change have copied tactics from previous tobacco campaigns, and again 'bought off' some scientists and paid them to argue in the media that there is no consensus on climate change science and therefore no reason to act (Oreskes and Conway 2010). Climate scientists responded to this by forming the Intergovernmental Panel on Climate Change (IPCC) in 1988. The IPCC, formed under the auspices of the World Meteorological Organization (WMO) and the UN Environment Programme (UNEP), produces major reviews on the consensus of the latest developments in the physical science of climate change as well as on how best to effectively adapt to and mitigate climate change. The IPCC is important because, among other things, it provides a process through which to achieve consensus by the clear majority of the world's climate scientists. The work of the IPCC discredits arguments from those who would contend that there is no consensus on climate change science. It also involves representatives from all the governments of the world who can help shape and review drafts. As it is a document of both scientists and governments, governments are far more likely to utilise its findings.

The IPCC assessments continue to affirm that climate change is occurring and, with still higher certainty, that it is human-caused. The IPCC's special report on keeping global warming below 1.5 °C argues that humanity only has 12 years to avoid risks of irreversible climate change and avoid crossing the 1.5 °C guardrail (IPCC 2018). The IPCC called on all governments to cut greenhouse gas emissions by at least 40% by 2020 below 1990 levels and achieve net zero emissions by 2050. The IPCC's work and reports continue to affirm that the costs of inaction on climate change were far greater than the costs of action.

24.3.1 Working with National and Subnational Governments

In the lead up to the Paris Agreement in 2015, and since, a range of NGO initiatives have been important for building political will for nations, regional sub-governments and cities adopting and implementing the Paris Agreement's greenhouse gas reduction targets in line with the latest climate science (Jacobs 2016).

The UN Deep Decarbonisation Pathways Project (Deep Decarbonisation Pathways Project 2015) developed detailed decarbonisation pathways for the 16 nations responsible for over 60% of global emissions, demonstrating how they could best contribute cost-effectively to global efforts to keep warming under 2 °C. The Under 2 Degrees Coalition – which includes over 200 subnational governments that are formally committed to keeping warming under 2 °C – comprises subnational governments that represent over 1.3 billion people and nearly 40% of the global economy (Under 2 Degrees Coalition n.d.). The Climate and Clean Air Coalition helps over 120 nations, cities and companies to focus on and share best-practice knowledge on how to reduce emissions from non-CO_2 (carbon dioxide) greenhouse gases (Climate and Clean Air Coalition n.d.). The Powering Past Coal

Alliance is a group of over 95 countries, cities, regions and corporations formally committed to phasing out of traditional or unabated coal power (Powering Past Coal Alliance n.d.).

Many of the countries most vulnerable to climate change do not have large economies and are therefore excluded from the G5, G7, G20 and other international forums to make their case for stronger and more urgent action on climate change. To address this problem, a new NGO was created, the Climate Vulnerable Forum, in the lead-up to the Paris Agreement. This forum is made up of 43 countries and was formed to help these nations be heard alongside the traditional groupings of small island states, least-developed countries and African countries.

These nations, progressive institutional investors, businesses and their NGO partners played a critical role in the resulting Paris Climate Change Agreement (Jacobs 2016), which formally adopted the target of keeping global warming closer to 1.5 °C than 2 °C, as well as the long-term goal of net zero emissions.

24.3.2 Working with Institutional Investors

Institutional investors, such as those managing workers' retirement funds (known as superannuation funds in some countries), manage investments over a long time frame. Over the last three decades, superannuation funds have gone from bit players to managing significant amounts of capital in many major economies, capable of wielding significant power (Hargroves and Smith 2005; Smith and Hargroves 2006). Moreover, some fund managers have increasingly adopted an ethical lens to their investment decision-making and, as evidence has mounted that climate-conscious companies were outperforming the market, large investment funds are increasingly divesting from fossil fuel investments, as shown in the previous chapter, and large investment funds and superfunds have been increasingly calling on CEOs and boards to better assess their risks from and act on climate change. This is illustrated in the formation of the International Investors Group on Climate Change (IIGCC), which represents over 230 institutional investors, with investments of over USD 40 trillion in the global economy, and which is calling on both CEOs and nations to take strong action on climate change.

In addition, as discussed in Chapter 23, 450 institutional investors have formed the Climate Action 100+ Initiative (Climate Action 100+ Initiative n.d.) to pressure the 160 largest carbon-emitting companies to properly assess and report on how they are addressing climate risk and developing plans for a carbon-constrained future. This has contributed to some of the largest mining companies agreeing to cap coal production and work to align their activities with the Paris Agreement targets. Institutional investors worth over USD 70 trillion, for instance, are also members of the NGO the Carbon Disclosure Project (CDP n.d.), which asks large businesses and cities every year to answer a series of questions to disclose their exposure to climate change risks and what they are doing to act on climate change. Further evidence that institutional investors are joining forces to collaborate to demand action on climate is shown by the formation in 2019 of the Coalition for Climate Resilient Investment, which includes institutional investors responsible for over USD 5 trillion in investment (Coalition for Climate Resilient Investment 2019).

24.3.3 Working with Progressive Business Actors

Historically, businesses, including fossil fuel and other emissions-intensive industries, have been identified as key climate solutions blockers. But with the growing consensus on climate change science, and the economics of climate change showing that action could increase prosperity, combined with the shift in views by institutional investors, business CEOs have demonstrated an increasing consciousness and support of climate change action (WEF 2018). While this may be a response to movements in civil society, it is also evident that businesses are seeing threats to their bottom lines from, for example, extreme weather events. They also see opportunities to reduce operational costs through energy efficiency and appreciate the benefits of relatively cheap renewable energy. Most businesses can also see opportunities to position themselves for fast-growing markets in energy-efficient low-carbon products and services against the 'old economy', promoting their interests in a business alternative (Newell and Paterson 1998: 694).

Since 2000, leading global corporations have begun to argue in public that strong climate policy is in the interests of business. To tackle the power of traditional blocking coalitions, the emerging climate change social movement and its allied NGOs are encouraging and building new coalitions among those businesses arguing for strong climate policy, most of whom have much to gain from a transition to a low-carbon future. These coalitions can lobby governments to adopt stronger climate change policies. One of the most prominent examples of this is the NGO We Mean Business. We Mean Business was launched in New York ahead of the September 2014 Paris UN summit:

We Mean Business is a coalition of organizations working with thousands of the world's most influential businesses and investors. These businesses recognise that the transition to a low carbon economy is the only way to secure sustainable economic growth and prosperity for all. To accelerate this transition, we have formed a common platform to amplify the business voice, catalyze bold climate action by all, and promote smart policy frameworks. (*We Mean Business n.d.*)

Now listing over 1000 major global corporations as partners with their ambitious climate change commitments (including over 230 major corporations committing to become 100% renewable), they are helping build momentum for stronger action on climate change globally.

In addition to We Mean Business, in recent years the NGO The Climate Group and its NGO partners have co-founded the 'Renewable Energy 100' (RE100), 'Electric Vehicle 100' (EV100) and 'Doubling Energy Productivity by 2030' (EP100) initiatives, which had, by 2020, successfully encouraged 290+ companies to commit to 100% renewable energy targets, over 100 companies to adopt electric vehicle (EV) fleet targets and over 100 companies to double their energy productivity by 2030 (RE100 n.d.; The Climate Group n.d.a, n.d.b).

At the national level, NGOs are organising similar coalitions of progressive local businesses. For instance, in the USA, the NGO Ceres has organised a US-based coalition of over 1000 progressive investors and businesses that want smart policies and strong action on climate change. These businesses and investors have called on the US Government administration to maintain climate change policies and stay within the Paris Climate Change Agreement (Ceres n.d.).

Other NGOs representing the interests of progressive business include, at the global level, the World Business Council for Sustainable Development, the World Green Building Council, the International Renewable Energy Agency and the International Fair Trade Association; alongside related national-level organisations, for instance, in Australia, the Business Council for Sustainable Development, the Green Building Council of Australia, Eco-Tourism Australia, the Council of Australian Recyclers, the Clean Energy Council, the Energy Efficiency Council and Fair Trade Australia/New Zealand, to name a few. There is an opportunity for policy think tanks or policy-focused universities to better work with the now large array of progressive 'green' business groups.

Corporate climate change advocacy and actions sometimes evoke scepticism and when companies engage in greenwashing – giving the appearance of climate concerns while persisting with practices inimical to the environment and greenhouse gas emissions reductions – such views are reinforced. However, corporations are an integral part of society and thus must be participants in action on climate change.

24.3.4 *Working with Cities and Mayors*

At the 2015 Paris Conference of Parties, mayors from almost 1000 cities committed to achieving either 100% renewable energy or 80% cuts to greenhouse gas emissions by 2050. This is a significant start, but many cities are aiming for even faster transitions, seeking to be net carbon neutral well before 2050. This is important in helping to build political will to adopt more ambitious targets at the national level. This surge in activity led by cities has been helped by the NGOs facilitating capacity-building and knowledge-sharing between cities, such as ICLEI and the Compact of Mayors (ICLEI Local Governments for Sustainability n.d.), as well as by other NGO initiatives, such as the C40 Cities low-carbon initiative.

24.4 **Climate Change Social Movement**

Civil society played an influential role in the UNFCCC Paris Agreement process, but it is clear that the Paris Agreement is not alone sufficient to achieve effective climate change policy across the world. With some nation states resisting the adoption of effective climate policy, civil society can play an influential role building pressure for change. A broad and diverse civil society movement will provide many avenues for driving change. Table 24.1 summarises just a sample of current civil society initiatives and partnerships on climate change. The breadth of the movement reflects the fact that climate change negatively affects everyone struggling for development and long-term job opportunities, women's rights, the rights of Indigenous people and other social and economic issues. Conversely,

Table 24.1. *Examples of organisations, sectors or communities in the social movement for climate action*

Organisation/community description	Examples of organisations and actions taken
Private sector and NGOs forming progressive business coalitions to encourage investors and business to change practices and 'walk the talk' on reducing emissions. Such coalitions are building climate change policy consensus within the business community and thereby contributing to climate change policy reform debates.	Examples of organisations include: the World Business Council for Sustainable Development, Climate-KIC Europe, Global Investors Coalition on Climate Change, Principles of Responsible Investment and the Carbon Disclosure Project. More recently: 2014: We Mean Business coalition launches (We Mean Business n.d.). 2017: The Climate Action 100+ Initiative (n.d.) – 370 investment funds with USD 35 trillion of assets under management demanding that the companies with which they invest properly assess climate risk and develop plans for a low-carbon future. 2013–2017: Ceres (USA) and its Business for Climate and Energy Policy (BICEP) coalition's Climate Declaration policy statement is supported by over 1000 businesses (Ceres n.d.). 2015: Global Reporting Initiative (GRI), the United Nations Global Compact and the World Business Council for Sustainable Development (WBCSD) join forces to mobilise the private sector as a key player in achieving the UN's Sustainable Development Goals (SDGs). 2014: RE100 initiative is launched. This now is a coalition of over 230 multinational corporations that have committed to achieving 100% renewable energy (RE100 n.d.). 2015: EP100 initiative launched. A coalition of over 100 major businesses formed to increase the number of large corporations formally committed to doubling energy productivity by 2030 to lower greenhouse emissions (The Climate Group n.d.a). 2017: EV100 initiative launched. A coalition of over 100 major multinationals including Unilever and Ikea committing to transitioning to 100% electric vehicle corporate fleets (The Climate Group n.d.b).
Labour movements: many unions across the world have published documents or run campaigns focused on the impact of climate change on workers, and suggesting solutions workers can be involved in.	2008: Australian Manufacturing Workers Union publishes *Making Our Future: Just Transitions for Climate Mitigation* (AMWU 2008). 2014: Australian firefighter union members launch Australian Firefighters' Climate Alliance to advocate for climate action.

Table 24.1. (*cont.*)

Organisation/community description	Examples of organisations and actions taken
	2015: International Trade Union Congress: 'No jobs on a dead planet' (ITUC 2014). 200 delegates from 50 unions across the world participate in the sign-up for climate action, calling on UN negotiations to ensure a just transition for workers from a high-emissions economy to sustainable jobs.
Indigenous movements: Indigenous communities across the world have existing campaigns and actions for land rights and justice. Many Indigenous communities have recognised the impact of climate change, and have integrated a call of climate justice into their struggles.	2010: Bolivia World People's Conference on Climate Change and Rights of Mother Earth. Indigenous representatives play lead role drafting Cochabamba People's Accord and Universal Declaration on Rights of Mother Earth (People's Agreement of Cochabamba 2010). 2015: Indigenous communities are resisting fossil fuel mining on their land, for example expulsion of oil companies exploiting oil from Sarayuku land in the Amazon (Goodman 2015). 2014: Founding of Seed Youth Indigenous Climate Network, fostering Indigenous youth leadership for climate action in Australia (Seed 2014). 2016: Standing Rock Sioux tribal council opposing the North Dakota Access Oil Pipeline, gathering support across the USA and the world (Petronizio 2016).
Faith movements: leaders of most faith traditions across the world have recognised climate change as a key problem that people of faith must act upon, to care for creation, humans and other species.	2015: Papal encyclical *Laudato Si Dialogue* calls on all people to enter dialogue on climate change and take action (Francis 2015). 2015: Council of Islamic leaders agree to the international Islamic declaration on climate change (UNFCCC 2015). 2015: The People's Pilgrimage (n.d.) international event for people of all faiths to highlight climate change and call for action in the lead-up to UN Paris climate change negotiations.
Decentralised groups calling for institutional divestment: united by groups like 350.org, institutional investors, teachers, union members, churches, communities and individuals increasingly choosing to divest their superannuation/pension funds of fossil fuels and invest in low-carbon technology companies.	2012: Unity College in Maine started the trend and by 2020, over 1000 institutions had voted to divest close to USD 8 trillion in investments in fossil fuel companies (McKibben 2018). This includes a wide variety of institutions: University campaigns that have led to university divestment from some companies: University of California, University of Glasgow, Oxford University, Australian National University; Government divestment: City of Moreland (Melbourne, Australia), San Francisco, USA;

Table 24.1. (*cont.*)

Organisation/community description	Examples of organisations and actions taken
	Faith institutions divestment: World Council of Churches, Church of England; Financial institution divestment: Norwegian Sovereign Wealth Fund (Ambrose 2019), Rockefeller Brothers (Goldenberg 2014), HESTA Super fund.
Social justice organisations: organisations committed to equality, poverty alleviation and development have recognised that the impacts of climate change could undermine all efforts to end poverty and inequality.	2015: Nations of the world adopt the UN's Sustainable Development Goals. The OECD, the UN Development Programme and major development organisations (for example, Oxfam) have recognised climate change as a major impediment to ending poverty.
Health professional advocates: since the International Society of Doctors for the Environment was formed in 1990, international networks of doctors and other health professionals have been running public campaigns to highlight that health is directly linked to a healthy environment.	2019: Australian Medical Association declares a climate emergency. 2019: UK Health Alliance on Climate Change, which represents 650 000 healthcare professionals in the UK, sends a letter to the UK Government calling for the adoption of a zero emissions target by 2050. 2019: In the USA, 70 public health groups issue a statement stating that the climate crisis is also a health emergency, calling on government and businesses to take urgent action.
Community organisations: common community organisations like sporting and youth activity groups have recognised the impacts of climate change.	2011: Surf Lifesaving Australia releases report on climate change impacts, outlining plans to support all clubs to adapt their coastal infrastructure to climate change and reduce emissions with renewable energy (Elrick and Kay 2011). 2010–2020: The World Organisation of the Scout Movement, an international partner with the World Wildlife Fund, encourage scouts to host Earth Hour events to raise awareness of climate change. They have now run this for a decade (Scouts 2010, 2018). 2019: More than 20 former fire and emergency chiefs from multiple states and territories in Australia issue a statement calling for greater preparedness for the risk of more intense bushfires due to drought conditions and climate change (Cox 2019).
Online advocacy communities: the internet has become a powerful way for millions to	2010: 350.org launches international network linking existing grassroots climate action groups (350.org n.d.).

Table 24.1. (*cont.*)

Organisation/community description	Examples of organisations and actions taken
communicate their support for climate action to decision makers in global campaigns.	2014: Avaaz, an online community with 14 million supporters, rolls out online petition and campaign calling on European leaders to set 'at least' 40% cuts to greenhouse gas emissions by 2030. When EU leaders agree, German Environment Minister recognises Avaaz for their role (Avaaz n.d.).
Cities and local governments: many local-scale governments, such as city councils, and state or province governments are setting strong climate policies and reducing emissions at the local scale.	2007: International Local Governments Climate Roadmap launched (ICLEI Local Governments for Sustainability n.d.). The aim is for local governments to show leadership on climate action and to stimulate agreement to climate action at the national and international scale. 2015: Compact of Mayors. At the Paris Conference of Parties in 2015, mayors from almost 1000 cities committed to achieving either 100% renewable energy or 80% cuts to greenhouse gas emissions by 2050.
Youth networks for climate action and schools for climate strike networks: young people will face more of the effects of climate change as they grow older. Youth organisations across the world have formed to call for climate action and intergenerational equity. They have been pivotal to the rapid spread of the schools' strike for climate action movement.	2002: South Asia Youth Environment Network (SAYEN n.d.) supports groups in many nations to run projects such education on climate change. 2015: Australian Youth Climate Coalition has grown to 120 000 supporters, and runs successful campaigns calling on banks to halt investments in coal mine infrastructure (AYCC 2015). 2018 onwards: Youth networks for climate change have been important for supporting Greta Thunberg's call for schools to strike for the climate. These have been further supported by teachers and teacher unions, parents' and grandparents' groups' calls for action on climate change. September 2019 saw the largest global school strike for action on climate change, involving over 1200 cities and 6 million students. It was the largest global youth strike in history.
Local campaigns against fossil fuel extraction and power generation: communities across the world, concerned about the impacts of fossil fuel extraction and use on issues such as health, water and land rights are campaigning to halt projects. Many campaigns use powerful direct action to physically halt or delay projects.	2007: James Hansen, retired NASA climate scientist, publicly calls for all new coal-fired power plants to be stopped. Soon after, CoalSwarm, an online information hub tracking coal-fired power plants across the world is formed to support thousands of grassroots campaigns to halt new coal plants.

Table 24.1. (*cont.*)

Organisation/community description	Examples of organisations and actions taken
	2010: Lock the Gate, an Australian alliance of communities standing against coal-seam-gas development, is founded (Lock the Gate n.d.). Communities in the alliance have mobilised to halt new projects and have won legislative change. The alliance now also includes communities working against coal projects.
	2011: USA Sierra Club throws support behind the Beyond Coal campaign, supporting local communities fighting coal plants in their area. Following the success of this model, Sierra Club launches Beyond Gas (Sierra Club n.d.a, n.d.b).

action on climate change can reduce poverty, empower women, create jobs and build more resilient low-carbon communities. The climate change social movement has been flexible, responding to local and global needs through, for instance, organising international days of action in the lead-up to major international climate meetings, or responding to the threat of the fossil fuel sector to local Indigenous communities.

24.5 Community Renewable Energy

The climate movement encompasses not only groups working to prevent corporations and governments from investing in further fossil fuel infrastructure projects, which risk locking in greenhouse emissions into the future, but also groups working to build zero-carbon resilient alternatives. Non-governmental organisations work with investors, sustainable businesses, sustainable designers, farmers, local and city governments and local communities to create concrete examples of decarbonised alternatives.

Local communities and 'green' businesses across the world are providing concrete examples of how people *can* provide all their day-to-day needs in ways that do not cause climate pollution.

Local communities are investing in renewable energy solutions to bring about an energy transition away from fossil fuels, one town, city or region at a time. Local community-owned solar farms allow individuals to invest in a solar project, allowing those without their own roof for solar photovoltaic (PV) panels, or without the financial means to install solar panels, to purchase energy. For example, Canberra SolarShare is setting up a community-owned solar farm in the Australian capital city (SolarShare Canberra n.d.). At the city government level, communities have led campaigns to drive increases in renewable energy in their local area. In September 2013, citizens of Hamburg, Germany, voted to buy back their energy grid, which they had earlier privatised. The 'Our Hamburg–Our Grid' alliance had campaigned in the community when private energy companies were seen to be stifling the move to renewable energy. Communities have been driving renewable energy investment in Germany, a country seen as a leader in renewable energy implementation, with 650 renewable energy community cooperative projects (Leidreiter 2013).

24.5.1 Transition Towns

Transition Towns is a global movement that unites participating localities that are 'transitioning' elements of their communities from climate pollution sources to low- or zero-emissions alternatives. These campaigns include projects to provide renewable energy, grow local food using local resources, increase public transport, build low-energy-consumption housing and develop community ownership models that employ and engage local people in the provision of services (Transition Network n.d.).

24.5.2 Community NGOs That Assist Households and Communities to Achieve Practical Change

The community group SEE-Change (Social, Economic and Environmental Change) is an alliance of groups from suburbs across Canberra, Australia, who undertake their own projects to reduce the emissions of their community (SEE-Change n.d.). Their projects include building resilient food systems, such as a city farm, local energy production through community solar power purchases and encouraging sustainable transport with bike servicing workshops and trailer hire. They also support local political campaigns that have secured legislation for a 40% reduction in emissions, a 90% renewable energy target that stimulated a solar farm in the area and investment in public transport infrastructure (SEE-Change n.d.).

24.6 Bottom-Up Community Leadership

From international networks to local, specific groups and issues, the climate change social movement extends to the actions each of us can take each day. The catch-cry 'think global, act local' is identified with many community-scale actions individuals can easily get involved with. Table 24.2 outlines examples of local community actions to tackle climate change.

24.7 The Value of Individual Leadership and Behaviour Change

This book has focused mainly on the benefits of national government and private-sector climate change leadership, underpinned by effective government climate change policy reform (Chapter 1), and the wide range of enabling technologies available to support it. This chapter has outlined how civil society can help, by building partnerships with the

Table 24.2. *Doing it ourselves: examples of local community actions to reduce emissions*

Community climate action	Example
Community-owned renewable energy projects.	SolarShare is a community project in Canberra, Australia, connecting people who want to invest in renewable energy but are unable to, due to financial limitations or lack of home ownership. It is creating a community solar farm which people can invest in (SolarShare Canberra n.d.).
Online carbon footprint reduction education.	Carbon Rally is a USA-based website that hosts teams to compete to reduce their personal emissions. The site lists a range of challenges the team can take on, educating participants about the emissions related to an everyday behaviour, and an alternative to try. Over 150 000 individuals have participated on the Carbon Rally site.
Reinvestment: following the success of the divestment movement, communities are now finding sustainable projects like renewable energy, for the divested money to be diverted towards.	The Our Power campaign is linked with the divestment movement, to identify community initiatives that are building sustainable alternative energy and other systems. Black Mesa in Arizona is a pilot reinvestment community, driven by the local Navajo community. It aims to bring sustainable, clean jobs to the area, where two coal mines have had impacts on the local environment.
Reducing consumption through reuse: community networks to help people swap second-hand goods are not a new idea, but many have now been associated with reducing your carbon footprint.	Swishing is a new term introduced by climate communications group Futerra Sustainability Communications (n.d.) in the UK. It means clothes swaps, usually a social event where friends get together and bring clothes they no longer wear to trade. Swishing events can be posted online for others to find. Other examples include libraries, toy libraries, second-hand shops, recycling centres or giveaway gifting groups.
Community 'greening' infrastructure: local faith groups and sports clubs (for example) are investing in solar panels or energy efficiency to reduce their footprint. Community bulk buys for emissions reduction.	Over 3700 schools in the USA have invested in solar PV systems for their rooftops. The Brighter Future report found the main reasons schools adopt solar power were: reducing electricity costs and reducing emissions, as well as educating students on science and technology issues. The UK Community Pathways group encourages pooling finances to purchase energy-efficient equipment of tools for their neighbourhood/community group.
Home retrofit programmes aimed at those most in need.	Transition Streets (n.d.), a programme that started in the UK, supports neighbours in a

Table 24.2. (*cont.*)

Community climate action	Example
	street to take an energy-saving challenge together. Transition Streets aims to increase social interactions to assist isolated and disadvantaged households reduce energy use. A street champion contacts neighbours in their street to meet and learn from a workbook how to reduce emissions and save energy through modifications to their homes.
Community education for sustainable living: people interested in reducing their environmental impact are building networks to share ideas, run working bees and support one another to learn how to live a low-emissions lifestyle.	Urban homesteading means finding ways to be more self-sufficient in food, energy and other goods, and to reduce your personal environmental footprint. The Canberra Urban Homesteaders group uses both online discussions and face-to-face workshops to socialise and share skills like gardening, preserving, keeping poultry and so on.
Low-carbon-footprint food: agriculture and food production is a major source of emissions driving climate change. Communities across the world are building alternative food systems that aim to regenerate landscapes, minimise inputs and reduce the distance food is transported from production to consumption.	Slow Food (n.d.) is an international movement that started in Italy in 1989 in response to the rise of industrial agriculture, fast food and the health impacts of poor diets. Slow Food networks have developed in 160 countries. Groups across the world support local food production, farmers' markets and food traditions like preserving. Slow Food advocates for environmentally sustainable food production and consumption, specifically reducing the agricultural practices linked to climate change (Slow Food n.d.).

institutional investment, business and government communities to support, knowledge-share and de-risk the transition to a zero-carbon resilient future. It is critical, before finishing, to acknowledge that all of this requires leadership from individuals with the confidence to collaborate with colleagues to achieve systematic change within those institutions and organisations: this confidence is often built over time by individuals successfully tackling smaller, but no less important, challenges and sustainability opportunities within local spheres of influence, such as through the local community initiatives overviewed above.

Individuals can further grow the confidence to effectively communicate and implement the ideas in this book (and others) by 'walking their talk': taking steps to reduce their carbon and ecological footprints through behaviour change and investing themselves in zero-carbon technologies at the household level. Everything consumed at an individual level has long supply chains, so individual consumers can cut both greenhouse gas emissions and costs significantly by shifting to on-site sustainable sources of household energy (i.e. solar PV systems) and water (i.e. rainwater harvesting and local usage), and embracing a healthy low-carbon diet and low-carbon personal transportation, as well as participating in the circular economy.

If individuals can demonstrate and articulate the benefits of such change in their own lives, they can be more effective at inspiring change in their community or at work. With the internet, such individual examples of leadership can be relatively easily shared to inspire action, locally, regionally and even internationally. Exemplar individual leadership and knowledge-sharing helps to de-risk new approaches and take-up of new technologies.

24.8 Conclusion

The technical means to mitigate climate change and adapt to its effects have been available for some time but they have not yet been implemented with the vigour required to meet global emissions targets of no more than 1.5 °C above pre-industrial levels (IPCC 2018). As a consequence, concerned communities of interest have formed to address the gap. These have ranged from local to national and global coalitions and have involved community, business, scientific, faith and other groups, as outlined in Table 24.1. The activities of civil society, the scientific community and some institutional investor and business interests were instrumental in bringing about the UNFCCC Paris Agreement and are now also key to ensure its intent is implemented (Jacobs 2016) at the scale and speed required to avoid the worst forms of dangerous climate change.

A critical element in the climate change 'debate' has been to change the narrative from one focused on the costs of change to one highlighting the opportunities offered by transition to and operation of a carbon-free economy at the macro- and microeconomic levels. Studies have shown both that the cost of change will be greater the longer it is delayed and that the economic opportunities offered by transition far exceed those that can be expected from BAU (Repetto and Austin 1997; Stern 2006; Smith et al. 2010; WRI 2014; Smith 2015; OECD 2017; Global Commission on the Economy and Climate 2018). Defenders of the status quo argue that the required changes will be expensive and will exert negative pressures on national economies. They assert, among other things, that the fossil fuel industries are essential to maintain current and future prosperity and to improve the lives of the poor in developing countries. Institutional bodies, some governments and civil society NGOs have undertaken relatively new significant analysis to counter these arguments and to build the case for action on climate change, evidencing that strong and smart action on climate can boost the global economy (OECD 2017) USD 26 trillion above BAU by 2030 (Global Commission on the Economy and Climate 2018). Scientific and government resources have been brought together in the IPCC, which has produced landmark reports evidencing how to technically and economically keep global warming under 1.5 °C (IPCC 2018). The UN Deep Decarbonisation Pathways Project has been a key initiative, both in highlighting the urgency of action and in defining technically and economically viable pathways to decarbonisation of 16 national economies (Deep Decarbonisation Pathways Project 2015).

Institutional funds, which have in recent years become significant repositories of capital, have adopted positions supporting action on climate change, resulting in institutional investors worth over USD 40 trillion forming the IIGCC and close to USD 8 trillion being divested out of fossil fuels. Over 450 institutional investors have formed Climate Action 100+, which is demanding action from the largest polluting businesses. This is one of many factors resulting in businesses increasingly recognising both that they face threats from global warming, such as from more intense extreme weather events, and that transition to a low-carbon economy offers business opportunities. Many large businesses have incorporated climate change considerations into their business models and have committed to achieving greenhouse gas reduction targets as part of the We Mean Business (We Mean Business n.d.), 'Renewable Energy 100' (RE100), 'Energy Productivity 100' (EP100) and 'Electric Vehicle 100' (EV100) initiatives (RE100 n.d.; The Climate Group n.d.a, n.d.b). By doing so, businesses are working to transition to 100% renewable electricity, transition their fleets to electric vehicles and double energy productivity by 2030 from a 2005 baseline.

This chapter has shown that this creates opportunities for civil society NGOs to form new partnerships, alliances and coalitions, speeding up the rate of decarbonisation and adaptation by de-risking the transition through knowledge-sharing, consensus-building and target-setting. This work is critical to enable ratcheting up of climate change mitigation and adaptation ambitions and implementation, in concert with national and subnational government efforts to review and improve their mitigation targets and ensure their delivery.

References

350.org (n.d.). History. *350.org*. Available at: https://350.org/about/#history.

Ambrose, J. (2019). World's biggest sovereign wealth fund to ditch fossil fuels. *The Guardian*. 13 June. Available at: www.theguardian.com/business/2019/jun/12/worlds-biggest-sovereign-wealth-fund-to-ditch-fossil-fuels.

AMWU (Australian Manufacturers Workers Union) (2008). *Making Our Future: Just Transitions for Climate Change Mitigation*. Australian Manufacturers Workers Union. Available at: www.aph.gov.au/DocumentStore.ashx?id=3a4a8f80-43f0-437e-89e9-fd7138fd0304&subId=460076.

Avaaz (n.d.). Avaaz is a global web movement to bring people-powered politics to decision-making everywhere. *Avaaz*. Available at: www.avaaz.org/en/about.php.

AYCC (Australian Youth Climate Coalition) (2015). *2015 Impact Report*. Australian Youth Climate Coalition. Available at: https://d3n8a8pro7vhmx.cloudfront.net/aycc/pages/77/attachments/original/1459922021/AYCC_Annual_Report_2015_w6.pdf?1459922021.

CDP (Carbon Disclosure Project) (n.d.). Carbon Disclosure Project. *CDP.net*. Available at: www.cdp.net/en.

CDP (2019). World's top green businesses revealed in the CDP A List. *CDP.net*. 22 January. Available at: www.cdp.net/en/articles/companies/worlds-top-green-businesses-revealed-in-the-cdp-a-list.

Ceres (n.d.). We are still in. *Ceres.org*. Available at: www.ceres.org/initiatives/we-are-still-in.

Climate Action 100+ Initiative (n.d.). *Climate Action 100+*. Available at: www
.climateaction100.org/.

Climate and Clean Air Coalition (n.d.). Who we are. *Climate and Clean Air Coalition*.
Available at: https://ccacoalition.org/en/content/who-we-are.

Coalition for Climate Resilient Investment (2019). Private-sector led Coalition for Climate
Resilient Investment brings together companies across the infrastructure investment
value chain with assets totalling USD 5 trillion. *Globe Newswire*. 23 September.
Available at: www.globenewswire.com/news-release/2019/09/23/1919391/0/en/
Private-sector-led-Coalition-for-Climate-Resilient-Investment-brings-together-com
panies-across-the-infrastructure-investment-value-chain-with-assets-totalling-USD-
5-trillion.html.

Cox, L. (2019). Former fire chiefs warn Australia unprepared for escalating climate threat.
The Guardian. 10 April. Available at: www.theguardian.com/australia-news/2019/
apr/09/former-fire-chiefs-warn-australia-unprepared-for-escalating-climate-threat.

Deep Decarbonisation Pathways Project (2015). *Pathways to Deep Decarbonization: 2015
Report*. Sustainable Development Solutions Network and the Institute for Sustainable
Development and International Relations. Available at: https://resources.unsdsn.org/
pathways-to-deep-decarbonization-2015-synthesis-report.

Dunlap, R. and McCright, A. (2015). Challenging climate change: The denial counter
movement. In R. Dunlap and R. Brulle, eds., *Climate Change and Society:
Sociological Perspectives*. Oxford: Oxford University Press.

Elrick, C. and Kay, R. (2011). *Impact of Extreme Weather Events and Climate Change on
Surf Life Saving Services*. Summary Report prepared for Surf Life Saving Australia.
Claremont, Western Australia: Coastal Zone Management Pty Ltd. Available at:
https://issuu.com/surflifesaving/docs/impact-of-extreme-weather-events-slsa.

Francis (2015). *Encyclical Letter Laudato Si' of the Holy Father Francis: On Care for Our
Common Home*. Rome: Vatican Press. Available at: http://w2.vatican.va/content/
dam/francesco/pdf/encyclicals/documents/papa-francesco_20150524_enciclica-lau
dato-si_en.pdf.

Futerra Sustainability Communications (n.d.). About. *Swishing*. Available at: http://
swishing.com/about_swishing/.

Global Commission on the Economy and Climate (2018). *Unlocking the Inclusive Growth
Story of the 21st Century: Accelerating Climate Action in Urgent Times*. Washington,
DC: The Global Commission on the Economy and Climate. Available at: http://
newclimateeconomy.report/2018/.

Goldenberg, S. (2014). Heirs to Rockefeller oil fortune divest from fossil fuels over climate
change. *The Guardian*. 23 September. Available at: www.theguardian.com/environ
ment/2014/sep/22/rockefeller-heirs-divest-fossil-fuels-climate-change.

Goodman, D. (2015). Deep in the Amazon, a tiny tribe is beating Big Oil. *YES! Magazine*.
13 February. Available at: www.yesmagazine.org/issues/together-with-earth/deep-in-
the-amazon-a-tiny-tribe-is-beating-big-oil.

Hargroves, K. and Smith, M. (2005). *The Natural Advantage of Nations: Business
Opportunities, Innovation and Governance in the 21st Century*. London: Earthscan.

ICLEI Local Governments for Sustainability (n.d.). *Local Government Climate Roadmap*.
Available at: http://old.iclei.org/climate-roadmap/home.html.

IPCC (2018). *Global Warming of 1.5 °C: An IPCC Special Report on the Impacts of
Global Warming of 1.5 °C Above Pre-Industrial Levels and Related Global
Greenhouse Gas Emission Pathways, in the Context of Strengthening the Global
Response to the Threat of Climate Change, Sustainable Development, and Efforts to
Eradicate Poverty*. Edited by V. Masson-Delmotte, P. Zhai, H.-O. Pörtner et al.
Cambridge: Cambridge University Press. Available at: www.ipcc.ch/sr15/.

ITUC (International Trade Unions Congress) (2014). Union leaders announce their commitments to the fight against climate change. *3rd ITUC World Congress*. 21 May. Available at: www.ituc-csi.org/union-leaders-announce-their.

Jacobs, M. (2016). High pressure for low emissions: How civil society created the Paris climate agreement. *IPPR: The Progressive Policy Think Tank*. 14 March. Available at: www.ippr.org/juncture/high-pressure-for-low-emissions-how-civil-society-created-the-paris-climate-agreement.

Leidreiter, A. (2013). Hamburg citizens vote to buy back energy grid. *Energy Transition*. 8 October. Available at: https://energytransition.org/2013/10/hamburg-citizens-buy-back-energy-grid/.

Lock the Gate (n.d.). *Lock the Gate Alliance*. Available at: www.lockthegate.org.au/.

Maddow, R. (2019). *Blowout: Corrupted Democracy, Rogue State Russia, and the Richest, Most Destructive Industry on Earth*. London: Crown Publishing.

McKibben, B. (2018). At last, divestment is hitting the fossil fuel industry where it hurts. *The Guardian*. 17 December. Available at: www.theguardian.com/commentisfree/2018/dec/16/divestment-fossil-fuel-industry-trillions-dollars-investments-carbon.

Newell, P. and Paterson, M. (1998). A climate for business: Global warming, the state and capital. *Review of International Political Economy*, 5, 679–703.

OECD (2017). *Investing in Climate, Investing in Growth*. Paris: OECD. Available at: www.oecd.org/economy/taking-action-on-climate-change-will-boost-economic-growth.htm.

Oreskes, N. and Conway, E. (2010). *Merchants of Doubt: How a Handful of Scientists Obscured the Truth on Issues from Tobacco Smoke to Climate Change*. Bloomsbury Publishing.

People's Agreement of Cochabamba (2010). People's Agreement of Cochabamba. World People's Conference on Climate Change and the Rights of Mother Earth. Available at: https://pwccc.wordpress.com/2010/04/24/peoples-agreement/.

People's Pilgrimage (n.d.). About the People's Pilgrimage. *People's Pilgrimage*. Available at: http://peoplespilgrimage.org/.

Petronizio, M. (2016). How young Native Americans used social media to build up #NoDAPL. *Mashable*. 8 December. Available at: http://mashable.com/2016/12/07/standing-rock-nodapl-youth/#7bxk4hKZdqqm.

Powering Past Coal Alliance (n.d.). *Powering Past Coal Alliance*. Available at: https://poweringpastcoal.org.

RE100 (n.d.). RE100 overview. *RE100*. Available at: http://there100.org/re100.

Repetto, R. and Austin, D. (1997). *The Costs of Climate Protection: A Guide for the Perplexed*. Washington, DC: World Resources Institute. Available at: http://pdf.wri.org/costsclimateprotection_bw.pdf.

SAYEN (South Asia Youth Environment Network) (n.d.). Projects. *SAYEN.org*. Available at: https://sayen.org/sayen/templates/default/page_tpl.php?p=30.

Scouts (2010). Scouts support Earth Hour 2010. *Scouts*. Available at: www.scout.org/es/node/6810.

Scouts (2018). WWF and World Scouting join forces for a healthy planet. *Scouts*. 12 March. Available at: www.scout.org/wwf-world-scouting-join-forces-to-mobilise-young-people-for-healthy-planet.

SEE-Change (n.d.). Working together to create sustainable change in Canberra. *SEE-Change*. Available at: www.see-change.org.au.

Seed (2014). Indigenous youth declaration for climate justice. *Seed Mob*. Available at: www.seedmob.org.au/indigenous_youth_declaration.

Sierra Club (n.d.a). We're moving beyond coal and gas. *Sierra Club: Beyond Coal*. Available at: https://coal.sierraclub.org/campaign.

Sierra Club (n.d.b). Beyond gas. *Sierra Club: John Muir Chapter.* Available at: www
.sierraclub.org/wisconsin/beyond-gas.

Sinclair, U. (1994). *I, Candidate for Governor: And How I Got Licked,* 1st ed. Berkeley,
CA: University of California Press.

Slow Food (n.d.). Climate change and the food system. Position Paper. Available at: www
.slowfood.com/sloweurope/wp-content/uploads/ENG-PAPER-climatechange.pdf.

Smith, M. (2009). Advancing and resolving the great sustainability debates and discourses.
PhD thesis, The Australian National University. Available at: https://openresearch-
repository.anu.edu.au/handle/1885/49387.

Smith, M. (2013a). *Assessing Climate Change Risks and Opportunities for
Investors: Property and Construction Sector.* Canberra: The Investor Group
on Climate Change (IGCC) and The Australian National University (ANU).
Available   at: https://igcc.org.au/wp-content/uploads/2016/04/Property-and-
Construction-1.pdf.

Smith, M. (2013b). *Assessing Climate Change Risks and Opportunities for Investors:
Mining and Mineral Processing Sector.* Canberra, Australia: The Investor Group
on Climate Change (IGCC) and The Australian National University (ANU).
Available at: https://igcc.org.au/wp-content/uploads/2016/04/mining_assessing-cli
mate_change_risks_for_investors.pdf.

Smith, M. (2013c). *Assessing Climate Change Risks and Opportunities for Investors: Oil
and Gas Sector.* Canberra: The Investor Group on Climate Change (IGCC) and The
Australian National University (ANU). Available at: https://igcc.org.au/wp-content/
uploads/2016/04/Oil-and-Gas.pdf.

Smith, M. (2014). *Assessing Climate Change Risks and Opportunities for Investors:
Industrials and Manufacturing Sector.* Canberra: The Investor Group on Climate
Change (IGCC) and The Australian National University (ANU). Available at: https://
igcc.org.au/wp-content/uploads/2017/06/Assessing-Climate-Change-Risks-and-
Opportunities-for-Investors.pdf.

Smith, M. (2015). *Doubling Energy and Resource Productivity by 2030: Boosting the
Global Economy by >$25 Trillion above BAU Whilst Transitioning to a Low Carbon
Economy.* ANU Discussion Paper. The Australian National University.

Smith, M. and Hargroves, K. (2006). $uper powers: How investment funds are driving
progress in an emerging 'sustainability' economy. *ECOS,* 132, 24–28.

Smith, M., Hargroves, K. and Desha, C. (2010). *Cents and Sustainability: Securing Our
Common Future by Decoupling Economic Growth from Environmental Pressures.*
London: Routledge.

SolarShare Canberra (n.d.). *SolarShare Canberra.* Available at: https://solarshare.com.au/.

Stern, N. (2006). *The Stern Review: The Economics of Climate Change.* Cambridge:
Cambridge University Press.

The Climate Group (n.d.a). EP100. *The Climate Group.* Available at: www
.theclimategroup.org/project/ep100.

The Climate Group (n.d.b). EV100. *The Climate Group.* Available at: www
.theclimategroup.org/project/ev100.

Transition Network (n.d.). About the movement. *Transition Network.org.* Available at:
https://transitionnetwork.org/about-the-movement/.

Transition Streets (n.d.). How Transition Streets works. *Transition Streets.* Available at:
www.transitionstreets.org.uk/about/.

Under 2 Degrees Coalition (n.d.). *Under2Coalition.org.* Available at: www
.under2coalition.org/.

UNFCCC (UN Framework Convention on Climate Change) (2015). Islamic declaration on global climate change. *UN Climate Change*. 18 August. Available at: https://unfccc .int/news/islamic-declaration-on-climate-change.

We Mean Business (n.d.). *We Mean Business Coalition*. Available at: www .wemeanbusinesscoalition.org/about.

WEF (World Economic Forum) (2018). An open letter from business to world leaders: 'Be ambitious, and together we can address climate change'. *World Economic Forum*. 29 November. Available at: www.weforum.org/agenda/2018/11/alliance-ceos-open-letter-climate-change-action/.

WRI (World Resources Institute) (2014). *Better Growth, Better Climate. The New Climate Economy Report*. The Global Commission on the Economy and Climate. Washington, DC: The Global Commission on the Economy and Climate, World Resources Institute. Available at: www.newclimateeconomy.report//2014/.

Index

Page numbers in italics indicate data contained in tables or figures.